MW00607512

CLIMATE

CLIMATE
History, Periodicity, and Predictability

Edited by

Michael R. Rampino
John E. Sanders
Walter S. Newman
L. K. Königsson

VNR VAN NOSTRAND REINHOLD
New York

Library of Congress Catalog Card Number 87-21559
ISBN 0-442-27866-7

Printed in the United States of America

Van Nostrand Reinhold
115 Fifth Avenue
New York, New York 10003

Van Nostrand Reinhold International Company Limited
11 New Fetter Lane
London EC4P 4EE, England

Van Nostrand Reinhold
480 La Trobe Street
Melbourne, Victoria 3000, Australia

Macmillan of Canada
Division of Canada Publishing Corporation
164 Commander Boulevard
Agincourt, Ontario M1S 3C7, Canada

15 14 13 12 11 10 9 8 7 6 5 4 3 2

Library of Congress Cataloging-in-Publication Data
Climate, history, periodicity, and predictability.
Bibliography: p.
Includes indexes.
1. Climatic changes. 2. Paleoclimatology. 3. Meteorology—Periodicity. I. Rampino, Michael R.
QC981.8.C5C57 1987 551.6 87-21559
ISBN 0-442-27866-7

Contents

PART III: SEA-LEVEL CHANGE AND CLIMATE

PART IV: SHORT-TERM CLIMATE (10–10^2 YR) AND PERIODICITY

PART V: LONG-TERM CLIMATE (10^3–10^7 YR) AND PERIODICITY

Preface

This book contains the papers presented at a meeting held May 21-23, 1984, at Barnard College, Columbia University, New York City, to honor Professor Emeritus Rhodes W. Fairbridge on his seventieth birthday. At the meeting, 80 specialists from 12 countries discussed various aspects of the Earth's climate. The papers included here present new data ranging from measurements of recent atmospheric temperatures to geologic evidence extending back several billion years. Some are devoted to sophisticated computer modeling and time-series analyses. Many participants discussed causal mechanisms for climate change that ranged from the purely terrestrial to those involving the sun, the moon, the dynamics of the solar system, and even the rhythms of the galaxy. A major emphasis includes climate cycles and the possibilities for predicting the trend of future climate.

The title of the conference, "Climate: History, Periodicity, and Predictability," reflects what we think are the essential elements for understanding the Earth's climate. In 1961, Professor Fairbridge organized a similar conference entitled "Solar Variations, Climatic Change, and Related Geophysical Problems" (Fairbridge, 1961). As organizers of the 1984 conference, one of our objectives was to reexamine the subject matter of the 1961 conference. We also hoped to present something other than just another symposium on climate, of which many have been held recently. Since 1961, the Milankovitch hypothesis—relating climate changes on Earth to the interaction between an essentially constant sun and well-known Earth-orbit and Earth-axial changes, and thus yielding periodic variations in climate on time scales of 10^4-10^5 yr—has become widely accepted. Given such acceptance, one might suppose the existence of a certain greater appreciation of the potential periodic climatic inputs from the orbital dynamics of other parts of the solar system, but this has been slow in coming. For example, in his monumental two-volume summary of climate and historical climate data, Lamb (1972, 1977) examined the subject of periodicities and, along with many others, remained skeptical as to whether the Earth's short-term climate is influenced by periodic input functions, especially by external periodic forcings. Such skepticism seems to be rooted in the lack of acceptable physical mechanisms linking the various factors that would have to be involved.

In planning for this symposium, we tried to bring together speakers who could assess the historical and geologic records of climates for demonstrated periodic features with those who would present evidence of periodicities within the solar system. If connections could be made between the two, then a basis might be established for using solar system periodicities to predict climate changes on the Earth. Papers 24, 25, and 26 address such possible relationships between various solar-orbital parameters and solar-terrestrial climatic effects.

The larger issue of the periodic machinery of the solar system and its possible effects on the sun, and through such effects, on the Earth's climate, continues to be the subject of debate. One approach to possible effects of solar system dynamics was charted by Professor Fairbridge's good friend, the late Clyde M. Stacey, an engineer who spent his active professional career designing gears for sewing machines and who spent his spare time, later full time in retirement, working out a new approach to planetary orbits. Stacey visualized the planetary orbits as following the same rules that govern a complex train of gears. His fundamental conclusion was that even the relatively short-term motions of the planets could affect the Earth's climate.

As emphasized by Milankovitch (1941), the essential element in orbital variations is their recurrence at predictable intervals. Therefore, if one is able to establish that any recurrent effects related to planetary orbits are indeed registered on the Earth, for example, then one is armed with a powerful predictive tool. But the nub of the problem is to establish the supposed connection and to provide a convincing causal mechanism(s).

Of the more than 50 presentations given at the conference, 26 are published here. If the mixture seems eclectic, it is because of Fairbridge's widespread interests and influence in climatological studies. Some conference presentations represented work in progress that will appear in more complete form in professional journals during the next few years. Several studies have already been published elsewhere. We highlight here only a few of the many aspects of the wide-ranging conference. On short time scales, several investigators reported climatic phenomena whose periods match possible solar and/or lunar signals. These phenomena include a 12.18-yr cycle in tree rings (E. Cook, unpublished), approximate 22-yr cycles from the Hudson Valley in recent temperatures (paper 3) and in varves of Late Pleistocene age (R. Pardi, unpublished), and 18.6- and 22.1-yr cycles in the minima of the record of Nile flow (paper 1) going back to A.D. 662 (Fairbridge, 1984).

Longer cycles (~100-1,000-yr scale) were reported in climate, sea-level, and various geomorphologic features. Holocene climate fluctuations are now rather well established and show up clearly in oscillations of wet/dry conditions that give rise to cycles of alluviation and dissection in arid and semiarid regions. R. Gerson (paper 17) in the Negev Desert, Israel, and R. Paepe (see Paepe et al., 1983) in Greece inferred cycles of about 1000 yr. Through studies of paleosols dated by tree ring, radiocarbon, and archeologic evidence, T. Karlstrom (unpublished) reported on drought cycles in the southwestern United States having periods of 550 yr and 275 yr.

Several papers discuss the excellent climate records in tree rings (paper 4) and in lake deposits in France (paper 5) and Africa (paper 6). Others search for causes in geomagnetic variations (paper 13), fluctuations in solar activity (paper 24), and various mechanisms of change both external and internal to the climate system (papers 14 and 15).

The Fairbridge Sea-Level Curve, first presented at the 1961 conference, is still being hotly debated. Paper 7 presents a brief summary by Professor Fairbridge of the Holocene sea-level situation as viewed from the perspective of more than 25 yr. Evidence from some areas, such as the southeastern United States (paper 8) and the Normandy coast (paper 12), seems to show low-amplitude Holocene sea-level fluctuations, whereas other regions show little evidence of such oscillations. Based on a compilation of more than 4,000 radiocarbon-dated sea-level indicators worldwide, W. S. Newman (see Newman et al., 1980) proposed that low-latitude coasts seem to display the Fairbridge-type wiggles, whereas in areas lying north of latitude 38°N, readjustments of the postglacial geoid largely mask the record of what may have been true oscillations of sea level.

Many conference speakers urged that researchers focus attention on long-term chronologies wherein events can be placed on a year-by-year basis, such as tree rings, ice cores, and varved sediments. Of particular interest is the possibility of establishing unique time markers of Late Quaternary age. Extreme events such as floods, unusually cold years, widespread tephra from known volcanic explosions, and other singularities can provide such marker horizons. For example, V. LaMarche (see LaMarche and Hirschboeck, 1984) showed that frost rings can be recognized in dendrochronological studies, and that some of these can be correlated with known volcanic eruptions and acidity spikes in cores of glacial ice. As noted by D. Schove (paper 21), such singularities can be very helpful in checking dubious correlations and in tying down floating segments of chronologic sequences.

Several papers were presented in which the cause of short-term terrestrial climate cycles was related to the regularities of the sun's orbit around the center of mass (barycenter) of the solar system, an orbit driven by the changing distributions of mass of the planets in their orbits. Using a simple model (an ellipse inscribed on a transparent plastic sheet, rotated about a pin at one focus representing the solar system barycenter, over another transparent plastic sheet laid over the glass of an overhead projector), J. E. Sanders (paper 26) demonstrated the predominant effects on the sun's orbit of Jupiter and Saturn. Stacey (1967) had earlier emphasized the importance of the Jupiter-Saturn lap cycle of 19.857 yr. Sanders reported solar orbit cycles ranging from about 400 days to a fundamental period of about 178 yr. Similar periodicities have been reported in various climatic indicators. Solar activity may be affected by changing planetary torque on the sun. T. Landscheidt (paper 25) has found wavelike variations in his calculations of the impulses of the torque related to the sun's orbit around the barycenter and has shown that these correlate with solar cycles and, to some extent, with changes in the Earth's climate. Cycles of solar flares have been found with periods of 9 yr, 2.25 yr, and 3.3 mo. Longer torque cycles display periods of 13.3 yr, 83 yr, and 391 yr.

The Milankovitch model for glacial/interglacial climate fluctuations was discussed with regard to the possible causes of the dominant 23,000, 41,000, and 100,000-yr cycles inferred from Quaternary climate records. Whatever the origin of such cycles, the sedimentary record of the high interglacial sea levels has been well documented on the mid-Atlantic coast of the United States (paper 9) and in Europe (paper 11). Based on this evidence, Kraft, Belknap, and Demarest (paper 10) predict that the current interglacial sea level will peak in 2,000-4,000 yr, thus allowing the sea to transgress inland up to 10 km from present coastlines.

On time scales of 10^7 yr and 10^8 yr, M. Rampino (see Rampino and Stothers, 1986) reported on a time-series analysis that showed that the geologic record has been dominated by two significant cycles at 33×10^6 and 260×10^6 yr. These periodicities appear in records of global sea level, various kinds of tectonic activities, and geomagnetic reversal frequency, as well as in biological mass extinctions. The same two periods dominate the record of dated impact craters on the Earth. From this, one can infer that impact energy could strongly affect terrestrial geologic processes. Rampino contends that these two basic cycles are related to the movements of the sun and planets through the Milky Way galaxy.

W. Donn (paper 20) makes an interesting case for polar wandering to explain the seeming paradox of evidence of warm climate at high latitudes during Mesozoic and early Tertiary times. This problem is of great interest to research groups trying to model the atmospheric and oceanic circulation patterns during those times of decreased latitudinal temperature gradients.

Prediction was a major focus of the symposium, and climatic forecasts were made on time scales ranging from a few years to a few million years. On short time scales, a major question relates to the expected greenhouse effect resulting from atmospheric pollution by CO_2 and other trace gases. A. Lacis described results from the climate model used at the Goddard Institute for Space Studies that predicts a global warming of 2.5-5°C (largely from positive feedback effects such as increased water vapor in the atmosphere) if atmospheric CO_2 doubles (Hansen et al., 1984), as has been projected for sometime in the next century. Since the mid-1800s, global mean temperature seems to have increased by 0.5°C (a finding also supported by tree-ring studies), and such an increase has been ascribed by some to the progressive increase in atmospheric CO_2. In contrast, two predictions based on the solar cycle (papers 23 and 25) suggest that we may be entering a natural cool period. Landscheidt (paper 25) forecasts an extreme minimum of solar activity beginning around 1990 and lasting until 2070, reaching its minimum around 2030. He foresees cool climate and advance of glaciers, as in the Little Ice Age.

Because climate modelers and others make projections about future climates in the next several decades based on a warm, CO_2-enriched atmosphere, we should not lose touch with the fact that any greenhouse effect is distinct from natural climatic cycles, in whichever direction these proceed. For example, if periodicities exist in the trends of surface temperatures, as has been inferred from many studies of tree rings, ice cores, and other proxy climate indicators, then knowledge of these natural cycles would certainly be required for any future forecasts and temperature predictions based on the atmospheric-pollution greenhouse effect.

For many years, Fairbridge has argued that Holocene climate has been characterized by significant oscillations, and we have no reason to expect that such oscillations have come to an end. With the Milankovitch cycles now well accepted, it seems worthwhile to ask what is causing the shorter-term fluctuations in our current interglacial epoch. The time also seems ripe for putting the same kind of concerted effort of data collection, analysis, and statistical methods that worked for the Pleistocene into an effort to understand the more immediate problems of short-term climate cycles. To aid in this, we have included, as an appendix, a selected bibliography of material pertaining to climatic periodicities and solar-planetary relationships.

If Fairbridge and others are right, and the climate on a scale of decades to centuries and perhaps even of millenia, is governed by predictable, yet complex, cycles driven in part by regular solar and/or solar-planetary configurations, then there is real hope that climate trends can be foreseen and planned for in ways that will make the results of climatic change more manageable for the human population.

We would like to thank the many individuals who made the conference and this book possible, especially Jean McCurry and her staff at Barnard College, Chuck Hutchinson for initiating the idea of a volume of the conference proceedings, and our ever-patient editor at Van Nostrand Reinhold, Bernice Pettinato. John Sanders composed the comprehensive subject index—an heroic task.

We note with sadness the deaths of four of our contributing authors: William L. Donn, Leona M. Libby, Derek Justin Schove, and Mireille Ters. We also regret the loss of our coeditor and close friend, Walter S. Newman. Without his help and enthusiasm, the "Fairbridge Conference," and this volume would never have come to fruition. He will be greatly missed.

REFERENCES

Fairbridge, R. W., conference editor, 1961, Solar variations, climatic changes and related geophysical problems, *New York Acad. Sci. Annals* **95:**1-740.

Fairbridge, R. W., 1984, The Nile floods as a global climatic/solar proxy, in N.-A. Mörner and W. Karlén (eds.), *Climatic Changes on a Yearly to Millenial Basis*, Berlin: D. Reidel, pp. 181-190.

Hansen, J.; Lacis, D.; Rind, D.; Russell, G.; Stone, P.; Fung, I.; Ruedy, R.; and Lerner, J., 1984, Climate sensitivity: Analysis of feedback mechanisms, in J. E. Hansen and T. Takahashi (eds.), *Climate Processes and Climate Sensitivity, Geophysical Monograph 29, Maurice Ewing Series No. 5*, Washington, D.C.: American Geophysical Union, pp. 130-163.

LaMarche, V. C., Jr., and Hirschboeck, K. K., 1984, Frost rings in trees as records of major volcanic eruptions, *Nature* **307:**121-126.

Lamb, H. H., 1972, *Climate, Present, Past, and Future*, vol. 1, London: Methuen and Company, Ltd., 613p.

Lamb, H. H., 1977, *Climate, Present, Past, and Future*, vol. 2, London: Methuen and Company, Ltd., 835p.

Milankovitch, M., 1941, *Canon of Insolation and the Ice-Age Problem*, Belgrade: Royal Serbian Academy. (English translation, 1969, Jerusalem: Israel Program for Scientific Translations.)

Newman, W. S.; Marcus, L. F.; Pardi, R., Paccione, J. A.; and Tomacek, S. M., 1980, Eustasy and deformation of the geoid: 1000-6000 radiocarbon years BP, in N.-A. Mörner (ed.), *Earth Rheology, Isostasy and Eustasy*, New York: Wiley, pp. 555-567.

Paepe, R.; Hatziotis, M. E.; Thorez, J.; Van Overloop, E; and Demarée, G., 1983, Climatic indexes on the basis of sedimentation parameters in geological and archaeological sections, in A. Ghazi (ed.), *Paleoclimatic Research and Models*, Dordrecht: D. Reidel, pp. 129-138.

Rampino, M. R., and Stothers, R. B., 1986, Geologic periodicities and the galaxy, in R. Smoluchowski, J. N. Bahcall, and M. S. Matthews (eds.), *The Galaxy and the Solar System*, Tucson: University of Arizona Press, pp. 241-259.

Stacey, C. M., 1967, Earth motions, in R. W. Fairbridge (ed.), *Encyclopedia of the Earth Sciences*, vol. 2: *Encyclopedia of Atmospheric Sciences and Astrogeology*, New York: Reinhold Publishing Corporation, pp. 335-340.

Contributors

Daniel F. Belknap, Department of Geological Sciences, University of Maine, Orono, ME 04469

A. Berger, Institut d'Astronomie et de Geophysique, Université Catholique de Louvain-la-Neuve, Chemin du Cyclotron 2, B-1348 Louvain-la-Neuve, Belgium

Jane B. Blizard, Department of Astrophysical, Planetary, and Atmospheric Sciences, Campus Box 391, University of Colorado, Boulder, CO 80309-0391

E. Bonifay, Laboratoire Géologie du Quaternaire, U.A. 692 du CNRS Luminy 907, 13288 Marseille Cedex 9, France

Mark J. Brooks, Department of Geology, and the South Carolina Institute of Archaeology and Anthropology, University of South Carolina, Columbia, SC 29208

Joël Casanova, Laboratoire Géologie du Quaternaire, CNRS, Case 907, Université de Luminy, 13288 Marseille Cedex 9, France

L. Casta, Laboratoire Géologie du Quaternaire, U.A. 692 du CNRS Luminy 907, 13288 Marseille Cedex 9, France

Donald J. Colquhoun, Department of Geology, University of South Carolina, Columbia, SC 29208

K. M. Creer, Department of Geophysics, University of Edinburgh, Edinburgh EH9 3JZ, U.K.

Thomas M. Cronin, U.S. Geological Survey, 970 National Center, Reston, VA 22902

Robert Guinn Currie, Laboratory for Planetary Atmospheres Research, State University of New York, Stony Brook, NY 11794

J. L. de Beaulieu, Laboratoire de Botanique Historique et Palynologie, Université d'Aix-Marseille III, 13397 Marseille, France

G. Delibrias, Laboratoire Mixte CNRS-CEA, Centre des Faibles Radioactivités, 91190 Gif-sur-Yvette, France

J. M. Demarest, Exxon Production Research Co., Houston, TX 77001

William L. Donn, Lamont-Doherty Geological Observatory, Columbia University, Palisades, NY 10964

J. C. Duplessy, Centre des Faibles Radioactivités, Laboratoire Mixte CNRS-CEA, Parc du CNRS, F-91190 Gif-sur-Yvette, France

David B. Ericson, Lamont-Doherty Geological Observatory of Columbia University, Palisades, NY 10964

Rhodes W. Fairbridge, Professor Emeritus, Columbia University, New York, NY 10027

Charles W. Finkl, Jnr., Center for Coastal Research, 355 West Rio Road, Box 8068, Charlottesville, VA 22906

Ran Gerson, Institute of Earth Sciences, The Hebrew University of Jerusalem, Jerusalem 91904, Israel

Vivien Gornitz, Lamont-Doherty Geological Observatory, Columbia University, and NASA, Goddard Space Flight Center, Institute for Space Studies, 2880 Broadway, New York, NY 10025
Sari Grossman, Institute of Earth Sciences, The Hebrew University of Jerusalem, Jerusalem 91904, Israel
Fekri A. Hassan, Geoarchaeology Laboratory, Department of Anthropology, Washington State University, Pullman, WA 99164
Claude Hillaire-Marcel, Départment des Sciences de la Terre, Université du Quebec a Montreal, C.P. 8888, Succ. "A," Montreal, P.Q. H3C 3P8, Canada
J. C. Kraft, Department of Geology, University of Delaware, Newark, DE 19716
Leona Marshall Libby (deceased), Environmental Science and Engineering, University of California, Los Angeles, CA 90024
Theodor Landscheidt, Schroeter Institute for Research in Cycles of Solar Activity, Im Dorfe 14, D-2804 Lilienthal-Klostermoor, F.R. Germany
Nils-Axel Mörner, Geological Institute, Stockholm University, S-106 91 Stockholm, Sweden and Gronby Independent Research Center, S-230 23 Anderslov, Sweden
Douglas Alan Paine, Biometeorology Unit, Cornell University, Ithaca, NY 14853
G. Perinet, Laboratoire Géologie du Quaternaire, U.A. 692 du CNRS Luminy 907, 13288 Marseille Cedex 9, France
P. Pestiaux, Institut d'Astronomie et de Geophysique, Université Catholique de Louvain-la-Neuve, Chemin du Cyclotron 2, B-1348 Louvain-la-Neuve, Belgium
A. Pons, Laboratoire de Botanique Historique et Palynologie, Université d'Aix-Marseille III, 13397 Marseille, France
M. Reille, Laboratoire de Botanique Historique et Palynologie, Université d'Aix-Marseille III, 13397 Marseille, France
John E. Sanders, Department of Geology, Barnard College, Columbia University, New York, NY 10027
D. J. Schove (deceased), St. David's College, Beckenham, Kent, U.K.
S. Servant, Laboratoire Géologie du Quaternaire, U.A. 692 du CNRS Luminy 907, 13288 Marseille Cedex 9, France and ORSTOM, Paris, France
G. Smith, Department of Geophysics, University of Edinburgh, Edinburgh EH9 3JZ, U.K.
Laszek Starkel, Polish Academy of Sciences, Institute of Geography, Department of Geomorphology and Hydrology, Krakow, Poland
Barbara R. Stucki, Department of Anthropology, Northwestern University, 2006 Sheridan Road, Evanston, IL 60201
Maurice Taieb, Laboratoire Géologie du Quaternaire, CNRS, Case 907, Université de Luminy, 13288 Marseille Cedex 9, France
Mireille Ters (deceased), Laboratory of Geomorphology, Sorbonne University, 191 rue Sainte Jacques, 75005 Paris, France
Jerome S. Thaler, Hudson Valley Climate Service, Clark Place, Mahopac, NY 10541
N. Thouveny, Laboratoire Géologie du Quaternaire, U.A. 692 du CNRS Luminy 907, 13288 Marseille Cedex 9, France
E. Truze, Laboratoire Géologie du Quaternaire, U.A. 692 du CNRS Luminy 907, 13288 Marseille Cedex 9, France

P. Tucholka, Laboratoire Géologie du Quaternaire, U.A. 692 du CNRS Luminy 907, 13288 Marseille Cedex 9, France and Department of Geophysics, University of Edinburgh, Edinburgh EH9 3JZ, U.K.
C. Vita-Finzi, University College, Gower Street, London WC1E 6BT, U.K.
John F. Wehmiller, Department of Geology, University of Delaware, Newark, DE 19716
Hurd C. Willett, Professor Emeritus, Massachusetts Institute of Technology, Cambridge, MA 02138
Goesta Wollin, 617 Jones Cove Road, Clyde, NC 28721

CLIMATE

Rhodes W. Fairbridge: An Appreciation

Charles W. Finkl, Jnr.

Rhodes Fairbridge has been called many things, depending on the point of view, but most epithets are complimentary, if not endearing. Variously referred to as an "outback Aussie with an Oxford polish" or, simply, "the Professor," the emphasis is almost always on the gentle nature of this academician who pursues the intellectual sport. A leader in the greatest game of all, the pursuit of knowledge, Rhodes brings to his students and colleagues a keen sense of honor and propriety. His ability to inculcate fairness in interprofessional relationships and unbiased interpretation of available evidence marks this man as a great researcher and a born teacher, although he chooses to regard himself as forever a student of the natural sciences. An ability to sift through mountains of data to get to the crux of the matter quickly and efficiently leads Rhodes to conclusions, often the obvious, that others have missed in their researches. One of his former students recently remarked that some newly trumpeted principle or paradigm was nothing new to him because "Fairbridge went over the whole thing with us in his lectures twenty years ago."

His keen powers of observation in the field and his love for the outdoors make Rhodes a natural leader and participant in major expeditions, field excursions, camps, and one- or two-day tours. Those who have been fortunate enough to spend time with Rhodes in the field appreciate his almost uncanny sense of direction and critical evaluation of extant geologic conditions. As Rhodes has often pointed out, the prepared mind knows what to look for and how to interpret the field conditions once they have been identified. These, and other abilities, often enable this geoscientist to draw new conclusions that are somewhat, if not completely, different from those long espoused by the establishment. This somewhat rebellious attitude did not help to make him a crony of that establishment. He was always impatient of red tape and detested what he regarded as the indignity of submitting grant proposals—"crawling on his belly," as he put it—to the pompous peers and petty bureaucrats.

The life of an iconoclast is not easy, but Rhodes deals with ardent supporters of dogma much as he does with family members: with candor and respect, and perhaps with a little paternal nudge toward new directions that he thinks might be the most enlightening and profitable. Life to Rhodes is, above all, fun, and every moment is to a very large degree well spent within this dedication of purpose, with his devotion to science and with a practicum in the classroom of planet Earth.

The ingredients that went into the mix that created this inveterate scholar are not unique, but there must have been a felicitous amalgam. Born in 1914 near Pinjarra, a small town ninety kilometers south of Perth, the capital of Western Australia, at a place known as the Fairbridge Farm School, Rhodes began his formal education at Bedales School in England in 1924. From 1932 to 1936 he studied for a bachelor of arts degree at Queen's University (Kingston, Ontario) and then went for a bachelor of science at Oxford University (Exeter College). His doctor of science degree was granted in 1944 from the University of Western Australia (Nedlands).

From 1943 to 1946 Rhodes served in the Royal Australian Air Force attached to the headquarters of General Douglas MacArthur, and he eventually rose to the position of deputy director of air intelligence for the Southwest Pacific region. There, in Brisbane, Rhodes met, and married, Dolores Collingwood Carrington. This wartime marriage has survived splendidly. They have a son, Kingsley, and now a grandson, Sean. Throughout his long career and multitudinous excursions and travels, Dolores has been at Rhodes's side, a tower of strength and provider of a wonderful home life. For some time she worked as an interior decorator, and her tastes in color and form are impeccable. Not an idle director, Dolores has always done a great deal of the work herself, drawing up the architectural plans, painting and repainting the house inside and out, puttying nail holes, and without her help, Rhodes would never have achieved his tremendous productivity. Her role as the social organizer has always made their nonscientific hours interesting and pleasant and a wonderful experience for visitors.

Immediately after World War II Rhodes returned to the department of geology at his alma mater in Perth, serving as lecturer from 1946 to 1953. There he taught stratigraphy and sedimentology and produced some of his first important papers dealing with coral reefs, sea levels, and paleoclimates.

Acceptance of a one-year (1953-1954) sabbatical appointment to the University of Illinois (Urbana campus) as assistant professor to teach structural and petroleum geology brought Rhodes to a land that he now calls home. The United States is also where he built his professional reputation. In 1954 Rhodes accepted a post at Columbia University where he remains today as professor emeritus (since 1982). In the three decades that Rhodes was at Columbia, his activities proliferated into many fields where he accomplished far more than one might reasonably expect. During this time he participated in endeavors that included teaching summer field courses at the University of Wyoming (1957 and 1958) and serving as visiting professor at the University of Paris (Sorbonne) in 1962 and at the University of Cologne (Köln), Germany, in 1976-1977.

Among some of his more noteworthy practical experiences outside of academic halls were the following appointments and expeditions: Iraq Petroleum Co., Middle East (1938-1941); Caltex (1941), northwest Australia; Hydro-Electric Commission (1947), Tasmania; Richfield Oil Company (1948), Western Australia; Snowy Mountains Hydroelectric Authority (1951), New South Wales; Scripps Institution of Oceanography (1952), "Capricorn Expedition" to the South Pacific; Pure Oil Company, Chicago (1955-1956), World Sedimentary Basin Analysis; Office of Naval Research, Washington, D.C. (1959-1960), geomorphology and sedimentology of Long Island Sound; UNESCO-U.S. National Science Foundation (1961), Nile expedition to Egypt and the Sudan; Indiana Museum (1965), archeology in coastal Brazil; Fairbridge Coral Reef Expedition (1969), New Guinea; French Institut du Pétrole Expedition (1970) to the central Sahara of Algeria; Bank for East Asian Development (1983), study of Bangladesh petroleum potential.

In 1961 Rhodes was made president of the Shorelines Commission of INQUA, the International Union for Quaternary Research, and filled this position for 8 yr. It became his passionate concern for the next two decades. This commission grew enormously under his tutelage, gaining regional subcommissions in different parts of the world and more than a hundred members. They organize field conferences and symposia, often jointly with their kindred organizations: the Neotectonics Commission and the Holocene Commission. Joint meetings were held with the International Geological Correlation Programme (IGCP) 61 and 200 (the shoreline dating and correlation projects). These gatherings educated Rhodes and his colleagues with a variety of coastal phenomena, so that gradually they learned to discriminate the local from the general and how to consolidate impressions into principles. Rhodes also served one 4-yr stint as president of the INQUA Neotectonics Commission.

A listing of some of these INQUA gatherings and his related travels makes an impression: Warsaw, Greece, Turkey, and Egypt (1961); Sudan, Libya, France, Netherlands, and Spain (1962); Florida, Bahamas, Virgin Islands, and Puerto Rico (1964); Portugal, Germany, and England (1964); Denver, Colorado, and New Mexico (1965); Brazil (1966); Trinidad and Barbados (1967); Louisiana and Guatemala (1967); East Germany and Baltic (1967); Czechoslovakia, Hungary, and Yugoslavia (1968); Mexico (1968); Papua, New Guinea, and Australia (1969); Singapore, Bangkok, Hong Kong, and Japan, (1969); Paris, Brittany, and Corsica (1969); Spain and Algeria (1970); England, southwestern Sweden, and West Germany (1971); Nova Scotia, Newfoundland, and Quebec (1972); Tahiti and New Zealand (1973); Sicily and southern Italy (1973); Jamaica (1974); Australia, Victoria, and New South Wales (1975); West Germany, Austria, Hungary, and Sweden (1976/1977); Dakar, Senegal (1976); Birmingham and Wales (1977); Brazil (1978); Tunisia and Italy (1979); Hudson Bay, Quebec (1979); Finland and northern Sweden (1979); Baja California Norte, Mexico (1979), Israel and Sinai (1980); South Carolina and Georgia (1981); Uppsala (Florin Jubilee), Sweden (1980); Gothenburg (Holocene stratotype), Sweden (1981); Bangladesh (1983); Stockholm (Nordic Paleoclimate Symposium); Sweden (1983); Hermosillo, Mexico, and Baja California (1984); Scotland, India, and Western Australia (1984).

The last was a sort of jubilee trip that included visits to sites of his early discoveries of Holocene sea levels at Point Peron, Western Australia. There also, on Rottnest Island, he revisited the old Pleistocene coral reef that Curt Teichert had discovered and where he realized that sea level repeatedly occupied the same levels in the past. To his delight, he found that the State Geological Survey, through the suggestion of one of his old students, Phillip Playford, had named the spot Fairbridge Bluff, and there it was on the map and neatly signposted.

Following some early work in the 1960s Rhodes became deeply interested in the exogenetic signal and the cyclic messages of geology that seemed to call for galactic, planetary, or solar-lunar forcing of climatically related features. A key event was the New York Academy of Sciences Symposium, "Solar Variations, Climate Change and Related Geophysical Problems," for which he was the convener and editor of the final volume (1961). He met H. Lamb, H. C. Willet, G. W. Brier, and D. Schove, and many others who influenced his thinking in that controversial area. For every person who demonstrated a relationship, there was someone else who proved—at least to his or her satisfaction—that it was a fallacy. Rhodes could not offer proofs, but he always said he trusted the rocks, and they said, "Cycles." Over the front door of Schermerhorn Hall at Columbia, a quotation from the Old Testament book of Job is engraved in stone: "Speak to the Earth, and it shall answer thee."

Rhodes received much encouragement from Keith Runcorn, Don Tarling, and others at Newcastle-upon-Tyne. They organized several NATO-sponsored paleoclimate symposia, where he had the chance of meeting Martin Schwarzbach, Lester King, H. Erhart, Herman Flohn, Andre Berger, and other experts. At the 1964 gathering he presented his general theory of Ice Age aridity, which was strongly opposed by the pluvialists—at first. The pre-Quaternary ice ages and climatic changes had begun to receive systematic attention with the advent of plate tectonics, and Newcastle published a two-volume collection on a conference (1973), where Rhodes spoke on glaciation and plate migration.

Rhodes returned his attention to the actualistic aspects of climate when he edited the *Encyclopedia of Atmospheric Sciences and Astrogeology* (1967). He had met an amateur astronomer, Clyde Stacey, at the 1961 symposium, and Clyde prepared a monograph, at his suggestion, for the New York Academy of Sciences (1963), entitled "Cyclical Measures"; further thoughts were included in the 1967 encyclopedia, including a revision of the Titius-Bode law. Stacey was convinced that the barycenter (center of mass) of the solar system, as defined by Isaac Newton, was the key to understanding terrestrial climatic dynamics, and he proposed an astronomic control for the Fairbridge Curve of Holocene sea level.

It is the study of the past 10,000 years of geologic time and the actualistic correlations of paleoclimatic data that has more occupied Rhodes in recent years. This interest has received more than a nudge from certain events and personalities. The work of Claude Hillaire-Marcel on the Hudson Bay 45-yr cycles, the recognition by John Sanders of Stacey's barycentric dynamics (incorporated in Sanders's textbook, *Physical Geology*, Wiley, 1981), the statistical work on the sunspot cycle and climate during the Maunder Minimum made possible by the special skills of Sultan Hameed, the increasing recognition of lunar cycles that followed his friendship with Robert Currie, and the computational skills brought by James Shirley all have helped to direct Rhodes into this essentially new field at the boundaries of geology, climatology, and astronomy.

FAMILY TRADITIONS AND BACKGROUND

From his earliest childhood Rhodes can recall the adventure stories told by his father, Kingsley. These stories were not made up; they were about real things and real people. He also loved the fantasy of Kipling's *Just So Stories* and did not find the trolls too scary in the tales of the Brothers Grimm.

Kingsley's father, Rhys, who was a mining surveyor, noticing that Kingsley, at the age of 10, didn't seem to be enjoying school very much, suggested he come up to central Africa with Rhys, do a bit of gold prospecting, and maybe shoot an elephant. Mother protested, of course, but Rhys was an impetuous romantic—his favorite book was Charles Kingsley's *The Water Babies*—and what boy could resist such an invitation? The trip had a momentous outcome. They never did shoot an elephant, we are glad to note. They were in what is now Zimbabwe, heading toward the Zambesi River, but the local Mashona and Matabele people were busy massacring one another, just as they do today. Kingsley loved the African people, but he realized that some tribes needed to be kept apart. He translated some of their folk stories, which were published later. When Kingsley was 12 years old and was preparing survey markers ahead of Rhys, he found himself in an area that had been totally decimated as a result of the tribal wars; all the animals had been driven away or killed and all the villages and foodstuffs burned. Kingsley and his loyal African help were starving. It took them about a week to get back. On the way, the boy had a vision. "Someday," he mused, "I would like to see this land dotted with farms. I will bring farmers and settle them on the land," a rich and fertile plateau.

Later, a Rhodes Scholarship took him to Oxford. How he managed the entrance exam after having no schooling since the age of 10, and only three months with a crammer, is a miracle. Then he went down with malaria on the first day of the exams and of course failed, but six months later he tried again and passed—including Greek and Latin. The story of those early adventures and days at Oxford is told in a remarkable book, *The Autobiography of Kingsley Fairbridge* (published by Oxford University Press, 1927).

At Oxford Kingsley went out of his way to gain support for his dream. He became a champion boxer because it was a one-man sport; either you made it or you didn't. Without money or position, sport was the only way to gain widespread acceptance. One of the Percys, later the duke of Northumberland, invited him to Alnwick Castle; Gerald Kearley, later Viscount Devonport, became one of his closest friends. Influential people were becoming interested in the colonial boy. He founded the Child Emigration Society, and Gerald paid the first subscription, just in front of Exeter College.

A few years earlier Kingsley had earned some money to take himself on a voyage around the world when he was 17, where he saw the slums and horrors of the poor in the great cities. He became convinced that the only real farmers were those who started young, so he decided to start a farm school and to help underprivileged children at the same time.

Then Kingsley met Ruby Whitmore at Edenbridge, in Kent, the well-protected daughter of a long line of west country farmers and industrialists. When Kingsley proposed, as Ruby reported later, she said she would follow him to the ends of the Earth, little knowing, perhaps, that she would indeed be called on to do just that.

They had decided, reluctantly, that Africa was not really the best place to start a farm school, and a grant of land in Western Australia, a state three times the size of Texas, confirmed the decision. A modest start was made there in 1912, and although the school nearly went under during World War I, by the 1920s it was truly established. Today, those once-underprivileged children—the "Old Fairbridgians"—realize that the old farm was the experience of a lifetime.

Rhodes spent most of his first ten years at the Fairbridge Farm School, learning about the land, about bushcraft in the undeveloped parts of their 3,000 acres, and how to ride a horse. New cottages were always being built, and Kingsley designed and constructed a family home (in a modified South African Dutch style) that is now on the Australian register of beautiful and historical buildings. To Rhodes, the hammer and saw were his favorite things. For his first years life was largely a matter of self-reliance, and everyone learned it on a sink-or-swim basis. He regards himself as lucky not to have worn shoes (other than sandals, when the ground was prickly) and equally fortunate not to have ruined his teeth on ice cream, candy, and "fast foods." The story of the farm school was told by Ruby in a book entitled *Pinjarra* (Oxford University Press, 1937). Other farm schools were later started in various parts of the world, but the Fairbridge Society, as the foundation is now called, has had to adjust to a worldwide trend away from farms, and it now concentrates more on city rescue and gives Fairbridge scholarships.

The Fairbridge family motto is *Discrimine salus* (safety in danger), which tells us something of a long tradition. It is reported that an ancestor, William, got his family name in the eleventh century from William the Conqueror for helping to defend the Norman king in Yorkshire. There is in fact a small river there called the Fair. It was from Middlesborough in Yorkshire that in 1824 Dr. William Fairbridge (with a freshly acquired M.D. from Edinburgh) set forth to the new colony at Cape Town. The first medical doctor in South Africa, he set up a free clinic for the black people there. It seems that dedication and pioneering were no accidents in the family tradition. The farm school at Pinjarra seems to have been a logical successor.

Thus, the Fairbridge Farm School in Pinjarra was home for Rhodes, the place where he grew in peace and tranquility. Rhodes's first trip from the farm school to the seaside was across the coastal sand plain (part of the Swan Coastal Plain) on a great wagon drawn by a team of horses that was borrowed from the nearby McLarty farm. Jock McLarty later became premier of the state. At the Mandurah Inlet Rhodes's father gave him his first swimming lessons—in the Indian Ocean. Rhodes fondly recalls, "I have always thought of that ocean as my own special swimming pool." He later came to appreciate this ocean as the one-time center of the ancient continent of Gondwanaland.

In 1919 Rhodes's father decided that they should all make a trip to England. Money had to be raised to continue the work of the farm school. Rhodes recalls that they had an old copy of the *Times Atlas*, one corner severely chewed by their dog, Jock (after *Jock of the Bushveld* by Sir Percy Fitzpatrick, a favorite for Rhodes's recreational reading in later years). They went to the mother country by ship, stopping in Durban, Cape Town, and the Canary Islands to take on coal. They were to return by way of the Suez Canal, Aden, and India. Rhodes recalls his father showing him the route in the atlas. The Cape peninsula was treated to an enlargement that showed the coast, roads, and small towns, Tracing out copies of all the appropriate maps, Rhodes now realizes that he loved the sensuality of the colors and shapes and that he enjoyed the cartographic adventure in an imagination already well fed by his father's tales of lions, antelopes, and monkeys. It must have been thrilling when they landed at Durban and saw monkeys swinging through the tree branches. At Cape Town their cousin, a newspaper editor, met them, and they set forth on a brief safari to the Cape of Good Hope. Rhodes was later told that at one turn in the road there was a discussion about the correct way to go. Out came his pencilled sketch and in a piping voice, the five-year-old said with total confidence, "To the right." And that was that.

Rhodes read maps as others would read a romance or a detective story. To this day Rhodes derives pleasure from the fantasies conjured up by these documents, as well as from the precise informational data that can lead to quite extraordinarily rich and dynamic interpretations. To Rhodes, multifaceted paleogeographic reconstructions are among the prime challenges, the objectives and rewards of geology. Deriving from this approach are the main trends of his life's work in paleoclimatology and paleoecology.

Rhodes has always thought of the planet Earth as one unit in the expanse of the universe. His farm was there in Western Australia. His father's parents lived over there in Africa. And his mother's family came from Britain, and one branch, the D'Ives, came from France. And there were cousins in Canada, New Zealand, and California, an uncle in India. After marriage he gained another cluster of relatives in Sweden, and after his son married he learned of distant kin in Mexico. Thus, for humankind, Earth is a habitat, and for science it is one system, a very complex integrated system.

EARLY INTEREST IN GEOLOGY

One might well ask, "How did Rhodes become so involved with geology, anyway?" How he entered into this love affair with the scientific study of the Earth, this interest that has sustained his loyalty even since he was quite young, is a long and involved story. It was in part the environment of his upbringing and in part family traditions. The environmental part begins with his birth, May 21, 1914, before the beginning of World War I, two months before the assassination of Archduke Ferdinand at Sarejevo, an act that triggered the eventual downfall of the British Empire and prepared the way for some political developments and gruesome events that debased the course of civilization through his entire career. Happily unaware of the details of the appalling global carnage, Rhodes listened to the chatter around him that made him aware of the strange and horrible drama unfolding in faraway places with strange-sounding names. As a youngster, he remembers the armistice of November 11, 1918, the eleventh hour, to be followed by the infamous Treaty of Versailles of 1919. His home at the time was a small farm in the southwest of the largest of the Australian states, Western Australia. Although he was far from the battles, he knew of neighboring menfolk who had been decimated in the Australian volunteer armies. Among his earliest memories are recollections of the womenfolk, with their cups of tea, talking about the latest casualty. Rhodes's father, Kingsley, was rejected as a volunteer due to a history of malaria and bilharzia (schistosomiasis), among other ailments. Tragically, Kingsley died in 1924, aged only 39.

The family—mother, sisters Barbara and Elizabeth, and younger brother Wolfe—moved to England, where for a while they lived with the maternal grandparents, the Whitmores. These were wonderful people, evidently putting up with the wild colonial children with remarkable equanimity. Those wild colonials were not easily tamed. They detested their English schools, English weather, English food, everything. Occasionally there were breaks in the clouds. Rich uncle Herbert would sometimes invite them to his stately home, and they had to learn the discipline required by a large household of servants and to appreciate the dilemma on going to the circus: "Shall we take the Rolls or the Bentley?"

Then to Rhodes came a stroke of extraordinarily good fortune. A new and experimental school called Bedales, near Petersfield, in Hampshire, which was coeducational and boarding and took children up to age 18, had been started in about 1900. It was a science-art-music-literature establishment, elitist certainly, but it tried to break away from the stultifying classics of traditional segregated schools. For a budding young scientist, this was a critical turning point. They all did chemistry, physics, biology, and geology as a matter of course, as well as music, painting, carpentry, and metal work. School rules were only those passed by the children's council.

Rhodes had learned to love maps and natural history in Pinjarra, but at Bedales this love became disciplined and organized. For the Oxford and Cambridge matriculation examinations they took, he earned straight *A*s, except Latin (languages did not appeal to him). Every year, they did a Shakespearean play and a Gilbert and Sullivan opera. Rhodes says all the fundamental human attributes, foibles, and situations are effectively and unforgettably dealt with by Gilbert, while Sullivan made them hummable. At Bedales they were blessed by a geography-geology-biology teacher named Bobby Browning, who had trained under Hawkes at the University of Reading, and Bobby would take his class out fossil hunting in the Cretaceous of the Weald. Soon Rhodes, on his bicycle, and often with a friend, was exploring and collecting fossils from every quarry within range. Every long vacation they took tents and went camping and geologizing in Dorset, Scotland, and in Wales, with more adventurous bike trips even to France, Belgium, Holland, and Germany. Then he was taking those fossils up to the Geology Museum, and the finest specimens were donated to the national collection. Many years later he found a fossil in the Cretaceous Gingin Chalk of Western Australia that to his great delight was called *Chlamys fairbridgei* (Feldtmann, 1963).

After Bedales, a university was in order, but scholarships were hard to come by in those days. Suddenly came an invitation from Canada: to Queen's University in Kingston, Ontario. It was definitely a hard-rock, Precambrian school, and Rhodes was much more concerned with sediments and fossils, but on arrival, he sat for the first two years' exams and promptly started in the third year of geology, and to round things off he took a double major, doubling up with biology, a decision for which he has had reason to be grateful for his entire life. The Canadian training was sound, and it opened some interesting windows.

For the first summer vacation, Rhodes and a friend, John Stephenson, bought bicycles and bedrolls and made a three-month excursion through the eastern United States. They explored the Appalachians as far south as West Virginia, where they encountered the friendly mountain people, listened to their music (violins played on the knee like a cello), and enjoyed their hospitality (beans for dinner—just beans; when you are hungry they are delicious). Then they headed for the Shenandoah Valley with its glorious spring colors and out to the coast to see Yorktown for history and the cliffs for fossil collecting—a profusion of Miocene, almost unknown in Great Britain. Next out to Kitty Hawk and Cape Hatteras—more history, for the airplane is really the symbol of the twentieth century. They encountered black people for the first time in North Carolina and were hospitably entertained with stories, laughter, and warm Coke. They turned back through Richmond (tobacco smell), Washington, D.C. (Greek temples), and New York—over the just-opened George Washington Bridge (wow). And then on again, as far as Bridgeport, Connecticut.

So far they had been extraordinarily lucky with the weather, but at Bridgeport it looked as if the monsoon had come in. Taking shelter at a gas station, the proprietor questioned their sanity, as many others had done, but they patiently explained that in spite of the Depression, they both had saved a little pocket money and bicycles did not need gas. The proprietor pointed out, rather obviously, that bicycles provided little shelter from the rain and that, as the two were headed north, a small investment in an internal combustion vehicle would be richly rewarded. Rhodes protested that this was (1) outside of their financial bracket, (2) he couldn't drive, and (3) he had no license. The year was 1933, the time of the Great Depression and Prohibition (so the only money-makers were at speakeasies). The salesman was determined. He was a friend of the police chief and could get them a temporary license without a test for $1; he would provide free driving lessons; and this incredibly beautiful Model-T Ford (1924) coupe could be theirs for $7.50. Rhodes is not easily persuaded, but he saw the logic of the argument. The rain beat down in an unending stream. The sale was made.

Off they went. Driving was a gorgeous feeling. The rains stopped, and they travelled through Providence, Boston, and eventually Calais, Maine, and to the Canadian marine lab on

Passamaquoddy Bay, where Stephenson's father was a marine biologist. There they slept in real beds for the first time in many weeks. The 45-foot tide on the Bay of Fundy was really something. Next they went north through New Brunswick and into Gaspe, where they had to dredge up some schoolboy French. The roads were not paved and they had lots of exercise fixing tires—perhaps 20 punctures on this sector. The Model-T bands wore out and they also found out how to fix them. So, down through Quebec and Montreal, and back to Kingston.

A car revolutionized the shy undergraduate's life. Social life began to open up, although the Model-T had certain shortcomings—for example, no side curtains—which meant that the snow blew in one side and out the other. It was a good test of whether or not the girlfriend of the evening was serious. A particularly good friend was Raymond Sternberg, son of the distinguished dinosaur man, Charlie Sternberg. Rhodes and Ray decided, after Christmas, to drive across the frozen Lake Ontario and go to the Geological Society of America convention at Rochester, New York. They arrived, half-frozen, to find there were no rooms, but some of the conventioneers were willing to share. Rhodes got Maxim K. Elias, the famous paleoecologist, a lucky draw. He was introduced to Romer, and to Schuchert and Dunbar, his textbook authors; they were *real* people! Geology suddenly became quite different. It was not all fossils and rocks but also full of warm exciting personalities.

The next summer (1934) Rhodes was ready for a job. Charlie Sternberg invited him to work on the dinosaurs at the National Museum in Ottawa. As a warm-up Charlie gave him some sacks full of Cretaceous turtle plates (many individuals). They all had to be reconstructed, and no cheating. One month later, Rhodes got his first hooded hadrosaur.

One day, on a steep hill in Quebec, the Model-T slid off the gravelly back road and into a tree—alas! He sold the wreck and hitchhiked to Montreal to pick up a 1926 Ford Model-A. It also gave good service until it was time to graduate.

Not all was roses in the academic line. Rhodes had a golden rule. Never open a textbook after 6 P.M. He failed a couple of courses, chemistry (of all things) and German. Summer school and makeups. For fun, he made a quick trip to England as a deck hand on a cattle boat, returning with a shipload of tinplate from Cardiff. On the dock a young lady of (mis)fortune offered personal services for a shilling (equivalent of 25 cents at that time). Rhodes, a shy boy, declined. She lowered the price to 10 pennies. Still it was no, but it also brought home to him in an unforgettable way that in the Great Depression some folks got pretty desperate.

Eventually came graduation, B.A. The trouble, however, with one degree is that one yearns for another. Someone suggested that Rhodes go see the chairman of the Rhodes Trustees at Oxford. His father had been a distinguished scholar, but Rhodes, having been born in Australia, and since he was in college in Canada, was not eligible. He saw the great man and was taken to lunch. He didn't spill his sherry or eat peas with his knife. "You'll be going to Exeter College, of course, won't you?" Nothing further to discuss.

Oxford was pure joy. He was sitting next to a young man during the admission procedures, and to break the ice, Rhodes remarked, "My father was here thirty years ago." The young man replied, "So was mine, and earlier on, those chaps up there," pointing to the paintings of his grandfather and great grandfather. Touché. A new friend was Michael Pitt-Rivers. His father had founded the Pitt-Rivers Museum (of archeology). Another friend was Christopher Cadogan, who was later killed in a commando raid on Rommel's headquarters in North Africa. After the war Rhodes tried to contact quite a lot of his old buddies from Bedales and Oxford, but alas, many of them had perished. One who survived was Harold Cox. A great train buff, he and Rhodes used to watch trains together, collecting locomotive numbers and photographing them. Later in life on field trips Rhodes rediscovered the wonderful old steam locomotives of Poland, Czechoslovakia, and Hungary and trains like the Chittagong Express in Bangladesh; the Chichuahua al Pacifico in Mexico, which used raw beef fat for bearings; and of course, the Sudan Railways' Wadi Halfa Express, which has toilet stops in the Sahara Desert.

At Oxford Rhodes's supervisor was W. J. Arkell, the ammonite specialist. Arkell explained that getting permissions to visit quarries on field trips in England was no problem. Driving up in his Rolls Royce demonstrated for all that Arkell was a V.I.P. and that deferential cap-touching was assured. Arkell knew every Jurassic outcrop in the land, down to the inch. For a thesis topic he pushed Rhodes to look abroad. "England," he said, "is all used up." By chance, Rhodes heard of an available summer job giving English conversation lessons to a young Hungarian, Baron Janos Radvanszky, who had a castle in the Carpathian Mountains—according to Dolores,"deep in Dracula country." The big spike over the castle gateway was where the head of an ancestral Radvanszky had been placed after a less successful tussle with neighbors a few centuries back.

The Radvanszkys were one of the royal families of Hungary. In the old days, kings were usually elected by vote, a selection from the best of the landed princes. The experience of suddenly coming to live in a largely feudal society was extraordinarily interesting, and the old Baroness Wanda taught Rhodes the history of the Holy Roman Empire, the Magyars, the Habsburgs, and the Turkish wars. The butler, Djuri, had once been a footman in the court of Archduke Ferdinand and spoke a beautiful courtly German, which left its mark on Rhodes's rather sparse knowledge of the local languages. He made some effort at Hungarian, but Slovak he felt was hopeless. For summer relaxation, the Radvanszkys liked to go for extended hikes in the mountains. Here was the solution for Rhodes's thesis topic.

A particularly fascinating feature of the Carpathians is the Klippe Zone, a narrow belt of Mesozoic limestone blocks set in a matrix of Late Cretaceous to Early Tertiary flysch. The latter was a soft sandy or shaley formation, commonly cloaked in forest or cultivated as green fields. Standing up dramatically every few miles was a giant crag, "like the dragons' teeth of the Argonauts," according to Rhodes. The blocks were all sizes, from a conglomerate or breccia up to tectonic units as much as perhaps 10 kilometers in length. He read avidly (most of the literature was in German or French) to discover that these exotic blocks were found also in the Alps and had been traced to Rumania, Turkey, and even to the Himalayas. Their emplacement became his special problem. The Klippe Zone was complicated by nappes and thrusting, but the blocks, Rhodes was convinced, were emplaced in a sedimentary way, by the synchronous action of the brow of the nappes, breaking off as they encountered the seafloor. This, of course, we can now recognize as a subduction zone, but in those days the term had not been invented. An Austrian, however, in the State Geological Survey, Otto Ampferer, had proposed the word *Verschluckungzone* (sliding-under zone) back around 1906, which really means 'subduction,' but it was very difficult to pronounce and few Western geologists read central European survey reports written in German. Rhodes was rewarded with a B.Sc. (at Oxford, a research degree). He had hoped for a D.Phil., but that would have meant another year or two, and funds were running out.

The Radvanszkys did not disappear from Rhodes's life. After World War II, he learned that they had used the castle as a staging base for escaping prisoners, downed airmen, and Jewish refugees, while the Nazis used the upstairs rooms as an officers' mess. Leftovers went down to the cellars, via a secret passage. But the eventual "liberators" were, alas, from the USSR, and Janos was soon on the hit list. Rhodes and Dolores organized a rescue, and soon the family was on its way to Australia where they gradually settled in. The migrant shock was mitigated by many good friends, and today Janos is a professor of philosophy—what else?

MIDDLE EAST AND SOUTHWEST PACIFIC

After two years at Oxford, funds were exhausted and Rhodes had to try to become an "honest man." He went, by invitation, to the Hague and spent a week looking at the Shell labs, where he met a senior medical man. Rhodes asked, "What happens to geologists who fall sick out in the jungles of Borneo?" The old Dutchman laughed and replied. "Nothing to worry about, my boy, we *always* send in an expedition to bring them out." That was a gentle hint, and Rhodes declined the job. A few years later in the war, Rhodes's unit rescued a Dutch geologist on the coast of Timor; he was not in great shape. A lot of the others died.

Rhodes returned to London and talked with G. M. Lees of the Anglo-Iranian Oil Co. and Norval Baker of the Iraq Petroleum Co. (IPC). Both offered exciting geology in arid terrain and 10% less salary than Shell. IPC was working in the classic lands of the Middle East, and that clinched it. In fall 1938, Rhodes arrived in Palestine using the Italian flying boat service that went by way of Trieste-Bari-Athens-Rhodes-Haifa. The IPC oil terminal was at Haifa, and after picking up a colleague, some equipment, a station wagon, and an Arab driver, Rhodes was on his way up the coast to Beyrouth, then under French mandate. Crossing the Lebanon mountains in a snow storm, the driver nearly killed them by skidding off a cliff, but eventually they arrived in Damascus. Now, this was the real Arab world, history and archeology back as far as you want. A friendly British consul showed Rhodes how in those seething crowds, every person was different. "Look at their sandals," he said, "the style of a Palmyra sandal is quite different from a Hedjaz or Baghdad sandal." Faces, headdresses, clothes—everything was distinctive. There were Armenians, Iraqi Jews, Copts, Maronites, Assyrians, and peoples you would think were extinct. But there they were, in Damascus.

With a new station wagon, Rhodes headed out across the desert to Palmyra. "If you can't find the hotel or any place to sleep, try the French Foreign Legion fort," the British Consul said. There was no road, just a lot of old tracks, and there had been rain. So there was some digging. It was pitch dark in Palmyra, but the Greco-Roman ruins, half buried in drifting sand, were silhouetted in the starlight. Here was the corner of a roof projecting from a sand dune and an old sign: Zenobia Hotel. "Aha! so that's it." A little later the legion fort was discovered and a place found to spread bed rolls.

The IPC had a concession for Syria, and Rhodes was given some anticlines along the Turkish border to map. At Deir-ez-Zor on the Euphrates, he picked up an experienced Swiss geologist, Heli Badoux, and tents, equipment, and a team of Arab workers. They set forth to the Djebel Abd-el-Aziz, made camp, and went out to start mapping. That was the tricky part. What the IPC had never bothered to mention was that this part of the world had never been mapped, by anyone. "We had to start with a clean sheet of paper," said Rhodes. To make matters worse, none of those fine university courses had ever mentioned the existence of the plane table or the alidade. Badoux asked, "When they interviewed you, did you lie and say you knew *all* about plane tabling?" Rhodes had to confess. Badoux chuckled, "Last year, I did too." But in one year one can learn a lot.

At the Djebel Abd-el-Aziz Rhodes learned not only the mysteries of the plane table but also how to entertain visiting bedouin chiefs, how to hunt for wild turkey, and one wild night in the snow, how to fight off an attack by a pack of hungry wolves that had come down from the Anatolian Mountains that lay to the north. One day, at the urging of their cook, he shot a gazelle, but its death left him without an appetite and never again did he shoot one. He had been on the rifle team at Bedales and in college and was a dead shot, but the idea of using that skill for sport was anathema.

The baselines for the mapping were finally laid out, and the anticline began to take shape. It had a Cretaceous core, exposed in some deep wadis, with a rim of early Tertiary limestones. One day, in among the Cretaceous chalk beds, he found trilobites and Paleozoic brachiopods. "Mon Dieu," said Badoux. Rhodes suddenly remembered the klippes of the Carpathians. Badoux recalled the Alpes Swisses Romandes and said that people would laugh at them. One doesn't find klippes out on the Arabian craton. Telegrams were sent. The regional chief came, the senior French geologist came, a man from the British Museum came, and the big boss from New York came. It was certainly extraordinary. The group came to the conclusion that exotic blocks, up to 500 m long, had slipped off an active submarine fault scarp. Thirty years later, Rhodes was listening to a student lecture at Lamont about the seismic evidence for the Anatolian microplate. Its southern boundary, and northern edge of the Arabian plate, runs E-W, through the Djebel Abd-el-Aziz.

Syria was more than an education. It was a total experience—people, history, geology. For the summer of 1939, Rhodes was shipped off to Cyprus to map the Kyrenia Range. Not only is Cyprus a gorgeous place to live in, but also the geology (on a plate margin) is infinitely challenging. But then one day at lunch time in the hotel in Nicosia, the radio brought the awful news. Hitler had attacked. History was repeating itself. The IPC ordered Rhodes back to Syria. But how? All communications, except radio, were cut. Eventually a Rumanian tramp steamer arrived, packed with Poles and Jews from Warsaw. There was absolutely no more room. The purser, however, on seeing a gorgeous white crinkly English £5 note, decided that one more passenger would surely not sink the ship; in fact, his own cabin could be shared.

That fall, Rhodes started mapping again in Syria, but the legion people were not so friendly, and gradually a horrible truth began to dawn. These were Vichy-type (pro-Nazi) Frenchmen for the most part. The British consul sent a message: "Run for it." Without ceremony, they did, "retiring to prepared positions," as military communiqués put it, which in this case meant Haifa. The IPC had concessions in Palestine and Jordan, so with a new associate, Eric Tiratsoo, they were soon off to the Negev beyond the borders of the Sinai Desert. At first they set up a base in Beersheba. The company said they'd find a police post there with telephone connections in case they needed anything. Bumping along the track from Gaza, Eric remarked that the telephone lines looked rather odd, hanging down on each side of every pole. They found the police station all right, but inside, the walls were blackened and there was blue sky overhead. Rhodes was comforted by having decided to purchase certain defensive hardware in Haifa. In Syria they had a platoon of the Foreign Legion as morale builders against bandits or whatever, but the Palestine authorities could not spare official bodyguards, saying the oil people could hire their own personnel. Eric came up with a brilliant scheme. He organized a feast: greasy mutton, mountains of rice, sugary drinks, and gold-tipped Balkan Sobranie cigarettes. All the Negev sheiks were invited. Eric made a speech (in Arabic). The oil people would, among other things, find places to drill for water, which was something they all wanted, and furthermore, there would be a job for the eldest (and bravest) son of each family as a bodyguard. (Previous exploring groups to the Negev had been used as target practice; some had been slaughtered to the last man.) One small detail: no money would be taken to the desert. At the end of the project, everyone would be paid *in gold* at Barclay's Bank in Jerusalem, the check being jointly signed by Eric and Rhodes, both to be present: both intact and in good health.

Thus, Eric and Rhodes had a fine survey in the Negev. Just to make sure that the locals knew they were armed, they would sometimes put up an old four-gallon kerosene tin on a hilltop and then pump it full of lead. No bandits ever cared to challenge them. Seeing is believing. On the modern Israeli geological maps, Rhodes has been told, they still use their old stratigraphic names. In the comfort of a modern airplane, Rhodes and Dolores flew down over the same terrain in 1980, but in 1940 it was rough going. The dust was choking. Rhodes eventually went down with pneumonia and had to be nursed back to health in the Gaza missionary hospital. There were complications, however: he developed permanent lung damage (bronchiectasis) that has dogged him since then, making him excessively susceptible to pulmonary infections. His sister, Elizabeth, by coincidence, at this precise time, was traveling by ship from Great Britain to Australia and, on hearing of his predicament, jumped ship at Suez and came up by train to Gaza. She nobly helped nurse him back to health.

Meanwhile, World War II was heating up in

the Middle East. The Nazis landed paratroops at Kirkuk, the principal IPC oilfields. Syria and Lebanon had gone over to the Vichy people. The Australians landed in Suez and, in a short but violent campaign, won back both Syria and Lebanon, so the Germans had to give up their strategic position on the oil wells. One day, Rhodes was sipping a noonday beer with the British intelligence chief on the front terrace of the Gat Rimon Hotel on the slope of Mt. Carmel. An unfamiliar rumbling sound came across the sea. "Sounds like aircraft," said the intelligence man. "Impossible," said Rhodes, "we don't have that many." It was the first joint Luftwaffe-Italian daylight raid. The exploding oil tanks made quite a spectacle.

Rhodes thought he ought to do something. Sister Elizabeth was immediately hired and went to Jerusalem and later to Teheran, where she had "something to do" with the famous Churchill-Roosevelt-Stalin meeting and the kidnapping of General Zahedi, who was planning a little monkey business to sabotage that gathering. Churchill's son Randolph came to Rhodes with a marvelous (but crazy) plan to go into Ethiopia, then occupied by the Italians. Rhodes signed off with the IPC and took the train to Cairo where he talked with Dr. (Col.) R. A. Bagnold, the sand dune expert, then with the Long-Range Desert Group. Rhodes's cousin Roger Thomas was also with them. He also spent a couple weeks in Alexandria, at the Royal Air Force HQ, where the Duke of Northumberland was the group captain. The Germans were bombing them nightly at this stage.

The powers-that-be suggested that Rhodes would be of more use to them if he went to a formal officer's school and into intelligence. Thus, in July 1940 he was in Suez, taking ship to Australia. It was one of the giant Cunarders, painted gray and given a number. To Rhodes's astonishment it was almost entirely filled with Italians, who had been captured in the western desert and were now on their way to Australia, where they were eventually to be paroled. They did invaluable work on the farms.

Back in Australia, Rhodes enlisted in the Royal Australian Air Force but was told to twiddle his thumbs for a while until there was a place in the training school. He filled in by registering for a D.Sc. at the University of Western Australia, where this advanced degree can be taken by thesis alone. Rhodes had been formulating an idea since his Abd-el-Aziz experience, reading and writing numerous chapters, whenever the opportunity came. Professor Leo Picard at the University of Jerusalem had played an important role in urging him on. And now he had a thesis: "Subaqueous sliding and slumped blocks" (parts of it appeared in 1946 in the *AAPG Bulletin* and in 1947 in the *American Journal of Science*). He was already in uniform by the time of the degree ceremony.

Rhodes had every intention of returning to the Middle East once he had the necessary training, but history decreed otherwise. In December 1941 there was Pearl Harbor, and after the Philippine collapse, General MacArthur escaped to Australia. Obviously well trained and experienced, he was made supreme commander, Southwest Pacific, over all allied forces. That included Rhodes and his air force colleagues. MacArthur arrived from the Philippines with no staff. Rhodes met him in Brisbane and became one of his intelligence people since he knew quite a bit about maps, photointerpretation, and the Middle East—but absolutely nothing about the Japanese or the Pacific. Ah well! We will return to some of his Pacific adventures in the section "Coral Reefs."

CONTINENTAL DRIFT AND ANCIENT CLIMATES

Rhodes's father studied forestry at Oxford, along with botany, geology, and various uplifting things such as how to behave at formal dinners. He was always showing his children little things related to bushcraft. He taught them to love the creatures of the wild; to delight in kangaroos, goannas, and kookaburras; and to take care of snakes (some in Western Australia, such as the dugite, had to be treated with great respect). But it seems that a visiting professor of zoology from the University of Western Australia showed Rhodes that a person could become a natural scientist. Professor Nicholls showed Rhodes the difference between a centipede and a millipede. Under every piece of bark there was another. In the small creek on the farm there was a menagerie of little creatures.

The professor also told Rhodes the "fairy tale" of Alfred Wegener and about his theory of continental drift. Nicholls had been studying fresh-water cave shrimp and found closely related species all across the Southern Hemisphere. The cave shrimp could not swim over the ocean, but by continental drift they could have been passengers on those floating life rafts. Rhodes says that Professor Nicholls may well have been wrong with his examples, but the seed was sown. Ten years later, in 1928, Rhodes read J. G. A. Skerl's translation of Wegener, and that seemed to make sense. Fortunately, Rhodes did not come into contact with Sir Harold Jeffreys of Cambridge until several decades later! According to Rhodes, Sir Harold was a lovable person but passionately dedicated to what he perceived as the established truth of geophysics and the immutable laws of physics. Little did he know that later Rhodes would promulgate an antithetical philosophy, particularly devised in the light of experiences in the earth sciences. Rhodes clearly perceived that in open systems like those of planet Earth and the solar system, all internal systems are inevitably complex and open-ended. Moreover, he knew that straightforward solutions that are both elegant and unique are also false.

In the 1940s, the distinguished economic geologist Blanchard once told Rhodes that his thoughts about continental drift should be kept to himself else he would never see a full professorship with tenure. At that time Rhodes had not made any discoveries that justified expressing these opinions, so the moral dilemma never arose. During and after World War II, Rhodes was back in Western Australia. Following demobilization after the war effort, he was appointed to the lectureship previously held by his old friend and mentor Curt Teichert, one of the world's most distinguished paleontologists. A refugee, born in Königsburg in what was then East Prussia, Teichert had almost single-handedly revolutionized the stratigraphic geology of Western Australia, discovering an almost continuous marine sequence along the coast from Ordovician to the present. Earlier, Teichert had adventured with Wegener on the Greenland ice cap. Wegener had tragically died, but Teichert had not been with him on that last and fatal traverse. In Western Australia Teichert was in a marvelous situation to check out one of Wegener's key reconstructions. The east coast of India had been placed side by side with that of Western Australia. This should be critical. But, alas, India's east coast contained none of the richly fossiliferous Western Australian Paleozoic formations, and the rest were indecisive. Wegener, their hero, seemed to have feet of clay! As far as Teichert was concerned, Wegener's hypothesis had been tested and found wanting. Thanks to information gleaned from field checks in the light of plate tectonic theory, we now know that eastern India was formerly situated alongside East Antarctica. Wegener's model has been adjusted accordingly. Rhodes is quick to point out, however, that one must not throw out the baby with the bath water. Wegener's basic drift concept is sound, but like some of Darwin's arguments about biological evolution, some bits and pieces were misplaced.

For Rhodes the moment of truth came during a field trip when he took some students to the Irwin River for their annual field course. They came across a magnificent Permian tillite, with a striated pavement and fossil proof of its age. So the inevitable question was put: "What do *you* think of Wegener's hypothesis?" Well, here they had certain glacial evidence at about 30° S latitude, and there was no way to explain it climatologically. After the candor of the moment, Rhodes wisely passed on Blanchard's advice. On another day, Joe Lord of the state geological survey had done some core drilling at Collie (a small coal-mining town south of Perth). They drilled through the Permian coal measures into a clayey, bluish-colored angular conglomerate. Joe wanted to know what Rhodes thought of this material. The underlying contact was clean as a whistle, on polished, fresh Precambrian granite-gneiss. They could not prevaricate: it had to be a Permian tillite. Rhodes reported it as such to the Gondwana Commission meeting at the next International Geological

Congress (Algiers, 1952). The Collie structure lay within the craton, previously reported simply as a graben, but it had to be a deep glacial scour. Several other related features were subsequently discovered.

Then, on another field trip to the Irwin River, Rhodes and his students discovered some yellow sandstones, later called the Yarragadee Formation. The beds were packed with gorgeous Mesozoic plant fossils, well preserved by silcrete. Samples went to the Sydney Museum showed that they were not familiar Australian flora. So Rhodes packed up another parcel of samples and sent them to Mrs. Jacob, a paleobotanist with the Geological Survey of India in Calcutta. The reply came back in a crumpled blue airletter. The fossils were all, genus and species, identical to those of the east coast Gondwanas of India. So, after all, Wegener was right.

Many years later (1979) Rhodes returned to this fascinating problem when he was asked to contribute to a *Festschrift* for a Dutch geologist, R. W. van Bemmelen, who had been a great inspiration to Rhodes, partly from his work on nickel deposits in the East Indies. At that time Rhodes's colleagues at Lamont and elsewhere had developed the idea that when continents rift apart, they do so along a series of hot-spot upbulges, which would split across and initiate drifting. The Red Sea, which split in the early Miocene, was supposed to represent the archetype. Hans Cloos, in Bonn, had sketched a persuasively clear three-dimensional model of it. But, alas! The historical record is a little different. Rhodes had fortunately studied the Red Sea paleogeography while with the IPC. At that time he deduced the following sequence of events. First, the site of the Red Sea had been a line of weakness back in the Late Precambrian-Early Paleozoic times. In the Carboniferous it developed as a marine graben, but open from the north end. Along the Jordan-Dead Sea strike-slip line there were precursory motions during the Late Cretaceous. Downwarping began in the Early Cenozoic, moving in from the southeast. A shallow-water marine embayment extended from the region to Somalia. Crustal thinning and subsidence preceded the mid-Tertiary rifting, not upwarping. Crustal stretching, in this, seems to be in advance of the uplift of the two marginal horsts.

To Rhodes this all had significance for the Western Australian basins. Teichert had found a marine embayment here, operative since the Ordovician. They both had discovered evidence of granitic clastics derived from the west, where the Indian Ocean is now located. In any case, there are slivers of a Precambrian granite-gneiss on the western side of the Leeuwin Block and north of Geraldton. It occurred to Rhodes that the western basins began as a Red Sea-like graben, evolving gradually by intense sediment loading through the entire Paleozoic and early Mesozoic. This was not an open geosyncline as visualized by Teichert but a rift basin. This kind of feature is what Rhodes's good friend Marshall Kay wanted (for excellent reasons) to call a "taphrogeosyncline." By analogy with *orogeny* and *epeirogeny, taphrogeny* was the term for block faulting. Unfortunately, Kay's contemporaries did not care for high-falluting Greek-based terminology.

In the van Bemmelen *Festschrift,* Rhodes developed his Red Sea model and pointed out that the breakup of Gondwanaland had been a progressive affair. The rifting went all the way back to the Precambrian, and it was only in the Jurassic that Australia's western borders drifted away to become eventually buried somewhere beneath Tibet. Rhodes further developed the idea in another *Festschrift,* this one for his dear friend Bruce Heezen, one of his former students. Rhodes vividly recalls Bruce's Ph.D. defense. Bruce had put it off for years, always being away on another expedition. But the axe had to fall. Now he was being examined by a blue-ribbon committee. The trouble was, as Rhodes recalls, Heezen knew far more about the subject than any of his examiners, who included Ewing, Kay, and others, none of them lightweights. Someone asked Bruce, innocently, if he could give references for some assertion. He looked rather sheepishly at his feet and mumbled, "I wrote them all."

One of the turning points in Rhodes's professional career began with a friendship that had been started with Armin K. Lobeck, who taught the Columbia College introductory geology course and who was famous for his brilliantly drawn field sketches (see his *Geomorphology*, McGraw-Hill, 1939) and his physiographic maps of the continents. The technique was later extended to the ocean basins by Bruce Heezen and Marie Tharp. Lobeck had visited Australia around 1950 and had commissioned Rhodes to help him do a physiographic map of Australia. Rhodes blithely agreed, little realizing what a herculean task it was to devise geomorphological units and map boundaries for a continent that had never been so treated before. During World War II Rhodes had watched Professor E. Sherbon Hills prepare a 1:1,000,000 scale physiographic model of the entire country (with the aid of a grant from the Army map service). The Lobeck map, however, was to be classificatory as well as visual. Rhodes consulted with his distinguished climatological colleague in Western Australia, Joe Gentilli. Joe helped Rhodes begin to appreciate the role of climate in geomorphology. Until then, Rhodes says that he had never heard it mentioned, but eventually *climatogeomorphology* was to become a full-fledged discipline.

Some years before, Sir Charles Cotton from New Zealand, who had been responsible for Rhodes's first readings in geomorphology, also wrote a book called *Climatic Accidents* (Whitcombe and Tombs, Auckland, 1947)—notably, deserts and glaciers, which reflected the attitude of William Morris Davis (Harvard). But in the 1960s Rhodes's French friends Jean Tricart and Andre Cailleux had outlined the principles of what was to be a new science in stencilled readings for their students in Strasbourg and Paris. Finally, in Germany, Rhodes's late colleague Julius Büdel (Würzburg) produced a glossy textbook *Klima-Geomorphologie* (1977), which soon ran into a second edition and an English translation.

GEOMORPHOLOGY AND THE CRATONIC REGIME

This concern with geomorphology that originated with the Australian physiographic map gradually developed, Rhodes recalls, into a lifelong fascination, especially after he went to Columbia. Arthur Strahler had been in charge of the geomorphology program there, but in the turbulent 1960s he took an early retirement and Rhodes inherited his courses. The *Encyclopedia of Geomorphology* (1968) was a notable upshot. Running to 1,295 pages, it nearly ruined his publisher, but it is widely regarded as the major publication in this field.

His interest in the general geomorphology of Australia and its paleoclimatic history was reawakened by another happy chance. On the trail of finding encyclopedia writers in the 1960s, Rhodes had written to Joe Gentilli asking his advice on a good soil man. He was guided to a young American who had taken his Ph.D. in soil science at the University of Western Australia, Charles Finkl, the author of this appreciation. From this meeting a long friendship began. One day, I asked Rhodes to look over my thesis, which dealt with paleosols on the southwestern edge of the Australian craton. I had discovered a series of ancient soil layers and alluvial deposits that contained gravels and boulders that became progressively coarser as one went back in time, until the oldest was reached and found to contain what looked like glacial erratics. The last glaciation on the Western Australian shield had been early Permian, at least 280 m.y. ago. Surely it was inconceivable that a fluvioglacial bedload should be preserved from erosion over such an immense time? But then we found that the paleovalleys, descending to about sea level or lower near the edge of the shield, ended in U-shaped gorges, albeit somewhat degraded. Henno Martin, the German explorer, then reported that these so-called fossil fjords also occurred in the Permian of South-West Africa (Namibia). Back in Western Australia, above the erratic-bearing alluvium, there were gravels and sands that could be traced, partly on the shoulders of the old glacial valleys, reaching seaward, to pass into fossiliferous Mesozoic formations. Still higher there were the alluvial deep leads (containing placer tin in places) that pass seaward into fossiliferous Cenozoic.

So it was that, with a nudge from Rhodes, I had stumbled onto an epoch-making discovery. The mean rate of downward post-Permian erosion on the Western Australian shield was calculated at around 10 cm/m.y. This value is some orders of magnitude less than the one that had hitherto been assumed for the continents. The explanation, as seen by Rhodes, seems to lie in the all-too-easy presumption that one could just take the average annual sediment discharge of the Mississippi River as an actualistic model. Rhodes emphasizes that researchers must be careful to consider whether the twentieth century is a fair measure. What of all the busy farmers and their overactive plows? And what about the glacial and periglacial "preparation" of the North American craton, cloaking it with loose till, loess, and colluvium? As Rhodes says, "Now we know!" the Holocene is extraordinarily unusual as a geological epoch. The surface of North America is also highly unusual as a standard continent. And the anthropogenic attack has vastly amplified the progressive erosion on an already unusually fragile landscape. Putting all reservations aside, Rhodes says that he considers this to be the most important discovery in geomorphology in the twentieth century.

When Rhodes arrived at the University of Western Australia in January 1946, fresh out of the air force, there was still food rationing and no civilian clothes in the shops. So, for a year or so he just wore ragged bits of uniform (insignia removed, of course). Rhodes then started to explore the neighborhood of Perth with his students, many of whom were his age and also recently demobilized. He jests that they were really all almost fellow students as he was often keeping just one chapter ahead of them, sometimes not even that.

Anyone looking at world structural maps when they came to Australia must have been struck by the rectilinear Darling Fault that cuts off the western edge of the Precambrian shield with a north-south scarp for 1,000 km or more. Near Perth the scarp has a relief of 300 m or so, and it looks like a classic fault-line scarp. Once when Rhodes remarked as such, the professor, who was a classic in the old-style Scottish tradition of cautious skepticism, asked, "Do you really think so?" To Rhodes and others the professor, the only one at the top of the departmental pyramid, was a most lovable New Zealander, E. de Courcy Clarke, known as Corky. In those days there was not much of a departmental hierarchy. Corky handled the big first-year classes (there were hundreds of students) and the administration. Rex Prider, a couple years older than Rhodes, and with a Cambridge degree, taught mineralogy and petrology, having a fundamental affection for hard rocks. Rhodes remembers being left with all the soft rocks: sedimentology, paleontology, stratigraphy, structure, geomorphology, Quaternary geology, and petroleum geology—a little bit of everything. He gave the first lectures on paleoecology to a mixed audience of biologists and geologists.

We must, however, return to Rhodes's story about the Darling Scarp. He continues by saying that Corky insisted that the scarp was really just an old sea coast (rectilinear? Precambrian cliffs?). Rex Prider had found subovate quartzite boulders up to 20 cm across along the foot of the scarp along with secondary, colluvial laterite. According to a principle established by the late André Cailleux, Rhodes thinks that marine, wave-rounded boulders tend to be subovate pillows or shingle in contrast to fluvial boulders that tend to become spherical. Rhodes agreed that clearly this was an old sea coast. He investigated the logs and samples from water borings along the foot of the scarp and on the coastal plain. They showed only sedimentary formations, with fossil evidence of ages back to the early Cretaceous. The geophysical evidence, gravity, was first surveyed by the Dutch pioneer, F. A. Vening Meinesz, when he stopped at Fremantle on one of his heroic expeditions that took the idea of gravity at sea level all over the world in a leaky old ex-German submarine. Before the invention of the modern marine gravity meter, it was possible to get a stable reading on the torsion balance only by submerging the submarine to below wave base, shutting off the engines, and forbidding anyone to move for half-an-hour or so while Vening Meinesz took his readings. Years later he sent to Rhodes all his reports and greatly encouraged Rhodes's interest in seafloor structures and eustatic history. Vening Meinesz's profile across the Darling Range suggested a vertical boundary fault of immense displacement, possibly more than 10,000 m. Rhodes says that he could hardly believe it but that subsequent geophysical work and extensive oil exploration borings reinforce the idea.

Later on, following my own work on the paleovalleys in the edge of the Darling Scarp and assessing our common interest in the breakup of Gondwanaland, Rhodes says that it became clear to him that the Darling Scarp was indeed a normal fault, a relict of a once gigantic graben, such as on the Black Forest side of the Rhine Graben or the Kenya side of the East African Rift. The west side had rifted away and drifted northward during India's migration away from Antarctic during the Late Jurassic and Cretaceous. With typical fairness, Rhodes is quite careful to point out that Corky and Rex were right too. The scarp had in fact been a seashore, on and off, for 450 m.y., since the Late Ordovician. It had, of course, been repeatedly buried by sediments and then periodically re-excavated or structurally revived. Rhodes doesn't think there has been any tectonism along it for at least 5 m.y., which adds another dimension to ideas about marine erosion. In the mid-nineteenth century A. C. Ramsay (1846) imagined marine erosion of continents to peneplains. Later, William Morris Davis developed the subaerial model but did not gain much appreciation from the process school of the late twentieth-century geomorphology. Today, however, the study of paleosols has convinced Rhodes and me that Davis was basically right.

CORAL REEFS, LIMESTONE EROSION, AND SEA-LEVEL CHANGE

Rhodes's introduction to the world of the Quaternary came quite late in his career. Until World War II he had been concerned primarily with problems in the Paleozoic and Mesozoic, and as he quite often remarks, his former professors left him without the slightest hint that geologic time did not come to an end with the close of the Pliocene epoch. The war brought him to the southwest Pacific and to the close-up study of coral reefs. The reefs were, at that time, seen as an intelligence problem of military logistics and operational planning. The scientific problems soon piqued his curiosity and led to his formulation of two basic and critical questions: (1) What was the mechanism of marine erosion on the horizontal intertidal limestone platforms of tropical seas, and (2) what is the explanation for the emerged steplike platforms or benches and multiple notches in the sea cliffs of tropical limestones? In the first case a problem exists because there are contemporary platforms and undercuts (notched cliffs) even in bays where there are no storm waves. The problem in the second case relates to the fact that benches and notches are often associated with raised beaches that are familiar in glacio-isostatic landscapes but not in the tropical Pacific. Both questions are distinct yet inextricably interrelated. Rhodes found solutions to the two problems that were both elusive and demanding. These same problems will probably continue to nourish research for others over future decades.

While at MacArthur's headquarters, Rhodes realized that their island-hopping strategy was fraught with risks unless they learned something more about the reef platforms. Curt Teichert knew more about reefs than anyone else, but Curt was technically an enemy alien and restricted to the neighborhood of the University of Western Australia. The local security people would not let him go, so Rhodes and his colleagues simply spirited him away in the dead of night and flew him to Brisbane. Curt was decked out in an imaginative uniform: part Dutch Air Force (to explain the accent), part U.S. Navy, and part Australian Army. With some V.I.P. passes they explored the barrier reefs and coral islets in great number. Weathering a hurricane on Low Isles, they learned a great deal about reef dynamics both from the air and, dodging the sharks, from below.

Later, after Rhodes was installed at the University of Western Australia, he followed Curt's advice and started to explore the coastal eolianites and their reef platforms. These are very peculiar rocks, scarcely mentioned in textbooks, formed of calcarenites, carbonate clastics with the texture of coarse sandstone but subject to solution like limestone. These formations, though puzzling, were ideal for preserving traces of former sea levels.

Rhodes found that there were three distinctive levels that appeared to be Holocene in age, because of their freshness, and with Curt, discovered there was also an older generation of emerged reefs and platforms that had been plastered over by the eolian formations and then retouched by Holocene intertidal erosion. In 1949 Rhodes announced "Multiple stands of the sea in postglacial times" at the Seventh Pacific Science Congress (New Zealand). The paper triggered a furor. The platforms were variously claimed to be products of storm-wave erosion (but what of the bays?) or differential weathering (in homogeneous formations?), or maybe they were really Pleistocene—mistaken for Holocene. As with many things in nature, there was some grain of truth in every hypothesis.

An analogous sequence of major strandlines, the Littorina beaches, had long been recognized in Sweden and dated by Baron de Geer's varve chronology. Rhodes tried a long-shot correlation, arguing that if the strandlines were eustatic in one place, some traces should be found worldwide. Of course, the correlation was regarded as outrageous, being halfway around the world from Sweden. But a few years later Ed Deevey from Yale offered to date the specimens by radiocarbon, Libby's newly discovered technique. The dates, with allowance for the secular error, were identical with the sidereal varve dates. Researchers, however, are still arguing about it.

Rhodes was disappointed, of course, that he was not able to convince the scientific world of what he considered his great discovery. He was never able to secure a grant that would pay for crucial radiocarbon dating and systematic fieldwork. Bit by bit, however, the picture has been filled in by others, and today the Fairbridge Curve of the Holocene eustatic fluctuations finds its way into countless textbooks. It is funny to note that the expression was first used by the opposition in a deprecating way, but the label stuck.

In a way it was a good thing that lack of grant money forced Rhodes into other areas, broadening his already wide interests, with many good results. While still in Western Australia he spent a lot of time studying reef ecology and the actual mechanism of reef-platform erosion. The pH meter became a key tool in this investigation, and an invitation by Roger Revelle to join the *Capricorn* expedition to the South Pacific in 1951 brought an opportunity to discuss the problem with experts.

Rhodes had long been puzzled as to how seawater, with a pH of 8 or higher, seemed to generate a chemical etching effect on raised coral reefs and the calcareous eolianites of Western Australia at around the air-water contact, often carving a deep undercutting. Phillip Kuenen from the Netherlands, and others, had noticed the same phenomenon in the East Indies and commented on this horizontal sawing action. It occurred just as frequently in bays, so it was not related to storm waves or mechanical abrasion, the textbook explanation. Rhodes had already discovered a pH range of 5 to 10 in the tide pools over 24-hour vigils.

When Expedition Capricorn got to Tahiti, Rhodes, Roger Revelle, and Walter Munk borrowed a dugout canoe and paddled out to the outer barrier reef early one morning at dead low tide. With a pH meter they measured the acidity/alkalinity relationships in the open water, in the lagoon, in tide pools, and under the banks of algae. That was it! In the microenvironments, in the early morning, the CO_2 production pushed the pH down toward 5, and at that value the carbonate is easily dissolved. Later, Rhodes tried thin-sectioning the limestones in the undercut shore platforms in Western Australia, and this confirmed also the presence of algal borers, which several earlier investigators had reported. Biochemical solution advances in several ways, and the boring algae explained the deep undercuts. When exposed to sunshine at low tides, the microenvironments of evaporating tide pools and open pore spaces swing over to a high pH, CO_2 is evaporated, and carbonates are reprecipitated, creating a hard, indurated crust that will ring dramatically to a hammer blow. The riddle of coastal limestone erosion is really just a question of day and night.

DISCOVERING THE UNITED STATES

When Rhodes finally left Western Australia for a sabbatical in the United States (in 1953), he was fortunate enough to get a Fulbright grant. Also, the University of Illinois invited him to a visiting professorship. It turned out to be an exhilarating experience after the long isolation in Western Australia, in spite of all his exciting discoveries made there. Rhodes found a new niche, one with audiences eager to hear his ideas. George White, his chairman at the University of Illinois, told him that he should not feel tied down to Urbana but to accept lecturing invitations wherever he was invited. He responded to the call and gave dozens of lectures, somewhat overextending himself, and at times he was totally exhausted. Some days, it was only with Dolores's loyal support that he was able to pull through. From his tours in the United States, Rhodes discovered very rapidly what a warm-hearted and marvelously receptive environment existed in North America. Innumerable speaking schedules and field trips took him to almost every state as well as Canada and Mexico.

After the sojourn in Illinois, Rhodes put in six months at Scripps Institution of Oceanography at La Jolla, California. One day, Marshall Kay at Columbia University in New York called. He wanted to know if Rhodes would consider coming there. Of all institutions in the world, Rhodes regarded Columbia as the leader in the earth sciences. Rhodes said that he would come, even as a junior lecturer. Marshall chuckled and said that Rhodes should have bargained a little, but what to bargain for? This was to be a permanent appointment for a full professorship with tenure, extended by a unanimous vote of the faculty. Rhodes got the largest office in Schermerhorn Hall, and over the years he filled it beyond capacity with his global collections.

While at La Jolla, Rhodes and Dolores enjoyed quite a social whirl, but not to the exclusion of very exciting scientific work. Joel Hedgpeth at that time was editing his monumental Geological Society of America (GSA) treatises on ecology and paleoecology; isn't it curious that the first comprehensive collections on ecology were published by a geological society? At the University of Western Australia, Rhodes was the first to give lectures on ecology and in fact had to define the word for the dean. Joel asked Roger Revelle and Rhodes to write the chapter on carbonates and carbon dioxide. It was an interest that has suddenly reblossomed in the 1980s, and now everybody is talking about CO_2—much of it nonsense, incidentally.

Also at La Jolla, Rhodes and Dolores met Francis and Elizabeth Shepard. Fran taught Rhodes how to use the scuba diving equipment, and Rhodes took the official training course. Fran had already prepared a eustatic curve, based mainly on his experiences on the Gulf Coast (which is a subsiding region). He drew a smooth sine curve through the available data points, and this became known as the "Shepard Curve." It has often been reproduced in textbooks, contrasted against the Fairbridge Curve, much to Rhodes's embarrassment. Obviously, he said, it was only a question of viewpoint. The Shepard Curve was simply a mathematical smoothing based on data generated from a subsiding region. Rhodes's wiggly curve recognized the multiple strandlines seen on nonsubsiding coasts. Also, as more dates emerged, it became clear that the negative swings corresponded to cold climatic cycles, something shown by records of glacier advances, by the negative spikes in ice cores, by sequences of narrow rings in dendrochronology, and by infusions of cold-climate pollen in the palynological diagrams. Rhodes argued that, in view of these converging proofs of major Holocene cooling incidents, it would be astonishing if sea level did not fluctuate.

Another subject discussed at La Jolla was

the question of steric change of sea level. J. Pattulo, W. Munk, R. Revelle, and E. Strong wrote a paper on it (*Journal of Marine Research*, v. 14, 1955). A steric change is a volumetric response of sea level to the expansion or contraction of the water column due to warming or cooling. They calculated that about 50% of the 3-m rise seen in the South Pacific coral reefs could be better explained as steric rather than eustatic. This was a helpful suggestion because large eustatic fluctuations would call for impossibly high rates of glacier melting or rebuilding.

Other shipmates on Expedition Capricorn at La Jolla included Bill Menard, Bob Fisher, and Harris Stewart. Together with Rhodes they had been on the R. V. *Horizon* (in January 1951) when they discovered the greatest deep in the Southern Hemisphere in the Tonga Trench. Later it was called the "Horizon Depth" (10,800-10,882 m).

Rhodes had become very interested in the morphology and terminology of deep-sea features and for a long time served on a committee about them organized by the International Association of Physical Oceanography. How exciting to be in on one of the major discoveries! In those days, with the newly developed PDRs (precision depth recorders), it was hardly possible not to discover something, even if only a seamount or two. Rhodes and Harris Stewart wrote a paper (in *Deep-Sea Research* v. 7, 1960) on the partly drowned coral reefs of the Alexa Bank. They and some other reefs are located on horst blocks along the northern margin of the Melanesian Plateau, a submerged complex that lies along the inner side of the main border trenches of the southwest Pacific. In those pre-plate-tectonic days it was all very mystifying. At least they could identify the physical features and give names to them. Rhodes was very flattered when the editors of the new *Times Atlas* (published in 1967) invited him to supply the proper names for the submarine features of both the Indian Ocean and the southwest Pacific. He had now come quite a way since tracing his first maps from an earlier edition of that wonderful atlas.

DIAGENESIS AND THE DOLOMITE PROBLEM

Studies of carbonates naturally led to an interest in dolomites, especially during his sabbatical at the University of Illinois where Rhodes enjoyed innumerable field trips into the hitherto unknown terrain of the mid-craton Paleozoic. With Harold Wanless, he returned to the microfacies potential for stratigraphic analysis that had earlier been introduced to him in the Middle East by Henson while they were at IPC. Examining the Paleozoic dolomites, which at that time were regarded as deep-water facies, it struck Rhodes that, in alternating limestone-dolomite formations, the ratio of terrigenous components always went up in the dolomites, suggesting that they were in fact nearshore. Then he met Don Graf, Keith Chave, and George Chilingar, who showed him analyses of contemporary lagoonal carbonates that were rich in the metastable high-magnesium calcites, which in turn are characteristically accumulated by the carbonate-secreting algae of shallow, warm environments.

Here was another "Eureka" experience! The dolomites must be warm, shallow-water, nearshore facies, not deep-water deposits. This led to a Society of Economic Paleontologists and Mineralogists' (SEPM) paper ("The dolomite problem," 1957) and a series of papers on the use of carbonates as paleoclimatic indicators. Later, some Shell geologists told Rhodes that this discovery completely changed their exploration strategy.

Studies of dolomitization led naturally to the question of diagenesis and then to two large chapters in volumes edited by Chilingar (1967). He recognized that diagenesis took place not only immediately after deposition ("syndiagenesis") but also during deep burial, which he called "anadiagenesis." Following uplift or emergence and re-exposure to low-salinity, oxygenated meteoric waters, a third regime is established; this is "epidiagenesis."

Rhodes traveled so much that it was a matter of pride that he was personally acquainted with at least half the people mentioned in his reference lists. Glancing at the bibliography of the 1983 revision of his diagenesis report, he marked out for me those he had actually met and who had materially helped in that work: Abelson, Adams, Alderman, Alimen, Allen (J. R. L.), Amstutz, Arrhenius, Ault, Avias, Baas Becking, Berner, Black, Chave, Chilingar, Cloud, Correns, Crook, Curtis, Damuth, Dansgaard, Dapples, Debelmas, Degens, Drever, Dunham, Dunoyer de Segonzac, Dury, Edwards, Erhart, Feely, Fischer, Folk, Friedman, Fyfe, Garrels, Ginsburg, Glover, Goldberg, Graf, Grim, Hallam, Hay, Heezen, Herman, Holland, Hough, Hunt, Jacobs, Kay, Keller, Kerr, Kirkland, Kolodny, Krauskopf, Krejci-Graf, Krumbein, Kuenen, Kvenvolden, Langford-Smith, Lees, Lloyd, Lobeck, Margaritz, Manheim, Mawson, Millot, Nagy, Newell, Parker, Purser, Revelle, Riedel, Rona, Ronov, Rutten, Sanders, Scheidegger, Schlanger, Schmalz, Selley, Shearman, Shinn, Siever, Sillen, Skinner, Smirnov, Steinitz, Stephens, Teichert, Termier (H. and G.), Ters, Trendall, Van Andel, Van der Lingen, Van Houten, Van Straaten, Veizer, Weiss, Wetzel, Zen, Zenger, and ZoBell. What an astonishing list! Rhodes, in the geographic isolation of Western Australia, began a correspondence with numerous people with kindred interests. He was always surprised and gratified to find how approachable and generous were the most famous scientists. In later years he has tried to be the same to the young and aspiring scientists. He felt this was partly a reflection of the smallness of our profession but also expresses the freedom and openness that comes from a common love of nature and from a shared curiosity with its mysteries.

GLOBAL RHODES

Soon after Rhodes had settled in to Schermerhorn Hall at Columbia, and Dolores, footsore after a month of search, had discovered an apartment nearby on Riverside Drive with a gorgeous view of the Hudson River, Rhodes was approached by two Chappaqua artists, Sam Berman and Ken Fagg. Would he like to be their consultant on a project to build a giant, 6-foot-diameter relief globe? What a challenge! The vertical scale was logarithmic so that the low elevations would be exaggerated enough to show up clearly while the high relief would be subdued to reasonable proportions. Everything had to be sculpted in detail. It was great fun, but the finished product had to be sold (as Geophysical Globes, Inc.) to Rand McNally, and the project was a financial disaster.

One happy outcome, however, was one blank globe delivered to the Geology Library in Schermerhorn Hall. Rhodes prepared template maps of the major geotectonic provinces of the world, and Dolores, with acrylics, painted the entire surface. Bruce Heezen and Marie Tharp supplied their first drafts of the seafloor physiographic charts that had been inspired by Lobeck's work for the continents. One day, Harold Urey chanced by. He studied the globe, fascinated, for about half-an-hour and then exclaimed, "How extraordinary! Fairbridge, how *can* the Earth be so complicated?" Rhodes replied that this merely represented a triumph of condensation. He was always astonished at the way physicists and chemists tend to imagine natural phenomena in oversimplified terms. On another day a distinguished alumnus who had become blind was brought in. He spent hours feeling the globe as if it were Braille. His face was illuminated with joy. "I've just been on a wonderful journey."

Besides the 6-foot "dinosaur," the Geophysical Globe company manufactured a low-cost 12-inch plastic relief model. Rhodes and Bruce took blank copies and tried out many schemes. One day the Smithsonian Institution called. They wanted a duplicate of the 6-foot geotectonic globe, and they gave Dolores the painting contract. They also wanted about a dozen of the 12-inch series, painted in various geophysical styles, including gravity, magnetism, volcanism, seismicity, ocean currents, and meterological features. As Dolores commented wryly, yes, it was nice to get some consulting money, but, of course, instead of being wisely invested in gilt-edged securities, the funds had to be ploughed right back into the next project.

ENCYCLOPEDIAS AND BENCHMARKS

Rhodes feels about books as a socialite feels about diamonds: You can't have too many of them. He had begun his collection in his teens, but after leaving Oxford, he put them in storage, only to have his library blown to bits by a Nazi bomb during World War II. When he started teaching at the University of Western Australia, he began with a single volume. He realized that lots of information was scattered in snippets throughout the scientific journals; somehow this scattered treasure must be gathered together. For his Australian students there was no such thing as a basic textbook on Australia stratigraphy, so he set about writing one, aided by extensive quotations from the regional journals. This classic was issued by the Western Australia Textbooks Board in 1949.

When Rhodes came to New York, it was natural he should be approached by innumerable publishers, but at first he wisely turned them aside. There were more pressing researches in hand. He agreed to serve, however, as a translation series editor for the Van Nostrand Co. in Princeton, New Jersey, and was instrumental in translating and issuing for the English-speaking audience some invaluable references: Schwarzbach's *Climates of the Past* (1963), Borchert and Muir's *Salt Deposits* (1964), and Termier's *Sedimentation*. He also gave a nudge to the translation of Guilcher's invaluable *Coastal and Submarine Morphology* (Methuen and Wiley, 1958).

He founded, in addition, a new journal of translations, with a grant from the American Geological Institute, entitled *International Geology Review*, which is devoted largely to Russian material. Arrangements were also made for the translation of V. V. Beloussov's book *Geotectonics*. Beloussov was a convinced verticalist with his background of work in the Urals and Caucasus, but on Rhodes's suggestion he visited the United States and even spent some weeks at Lamont, so he was well-exposed to the evidence of plate tectonics. Nevertheless, one day he put a fatherly arm over Rhodes's shoulder and said, "My boy, why don't you give up this stupid drift theory—then you will leap ahead." After the 1968 debacle in Prague (the Russian invasion, which coincided with the International Geological Congress), Rhodes never felt quite the same about the socialist world and was greatly saddened for sure.

A crucial point in his editorial activities came one day when the Reinhold publishing people (later merged with Van Nostrand) came to him with a proposal that he should edit an encyclopedia of earth sciences. Rhodes was delighted with the idea, but as with many projects, he failed hopelessly in anticipating what it would lead to. The important thing was that it needed to be done. It would be unique and invaluable for the profession. It was all to be condensed into one volume. That was the rub. It wouldn't fit. So they worked out a revised plan: each branch of the subject should get one volume. *Oceanography* (1966) was first, a smashing success, selling over 12,000 copies. *Atmospheric Sciences and Astrogeology* (1967) was next, then *Geomorphology* (1968), then a pause and *Geochemistry and Environmental Sciences* (1972). Rhodes did not want to be tied down and was able to persuade a student, Joanne Bourgeois, jointly to take on *Sedimentology* (1978), and David Jablonski *Paleontology* (1979). I took on *Soil Science* (1979), *Applied Geology* (1982), and most recently, *Field and General Geology* (1987); Keith Frye did *Mineralogy* (1981); Maurice Schwartz, Rhodes's old teaching assistant, did *Beaches and Coastal Environments* (1982); and John Oliver did *Climatology* (1986). Rhodes handled *World Regional (Western Hemisphere)* (1975), but the publishers, after several shufflings and takeovers, began to lose interest in a project that was not a great money-maker.

Another publishing venture involved the Benchmark Papers in Geology, an idea originated by Milo Dowden, one of a partnership of Dowden, Hutchinson & Ross, which was formed after an early Van Nostrand Reinhold shakeup. Chuck Hutchinson, who had a degree in geology, played a major role in encouraging Rhodes's editorial interests. This series generated a remarkable number of very fine volumes, over 90 in all. Volume 68 was *Sunspot Cycles* (Schove,

1983), and Volume 88 was *Continental Drift* (Shea, 1985), so the series in part evidently reflects Rhodes's broad outlook.

VOYAGE UP THE NILE

One day Rhodes's friend from archaeology, Ralph Solecki, dropped in and remarked that, by some administrative bungle in Washington, D.C., he had gotten two concurrent research grants, one to the Nile to help the UNESCO Save the Nubian Monuments campaign, the other to Iran to dig for Neanderthal skeletons at Shanidar. Evidently that was where Ralph's heart lay, so would Rhodes care to save the monuments? This task would involve doing the geology of the Nile River deposits. Who could resist?

"The trouble," Rhodes explained to me, "was that I knew nothing whatever about the Nile or fluvial sedimentation in general. Not a clue. . . ." But he knew that his old friend Richard Foster Flint had recently been to Africa and that "Rocky" Flint knew everything there was to know about the Quaternary. Rhodes phoned him: "Dick, what do I have to do, what to read?" The reply was somewhat alarming: "Rhodes, everything that's been written so far about the Nile is nonsense. Don't look at the publications; they'll only distract you. Just use your instincts and you'll know what to do."

Actually, Rhodes already had some old, more or less archeological, papers by his friends W. J. Arkell and K. S. Sandford, who had been his instructors at Oxford. He had Hurst's excellent book on the Nile hydrology and J. Ball's interesting *Contributions to the Geography of Egypt* (1939), which he had picked up from Ball himself in between doing a bit of intelligence work in Cairo in 1940.

Rocky Flint was right, however. The early reports were not very informative and, in many of their interpretations, turned out to be dead wrong. Rhodes had already done some general reading about equatorial and subtropical climates during the Quaternary and had become aware of the standard dogma: that glacials equal pluvials (in the tropics) and that interglacials equal arid phases. He had, however, been talking to Dave Erickson and Cesare Emiliani about ocean temperature during the last glacial. Both agreed—one from the foraminifera, the other from oxygen isotopes—that middle- and low-latitude surface ocean waters were 3-5°C cooler than they are today. Surely, he reasoned, the cooling of the glacial-stage ocean surface would greatly reduce the evaporation, and then we should expect to see a greatly weakened monsoon and a vast increase of deserts.

Geological data from Africa were unfortunately not supported by a reliable chronology, but Rhodes found the answers in French archeological papers, first by L. Balout and later by Miss H. Alimen, both competent specialists. The last glacial maximum was a time of total aridity in the Sahara, with dunes extending well to the south of their present border. The last pluvial phase corresponded to the Neolithic cultures of the early to mid-Holocene, and this was proven by radiocarbon dating.

On their way to Cairo, Rhodes and Dolores stopped in Rome, where a Paleoclimate Symposium was being organized by FAO/WMO (the United Nations Food and Agriculture Organization, jointly with the World Meteorological Organization). Rhodes presented a couple of papers. One paper dealt with sea level, climate, and postulated solar radiation changes; the other with his suggested reinterpretation of the tropical climates of the glacial stages. Both papers were violently attacked by the establishment. Rhodes, as ever, was not dismayed, although mildly puzzled at his total lack of ability to convince people whose minds are already made up.

In a few days they were in Cairo and heading up to Aswan. After a week of traditional tomb explorations, they were joined by their son Kingsley (then aged 16) and Tony Marks, one of Solecki's archeology students. Those two had picked up a Jeep from a ship at Alexandria, bribed their way through customs ("Never let them see more then 100 piasters," counseled Papa Rhodes), and heroically drove 1,000 km up to Aswan, arriving dusty and tired but on schedule.

Shortly afterward, they were comfortably installed on a Nile River steamer, the *Zahra*, heading for Wadi Halfa in the Sudan, following the age-old route of the pharaohs, the Napoleonic troops, the Kitchener campaign, and Agatha Christie. There, installed in a bare but recently constructed Nubian-style house, they hired a marvelous cook, Bisri Ali (who was "illiterate in five languages"), and set out to explore the region. They visited Abu Simbel (before it was chopped up and shifted) and pushed up river as far as Khartoum where the Blue and White Niles merge. They visited numbers of archeological digs, English, French, Swedish, Polish, but none of them had a geologist along, until after a couple of months they were joined by Jean de Heinzelin and his student Roland Paepe, both fresh in from the Congo. Roland was suffering from dysentery and was pitifully thin, but Dolores nursed him back to health.

To understand the Nile, Rhodes reasoned, one should study it on an actualistic basis. What has happened recently? Every year in midsummer the river comes down in floods, but the interannual variance sometimes exceeds 50% of total discharge, and even in the twentieth century there have been large cyclic fluctuations. During each low-water stand, sand banks emerge and the silt leaves another varvelike layer on the floodplain. Behind the floodplain there were silt terraces up to 40 m higher, and the oldest of them contained artifacts of the last glacial time. Since then, there has been a history of cut and fill, with progressively lower amplitudes. The silts carried layers of a charming little gastropod shell called *Cleopatra* and included desiccation surfaces with artifacts and fireplaces with charcoal. Both were collected for radiocarbon dating. Even with the latter, however, the dynamics appeared to be clear. During desiccation phases the middle Nile valley would choke up with silt, which would be carved out again during the high-discharge cycles.

Later, when the ^{14}C dates came in, Rhodes was able to set an approximate chronology to the story that was published in *Nature* (1962). The astonishing thing was that the fluctuations were in phase with the sea-level record, although above the cataracts there was naturally no possibility of eustatic ("thalassostatic") control. What became clear was that the monsoonal rains reflected the Indian Ocean evaporation and tropical water temperatures, which were in phase with the ups and downs of sea level. The latter in turn had already been shown to be in phase with high-latitude climatic oscillations. So the global climate system, at least on the scale of 500- to 1,000-year cycles, was responding to a synchronous beat!

Rhodes developed this theme in a number of papers—for example, at the NATO Paleoclimate Symposium in Newcastle-upon-Tyne in 1963 (reported in a volume edited by A. E. M. Nairn in 1964 entitled "African Ice-Age Aridity") and in two articles in *Quaternary Research* (1972, 1976).

A trip to South America in 1966 brought him in contact with John Bigarella (Curitiba, Brazil) who showed him how the Brazilian rain forests in the coastal ranges are actually growing on sediments of the sort that today can be seen forming in Arizona. The colluvial wash showed spots of modern kaolinite in the shape of feldspar clasts. These wash sheets could be traced in bores to 100 m below sea level and were thus evidently representative of glacial climates. Describing this situation one day in the classroom, one of the students, Jed Damuth, mentioned he was studying some *Vema* cores taken offshore from Brazil. Did Rhodes think the glacial stages would be marked in the Amazon submarine fan by fresh feldspars? He certainly did. And Jed found indeed it was so, and that made a nice joint paper for the *GSA Bulletin* (1970).

It seems now that during the coldest epochs there was a global decrease in evaporation (expansion of sea ice, reduction of continental shelf waters, cooling of ocean surface waters in general). The old idea that climatic belts would just be shifted equatorward has only very limited validity. Thus, what started with a simple voyage up the Nile gradually blossomed into a global picture of tropical climate change. Rhodes believes that when the postglacial (Neolithic) pluvial stage came to an end about 4000 B.C., and the former hunters or herders were driven out of the erstwhile lake-studded savannas of what is now the Sahara and forced to settle in the Nile valley and elsewhere, this was really the beginning of urban civilization as we know it. Humans had to settle down, organize communities, and devise systems of water supply, irrigation engineering, and administration.

CYCLES: GALACTIC, SOLAR, LUNAR

One day in La Jolla, Hans Suess said, "But I'm not a cycle freak, like you, Rhodes." His grandfather, Eduard Suess, of Vienna, had created the expression *eustatic change* for long-term cyclic variations of sea level, so Hans already had a tradition, but perhaps he was just being cautious. For Rhodes, a critical experience emerged from writing an article on sea levels for *Scientific American* (1960). He mentioned the apparently cyclic long-term pattern of sea levels that seemed to go hand-in-hand with ice ages, except that there was one missing, in the mid-Paleozoic. The editor said we must have a schematic curve; is it O.K. to show that one with a question mark? Rather reluctantly Rhodes agreed. Nine years later, he noticed that according to the newest paleomagnetic data from Australia, one could trace back a south pole (a "Gondwana pole") into the Sahara Desert of western Africa for the Late Ordovician. A two-page note was dispatched to *GSA Bulletin* (1969).

Soon after, Rhodes was at a meeting of the International Association of Sedimentology (Reading and Edinburgh) and was approached by a young French oil geologist. Would Rhodes like to see some color photos taken in the central Sahara? Most certainly. They were demonstrably Upper Ordovician, with distinctive marine fossils, trilobites, and brachiopods. Differential weathering disclosed remarkably grooved surfaces. In places there were large erratics, some of them nicely faceted. They looked like typical tillites. Would "Monsieur le professeur" like to join some colleagues on an expedition to the Algerian Sahara next January? The Institut Française du Pétrôle would happily finance the thing and organize a small symposium in Algiers to follow.

It was an unforgettable trip, made especially enjoyable by charming French and Algerian hosts (B. Bijou-Duval, Pierre Rognon, Yvonne Gubler, Jean Dresch, André Vatan, A. Benamar, and A. Bennacef) and the distinguished international colleagues (Paul Potter, United States; Alexei Bogdanoff, USSR; Percy Allen, United Kingdom; Stefan Rozycki, Poland; Adolf Seilacher, West Germany; L. M. J. U. van Straaten, the Netherlands; J. J. Bigarella, Brazil; Nils Spjeldnaes, Denmark; and Anders Rapp, Sweden). With balloon-tired Land Rovers and desert trucks, the expedition visited all the finest localities. They eventually came to a spot east of the Hoggar Mountains where glacial grooves could be followed in these silicified flat-lying sediments, extending almost to the horizon. The proof was overwhelming. At lunch time Bijou-Duval called the international panel together. Well, were they convinced? Totally. Rhodes said the evidence was clearer for the Sahara Ordovician than it was for the Pleistocene of Canada. He had seen also the Permo-Carboniferous and the Upper Precambrian glacial deposits. So that was the last word, and Bijou called for champagne. The young devil had secreted a whole case in the refrigerator truck, and out it came, glassware and all! They drank to the Great Sahara Glaciation. When they reached home, Rhodes gave the story to Walter Sullivan, and it got the front page of *The New York Times*.

They didn't get away with it without protest. A Dutch specialist, L. J. G. Schermerhorn (some distant relative of the donor of Columbia University's Schermerhorn Hall?) wrote lengthy papers for the *American Journal of Science* and the *GSA Bulletin*, to which Rhodes replied at length. No, he explained it would really be difficult to generate submarine landslides and pebbly mudstones right out there in the middle of the African craton.

The discovery opened a floodgate for more finds of the Ordovician glaciation, ranging from Sierra Leone on the Atlantic shore, all the way across north Africa, to Ethiopia, Yemen, and Saudi Arabia. It was truly the largest glaciation that has been documented. Much later, Rhodes was one day idly reading a rare volume by A. W. Grabau, published in Peking in 1940. And there, to his amazement, was a paleogeographic reconstruction of the Late Ordovician, with a great ice sheet bang in the middle of west Africa! How could Grabau, living in China, possibly have known anything about the Sahara glaciation? He couldn't. He was a superb paleontologist, however, and using Wegener's drift reconstruction, he deduced from the evidence of climatic zonation (from the morphology of the fossils) that there had to have been a great ice sheet there. And it was based purely on *deduction* from observed proxies. Rhodes remarked that Grabau, long before, had also been a Columbia professor.

As to the cause of the great ice-age cycles, Rhodes had long favored the galactic idea, but an accurate chronology was difficult to establish. It gave him great pleasure when the idea was backed by Mike Rampino, together with another friend and colleague, Dick Stothers (both at NASA's Goddard Institute for Space Studies, where Rhodes now spends a lot of time). A secondary modulation, with a 30-32 m.y. period, is apparently introduced by the wavelike motion of the solar system above and below the mean galactic plane, which seems to result in periodic streams of cometary impacts. In turn, these events seem to match up to the key extinctions and climatic fluctuations. Grabau had also identified this periodicity in the stratigraphy and paleontology.

During the 1970s and 1980s, an important turning point was reached in the philosophy of climatology: the demonstration (mainly by the marine geologists, many of them from Columbia's Lamont-Doherty Geological Observatory) that the Milankovitch theory of insolation seemed to be perfectly synchronous with paleoclimatic, sedimentological, geochemical, and paleontological changes in Quaternary time sequences. It amounted to the recognition that climate changes were forced by factors outside the normal terrestrial atmosphere and its dynamic systems. This is exogenetic determinism, and Rhodes feels that there has until now been a psychological-philosophic roadblock opposing such reasoning. Thus, climatic change on a second time frame (10^4-10^5 years) joined the galactic phenomena (10^6-10^8 years) in the recognition of external forcing.

A third category (10^1-10^3 years) of exogenetic control was soon to join the longer-term cycles, although many members of the meteorological establishment still appear to fight the concept tooth and nail. Rhodes argues patiently that if the geological evidence of millennia is sitting there, like Mt. Everest, we can't just pretend it is not there or prove by fancy statistics that it is a figment of an overheated imagination. Already, before the 1961 New York Academy of Sciences symposium on solar radiation and climate, Rhodes was clearly thinking about this question. Like a good student of Lyell (Sir Charles had been at the same Oxford college a century earlier), Rhodes is a passionate believer in actualistic reasoning. Uniformitarianism, however, he feels, was overdone by Giekie and others. The argument for a paleoevent must be supported by evidence of a recent one where at all possible.

At the New York Academy meeting, he met the late Clyde Stacey (an amateur astronomer, to whom Chapter 26, by Fairbridge and Sanders is dedicated). Clyde made Rhodes and John Sanders aware of the importance of planetary patterns and dynamics in dictating the Sun's barycentric inertial motions.

From the seeds sown by Stacey, Rhodes and John were drawn more into planetary motion studies. Celestial mechanics is a highly specialized branch of astronomy, handled rather superficially by the observer fraternity but treated very seriously in celestial navigation for space exploration. Rhodes was convinced that if NASA was smart enough to get an observer package to Jupiter or Uranus, they must have an ephemeris oriented to the solar system's barycenter. Indeed they did! It had been prepared at the Jet Propulsion Lab (JPL) in Pasadena, and after some negotiations a printout of more than a millennium of planetary positions was received. Rhodes spent a summer plotting orbital paths for the sun. Sir Isaac Newton had pointed out the principle that the sun moved "every which way" with respect to the center of mass. Rhodes discovered that, although no two orbits are alike, they fall into distinctive patterns. The orbital path is called an *epitrochoid* (a big circle continuous with a little ring nestling asymmetrically inside it), but as with his early experience in drawing fossils like foraminifera, he could identify a simple axis of symmetry. He established eight standard forms, to which the sun's path returns from time to time. Each class is associated with a distinctive constellation of the major planets. Rhodes noticed that now and then there was a distinctive flutter in the cycle of change in radial distance from barycenter to heliocenter. These events were accompanied by dramatic decreases in sunspot activity, the latter in turn accompanied on Earth by catastrophic cooling cycles of the little ice age type. Aha! So there was support for the gut feeling that when things happen on the sun it seemed intuitively reasonable to expect some effects on planet Earth.

What is the mechanism exactly? People are still working on it. One of Rhodes's favorite aphorisms is that, in open systems like natural sciences, any solution that is straightforward, clear, and simple is automatically false, or at least it should be suspect. Complex interrelationships are the norm. Sun-Earth relations are gorgeously complex, modulated by the planets and our moon. Solar emissions are variable in amplitude, often apparently relating to planetary constellations (although some doubting Thomases, of course, protest), which are variable as to type. Both Maxwellian electromagnetic radiations and "corpuscular" or charged particles reach the Earth in pulses of the solar wind. The principal variance in the first category is in the very short-wave ultraviolet that modulates our stratospheric ozone (most effective above the equator), and in the second category we receive aperiodic showers of high-energy particles that interact with our atmosphere, mainly in the high latitudes around the magnetic poles. Rhodes says it operates like the "push me-pull you," an apocryphal beast—a donkey with two heads pointing in opposite directions— invented by Dr. Doolittle, who wrote children's books.

* * * * * * * * *

This vignette of a displaced Western Australian summarizes some of the high points in the unusual career of Rhodes Fairbridge. Although his various technical accomplishments are well known to many, Rhodes has managed to maintain a remarkably modest decorum in his personal life. I hope that the anecdotes and historical events marshalled here not only will provide some insight into Rhodes the man but also may serve to inspire and encourage younger workers to strive for greater understanding of natural processes in the world about them and to work toward a greater appreciation of the impacts of exogenic forces that can profoundly affect the face of planet Earth. This eclectic recounting of some interesting facets in one man's life does not provide a how-to-do-it, but it certainly must suggest, at least to some, degrees of professional involvement that provide great personal satisfaction.

My brief encounters with mature studies of the Earth have shown the way to proceed from simple beginnings. There is so much to be learned and so little time to research the different possibilities and to consider applications of our knowledge, the nature of which we have not even dreamed. Rhodes is a visionary who has seized his opportunities, a geoscientist able to put pieces together in the great jigsaw puzzle of Earth. He is in particular, I think, an inspiration to all professionals, but especially to young scientists just getting started. Rhodes is not only the author of more than 1,000 scientific articles, the editor of the Benchmarks in Geology Series with more than 90 volumes in print, and general editor of the Fairbridge Encyclopedias of Earth Science but also a living, walking, talking encyclopedia of geological knowledge. His global concepts and imaginative powers have not diminished in retirement. On the contrary, his colleagues can expect to hear still more. "Even when I am wrong," says Rhodes, "I think it is useful to occasionally play the role of the gadfly that shakes people up, provokes them to think, to investigate, to explore—to discover!"

ANNOTATED BIBLIOGRAPHY OF RHODES W. FAIRBRIDGE

Many abstracts, reviews, and encyclopedia entries have been omitted.

1946. Coarse sediment on the edge of the continental shelf, *Am. Jour. Sci.* **245:**146-153. (First paper, extracted from D.Sc. thesis, indicating eustatic origin of littoral sediments on shelf margin and possible role of slumping.)

1946. Submarine slumping and location of oil bodies, *Am. Assoc. Petroleum Geologists Bull.* **30:**84-92. (Based on part of D.Sc. thesis, shows how gravitational slides and/or tectonics can bury petroliferous sediments.)

1947. Causes of intraformational disturbances in Carboniferous varve rocks of Australia, *Proc. Royal Soc. New South Wales* **81:**99-121. (Describes structural features of glacial lake sediments.)

1947. (with E. D. Gill). The study of eustatic changes of sea-level, *Australian Jour. Sci.* **10:**63-67. (First attempt at guidelines and datum selection for eustatic research.)

1947. A contemporary eustatic rise in sea level? *Geog. Jour.* **109:**157. (Evidence of upward growth of reef-flat coral, in Abrolhos Islands, Western Australia, suggesting twentieth-century sea-level rise.)

1947. (with C. Teichert). The rampart system at Low Isles, 1928-45, *Great Barrier Reef Comm. Reports*, vol. 6, pp. 1-6. (Fieldwork showing rapid buildup of coral debris during a hurricane.)

1948. Notes on the geomorphology of the Pelsart Group of the Houtman's Abrolhos Islands, *Jour. Royal Soc. Western Australia* **33:**1-43. (Geomorphic features due to sea-level changes.)

1948. (with C. Teichert). The Low Isles of the Great Barrier Reef: A new analysis, *Geog. Jour.* **111:**67-88. (Geomorphic effects of hurricanes.)

1948. Problems of eustatism in Australia, in International Geographical Union, *Problèmes des Terrasses* (6th Comm. Rept.), Louvain, pp. 47-51. (Review of eustatic problem for international audience.)

1948. The juvenility of the Indian Ocean, *Scope (Journal of Science Union, University of Western Australia)* **1**(3):29-35. (Discussion of the Mesozoic opening of Indian Ocean: Stille versus Wegener.)

1948. (with C. Teichert). The Low Isles of the Great Barrier Reef: A new analysis, *Geog. Jour.* **111:**67-88. (Detailed geomorphic analysis of a well-studied platform reef and its islets, including mangrove swamp.)

1948. (with C. Teichert). Some coral reefs of the Sahul Shelf, *Geog. Rev.* **38:**222-249. (Air photo interpretation of varied reefs and islets northwest of Australia, outlining tectonic and eustatic alternatives.)

1948. Discoveries in the Timor Sea, North-West Australia, *Royal Australian Hist. Soc. Jour. and Proc.* **34:**193-218. (Revised interpretation of Dampier's voyage.)

1948. Gravitational tectonics at Shorncliffe, S.-E. Queensland, *Royal Soc. Queensland Proc.* **59:** 179-201. (Field evidence from block-faulted Triassic sediments showing slide structures analogous to some Alpine nappes.)

1949. Geology of the country around Waddamana, central Tasmania. *Royal Soc. Tasmania Papers and Proc.* (for 1948), pp. 111-149. (Regional survey, emphasizing lineaments and fracture patterns in Triassic dolerite-sill terrain.)

1949. Antarctica and geology, *Scope (Journal of Science Union, University of Western Australia)* **1**(4):25-30. (Review pointing out key position in ancient Gondwanaland, evidence of drift?)

1950. The geology and geomorphology of Point Peron, Western Australia, *Jour. Royal Soc. W. Australia* **34**(3):35-72. (Chronology of Holocene sea-level changes.)

1950. Recent and Pleistocene coral reefs of Australia, *Jour. Geology* **58:**330-401. (Growth of coral reefs and sea-level features of continent under changing world climate controls.)

1950. Precambrian algal limestones in Western Australia, *Geol. Mag.* **87:**324-330. (First discovery of Precambrian fossils in Western Australia, showing warm climate comparable to present.)

1950. Problems of Australian geotectonics, *Scope (Journal of Science Union, University of Western Australia)* **1**(5):22-28. (Reviews continental structure in the light of Stille's consolidation theory.)

1950. Landslide patterns on oceanic volcanoes and atolls, *Geog. Jour.* **115:**84-88. (Using wartime air photography, shows pattern of slide scars on insular volcanoes and a possible explanation of star-shaped atoll form.)

1951. (with E. de C. Clarke, K. L. Prendergast, and C. Teichert). Permian succession and structure in the northern part of the Irwin Basin, Western Australia, *Jour. Royal Soc. W. Australia* **35:**31-84. (First detailed survey of Permian glacial formations in this part of Australia.)

1951. Some recent advances in the geology of Western Australia, *Geol. Rundschau* **59**(1):282-292. (Permian glacial material.)

1951. The Aroe Islands and the continental shelf north of Australia, *Scope (Journal of Science Union, University of Western Australia)* **1**(6):24-29. (An air-photo and interpretive Quaternary history of some peculiar islands.)

1952. Multiple stands of the sea in post-Glacial time, *7th Pacific Sci. Congr. (New Zealand, 1949) Proc.* **3:**345-347. (Triple high sea levels are basically eustatic.)

1952. Marine erosion. *7th Pacific Sci. Congr. (New Zealand, 1949) Proc.* **3:**347-359. (Historical study of the coast erosion shows neglect of chemical process.)

1952. Permian of southwestern Australia, *19th Internat. Geol. Congr.* (Alger, 1952, Gondwana Symposium). (First discovery of Permo-Carboniferous glacial deposits beneath coal formations located in deep scoured depression.)

1952. The geology of the Antarctic, in *The Antarctic Today*, Wellington, N.Z., pp. 56-1010. (Review of the continent, seen as component of Gondwanaland.)

1953. (with C. Teichert). Soil horizons and marine bands in the coastal limestones of Western Australia, *Jour. Royal Soc. New South Wales* **86:**68-87. (Alternating wet and dry climates identified.)

1953. The Sahul Shelf, Northern Australia, its structure and geological relationships, *Jour. Royal Soc. W. Australia* **37:**1-33. (Glacial landbridge, marked by river systems in arid landscape.)

1953. *Australian Stratigraphy*, Nedlands: University of Western Australia Text Books Board, 516p. (University textbook, developing the Stille method of progressive cratonic consolidation, with cyclic phasing.)

1953. Australia in the Indian Ocean? *Walkabout (Australian Geog. Mag.)* **19**(9):10-11. (Review of problem of ocean boundary definition with historical examples.)

1953. Niveaus multiples de la mer à l'age postglacial, in *Actes du IV Congrès International du Quaternaire*, Rome-Pisa: Instituto Italieno di Paleontologia Umana, pp. 1-7. (References to new work at Point Peron, Abrolhos and Rottnest islands, and classic eustatic interpretations going back to Ehrenberg in 1834.)

1954. (with M. A. Carrigy). Recent sedimentation, physiography and structure of the continental shelves of Western Australia, *Jour. Royal Soc. W. Australia* **38:**65-95. (A synthesis of modern shelf sedimentation in an atectonic setting.)

1954. Stratigraphic correlation by microfacies, *Am. Jour. Sci.* **252:**683-694. (Practical utility of thin-section appearance, discussing paleoclimatic and geotectonic control of universality and rhythmic repetition.)

1954. Oceanographic research in the Indian Ocean, *Deep-Sea Research* **1:**185-186. (Discussion of nature of deep-sea floor.)

1954. (with V. N. Serventy). The Archipelago of the Recherde, 1b: Physiography, *Australian Geog. Soc. Rept.* **1:**9-28. (Examines anomalous drowned-inselberg islands on the continental shelf, slow rate of erosion, and eolianite production.)

1955. Some bathymetric and geotectonic features of the eastern part of the Indian Ocean, *Deep-Sea Research* **2:**161-171. (Analyzes key structures and possible nature of "thalassocratonic" crust.)

1955. Report on the limits of the Indian Ocean, *Pan Indian Ocean Sci. Congr. (Perth, 1954) Proc.*, 18-28. (Detailed historical view of the nomenclature and boundaries of the ocean.)

1955. Warm marine carbonate environments and dolomitization, *Tulsa Geol. Soc. Digest* **23:**39-48. (First draft of new interpretation of an environmental indicator.)

1957. The dolomite question, in R. J. Le Blanc and G. J. Greeding (eds.), *Sociey of Economic Paleontologists and Mineralogists Special Paper No. 5*, pp. 125-178. (First recognition of dolomite as a warm shallow-water paleogeographic indicator.)

1957. (with R. Revelle). Carbonates and carbon dioxide, *Geol. Soc. America Mem.* **67:**239-296. (Review of the most useful paleoclimatic and paleoecologic geochemical indicators.)

1957. Continental margin of the South-West Pacific: Advancing or retreating? *Mining Gazette (Thailand)* **2**(11):C/13-14 (9th Pacific Science Congress Abstracts). (Indicates widespread evidence of block-faulting and differential vertical motion, as well as long-term history of lateral expansion.)

1958. Dating the latest movements of the Quaternary sea level, *New York Acad. Sci. Trans.* **20:**471-482. (First, and very simplified, version of the Fairbridge Curve of Holocene fluctuations of mean sea level.)

1958. What is a consanguineous association? *Jour. Geology* **66**(3):319-324. (Discussion of lithofacies, bathymetry, and paleogeography of flysch, molasse, and paralic facies.)

1959. Statistics of non-folded basins, *Publications, Bureau Centrale Seismologique International, Ser. A* **20:**419-440. (Summary of global study, made for Pure Oil Co., showing how each of standardized basin types, following Kay or Weeks, possessed predictable shape and thickness parameters.)

1959. Periodicity of eustatic oscillations *(Periodizitaet der eustatischen Oszillationen)*, International Oceanographic Congress, Proceedings (Washington, D.C., Am. Assoc. Advancement Sci.), pp. 97-99. (Cyclicity of sea-level fluctuations seen on varied time scales, throughout geologic time.)

1960. The changing level of the sea, *Sci. American* **202**(5):70-79. (First presentation of the Holocene Fairbridge Curve of eustatic sea level. Uplifted Pleistocene terraces illustrated from New Guinea. Galactic period of sea level and glaciations, e.g., anticipating discovery of Late Ordovician in Africa.)

1960. (with W. S. Newman). Glacial lakes in Long Island Sound, *Geol. Soc. America Bull.* (abstract) **71:**1936. (Discovery of varved clays upthrust in Long Island moraine, taken with seismic profile evidence, demonstrates former lake, "Glacial Lake Lougee.")

1960. (with W. S. Newman). Active subsidence in the New York area, *Geol. Soc. America Bull.* **71:**2107-2108. (Geodetic evidence of differential warping, supported by long-term data.)

1961. Eustatic changes in sea-level, in L. H. Ahrens, K. Rankama, F. Press, and S. K. Runcorn (eds.), *Physics and Chemistry of the Earth*, vol. 4, London: Pergamon Press, pp. 99-185. (Definitive work on sea-level change, identifying multiple mechanisms, climatic, steric [thermal], displacement, tectonic, geoidal, dynamic [Coriolis effect], etc.; first radiocarbon-dated proof of Milankovitch insolation prediction with retardation effect.)

1961. Radiation solaire et variations cyclique du niveau de la mer, *Rev. Géographie Physique et Géologie Dynam.* **4:**2-4. (Relationship of recent sea-level fluctuation to global climate trends and solar cycle.)

1961. (Editor). Solar variations, climatic change, and related geophysical problems, *New York Acad. Sci. Annals* **95:**1-740. (Collected proceedings of a major international conference on sun-Earth relations, with key articles by the major world experts.)

1961. Convergence of evidence on climatic change and ice ages, *New York Acad. Sci. Annals* **95**(1): 542-579. (An "Eclectic Theory of Ice Ages," recognizing polar wandering or crustal shift; also a now-discarded "Polar Coincidence Theory." Significant factors include: closure of seaways, eustatic controlled continentality, Milankovitch climatic-insolation cycles leading to rapid glacial response in high-elevation sites in "sensitive latitudes" with feedback acceleration; secondary cycles during past 1,000 yr relate climate and sea level to radiocarbon flux and solar particulate radiation.)

1961. La base eustatique de la géomorphologie, *Ann. Géogr.* **70:**486-492. (Relates geomorphic processes to eustatic record, the ultimate control of fluvial base level.)

1961. The Melanesian Border Plateau, a zone of crustal shearing in the S.W. Pacific, *Publications Bureau Centrale Seismologique International, Ser. A* **22:**137-149. (Outcome of the Scripps Expedition "Capricorn," which paved the way to the understanding of crustal subduction in the Tonga Trench, where the deepest sounding was named "Horizon Depth" after our ship.)

1961. Sea-level and climate, *New Scientist* **15:**242-245. (Semipopular review, drawing attention to the role of geodetic adjustment.)

1962. (with W. S. Newman). Postglacial subsidence of coastal New England, *Geol. Soc. America Spec. Paper 68*, pp. 239-240. (Regional analysis of dated sea levels discloses a crustal rebound pattern for coastal belt.)

1962. (with W. S. Newman). Postglacial sea level, coastal subsidence and littoral environments in the Metropolitan New York City area, *First Natl. Coastal and Shallow Water Conf. Proc.*, Natl. Sci. Foundation and Office of Naval Research, pp. 188-190. (New radiocarbon dates help establish crustal adjustment to postglacial events.)

1962. World sea level and climatic changes, *Quaternaria* **6:**111-134. (Review of Quaternary climate history and evidence of sea-level fluctuation.)

1962. (with O. B. Krebs, Jr.). Sea level and the Southern Oscillation, *Geophysical Jour. (London)* **6**(4):532-545. (Analysis of global tide-gauge records, identifying global eustatic effect, parallel to mean sunspot maxima, and regional but inversely related trends such as S.O.)

1962. New radiocarbon dates of Nile sediments, *Nature* **196:**108-110. (First radiocarbon-dated proof of very large fluctuations of Nile River discharge, with minimum in last glacial phase. Present Sahara desiccation began before 3,000 B.P.)

1963. Geology of the world (map), *Great World Atlas*, New York: Reader's Digest, plate 139. (First global map combining geotectonics with Heezen's new seafloor structural physiography and fracture zones.)

1963. Mean sea level related to solar radiation during the last 20,000 years, *UNESCO-WMO Symposium (Rome)*, Paris: UNESCO, pp. 229-242. (Relates sea level and climate to solar control; predicts that the high-pressure tropics would be much drier during glacial phases.)

1963. Sedimentation above Wadi Halfa during the last 20,000 years, *Kush* **11:**96-107. (Field details of sedimentation during decreased discharge building up the Nile bed, in contrast to conventional model equating siltation with high flood. At times, the flow in Nubia almost ceased.)

1964. The importance of limestone and its Ca/Mg content to palaeoclimatology, in A. E. M. Nairn (ed.), *Problems in Paleoclimatology*, London: Wiley-Interscience, pp. 431-530. (Threshold theory of atmospheric evolution, analogous to punctuated theory developed later by paleontologists, recognizing five major revolutions in evolution of the Earth's atmosphere (and ocean); calcium/magnesium ratios in rocks can serve as paleoclimatic proxies.)

1964. African Ice Age aridity, in A. E. M. Nairn (ed.), *Problems in Palaeoclimatology*, London: Wiley-Interscience, pp. 356-363. (Attack on conventional view that "pluvial" deposits in tropical Africa reflect heavy precipitation during glacial stages, presenting evidence of global oceanic cooling and general desiccation.)

1964. Eiszeitklima in Nordafrika, *Geol. Rundschau* **54:**399-414. (German version of earlier North African and Nile papers, combined with new illustrations.)

1964. Thoughts about an expanding globe, in *Advancing Frontiers in Geology and Geophysics, Krishnan Volume*, Hyderabad: Osmania University Press, pp. 59-88. (Consideration of Wegener's drift theory and the new evidence of youthful seafloors; mechanisms to trigger global tectonics are multiple.)

1965. The Indian Ocean and the status of Gondwanaland, *Problems in Oceanography*, vol. 3, Oxford: Pergamon Press, pp. 83-136. (Examination of geologic and climatic history of Gondwanaland showed that Indian Ocean is a new ocean, in contrast to Pacific Ocean.)

1965. (with H. G. Richards). *Annotated Bibliography of Quaternary Shorelines*, Philadelphia: Academy of Natural Science. (Product of global collaboration by INQUA Shorelines Commission under R. W. Fairbridge, president; worldwide evidence of Holocene sea-level fluctuations, although chronology needs strengthening. Regular supplements to this bibliography were prepared by Richards up to 1986.)

1965. (with W. S. Newman). Sea level and the Holocene boundary in the eastern United States, in *Report of the Sixth International Congress on Quaternary, Warsaw, 1961*, vol. 1, *Subcommission on the Holocene*, pp. 397-418. (Urge use of term *Holocene*, bounded by pollen zone III/IV, dated about 10,500 B.P. by radiocarbon, and marked by shelf unconformity about −40 m.)

1966. Endospheres and interzonal coupling, *New York Acad. Sci. Annals* **140:**133-148. (Emphasizes interzonal coupling, mechanical and chemical, between the Earth spheres beginning with solar modulation—e.g., ultraviolet flux, ozone, and stratospheric greenhouse effect. Mass transfer on and within the Earth ensures geodynamic change, modified by cyclic changes and by irreversible secular change and possible Earth expansion.)

1966. (Editor). *The Encyclopedia of Oceanography,* New York: Reinhold, 1,021p. (Articles of climatic interest include Amphidromic Point, Calcium Carbonate Compensation Depth, Mean Sea Level Changes, Rossby Number, Storm, Storminess Waves as Energy Sources. The first volume in the Encyclopedia of Earth Sciences Series.)

1967. Phases of diagenesis and authigenesis, in G. Larsen and G. V. Chilingar (eds.), *Diagenesis in Sediments,* Amsterdam: Elsevier, pp. 19-89. (Fundamental historical review of principles of diagenesis, as temporal and spatial processes, proposing new terms *anadiagenesis* and *epidiagenesis.*)

1967. (Editor). *The Encyclopedia of Atmospheric Sciences and Astrogeology,* New York: Reinhold, 1,200p. (Articles include Earth, Galaxy, Ice-Age Theory, Zonal Circulation. A volume in the Encyclopedia of Earth Sciences Series.)

1967. Geological and cosmic cycles, *New York Acad. Sci. Annals* **138:**433-439. (Discusses scale of cycles, from human heartbeat, about 1 sec, to galactic cycle, 7×10^{15} sec; emphasizes 11-22-44-yr solar activity, Milankovitch insolation, and effect on human history.)

1967. Carbonate rocks and paleoclimatology in the biogeochemical history of the planet, in R. W. Fairbridge, C. V. Chilingar, and H. J. Bissell (eds.), *Carbonate Rocks,* Amsterdam: Elsevier, pp. 399-432 (also in Russian translation, Moscow: MIR, 1970). (Uses carbonate sediments, Ca/Mg and Ca/Sr ratios, and paleoecology to postulate that in the nonglacial past the equatorial latitudes were 3-5°C warmer, poles 20-40°C warmer, and deep sea 8-10°C warmer than today, corresponding to a high-sea-level, thalassostatic globe.)

1968. (Editor). *The Encyclopedia of Geomorphology,* New York: Reinhold, 1,295p. (Articles include Denudation, Tafoni, Terraces, Limestone Coasts, Quaternary, Holocene. A volume in the Encyclopedia of Earth Sciences Series.)

1968. (with W. S. Newman). Postglacial crustal subsidence of the New York area, *Zeitschr. Geomorphologie,* **12**(3):296-317. (Neotectonic warping proved by geodetic releveling).

1969. Early Paleozoic South Pole in northwest Africa, *Geol. Soc. America Bull.* **80:**113-114. (From recent Ordovician paleomagnetic data, deduces that traces of a former glaciation would be anticipated somewhere in northwest Africa. Interpretation of conglomerates as tillites.)

1970. World paleoclimatology of the Quaternary, *Rev. Géographie Phys. Géologie Dynam.* **12**(2): 97-104. (Ice-age glacial cooling of sea-surface temperature leads to reduced evaporation and desiccation of land, augmented by sea-ice growth and eustatic fall; "cold pluvials" or two wet and two dry cycles per 100,000 yr in some Mediterranean lands. Lists laws and principles of paleoclimate proxy interpretation.)

1970. (with J. E. Damuth). Equatorial Atlantic deep-sea arkosic sands and ice-age aridity in tropical South America, *Geol. Soc. America Bull.* **81:**189-206. (Deep-sea cores offshore from Brazil show thick clastic sections corresponding to glacial stages with unweathered arkosic sands, suggesting mechanical, arid erosion.)

1970. South Pole reaches the Sahara, *Science* **168:**878-881. (French petroleum geologists with an international group identified Ordovician glacial formations in Central Sahara, as predicted in *Sci. American* paper of 1960.)

1970. (with H. G. Richards). Eastern coast and shelf of South America, *Quaternaria* **12:**47-55. (Regional review of recent Quaternary studies.)

1970. An ice-age in the Sahara, *Geotimes* **15:**18-20. (Popular account of expedition.)

1971. (with W. S. Newman and S. March). Marginal subsidence of glaciated areas: United States, Baltic and North Seas, in M. Ters (ed.), *Etudes sur le Quaternaire dans le Monde* (VIII Congr. INQUA), Paris, pp. 795-801. (A comparison of marginal bulge areas.)

1971. Upper Ordovician glaciation in Northwest Africa? Reply, *Geol. Soc. America Bull.* **82:**269-274. (Defends glacial evidence versus landslide and "pebbly mudstone" hypothesis.)

1971. Quaternary shoreline problems at INQUA 1969, *Quaternaria* **15:**1-18. (Discussion of nonglacio-eustatic Pleistocene regression; local importance of flexure; ice-cut "marine abrasion" platforms; high-level ice-borne erratics; periglacial gravel trains; eolianites at regressions; pre-Quaternary hard-rock relief; stability of hardrock coasts and shelves; geoid adjustment versus eustasy.)

1972. Climatology of a glacial cycle, *Quaternary Research* **2:**283-302. (Ice-age preparation due to continental drift, oceanic blocking, and orogeny. Modulated by Milankovitch insolation with varied feedbacks under anaglacial, pleniglacial, and kataglacial asymmetric patterns; diachronism from low to high latitudes; sea-ice erosion at anomalous elevations. Expansion and contraction of arid zones. Paleowind maps of Africa and South America.)

1972. Quaternary sedimentation in the Mediterranean region controlled by tectonics, paleoclimates and sea level, in D. J. Stanley (ed.), *The Mediterranean Sea,* Stroudsburg, Pa.: Dowden, Hutchinson

& Ross, pp. 99-113. (New interpretation of interglacial sea levels in framework of Milankovitch and magnetochronology. Quaternary cycle effects appear to increase progressively in amplitude, superimposed on secular rise of mean sea level.)

1972. (Editor). *The Encyclopedia of Geochemistry and Environmental Sciences,* New York: Van Nostrand Reinhold, 1,321p. (Vol. 4A of the Encyclopedia of Earth Sciences Series. A companion volume, 4B, *The Encyclopedia of Mineralogy,* was edited by K. Frye in 1981.)

1973. Glaciation and plate migration, in D. H. Tarling and K. Runcorn (eds.), *Implications of Continental Drift to the Earth Sciences,* vol. 1, London: Academic Press, pp. 503-515. (Phanerozoic ice ages are discontinuous, probably reflecting galactic cycle, but require continental plates in high latitudes—"ice-age preparation.")

1973. Friends of the Mediterranean Quaternary visit type sections, *Geotimes* **18**(11):24-26.

1974. INQUA in New Zealand, *Geology* **2:**505-506. (Congress report listing 33 international commissions and subcommissions, with names of presidents and secretaries.)

1974. Articles *Holocene,* etc., *Encyclopaedia Britannica,* 15th edition. ("Holocene" is the first specifically designated article for this chronostratigraphic epoch. Its major subdivisions and dates are tabulated.)

1974. Glacial grooves and periglacial features in the Saharan Ordovician, in D. R. Coates (ed.), *Glacial Geomorphology,* Binghamton: State University of New York, pp. 317-327. (Geomorphological criteria for an ancient glaciation, exceptionally well preserved in Sahara.)

1975. Epidiagenetic silification, *IXme Congrès Internat. de Sédimentologie,* Nice, pp. 49-54. (Silica concentration in semiarid soil formation leaves almost indestructible paleosols, and therefore a most valuable paleoclimate proxy for high-pressure subtropics.)

1975. Continental margin sedimentation, *IXme Congrès Internat. de Sédimentologie,* Nice, pp. 109-121. (Illustrates systematic distinctions between trailing-edge or passive continental margin sedimentary facies and those of subductive and collision type.)

1975. Evoluzione del clima nel passato, Milano, A. Mondadori (ed.), *Scienza & Tecnica 75,* pp. 299-316. (Italian-language review of paleoclimate record of past 500 m.y.)

1975. (with A. Rice). Thermal runaway in the mantle and neotectonics, in *Recent Crustal Movements,* N. Pavoni and R. Green (eds.), *Tectonophysics* **29**(1-4):59-72. (Hypothesis of discontinuous convective transport in the mantle and its relationship to periodic acceleration of plate tectonics and seafloor spreading.)

1975. (Editor). Contributions to coastal geomorphology, *Zeitschr. Geomorphologie* **22**(Supp.):1-170. (A collective volume dedicated to coastal processes and classification.)

1975. (Editor). *The Encyclopedia of World Regional Geology, Part 1: Western Hemisphere,* Stroudsburg, Pa.: Dowden, Hutchinson & Ross, 704p. (Regional and country-by-country, multi-authored collection. A volume in the Encyclopedia of Earth Sciences Series.)

1976. The search for a boundary-stratotype of the Holocene, *Boreas* **5:**194-197. (Outlines international stratigraphic commission criteria for stratotype selection and options available.)

1976. Effects of Holocene climate change on some tropical geomorphic processes, *Quat. Research* **6:**529-556. (Human interference with landscape processes creates a "para-glacial" agency, paralleling the worldwide destruction found in glacial cycles, but short natural fluctuations in climate of natural origin antedate human influence.)

1976. Shellfish-eating preceramic Indians in coastal Brazil, *Science* **191:**353-359. (Aboriginal shell middens confirm short-lived cold cycles with wind reversals and sea-level fall. Revised version of Fairbridge Curve presented on basis of new radiocarbon dating.)

1976. (with G. F. Adams). North American cratonic warping since the Cretaceous. *EOS (Am. Geophys. Union Trans.)* (abstract) **57**:325. (Summary of a government-funded contract involving reconstruction of major erosional surfaces.)

1977. Global climate change during the 13,500-B.P Gothenburg geomagnetic excursion, *Nature* **265:**430-431. (Paleomagnetic event may mark the abrupt rise of sea level and global climate change at end of last glaciation.)

1977. (with C. Hillaire-Marcel). An 8,000-yr paleoclimatic record of the "Double-Hale" 45-yr solar cycle, *Nature* **268:**413-416. (First recognition, with radiocarbon dating, of a 45-yr cycle of a terrestrial climate proxy continuously recorded over 8,300 yr; solar correlation proposed.)

1977. Rates of sea-ice erosion of Quaternary littoral platforms, *Studia Geol. Polonica* **52:**135-141. (Evidence for sea-ice erosion of coastal rock platforms in regions with no sea ice today; such platforms absent from tropics. Anomalous elevations suggest geoid shift.)

1977. Conference report: The IGCP and sea levels in Senegal, Dec. 1976, *Zeitschr. Geomorphologie* **21:**228-235 (also in Chinese: *Geo-Science Translation,* pp. 1-2, 1978). (IGCP Project 61, led by Arthur Bloom, met in Senegal and carried out several field trips; the submission of an album of sea-level curves for the last 15,000 yr showed strong local variations, attributed partly to hydroisostasy, neotectonics, and other variables.)

1977. Discussion: Recent crustal movements, *Zeitschr. Geomorphologie* **21**(2):236-238. (Symposium of IUGG held at Grenoble in 1974; new lunar ranging equipment offers potential for monitoring plate motion and neotectonic fluctuations.)

1978. New results of landscape evolution, The Tenth INQUA Congress, Birmingham, England, 1977, *Zeitschr. Geomorphologie* **22**(2):211-222. (Notes on geomorphic items of interest at INQUA, commission leaders; R. W. Fairbridge became president of Neotectonics Commission; paleomagnetics; boundary problems.)

1978. (with C. W. Finkl, Jnr.). Geomorphic analysis of the rifted cratonic margins of Western Australia, *Zeitschr. Geomorphologie* **22**(4):369-389. (Alluvial deposits increase in coarseness, back through time, to the Permian, which carries glacially faceted boulders.)

1978. Exo- and endogenetic geomagnetic modulation of climates on decadal to galactic scale, *EOS (Am. Geophys. Union Trans.)* (abstract) **59**(4):269. (Summary of Florida American Geophysical Union meeting paper indicating the rising volume of evidence and improving chronology to suggest solar-planetary and lunar periodicities in terrestrial processes, as forcing functions.)

1978. Spaltet der Rheingraben Europa? Beispiel aus der geologischen Vergangenheit, Gondwana—das Auseinanderbrechen eines Urkontinents, *Umschau* **78:**69-75. (German-language outline of the manner and history of the Gondwanaland breakup, with comparison to the present-day Rhine Valley rifting.)

1978. (Editor with J. Bourgeois). *The Encyclopedia of Sedimentology,* Stroudsburg, Pa.: Dowden, Hutchinson & Ross, 901p. (A multi-authored, alphabetic volume in the Encyclopedia of Earth Sciences Series.)

1978. (with C. Hillaire-Marcel). Isostasy and eustasy of Hudson Bay, *Geology* **6:**117-122. (Further description, following 1977 *Nature* announcement, analyzing the isostatic uplift, which can be detrended to furnish a Holocene eustatic and climate curve.)

1978. Global cycles and climate, in A. B. Pittock et al. (eds.), *Climatic Change and Variability—A Southern Perspective,* Cambridge: Cambridge University Press, pp. 200-211. (Review of 4-b.y. climate history of Earth, emphasizing secular but punctuated changes in atmosphere, increase in oceanic alkalinity, increasing continentality, decreasing continental shelf areas, and increasing speciation in biosphere.)

1979. (with M. R. Rampino and S. Self). Can rapid climate change trigger volcanic eruptions? *Science* **206:**826-829. (Many major volcanic eruptions in the historical period occurred during cold cycles, and thus the dust veil was only a secondary cause of planetary cooling.)

1979. (Editor with D. Jablonski). *The Encyclopedia of Paleontology,* Stroudsburg, Pa.: Dowden, Hutchinson & Ross, 886p. (An alphabetic, multi-authored autonomous encyclopedia in the Encyclopedia of Earth Sciences Series.)

1979. (with C. W. Finkl, Jnr.). Paleogeographic evolution of a rifted cratonic margin: S.W. Australia, *Palaeogeography, Palaeoclimatology, Palaeoecology* **26:**221-252. (The western continental margin was initiated with an aulacogen that opened from the north in Ordovician time. Permian glaciers cut fjords in the scarp, but the seaward side only broke away when seafloor spreading began in Jurassic time.)

1979. Traces from the desert: Ordovician, in B. S. John (ed.), *The Winters of the World,* Newton Abbot, U.K.: David and Charles, pp. 131-153. (Popularization of the Ordovician ice age; discussed postglacial 300-500 m sea-level rise in Silurian, possibly caused by pole shift and geoid adjustment.)

1979. Vertical crustal movements and the rifting of continents, *Geologie en Mijnbouw* **58**(2):273-276. (Heat flows associated with continental rifting cause rim uplift, triggering rapid erosion; entry of ocean waters into Paleozoic grabens or aulacogens provided moisture for Gondwana glaciations and rifts became sites of fjords prior to Mesozoic breakup and seafloor spreading.)

1980. Prediction of long-term geologic and climatic changes that might affect the isolation of radioactive waste, *Underground Disposal of Radioactive Wastes* **2:**285-405 (Vienna: International Atomic Energy Agency, IAEA-SM-243/43). (Analysis of global geodynamic and climatic processes that can disturb the crust and so-called safe disposal sites. Appendix supplies synthesis of four fundamental Earth laws and principles, commonly unknown outside geology.)

1980. Review: *Earth Rheology, Isostasy and Eustasy,* a new book edited by N.-A. Mörner. *Zeitschr. Geomorphologie N.F.* **24**(4):472-478. (Discussion of the importance of Scandinavian data in our understanding of these interrelated processes.)

1980. (with C. W. Finkl, Jnr.). Cratonic erosional unconformities and peneplains, *Jour. Geology* **88:**69-86. (Demonstrates repeated occupation of cratons by shallow seas, followed by regression and re-exposure, but resulting in very little erosion over very long intervals.)

1980. Thresholds and energy transfer in geomorphology, in D. R. Coates and J. D. Vitek (eds.), *Thresholds in Geomorphology,* London: Allen & Unwin, pp. 43-49. (Discussion of episodic nature of surficial processes.)

1980. The estuary: Its definition and geodynamic cycle, in E. Olaussen and I. Cato (eds.), *Chemistry and Biogeochemistry of Estuaries,* New York: J. Wiley & Sons, pp. 1-35. (Estuary redefined in terms of both tides and mixing analyzed in light of 6,000-yr average age, disclosing Davisian cycle, modulated by climatic/eustatic dynamics.)

1980. (Editor). *Plianeta Mare: Enciclopedia di Scienza,* Milano: Fabbri. (Also contributor of many sections dealing with oceanography and exploration of the seas.)

1981. *Grande Enciclopedia delle Natura,* Milano: Fabbri. (Contributions for each major period of geologic time.)

1981. The concept of neotectonics—An introduction, *Zeitschr. Geomorphologie,* **40**(Suppl.):vii-xii. (Outline of history and meaning of term, with appreciation of its present-day importance.)

1981. Holocene wiggles, *Nature* **292:**670. (Conference report of London INQUA commission meeting, treating ice cores, varves, and tree rings.)

1981. Holocene sea-level oscillations, in L. K. Königsson and K. Paabo (eds.), *Striae, Florilegium Florinis Dedicatum,* vol. 14, Uppsala, pp. 131-141. (History of Fairbridge Curve and its Swedish chronology. Examines Holocene climate record in light of exogenic, solar-planetary determinism.)

1982. The fracturing of Gondwanaland, in R. A. Scrutten and M. Talwani (eds.), *The Ocean Floor,* Chichester/New York: J. Wiley & Sons, pp. 229-235. (The megacontinent developed lineaments and aulacogens in the Precambrian, leading to Proterozoic intracontinental geosynclinal belts. Repeated fracturing occurred during the Phanerozoic, with seafloor spreading only during the later Mesozoic and Cenozoic.)

1982. The Holocene boundary stratotype: Local and global problems, *Sveriges Geol. Undersökning Arsb.* **794c:**281-286. (Review of type area and desirable features for the stratotype selection.)

1982. (with S. Hameed and F. Hassan). An 1112-yr cycle analysis of Nile floods, auroras and sunspots, *Geol. Soc. America Abs. with Programs* **14**(7):486. (First report on long-term solar system cycles applied to a terrestrial monsoon area.)

1983. (with S. Hameed). Phase coherence of solar cycle minima over two 178-year periods, *Astron. Jour.* **88**(6):867-869. (The so-called 11-yr cycle undergoes a systematic change in length that is twice repeated.)

1983. Syndiagenesis—Anadiagenesis—Epidiagenesis: Phases in Lithogenesis, in G. Larson and G. V. Chilingar (eds.), *Diagenesis in Sediments,* Amsterdam: Elsevier, pp. 17-113. (Fundamental sedimentological processes that vary with environment, burial, and exposure; long-term variations in storage or recycling with burial of elements normally associated with hydrosphere and atmosphere.)

1983. The Pleistocene-Holocene boundary, *Quaternary Sci. Rev.* **1:**215-244. (Report of stratigraphic commission decision with discussion of methods for improving chronology, including solar system and lunar determinism.)

1983. Himalayan uplift and growth of Ganges-Brahmaputra delta, *Geol. Soc. America Abstracts with Programs* **16**(6):569. (Brief announcement of sedimentologic and palynologic demonstration of time and rate of mountain uplift.)

1983. Isostasy and eustasy, in D. Smith and A. Dawson (eds.), *Shorelines and Isostasy,* London: Academic Press, pp. 3-25. (Analysis of Hudson Bay climate. Analysis of isostatic/eustatic interactions on Hudson Bay, Mississippi delta and Pacific atolls.)

1984. (with C. W. Finkl, Jnr.). Tropical stone lines and podzolized sand plains as paleoclimatic indicators for weathered cratons, *Quaternary Sci. Rev.* **3**(1):41-72. (Paleoclimatic catastrophism and secular change in generating widespread geomorphic-pedologic features.)

1984. Planetary periodicities and terrestrial climate stress, in N. A. Mörner and W. Karlén (eds.), *Climatic Changes on a Yearly to Millennial Basis,* Dordrecht: Reidel Publ. Co., pp. 509-520. (Planetary orbital stress, mainly by Jupiter and Saturn, modulates spin rates and orbital velocities of the Sun, affecting solar radiation, and of the other planets, including the Earth/Moon pair, affecting climatic processes subject to gravitational and geomagnetic modulation.)

1984. The Nile Flood as a global climatic/solar proxy, in N. A. Mörner and W. Karlén (eds.), *Climatic Changes on a Yearly to Millennial Basis,* Dordrecht: Reidel Publ. Co., pp. 181-190. (Since A.D. 622 flow data provide world's longest annual climate proxy, disclosing both lunar and solar input and singularities calling for individual explanations such as volcanic eruptions.)

1985. (with R. G. Currie). Periodic 18.6-year and cyclic 11-year induced drought and flood in northeastern China and some global implications, *Quaternary Sci. Rev.* **4:**109-134. (A 500-yr proxy record discloses very large fluctuations in the east Asian monsoon and zonal circulation. Both solar and lunar periodicities suggest exogenetic forcing.)

1986. (Editor with J. E. Oliver). *The Encyclopedia of Climatology,* New York: Van Nostrand Reinhold, 986p. (A multi-authored, alphabetic, autonomous volume in the Encyclopedia of Earth Sciences Series.)

1986. (with W. S. Newman). The management of sea-level rise, *Nature* **320:**319-321. (The hydrologic cycle is regulated by dam construction and can be much further controlled if the need arises but at a considerable economic and environmental cost.)

I

Historical Climate Change

1: Nile Floods and Climatic Change

Fekri A. Hassan
Washington State University

Barbara R. Stucki
Northwestern University

Abstract: Analysis of the cumulative deviation of annual flood maxima and minima from 100-yr trend values for the past 1,300 yr reveals distinct episodic variations in the magnitude of flood discharge. Statistical analysis of the differences between the deviations of flood maxima and minima and of the climate of the Nile basin since 1650 indicates that the variations are caused by differences in the contribution from equatorial Africa relative to that from Ethiopia. Episodes of high Nile flood discharge correspond with an increase in winter-early spring and late fall precipitation in eastern Africa. These episodes also coincide with warmer climate in northern Europe, high lake levels in eastern Africa, and greater rainfall in the Sahara.

HISTORICAL NILE FLOOD DISCHARGE AND CLIMATIC CHANGE

The rise of the Egyptian civilization and the continued prosperity of Egypt until modern times has rested on an agricultural economy based on irrigation by Nile water. The Nile in Egypt is an allogenous river, which derives its water from distant sources in Ethiopia and the Lake Plateau of equatorial Africa. At present, the Nile north of Atbara in the Sudan is fed mostly by the water delivered by the Ethiopian tributaries, the Atbara and the Blue Nile. Compared with an average discharge of 820 m^3/s from the White Nile, which receives its water from the Lake Plateau, the Blue Nile delivers an annual average of 1,620 m^3/s and the Atbara 380 m^3/s. The average contribution from the Ethiopia tributaries thus amounts to about 75% of the total discharge. Because of differences in the timing of rainfall over the Lake Plateau and Ethiopia, the main Nile experiences its major flood surge during the summer when the contribution from Ethiopia peaks to 90-95% of the total discharge. During the period of low Nile (March-June), the contribution from Ethiopia is very low. At that time, the contribution from the White Nile rises to about 75% of the total discharge (Willcocks and Craig, 1913; Hurst, 1957).

Cognizant of the relationship between Nile flood heights and agricultural productivity, the Egyptians began, apparently well before the rise of the unified Egyptian state circa 3150 B.C., to keep records of the Nile flood heights. The earliest known record, however, dates to the Early Dynastic period (Bell, 1970). This record was carved on a large stone stele during the Fifth Dynasty and includes the height of the flood for every year back to the reign of King Djer (about 3090 B.C.). The measurements of the Nile flood heights were made either by a portable measuring device or by Nilometers consisting of a staircase built into a well or quay with a fixed scale. The measurements were made in cubits and fingers. The cubit measured about 54 cm and consisted of 24 fingers. Traces of Nilometers from ancient Egypı are mostly from the Greco-Roman times. The Roda Nilometer near Cairo was first built in A.D. 715 and rebuilt in A.D. 861, with numerous repairs and modifications since then (Ghaleb, 1951).

The history of the Nile was of particular fascination to one member of Egypt's ex-royal family, Prince Omar Toussoun, to whom we are indebted for a comprehensive compilation of the record of Nile flood maxima and Nile flow minima from A.D. 622 to 1921. The data are mostly from the measurements taken at the Roda Nilometer and were gleaned from original Arabic texts. Toussoun also attempted to convert the Nile flood heights to elevation above sea level, but his results, which provide the basis for the plots by Jarvis (1936), suffer from several inadequacies. Excavation of the Roda Nilometer for restoration by Ghaleb (1951) on behalf of the ministry of public works and a critical examination of the records by Popper (1951) revealed that Toussoun ignored changes in the basal level of the Nilometer above sea level, variations in the length of a cubit, and changes in the number of fingers per cubit. Taking into consideration these sources of error, one of us (Hassan) began to prepare a table of corrected values of the height of Nile flood maxima. With the assistance of B. Stucki, the conversions were later computerized and a set of tables of corrected flood maxima and minima were prepared. In 1981, Hassan provided a first approximation of the episodic variation in Nile floods based on an analysis of centennial averages of flood maxima and using three different statistical techniques. In this chapter, the results of an analysis of the annual Nile flood maxima and minima by the Kraus method (1955) provide the basis for an examination of the link between Nile floods and climatic change.

The corrected flood heights used here are based on the conversion scale proposed by Popper (1951). Comparison between the corrections from Popper with those from Ghaleb (1951) revealed no significant differences in the pattern of variation (Fig. 1-1A-1-1E). The main difference lies in the absolute height, especially from A.D. 1520 to 1700 (Fig. 1-1D).

The plots in Figure 1-2A to E represent the normalized cumulative deviation. To obtain the value of this parameter for individual years, flood heights were converted to discharge by a formula based on a strong correlation ($r = 0.93$) between flood height *(H)* and discharge *(V)* based on data by Willcocks (1889):

$$V = 780.63 + 57.49H.$$

It was also deemed necessary, considering that flood heights show a progressive variation through time as a result of siltation, to estimate the trend value using a 50-yr moving average. The deviation of flood height *(H)* from the trend value of flood height *(dH)* is then calculated and converted to a deviation of discharge *(dV)* from the formula

$$dV = 57.49 \times dH.$$

The normalized cumulative deviation (see Fig. 1-2) is then obtained from the cumulative sum of deviations divided by the trend value of discharge.

Although the cumulative deviations of minima and maxima generally correspond, some occasional differences do appear. The differences between the flood minima and maxima result from the difference in the source of the water during the low flow stage and the maximum flood stage. During the period of maximum discharge, almost 100% of the water is from the Ethiopian tributaries (mostly the Blue Nile and the Atbara). During the period of minimum flow stage, by contrast, about 75% of the water is from equatorial Africa (the White Nile and its tributaries). To assess the long-term differences between the contribution from Ethiopia relative to that from eastern Africa, the cumulative difference between the deviations of flood minima and maxima was calculated (Fig. 1-3), using the formula

$$I = (dV_{\min_i} - dV_{\max_i}),$$

where $dV_{\min_i}$ and $dV_{\max_i}$ are the deviations from average minimum and maximum discharge for the *i*th decade, respectively. A value of *I* greater than zero would indicate greater contribution from the equatorial African sources relative to

Figure 1-1. Nile flood maxima (in meters) showing calibrations based on both Ghaleb's (*) (1951) and Popper's (x) (1951) corrections.

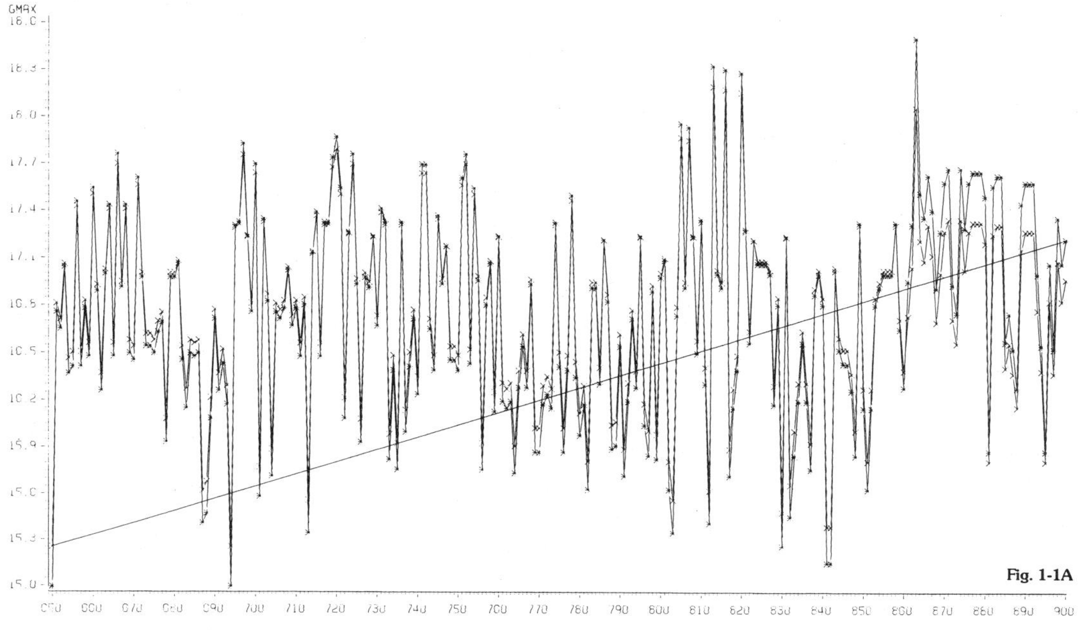

Year A.D. 650 to 900

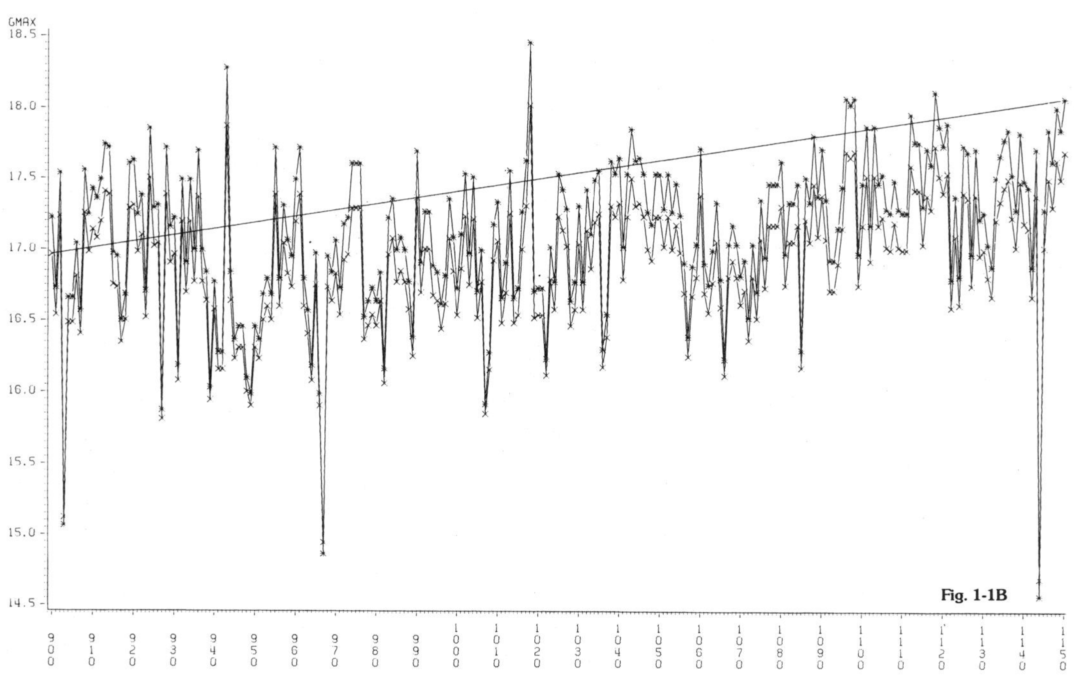

Year A.D. 900 to 1150

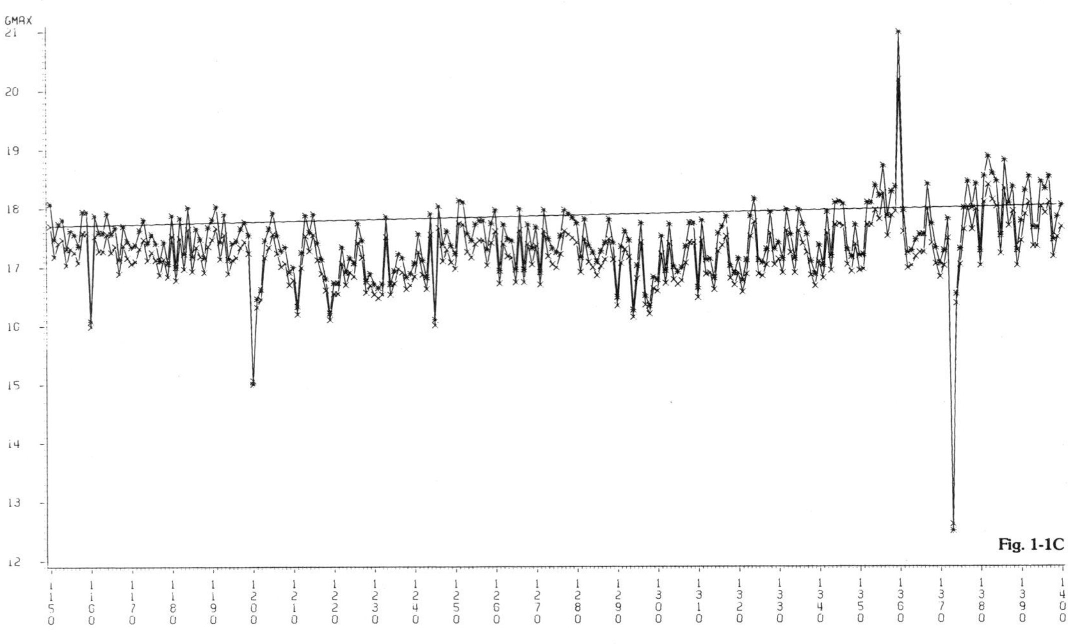

Fig. 1-1C

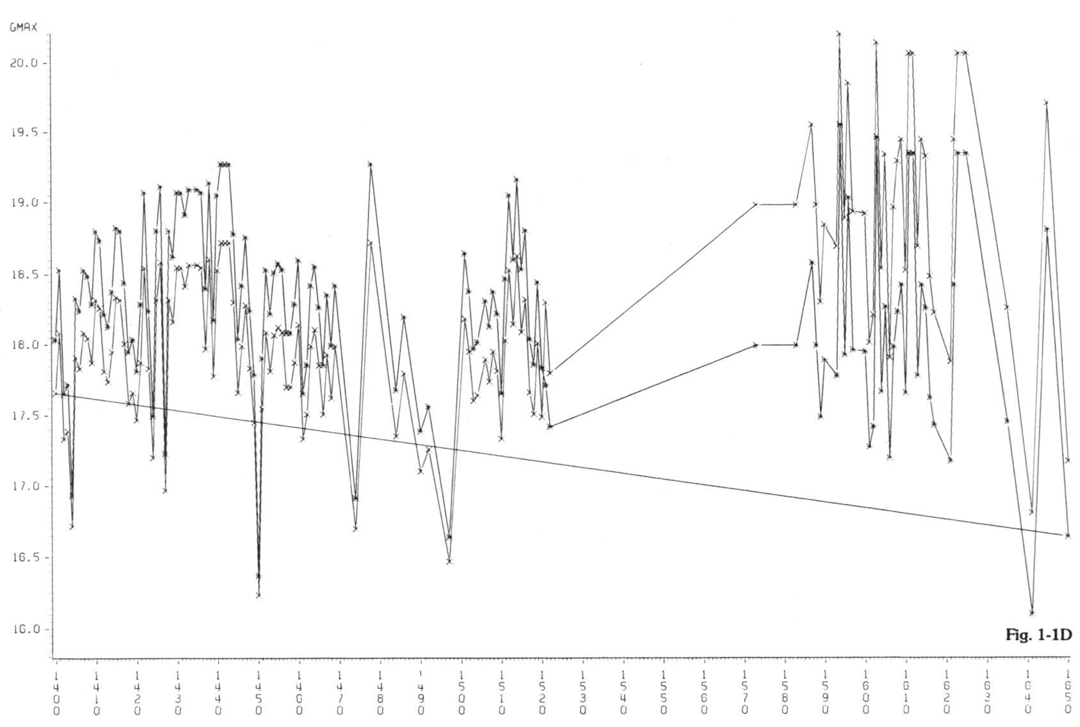

Fig. 1-1D

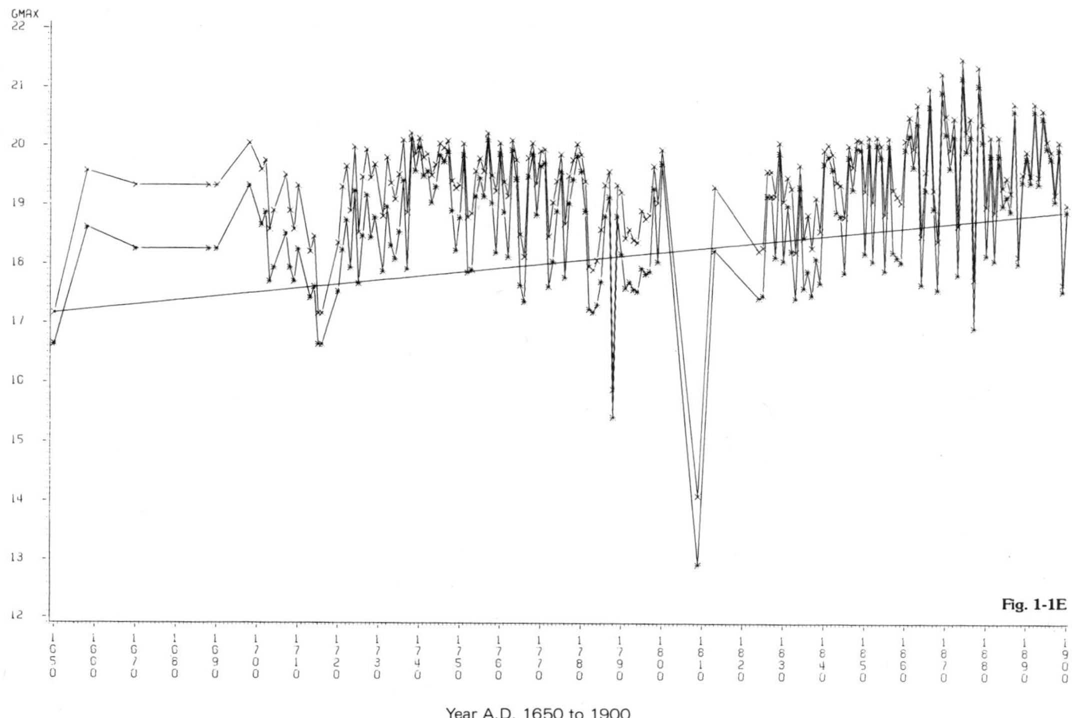

Figure 1-2. Normalized cumulative deviations of Nile flood minima (+) and maxima (x) from 50-yr moving averages. Flood heights used were calibrated following Popper (1951) corrections.

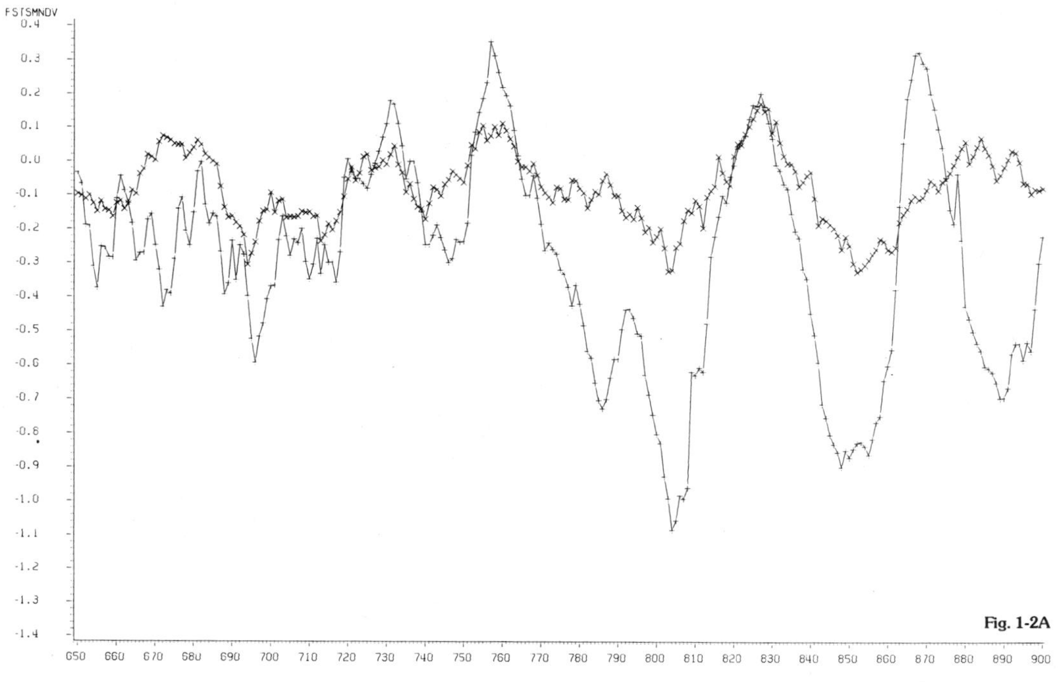

Year A.D. 650 to 900

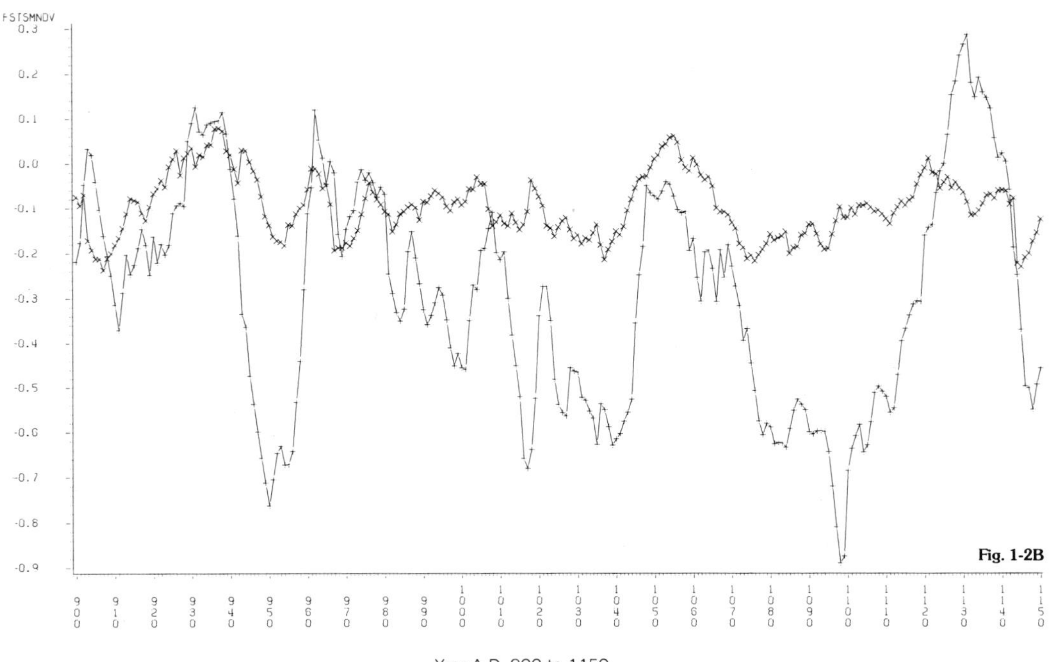

Fig. 1-2B

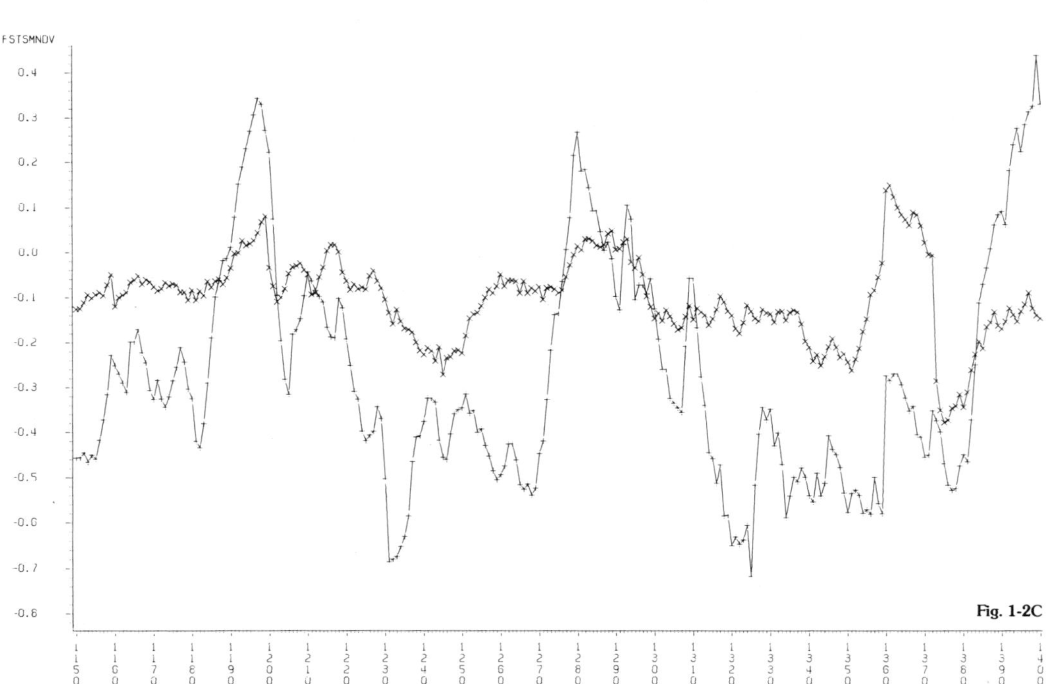

Fig. 1-2C

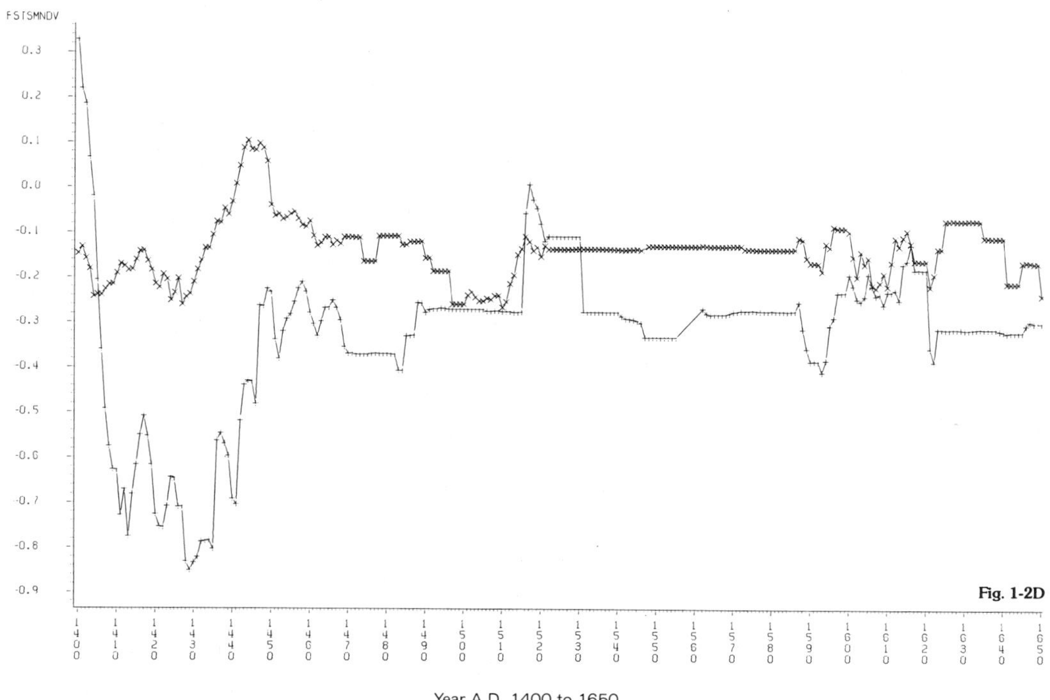

Fig. 1-2D

Year A.D. 1400 to 1650

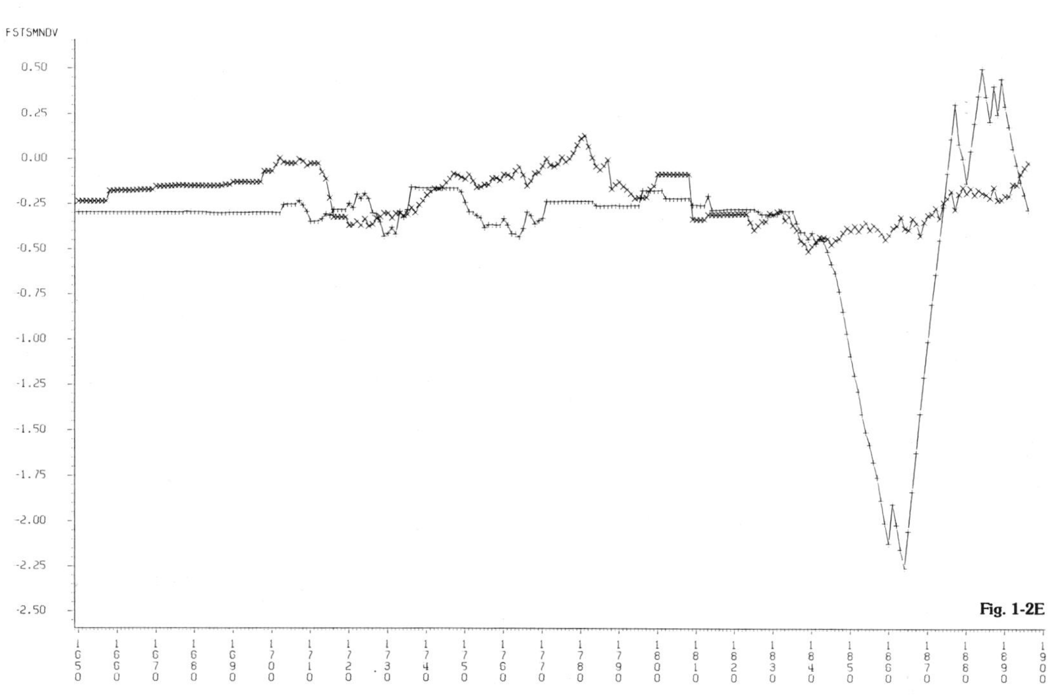

Fig. 1-2E

Year A.D. 1650 to 1900

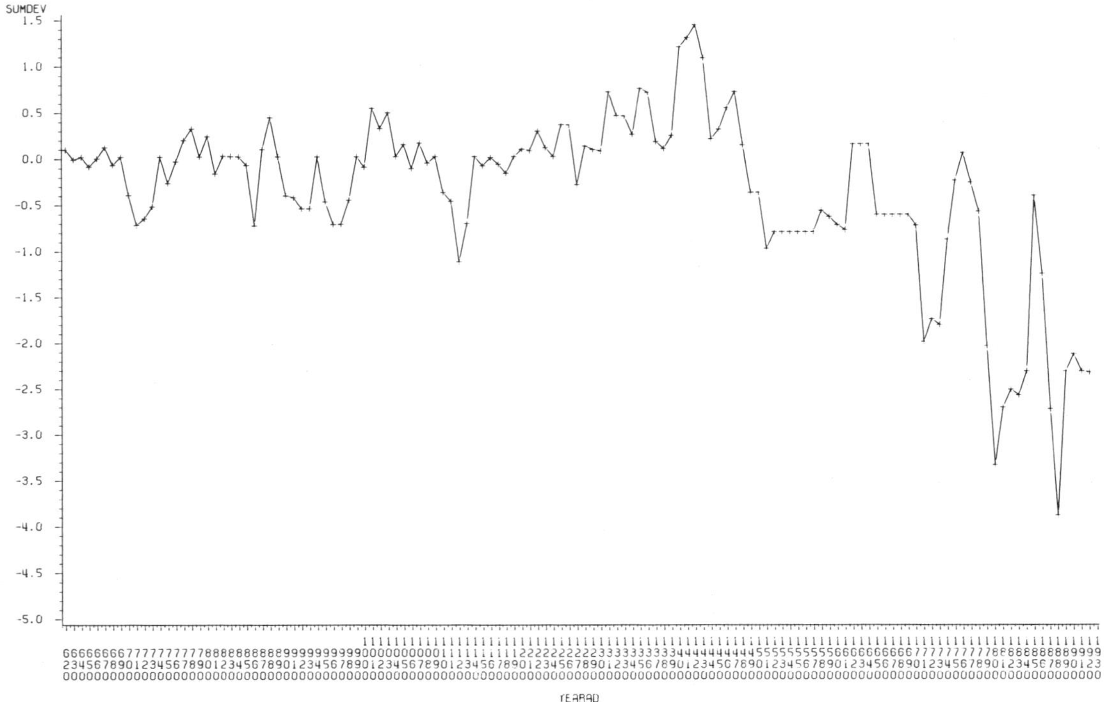

Figure 1-3. Cumulative differences between the deviations from decadal flood averages of flood maxima and minima. Peaks indicate greater contributions from the equatorial sources of the Nile relative to those from Ethiopia. (From Popper, 1951)

those from Ethiopia (independent of the magnitude of discharge). The graph shows that, at the scale of decades, the value of *I* varies markedly. In addition, the graph discerns three main episodes: (1) a period of more or less equitable contribution from both sources from 650 to 1250, (2) a period of greater contribution from equatorial Africa from 1250 to 1480, and (3) a period of lower discharge from the equatorial sources from 1480 to 1870.

NILE RIVER DISCHARGE AND CLIMATE: 1650–1960 A.D.

The significance of these changes in the sources of water to the Nile may best be examined by a detailed investigation of the Nile hydrology from 1650 to 1960, with special emphasis on the period from 1870 to 1960 when data from various sources are available on the discharge of Nile tributaries and climate at many places in the world.

The cumulative deviations of the flood maxima show a low Nile discharge from 1710 to 1745 and again from 1800 to 1880 (see Figs. 1-2E and 1-3). From 1880 to 1945 Nile discharge was high, subsequently declining with a minor episode of increased discharge in the 1960s. The variation in maximum Nile discharge closely parallels that of the level of Lake Chad (Hassan, 1981; Maley, 1973, 1981). The reduction in maximum Nile discharge since 1880 coincides with a reduction in rainfall in a number of stations in tropical regions. The effect is most pronounced in the rainfall data from Sierra Leone and Freetown (Bell, in Bell and Menzel, 1972). Bell also examined the hydrographic changes of the major Nile tributaries (Atbara, Blue Nile, Sobat, White Nile) from the beginning of their records early in this century. Her results clearly show that the increase in the discharge of the main Nile in the 1960s corresponds to an increase in the flow of the White Nile by as much as 150-250% of the 1911-1960 mean. The flood levels of the Blue Nile and Sobat do not show a significant increase in flow during the winter and early spring when they are normally low. Lake Victoria in eastern Africa also showed a rise in its level of 1-2 m about its twentieth-century norm through the 1960s (Schove, 1977). River runoff in the Sahel also showed a marked increase in the 1960s (Faure and Gac, 1981).

Bell also investigated the rainfall records of 183 stations in eastern Africa. She found that 167 stations showed wetter conditions in the period 1961-1970 compared with their mean from the 1931-1963 period. Only 16 stations recorded a reduction in rainfall. These stations were concentrated mostly in the dry region of the coastal area between Mombassa and Dar es Salaam. The increased rainfall extended into northern Zambia to about 12°N and into southern Sudan to about 8°N. Throughout eastern Africa the greatest and most consistent increase in rainfall occurred in November, October, February, and March.

From these investigations it is clear that the increase in Nile flood discharge is a result of an increase in winter and early spring and late fall over eastern Africa and Ethiopia. These rainfall changes are reflected in relatively greater discharge from the equatorial sources than the norm, in agreement with previous findings by Hassan (1981, p. 1,144).

Reduction in Nile discharge also correlates with lowering of temperature in northern Europe. The correlation is clear from a comparison of the records of a number of stations where there are long instrumental records (Barry and Chorley, 1974, p. 321). The temperature records show that the low Nile flood discharge from 1800 to 1880 corresponds with temperature lower than that recorded for the period 1890-1940 when the Nile discharge was high. Kukla (1969) also notes a worldwide rise in mean winter temperature and a consequent rise of sea level of about 1 mm/yr from 1870 to 1940. This correspondence between periods of reduced Nile flow and periods of lower temperature in Europe during the Holocene as suggested by Fairbridge (1972) and Hassan (1981) is thus in agreement with the historical data.

Periods of low Nile discharge also correspond with arid episodes in northern Africa. In a pioneer study of Nile sedimentation and climate, Fairbridge (1963) concluded that high Nile discharge resulted from an increase in equatorial and monsoonal rains. Nicholson and Flohn (1980) provide an explanatory climatic model for the increase in rainfall over eastern and northern Africa during the 10,000-8,000 B.P. period. They noted that rainfall in the southern Sahara develops from the interaction of troughs in the upper-level westerlies with humid air south of the Intertropical Convergence Zone (ITCZ). Nicholson and Flohn hypothesize that a re-establishment of the humid southwesterly monsoons in western Africa and a northward shift of the ITCZ must have led to greater interaction between the westerlies and the humid air south of the ITCZ, accounting for a greater rainfall in the southern Sahara. Such a climatic regime would also lead to an increase in summer monsoonal rain in northern Africa. Rognon and Williams (1977) and Bryson and Murray (1977, pp. 104-106) also attribute climatic change over northern Africa to the effect of the jet stream of westerlies, which influences the position of the subtropical anticyclone, which in turn affects the poleward movement of the ITCZ. The climatic changes appear to correspond with increased solar heating and consequent changes in albedo (Tanaka et al., 1975).

The variations in the cumulative difference between the deviations of flow minima and maxima may thus serve as an indicator of the strength of the southwesterly monsoons and the poleward shift of the ITCZ. As noted earlier, the period since A.D. 1480 represents a major reduction in the contribution from equatorial Africa relative to that from Ethiopia compared with the period from A.D. 650 to 1480. This change coincides with the onset of the Little Ice Age in Europe from 1430-1550 to 1850-1920 (Lamb, 1972; Bryson and Padoch, 1981, p. 12).

The pattern of variation of the cumulative differences between flood maxima and minima also corresponds well with the decadal mean annual temperature in Iceland and other records of temperature (Bryson, 1975), with an increase in temperature matched with relatively greater contributions for equatorial Africa. This finding is in good agreement with the strong correlation in tropical Africa between warm periods and heavy rainfall extending well back into the Pleistocene (Rossignol-Strick, 1983). The paleoclimate evidence of strong monsoons between 10,000 yr and 5,000 yr ago in Africa also seems to have resulted from an increase in global radiation (Kutzbach, 1981).

REFERENCES

Barry, R. G., and Chorley, R. J., 1974, *Atmosphere, Weather, and Climate,* London: Methuen (first published 1968), 379p.

Bell, B., 1970, The oldest records of the Nile floods, *Geog. Jour.* **136:**569-573.

Bell, B., and Menzel, D. H., 1972, Toward the observation and interpretation of solar phenomena, AFCRL F19628-69-C-0077 and AFCRL-TR-74-0357, Air Force Cambridge Research Laboratories, pp. 8-12.

Bryson, R. A., 1975, The lessons of climatic history, *Environmental Conservation* **2**(3):163-170.

Bryson, R. A., and Murray, T. J., 1977, *Climates of Hunger,* Madison: The University of Wisconsin Press, 171p.

Bryson, R. A., and Padoch, C., 1981, On the climates of history, in I. Rotberg and T. K. Rabb (eds.), *Climates and history,* Princeton, N.J.: Princeton University Press, pp. 3-17.

Fairbridge, R. W., 1963, Nile sedimentation above Wadi Halfa during the last 20,000 years, *Kush* **11:**96-107.

Fairbridge, R. W., 1972, Quaternary sedimentation in the Mediterranean region controlled by tectonics, paleoclimate, and sea level, in D. J. Stanley (ed.), *The Mediterranean Sea: A Natural Sedimentation Laboratory,* Stroudsburg, Pa.: Dowden, Hutchinson & Ross, pp. 99-113.

Faure, H., and Gac, J.-Y., 1981, Will the Sahelian drought end in 1985? *Nature* **291:**475-478.

Ghaleb, K. O., 1951, Le Mikyas ou Nilométre de l'Île de Rodah, *Inst. Desert. Egypte,* Mém. **54,** 182p.

Hassan, F. A., 1981, Historical Nile floods and their implications for climatic change, *Science* **212:**1142-1145.

Hurst, H. E., 1957, *The Nile,* London: Constable, 331p.

Jarvis, C. S., 1936, Flood-stage records of the River Nile, *Am. Soc. Civil Engineers Trans.* **101:**1012-1071.

Kraus, E. B., 1955, Secular changes of tropical rainfall regimes, *Royal Meteorol. Soc. Quart. Jour.* **81:**198-210.

Kukla, G., 1969, The causes of the Holocene climatic change, *Geologie en Mijnbouw* **48**(3):307-334.

Kutzbach, J. E., 1981, Monsoon climate of the early Holocene: Climate experiment with the Earth's orbital parameters for 5000 years ago, *Science* **214:**59-61.

Lamb, H. H., 1972, *Climate: Present, Past, and Future,* London: Methuen, 613p.

Maley, J., 1973, Mechanisme des changements climatiques dans basses latitudes, *Palaeogeography, Palaeoclimatology, and Palaeoecology* **14:**193-227.

Maley, J., 1981, Etudes palynologiques dans le bassin du Tchad et paleoclimatologie de l'Afrique nord-tropicale de 30,000 ans à l'epoque actuelle, *Travaux et Document de l'Orstom No. 129,* pp. 1-586.

Nicholson, S. E., and Flohn, H., 1980, African environment and climatic changes and the general atmospheric circulation in late Pleistocene and Holocene, *Climatic Change* **2:**313-348.

Popper, W., 1951, *The Cairo Nilometer,* Berkeley: University of California Press, 269p.

Riehl, H., and Meitin, J., 1979, Discharge of the Nile River: A barometer of short-period climate variations, *Science* **206:**1178-1179.

Riehl, H.; El-Bakry, M.; and Meitin, J., 1979, Nile River discharge, *Monthly Weather Rev.* **107:**1546-1553.

Rognon, P., and Williams, M. A. J., 1977, Late Quaternary climatic changes in Australia and North Africa: A preliminary interpretation, *Palaeogeography, Palaeoclimatology, Palaeoecology* **21:**285-327.

Rossignol-Strick, M., 1983, African monsoons and immediate climatic response to orbit insolation, *Nature* **304:**46-49.

Schove, D. J., 1977, African droughts and the spectrum of time, in D. Dalby, R. O. Harrison Church, and F. Bezaz (eds.), *Drought in Africa,* Afrenviron Special Report No. 6, London: International African Institute, pp. 38-53.

Tanaka, M.; Weare, B. C.; Navato, A. R.; and Newell, R. E., 1975, Recent African rainfall patterns, *Nature* **255:**201-203.

Willcocks, W., 1889, *Egyptian Irrigation,* London: Spon, 448p.

Willcocks, W., and Craig, J. I., 1913, *Egyptian Irrigation,* 3rd Edition, Vol. 2, London: Spon, pp. 449-884.

2: Climatic Consequences of Anthropogenic Vegetation Changes from 1880–1980

Vivien Gornitz
Lamont-Doherty Geological Observatory
Columbia University
and
Goddard Space Flight Center

Abstract: Anthropogenic modification of the land cover can influence climate directly through release of CO_2 into the atmosphere, through changes in the surface albedo, and indirectly through disruption of the hydrologic cycle and in generation of wind-blown dust and aerosols. This chapter focuses on the albedo changes associated with land-use alterations from the end of the nineteenth century to the present and the ensuing climatic implications.

The expansion of cultivated land provides one of the few documented records of land-use change, yet estimates of conversion to agriculture since the end of the nineteenth century range between 725 million ha and 931 million ha. Since the 1950s, the global forest area has remained close to 4 billion ha, but poor statistics in older surveys and some temperate-land reforestation mask any anticipated decrease. At present, deforestation affects between 9 million ha and 19 million ha per year (Seiler and Crutzen, 1980; Lanly, 1982).

A re-examination by Henderson-Sellers and Gornitz (1984) of Sagan, Toon, and Pollack's (1979) calculation of albedo changes associated with anthropogenic vegetation changes during a 30-yr period showed an increase in albedo of only around half that reported by Sagan and co-workers. These albedo changes would result in a temperature decrease of only 0.06-0.09 K.

Western Africa, as a microcosm of vegetation changes in the tropics, was selected for more detailed study. Land-use changes in western Africa were reconstructed from historical records, censuses, atlases, and descriptive reports. Agricultural and population censuses suggest that since 1930, permanently cultivated land has increased by about 40 million ha (44 million ha extrapolated to 1910). If fallow land is included, up to 115 million ha may have been cleared since 1910. The forest zone in the twentieth century has been reduced by around 60-70%. A conservative estimate suggests removal of up to 123 million ha of forest and woodland.

Albedo changes corresponding to these land-use alterations were calculated. The mean regional albedo increased from 17.3% (natural vegetation only) to 17.4% in 1880 and 17.8% in 1980, or an increase of 0.4% since the late nineteenth century and 0.5% since agriculture began.

Precipitation shows no apparent secular decrease linked to vegetation clearing, as suggested by some climatic models. However, disruption of the hydrologic cycle as a result of devegetation may have led to reductions in soil moisture, placing plants under stress and simulating a climatic desiccation.

INTRODUCTION

The role of humans as a geologic agent in altering the natural environment is now accepted. One of the most pervasive means of environmental alteration has been by clearing the natural vegetation for agriculture. The use of fire to remove forests and woodlands for improved hunting, grazing, farmland, and soil fertility may have been the earliest form of environmental modification (Stewart, 1956; Bartlett, 1956; Kozlowski and Ahlgren, 1974). Much of the Earth's grasslands and savannas may have been created or enlarged artificially because of burning (Stewart, 1956; Bartlett, 1956; Aubreville, 1949; Looman, 1983).

Modification of the vegetation cover has led to destabilization and erosion of soils, and changes in the hydrologic cycle and has created a potential for global climate change. Anthropogenic disturbance to the vegetation cover impacts climate directly by altering the atmospheric CO_2 level and the surface albedo, thus affecting the Earth's radiation balance. Burning to clear land for agriculture and the generation of wind-borne dust from bare fields or deforested slopes indirectly introduce aerosols and particles into the troposphere, which in turn interfere with the absorption of sunlight and the condensation of moisture. Another indirect influence on climate is the disruption of the hydrologic cycle by altering stream flow, water infiltration into soils, and evapotranspiration.

This chapter briefly reviews the various ways in which land-use modification can influence climate; however, the impact of surface albedo changes receives the major emphasis. To estimate the magnitude of the albedo change, it is necessary to reconstruct more accurately the history and areal extent of land conversion, primarily for agriculture. Two recent conferences on global deforestation have begun to address this issue (Tucker and Richards, 1983; Richards and Tucker, 1987), but only limited regional coverage has been completed to date.

Another section summarizes global land-cover changes since 1880. Estimates of land conversion from agricultural expansion and deforestation are presented. Western Africa has been examined in greater detail than other areas (Gornitz, 1985). This region, comprising 17 countries, offers a good sample of the range of land-use transformations affecting the tropical zone. Western Africa spans a variety of vegetation zones, grading latitudinally northward from rain forest to desert scrub. The region encompasses both problems of desertification in the Sahel and extremely rapid deforestation of the tropical forest zone (e.g., Ivory Coast, Nigeria, Ghana). The principal processes of land-cover modification during the twentieth century include clearing of natural vegetation for agriculture, grazing, and logging and degradation of semiarid to arid vegetation under the combined influence of drought and overgrazing. Albedo changes associated with these land-use conversions form the basis of the last section. In addition, it discusses the climatic effects of albedo change.

REVIEW OF THE INFLUENCE OF ANTHROPOGENIC VEGETATION CHANGES ON CLIMATE

Atmospheric CO_2

The CO_2 concentration of the atmosphere affects the heat balance because CO_2 is transparent to incoming solar radiation but absorbs outgoing infrared radiation, reradiating it back toward the Earth's surface, thus leading to global warming (the so-called greenhouse effect). Trace gases such as N_2O, CH_4, and chlorofluoromethanes may reinforce this effect (Hansen et al., 1981).

The best estimates of pre-industrial atmospheric CO_2 concentration lie in the range of 260-280 ppmv, whereas back-extrapolation of the Mauna Loa data to 1860 (assuming that the increase was due solely to fossil fuel combustion) yields a value of 294 ppmv (Bojkov, 1983). Thus, the present atmospheric value of 340 ppmv CO_2 represents an initial level of around 270 ppmv, plus contributions from both fossil fuel combustion and the biosphere through deforestation and decay of exposed soil humus (Kerr, 1983). Although researchers agree that the biosphere has been a source of CO_2 during the 1900s, the total net release and its disposition are still widely debated (for a recent summary, see Clark, 1982). Woodwell et al. (1983) find a cumulative C release of around 180×10^{15} g, between 1860 and 1980, with a 1980 release of $1.8\text{-}4.7 \times 10^{15}$ g. In contrast, Richards et al. (1983) estimate a C release of only 62×10^{15} g over the same period, corresponding to an annual average release of 0.5×10^{15} g. The discrepancy among these figures arises largely from uncertainties in the history of forest clearing, particularly in the tropics; the decay of organic soil matter; and the biomass. Richards et al. may underestimate total vegetation clearance since they omit activities such as logging, grazing, firewood collection, and shifting cultivation.

Although the ocean is a potentially large sink for CO_2, since 1958 only 37% of the CO_2 released by fossil fuel combustion appears to be taken up by the oceans (Broecker et al., 1979). Adding this figure to the airborne fraction (52%) and organic burial accounts for a total of 91% of the observed increase in CO_2. Therefore, the present biospheric contribution to the total CO_2 increase should be relatively small.

Albedo

Processes that influence surface albedo are reviewed by Henderson-Sellers and Wilson (1983). In general, human-induced land-use changes tend to increase albedo. Results from several climate models suggest that the higher albedo over devegetated or overgrazed terrain leads to a net radiative loss, increased subsidence, and decreased rainfall (Charney, 1975; Charney et al., 1977; Berkofsky, 1976; Chervin, 1979; Sud and Fennessy, 1982). Lettau et al. (1979), however, find opposite results for Amazonian deforestation. Contradictory results are also reported for the relationship between albedo change and temperature. Otterman (1974, 1977) suggests that the higher albedos caused by overgrazing in the Sinai desert lead to lower surface temperatures, whereas Jackson and Idso (1975) and Idso (1977) find that, on denuded soil, the reduced evapotranspiration exceeds the cooling due to higher albedo, producing a net warming of the surface. Sagan and co-workers (1979) estimate a global cooling of 1 K from anthropogenic albedo changes over several millennia, whereas Potter et al. (1981), using the same assumptions with a different climate model, calculate a decrease of only 0.2 K. Henderson-Sellers and Gornitz (1984), updating the albedo data of Sagan, Toon, and Pollack (1979), find an albedo change since the 1950's of only half of that computed by Sagan and co-workers (1979), which using the sensitivity results of Hansen et al. (1981) (see the section on albedo changes caused by anthropogenic modifications), corresponds to a temperature decrease of around 0.1 K.

Hydrologic Cycle

The energy and hydrologic cycles are closely interrelated. Disturbance of the vegetation cover affects the atmosphere's heat balance by increasing surface albedo and by reducing the amount of energy absorbed at the ground—hence, cooling it. However, vegetation removal can also be expected to affect the surface atmosphere energy exchange by altering the relationship of sensible to latent heat (the Bowen ratio), largely through changes in evapotranspiration (Dickinson, 1980).

Denudation of the vegetation cover usually increases the surface runoff, and thus soil erosion, and decreases water infiltration into the ground (Pereira, 1973; Goudie, 1982). Deforestation is traditionally believed, at least at a local or regional scale, to lead to a decrease in precipitation (Aubreville, 1949; Thompson, 1980). Results of several recent climate model studies tend to support this viewpoint (Charney, 1975; Charney et al., 1977; Chervin, 1979; Sud and Fennessy, 1982). In a simulated deforestation of Amazonia, Henderson-Sellers and Gornitz (1984) find local decreases in evapotranspiration, precipitation, and cloud cover (see the section on anthropogenic modification). Amazonian rain forests may recycle up to 50% of the rainfall they receive through evapotranspiration (Marques et al., 1977; Salati and Vose, 1984). Thus, large-scale forest clearance could lead to a significant decrease in evapotranspiration and precipitation.

Conversely, the effect of vegetation disturbance on evapotranspiration depends on factors such as the areal extent and nature of the vegetation change, the amount and seasonal distribution of precipitation, and the rate of desiccation of bare soil (Dickinson, 1980). Vegetated and bare surfaces differ in the way they absorb heat (Wendler and Eaton, 1983): bare soil shows an inverse relationship between surface albedo and ground temperature, whereas these two parameters covary for vegetated surfaces. In the latter case, evapotranspiration overcompensates for the expected cooling due to higher albedo (Jackson and Idso, 1975). Because of these counteracting factors, it is difficult to predict the net effects of deforestation on climate.

Atmospheric Particulates and Aerosols

Overgrazing and slash-and-burn agriculture (Seiler and Crutzen, 1980) tend to increase the atmospheric aerosol and dust loading. An increase in wind-blown soil and dust affects the surface radiation and the stability of the overlying atmosphere. Little is known of the cumulative effects on climate, but it is believed to be less significant than the other processes described earlier (Kellogg, 1978).

Aerosols both scatter and absorb sunlight. Upward-scattered radiation will be lost to space. Although the reduction in sunlight reaching the ground will increase the net system albedo, producing a net cooling, absorption by particles and surrounding atmosphere may counteract this cooling by reducing the net albedo. The ultimate effect on temperature depends on the ratio of particle absorption to its back scatter and the underlying ground albedo (Kellogg, 1978). Aerosols of a given absorption/back scatter ratio located over a dark surface (e.g., ocean) are more likely to increase the net albedo than if they are over a bright surface (e.g., snow or desert sand).

Although most anthropogenic aerosols remain over land where they originate, winds can carry the dust over the ocean. The average residence time for aerosols before being washed out by rain may not be much longer that 1-2 wk. A threefold increase in lower troposphere aerosols between 1965 and 1976 in the equatorial Atlantic may have been related to desertification and drought in the Sahel (Prospero and Nees, 1977). Dust-storm activity in the Sahel increased by a factor of 5 to 6 times between 1968 and 1978 over pre-drought values (Middleton, 1985). An anthropogenic dust veil caused by excessive cultivation in Rajasthan, India, is believed to have caused atmospheric subsidence, cooling, and suppression of rainfall (Bryson and Baerris, 1967).

Aerosols can also act as condensation or freezing nuclei, hastening the formation of water or ice droplets. Although the climatic effects are still unclear, the enhanced cloud cover could locally increase albedo. Aerosols may enhance precipitation downwind of certain urban areas, such as the Chicago, Illinois-Gary, Indiana, conglomeration (Goudie, 1982).

ANTHROPOGENIC CHANGES IN LAND COVER BETWEEN 1880 AND 1980

Global Survey

Most human-induced transformations of the Earth's plant cover have gone undocumented. However, two major sources of data include the expansion of agriculture and forestry surveys.

Growth of Agriculture. During the second half of the nineteenth century, a major transformation of the Earth's surface occurred as agriculture expanded rapidly into the Earth's grasslands and steppes in the western United States and Canada, the Asiatic USSR, Australia, South Africa, and South America. This process has been termed the *pioneer agricultural explosion* (Wilson, 1978).

The Great Plains of the United States and Canada were brought under cultivation as recently as the end of the nineteenth to early twentieth centuries. Between 1880 and 1910, the main expansion in the U.S. agricultural area has occurred in the Great Plains states (Krause, 1970). Around 93 million ha were cleared between 1880 and 1900, in roughly 2:1 proportion of grassland to forest (Primack, 1962). Further growth took place between 1910 and 1920 in parts of Texas, Oklahoma, and the western Great Plains states and Montana. Total area of cropland (excluding pasture) peaked by 1940 and has declined slightly since (Krause, 1970).

Cultivated area reached a maximum by the early nineteenth century in northwestern Europe and by the late nineteenth century in southeastern Europe. The period of agricultural growth ceased after industrialization. Since the 1950s abandonment of farmland in western Europe has resulted in a decrease of around 4.6 million ha (3.2% total cultivated area), especially in Italy, France, and Germany (Commission of European Communities, 1980). The total agricultural area of Europe has remained relatively constant since 1870 (Table 2-1).

The expansion of Soviet agriculture into the virgin lands of the semiarid steppes has doubled the arable land since 1870 (Table 2-1). In mainland China, in spite of intensive efforts, the presently cultivated area has not increased significantly over prewar figures (Buck et al., 1966; FAO Production Yearbook, 1980), although considerable expansion took place in the nineteenth century (Table 2-2). India also experienced major agricultural growth in the nineteenth century (Blyn, 1966; Tucker and Richards, 1983). The agricultural area of Japan increased prior to 1910 but has remained fairly constant between 1910 and 1970, showing a decline in recent years (see Table 2-1). The loss of farmland in Japan follows the Western pattern and is probably caused by urbanization and industrialization. The other pioneer agricultural lands (Argentina, Uruguay, and Australia) show increases of arable land through the 1930s, which have subsequently stabilized.

In contrast to the developed countries, exploitation of natural resources has intensified in the developing countries under pressures of modernization and the population explosion especially since World War II. Consequently, the major conversion in vegetation cover from forests to farms and grassland as well as desertification are currently occurring in the Third World. Western Africa, as an example of the developing world, is examined later in greater detail.

Agricultural data for the last 100 years are unavailable for most developing countries, but *FAO Production Yearbooks* cover the period since the 1950s (although 1950 data are less comprehensive and less accurate than 1980 figures). Cropland in Latin America has doubled in that period; Asian arable land increased by 31%; but Africa shows an anomalous 20% decrease, apparently due to exclusion of fallow land in the most recent reports.

The arable area globally has doubled between 1870 and 1979 (see Table 2-1), but it increased by only 13.9% from 1951 to 1980 (*FAO Production Yearbooks,* 1951, 1980). Considerable uncertainty remains, however. A comparison of net changes in cultivated area with other compilations still shows a wide range between 725 million ha and 930.5 million ha (see Table 2-2).

Forestry Data. Global forestry inventories span a period of 60 yr (Table 2-3). The reported global total forest area has remained close to 4 billion ha since the 1950s, partly because of incomplete coverage in older surveys, resulting in underestimates, and partly because tropical removal has been compensated to some extent by temperate reforestation (Armentano and Ralston, 1980; FAO, 1976). The large uncertainties in the forestry data conceal any trend over time, although a decrease in area is widely accepted. The proportion of cultivated to forested land has remained fairly constant in the developed countries during the twentieth century. The area of temperate and boreal forests has stabilized in this century and may even be increasing slightly in North America and Europe (Armentano and Ralston, 1980). Forestry data for the USSR show wide differences (Sutton, 1975). The small increase in temperate forests can be attributed to the abandonment of marginal farmland and reversion of the fields to native vegetation.

Recent estimates for total tropical forest areas range between 935 million ha and nearly 2 billion ha. The totals fall into two discrete groups,

Table 2-1. Expansion of the Arable Area 1870–1980 (million ha)

Region	*1870*	*1900*	*1910*	*1920*	*1930*	*1940*	*1950*	*1960*	*1970*	*1980*
North America	80		154.5	182.9	215.5	237.6	219.8	205.8	218.3	223.7
Argentina, Uruguay	0.4		29		33		33.3	30.3	33.5	37
Australia	0.4	3.6	4.4	6.1	11.1	8.5	9.7	12	16.2	17.4
USSR	102		160	(160)	160		200	220.7	232.6	231.9
Europe	141		147	(148.5)	150		148	152.5	147.5	141.3
China	81		91		100		94	110.1	110.2	99.3
India, Pakistan, Bangladesh	68	73.3	82.7	72	95.1	82.2	157	151	193	199
Japan	3.2		5.4		6		5.4	6.1	5.6	4.9
Java	5.6		7		8.4			8.8	—	—
Southeast Asia (rice land only)	3		9		12		13	15	17	(33.9)
Total, accounting for 67% of world agricultural area (1950-1970)	485		690		791.1		880	912	974	988
Global	724		1,030		1,181		1,272	1,405	1,424	1,449
Total (est., prior 1950)										

Sources: Data from Grigg, 1974; Buck et al., 1966; Blyn, 1966; historical statistics of the United States, colonial times to 1970; *Canada Yearbook*, 1950, 1960, 1970; *Official Yearbook of Australia 1975-1976*, 1981; *FAO Production Yearbooks*, 1951, 1960, 1970, 1980.

Table 2-2. A Comparison of Net Changes in Cultivated Areas, 1860–1978 (areas $\times 10^6$ ha)

Region	*1860-1920*	*1920-1978*	*Total Change 1860-1978*	*Reference*
North America	95 (1870-1920)	43 (1920-1970)	138	1
		14.8 (1950)		2
	161.2	−1.5	159.7	3
			138.1	6
Europe	7.5 (1870-1920)	−1.5 (1920-1970)	6	1
		−5.9 (1950-1978)		2
	20.6	1.1	21.7	3
			−1.6	6
Australia	5.6	11.3	16.9	4
	7.6 (1870-1920)	11 (1920-1970)	18.6	1
	7.2	36.2	43.4	3
USSR	58 (1870-1920)	71.8	129.8	5
	88	62.9	150.9	3
			134.7	6
China	14.5 (1870-1920)	4.1	18.6	5
	12.6	6	18.6	3
			55.8	6
India (including Pakistan and Bangladesh)	4 (1870-1920)	127	131	5
	49.1	65.2	114.3	3
Global	381.5 (1870-1920)	343.5	725	5
	432.2	419.3	851.5	3
			930.5	6

References: (1) Grigg, 1974; (2) *FAO Production Yearbooks,* 1951, 1970, 1979; (3) Richards et al., 1983; (4) *Official Yearbook of Australia 1975-1976,* 1981 (excludes cultivated grassland); (5) from Table 2-1; (6) Houghton et al., 1983.

however, averaging around 1.1 billion ha and 1.9 billion ha respectively, which correspond to closed forests and total forest land, including open woodland and, often, savanna (Henderson-Sellers and Gornitz, 1984). Estimates indicate that clearing of tropical forests for shifting cultivation and new permanent agriculture, cattle grazing, logging, and fuel-wood consumption have completely removed between 9 million ha and 15 million ha/yr (Seiler and Crutzen, 1980; Lanly, 1982). Myers (1980), who considers the total area of degraded or only partly destroyed forests, finds a recent deforestation rate of 24.5 million ha/yr. Although the present rate of clearing of all tropical forests is around 0.6%/yr of the existing closed forest area (Lanly, 1982), this rate varies widely from country to country and even between regions within a country. However, the historical trend of deforestation usually cannot be reconstructed from available published sources, except for limited areas, because of incomplete data and poor sampling in older surveys.

Western Africa

European colonialism contributed to a major modification of the natural vegetation of western Africa after 1890 because of the vast expansion of commercial agriculture and logging and the rapid growth in population that was stimulated by the introduction of modern medicine and hygiene. These developments intensified the need for land, thus extending cultivation into previously unoccupied or sparsely settled areas, and shortening the length of fallow, contributing to a loss in soil fertility. Deforestation of the forest zone accelerated following a southward population migration from traditional trade cen-

Table 2-3. World Forest Resources—Regional Summary of Data from Various Sources (10^3 ha)

Region	*Zon and Sparhawk, 1923*	*FAO, 1946*	*FAO, 1948(50)*	*FAO, 1954*	*Haden-Guest et al., 1956*	*Paterson, 1956*	*FAO, 1960*	*FAO Production Yearbook, 1963*	*Persson, 1974*	*FAO Production Yearbook, 1979*	*FAO Production Yearbook, 1980*
Europe	127,687	132,600	1,046,000 (est.)	135,700	128,633	128,879	141,000	138,000[c] (144,000)	140,000[c] (170,000)	154,670	154,585
USSR	646,433	960,000		742,600	1,068,600	920,000	836,000	738,117 (910,009)	765,400 (914,900)	920,000	920,000
North America	511,167	635,100	653,251	656,413	690,183	650,960	733,000	713,059 (750,219)	630,000	616,229	616,889
Central America	69,513	88,800	74,822 (est.)	65,481	—	65,543	74,000	71,000 (76,000)	60,000 (65,000)	102,263	73,510
South America	850,534	726,100[a]	755,000 (est.)	861,000	860,770	754,465	957,000	830,000 (890,000)	530,000 (730,000)	918,396	946,340
Asia	416,479[b]	517,600	520,000 (est.)	535,827	433,292[b]	486,252	520,000	500,000 (550,000)	400,000 (530,000)	553,050	545,551
Africa	379,935[b]	536,100	849,000 (est.)	801,600	282,999[b]	794,152	753,000	700,000 (710,000)	190,000 (800,000)	636,608	699,882
Oceania	108,208	53,700	80,000 (est.)	116,660	93,568	72,752	96,000	207,267	80,000 (190,000)	155,457	151,227
Global total	3,109,956[b]	3,650,100	3,978,000 (est.)	3,915,300	3,557,435[b]	3,873,000	4,110,000	3,792,000 (4,126,000)	2,800,000 (4,030,000)	4,056,673	4,107,984

Notes: [a]No data for Bolivia.
[b]Incomplete data.
[c]Figures in parentheses represent total forest land, including unstocked areas. Lesser figures are for closed forest. Mean (total forests) 3,864,950 ± 312,631

ters (e.g., Timbuktu, Mali; Oyo, Nigeria) during the nineteenth century, triggered partly by tribal warfare and partly by better economic opportunities near the coast and the development of commercial crops (Morgan and Pugh, 1969, p. 313; Allison, 1961; Morgan, 1959). Although difficult to document, it is possible that the southward spread of the so-called derived savanna coincided with this population shift (Allison, 1962).

Clearing of the western African rain forest began in the 1840s in Sierra Leone (Dorward and Payne, 1975). By the 1930s, only 6% of Sierra Leone remained forested (Sierra Leone, 1937, 1939). Widespread planting of rice in coastal mangroves began in northern Sierra Leone in the 1880s but intensified since the 1930s (Kaplan et al., 1976; Church, 1974, pp. 312-313; Morgan and Pugh, 1969, p. 655).

Growth of the oil palm industry by the mid-nineteenth century also greatly affected the forest zone. Oil palms were originally grown near Old Calabar, Niger Delta, in the 1830s but spread shortly thereafter to Dahomey, Togo, and Ghana. Cocoa, another important tree crop, was first cultivated in southeast Ghana, but by the 1890s cultivation had spread into central and western Ghana, southwestern Nigeria, Ivory Coast, and Cameroon (Morgan and Pugh, 1969, p. 474; Agboola, 1979). By the 1950s, the cocoa belt in Ghana had reached the Ivory Coast border (White and Gleave, 1971, p. 115). Coffee, another forest zone cash crop, has been grown extensively in the Ivory Coast since the 1930s (Church, 1974, p. 349). Rubber was introduced into southern Nigeria at the turn of the century and now occupies large areas of Calabar Province and near Benin City. In Liberia, the area planted in rubber grew from 800 ha in 1910 to 81,000 ha in 1960 (Morgan and Pugh, 1969, p. 480).

Ivory Coast has had a long history of deforestation. By 1980, around 70% of the forest present in 1900 had been cleared (Table 2-4). The coastal forests of Dahomey and Togo were largely cleared by the 1930s (Aubreville, 1937). In Ghana, by the 1950s, less than half the forest zone remained under tree cover (Table 2-5). The rain forests of Abeokuta province, western Nigeria, were largely destroyed by 1920 (Unwin, 1920), and further clearances occurred in the 1950s and early 1960s (White and Gleave, 1971, p. 158). Liberia, by 1980, had less than 1/3 of the closed forest cover present in 1920 (Table 2-6).

A widespread expansion of peanut cultivation took place in the Sudanian zone of Senegal (Atlas National du Senegal, 1977) and northern Nigeria and also of cotton in the latter country early in the twentieth century (White and Gleave, 1971, pp. 122-124; Church, 1974, p. 219). Irrigation has transformed the natural vegetation in the inland delta of the Niger River, Mali, starting in the 1920s. Main crops are rice

Table 2-4. Deforestation History of Ivory Coast

Year	*Area of Closed Forests* ($\times 10^3$ *ha*)	*Reference*
~1900	15,000 (14,500) (moist forest)	FAO, 1981
1920	12,140	Zon and Sparhawk, 1923
1933/1934	13,000	Ann. Stat. A.O.F., 1933/34
1950	7,000 (seems low)	Haden-Guest et al., 1956
1955	11,800 (moist forest)	FAO, 1981
1958/1963	8,000	FAO, 1963
1965	8,983 (moist forest)	FAO, 1981
1966	9,000	Persson, 1977
1973	6,200 (moist forest)	FAO, 1981
1980	4,458 3,993 (moist forest)	

Table 2-5. Deforestation History of Ghana (10^3 ha)

Year	Area of Closed Forests	Area of Woodland and Open Forest	Total Forest Area	Reference
1920	9,871 "forest area"			Zon and Sparhawk, 1923
1937/1938	4,789	11,111	15,900	Gold Coast Report on the Forestry Dept. for the year
1946/1947	4,375			Gold Coast Report . . .
1948/1949	4,236	11,085	15,321	Gold Coast Report . . .
1950/1951	4,087	11,045	15,132	Gold Coast Report . . .
1953	2,810			Charter, 1953
1958	2,493	11,331	13,824	Ghana Ann. Rept. Forestry Dept. 1958
1960/1961	2,424	10,687	13,111	1962 Ghana Stat. Yearbook
1968	2,207	9,711	11,918	1967-1968 Ghana Stat. Yearbook
1980	1,718	6,975	8,693	FAO, 1981

Table 2-6. Deforestation History of Liberia (10^3 ha)

Year	Area of Closed Forests	Total Forest Area	Reference
1920	6,475	6,475	Zon and Sparhawk, 1923
~1950	5,520		Haden-Guest et al., 1956
1968	2,500		Persson, 1977
1980	2,000	2,040	FAO, 1981

and cotton (Morgan and Pugh, 1969, pp. 645-654). The creation of large artificial lakes (Lake Volta, Ghana; Lake Kossou, Ivory Coast) has inundated vast areas of forest and savanna.

In the Sahelian zone, since the beginning of this century, sedentarization of nomads has accelerated; cultivation has extended farther north into former grazing land marginal to agriculture, and the number of well and boreholes has increased (Ware, 1977; Baier, 1980; Dresch, 1959). Even prior to the 1970s drought, the rapid expansion of cattle herd size, together with areal reduction of grazing lands, and the concentration of herds around boreholes has lead to severe overgrazing around wells and towns (Mabbutt and Floret, 1980).

Figure 2-1 provides a schematic map of deforestation and vegetation degradation in west Africa. It is incomplete in certain areas due to insufficient data. The reliability is greater for Senegal, Ivory Coast, Ghana, and southern Nigeria than for other countries.

Estimates of the Extent of Anthropogenic Vegetation Changes in Western Africa

Changes in the area of cropland may serve as an indirect measure of vegetation change. Although exports (tonnages) of commercial crops have been documented since around 1910, records of the cultivated area cover only the period since the 1930s or later. Permanently cultivated land has increased by 12.9 million ha for eight former French west African countries between 1930 and 1979 (International Institute of Agriculture, 1939; *FAO Production Yearbook*, 1987). Applying the same proportion of growth to all of western Africa would result in an increase in permanent cultivation of around 42.7 million ha over this period.

Population growth can be used as a surrogate for cropland expansion and, hence, indirectly, changes in the natural vegetation, in

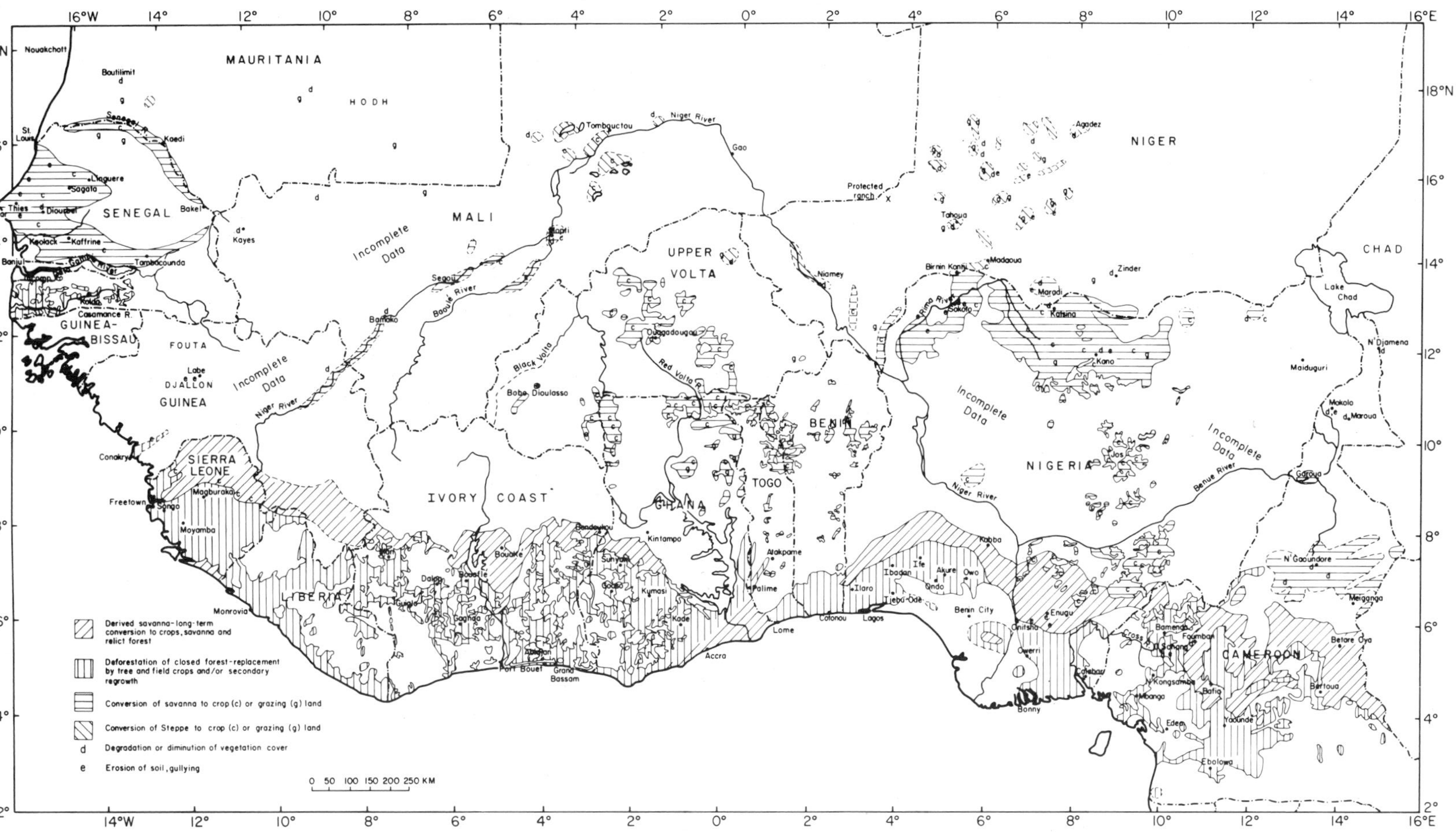

Figure 2-1. Schematic map of anthropogenic vegetation changes in western Africa during the last century.

predominantly rural regions in particular where traditional agricultural methods still prevail. The assumptions and data used for calculating the expansion of agriculture based on population growth are summarized in Table 2-7. This table shows that the permanently cultivated area for former French west Africa has expanded by 10.5 million ha between 1930 and 1980, as compared with 12.9 million ha from IIA (1939) and *FAO Production Yearbook* (1980). Over this same period, the permanently cultivated area for all of western Africa increased by 39.4 million ha, which is close to the 42.7 million ha derived from agricultural surveys. Between 1910 and 1980, the permanently cultivated land increased by 44.3 million ha and total cultivated area (including fallow) by 114.9 million ha.

Although on a regional level available forestry data are inadequate to reconstruct the deforestation history for all of western Africa, data for a few countries (Ghana, Ivory Coast, Liberia) provide a reasonable estimate of deforestation (see Tables 2-4, 2-5, and 2-6). The area of closed forest in Ivory Coast has decreased by 70.3% in 80 yr (Table 2-4). The current rate of deforestation of closed forest (7%/yr) is the highest in the tropics (FAO, 1981). In Ghana, closed forest shrank by 64.1% and open forest and woodland by 37.2% in 43 yr (Table 2-5). The present rate of deforestation is 1.6%/yr (FAO, 1981). Liberian (Table 2-6) closed forests decreased by 69% since 1920 (Zon and Sparhawk, 1923; FAO, 1981).

However, deforestation in these three countries may be higher than for western Africa as a whole. Phillips (1974) has estimated that 57% of the closed forest zone in the Guinean and Nigerian blocks (excluding Cameroon) has been lost (over an unspecified period). Vegetation maps indicate that tropical forests should cover around 83.6 million ha in western Africa (Matthews, 1983); however, existing closed forest occupies only 36.6 million ha (FAO, 1981). This represents a 56.3% reduction, which agrees remarkably well with Phillips's estimate, although being somewhat lower than the data from Ghana and the Ivory Coast. Using a conservative figure of 56% reduction in closed forests since the early 1900s, 46.8 million ha have been cleared. If the decrease in open woodland and savanna in Ghana is representative of the whole region, 37%, or 76 million ha, could have been degraded and/or converted to agriculture. Adding these two figures, up to 122.8 million ha of forest, woodland, and savanna may have been cleared.

Albedo Changes Due to Anthropogenic Modification

The anthropogenic role in affecting climate through surface albedo changes has been examined in a number of papers. Otterman (1974, 1977) finds a local cooling associated with an increase in albedo caused by overgrazing. Charney (1975) proposes a biogeophysical feedback mechanism for the persistence of Sahelian drought, in which the higher albedo resulting from removal of vegetation increases the reflection of solar radiation to space, cooling the air above. This cooler air descends, compresses adiabatically, and decreases the relative humidity, thereby enhancing the dryness of the desert. While the initial Charney study has been criticized for neglecting the role of evapotranspiration (Ripley, 1976; Idso, 1977; Jackson and Idso, 1975), subsequent studies, with improved specification of albedo and ground hydrology, have provided additional support for the Charney mechanism (Charney et al., 1977; Walker and Rowntree, 1977; Chervin, 1979; Sud and Fennessy, 1982).

Sagan, Toon, and Pollack (1979) calculate a global cooling of 1 K from anthropogenic albedo changes over several millennia and of 0.2 K since around 1960. Potter et al. (1981), using the same assumptions of albedo change as Sagan and co-workers but a different climate model, including cloud and ice albedo feedbacks, find a global temperature decrease of only 0.2 K for the cumulative vegetation change. More recently, Henderson-Sellers and Gornitz (1984) have re-evaluated the albedo changes of Sagan and co-workers and find a total global albedo increase since the 1950s of between 0.00033 and 0.00064, or roughly half of Sagan and co-workers' estimate of 0.001. The main differences from these workers lie in a smaller

Table 2-7. Growth of Cultivated Land from Population Data

	1910	*1920*	*1930*	*1940*	*1950*	*1960*	*1970*	*1980*
Former French West Africa								
Population × 10^3	11,097	12,400	14,576	15,955	18,982	23,388	31,968	42,671
Percent in agriculture	95	95	95	95	93	88	82	77
Agricultural population × 10^3	10,542	11,780	13,847	15,157	17,653	20,584	26,214	32,857
Average area cultivated per agriculturalist ha	0.55	0.55	0.55	0.55	0.55	0.55	0.55	0.55
Area permanently cultivated × 10^3 ha	5,798	6,479	7,616	8,337	9,709	11,320	14,418	18,071
Ratio of cultivated to fallow	1:6.5	1:6	1:5.5	1:5	1:4.5	1:4	1:3	1:3
Total area cultivated (including fallow) × 10^3 ha	43,487	45,353	49,504	50,019	53,401	56,599	57,670	72,285
				Permanent agriculture			Total, including fallow	
				△ 1930-1980	10,455		22,781	
				△ 1960-1980	12,273		28,798	
All West Africa								
Population × 10^3	34,171	39,237	43,559	49,318	65,605	87,778	113,274	152,831
Percent in agriculture	95	95	95	95	90	85	79	74
Agricultural population	32,462	37,275	41,381	46,852	59,044	74,611	89,486	113,095
Area permanently cultivated[a]	17,854	20,501	22,760	25,769	32,474	41,036	49,218	62,202
Total area cultivated[a]	133,907	143,510	147,937	154,612	178,610	205,181	196,870	248,809
				Permanent agriculture			Total, including fallow	
				△ 1930-1980	39,442		100,872	
				△ 1910-1980	44,348		114,902	

Source: Data from Gornitz, 1985.

[a] Average area cultivated and ratio of cultivated to fallow land assumed same as in former French West Africa.

albedo change associated with tropical deforestation, a more conservative estimate of permanently desertified land and allowance for partial recovery of such land following the end of drought, a significantly lower estimate of the effect of salinization, and inclusion of arid-land irrigation and dam building, both of which tend to reduce albedo (Table 2-8). These albedo changes correspond to surface temperature decreases of between 0.06 and 0.09 K (Henderson-Sellers and Gornitz, 1984). The same paper estimated the maximum impact of tropical deforestation, in a hypothetical total replacement of tropical rain forest in Amazonia (albedo 11%) by a grass/savanna cover (albedo 19%) (Fig. 2-2), using the GISS Model II GCM (Hansen et al., 1983). The simulated deforestation, locally, decreased precipitation by 0.6 mm/day, and evapotranspiration by 0.4-0.5 mm/day, but caused no significant change in temperature in spite of an 8% increase in surface albedo. A rise in temperature from a reduction in evapotranspiration may just offset the decreased temperature caused by the increased surface albedo (Jackson and Idso, 1975; Idso, 1977). However, the simulated Amazonian deforestation did not have a significant impact on global climate.

A more realistic estimate of the extent of albedo change has been calculated for western Africa (q.v.) from historical records (Gornitz, 1985). Present-day cultivation and land-use data from atlases and maps have been digitized on a $1° \times 1°$ grid, using the vegetation map and land-use intensity scale of Matthews (1983). The land-use classification differs from Matthews (1983) chiefly in assigning cells of greater than 70-100 people/km^2 to cultivation intensity class 4; population density over 30-40 people/km^2 to class 3; over 10-20/km^2 to class 2 (traditional subsistence economy), and less than 10-20/km^2 to class 1. Land-use intensities of the late 1800s are much more difficult to estimate than current ones since very few atlases present such data. However, a historical pattern of land-use change since that time has been reconstructed from government censuses, atlases, and reports of forestry and agriculture ministries, supplemented by descriptive reports of forest clearance by logging or cultivation, and the expansion of both traditional and commercial agriculture taken from historical, economic, and geographic sources (Fig. 2-1; Gornitz, 1985).

Albedos associated with given vegetation types were adapted from Matthews (1984). Albedos for various west African crops were taken from Oguntoyinbo (1970*a*, 1970*b*). Where known, the average albedo of the three to four dominant crop types present in a grid cell was used; otherwise, the mean value of 17% for mixed crop was taken. The albedo for each cell is calculated as an area-weighted average of the crop and the natural vegetation albedos, using the digitized land-use intensity scale as a measure of the percentage of area cultivated. Maps of the *difference* in albedo between 1880 and 1980 (100-yr change) and that between the original vegetation in the absence of agriculture and that of 1980 (cumulative change) are presented in Figures 2-3 and 2-4.

Albedo changes were also calculated, with several modifying assumptions. An albedo of 11% for tropical rain forest and seasonal forest may be too low. Studies show a value closer to 13% (Oguntoyinbo, 1970*b;* Pinker et al., 1980). Albedos of tropical forests lying between 2° and 8°N were thus changed from 11% to 13%. The second modifying assumption involves estimating the effect of desertification due to overgrazing on albedo. The albedo ratio between a protected exclosure and its surroundings provides a measure of the albedo change associated with this process. A protected ranch in Niger, near the Mali border (15°54′N, 4°7′E), has an albedo of 34.2% as compared to 42.3% for nearby overgrazed terrain (Otterman and Frazer, 1976; Otterman, 1981), yielding an albedo contrast ratio of 1.24. A figure of around 10% potentially heavily grazed land in the Eghazer-Azaouak region of Niger (Mabbutt and Floret, 1980) is considered representative for the entire Sahel. The increase in albedo due to desertification was recalculated by multiplying the previously calculated 1980 albedo for each Sahelian cell by 1.24 and assigning this desertified albedo an areal weight of 0.1 for cells lying between 14° and 17°N, except for the agricultural areas of Senegal (14°-16°N, 15°-17°W) and unin-

Table 2-8. Global Albedo Changes Due to Anthropogenic Modifications

Process	*Land-Type Change*	*Δrs, Surface Albedo Change (Integrated)*	*A, Change in Area, over 30 yr, Relative to Earth's Surface Area*	*F, Cloud Cover Fraction*	*I, Insolation Factor*	*ΔR, Fractional Change in Global Albedo over 30 yr*
Deforestation of tropical forests	(1) Forest → savanna (max.) grassland	(1) 0.11 → 0.20 = 0.09	0.00647	0.5	1.2	(1) 4.59×10^{-4}
	(2) Forest → clearings (min.) (Pinker et al., 1980)	(2) 0.13 → 0.16 = 0.03				(2) 1.53×10^{-4}
Deforestation of temperate forests	Forest → field pasture	0.13 → 0.18 = 0.05	Small	0.5	0.8	Small
Dam building	Field → water	0.18 → 0.04 = −0.14	0.00059	0.25	1	-0.68×10^{-4}
Salinization	Field → saline field	0.18 → 0.24 = 0.06	0.000074	0.25	1	0.04×10^{-4}
Irrigation of arid land	Desert soil → field	0.35 → 0.18 = −0.17	0.00072	0	1	-1.23×10^{-4}
Urbanization	Field, pasture → urban Forest → urban	0.18 → 0.13 = −0.05 0.13 → 0.13 ≅ 0.0	0.00059 0.00059	0.5 0.5	1 1	-0.19×10^{-4}
Desertification	(i) Scrub/shrub → desert soil	0.23 → 0.35 = .12	(a) 0.00318 (b) 0.00159	0 0	1 1	(a) 3.81×10^{-4} (b) 1.91×10^{-4}
	or (ii) Protected exclosure → desert soil	0.39 → 0.44 = 0.05	(c) 0.00318 (d) 0.00159	0 0	1 1	(c) 1.59×10^{-4} (d) 0.79×10^{-4}

Total resulting planetary albedo change

		Case (1) Maximum tropical deforestation albedo change = 0.09		*Case (2)* Minimum tropical deforestation albedo change = 0.03
Desertification	(i)	(a) 6.35×10^{-4} (b) 4.45×10^{-4}	(i)	(a) 3.29×10^{-4} (b) 1.39×10^{-4}
	(ii)	(c) 4.13×10^{-4} (d) 3.33×10^{-4}	(ii)	(c) 1.07×10^{-4} (d) 0.27×10^{-4}

Note: Cases (a) and (c) use the UN (1977) estimate of the area of desertification while cases (b) and (d) use half this value.

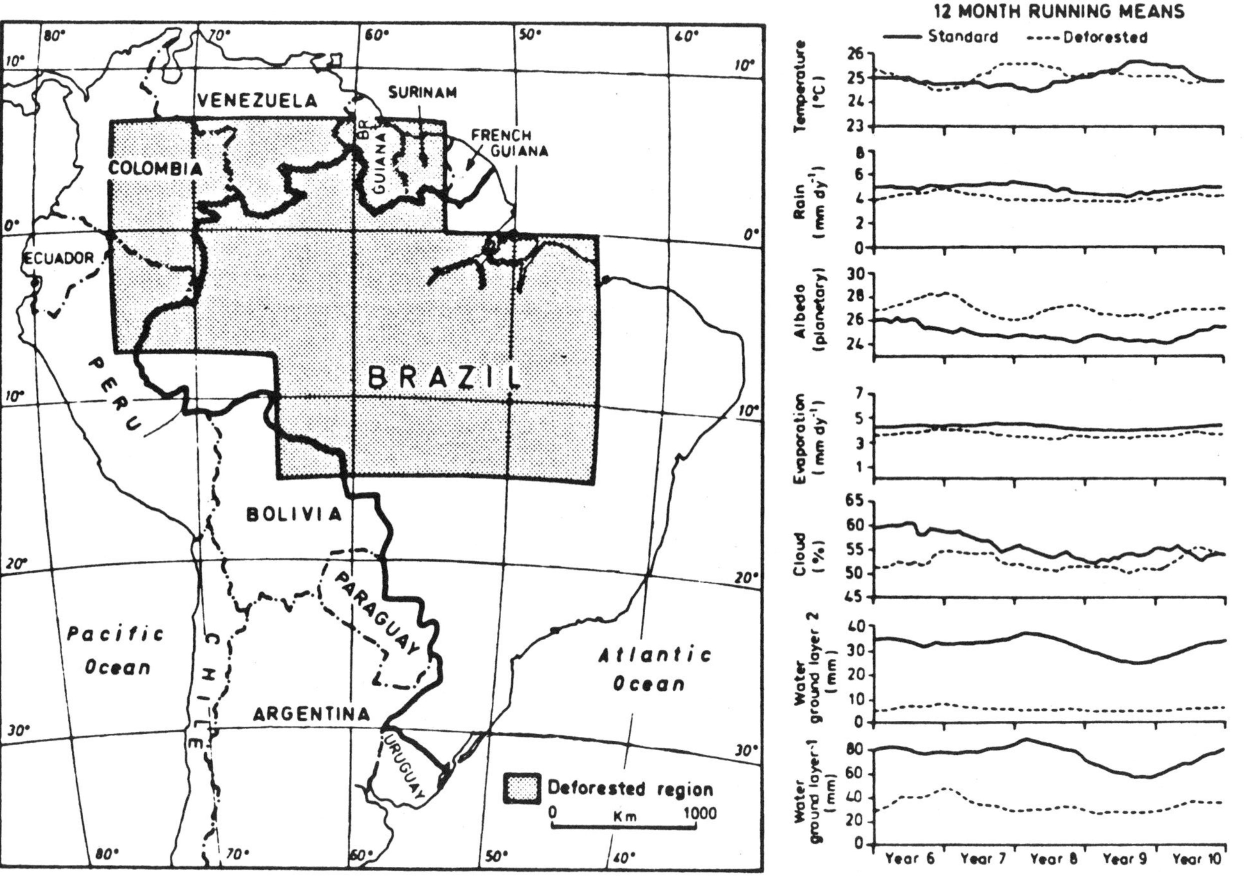

Figure 2-2. GISS GCM simulation of Amazon deforestation. *left:* Region deforested, *right:* 12-mo running means of selected climatic parameters. (After Henderson-Sellers and Gornitz, 1984)

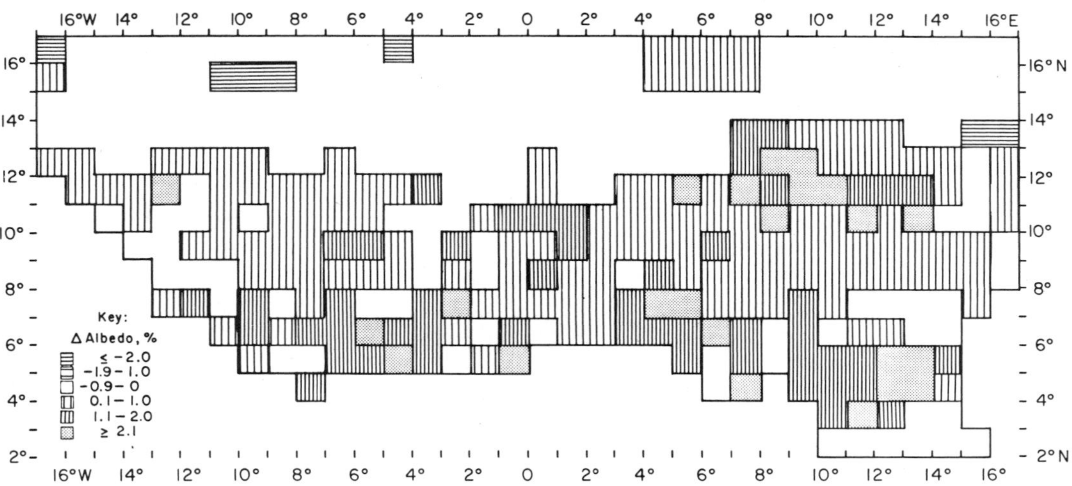

Figure 2-3. Map of albedo changes, 1880-1980.

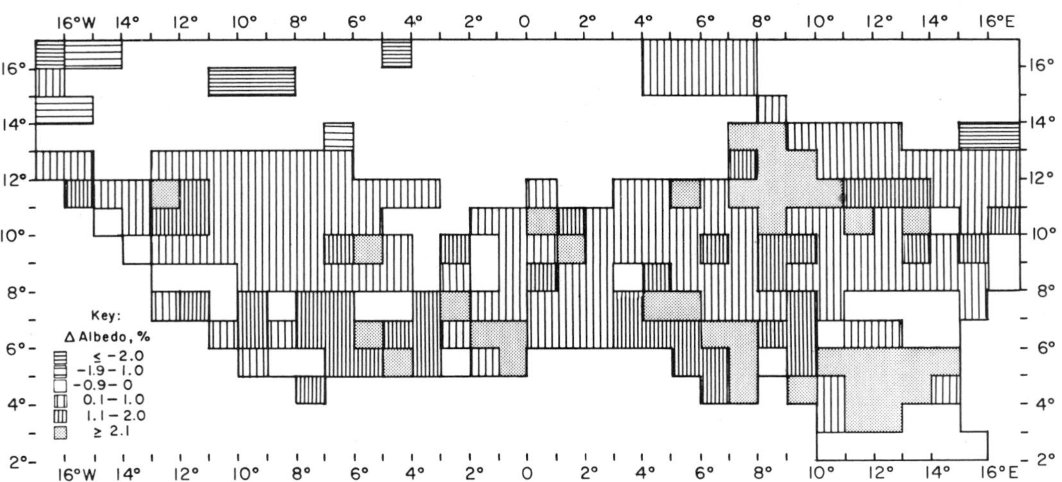

Figure 2-4. Map of albedo changes, pre-agriculture to present.

habited desert areas of Niger (15°-17°N, 9°-17°E).

In a separate case, the albedo calculation has been repeated, using the extreme assumption of 100% desertification between 14° and 20°N (with the exceptions noted previously and an albedo contrast ratio of 1.74 for 10% of the land immediately surrounding boreholes and 1.5 to the remaining 90%.

The average regional albedo of natural vegetation (spanning a range from rain forest in the south to desert scrub in the north) in western Africa is 17.3. The mean area-weighted regional change in albedo is 0.41% (100 yr) and 0.54% (cumulative change). Thus, the regional albedo increases from 17.3% to 17.4% by 1880 and to 17.8% by 1980. With the additional assumptions of a higher albedo for tropical forests and effects of moderate overgrazing, the regional mean albedo change is 0.42% (100 yr) and 0.54% (cumulative), which are nearly the same as before. The albedo change in the forest zone is now less than before because of a lower albedo contrast between forest and the replacing crops. This decrease is nearly equally counterbalanced by an increase in Sahelian albedo due to estimated desertification. With the extreme assumption of total desertification, the regional mean albedo increases by around 4% above that of natural vegetation for both the 100-yr and cumulative periods.

The maximum zones of increased albedo are concentrated in the forest zone between 4° and 8°N and in the savanna and southern Sahel between 10° and 12°N, which correspond to zones of maximum agricultural and population growth (see Figs. 2-3 and 2-4). North of 13°N, the albedo change is small or negative, due to a combination of less intensive land utilization and the replacement of an open vegetation cover on sandy soil by irrigated crops with lower albedo, such as along the Niger River (e.g., Morgan and Pugh, 1969, pp. 646-653; Church, 1974, pp. 250-251).

In addition to errors based on incorrect land-use intensities or improperly specified crop or vegetation albedos, this procedure may overestimate albedo change since 1880 relative to the cumulative change but may underestimate the cumulative vegetation change because the role of anthropogenic fire in expanding savannas has not been considered (see the introduction). It is widely believed that the climatic climax of dry deciduous woodland in western Africa has been gradually replaced by a fire climax of wooded savanna. The degradation of moist closed forest to a mosaic of secondary forest, tree crops (i.e., oil palm, cocoa, coffee), and bush fallow derived savanna (see Fig. 2-1) has accelerated since 1880 (Aubreville, 1949; Allison, 1962; Morgan, 1959). Conversely, the inference of a once more widespread densely wooded vegetation (Phillips, 1974; Aubreville, 1949) may be based partly on forest relicts dating back to a formerly more humid climate, possibly the last climatic optimum, or Hypsithermal (Aubreville, 1932, 1937, 1939).

One important finding is that the actual albedo change due to land-use transformation is likely to be much less than the extremes taken for climate modeling (e.g., 14-35%, a change of 21%; Charney et al., 1977), and therefore, the anticipated climatic impact should be correspondingly less. Wendler and Eaton (1983) find albedo differences of approximately 10% between protected and unprotected Sahelian vegetation. The largest increase in albedo for any given cell in this study was only 5.3% (Fig. 2-4). Other cells showed little or no albedo change due to either a low land-use intensity or replacement of natural vegetation by crops of similar albedo (as in western Senegal).

Furthermore, consideration of albedo changes alone neglects the role of the water balance, which under the right circumstances can counteract the effect of albedo change (see the section on the hydrologic cycle). A more thorough investigation into the effects of evapotranspiration is beyond the scope of this chapter. Turning to the historical climate records for western Africa, however, what evidence exists for any impact of the anthropogenic changes (outlined in the section on western Africa, Figs. 2-1, 2-3, Tables 2-4 through 2-7) on the hydrologic cycle, and thus indirectly, on climate? Three major droughts have occurred in the twentieth century: from 1910-1920, in the 1940s, and 1968-1974 (Nicholson, 1980*a*). [The 1970s

Sahelian drought may not have actually ended; the early 1980s have been as dry as 1972 and 1977 (Hare, 1983). It is much too early to conclude whether the most recent drought is part of a longer-term cycle or the beginning of climatic desiccation.] Although biogeophysical feedback mechanisms (Charney, 1975; Chervin, 1979) may help exacerbate drought, the geological record indicates that major climatic changes have occurred since the end of the last glaciation (Nicholson and Flohn, 1980; Talbot, 1980).

Over the last few centuries, the Little Ice Age (sixteenth to eighteenth centuries) in western Africa was, on the whole, wetter than at present (Nicholson, 1980*b*). Lake levels and historical records indicate a drying trend during the late eighteenth-early nineteenth centuries, followed by a partial return to a more humid climate between 1875 and 1895 (Nicholson, 1980*b*). Descriptions of lusher vegetation in the Sahel reported by nineteenth-century European travelers (Depierre and Gillet, 1971) may correspond to this moister period. Early twentieth-century references to the desiccation of Africa (Hubert, 1920; Migeod, 1922; Stebbing, 1935) could represent the effects of the 1910-1920 drought.

On the one hand, fluctuations in stream flow during the twentieth century (Sircoulon, 1976; Faure and Gac, 1981) correlate well with precipitation variations (Nicholson, 1980*a*; Palutikof et al., 1981), but there is no definitive evidence for a regional secular decrease in precipitation such as that predicted by several climate models (Charney, 1975; Chervin, 1979; Sud and Fennessy, 1982), although continued persistence of the current drought into the future may provide the first indication. On the other hand, human-induced devegetation has locally disrupted the hydrologic cycle, particularly in semiarid environments (Ledger, 1961; Delwaulle, 1973; Chevalier, 1950; Morgan and Pugh, 1969, p. 235), leading to increased runoff, soil erosion, siltation, and reduced water infiltration capacity. Anthropogenic vegetation disturbances, therefore, can produce effects similar to a climatic desiccation (so-called desertification). It has been suggested that the devegetation-hydrology feedback may be more important for surface properties than the albedo-precipitation (or temperature) feedback (Verstraete, 1981).

SUMMARY AND CONCLUSIONS

1. The expansion of cultivated land provides one of the few documented records of land-use change. The global arable area has doubled between 1870 and 1979, increasing by 725 million ha (see Table 2-1). A considerable degree of uncertainty exists in this figure, however; other compilations indicate increases of up to 930.5 million ha (see Table 2-2).

2. The global forest area appears to have remained close to 4 billion ha since the 1950s, but poor statistics in older surveys and some temperate land reforestation mask any anticipated decrease (see Table 2-3). Clearing of forests for agriculture, grazing, and logging, among other uses, is estimated to destroy between 9 million ha and 15 million ha/yr at present (Seiler and Crutzen, 1980; Lanly, 1982) and could degrade up to 24.5 million ha/yr (Myers, 1980).

3. A reappraisal of Sagan, Toon, and Pollack's (1979) assumptions of albedo change has found only half the increase in albedo calculated by those authors (see Table 2-8). These revised albedo changes could account for a temperature drop of only 0.6-0.09 K over a 30-yr period, or less than half that initially reported (Henderson-Sellers and Gornitz, 1984).

4. Western Africa, as a microcosm of land-use transformations in the tropics, was studied in greater detail. Agricultural and population censuses suggest an increase in permanent cultivation of around 40 million ha since 1930, and 44 million ha when extrapolated to 1910. If fallow land is included, up to 115 million ha may have been cleared (see Table 2-7). Forestry records, although incomplete and imprecise, suggest a loss in the forest zone of between 56% and 70% since the early 1900s (see Tables 2-4 through 2-6). Using the more conservative figure of 56% for the forest zone and 37% for woodland and savanna, up to 123 million ha of natural vegetation may have been cleared or converted to agriculture, as compared with 115 million ha from population data.

5. Albedo changes derived from a reconstruction of land-use changes in western Africa were digitized and mapped on a $1° \times 1°$ grid (see Figs. 2-3 and 2-4). The mean regional albedo increased by 0.41% over a 100-yr interval, and by 0.54% since agriculture began, from an initial regional mean value of 17.3%. Modifying assumptions including a higher albedo assignment for tropical forests and increased albedo due to desertification produced no significant change. Only the extreme assumption of total desertification yielded an increase in albedo of around 4%. With these assumptions, lower albedo increase in the forest zone due to reduced albedo contrast between forests and crops is nearly exactly compensated for by an increase in Sahelian albedo due to desertification.

6. The anthropogenic vegetation changes in western Africa do not appear to have affected the regional climate, as seen from historical records (Nicholson, 1980*a*). However, disruption of the hydrologic cycle caused by devegetation may have led to reductions in soil moisture, placing plants under stress and simulating a climatic desiccation. Desertification is more likely the consequence of altered soil hydrology than of diminished precipitation.

ACKNOWLEDGMENTS

The author thanks the following people for their helpful comments and suggestions: Dr. J. Hansen, Dr. A. Lacis, Ms. E. Matthews, and Dr. W. Rossow, GISS; Professor J. Otterman, Tel Aviv University; and Dr. A. Henderson-Sellers, University of Liverpool. This research was supported by NASA cooperative agreement NCC5-29, task A.

REFERENCES

Agboola, S. A., 1979, *An Agricultural Atlas of Nigeria*, Oxford: Oxford University Press, 248p.

Allison, P. A., 1962, Historical inferences to be drawn from the effect of human settlement on the vegetation of Africa, *Jour. African History* **3:**241-249.

Annuaire statistique de l'Afrique Occidentale Française, 1933-1934, Gouvernement génerale d l'Afrique Occidentale Française, Paris.

Armentano, T. V., and Ralston, C. W., 1980, The role of temperate forests in the global carbon cycle, *Canadian Jour. Forest Research* **10:**53-60.

Atlas National du Senégal, 1977, Inst. Géog. Natl., Paris.

Aubréville, A., 1932, La Fôret de la Côte d'Ivoire, *Com. d'Études Hist. et Sci. de l'A.O.F. Bull.* **15:**205-249.

Aubréville, A., 1937, Les Fôrets du Dahomey et du Togo, *Com. d'Études Hist. et Sci. de l'A.O.F. Bull.* **20:**1-112.

Aubréville, A., 1939, Fôrets reliques en Afrique Occidentale Française, *Rev. Botany Appl.* **19:**479-484.

Aubréville, A., 1949, *Climats, fôrets et désértification de l'Afrique tropicale*, Paris: Societé d'éditions géographiques, maritimes et coloniales, 351p.

Baier, S., 1980, *An Economic History of Central Niger*, Oxford: Oxford University Press, 325p.

Bartlett, H. H., 1956, Fire, primitive agriculture, and grazing in the tropics, in W. L. Thomas, Jr. (ed.), *Man's Role in Changing the Face of the Earth*, Chicago: University of Chicago Press, pp. 692-720.

Berkofsky, L., 1976, The effect of variable surface albedo on the atmospheric circulation in desert regions, *Jour. Appl. Meteorology* **15:**1139-1144.

Blyn, G., 1966, *Agricultural Trends in India, 1891-1947; Output, Availability, and Productivity*, Philadelphia: University of Pennsylvania Press, 370p.

Bojkov, R. D. (ed.), 1983, Report of the WMO Meeting of experts on the CO_2 concentrations from pre-industrial times to I.G.Y., *WMO Project on Research and Monitoring of Atmospheric* CO_2, *Report 10*, Geneva.

Broecker, W. S.; Takahashi, T.; Simpson, H. J.; and Peng, T. H., 1979, Fate of fossil fuel carbon dioxide and the global carbon budget, *Science* **206:**409-418.

Bryson, R. A., and Baerris, D. A., 1967, Possibilities of major climatic modification and their implications: Northwest India, a case for study, *Am. Meteorol. Soc. Bull.* **48:**136-142.

Buck, J. O.; Dawson, D. L.; and Wu, Y. L., 1966, *Food and Agriculture in Communist China*, New York: F. A. Praeger, 171p.

Canada Yearbook, 1950, 1960, 1970, Ottawa: Government Printing Office.

Charney, J. C., 1975, Dynamics of deserts and drought in the Sahel, *Royal Meteorol. Soc. Quart. Jour.* **101:**193-202.

Charney, J. C.; Quirk, W. J.; Chow, S. H.; and Kornfield, J., 1977, A comparative study of the effects of albedo change on drought in semi-arid regions, *Jour. Atmos. Sci.* **34:**1366-1388.

Chervin, R. M., 1979, Response of the NCAR general circulation model to changed land surface albedo, *GARP Publ. Series No. 22*, pp. 563-581.

Chevalier, A., 1950, La progression de l'aridité, du desséchement et de l'ensablement et de la décadence des sols en Afrique Occidentale Française, *Acad. Sci. Comptes Rendus* **230:**1550-1553.

Church, H. R. J., 1974, *West Africa* (7th ed.), London: Longman Group, 526p.

Clark, W. C. (ed.), 1982, *Carbon Dioxide Review: 1982*, New York: Oxford University Press, 469p.

Commission of European Communities, 1980, Effects on the environment of the abandonment of agricultural land, *Information on Agriculture No. 62*.

Delwaulle, J. C., 1973, Désértification de l'Afrique au sud de Sahara, *Bois et Fôrets des Tropiques* **149:**3-20.

Depierre, D., and Gillet, H., 1971, Désértification de la zone sahelienne au Tchad, *Bois et Forêts des Tropiques* **139:**3-25.

Dickinson, R. E., 1980, Effects of tropical deforestation on climate, in *Blowing in the Wind: Deforestation and Long-range Implications*, Publ. No. 14, Studies in Third World Societies, Williamsburg, Va.: Department of Anthropology, College of William and Mary, pp. 411-441.

Dorward, D. C., and Payne, A. I., 1975, Deforestation, the decline of the horse, and spread of the tsetse fly and trypanosomiasis (nagana) in 19th century Sierra Leone, *Jour. African History* **16:**239-256.

Dresch, J., 1959, Les transformations du Sahel Nigerien, *Acta Géog. Lovaniensia* **30:**3-12.

Faure, H., and Gac, J. Y., 1981, Will the Sahelian drought end in 1985? *Nature* **291:**475-478.

Food and Agricultural Organization (FAO) Production Yearbooks, 1951, 1960, 1963, 1970, 1979, 1980 (annual), Rome: United Nations.

FAO, 1946, *Forestry and Forest Products, World Situation 1937-1946,* Rome: United Nations, 93p.

FAO, 1948, Forest resources of the world, *Unasylva* **2:**161-182.

FAO, 1950, Forest resources of the world, *Unasylva* **4:**57-59 (supplement to 1948 data).

FAO, 1954, Forest resources of the world—A summary of the 1953 World Forest Inventory, *Unasylva* **8**(3):129-144.

FAO, 1960, The world's forest resources—A summary of the "World Forest Inventory, 1958," *Unasylva* **14:**131-150.

FAO, 1976, European timber trends and prospects 1950 to 2000, *Timber Bull. for Europe* **24**(Suppl. 3):308p.

FAO, 1981, *Forest Resources of Tropical Africa, Part I: Regional Synthesis,* J. P. Lanly and J. Clement; *Part II: Country Briefs,* Rome: Tropical Forest Resources Assessment Project.

Gornitz, V., 1985, A survey of anthropogenic vegetation changes in West Africa during the last century—Climatic implications, *Climatic Change* **7:**285-325.

Goudie, A., 1982, *The Human Impact; Man's Role in Environmental Change,* Cambridge, Mass.: MIT Press, 316p.

Grigg, D. B., 1974, The growth and distribution of the world's arable land 1870-1970, *Geography* **59:**104-110.

Haden-Guest, S.; Wright, J. K.; and Teclaff, E. M. (eds.), 1956, A world geography of forest resources, *Special Publication No. 33*, New York: American Geographical Society.

Hansen, J.; Johnson, D.; Lacis, A.; Lebedeff, S.; Lee, P.; Rind, D.; and Russell, G., 1981, Climate impact of increasing atmospheric carbon dioxide, *Science* **213:**957-966.

Hansen, J.; Lacis, A.; Rind, D.; Russell, G.; Stone, P.; Fung, I.; Ruedy, R.; and Lerner, J., 1984, Climate sensitivity: analysis of feedback mechanisms, in J. E. Hansen and T. Takahashi (eds.), *Climate Processes and Climate Sensitivity, Geophysical Monograph 29,* Maurice Ewing Series 5, Washington, D.C.: American Geophysical Union, pp. 130-163.

Hansen, J.; Russell, G.; Rind, D.; Stone, P.; Lacis, A.; Lebedeff, S.; Ruedy, R.; and Travis, L., 1983, Efficient three-dimensional global models for climate studies: Models I and II, *Mon. Weather Rev.* **111:**609-662.

Hare, F. K., 1983, *Climate and Desertification: A Revised Analysis,* WCP-44, World Climate Applications Program, Geneva: WMO, 149p.

Henderson-Sellers, A., and Gornitz, V., 1984, Possible climatic impacts of land cover transformations, with particular emphasis on tropical deforestation, *Climatic Change* **6:**231-257.

Henderson-Sellers, A., and Wilson, M. F., 1983, Surface albedo data for climatic modeling *Rev. Geophysics and Space Physics* **21:**1743-1778.

Houghton, R. A.; Hobbie, J. E.; Melillo, J. M.; Moore, B.; Peterson, B. J.; Shaver, G. R.; and Woodwell, G. M., 1983, Changes in the carbon content of terrestrial biota and soils between 1860 and 1980: A net release of CO_2 to the atmosphere, *Ecology Mon.* **53:**235-262.

Hubert, H., 1920, Le desséchement progressif en A.O.F., *Com. d'Études Hist. et Sci. Bull.* **4:**401-467.

Idso, S. B., 1977, A note on some recently proposed mechanisms of genesis of deserts, *Royal Meteorol. Soc. Quart. Jour.* **103:**369-370.

International Institute of Agriculture (IIA), 1939, *The First World Agricultural Census (1930),* vol. 5, Rome.

Jackson, R. D., and Idso, S. B., 1975, Surface albedo and desertification, *Science* **189:**1012-1013.

Kaplan, I.; Dobert, M.; McLaughlin, J. L.; Marvin, B. J.; and Whitaker, D. P., 1976, *Area Handbook for Sierra Leone,* Washington, D.C.: U.S. Government Printing Office.

Kellogg, W. W., 1978, Global influences of mankind on the climate, in J. Gribbin (ed.), *Climatic Change,* Cambridge: Cambridge University Press, pp. 205-227.

Kerr, R. A., 1983, The carbon cycle and climate warming, *Science* **222:**1107-1108.

Kozlowski, T. T., and Ahlgren, C. E. (eds.), 1974, *Fire and Ecosystems*, New York: Academic Press, 542p.

Krause, O. E., 1970, Cropland trends since World War II, regional changes in acreage and use, *Agriculture Economics Report No. 177*, Washington, D.C.: U.S. Dept. of Agriculture, Econ. Res. Service, 15p.

Lanly, J. P., 1982, Tropical forest resources, tropical forest resources assessment, *FAO Forestry Paper 30, Technical Report 4*, Rome: FAO/UNEP, 106p.

Ledger, D. C., 1961, Recent hydrological change in the Rima Basin, northern Nigeria, *Geog. Jour.* **127:**477-487.

Lettau, H.; Lettau, K.; and Molion, L. C. B., 1979, Amazonia's hydrologic cycle and the role of atmospheric recycling in assessing deforestation effects, *Monthly Weather Reviews* **107:**227-238.

Looman, J., 1983, Grassland as natural or seminatural vegetation, in W. Holzner, M. J. A. Werger, and I. Ikusima (eds.), *Man's Impact on Vegetation*, The Hague: Dr. W. Junk, Publ., pp. 173-184.

Mabbutt, J. A., and Floret, C. (eds.), 1980, *Case Studies on Desertification*, Paris: UNESCO, 279p.

Marques, J.; dos Santos, J. M.; Villa Nova, N. A.; and Salati, E., 1977, Precipitable water and water vapor flux between Belem and Manaus, *Acta Amazonica* **7**(3):355-362.

Matthews, E., 1983, Global vegetation and land use: New high-resolution data bases for climate studies, *Jour. Climate and Appl. Meteorol.* **22**(3):474-487.

Matthews, E., 1984, Vegetation, land-use and seasonal albedo data sets: Documentation of archived data tape, *NASA Tech. Memo. 86107*.

Middleton, N. J., 1985, Effects of drought on the dust production in the Sahel, *Nature* **316:**431-434.

Migeod, F. W. H., 1922, Some notes on the Lake Chad Region in British Territory, *Geog. Jour.* **60:**347-359.

Morgan, W. B., 1959, The influence of European contacts on the landscape of Southern Nigeria, *Geog. Jour.* **125:**48-64.

Morgan, W. B., and Pugh, J. C., 1969, *West Africa*, London: Methuen, 788p.

Myers, N., 1980, *Conversion of Tropical Moist Forests*, Washington, D.C.: National Academy of Science, 205p.

Nicholson, S. E., 1980*a*, Saharan climates in historic times, in M. A. J. Williams and H. Faure (eds.), *The Sahara and the Nile*, Rotterdam: A. A. Balkema, pp. 173-200.

Nicholson, S. E., 1980*b*, The nature of rainfall fluctuations in subtropical West Africa, *Monthly Weather Reviews* **108:**473-487.

Nicholson, S. E., and Flohn, H., 1980, African environmental and climatic changes and the general atmospheric circulation in late Pleistocene and Holocene, *Climatic Change* **2:**313-348.

Official Yearbook of Australia 1975-1976, 1981, Canberra.

Oguntoyinbo, J. S., 1970*a*, Surface measurements of albedo over different agricultural crop surfaces in Nigeria and their spatial and seasonal variability, *Nigerian Geog. Jour.* **13:**39-55.

Oguntoyinbo, J. S., 1970*b*, Reflection coefficients of natural vegetation crops and urban surfaces in Nigeria, *Royal Meteorol. Soc. Quart. Jour.* **96:** 430-441.

Otterman, J., 1974, Baring high albedo soils by overgrazing: A hypothesized desertification mechanism, *Science* **186:**531-533.

Otterman, J., 1977, Anthropogenic impact on the albedo of the earth, *Climatic Change* **1:**137-155.

Otterman, J., 1981, Satellite and field studies of man's impact on the surface in arid regions, *Tellus* **33:**68-77.

Otterman, J., and Fraser, R. S., 1976, Earth-atmosphere system and surface reflectives in arid regions from LANDSAT MSS data, *Remote Sensing Environment* **5:**247-266.

Palutikof, J. P.; Lough, J. M.; and Farmer, G., 1981, Senegal River runoff, *Nature* **293:**414.

Paterson, S. S., 1956, *The Forest Area of the World and Its Potential Productivity*, Goteborg, Sweden: Royal University of Goteborg, 216p.

Pereira, H. C., 1973, *Land Use and Water Resources in Temperate and Tropical Cimates*, London: Cambridge University Press, 246p.

Persson, R., 1974, *World Forest Resources*, Roy. Col. Forestry, Stockholm, Report No. 17, 261p.

Persson, R., 1977, *Forest Resources of Africa: Part II*, Roy. Col. Forestry, Stockholm, Report No. 22, 224p.

Phillips, J., 1974, Effects of fire in forest and savanna ecosystems of sub-Saharan Africa, in T. T. Kozlowski and C. E. Ahlgren (eds.), *Fire and Ecosystems*, New York: Academic Press.

Pinker, R. T.; Thompson, O. E.; and Eck, T. F., 1980, The albedo of a tropical evergreen forest, *Royal Meteorol. Soc. Quart. Jour.* **106:**551-558.

Potter, G. L.; Ellsaesser, H. W.; MacCracken, M. C.; and Ellis, J. S., 1981, Albedo change by man: Test of climatic effects, *Nature* **291:**47-49.

Primack, M. L., 1962, Land clearing under 19th century techniques, some preliminary calculations, *Jour. Econ. History* **22:**484-497.

Prospero, J. M., and Nees, R. T., 1977, Dust concentrations in the atmosphere of the equatorial North Atlantic: Possible relationship to the Sahelian drought, *Science* **196:**1196-1198.

Richards, J. F.; Olson, J. S.; and Rotty, R. M., 1983, Development of a data base for CO_2 releases resulting from conversion of land to agricultural uses, Research Memo, Inst. for Energy Analysis, Oak Ridge, Tenn.: Oak Ridge Assoc. Universities, ORAU/IEA-82-10(M).

Richards, J. F., and Tucker, R. P. (eds.), 1987, *World*

Deforestation in the 20th Century, Durham, N.C.: Duke University Press.

Ripley, E. A., 1976, Drought in the Sahara: Insufficient biogeophysical feedback? *Science* **191:**100.

Sagan, C.; Toon, O. B.; and Pollack, J. B., 1979, Anthropogenic albedo changes and the earth's climate, *Science* **206:**1363-1368.

Salati, E., and Vose, P. B., 1984, Amazon Basin: A system in equilibrium, *Science* **225:**129-138.

Seiler, W., and Crutzen, P. J., 1980, Estimates of gross and net fluxes of carbon between the biosphere and the atmosphere from biomass burning, *Climatic Change* **2:**207-247.

Sierra Leone, 1937, *Colonial Reports: Annual No. 1873,* Freetown: Govt. Printer.

Sierra Leone, 1939, *Forestry Report,* Freetown: Govt. Printer.

Sircoulon, J., 1976, Les données hydropluviométriques de la sécheresse récente en Afrique intertropicale, comparaison avec les sécheresses "1913" et "1940," *Cahiers Orstom Ser. Hydrologique,* **13:**75-174.

Stebbing, E. P., 1935, The encroaching Sahara: The threat to the West African colonies, *Geog. Jour.* **85:**506-524.

Stewart, O. C., 1956, Fire as the first great force employed by man, in W. L. Thomas, Jr. (ed.), *Man's Role in Changing the Face of the Earth,* Chicago: University of Chicago Press, pp. 115-133.

Sud, Y. C., and Fennessy, M., 1982, A study of the influence of surface albedo on July circulation in semi-arid regions using the GLAS GCM, *Jour. Clim.* **2:**105-125.

Sutton, W. R. J., 1975, The forest resources of the U.S.S.R., their exploitation and their potential, *Commonwealth Forestry Rev.* **54:**110-138.

Talbot, M. R., 1980, Environmental responses to climatic change in the West African Sahel over the past 20,000 yrs, in M. A. J. Williams and H. Faure (eds.), *The Sahara and the Nile,* Rotterdam: Balkema, pp. 37-62.

Thompson, K., 1980, Forests and climate change in America: Some early views, *Climatic Change* **3:**47-64.

Tucker, R. P., and Richards, J. F., 1983, *Global Deforestation and the Nineteenth Century World Economy,* Durham, N.C.: Duke University Press, 210p.

United Nations Conference on Desertification, 1977, *Desertification: Its Causes and Consequences,* Oxford: Pergamon Press, 448p.

Unwin, A. H., 1920, *West African Forests and Forestry,* London: T. Fisher Unwin Ltd., 527p.

Verstraete, M. M., 1981, Some impacts of desertification processes on the local and regional climate, in A. Berger (ed.), *Climate Variations and Variability: Facts and Theories,* Dordrecht, Holland: D. Reidel Publ. Co., pp. 717-721.

Walker, J., and Rowntree, P. R., 1977, The effect of soil moisture on circulation and rainfall in a tropical model, *Royal Meteorol. Soc. Quart. Jour.* **103:**29-46.

Ware, H., 1977, Desertification and population: Sub-Saharan Africa, in M. H. Glantz (ed.), *Desertification,* Boulder, Colo.: Westview Press, pp. 165-202.

Wendler, G., and Eaton, F., 1983, On the desertification of the Sahel zone, *Climatic Change* **5:**365-380.

White, H. P., and Gleave, M. P., 1971, *An Economic Geography of West Africa,* London: G. Bell and Sons, 322p.

Wilson, A. T., 1978, Pioneer agriculture explosion and CO_2 levels in the atmosphere, *Nature* **273:**40-41.

Woodwell, G. M.; Hobbie, J. E.; Houghton, R. A.; Medillo, J. M.; Moore, B.; Peterson, B. J.; and Shaver, G. R., 1983, Global deforestation: Contribution to atmospheric carbon dioxide, *Science* **222:**1081-1086.

Zon, R., and Sparhawk, W. N., 1923, *Forest Resources of the World, vol. 2,* New York: McGraw-Hill, 997p.

3: Hudson Valley Reconstructed Temperature Data Sets: History and Spectral Analysis

Jerome S. Thaler
Hudson Valley Climate Service
Mahopac, NY

Abstract: Time-series analyses (both standard power spectral analysis and maximum entropy spectral analysis) reveal a predominant 10.2-yr signal in winter temperatures in seven combinations of Hudson Valley stations. The predominant summer temperature signal is found at 12.9 yr. The strongest signal in the annual temperatures lies at 10.4 yr, with a secondary peak at 17.3 yr. These signals may result from solar and lunar-nodal forcing. A shorter 6-yr signal is seen in the winter and annual temperature spectra; this may be a harmonic of solar-lunar tidal forcing, or it may be related to the dynamics of the inner planets. A longer signal of around 22 yr is found in winter temperatures in Central Park records; when these records are combined with other Hudson Valley stations, a signal is found at about 28 yr.

INTRODUCTION

The purpose of this study is to determine whether cyclical temperature patterns reported for various areas by Spar (1954; Spar and Mayer, 1973), Mock and Hibler (1976), Hancock and Yarger (1979), and Currie (1979, 1981) are also revealed in temperatures in the Hudson Valley, New York, United States. This chapter goes beyond the earlier Hudson Valley filter studies of Thaler (1979, 1983) by employing spectral analysis methods on winter, summer, and annual temperatures. It also seeks to bring the earlier work of other investigators up to date and to determine whether signal periodicities exist that might be related to sunspot (10-11 yr), the solar-lunar tidal (54, 13.5 yr), and the lunar-nodal (18.6 yr) forcing agents, as some have claimed.

Currie (1979) found 10.4- and 10.1-yr temperature signals in the New York City (1869-1960) and Albany (1874-1960) annual records. In a more recent study Currie (1981) reported a 10.2- and 10.4-yr temperature signal for New York City and Albany in a 140-yr time span ending in 1960. Amplitudes of 0.25°C and 0.37°C were found in the 1979 study and 0.04°C and 0.14°C in the 1981 paper. A weaker signal of 18.4 yr was found for 54 stations in the northeastern quadrant of North America. Currie used maximum entropy spectral analysis (MESA), a relatively new and somewhat controversial technique that is claimed to be superior in sensitivity to classical spectral analysis in that it can identify and resolve weak signals to a far greater degree.

Currie (1979) claims that the solar-cycle signal (10-11 yr) that he found in 42 worldwide stations equal or greater in length than 85 yr was undetectable by conventional spectral analysis. However, it has been pointed out that the amplitudes of such signals are uncertain (Mitchell and Cook, personal communication). Currie has questioned the amplitudes found using the

MESA technique but notes that even a signal of as small as 0.1°C over a large area could significantly affect climate.

Borisenkov and co-workers (1983) identified an approximate 18-yr insolation period for summer and winter solstices in the high latitudes of 60°N and 80°N. For lower latitudes (20°N), cyclical components of 11.9, 5.9, 4, and 2.7 yr were identified. Related forcing mechanisms for these were attributed by Borisenkov and co-workers to the sidereal period of revolution of the inner planets (5.9, 4, and 2.7 yr). The latitudes of interest in this study of the Hudson Valley range from 40°.47′N to 42°.45′N; therefore, the shorter periods reported are of greatest interest here.

DATA SET HISTORY AND PREPARATION

The Hudson Valley climate history contains temperature data sets that go back to the early 1820s. The station locations at New York City, West Point, and Albany provide a representative portrait of the temperature profile of the lower, mid-, and upper Hudson Valley. These mean temperature time-series data sets (partly reconstructed for this chapter), together with data from New Haven, Connecticut (1780-present), reconstructed New Brunswick, New Jersey, seasonal temperatures (1831-present), and reconstructed seasonal temperature records from Philadelphia, Pennsylvania (1738-present), constitute the oldest and longest continuous temperature series in North America.

Reconstructions of climate data are subject to estimations, errors, and educated guesses. The assumptions made here in the combination of nineteenth- and twentieth-century data into a continuous time series are revealed in the three station histories (Table 3-1). The uncertainties of daily mean determination are easily seen in unspecified reading times of "before sunrise" and "after sunset," unspecified locations with no coordinates, and no lateral distances given in station relocation. Monthly mean determinations using both the average of three readings per day, and the minimum plus the maximum, divided by two, show significant statistical differences. The data have not been corrected for these differences at the three stations. In a previous paper, Thaler (1983) discussed the problems of such flawed data in greater detail. Reconstruction of time spans noted in Table 3-1 had to be made by extrapolation using overlapping periods so that a mean difference could be used to estimate missing months. A climatic homogeneity test (Reiss, Groveman, and Scott, 1980) was made to verify whether such comparisons were legitimate. The comparisons—Fort Columbus with Central Park, West Point with Carmel, and Albany (city) with Albany Airport—all show homogeneity constants for all months of the year that are below the critical ratio value, above which extrapolation would be invalid.

PROCEDURE

In this spectral analysis, both the standard (Blackman-Tuckey) power spectral analysis and the MESA are employed, not so much to reveal differences between the two methods but to determine whether both methods show agreement of the predominant periodicities. In this study, the combined Albany and West Point temperature record was used for winter, summer, and annual temperature. Central Park, West Point, and Albany together were used for winter temperature. This method was determined to be the best way of minimizing observational inaccuracies, lessening errors in missing monthly estimations, and obtaining a more representative temperature profile of the Hudson Valley.

RESULTS

Results of spectral analysis for winter temperatures (December to March) are as follows: For Central Park (Fig. 3-1) common peaks in standard power spectrum and MESA analyses are found at about 22.2 yr, 10.4 yr, 6 yr, 4.1 yr, and

Table 3-1. Station Histories

Year	Observation Time	Station Location Changes
Albany		
1820-1825	7A.M., 2 P.M., 9 P.M.	
1826-1849	Before sunrise, 3 P.M., after sunset	
1850-1852	6 A.M., 2 P.M., 10 P.M.	
1853-1861	7 A.M., 2 P.M., 9 P.M.	No known elevation; 1858 readings taken locally at different site
1862-1873	8 A.M. and 7 P.M.	No known elevation
1874-1879/1884	Minimum/Maximum thermometer	Elevation changed 67 m (no distance given)
October 1884-1938		Station relocated to 6 m elevation (nearer Hudson River)
March 1938-1970		Concurrent record with Albany Airport
1971-present		Albany closed, extrapolated from Albany Airport
New York City (Central Park)		
1822-1868	7 A.M., 2 P.M., 9 P.M.	Extrapolated from the Ft. Columbus period in common (1869-1888) 10 km to the south
1869- present	Minimum/Maximum thermometer	
West Point		
1821-1839	7 A.M., 2 P.M., 9 P.M.	1820, 1821, 1822, some months estimated
1840-1854	Sunrise, 9 A.M., 3 P.M., 9 P.M.	
1855-1869	7 A.M., 2 P.M., 9 P.M.	
1870-1890	7 A.M., 2 P.M., 9 P.M., also minimum/maximum thermometer	
1891-1899	Minimum/Maximum thermometer	March 1890 moved 305 m east from 167 ft to 160 ft elevation
1900-March 1905	No observations, extrapolated from Carmel 15 km to east	
April 1905-present	Minimum/Maximum thermometer	June 1946 moved west 110 m elevation to courtyard

2 yr. For common prominent peaks, an average has been taken of the peaks derived by the two methods of analysis. For the combined Albany, West Point, and Central Park (Fig. 3-2), common peaks are found at 28 yr, 10.2 yr, 6 yr, 3.6 yr, and 2.1 yr. For the combined Albany and West Point (Fig. 3-3), common peaks are found at 27.9 yr, 10.1 yr, 6 yr, and 2.3 yr. The power spectral analysis shows a peak with greater than 95% confidence level only at 9.8 yr. For comparison, Mitchell (personal communication), using standard power spectral analysis, finds a spectral signal at 27 yr, 18 yr, and 10 yr.

These signals may be compared with a study

(Landsberg, Yu, and Huang, 1968) on reconstructed winter (December to February) temperatures of Philadelphia, Pennsylvania (1738-1967), in which variance spectral analysis shows a peak between 23 yr and 25 yr. Shorter periodicity winter signals are found in the present study at 6 yr, 2.3 yr, and at approximately 2.1 yr. Confidence levels of 95% are only found in peaks at 2 yr and 2.1 yr. There is also a signal ranging from 3.6 yr to 4.2 yr, which has a confidence level of 95% for a 4-yr signal in one of five spectral analyses.

Spectral analysis for summer temperatures (June to September) are as follows: For Central Park (Fig. 3-4) a common peak is only found at about 12.9 yr. For Albany and West Point (Fig. 3-5) common peaks are found at 53.8 yr and 13.2 yr, 7.6 yr and 4 yr. Mitchell (personal communication), using standard power spectral analysis, shows signals at 60 yr, 13.3 yr, 7.5 yr, and 3.9 yr. None of the peaks in Figures 3-4 and 3-5 is significant at the 95% confidence level. Landsberg and co-workers' (1968) Philadelphia study for summer (June to August) shows a peak between 23 yr and 25 yr (not seen here) plus one between 11.6 yr and 12.1 yr. Short-periodicity summer signals were found in the present study at 6 yr and from 3.9 yr to 4.3 yr.

Results of spectral analysis for annual temperatures are as follows: For Central Park (Fig. 3-6) a common peak is found at 17.5 yr and 10.4 yr. For Albany and West Point (Fig. 3-7) common peaks are found at 13.2 yr, 10.6 yr, and 6.4 yr. Mitchell (personal communication) finds annual signals at 17.1 yr, 10 yr, and 6 yr. Landsberg, Yu, and Huang's (1968) Philadelphia study for the annual series revealed peaks at 10 yr and between 7.5 yr and 8.2 yr.

In Figure 3-8 power spectral analysis is shown for Albany and West Point combined January mean temperatures. Previous work by Spar (1954; Spar and Mayer, 1973), Mock and Hibler (1976), and Hancock and Yarger (1979) noted a near 20-yr January temperature oscillation or harmonic. While spectral-amplitude 20-yr peaks are noted for the 1900-1984 and 1930-1984 periods, the predominant periodicity for the 1821-1900 period is 10 yr, and the entire 1821-1984 period shows no predominant spectral amplitudes. The analysis was not checked for confidence levels; it is possible that the periods are harmonically related. Just how this 20- and 10-yr periodicity might be related to the solar 10-11-yr cycle or the Hale double sunspot cycle is unclear.

DISCUSSION

While an 11.9-yr signal was not found in this study, the shorter periodicity signals noted here in general confirm the work of Borisenkov, Tsvetkov, and Agaponov (1983). Additional studies are needed to resolve the 2.7-yr and 2.1-yr discrepancy and to relate potential forcing mechanisms to these shorter-length periodicity signals. The 7-yr and 8-yr periodicities found in this study have no ready explanation.

The validity of spectral analysis on monthly and seasonal data has been called into question. Currie (1981) has emphatically pointed out that spectral analysis of monthly data such as the preceding January study is subject to aliasing errors that make the results of this type of investigation invalid. To some extent there could also be some problems in the seasonal analysis. Mitchell (personal communication) states that while aliasing could be a problem, analysis using other methods has not shown that the conclusions reached here are invalid. At the moment the question is open and requires additional investigation.

CONCLUSION

The common periodicity signals found using MESA and standard spectral analysis are as follows: The predominant winter-temperature signal is the 10.2-yr signal shown by seven combinations of Hudson Valley stations. The predominant summer temperature signal is seen at about 12.9 yr and is closest to the solar-lunar-tidal harmonic at 13.5 yr.

The predominant annual temperature signal is seen at 10.4 yr; this is the closest peak to the solar sunspot period, with a secondary signal of possible lunar-nodal origin at about 17.3 yr. These findings to 1983 confirm the findings of Currie (1979) regarding an annual signal of 10.1 yr and 10.4 yr for Albany and New York City (to 1960). An additional approximate 6-yr signal is seen in the winter and annual temperature spectrum This may be a harmonic of the solar-lunar-tidal forcing or of the inner-planet sidereal period of revolution identified by Borisenkov and co-workers (1983). The previously reported warming trend is confirmed in both the winter and summer data and in the annual spectra, which show the greatest power in the longest time intervals. This trend, however, might be the result of the urbanization of the area.

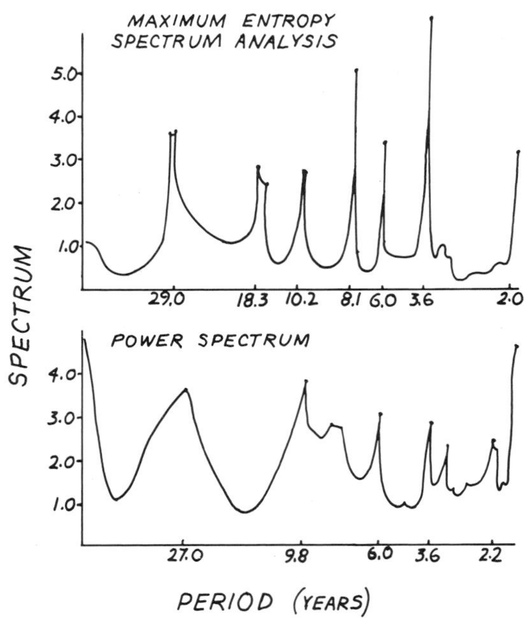

Figure 3-2. Winter temperatures, Albany, West Point, and Central Park, 1821-1982. No periods show a confidence level of 95%.

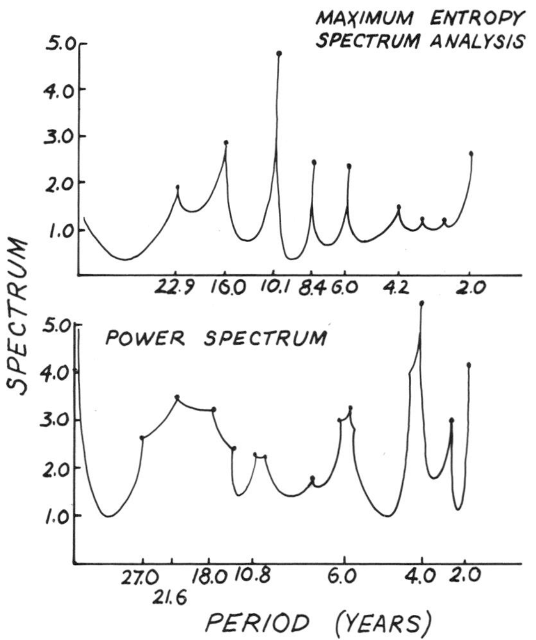

Figure 3-1. Winter temperatures, Central Park, 1821-1822 to 1982-1983. Only periods of 2 yr and 4 yr show a confidence level of 95%.

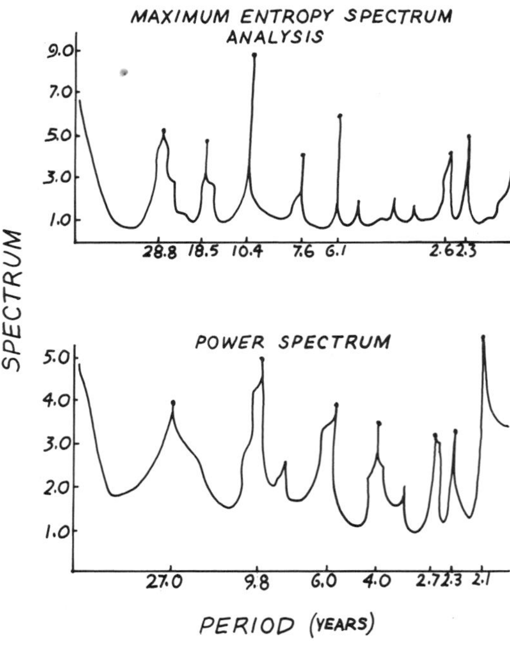

Figure 3-3. Winter temperatures, Albany and West Point, 1821-1822 to 1982-1983. Periods of 9.8 yr and 2.1 yr show a confidence level of 95%.

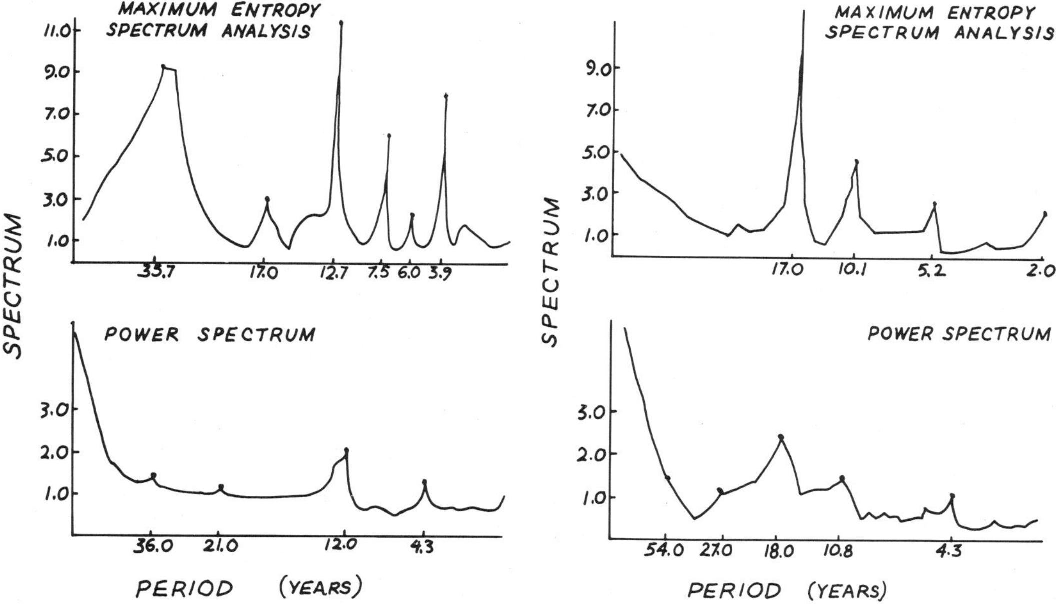

Figure 3-4. Summer temperatures, Central Park, 1822-1983. No periods show a confidence level of 95%.

Figure 3-6. Annual temperatures, Central Park, 1822-1983. No periods show a confidence level of 95%.

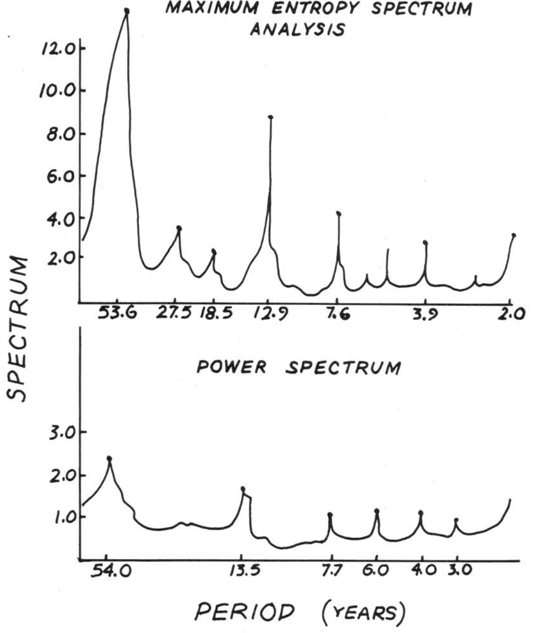

Figure 3-5. Summer temperatures, Albany and West Point, 1821-1983. No periods show a confidence level of 95%.

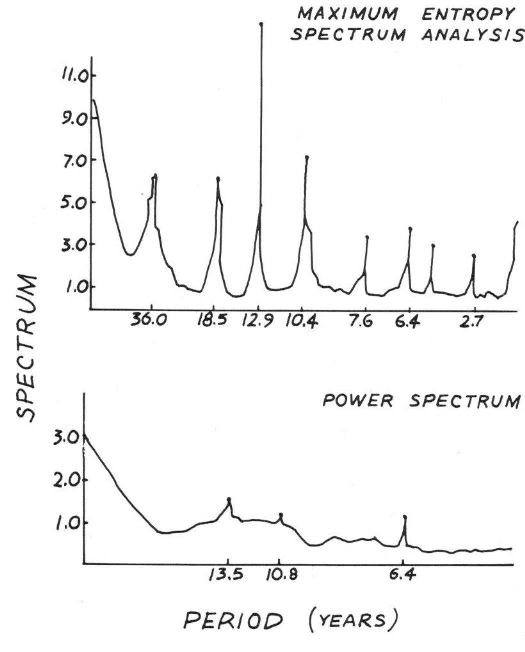

Figure 3-7. Annual temperatures, Albany and West Point, 1821-1983. No periods show a confidence level of 95%.

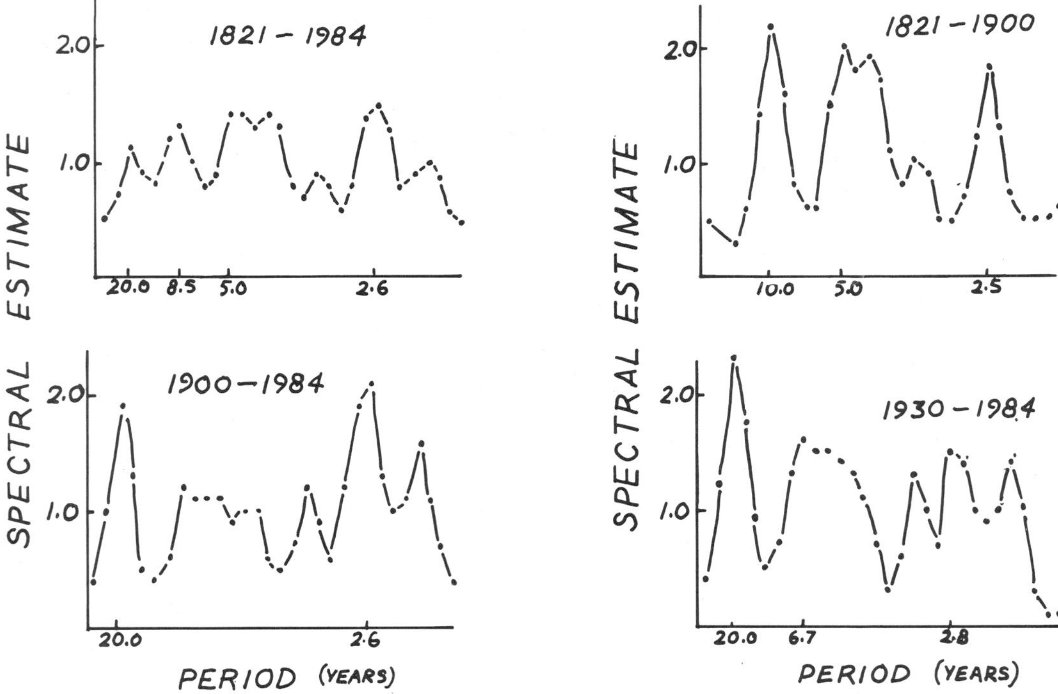

Figure 3-8. Mean temperatures during January for Albany and West Point.

ACKNOWLEDGMENTS

The author thanks Edward Cook, Murray Mitchell, Jr., and Benson Sundheim for their gracious assistance in providing computer time and expertise in the many aspects of power spectrum analysis. Without their help this study could not have been made.

REFERENCES

Borisenkov, Y. E.; Tsvetkov, A. V.; and Agaponov, S. V., 1983, On some characteristics of insolation changes in the past and future, *Climatic Change* **5:**237-244.

Currie, R. G., 1979, Distribution of solar cycle signal in surface air temperature over North America, *Jour. Geophys. Research* **84:**753-761.

Currie, R. G., 1981, Solar cycle signal in air temperature in North America—Amplitude, gradient, phase and distribution, *Jour. Atmos. Sci.* **38:**808-818.

Diaz, H. F., and Myers, E. F., 1978, *Index of Historical Surface Weather Records—New York*, Asheville, N.C.: National Climate Center.

Hancock, D. J., and Yarger, D. N., 1979, Cross-spectral analysis of sunspots and monthly mean temperature and precipitation for the contiguous United States, *Jour. Atmos. Sci.* **36:**746-753.

Landsberg, H. E.; Yu, C. S.; and Huang, L., 1968, Preliminary reconstruction of a long time series of climatic data for eastern United States, *Technical Note BN 571*, College Park, Md.: University of Maryland, Institute for Fluid Dynamics and Applied Mathematics.

Mock, S. J., and Hibler, W. D. III, 1976, The 20-year oscillation in eastern North American temperature records, *Nature* **261:**484-486.

National Archives of the United States, Climatological Records, 1952, Microfilm reels CL 875, 876, *West Point Monthly Observer's Reports, 1821-1893.*

Reiss, N. M.; Groveman, B. S.; and Scott, C. M., 1980, Construction of a long time-series of seasonal mean temperatures for New Brunswick, New Jersey, *New Jersey Acad. Sci. Bull.* **25:**1-11.

Spar, J., 1954, Temperature trends in New York City, *Weatherwise* **7:**149-151.

Spar, J., and Mayer, J. A., Temperature trends in New York City: A postscript, *Weatherwise* **26:**128-130.

Thaler, J. S., 1979, West Point—152 years of weather records, *Weatherwise* **32:**112-115.

Thaler, J. S., 1983, A cyclical pattern in 19th and 20th century Hudson Valley winter temperatures, in B. M. McCormac (ed.), *Weather and Climate Responses to Solar Variation Symposium*, Boulder, Colo.: Colorado Associated University Press, pp. 189-196.

U.S. Department of Agriculture, Weather Bureau, 1934, *Climatic Summary of the United States, Section 83, Eastern N.Y., 1820-1930*, Washington, D.C.: U.S. Gov't. Printing Office.

U.S. Department of Agriculture, 1894-1897, 1901-1938, *Climatological Data, New York Section*, Asheville, N.C.: National Climate Center.

U.S. Department of Commerce, 1956, *Substation History, New York*, Asheville, N.C.: National Climate Center.

U.S. Department of Commerce, 1958, *Climatography of the U.S., No. 40-30, Climatic Guide for New York City, N.Y. and Nearby Areas*, Washington, D.C.: U.S. Gov't. Printing Office.

U.S. Department of Commerce, 1963-1983, *Climatological Data, New York*, Asheville, N.C.: National Climate Center.

U.S. Weather Bureau, 1939-1962, *Climatological Data, New York*, Asheville, N.C.: National Climate Center.

II

Proxy Climate Indicators

4: Evaluation of Historic Climate and Prediction of Near-Future Climate from Stable-Isotope Variations in Tree Rings

Leona Marshall Libby
University of California, Los Angeles

Abstract: Trees are recording thermometers because they grow from water and atmospheric carbon dioxide. Isotope variations in hydrogen and oxygen in their rings are climate indicators because the isotope composition in rain varies with temperature. The temperature of relevance is the sea-surface temperature from which the rain distilled and the temperature local to the site of rainfall. This hypothesis has been proven in four different trees from contrasting locales by comparing variations of the measured isotope ratios with the mercury thermometer records of air local to the trees. The longest mercury thermometer record is from Germany, 1690 to present. This is compared with stable-isotope variations measured in a German oak from 1700 to 1950. The agreement of the variations in the two records is excellent, showing the cold of the second Little Ice Age, the sudden warm period 1710-1740, followed by cold extending to 1800, after which there was a long slow warming peaking in 1930. The longest record measured so far is the 2,000-yr sequence of a Japanese cedar. The climate periods indicated by the isotope record have been evaluated by Fourier transform, and the near-future climate variations have been predicted using these periods in the transform.

INTRODUCTION

Trees are seen as recording thermometers in that they store a chronologic record of climate changes in the form of variations in ratios of stable isotopes in their tree rings. Knowledge of changes in past climates is relevant to understanding how humans have evolved, how climatic change has influenced their way of living, and the developing technology by which they defended themselves against climate exigencies. The method of tree thermometry depends on measuring the ratios of the stable isotopes of hydrogen and oxygen in tree rings. These elements derive from rain and snow, which in turn originate as water vapor distilled from the surface of the seas. Variations in sea-surface temperature cause variations in the stable-isotope ratios in the water vapor and therefore changes in their ratios as they are stored year by year in tree rings. By evaluating the changes in these ratios in tree rings, we are able to determine climate changes of the past, as far back in time as tree rings may be secured. By applying the method of Fourier analysis to the stable-isotope measurements in trees, we can predict climate change for the near future. Such predictions have not been tested, but they should certainly be studied for their potential influence on world food and energy supplies.

The analytic chemistry used in measuring stable-isotope ratios in wood is described with care so that other investigators can use it to measure trees from all over the world. In this way we may learn if climate changes have been cosynchronous in both hemispheres. Measurements of carbon-isotope ratios are described, but they are not as useful because carbon does not derive from seawater (Libby, 1983).

PRINCIPLES

The onset of testing of hydrogen bombs in the atmosphere led to an understanding of isotope fractionation in the water vapor that is distilled from the ocean surfaces. This information is derived from the establishment of a global network of 155 collecting stations in 65 countries, beginning in 1953 and continuing to the present, by the International Atomic Energy Agency (IAEA) and the World Meteorological Organization. Monthly meteorological data (amount of precipitation and temperature) were reported, and monthly samples of precipitation were measured for deuterium-to-hydrogen ratio and oxygen-18-to-oxygen-16 ratio.

The measured ratios are expressed as δ_D and δ_{18}:

$$\delta_D = \left\{[(D/H)_s - (D/H)_{std}]/(D/H)_{std}\right\} \times 10^3 \text{‰}$$

$$\delta_{18} = \left\{[(^{18}O/^{16}O)_s - (^{18}O/^{16}O)_{std}]/(^{18}O/^{16}O)_{std}\right\} \times 10^3 \text{‰},$$

where subscript s refers to the sample, subscript std refers to standard mean ocean water (SMOW), and ‰ means parts per thousand.

The error of measurement of δ_D is about 2‰; of δ_{18}, it is about 0.2‰. The plot of world data from 1953 to 1963, of δ_D versus δ_{18}, shows that all the measurements for terrestrial surface waters lie on a line with a slope of 8, characterizing Rayleigh distillation of water vapor from the sea surface to form atmospheric precipitation. The slope of 8 was originally demonstrated by Harmon Craig (1961). The line is expressed by the relation $\delta_D = 8\delta_{18}$ + constant, where the slope of 8 can be computed from the measured temperature coefficients for $[(D/H)_{liquid}/(D/H)_{vapor}]$ and for $[(^{18}O/^{16}O)_{liquid}/(^{18}O/^{16}O)_{vapor}]$, both in hydrogen and in oxygen measured in tropical precipitation distilled from warm oceans. Points between have been measured in middle latitudes. The same slope of 8 is shown in our isotope measurements in tree rings.

The IAEA monthly measurements show seasonal variations in that the heavy isotopes are depleted in precipitation when water vapor is distilled from cold oceans in the winters and enriched in precipitation when water vapor is distilled from warm oceans in the summers.

Wood is composed approximately of cellulose and lignin. Cellulose is a multiple alcohol of schematic formula $(H\text{-}C\text{-}O\text{-}H)_n$ so that the reaction for the formation of cellulose may be written:

$$CO_2 + H_2O \rightarrow (H\text{-}C\text{-}O\text{-}H)_n + O_2.$$

Lignin contains interconnected aromatic and aliphatic rings and aliphatic chains containing about 30% oxygen by weight in the form of ether, carbonyl, and hydroxyl bonds. Wood is approximately 25% lignin. This percentage varies somewhat from spring wood to summer wood, and it is also possible that its percentage may vary somewhat from ring to ring, so in principle variation of lignin might affect the temperature coefficient of wood formation.

Assuming the principle of thermodynamic equilibrium to hold in the formation of wood, a very slow process, we have estimated what effect as much as 10% variation in the percentage of lignin may have on the temperature coefficient for the formation of wood and find it to be about 1.5%. Since we judge that a 1.5% uncertainty is tolerable within the limits of other errors inherent in the method, we have always analyzed whole wood in our study of isotope variations in tree-ring sequences.

In analyzing whole wood, one is confronted by the question of whether to use wet or dry chemistries. Of course, if one decides to separate cellulose from lignin, one is forced to use wet chemistries. Only in whole-wood analysis is dry chemistry possible. With wet chemistries, performed necessarily with hydrogen- and oxygen-containing solvents, there is the risk of isotope exchange with the solvent. Intimate exchange of hydroxyl oxygen (-O-H) with carbonyl oxygen (-CO-OH) occurs under all conditions of acidity and alkalinity in liquids such as water, ketones, aldehydes, and alcohols. Furthermore, the exchange of cellulose and whole-wood hydrogens with hydrogen in water is rapid and effective, leading one to expect similar exchanges with other solvents containing OH groups. The exchange of hydrogen in cellulose with water is 50% in 1 hr at 25°C. With dry chemistry there is no possibility for isotope exchange with reagents, and for this reason we have always used dry chemistry.

Thus, in our study of isotope variations in lengthy dendrochronological sequences, we are evaluating temperature fluctuations in the sea surface, the source of the precipitation that nourished the trees. The sea-surface temperatures, in turn, are affected by variations in the ultraviolet spectrum and other factors relating to the Sun. The climate record from variations in sea-surface temperatures is also stored in the form of variations of organic carbon and uranium in marine sediments, in turn influenced by the amount of living matter growing in the sea; this matter becomes more abundant as the sea surface grows warmer and, after death, drops to the bottom to form sediments. The uranium ions always present in seawater attach themselves readily to this organic matter during sedimentation and vary with the abundance of organic matter. The dry-chemistry technique is a new technology that may offer the first advance in the evaluation of solar physics since the invention of the solar coronagraph.

APPLICATIONS

By measuring stable-isotope ratio variations in tree-ring sequences, we hope to set limits on solar changes and evaluate their periodicities as far back in time as tree-ring sequences exist (about 8,000 yr). In so-called floating sequences in ancient trees not fitted to present-day sequences but dated by radiocarbon, we have another data base where evaluation of solar variations becomes possible for the more distant past. Marine cores and varves also offer data bases that should in principle allow interpretation of the history of the local sea-surface temperature.

Any of these data banks could have had their stable-isotope ratios perturbed during glacial intervals because the ice depletes the oceans in the light isotopes and therefore significantly enriches the sea in the heavy isotopes. Thus, marine sediments and continental precipitation, rain and snow, reflect this perturbation as well as variations caused by temperature changes alone.

We have obtained tree-ring sequences from Bernd Becker of the Universität Hohenheim, Stuttgart; Dieter Eckstein of the University of Hamburg; Paul Zinke of the Forestry Department at the University of California at Berkeley; Henry Michael of the University Museum of the University of Pennsylvania; U.S. Forest Service in Sequoia, Kings Canyon National Park, California; and K. Kigoshi, Gukashuin University, Tokyo. Sequences of bog oak have been prepared from the ancient oaks dug out of the bogs of England and Ireland by laboratories in those countries and also by Bernd Becker, Dieter Eckstein, and V. G. Sienbenlist in Germany. We hope that tree-ring sequences will also be prepared from trees of the Southern Hemisphere, from which one could learn whether climate changes have been simultaneous in both hemispheres.

TECHNIQUE

We considered whether to measure whole wood, lignin, or cellulose, wood being about 25% lignin on the average. But isotope fractionation at climatic temperatures is a function of the frequencies of the chemical bonds. We quote from Gerhard Herzberg (1945, p. 192) as follows: "One would expect the -C-H bond to have essentially the same electronic structure and therefore the same force constant in different molecules, and similarly for other bonds. This is indeed observed." For the -C-H bonds the vibrational frequencies in lignin and in cellulose are almost equal, but in fact they differ by 6% because cellulose is a multiple alcohol $(H\text{-}C\text{-}O\text{-}H)_n$ and lignin is a polymer containing both aromatic and aliphatic carbons connected to hydrogen. Therefore, assuming thermodynamic equilibrium, variations of the lignin concentration in tree rings might affect the hydrogen-isotope ratio by as much as 25% of 6%—namely, 1.5%.

Likewise, for the -C-O-H bonds, the vibrational frequencies are equal in lignin and in cellulose within a few percent. But lignin, different from cellulose, also contains ether linkages, -C-O-C-. The -C-O-H linkage has a C-O bond distance of 1.427 Å, and the ether linkage has a C-O bond distance of 1.43 Å. Therefore, the presence of lignin, containing 14% oxygen, of which 16% is ether-linked, might affect the isotope ratio of oxygen in whole wood by $0.3\% \times 25\% \times 14\% \times 60\%$ equal to $6 \times 10^{3}\%$, by variation from its average concentration of 25%. On these numerical arguments and the necessity to avoid isotope exchange with liquids, we based our decision to measure stable-isotope ratios in whole wood.

The next problem was concerned with which trees to measure. Many tree-ring sequences can be counted with an accuracy of 1 yr. Those that are not tied to modern sequences by overlapping ring patterns can be dated in favorable cases with an accuracy of about 30 yr by radiocarbon, depending on the age and the number of radiocarbon measurements made.

To prove our hypothesis that trees are thermometers, however, we needed to compare our measurements of stable-isotope ratios in the tree rings with mercury thermometer records from near where the trees grew. Thus, we could not use bristlecone pines because there is no lengthy temperature record for hundreds of miles of their location in the White Mountains of California.

Because the longest temperature records are in Europe, we obtained a German oak from the laboratory of Bruno Huber in Munich, where the oak rings were counted and labeled with the number of the years in which they grew. More recently his students, Bernd Becker and Dieter Eckstein, who have established tree-ring laboratories in Stuttgart and Hamburg respectively, have sent us additional sequences of German oaks in which they have counted and labeled the rings.

From K. Y. Kigoshi in Tokyo, we obtained a 2,000-yr ring sequence of a cedar from the southern tip of Japan, in which Kigoshi had counted and labeled the rings; in addition, he verified his dates by making 50 radiocarbon measurements in its wood. Although the accuracy of radiocarbon dating is only about 40 yr, by making 50 measurements the verification achieves an accuracy of $40/(50)^{1/2}$, or about 6 yr.

The temperature records needed for comparison with the German oaks exist at nearby Basel and Geneva, extending back more than two centuries, and in central England, extending back three centuries. Temperature records for the cedar have been made at Miyazaki, Japan, since 1890. In addition, there are proxy climate records for the Far East (Libby et al., 1976).

The reason for our use of local air temperature records is that sea-surface temperature records are not readily obtainable except in the commercial sea-lanes. We know, however, that oceans comprise nearly 71% of the surface of the world and determine the air temperatures of the continents. From the evidence of tritium in rainwater, we know that rain makes three or four hops, raining out and evaporating in crossing a continent, so that rainwater retains the signature of the isotope content caused by the temperature of evaporation from the sea surface, and this value is intimately tied to local air temperatures where the tree grew.

We had an idea of the magnitude of the oxygen- and hydrogen-isotope variations we could expect to find in these trees because, since 1953, the IAEA in Vienna has measured and published the stable-isotope variations in rain and snow versus air temperature month by month for 155 worldwide weather stations, including those in Germany, Austria, and Japan.

We considered whether old heartwood could exchange isotopes with modern sapwood. On the contrary, there is compelling evidence that when sapwood passes into heartwood it becomes sealed against sap and therefore against isotope exchange with sap, at least in species having tight rings. For example, Huber (1960) has shown, using biological dyes in many tree species, that water conduction remains limited to the outermost annual ring, as is true for radiocarbon sugars, injected or ingested. Furthermore, we have observed that, when an intact block of California redwood was soaked for a month in an atmosphere saturated with water previously labeled with $\delta D_{\mathrm{smow}} = 1{,}170‰$, no deuterium exchange with wood was observed. This is reasonable to expect because wood is a remarkable polymer, containing very large molecules, cross-linked internally and to each other with hydrogen bonds. Hence, we concluded that the climate record in heartwood cannot be modified or perturbed by the sap in the outermost ring of the current active year.

For our first tree sequence, we measured *D/H* by reacting sawdust with uranium to produce H_2, 99% quantitatively. For measurement of $^{18}O/^{16}O$, we modified the method of Rittenberg and Pentecorvo (1956) by carrying it out at very high temperatures, 99% quantitatively. The temperature must be 525°C. If it is lower, the reaction is not quantitative. Whether the oxygen in tree rings comes from water or from CO_2 is not in question, because M. Cohn and H. Urey (1938) showed that isotopic equilibrium between the two substances is obtained in a damp atmosphere within a few hours at room temperature.

For the first measurement of the oaks we used, perforce, a mass spectrometer of somewhat low accuracy, and we achieved the accuracy to demonstrate that trees are thermometers by making many measurements on each sample. On the tree sequences that we measured later, we used high-precision spectrometers with accuracies of ±0.1‰ for $^{18}O/^{16}O$ and $^{13}C/^{12}C$ and ±2‰ for *D/H*. The measurements are expressed in terms of δ_D and δ_{18}. In an intermediate stage, we used the original mass spectrometer built after World War II by Harold Urey at the University of Chicago, and we are grateful that Harold Urey and Kurt Marti allowed us access to it.

SAMPLE PREPARATION

We milled a groove perpendicular to the tree rings—that is, along the radius of the tree; we collected sawdust from each few rings into an individual vial with a camel's hair brush. The vials were dried at 50°C and capped off to protect the dried sawdust from damp air. The chemistry used is as follows.

To evolve CO_2 for measurement of $^{18}O/^{16}O$, pump for 4 hr on 3 mg of sawdust mixed with 120 mg of $HgCl_2$ in vacuo. Seal the container. Heat at 525°C for 4 hr; if the temperature is less than 525°C, production of CO_2 does not quantitatively remove oxygen. React with triple-distilled quinoline at boiling temperature until the quinoline turns yellow. Freeze in a slurry of ethanol and dry ice at −120°C. Pass the gas through two traps of dry ice and acetone.

To evolve H_2 for measurement of *D/H*, burn 5 mg of dry sawdust in 1 atm of O_2 in a cupric oxide furnace at 750°C. Use oxygen purified over silica gel and cupric oxide to ensure that the O_2 is hydrogen-free. Freeze out H_2O and CO_2 in a liquid oxygen trap. Release CO_2 at dry ice temperature. React H_2O vapor on clean uranium shavings at 950°F, thus producing H_2 quantitatively.

THERMOMETER RECORDS

To show that trees can be used as thermometers, we needed to measure modern trees whose rings have been correctly counted and to obtain lengthy mercury thermometer records from places near where the trees grew, preferably from the same altitude. The oldest European temperature records in the World Weather Records (1966) are for Basel and Geneva, each at about a 300-m altitude. As mentioned earlier, we fortunately obtained from Huber's laboratory a German oak grown in central Germany at an altitude of about 300 m, which had been correlated with the fiducial oak-tree-ring sequential pattern developed by Huber (1960). Later, we obtained samples of similarly calibrated German oaks from Becker and Eckstein.

So far, we have measured isotope ratios in four modern trees of four different species at four different altitudes, latitudes, and longitudes and compared them with local temperature records from mercury thermometers: German oak, *Quercus petraea;* Bavarian fir, *Abies alba;* Japanese cedar, *Cryptomeria japonica;* and California redwood, *Sequoia gigantea.* For the parts of the oak and the cedar extending beyond the beginning of mercury thermometer records, we compared the measured isotope ratios with proxy evidence of climate change, such as lateness of cherry tree bloomings, number of

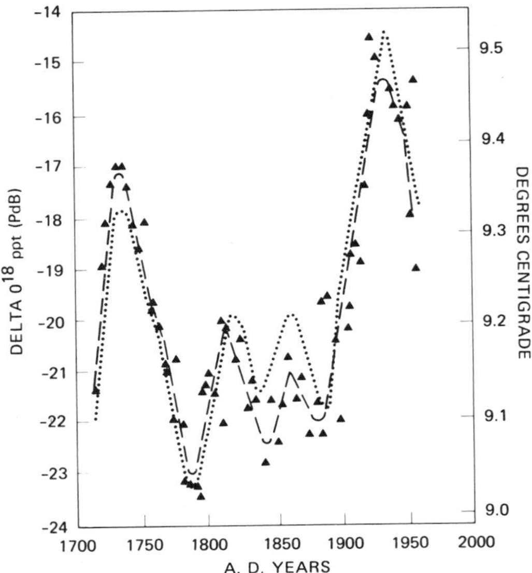

Figure 4-1. ^{18}O ratio in German oak, *Quercus petraea* (▲, seven samples, approximately 30-yr, running average, dashed line) compared with the annual average temperature in England (dotted line). (From Libby, 1983)

snowy days per year, and number of days per year when lakes were frozen. We found significant correlations.

In Figure 4-1 we compare oxygen-isotope measurements for a German oak with English temperature records back to the time of the invention of the mercury thermometer in the seventeenth century. Temperature records of Basel and Geneva from 1750 resemble those of central England and thus fit the isotope ratios well, but we show the English thermometer records here because they extend farther into the past.

CLIMATIC PERIODICITIES

Little is known about fluctuations of the solar constant, and even less is known about fluctuations of the ultraviolet part of solar radiation. Until now, periodicities of the sun have been evaluated solely from the sunspot cycles, observed over 350 yr (except for scattered observations in ancient China), but the sun has many ways to vary besides sunspots.

As was concluded by the U.S. Department of Transportation's Climatic Impact Assessment Program (CIAP) (Grobecker, 1975),

> All that is really known about the solar constant is that it has not changed by as much as a factor of two throughout the ecological history of the earth, that over the past 50 years it has not changed by as much as 10%, and that over the last 10 years, it has not changed by as much as 1%. These known limitations do not rule out long-period (hundreds of years or longer) changes of, say, less than 30% over the entire solar spectrum. Neither do they rule out short-period changes in the near UV nor UV flux (less than 2,500 Å) of the order of a factor of two.

We have therefore made Fourier transforms to deduce the power spectra of tree-ring periodicities. The same periods are found in deuterium and oxygen in the Japanese cedar, within experimental error. Our samples consist of wood from about 5 yr each, so we cannot expect to find meaningful evidence for periods of less than about 40 yr in these data. The Japanese cedar spans about 1,800 yr, so we cannot ask for meaningful evidence for periods of more than about 250 yr.

In making the Fourier transforms, we have used the deviations of the isotope ratios from the long-term slopes; that is, we have corrected for the slopes. We have tested the meaningfulness of the periods by manufacturing data sets consisting of a number taken from the least-squares fit to the data—namely, the straight-line fit—plus a random number varying over the numerical range of the deviations from the line. Each set of manufactured data was subjected to Fourier transform. In each case, in the transforms for 30 such manufactured data sets, no significant peaks were generated, indicating a confidence level for the periods of better than 96%.

The Fourier transforms were performed in the standard manner. No smoothing or filtering was employed. Subtraction of the data from the least-squares fit removes the constant or linear term characterizing a Markovian process. Fourier transform of the differences from the linear fit suppresses the enhancement of both the power and the amplitude of spectra at low frequencies.

The periods derived from the transform of the oxygen-isotope ratio in the *Sequoia gigantea* appear to agree with those in the *Cryptomeria japonica*, which are shown in Table 4-1. In the *Sequoia*, the samples were taken for as few as 2 yr at a time, so it is possible to have confidence that the 33-yr period found in it is meaningful. Because the tree has been analyzed for only about a 225-yr span, however, we cannot ask for meaningful evidence of periods of more than 143 yr, and although it appears to agree with the period of 156 yr found in the *Cryptomeria japonica*, and although in our experiments with artificially generated isotope ratios in the Fourier transforms no such large amplitudes were produced, one must still exercise caution.

In the Fourier transforms, amplitudes and phases are generated for each period. We have put the periods, amplitudes, and phases back into the summation of transcendental functions that represents the function of the data. By running this function into the future, we have made a prediction of the climate to be expected in King's Canyon: the climate will continue to cool on the average. Superimposed on the long-term decay will be a temporary warming up followed by a greater rate of cooling off. Naturally, a complete analysis of the 3,000-yr span of the *Sequoia gigantea* that we have in hand will yield a more reliable prediction of the future climate for Kings Canyon.

Five of the periods found in the oxygen- and hydrogen-isotope ratios of the Japanese cedar are also evidenced in variations of the oxygen-isotope ratio versus depth in the Greenland ice. The remarkable agreement between our tree records of oxygen and hydrogen isotopes and the ice record of oxygen isotopes is evident in yet another way: we have found the D/H and $^{18}O/^{16}O$ ratios for the Japanese cedar to be significantly correlated in opposite phase to the ^{14}C variations in bristlecone pines of southern California, as measured by Suess (1968). Similarly, Johnson and his colleagues (1971) found the oxygen isotope record in Greenland ice to be significantly correlated in opposite phase with bristlecone ^{14}C. Furthermore, the Fourier transform of the ^{14}C variations shows two of the same periods as in our tree, and one in the Greenland ice (see Table 4-1).

We conclude from the correlations of these four sets of data that the calculation of Johnson et al. (1971) of the age of ice versus depth in the Greenland ice cap seems to be correct with an error of not more than a couple of years, at least over the past 800 yr. We conclude that the climate variations in Greenland, southern Japan, and southern California have had the same periodicities for the past 800 yr or more.

A logical explanation for the global nature of these correlations is that they are all related to variations of the sun, which cause variations in the temperature of the sea surface, thus causing variations in the isotopic composition of

Table 4-1. Periods Yielded from Several Sets of Measurements by Fourier Transform (Years)

Cryptomeria japonica		*Santa Barbara Marine Core* [a]		*Bristlecone Pine* [b]	*Greenland Ice Core* [c]
D/H	$^{18}O/^{16}O$	*Organic Carbon*	*Uranium*	^{14}C	$^{18}O/^{16}O$
156	156	161	156	162	179
110	124	121	118	108	—
97	97	95	95	—	100
86	88	82	81	—	78
65	70	71	70	—	68
58	55	55	53	—	55

Source: From Libby, 1983, p. 45.
[a] Kalil, 1976.
[b] Suess, 1968.
[c] Johnson et al., 1971.

water vapor that is distilled from the sea and is stored as wood in trees and also forms the annual layers of the ice cap. The variations of the sun are furthermore related to the flux of solar neutrons in the Earth's atmosphere and so caused small variations in the ^{14}C content of the bristlecones. During times of a quiet sun, the average ^{14}C production is about 25% greater than when solar activity is high (Suess, 1968).

In two additional data sequences (Kalil, 1976) versus age, the same periodicities are revealed —namely, in the organic carbon and uranium concentrations in a deep-sea core from the Santa Barbara Basin off California. Preservation of annual varves in this anoxic environment provides a record of the age of the sediments, there being one varve, or distinct layer, for each year. The concentrations versus depth of organic carbon and uranium were measured in a continuous sequence of samples, each containing 7 yr of sediment, in core PT-8G, spanning the years 1264 to 1970. The age of sediment versus depth in the core was determined by comparing its varves with those in core 214, in which the varves had been counted (Doose, 1978; Soutar and Isaacs, 1969).

The age determination allowed Fourier transforms to be made, transforming concentration versus depth into signal power and amplitude versus the period expressed in years. The periods found are listed in Table 4-1. The deep-sea core spans over 700 yr, with each sample containing 7 yr of sediment, so evidence for periods between 40 yr and 200 yr should be meaningful.

The six periodicities found in uranium and organic carbon variations in the marine core are also found in stable isotopes of hydrogen and oxygen in the tree and in oxygen-isotope variations in the ice, enriching the interpretation of climate variations on a global scale and the attribution of these periodicities to variations of the sun as causing changes in sea-surface temperature. The sea-surface temperature determines the abundance of life and therefore the abundance of organic matter that falls to the ocean bottom and binds uranium ions in the seawater to it as it falls and accumulates. (It is well known that in uranium ores the amount of organic carbon is proportional to the amount of uranium.)

The CIAP study of the stratosphere (Grobecker, 1975) indicated mechanisms by which the sun's variation may influence temperature on the Earth's surface. In particular, solar energy absorbed in the stratosphere is rapidly converted by chemical reactions, producing a variety of chemical species that are redistributed both to space and to the ground. Thus, climate on the ground is sensitive to variations in stratospheric chemical components. The solar electromagnetic radiation is known to vary by factors of two and three in intensity at short wavelengths, affecting the concentrations of those species. Reradiation to the ground is especially sensitive to ozone and water concentrations in the stratosphere, and those variations are only beginning to be assessed. Studies of stable isotopes in trees and of uranium and organic carbon in marine sediments as a function of time in the past are thus seen as a powerful way of examining past behavior of the sun.

REFERENCES

Cohn, M., and Urey, H. C., 1938, Oxygen exchange reactions of organic compounds and water, *Am. Chem. Soc. Jour.* **60:**679-687.

Craig, H., 1961, Isotope variations in meteoric waters, *Science* **133:**1702-1703.

Doose, P. R., 1978, Ph.D. dissertation, Geochemistry Department, University of California at Los Angeles.

Grobecker, A. J. (ed.), 1975, *The Natural Stratosphere of 1974*, Climate Impact Assessment Program, CIAP Monograph No. 1, Washington, D.C.: Department of Transportation.

Herzberg, G., 1945, *Infrared and Raman Spectra of Polyatomic Molecules*, New York: Van Nostrand Reinhold.

Huber, B., 1960, Dendrochronologie, *Geol. Rundshau* **49:**120-131.

Johnson, S. J.; Dansgaard, W.; Clausen, H. B.; and Langway, C. C., 1971, Climatic oscillations 1200-200 AD, *Nature* **227:**482-483.

Kalil, E. K., 1976, Ph.D. dissertation, Geochemistry Department, University of California at Los Angeles.

Libby, L. M., 1983, *Past Climates: Tree Thermometers, Commodities, and People*, Austin: University of Texas Press, 143p.

Libby, L. M.; Pandolfi, L. J.; Payton, P. H.; Marshall, J.; Becker, B.; and Sienbenlist, V. G., 1976, Isotopic tree thermometers, *Nature* **261**:284-288.

Rittenberg, D., and Pontecorvo, L., 1956, A method for the determination of the ^{18}O concentration of the oxygen of organic compounds, *Internat. Jour. Appl. Radiat. Isotopes* **1**:208-214.

Soutar, A., and Isaacs, J. D., 1969, *History of Fish Populations Inferred from Fish Scales in Anaerobic Sediments off California*, La Jolla: Scripps Institution of Oceanography.

Suess, H. E., 1968, Climatic changes, solar activity, and the cosmic-ray production rate of natural radiocarbon, *Meteorological Monographs* **8**:146-150.

World Weather Records, vol. 2, Europe, 1966, Washington, D.C.: U.S. Department of Commerce, Environmental Services Administration.

5: Study of the Holocene and Late Würmian Sediments of Lac du Bouchet (Haute-Loire, France): First Results

E. Bonifay
Université de Luminy, Marseille

and

K. M. Creer
University of Edinburgh

with

J. L. de Beaulieu*, L. Casta†, G. Delibrias‡, G. Perinet+, A. Pons*, M. Reille*, S. Servant†§, G. Smith‖, N. Thouveny†, E. Truze†, and P. Tucholka†‖

Abstract: A comprehensive multidisciplinary research program on the sediments of Lac du Bouchet (44.9°N, 3.8°E) was begun in 1982, and since then about 50 Mackereth-type cores from 1.5 m to 8 m long have been collected. In this first report, results from three 6-m cores on which sedimentological, geochemical, palynological, diatom, radiocarbon, and paleomagnetic studies have been completed are discussed. The data obtained have been synthesized to define a zonation for the late part of the Würm period to the present time, and good agreement has been found among the pollen analytical zonation, the diatom zonation, and the sedimentological data. The quality of the paleomagnetic record recovered varies inversely as the organic content (OC) of the sediments and is exceptionally good through the late glacial for which the OC is only a few percent. Throughout the whole of the period covered—that is, from about 24,000 B.P. to the present—declination and inclination variations consist of oscillations of up to about 30° to 40° amplitude: there is no suggestion of any abnormal large-amplitude short-lived excursions either at the end of the glacial period corresponding to the Gothenburg Excursion or near the start of the record (ca. 24,000 B.P.) corresponding to the Mono Lake Excursion. The sense of rotation of the motion of the geomagnetic vector for different degrees of smoothing has been studied, and spectral analyses have been carried out on the declination and inclination curves. Studies of nearly 50 9-m cores are now nearing completion and are awaiting adequate dating control before publication.

*Université d'Aix-Marseille
†Laboratoire Géologie der Quaternaire, CNRS, Luminy
‡Laboratoire Mixte CNRS-CEA, Gif-sur-Yvette, France
§ORSTOM, Paris
‖University of Edinburgh

THE PALEOENVIRONMENT

A study of the recent sediments of the lakes of the French Massif Central was begun in 1982 under the direction of the first two authors of this chapter as part of a research program on the lakes and paleolakes of the Velay and the Auvergne (Bonifay, 1982). Three coring campaigns were undertaken in 1982 and two more in 1983 when the bottom sediments were sampled using Mackereth-type corers from the geophysics department, University of Edinburgh, in the lakes of Tazenat (one core of nearly 6 m length), d'Issarles (two 6-m cores), St. Front (three 6-m cores) and most important, Le Bouchet (more than fifty cores of length from 1.5 m to 9 m).

We report here on results only from the first investigations at the most important site, Lac du Bouchet, which is particularly favorable for investigations of the dynamics of sedimentation in maar-type lakes and of the paleoenvironment —for example, paleoclimatology, paleobiology, and paleomagnetism (Bonifay, 1982).

The study of cores collected in 1982 and in 1983 from Lac de Bouchet is not complete. The results presented here constitute only a first step toward our ultimate goal of studying the complete sedimentary section deposited in the lake, which amounts to at least 50 m. Such a study will provide paleoenvironmental and paleomagnetic data through much of the Brunhes Chron.

General Data on Lac du Bouchet

Situated at latitude 44°55′N, longitude 3°47′E, at an altitude of 1,205 m, Lac du Bouchet has a diameter of 800 m and a maximum depth of about 30 m (Delbeque, 1898). It is contained in a volcanic structure of the maar type (explosion crater of phreatomagmatic origin) (Bout, 1978), which gives it an almost perfect circular shape.

Lac du Bouchet has historically maintained a remarkably stable water level, the seasonal variations never having exceeded 0.5 m through several centuries. However, the discovery, in August 1983, of a Pre-Würmian paleobeach situated about +12 m above the actual water level shows that important variations in lake level occurred during the Pleistocene (Bonifay and Truze, 1984). Other minor variations of the order of a meter occurred at the end of the Late Würm (level down at about −1.5 m) and at the beginning of the Holocene (level up at about +2 m).

Of a complex structure, the volcano of Le Bouchet underwent effusive activity (basalt flows and scoriaceous cones) before entering an explosive phase that gave rise to the birth of the maar. The volcano belongs to the Deves Volcanic Complex, of Pleistocene age, which covers a Hercynian granite intrusive and contains a large number of scoriaceous cones and maar-type craters (de Goer de Hervé and Mergoil, 1971) (Fig. 5-1) that gave rise to numerous existing (Le Bouchet) and fossil lakes.

From the morphological point of view, Lac du Bouchet is remarkable for the small surface area of its watershed and by the absence of inflowing and outflowing streams. Seismic reflection profiles obtained in 1983 show the existence of a quite shallow (0-3 m) annular littoral prism, several dozens of meters in width, composed of highly unsorted littoral sediments. Its extent was limited by the steep internal slope that leads on to an almost flat base. All our cores were taken from this central part of the lake in water depths between 25 m and 30 m. We have not attempted to core from the littoral zones because of the large and variable grain size of the sediments, because of the steep slope, and because the shallow depths are unfavorable for operation of the corer.

SEDIMENT STRATIGRAPHY AND DYNAMICS

All the cores on which detailed stratigraphic studies have been made show the same succession of facies:

Assemblage A: An initial, superficial layer about 1 m thick, assemblage A consists of mud that is very fluid near the water-sediment interface, becoming more compact toward the base. It is

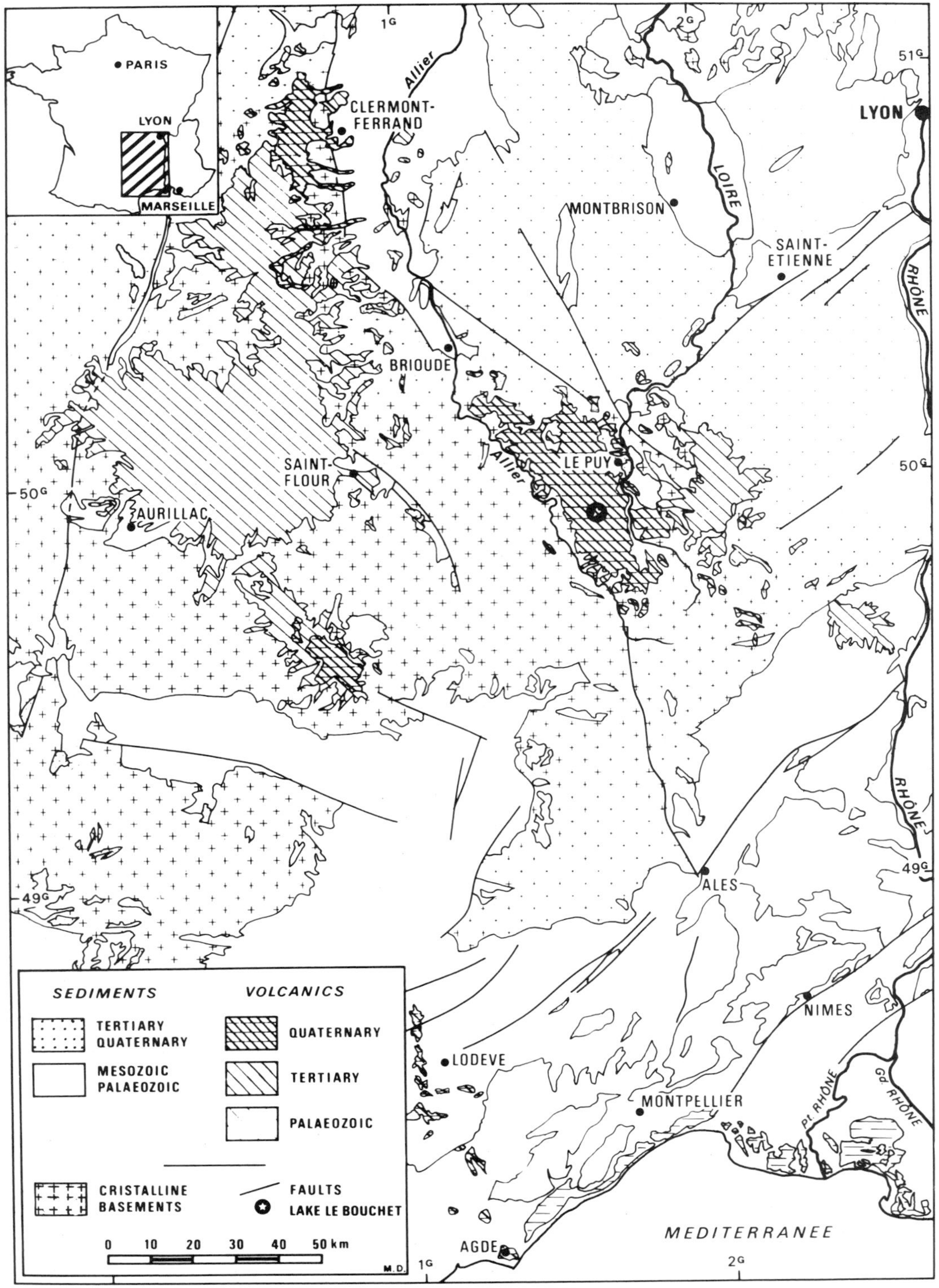

Figure 5-1. Structural map of the French Massif Central.

very rich in organic material (25-40%), reddish in color at the top and brown to blackish toward the base. A weak lamination is apparent and evidence of slumping rare or absent.

Assemblage B: Below assemblage A, this contains more compacted muds of blackish color, the thickness of which varies from 0.1 m to 1 m from place to place over the lake, very rich in organic matter (40-70%).

Assemblage C: Under assemblage B, 1-2 m of compact gray and brownish muds are found. They are poor in organic matter (2-4%), quite well laminated with intercalations of sandy levels resulting from underwater slumping.

Assemblage D: Under assemblage C there is about 1 m of clearly laminated, more or less gravelly gray muds that carry a large number of sandy or stony horizons resulting from the effects of slumping and also a few gravelly horizons (2-4% of organic matter).

Assemblage E: A less gravelly, well-laminated facies with a higher proportion of organic matter (6-11%) is found at the bases of the longer cores.

Examination of the different cores studied up to now shows that the fine-grained sediments deposited in Lac du Bouchet have resulted from decantation processes. The coarser-grained sediments, resulting from solifluction and inflow from streams, are restricted to the marginal prism that progressively spreads out toward the center of the lake. Any sediment overload deposited on the internal part of this marginal wedge produces disequilibrium, which initiates slumping. The coarser fractions of the sediment are then redeposited on the underwater slope and on the margins of the central plain, forming bosses of 1-3 m height, while the finer fractions remain in suspension and are spread out over the whole lake. The role of slumping appears to be an essential factor in the sedimentation processes in maar-type lakes. It may be supposed that the frequency of underwater slumping events should have been greater during cold climates due to factors such as diminution of vegetal cover and augmentation of erosion.

DATING CONSIDERATIONS

The pollen-analytic, paleomagnetic (by comparison with other dated cores), and isotopic data allow us to assign stratigraphic assemblages A and B and the upper part of assemblage C to the Post-Würm, the middle and lower part of assemblage C to the Late Würm, and assemblage D to the Pleni-Würm. The most complete datings, those for core B5, show:

1. an age of 2,280 ± 90 yr B.P. (Gif 5944) for the middle part of assemblage A at 45-55 cm;
2. an age of 5,500 ± 100 yr B.P. (Gif 5940) for the top of assemblage B at 90-92 cm;
3. an age of 8,340 ± 150 yr B.P. (Gif 5941) for the upper part of assemblage C at 160-170 cm;
4. an age of 15,800 ± 900 yr B.P. (Gif 5942) for the base of assemblage C at 280-300 cm;
5. an age of 19,400 ± 1,300 yr B.P. (Gif 5943) for the base of assemblage D at 530-547 cm;
6. the base of core B21 (equivalent depth along core B5 is ~700 cm; see "Intercore Correlation") has an age equal to or greater than 25,000 yr B.P. (Gif 6200).

SEDIMENTOLOGY

The sedimentological study is incomplete: a rather complete study has been made on core B21, and cores B1, B5, B12, B16, and B53 have been partly studied (Truze, 1983). These studies allow the definition of the following characteristics (Fig. 5-2) (Truze, 1983):

1. The proportion of organic material varies directly as the water content of the sediment. The water content does not depend only on the degree of compaction of the sediment because the organic matter plays a role in the retention of water.

2. The proportion of organic matter appears to depend on the climate, being very low in cold climates (Pleni-Würm and Late Würm) and becoming very important in the Post-Würm, probably resulting from the development of vegetal cover in the surrounding basin and from an increase in biological activity in the superficial lacustrine waters (development of plankton in the warm superficial water in summer).

3. The granulometry of the mineral phase varies with the climatic conditions and the frequency of slumping events. In the central part

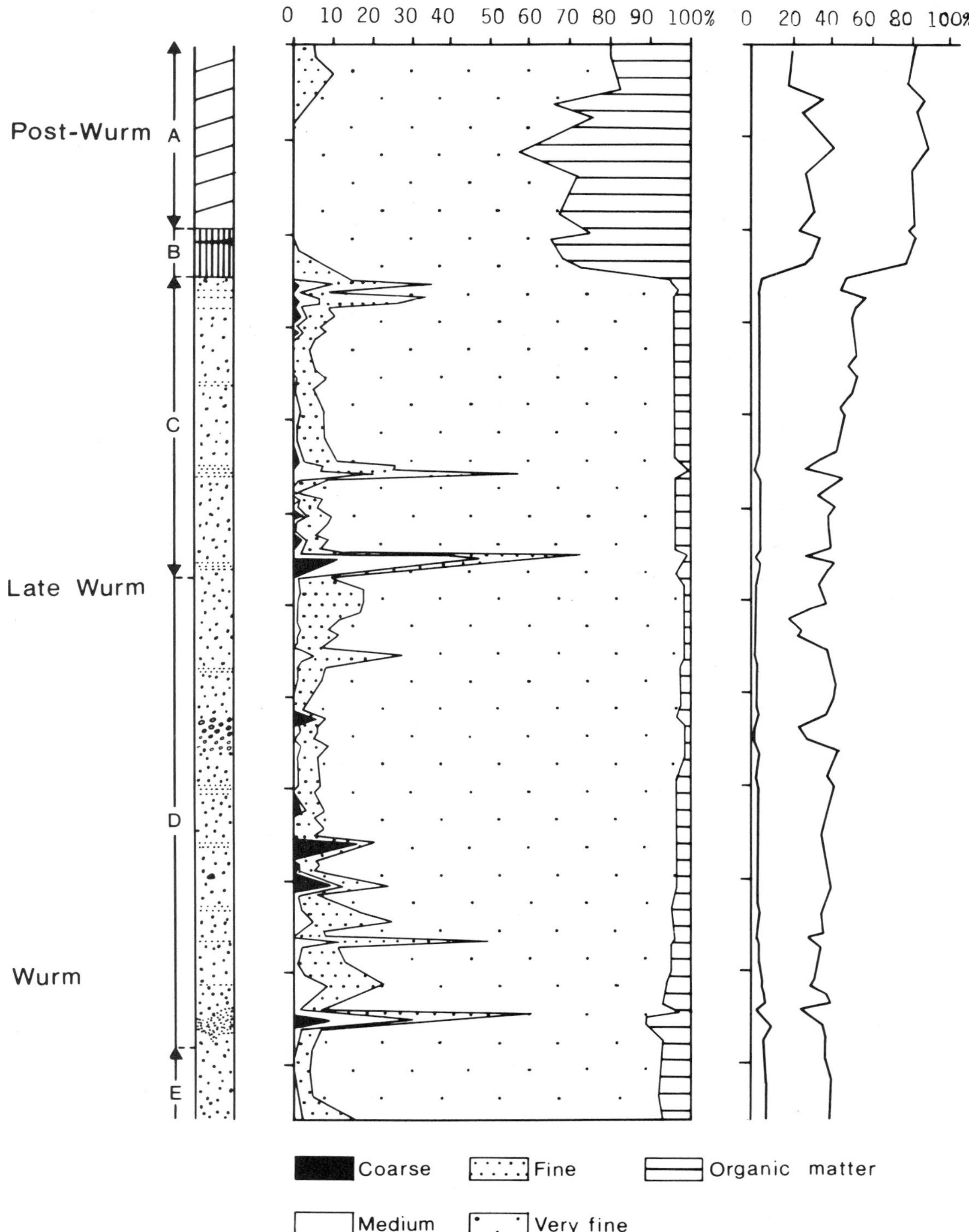

Figure 5-2. Sedimentological logs for core B5. *(a)* Stratigraphy—lithological assemblages A, B, C, D, and E are described in the text in the section on sediment stratigraphy and dynamics, *(b)* grain-size distribution, and *(c)* proportion of organic matter (left) and water content (right).

of the lake, the most important fraction at all levels in the core is fine-grained, less than 50 μm. The intermediate-sized fraction (50-315 μm) is a good recorder of turbidites (caused by slumping) and is almost completely absent from the Post-Würmian levels except in assemblage A, due to erosion caused by the action of humans. The coarsest fraction (greater than 315 μm) occurs only at the levels of the most important turbidites or in the gravelly beds of the Pleni-Würm (action of floating ice?). In core B21, the frequency of turbidites shows two maxima, which represent perhaps the two major cold phases of the recent Würm.

4. The proportion of organic matter is generally lower at the level of the turbidites, demonstrating their littoral origin (Bonifay and Truze, 1984).

5. The clay minerals, identified by X-ray diffractrometry, show a dominance of illite and chlorite in the lower levels of core B21, below 2.20 m (Pleni-Würm and Late Würm), while the upper levels (Post-Würm) are dominated by an association of smectites and kaolinite. The first association consists of primary minerals derived from decomposition of rock under the influence of a cold climate, while the second results from alteration in wet conditions, corresponding to the installation of the temperate Post-Würm climate.

6. The mineralogical composition of the sediments of Le Bouchet is still poorly known; the proportions of the diverse constituents seem very variable from level to level. Quartz is typically dominant, with feldspars, microscorias, and fragments of volcanic glass. Olivine and micas are also abundant in variable proportions.

7. The morphology of the quartz grains, equally poorly known, suggests a strong eolian influence; numerous unworn quartz grains of small size suggest the possibility of eolian transport of loessic type. The percentage of eolian grains (irregular-matt and round-matt) is also very high. Finally, polished grains originate from the lake shores where they acquired their shape.

Taking account of the proportion of organic matter in the sediments and of their degree of compaction (proportion of water), one can estimate to a first approximation that the rate of deposition of the mineral phase of the Le Bouchet sediments was extremely slow, from abut 0.25 mm/yr to 0.33 mm/yr in the lower parts of the cores that were deposited in a cold climate, falling to 0.05—0.07 mm/yr in the upper (Post-Würm) levels (Truze and Bonifay, 1984).

POLLEN ANALYSES

Pollen analyses have been made on subsamples taken from the smaller segments of core B5 opposite to the larger segments, which were subsampled for paleomagnetism. Seventy-nine analyses have been made. Samples from above 435 cm, after treatment in heavy liquids (Goeury and de Beaulieu, 1979), are sufficiently rich in sporopollonic material to permit the establishment of pollen spectra on the basis of counts greater than 400, including all the grains of pollen and vascular vegetal spores. In the lower part, valuable spectra could be established only at two levels, 510 cm and 520 cm. All the information collected is presented on a classic pollen diagram.

The zonation of this diagram (Fig. 5-3) was obtained from all the data about the vegetation history of the region (Couteaux, 1978, 1984; de Beaulieu, Pons, and Reille, 1982, 1984). The sequence of stages of the vegetation dynamics, named following the classical terminology, serves only as a chronozonation. The terms of these stages are provisionally fixed, as a function of the data actually available, in the following way (de Beaulieu, Pons, and Reille, 1982): start of Bolling, 13,000 B.P.; start of Allerod, 12,000 B.P.; start of Recent Dryas, 10,700 B.P.; start of Pre-Boreal, 10,300 B.P.; start of Boreal, 9,000 B.P.; start of Atlantic, 8,000 B.P.; start of Subboreal, 4,700 B.P.; start of Subatlantic, 2,600 B.P.

Final Pleni-Würm (Pollen Zones a and b)

The base of core B5 is extremely poor in pollen and even sometimes sterile. The very high percentages of *Pinus*, which is the only tree

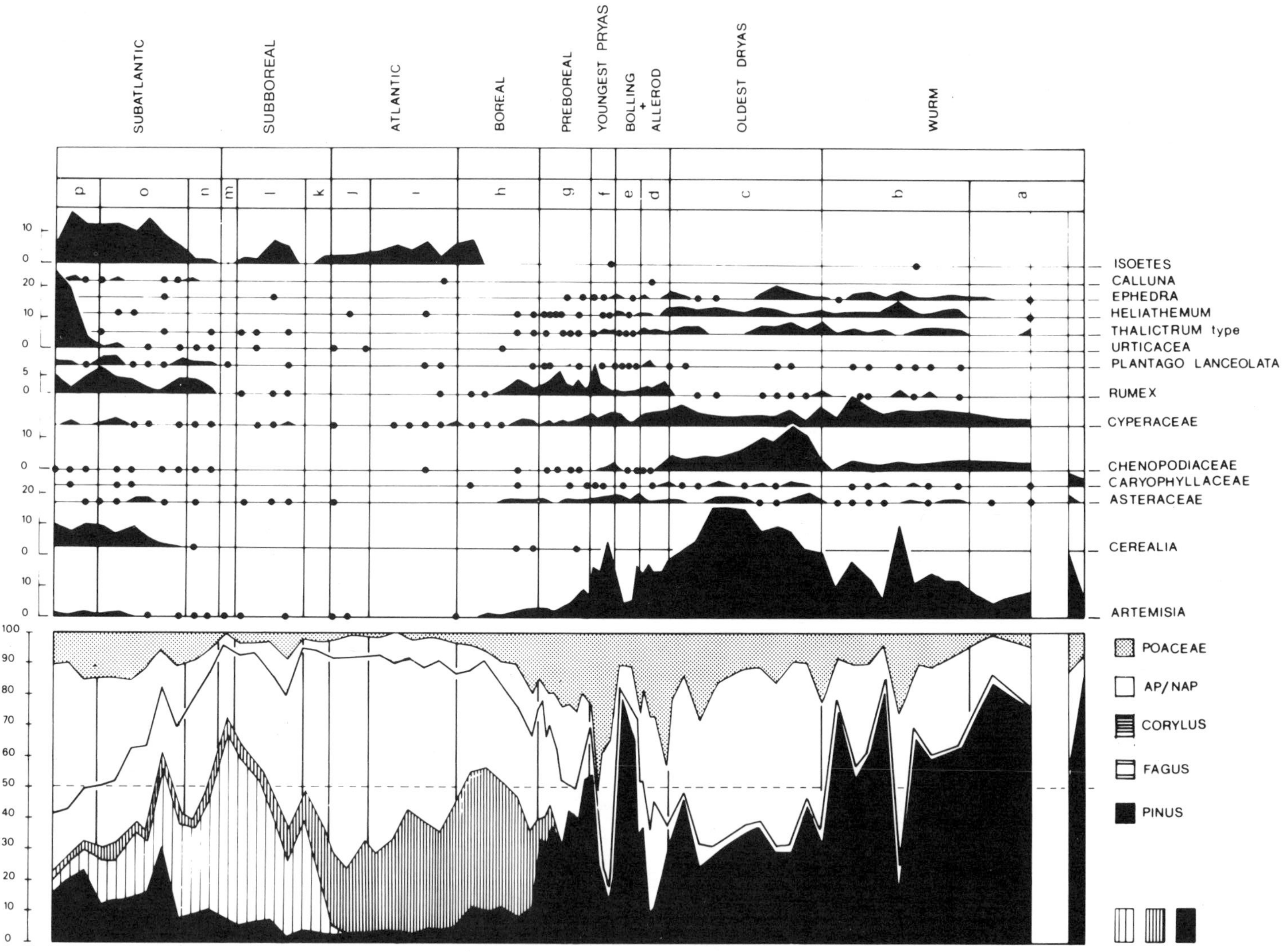
SUBATLANTIC
SUBBOREAL
ATLANTIC
BOREAL
PREBOREAL
YOUNGEST PRYAS
BOLLING
+
ALLEROD
OLDEST DRYAS
WURM
p
o
n
m
l
k
j
i
h
g
f
e
d
c
b
a
ISOETES
CALLUNA
EPHEDRA
HELIATHEMUM
THALICTRUM type
URTICACEA
PLANTAGO LANCEOLATA
RUMEX
CYPERACEAE
CHENOPODIACEAE
CARYOPHYLLACEAE
ASTERACEAE
CEREALIA
ARTEMISIA
POACEAE
AP/NAP
CORYLUS
FAGUS
PINUS
100
90
80
70
60
50
40
30
20
10
0

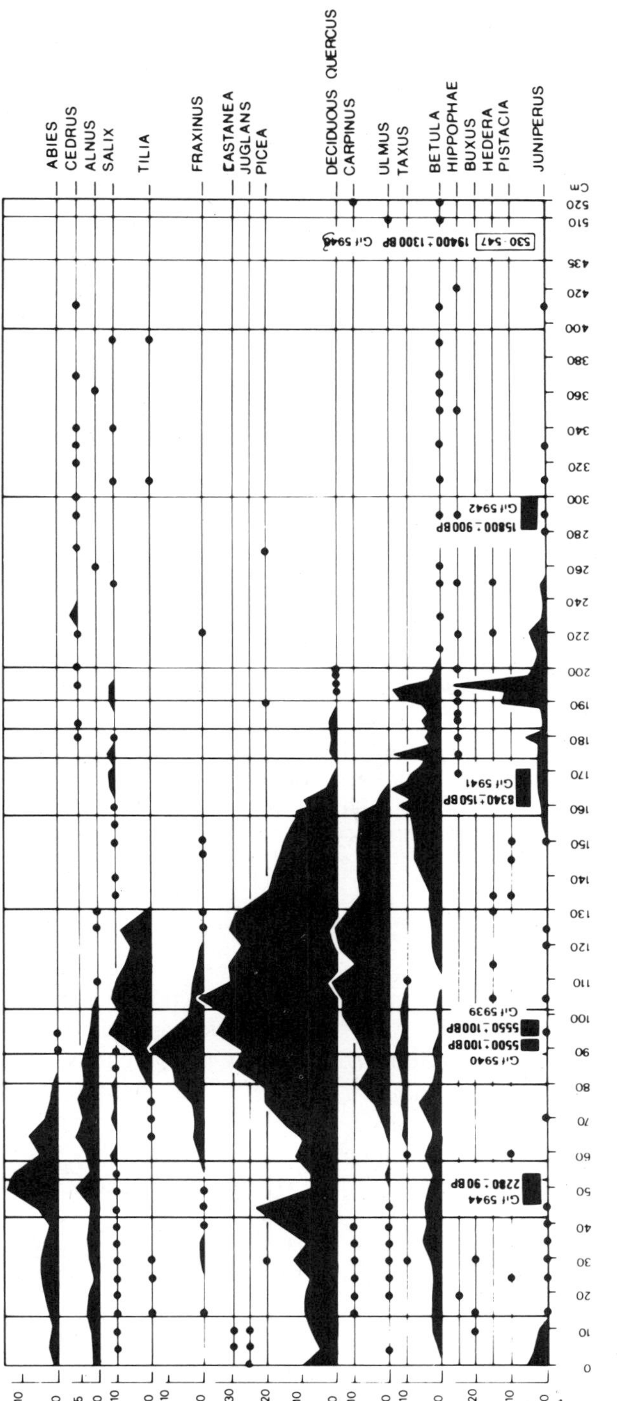

Figure 5-3. Set of pollen diagrams for core B5.

represented, are due to long-range transport in a desert environment. This preeminence of *Pinus*, associated with a very marked paucity in pollen in Pleni-Würm sediments, is well established (de Beaulieu and Reille, 1984*a*, 1984*b*; de Beaulieu, Pons, and Reille, 1984; Couteaux, 1984). Zone b is distinguished from zone a by an enrichment in the regional flora in taxons (above all, *Helianthemum*, *Ephedra*, *Thalictrum*).

Lower Ancient Dryas (Pollen Zone c)

The regional expansion of steppe (zone c) is recorded. Formations of *Artemisia* and Chenopodiaceae are first represented; then *Artemisia* attains its optimum when the Chenopodiaceae decline. Such a succession is observed at other sites as varied as Les Echets (de Beaulieu and Reille, 1984*a*) and Pelleautier (de Beaulieu and Reille, 1984*b*) where it has been shown to correspond to a dynamic stage of colonization of bare soils formed during the very first Late Würmian climatic improvement. At Les Echets, the increase in the *Artemisia* curve is placed at about 15,250 ± 290 B.P.; at Pelleautier, before 14,700 B.P. A date of 15,800 ± 900 B.P. (Gif 5942), obtained for the layer at 280-300 cm, confirms these results.

The Bolling–Allerod Complex (Pollen Zones d and e)

Juniperus *and* Betula *Phase (Pollen Zone d).* *Betula* is developed shortly after *Juniperus*, while the steppes persisted, while *Rumex* developed, and while the Gramineae attained their first optimum.

Pinus *Maximum (Pollen Zone e).* The definitive regression of *Juniperus* was followed by a maximum in *Pinus*, demonstrating the existence of the pollen source in the crater.

Recent Dryas (Pollen Zone f)

This zone is particularly notable for a brief recurrence of *Artemisia*, Chenopodiaceae, and Gramineae, which demonstrate the almost deforested state of the Velay, occurring in the Recent Dryas (frequencies of arboreal pollen below 25%).

Preboreal (Pollen Zone g)

This zone is characterized by the disappearance of steppe conditions, the local support of megaphorbic-type vegetation (with *Rumex* and *Filipendula*) around humid zones, and the local extension of *Betula* and especially of the *Pinus* taxon, proving that this source was not eliminated from the crater during the Recent Dryas.

The spectra for the second half of the Preboreal show an increase of *Quercus*, then of *Ulmus* and *Corylus*, which exceed *Betula* and *Pinus*, whose percentages show a steady decline. These spectra provide evidence of closure of tree cover, except at the borders of the site.

A date of 8,340 ± 150 B.P. (Gif 5941), which corresponds to the beginning of the extension of *Corylus*, must be considered as too young by about a millennium in view of the pollen results obtained in this region (de Beaulieu and Reille, 1984*a*).

Boreal (Pollen Zone h)

This zone is represented in all the Velay sites only by a single pollen zone characterized by the classic dominance of *Corylus*, the increase of the *Quercus* curve to a maximum, and the virtual disappearance of *Pinus*.

Atlantic (Pollen Zones i and j)

The Atlantic, whose onset is marked by a steady *Tilia* curve and the decline in the abundance of *Corylus*, can be divided into two periods:

1. The first takes account of the existence of a single type of forest vegetation with *Quercus*, *Tilia*, *Ulmus*, and *Fraxinus*, yet still rich in *Corylus*, which attained a long optimum.
2. The second shows a slow decline of *Quercus* and the occurrence of new arboreal taxons (*Alnus*, *Taxus*, *Abies*, *Fagus*).

Subboreal (Pollen Zones k, l, and m)

The extension of *Fagus* characterizes the onset of the Subboreal. After a long period of supremacy of *Fagus, Abies* exhibited a late development. The two dates of 5,550 B.P. and 5,500 B.P. (identical to within nearly 50 yr) precede the rise of the *Fagus* curve, which has been dated in the region at about 4,700 B.P. and should be considered as too old.

Subatlantic (Zones n, o, and p)

This zone is recognized by the decline of *Fagus* and the appearance of the Ruderosegetales (in particular, *Plantago lanceolata*), which briefly precedes that of the cereals. The marked abundance of Urticales suggests the culture of *Cannabis* on the lakeside. The optimum phase of *Abies* has been dated at 2,280 ± 90 B.P. (Gif 5944), which is in agreement with the date of 2,360 ± 100 B.P. (immediately subjacent to this feature) in the Forez (Janssen and van Straten, 1982).

Discussion

The pollen diagrams obtained from core B5 conform fully to our knowledge of the history of the regional vegetation as observed in peat bogs, at least since the Late Glacial. This finding indicates that the pollen sedimentation, and in consequence, the mineral sedimentation, in Lac du Bouchet have not been subjected to any large-scale sliding (the contrary has been observed in the neighboring maar of Issarles) (Couteaux, 1984). A similar absence of any perturbation certainly persisted for a much longer period, since an event that occurred at around 15,000 B.P. at Le Bouchet is identical to one that is known in the glacial lakes of Pelleautier (Hautes-Alpes) and Les Echets (Ain).

DIATOMOLOGY

Certain horizons of the recent sediments of Lac du Bouchet (core B5) contain a variety of microorganisms, assemblages A and B being rich in Ostracods. However, only the diatoms have been made the object of this preliminary examination: from an analysis of 32 samples, they are seen to be abundant and well conserved in the upper half of the series, but they are scarce and very fragmented in the lower half (Fig. 5-4).

Group 1: Samples 437 to 145

Levels 315 to 295, 247, and 175 are sterile. The other levels contain fragments of *Surirella* aff. *biseriata, Campylodiscus noricus,* and especially large-sized fragments of *Pinnularia* and also unidentifiable fragments.

Group 2: Sample 135

The quantity of frustules increases and the species are even more fragmented. *Pinnularia major* (Kütz.) Cleve, *Campylodiscus noricus* var. *hibernicus* (Ehr.) Grun., *Fragilaria brevistriata* Grun., and *Melosira arenaria* Moore are found, in order of increasing abundance. The latter species has been observed only at this level. *M. arenaria* lives in streams and in the middle of submerged bogs (Germain, 1981) with *C. noricus,* which lives equally well in streams in a benthic state, suggesting a fluviatile environment with soft, cold running water.

Group 3: Sample 125

The flora are abundant, well preserved, and especially varied. *Cyclotella kutzingiana* Thw., *C. ocellata* Pant., *Fragilaria brevistriata* Grun., *Diploneis ovalis* (Hilse) Cleve, *Navicula pseudoscutiformis* Hust., *N. scutelloides* W. Smith, and numerous other species are found. The presence, in very great abundance, of *N. farta* Hust., a little species widely distributed in the northern regions of Europe and abundant in Post-Glacial sediments in the same regions, makes this level unique and indicates a brief recooling of the surface layer of waters, relative to sample 1, in which *N. farta* is absent.

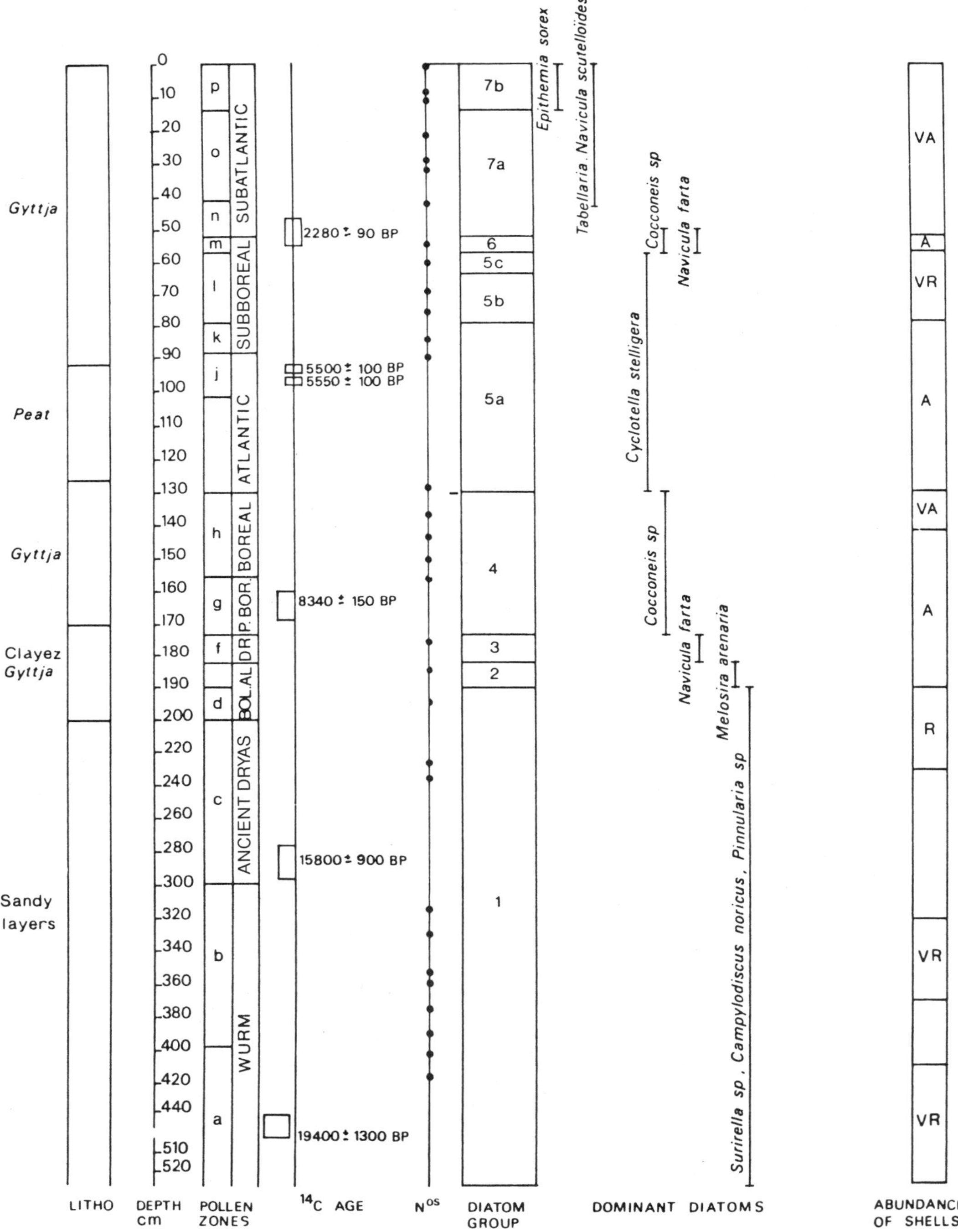

Figure 5-4. Diatom zones for core B5. In the abundance column on the right, VA = very abundant, A = abundant, R = rare, and VR = very rare. Radiocarbon ages, pollen zones, and the lithological log are also shown.

Group 4: Samples 109 to 91

Diatom frustules are abundant. The flora consists of a large number of species: *Stauroneis phoenicenteron* (Nitzsch) Ehr., *Navicula radiosa* Kütz., *Pinnularia acuminata* W. Smith, *P. interrupta* W. Smith, *Fragilaria construens* (Ehr.) Grun., and *Caloneis bacillum* (Grun.) Mereshk. These flora are cosmopolitan, oligohaline, indifferent to pH, eurythermal, littoral, and benthic.

The dominant species is *Cocconeis* sp. (The hypovalve is like *C. disculus,* but the epivalve is characterized by an exceptional enlargement of the pseudoraphe.) The frequency of this epiphytic species indicates that there was an extension of the vegetation belt on the marginal bank that was probably caused by a limited fall in the lake level.

Group 5: Samples 85 to 55

Group 5a: Samples 85 to 73. *Cocconeis* sp. disappears completely and is replaced by a planktonic species, (C1) *Cyclotella stelligera* Grun., which is of small size (diameter 5-6 μm) with barely visible ornamentation. The appearance of this species is accompanied by an increase in Chrysostomaceaen cysts. There is a clear decrease in the number of species in relation to group 4.

Group 5b: Samples 67 to 61. There is an impoverishment in the diatom flora to the advantage of Chrysostomaceaen cysts. Two types are particularly abundant: *Chrysostomum simplex* and *Clericia elegantissima.* The occurrence of Chrysostomaceaen cysts has already been reported in the lacustrine sediments of the Massif Central (Andrieu, 1937). Planktonic algae, which prefer to live in cold water, are present in the superficial layers of Lac du Bouchet.

Group 5c: Sample 55. The cysts persist but in smaller numbers. *Cyclotella stelligera* becomes the dominant diatom species. The individuals are larger in size (10 μm) and their ornamentation coarser than those in groups 5a and 5b. It seems that the abundance and the diversity of the diatom flora varies inversely with the Chrysostomaceae. Genera *Pinnularia, Eunotia, Gomphonema,* and *Tabellaria* are found in this sample.

The abundance of a planktonic species in this part of the core suggests a limited elevation of the lake level. This elevation caused a reduction in the vegetation band on the marginal slopes and an extension of the free water surface favorable to plankton.

Group 6: Sample 49

The assemblage in this sample is characterized by the reappearance of *Cocconeis* sp. (as found in group 4). These are accompanied by *Fragilaria strangulata* Hust., *Achnanthes* clevei Grun., *Diploneis ovalis* (Hilse) Cleve., *Navicula scutelloides* W. Smith, *Amphora lybica* (Ehr.) Cleve., and most important, the reappearance of *Navicula farta* (the dominant species in sample 125), which might relate to a decrease in the water temperature.

Group 7: Samples 37 to 1

Group 7a: Samples 37 to 19. The flora are abundant but not so well preserved. Many of the diatom frustules are fragmented. The Chrysostomaceaen cysts are abundant and remain so until the peak of the series. *Cocconeis* disappears completely and is replaced by *Fragilaria strangulata* Hust. in sample 37 and in the others by *Tabellaria flocculosa* (Rabh.) Kütz and *T. fenestrata* (Lyngb.) Kütz. These two species are associated with *Pinnularia* and *Eunotia.* They indicate an oligotrophic and oligohalobic medium with weak pH; they are abundant in the oligotrophic lakes of Sweden and Norway at the present time where the pH lies between 6 and 7 and the temperature between 5°C and 15°C. Conditions approaching those of the present became established in the basin after sample 31 (pollen zone o).

Group 7b: Samples 13 to 1. The species from group 7a persist in these samples. This subdivision is characterized by the appearance of *Epithemia sorex* Kütz.: in all other respects it is identical to group 7a. This species is very common in France, living equally well in fresh water and brackish water (Germain, 1981). It is often attached to the filaments of chlorophyceaen algae. It is an alkaliphile and could thus indicate an increase of pH and a modification of the trophic status of the water.

Conclusions on the Diatoms

1. The large majority of species belong to oligohalobic, oligotrophic, acidophilic to indifferent species, indicating that the paleolake of Le Bouchet was a fresh-water lake whose pH was less than 7. The lake was low in nutritive materials during a large part of its history except the Boreal period.
2. During the Preboreal, Boreal, and the end of the Subboreal, low lake levels are indicated by diatom associations that are composed essentially of epiphytic or benthic species. The abundance of the former type indicates an intensive colonization of the borders by a submerged aquatic vegetation; the latter type indicates that the waters of the lake were sufficiently clear to permit the penetration of light to depths of scores of meters. During the Subatlantic (pollen zone o), and particularly the Atlantic and the Subboreal, the predominance of planktonic species suggests episodes of high lake-water level.
3. The flora that developed from the end of the Allerod to the Subboreal, and especially to the Recent Dryas, indicates a lower water temperature than at the present time.
4. Conditions approaching the present state developed in the basin during the Subatlantic, at the end of which modification of pH and the trophic state (status) of the basin may have occurred (pollen zone p).

EVOLUTION OF THE PALEOENVIRONMENT OF LAC DU BOUCHET

The sedimentology, diatomology, and pollen analyses provide information about the evolution of the paleoenvironment of Lac du Bouchet through the Recent Würm and the Post-Würm. It must be noted that the different data agree well with a chronological framework whose precision rests on the very strong regional framework of analytical data.

The facies found in the cores taken from the center of the lake show identical sequences, permitting satisfactory stratigraphic correlations among them. These facies trace the evolution of the environmental landscape from the Pleni-Würm through the modern epoch.

During the Pleni-Würm, the rates of sedimentation were relatively high. The frequency of slumping events, the importance of apparently eolian action, and the low proportion of organic matter in the sediments and the pollen analyses suggest a desert landscape subjected to violent winds under periglacial-type conditions for the Deves Plateau. The lake was covered with ice for a large part of the year, and erosion and solifluction processes were intense on the internal slopes of the volcano. It was probably during this epoch that a small rocky glacier was temporarily installed in the southwest quarter of the volcano. The paucity of vegetal remains (pollen, diatoms) in the sediments confirms these observations.

In the Late Würm, the frequency and importance of sublake slumpings diminished, demonstrating a relaxation in the erosion processes and a decrease in detrital material transported into the lake due probably to the formation of the first soils on the internal slopes of the volcano and the colonization of the first vegetation at about 16,000 B.P. The climatic conditions were still severe, and the biomass produced in the lake and its basin remained extremely small (the amount of organic matter was still very low). The surface of Lac du Bouchet remained frozen throughout a large part of every year.

Between 15,000 B.P. and 13,000 B.P., a climatic amelioration led to more hydrolizing conditions, at least seasonally (warm seasons more differentiated?). These conditions, in turn, led to the onset of pedogenesis on the internal slopes of the volcano and to a change in the dominant associations of clay minerals deposited in the lake. Nevertheless, this climatic amelioration did not lead to an immediate floristic change but induced an increase in the vegetal cover.

The development of arbustive and arboreal

formations began between 13,000 B.P. and 10,700 B.P. (Bolling and Allerod) and coincided with a notable proliferation of diatoms and changes in sediment facies. Sublacustrine slumpings almost completely ceased from this time, implying a stabilization of the internal slopes of the volcano by the vegetation and an end to the supply of coarse detrital material to the marginal prism. In the Recent Dryas, a new climatic deterioration produced a clear increase in the importance of cold diatoms (zone 3). The extension of forest cover from the Preboreal until the end of the Atlantic (from 10,300 B.P. to 4,700 B.P., following the pollen chronology) caused a spectacular increase in the proportion of organic matter in the sediments (stratigraphic assemblage B), but during the Subboreal and the Subatlantic, a later decrease in forest cover is connected perhaps in part with anthropogenic activity. The impoverishment of diatom flora in the lake since about 8,000 B.P. and the presence of cold species only just before the present era (diatom zone 6 corresponds to the end of the Subboreal and the onset of the Subatlantic) could be explained by the climatic deterioration that has been recognized elsewhere for the same epoch.

Finally, the suggestion of variations of lake level during the Late Würm and the Post-Würm provided by the diatoms can be verified on the ground. These variations are weak in amplitude and do not appear to have exceeded about 1 m or 2 m up or down relative to the present lake level, but they have probably been sufficient to modify the bathymetry of the sublacustrine littoral plateau (summit of the marginal prism) and to perturb the vegetation population that was installed there.

The data on the paleoenvironment of Lac du Bouchet for the Pleistocene and the Holocene should be improved when the study of the other cores is more advanced. It will be very interesting to correlate the paleobiological and sedimentological data more precisely with the large changes in climate. In the passage from the Würm to the Post-Würm, it is evident that chronological correlations existed between climatic changes (end of sublacustrine slumping), the onset of pedogenesis (changes in the proportions of the different clay minerals), the colonization of soils by arbustive and arboreal vegetation, and the consequent extension of forests (augmentation of the biomass). These correlations should be precisely evaluated, and an exact calculation of the biomass produced within and outside the lake should be possible. Similarly, the older series of Lac du Bouchet will carry important information about the chronology of the upper Pleistocene (chronological limits and climates in stratigraphy of the last interglacial, e.g.) when deeper cores (50-200 m) can be obtained.

PALEOMAGNETISM

Subsampling of Cores

In the laboratory, cores were first cut into sections of 1.5-m length, and these sections were split into two equal segments through the plane containing the core axis. Subsamples were taken from cores B1, B3, and B21 by pushing almost square-shaped plastic boxes of ~22-mm internal side into the plane face of the exposed sediment, keeping the sides parallel to the sides of the core tube. The boxes were placed adjacent to one another to sample continuously down the central axis. On core B5, to obtain an increased sampling density, the boxes were inserted in a zigzag fashion down the core, each subsample overlapping adjacent ones. Arrows marking the upcore direction were then drawn before the boxes were gently extracted from the sediment core. In the first method, only sediment from close to the central axis of the cores was sampled, but in the second, sediment had to be sampled from closer to the edge of the core. Even though the action of the Mackereth corer (Mackereth, 1958) is smoother than that of other types of corers, some physical disturbance of the sediment fabric near the outside of cores is inevitable due to friction when the coring tube slides by as it penetrates the lake bed. The resulting slight increase in scatter of natural remanent magnetization (NRM) directions associated with the use of the second method is, however, compensated for by the increased density of sampling points. Verification of individual core records was obtained by intercore comparison.

Method

Subsamples (either in the large square boxes or in the small cylindrical boxes) were stored in the laboratory in a zero magnetic field environment for at least a month prior to measurement of NRM. A three-axis (Cryogenic Consultants, Ltd, London) CCL cryogenic magnetometer was used at Edinburgh, and a Digico fluxgate magnetometer was used at Marseille. A full discussion of methods and techniques is given in *Geomagnetism of Baked Clays and Recent Sediments* (Creer, Tucholka, and Barton, 1983).

NRM Results

As an example of the quality of the paleomagnetic results obtained from the Lac du Bouchet cores, a complete set of NRM logs and the lithological log for core B5 are shown in Figure 5-5. A highly organic layer (assemblage B) between 95 cm and 125 cm depth was not subsampled, as reflected by the break in the illustrated logs. Corresponding logs for the other cores studied verify the major features of the patterns of core B5. A preliminary account of the results for cores B1, B3, and B5 has been presented (Thouveny, 1983; Thouveny et al., 1985).

NRM Intensities and Susceptibilities. Throughout sediment assemblage A (which has ~20% organic matter) and also through the top part of assemblage C down to about 175 cm depth, NRM intensities *(J)* are very low (~10 μG or less). Between 175 cm and 200 cm, still within assemblage C, intensity magnitudes increase rapidly, attaining values peaking at ~600 μG. Three maxima and minima (~400 μG peak to peak) are observed between ~200 cm and ~450 cm, within the lower levels of assemblage A and the upper levels of assemblage D. In the lower part of assemblage D and in assemblage E, down to the bottom of the core, intensities decrease to ~200 μG. The magnitude of the NRM intensity depends on the organic carbon content (see Fig. 5-2), low

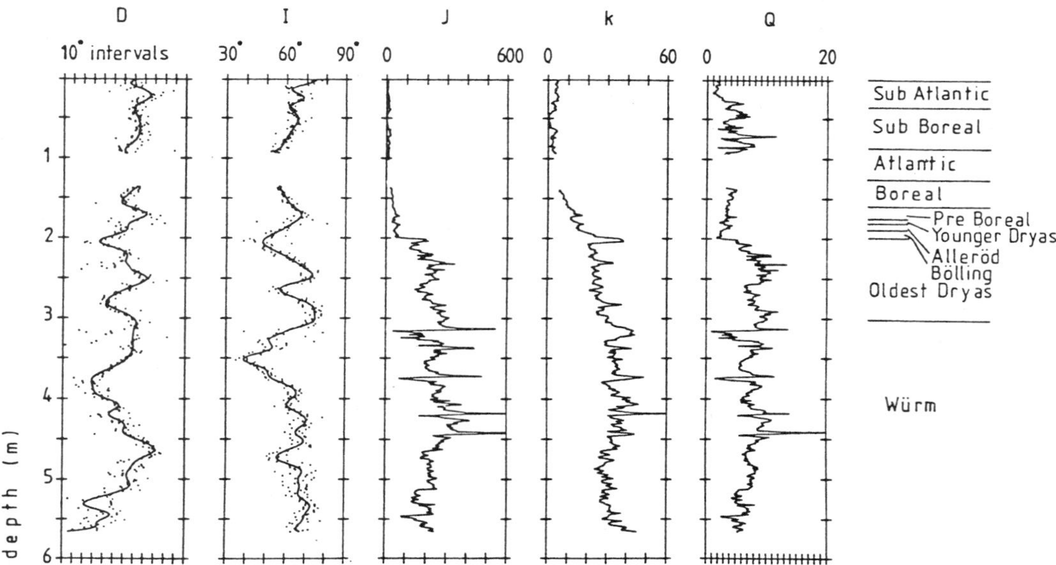

Figure 5-5. Logs of natural remanent magnetization (NRM) for core B5 (*D* = declination, *I* = inclination, *J* = intensity (μG), *k* = susceptibility (μG/oe), and *Q* = *Q*-ratio). Cubic spline curves are fitted with 50 equally spaced knots for *D* and *I*; *J*, *k*, and *Q* points are connected by a simple line.

values being associated with high organic content and vice versa.

NRM Directions through Assemblage A. The paleomagnetic signal recovered from assemblage A (which covers the Sub-Boreal and Sub-Atlantic pollen zones) is poorly defined, largely because of the high organic content (see Fig. 5-2) that results in weak NRM intensities. No signal could be recovered from sediment assemblage B where the organic matter is highly fibrous.

NRM Directions through Assemblage C. The top of assemblage C is associated with the top of the Boreal pollen zone which places it at about 8,000 B.P. Here, the organic content is only about 5%, and the high proportion of detrital grains, derived from the volcanic rocks forming the crater walls surrounding the lake, results in a strong paleomagnetic signal with well defined maxima and minima in both declination and inclination with amplitudes of about 40°,and 25° respectively.

NRM Directions through Assemblage D. Assemblage D sediments are placed in the Older Dryas pollen zone. The content of organic matter is typically less than 5%. Two complete declination oscillations occur in assemblage D. The upper one is accompanied by a single, well formed inclination oscillation while the lower declination oscillation is accompanied by smaller wavelength inclination oscillations. The larger amplitude inclination maximum coincides with an intensity maximum, and the underlying inclination minimum coincides with an intensity minimum.

Alternating Field Demagnetization

A weak viscous magnetic component was removed from pilot samples by alternating field (AF) demagnetization in 50 oe (oersteds). On further AF demagnetization, the median destructive fields (MDF) of subsamples from different levels were found typically to lie in the range of 150 oe to 200 oe. After removal of the viscous component, the remanent magnetization (RM) directions were very stable to demagnetizing fields of up to 600 oe at which 90% of the NRM had been removed (Fig. 5-6). As a result of these experiments, it was decided to use a peak field of 100 oe for bulk demagnetization.

Intercore Correlation

Paleomagnetic logs for cores B1 and B3 were replotted on the depth scale of core B5 by using transformation functions based on (in order of increasing detail) first, the depths at which the boundaries between the five lithological assemblages A-E had been identified (though comprehensive sedimentological analyses had been carried out only on core B5, with less thorough studies on cores B1 and B3); second, by correlating specific long-wavelength features characterizing the intensity and *Q*-ratio logs; third, by correlating the shorter-wavelength features exhibited by the susceptibility logs; and fourth, having satisfied ourselves that peaks of the oscillations in declination and inclination were in good gross agreement with the correlation so far established, matching the shapes of the profiles of the oscillations in those two parameters to perform a further fine tuning of the correlation.

Cleaned RM Results

Figure 5-7 shows declination and inclination logs for cores B5, B1, and B3, plotted on the depth scale of B5, after AF demagnetization in 100 oe. At this stage about 25% of the NRM signal had been removed (compare Fig. 5-7 with Fig. 5-5). The intensity, inclination, and declination records from all four cores show very good agreement, though there are clearly some slight differences in detail. In particular, the following common features are noted:

1. The break in the records corresponding to a large part of layer B, which was found to contain too much organic material to allow subsampling, is common to all four cores. The organic content is ~40% or more, about twice as high as in

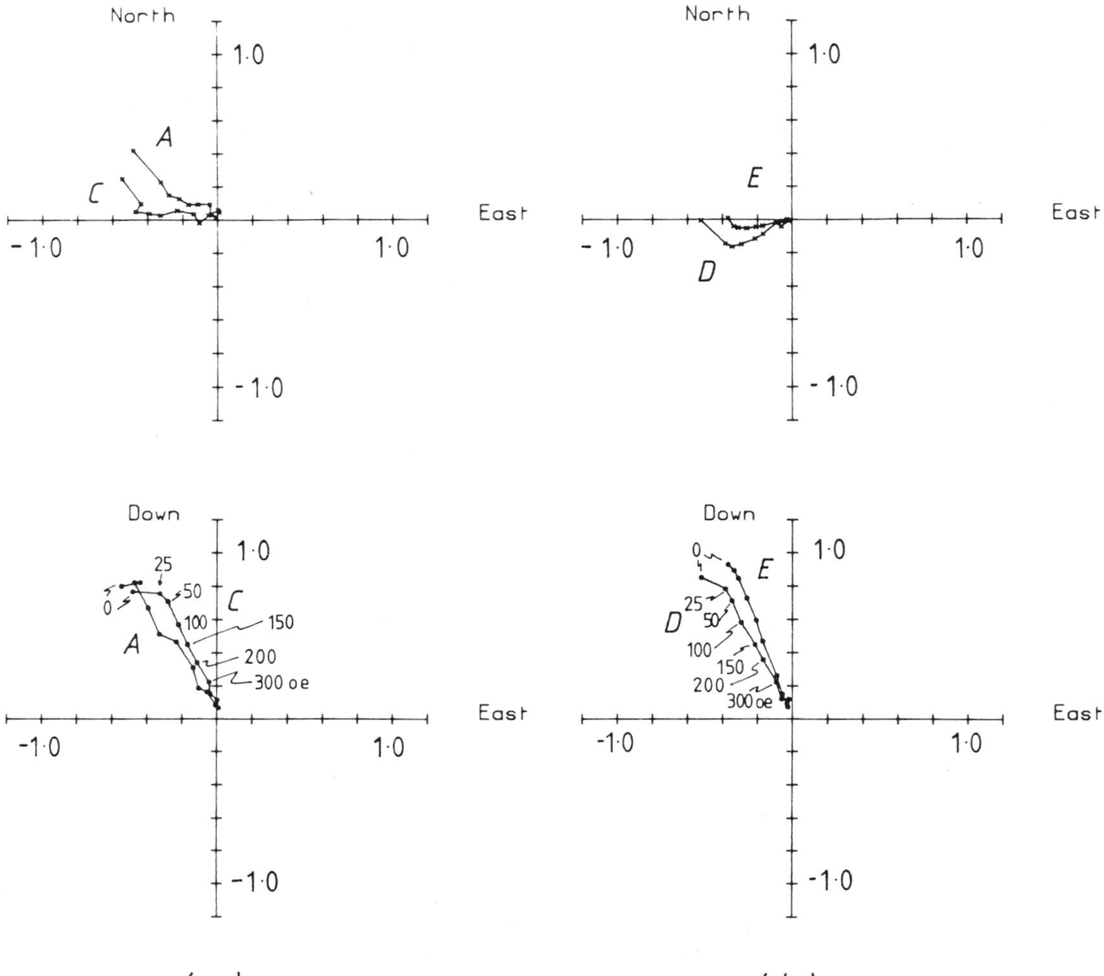

Figure 5-6. Alternating field demagnetization plots.

The remanent magnetization (RM) vector is projected on to the horizontal plane in the top row of plots (axes labeled *north* and *east*) and on to the vertical plane running east-west in the bottom row of plots (axes labeled *down* and *east*). The pair of plots on the left *(a)* illustrate the behavior of samples from lithological horizons A and B (see text), while the pair on the right *(b)* are for samples from lithological horizons C and D. Ten points are plotted for each sample showing the strength and orientation of the RM vector before demagnetization (label 0) and after demagnetization in alternating fields of peak value 25 oe, 50 oe, 100 oe, 150 oe, 200 oe, 300 oe, 400 oe, 500 oe, and 600 oe.

assemblage A. It was possible to obtain a few results from the middle part of assemblage B in core B1. The high organic content of assemblages A and B is accompanied by very low concentrations of detrital grains, explaining the low NRM intensities encountered in them.

2. The longer-wavelength oscillations in intensity correlate remarkably well along all three cores.
3. The zone with the very well-defined set of declination and inclination swings labeled Θ to Λ and μ to π respectively is followed by a zone where the declinations are 90° out of phase with the inclinations M to Π and ρ to τ, respectively (see Fig. 5-9). The intensities in the latter zone increase and appear broadly to follow the inclination variations.

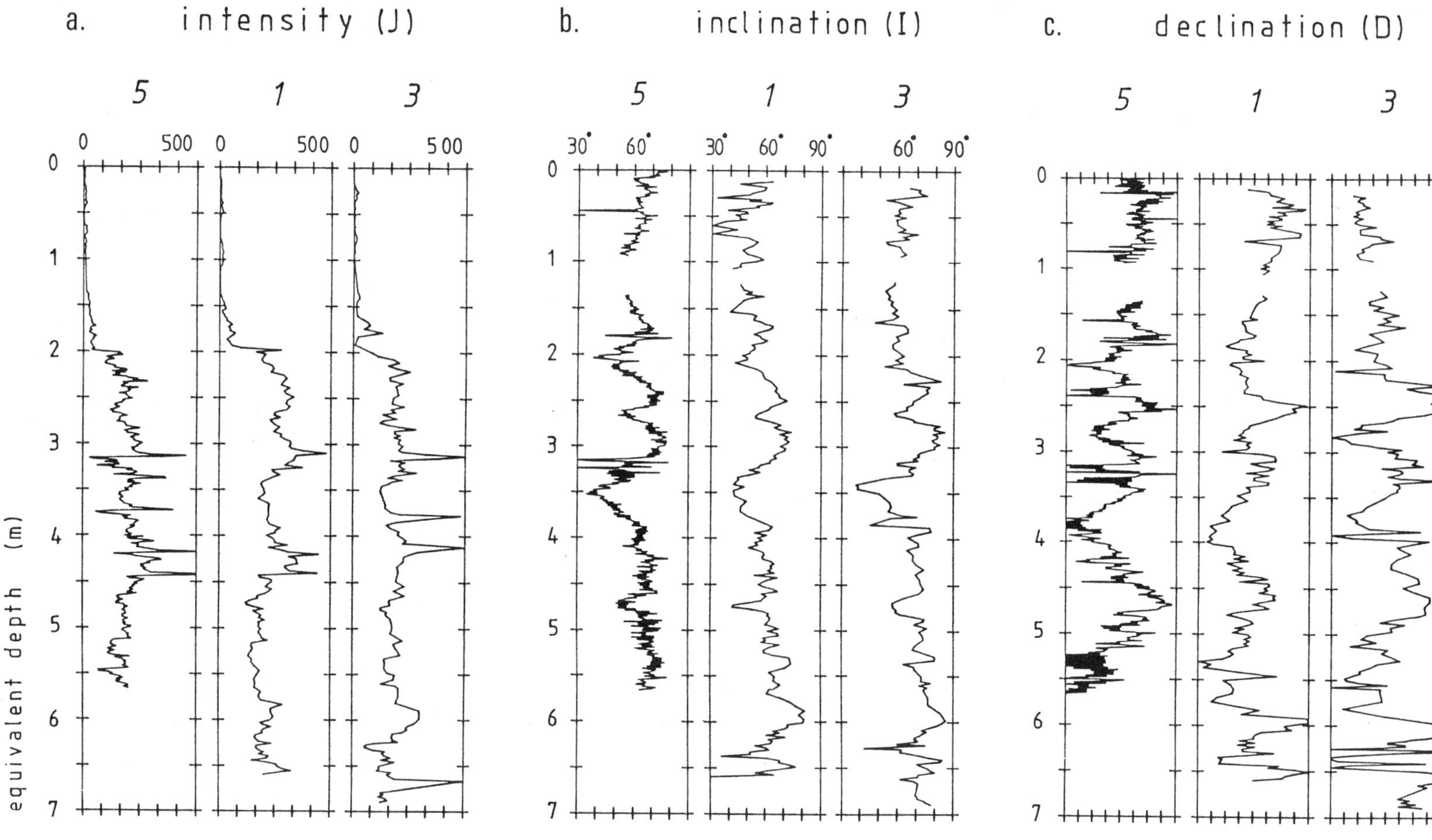

Figure 5-7. Paleomagnetic logs after AF cleaning in 100 oe peak field for cores B5, B1, and B3: *(a)* intensity, *(b)* inclination, *(c)* declination. The depth scales of cores B1 and B3 have been transformed to the depth scale of B5 so that the patterns line up (see "Intercore Correlation").

4. There are two discontinuities in the declination record, at 2.30 m and at 3.15 m. The latter is found to a lesser extent in other cores, and because it is associated with a pronounced discontinuity in intensity and a sharp peak in susceptibility, it is probably a direct effect of sedimentation. The upper discontinuity is not found in other cores, and therefore, it must originate from handling in the laboratory rather than from sedimentological or geomagnetic phenomena.
5. In all cores, the character of the inclination variations changes at about 380 cm depth (B5 scale). Below this level they are characterized by substantially reduced peak-to-peak amplitudes (less than 20°), with a noise level of as much as 12-15° in places. Could this be due to the increased compression at these depths, or is it a real geomagnetic phenomenon?

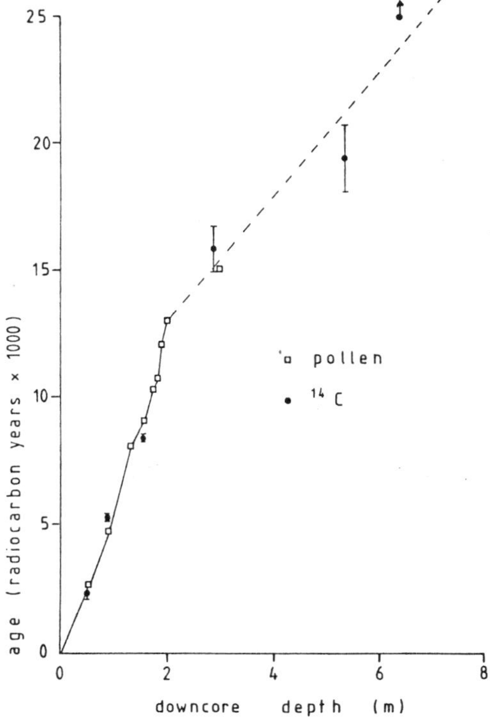

Figure 5-8. Depth/time transformation curve for core B5. Pollen zone boundaries represented by □, radiocarbon dates by ● with error bars (see text).

Transformation to Time Scale

Depth/Time Transfer Function. Radiocarbon age determinations carried out on core B5 are discussed in the section on dating considerations and illustrated together with the palynological results in Figure 5-3. The ages assigned to the pollen zone boundaries are given also in an earlier section. Figure 5-8 gives a graphical representation of these age controls.

We believe the ages inferred from the pollen studies to be more reliable than the radiocarbon ages that, for sediments, can frequently be subject to systematic error (Hedges, 1983). We have used the pollen-based control points to construct our depth-time transform function, ignoring two of the radiocarbon age determinations—viz. Gif 5941 (8,300 B.P.), which is thought to be too young (see section on the Preboreal), and Gif 5940 (5,500 B.P.), which is thought to be too old (see section on the Subboreal). Gif age 5944 (2,280 B.P.) is in good agreement with the palynology.

The final point, at >25,000 B.P., is a radiocarbon date for the bottom of core B21. This core has been the subject of comprehensive sedimentological studies (not reported here), and on these grounds, its base is estimated to be about 1 m below the base of core B5. Paleomagnetic measurements carried out on material remaining after the sedimentological samples had been taken confirm this correlation. A straight line has been fitted through the pollen ages at 13,000 B.P. and 15,000 B.P. and through radiocarbon ages at 15,800 B.P. and 19,400 B.P. on core B5 and 25,000 B.P. for core B21. Independent dating of the other cores is now being carried out.

Discussion of Results. The set of logs for core B5 after AF demagnetization in 100 oe, plotted on our time scale defined earlier, is illustrated in Figure 5-9, where the declination and inclination maxima and minima are labelled with Greek letters, upper and lower case, respectively, following the scheme described for the United Kingdom type curves by Creer (1985). The following points are noted:

1. The results for lithological assemblage A are stretched out on the time scale relative to the depth scale. We have already noted the low rate of deposition of this lithological unit. Inclination features α, β, γ, and δ are clearly seen, also declination features A, B, Γ, and Δ.
2. The break in the paleomagnetic record spanning assemblage B, which has been assigned to

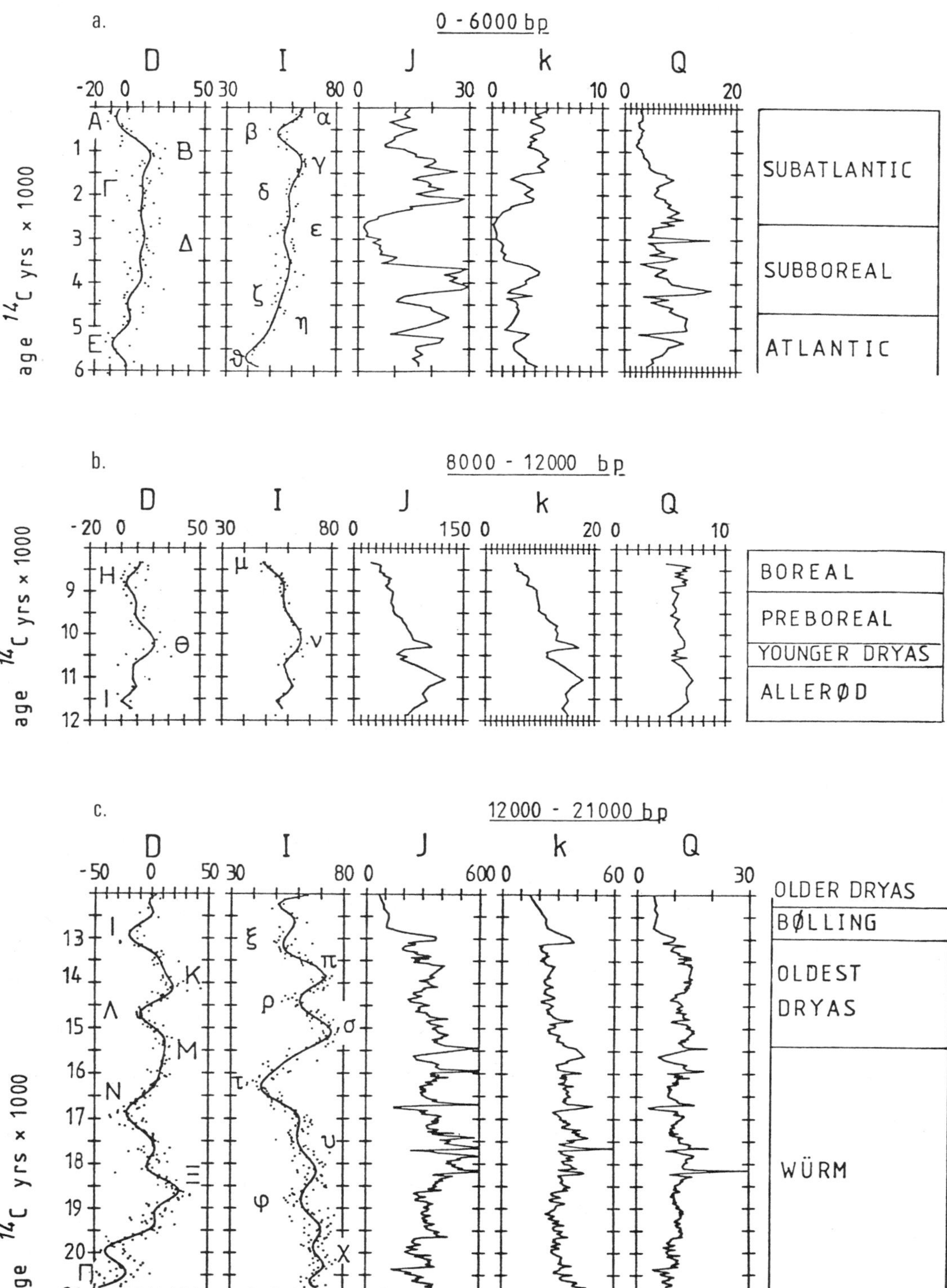

Figure 5-9. Paleomagnetic logs for core B5 transformed to a time scale. *(a)* 0-5,000 B.P., *(b)* 8,000-12,000 B.P., *(c)* 12,000-21,000 B.P. Cubic spline curves with 50 equally spaced knots fitted to *D* and *I* logs. Points along the *J*, *k*, and *Q* logs are connected by a simple line.

the Atlantic zone (see Fig. 5-3), is of some 3,000 yr duration.
3. A satisfactory coverage of the Boreal, Preboreal, and Younger Dryas zones (8,000-10,700 B.P.) has been obtained. Inclination maximum ν is centered at ~10,000 B.P. Declination peaks H (westerly) and Θ (easterly) are recorded.
4. Our coverage of the Allerød and Bølling zones is inadequate to reveal any significant secular variation features.
5. Dense coverage has been obtained for the Ancient Dryas zone (13,000-15,000 B.P.). A strong and well-defined paleomagnetic signal reveals two inclination minima ξ and ρ with maximum π in between and two westerly declination peaks I and Λ with easterly peak K in between.
6. Similarly, a strong signal is recorded through the Würm from 15,000 B.P. back to 22,000 B.P. at the bottom of the core. A large inclination minimum τ occurs at ~16,500 B.P. and is preceded by smaller-amplitude oscillations labelled υ, φ, and χ, each of which consists of two or three minor oscillations. Large-amplitude declination oscillations with two easterly peaks labeled M and Ξ and two westerly peaks N and Π are recorded.
7. A marked increase in NRM intensity occurs at the Bolling/Ancient Dryas boundary.

Results from Other Cores

Several additional 6-m cores are still under study at the time of writing, and the collection of a set of 9-m cores is planned (*Note added in proof:* The paleomagnetic results from these cores have since been written up and published in Thouveny et al., 1985; Creer et al., 1986; and Smith and Creer, 1986.)

PALEOMAGNETIC DATA ANALYSES

Representation of Logs by Cubic Spline Fits

The NRM logs for core B5, after AF demagnetization, have been fitted with cubic splines with knots equally spaced on the depth scale. Curve fitting was carried out at this stage rather than after transformation to the time scale because subsamples had been taken at uniform increments of depth.

We used the cross-validation technique (Clark, 1983) in an attempt to choose the optimum degree of smoothing, but we found that this method was insensitive. We have also noted that the fitted curves are quite similar for a wide range of knot spacings (see "Epilogue," Chapter 5, in Creer, Tucholka, and Barton, 1983). The fitted curves, illustrated in Figure 5-9, were computed using 50 knots, equally spaced along the depth scale—that is, one knot approximately every 10 cm, or ~10 subsamples. The aberrant directions referred to in item 4 of "Cleaned RM Results" were eliminated from the data set used for curve fitting.

The intensity and susceptibility logs exhibit many quite sharply defined short-wavelength features, usually of small amplitude, which are related with sudden changes in lithology. These data points cannot adequately be represented by a cubic spline fit, and therefore, Figure 5-9 shows only the individual data points.

Spectral Analyses

An extensive series of both Fourier and maximum entropy spectral analyses (MESA) have been carried out on our data from the Le Bouchet cores. This section describes the results of Fourier analysis of the 50-knot cubic spline curve fitted to the declination and inclination data for core B5. For the interval 0-5,000 B.P. (spectra not illustrated), peaks are found at 1,250 yr and 700 yr for declination and at 1,650 yr and 300 yr for inclination, but no longer periods could be identified due to the short length of this part of the record (Fig. 5-9*a*). For the interval 8,000-22,000 B.P. (record shown in Fig. 5-9*b* and *c*), peaks were found at 4,600 yr, 1,700 yr, 1,250 yr, 725 yr, and 1,000 yr in inclination (Fig. 5-10*a*), and at 4,600 yr, 1,950 yr, 2,750 yr, 1,250 yr, and 1,000 yr in declination (Fig. 5-10*b*). The periods are listed in order of decreasing power. The 4,600-yr, 1,250-yr, and 1,000-yr peaks are common to declination and inclination.

We have also tried analyses using maximum entropy (MEM) (Barton, 1983*a*, 1983*b*), but

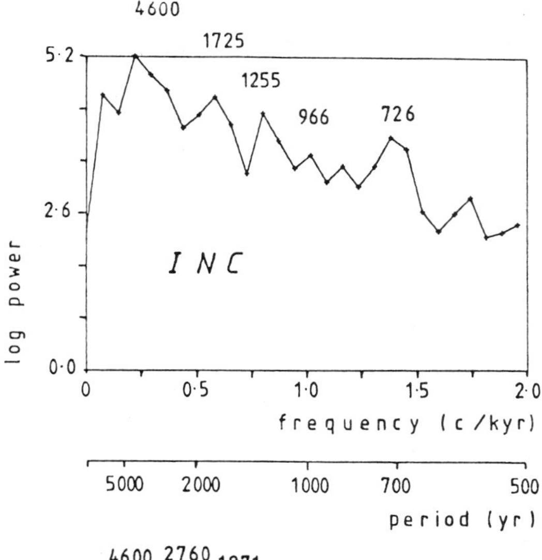

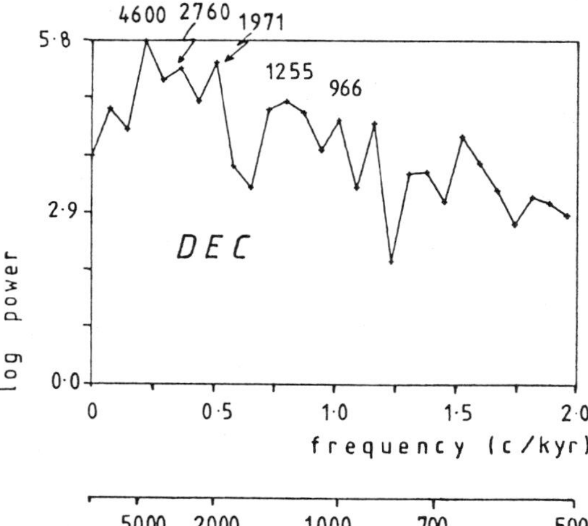

Figure 5-10. Spectra for 50-knot cubic spline curves fitted to *(a)* inclination and *(b)* declination (see "Spectral Analyses").

our experience has been that, to choose the appropriate prediction error filter (pef), one first needs the results of Fourier analysis (see Berryman, 1978; Ulrych and Bishop, 1975). If the pef is not correctly chosen, the results of the MEM analysis will not be correct. We have found, at least as far as our lake sediment data are concerned, and in agreement with Swingler (1980), that MEM analyses do not improve on Fourier analyses already undertaken, although it is useful to be able to show that the spectral peaks identified can be confirmed using a different method. Additional results, obtained since this chapter was written, are presented in Creer et al. (1986) and Smith and Creer (1986).

Curvature of Virtual Geomagnetic Pole Paths

It is useful to observe the pattern of movement of the geomagnetic vector with time rather than the individual movements of the declination and inclination curves. This observation can be achieved by using a stereographic projection, preferably centered on the axial dipole field (ADF) direction for the particular geographical site, or just by plotting declination against inclination as in a classic Bauer diagram. However, we have preferred to adopt the more commonly used procedure of computing virtual geomagnetic poles (VGPs) for successive declination-inclination pairs. We have done this computation for equal time increments along the 30-knot cubic spline curves fitted to the axial field demagnetized results for core B5. Overall, the path traced by the vector is predominantly of clockwise curvature. Between 0 B.P. and 5,000 B.P., however, there is substantial (~43% overall) anticlockwise curvature that is mainly concentrated through the past ~2,500 yr (Fig. 5-11*a*). This result has been found for other European data. Between 8,000 B.P. and 13,000 B.P., an elongated loop is traced out (Fig. 5-11*b*), with ~80% showing clockwise curvature. Between 13,000 B.P. and 18,000 B.P. (Fig. 5-11*c*) there is very open looping, essentially clockwise, with only ~27% consisting of minor weakly anticlockwise episodes. Between 18,000 B.P. and 22,000 B.P. (Fig. 5-11*d*), the VGP path traces out smaller clockwise loops with short lengths (comprising ~10% in total) of weak anticlockwise curvature.

VGP plots have been constructed for different degrees of cubic spline smoothing. These curves are very similar to one another, showing an increase in detail as the knot spacing decreases, as would be expected. The fewer the number of knots, the smoother the VGP plot,

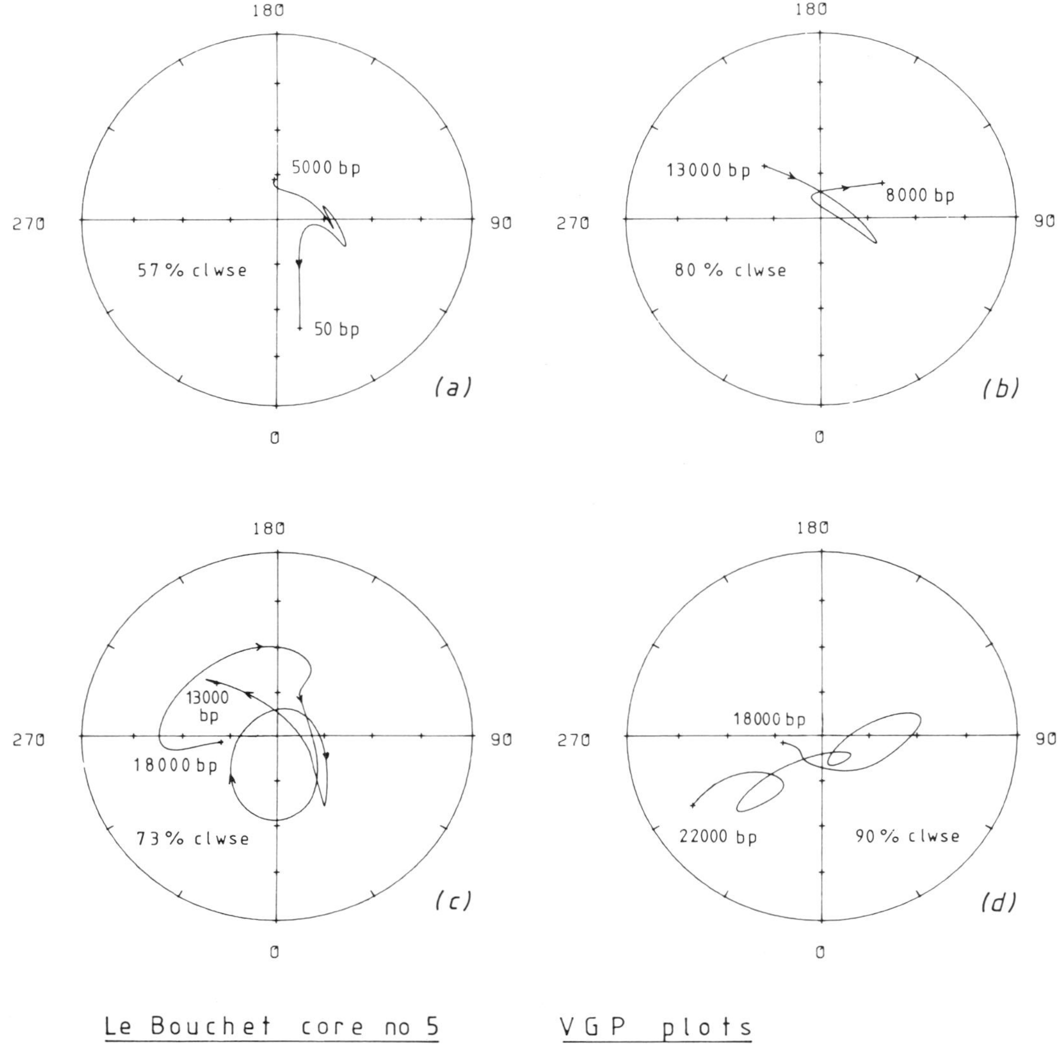

Figure 5-11. VGP paths corresponding to declination/inclination pairs along the 50-knot cubic spline curves. *(a)* 0-5,000 B.P., *(b)* 8,000-13,000 B.P. *(c)* 13,000-18,000 B.P., and *(d)* 18,000-22,000 B.P.

with a tendency for much more open looping, as indicated by broader peaks.

Open clockwise rotation would be expected from westward-drifting nondipole field anomalies and vice versa (Skiles, 1970), though this interpretation is not unique (Creer, 1983). Figure 5-12 shows more clearly the effects on curvature of cubic spline smoothing. There is a clear tendency for the episodes of anticlockwise looping to decrease with increased smoothing; that is, the longer periodicities appear to have clockwise rotation exclusively, except for the past few thousand years.

GEOMAGNETIC SECULAR VARIATIONS AND EXCURSIONS

Secular Variations

The Le Bouchet cores record irregular oscillations in declination and inclination of ampli-

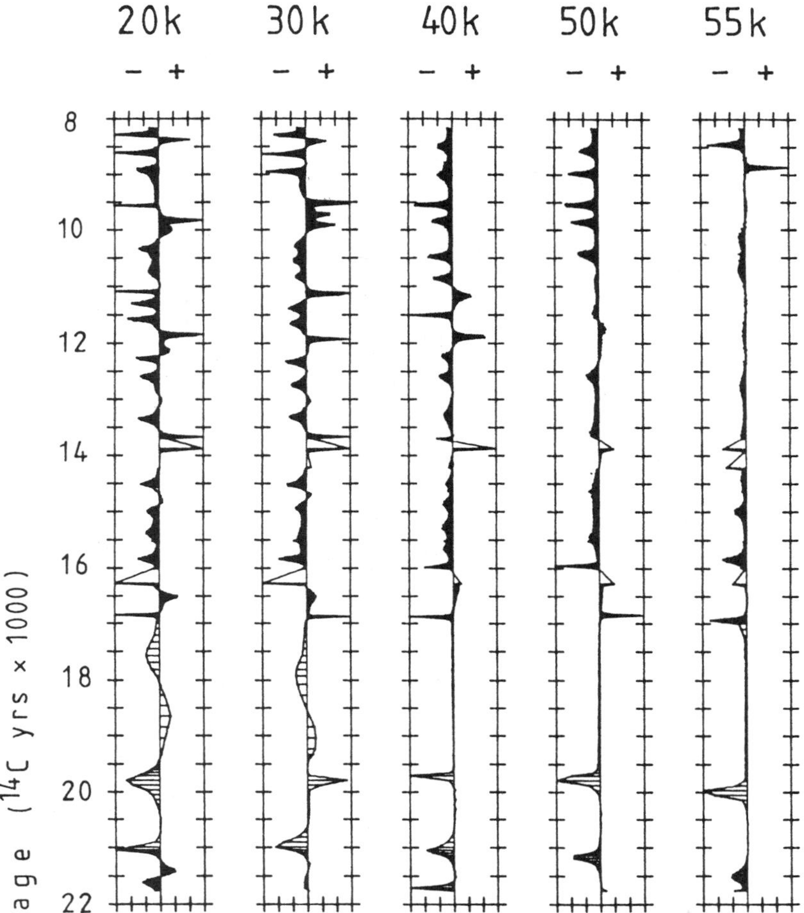

Figure 5-12. Plots illustrating sense of curvature of VGP paths (clockwise = positive and anticlockwise = negative) as a function of time for different degrees of smoothing achieved by fitting cubic splines with 20 knots, 30 knots, 40 knots, 50 knots, and 55 knots.

tude up to ~40° and ~25° respectively. Such oscillations are characteristic of spatial and temporal variations of a nondipole field with a pattern and intensity (relative to the main dipole field) similar to that computed for historic epochs during the past century or so. We therefore describe the Le Bouchet results as recording *regular* secular variations of the geomagnetic field; that is, there is no evidence of any geomagnetic changes of unexpectedly large amplitude, such as might have occurred if an unsuccessful or aborted attempt at polarity reversal had taken place.

Excursions

Geomagnetic excursions constitute departures of the geomagnetic field vector from its axial dipole configuration, well beyond the range of secular variations in direction and/or intensity observed for the historic geomagnetic field. The Laschamp excursion (or event, as it was originally described by Bonhommet and Babkine, 1967) constitutes reversed paleomagnetic directions recorded by lava flow of the Puy de Laschamp and by the lower flow of the Puy de Barme at Olby. These two volcanoes form part of the Chaine des Puys, which consists of about 80 volcanoes. The average paleomagnetic direction for these flows corresponds to a VGP located only about 7° from the present geographic pole (Bonhommet, 1970). Thus, the results obtained for the Laschamp and Olby flows are not typical of the bulk of the results obtained from the other volcanoes of the region. The NRM of these reversed flows, however, responds to the axial field and thermal demagnetization in an identical manner to the normally magnetized flows, and on these grounds Bonhommet (1970) maintained that they record a short-lived polarity reversal of the geomagnetic field during the Brunhes epoch of normal polarity. It is of particular interest to us because this area is less than 100 km to the north of Lac du Bouchet.

The reality of the Laschamp excursion has recently been questioned by Heller (1980) and by Heller and Petersen (1982) on the grounds that the excursion does not record a geomagnetic phenomenon but that it has been caused by a slow oxidation of unexolved titanium-rich titanomagnetites contained in these specific lava flows. These authors arrived at this conclusion because they observed complete or partial self-reversal of NRM in many Olby and in some Laschamp samples. Thus, it is particularly important to search for evidence of this excursion in the Lac du Bouchet sediments that acquired their NRM by a quite different physical process from the lavas. If it turns out that the NRM is not recorded in sediments of the same age from the same geographical region, then Heller and Petersen must be correct.

A series of attempts to date the Laschamp and Olby lava flows have been made by many investigators since 1969 using several different methods. Briefly, the age of the Laschamp flows can now be placed at about 32,500 B.P., using thermoluminescence measurements on plagioclase feldspars, while the Olby flows have been dated at about 37,300 B.P. (Guerin and Vallades, 1980). K-Ar ages are somewhat older, about 43,000 B.P. for the Laschamp and 50,000 B.P. for the Olby flows (Gillot et al., 1979).

We estimate the age of the sediments at the base of our longest cores (~8.5 m) to be about 30,000 B.P. by extrapolating our depth/time transformation curve of Figure 5-8. This is just younger than the time when the Laschamp and Olby lava flows were extruded. With our new single-piece 12-m corer, we would expect to reach sediments from about 37,000 B.P., on the assumption of linear extrapolation.

COMPARISON WITH OTHER GEOGRAPHICAL REGIONS

Records covering the Late Glacial have been recovered from only two other sites in Europe. These are Lac de Joux in Switzerland (Creer et al., 1980) and the eastern part of the Black Sea (Creer, 1974). The former record was derived from only two cores that penetrate through into the ancient Dryas: the age control is entirely palynological. The latter record was derived from a single 11-m core (no. 1474, taken by the Woods Hole Oceanographic Institution during cruise 49 of the Atlantis II in 1969). Seven radiocarbon age determinations were made on this core, and it yielded inclination and declination records extending back to ~24,000 B.P.

The de Joux and Black Sea records have been updated, and good agreement in both inclination and declination with the Le Bouchet results has been demonstrated (Creer, 1985). Since the Black Sea records had been derived from only a single core, not much credence had been attached to them, but they have now acquired some importance because they provide corroboration of the Le Bouchet results from a site located at about the same geographic latitude but some 30° to the east (see Fig. 5 of Creer, 1985). The ages of declination features Ξ and Π and inclination features υ and φ are about 1,000 yr older along the Black Sea records than along the Le Bouchet records. This age difference is likely due (at least in part) to insufficiently tight age control over the records from both regions rather than to drifting sources.

CONCLUSION

1. The record obtained for the Post-Glacial runs from the present through the Subatlantic and Subboreal zones back to ~5,000 B.P. Then there is a break spanning the 3,000 yr of the Atlantic zone. The record continues back through the Boreal and Preboreal zones. Due to the slow rate of deposition and the high organic content of the Le Bouchet Post-Glacial sediments, they do not provide a well-defined paleomagnetic record, though the major features of the United Kingdom type curve can be recognized.

2. The Late Glacial sediments carry a strong and stable remanence that provides a well-defined paleomagnetic record. The younger part, covering the Younger Dryas, Allerod, Bolling, and Ancient Dryas zones (<~15,000 B.P.), agrees well and provides some considerable improvement on published records from Lac de Joux (Creer et al., 1980) and the Black Sea (Creer, 1974).

3. The record through the Würm is well defined and shows good agreement with the Black Sea record in both declination and inclination back to ~22,000 B.P.

4. The sense of rotation of the geomagnetic vector is essentially clockwise. Short episodes of anticlockwise rotation seem to be identified with the joins between the larger loops and disappear when the shorter periods are filtered out. However, anticlockwise rotation seems to be important for the short part of the record that runs from 0 B.P. to 5,000 B.P. A possible, though not unique, interpretation of clockwise rotation is in terms of westward-drifting nondipole geomagnetic sources in the Earth's core (e.g., see Creer, 1983).

5. The main spectral frequencies common to both declination and inclination are ~4,600 yr, ~1,250 yr, and ~1,000 yr. Important additional spectral peaks occur in declination at ~1,950 yr and ~2,750 yr and in inclination at ~1,700 yr.

ACKNOWLEDGMENTS

The authors thank the organizations and individuals who gave permission for the field campaigns, particularly the General Council of the Haute Loire Department and the Municipalities of Cayres and le Bouchet. The work was cofinanced by the Centre National de la Recherche Scientifique (CNRS), the DGRST, and the Ministère de la Recherche (France) and by the National Environmental Research Council (NERC) and the University of Edinburgh (United Kingdom).

REFERENCES

Andrieu, B., 1937, Les chrysostomatacées d'Auvergne, I. Dépôt de Verneuge (Puy de Dome), *Soc. Française Microscopie Bull.* **6:**49-58.

Barton, C. E., 1983*a*, Analysis of palaeomagnetic time series and applications, *Geophys. Surveys* **5:**335-368.

Barton, C. E., 1983*b*, Spectrum analysis, in K. M. Creer, P. Tucholka, and C. E. Barton (eds.), *Geomagnetism of Baked Clays and Recent Sediments,* Amsterdam: Elsevier, pp. 262-266.

Berryman, J. G., 1978, Choice of operator length for Maximum Entropy Method spectral analysis, *Geophysics* **43:**1384-1391.

Bonhommet, N., 1970, Discovery of a new event in the Brunhes Period at Laschamp (France), in S. K. Runcorn (ed.), *Palaeogeophysics,* London: Academic Press, pp. 159-163.

Bonhommet, N., and Babkine, J., 1967, Sur la présence d'aimantations inversèes dans la chaine des puys, *Acad. Sci. Comptes Rendus* **264:**92-94.

Bonhommet, N., and Zahringer, J., 1969, Palaeomagnetism and potassium argon age determinations of the Laschamp geomagnetic polarity event, *Earth and Planetary Sci. Letters* **6:**43-46.

Bonifay, E., 1982, Etude des Maars du Velay et de leurs remplissages volcano-sédimentaires, in *Programme Géologie Profonde de la France,* Colloque National, Paris, June 1982, Document no. 39, Paris: Bureau de Recherche Geologique et Miniere, pp. 186-191.

Bonifay, E., and Truze, E., 1984, Paléorivages et évolution des lacs et paléolacs du Velay, *10th Réunion annuelle de Sciences de la Terre,* Bordeaux, April, p. 72.

Bout, P., 1978, *Problèmes du volcanism en Auvergne et Velay,* Brioude, France: Imprimerie Watel, 326p.

Clark, R. M., 1983, Statistical analysis of paleomagnetic data, in K. M. Creer, P. Tucholka, and C. E. Barton (eds.), *Geomagnetism of Baked Clays and Recent Sediments,* Amsterdam: Elsevier, pp. 249-261.

Couteaux, M., 1978, Analyses polliniques à Peyrebeille, Mezillac et Mazan (Ardèche), *Pollen et Spores* **20**(4):485-495.

Couteaux, M., 1984, Bilan des recherches pollen-analytiques en Ardèche (France), *Soc. Royale Bot. de Belgique Bull.* **117:**181-196.

Creer, K. M., 1974, Geomagnetic variations for the interval 7000-25000 yr B.P. as recorded in a core of sediment from Station 1474 of the Black Sea cruise of "Atlantis II," *Earth and Planetary Sci. Letters* **23:**34-42.

Creer, K. M., 1983, Computer synthesis of geomagnetic palaeosecular variations, *Nature* **304:**695-699.

Creer, K. M., 1985, Review of lake sediment palaeomagnetic data (Part I), *Geophys. Surveys* **7:**125-160.

Creer, K. M.; Hogg, T. E.; Readman, P. W.; and Reynaud, C., 1980, Palaeomagnetic secular variation curves extending back to 13400 years B.P. recorded by sediments deposited in Lac de Joux, Switzerland: Comparison with U.K. records, *Jour. Geophysics* **48:**139-147.

Creer, K. M.; Tucholka, P.; and Barton, C. E. (eds.), 1983, *Geomagnetism of Baked Clays and Recent Sediments*, Amsterdam: Elsevier, 324p.

Creer, K. M.; Bonifay, E.; Thouveny, N.; Truze, E.; and Tucholka, P., 1986, A preliminary palaeomagnetic study of the Holocene and Late Würmian sediments of Lac du Bouchet, *Royal Astron. Soc. Geophys. Jour.* **86:**943-964.

de Beaulieu, J. L., and Reille, M., 1984*a*, A long upper Pleistocene pollen record from les Echets near Lyon, France, *Boreas* **13:**111-132.

de Beaulieu, J. L., and Reille, M., 1984*b*, Paléoenvironement des marais de Pelleautier et Siguret (hautes Alpes). Analyses polliniques, *Ecologia Mediterranea* **9**(3-4):19-36.

de Beaulieu, J. L.; Pons, A.; and Reille, M., 1982, Recherches pollenanalytiques sur l'histoire de la végétation de la bordure Nord du massif du Cantal (France), *Pollen et Spores* **24**(2):251-300.

de Beaulieu, J. L.; Pons, A.; and Reille, M., 1984, Recherches pollenanalytiques sur l'histoire de la végétation des monts du Velay (Massif central, France), *Diss. Bot.* **72:**45-70.

de Goer de Hervé, A., and Mergoil, J., 1971, Structure et dynamique des édifices volcaniques Tertiaries et Quaternaires, in *Géologie, Géomorphologie et Structure du Massif Central,* Symposium J. Jung, pp. 345-370.

Delbecque, A., 1898, *Les lacs Français,* Paris: Chamerot et Renouard Ed., 436p.

Germain, H., 1981, *Flore des Diatomées D'eaux Douces et Saumatres,* Paris: Boubee, 444p.

Goeury, C., and de Beaulieu, J. L., 1979, A propos de la concentration du pollen à l'aide de la liqueur de Thoulet dans les sédiments minéraux, *Pollen et Spores* **21**(1-2):239-251.

Gillot, P. Y.; Labeyrie, L.; Laj, C.; Valladas, G.; Guerin, G.; Poupeau, G.; and Delibrias, G., 1979, Age of the Laschamp polarity magnetic excursion revisited, *Earth and Planetary Sci. Letters* **42:**444-450.

Guerin, G., and Valladas, G., 1980, Thermoluminescence dating of volcanic plagioclases, *Nature* **286:**697-699.

Hedges, R. E. M., 1983, Radiocarbon dating of sediments, in K. M. Creer, P. Tucholka, and C. E. Barton (eds.), *Geomagnetism of Baked Clays and Recent Sediments*, Amsterdam: Elsevier, pp. 37-44.

Heller, F., 1980, Self-reversal of natural remanent magnetization in Olby-Laschamp lavas, *Nature* **284:**334-335.

Heller, F., and Petersen, N., 1982, The Laschamp excursion, *Royal Soc. London Philos. Trans.* **A306:**169-177.

Janssen, C. R., and van Straten, R., 1982, Premiers résultats des recherches palynologiques en Forez, Plateau Central, *Acad. Sci. Comptes Rendus* **294** (Series II):83-86.

Mackereth, F. J. H., 1958, A portable core sampler for lake deposits, *Limnology and Oceanography* **3:**181-191.

Mardones, M., 1982, Le pléistocène supérieur et l'Holocène du Piemont de Lourdes: le gisement de Biscaye (Hautes Pyrenées, France), These de 3e cycle, Université Toulouse-Le Mirail, 96p.

Skiles, D. D., 1970, A method of inferring the direction of drift of the geomagnetic field from paleomagnetic data, *Jour. Geomagnetism and Geoelectricity* **22:**441-461.

Smith, G., and Creer, K. M., 1986, Analysis of geomagnetic secular variation 10,000-3,000 years BP, Lac du Bouchet, France, *Physics of the Earth and Planetary Interiors* **44:**1-14.

Swingler, D. N., 1980, Burg's maximum entropy algorithm versus the discrete Fourier transform as a frequency estimator for truncated real sinusoids, *Jour. Geophys. Research* **85:**1435-1438.

Thouveny, N., 1983, Etude paléomagnetique de formations du Plio-Pleistocène et de l'Holocène du Massif Central et de ses abords, These de Doctotat de 3e cycle, Université d'Aix-Marseille.

Thouveny, N.; Creer, K. M.; Smith, G.; and Tucholka, P., 1985, Geomagnetic oscillations and exclusions and Upper Pleistocene chronology, *Episodes* **8:**180-182.

Truze, E., 1983, *Etude préliminaire de la sédimentation dans les lacs de maars su Devés. Le Lac du Bouchet,* Marseille: Memoire de Diplome d'Etudes Approsondies, 54p.

Truze, E., and Bonifay, E., 1984, Sedimentation dans un lac de maar actuel: le Lac du Bouchet (Cayres, Haute Loire), *10th Réunion annuelle de Sciences de la Terre,* Bordeaux, April.

Ulrych, T. J., and Bishop, T. N., 1975, Maximum entropy spectral analysis and autoregressive decomposition, *Rev. Geophysics and Space Physics* **13:**183-200.

6: Isotopic Age and Lacustrine Environments during Late Quaternary in the Tanzanian Rift (Lake Natron)

Claude Hilaire-Marcel
Université du Québec à Montréal

Joël Casanova and Maurice Taieb
Université de Luminy

Abstract: Lake Natron is a very shallow (<1 m) water body, some 20 km wide and 60 km long, with extensive trona crust, located at the southern tip of the Gregory Rift Valley. The depression was filled by a larger water body on several occasions during the Late Pliocene and Pleistocene times. Two well-developed volcano-sedimentary formations, intensively faulted (the Humbu and Moinik formations), are observed on the west bank. Preliminary paleomagnetic measurements indicate that the Olduvai (1.87-1.67 m.y.) and possibly the Jaramillo (0.97-0.90 m.y.) events are recorded in these deposits. The Humbu and Moinik formations represent the oldest lacustrine episodes known in the area. High lake levels also existed in more recent times, as shown by the occurrence of an almost continuous belt of biogenic calcareous concretions, ranging from 40 m to 50 m above present lake level. Detailed morphologic and petrographic examinations of the stromatolites demonstrated several phases of formation: quite often, the more recent ones cover pebbles and boulders eroded from the older ones. The first generations of stromatolites are dated beyond the limits of the ^{14}C method (UQ-571 >40,000 B.P.; UQ-638 >31,000 ± 900 B.P.). Preliminary Th/U measurements indicate a Late Middle-Pleistocene age for these generations. Following these high lacustrine stands, an erosional episode took place: fluvial-type paleosurfaces are often covered by a layer of gravel and pebbles, partly of exotic origin. This paleosurface has been fossilized by the recent generation of stromatolites. Twelve samples have been radiocarbon dated. The ages range from 11,640 ± 100 B.P. (UQ-587) to 9,900 ± 150 B.P. (UQ-673). Some 200 ^{13}C and ^{18}O measurements of the stromatolites have been made. $\delta_{PDB}{}^{18}O$ values ranges from about 0‰ to 7‰. Granting a reasonable temperature range of 25 ± 5°C for the paleolake water during the episode and precipitation of calcite in equilibrium conditions, one may extrapolate $\delta_{SMOW}{}^{18}O$ values between about −2 ± 1‰ and about +5 ± 1‰ for the lake water. Such isotopic changes reflect variable residence times of the lake water in relation to the evaporation/precipitation balance, inasmuch as the hydrological setting corresponds mainly to runoff inputs. The ^{13}C values obtained on fossil stromatolites range from 2.6‰ to 8‰. The enrichment in heavy carbon is interpreted as the result of the metabolic activity of algae. The high Lake Natron episode may be equally interpreted as the consequence of more abundant precipitation or as the effect of change in seasonality; that is, the same amount of precipitation during a longer period would correspond to an increase in the average annual relative humidity and therefore to a decreasing evaporation rate. A positive balance (precipitation/evaporation) without any real increase in the water input would be sufficient to fill most depressions of the rift. Consequently, the recent paleoclimatic changes in eastern equatorial Africa could simply have resulted from less marked seasonally contrasted hydrological regimes.

INTRODUCTION

The late Quaternary hydrological changes in the East African Rift have been intensively studied during the 1970s and 1980s (Street and Grove, 1976; Gasse, Rognon, and Street, 1980; Richardson and Richardson, 1972; Nicholson and Flohn, 1980). Several paleontological (Gasse, 1975; Johnson, 1974; Holdship, 1976; Lezine and Bonnefille, 1982) and geochemical (Hay, 1967; Eugster, 1980; Fontes and Pouchan, 1975; O'Neil and Hay, 1973) techniques have been applied in the examination of paleolake deposits to reconstruct the Late Pleistocene and Holocene paleoclimatic changes (humid versus dry episodes). Correlations from one site to the other are difficult, however, because of several factors—for example, the limits of the ^{14}C dating method (type of material dated, apparent ^{14}C ages of paleolake waters, isotopic fractionation effects) and the hydrological settings of lakes (groundwater-fed lakes, terminal lakes, pluviometer lakes) (Fontes and Hillaire-Marcel, 1982). Moreover, paleoclimatic inferences are subject to controversy inasmuch as field and laboratory data essentially lead to the reconstruction of local paleohydrological changes. This chapter presents evidence for two high levels in Lake Natron (northern Tanzania) during the Late Quaternary, which are indicated by a belt of stromatolitic constructions some 50 m above present lake level (Fig. 6-1). Detailed results will be published later. In this chapter we discuss the paleohydrological conditions that prevailed during the most recent high lake level episode and its age from isotopic (^{14}C, ^{13}C, ^{18}O) measurements of fossil stromatolites.

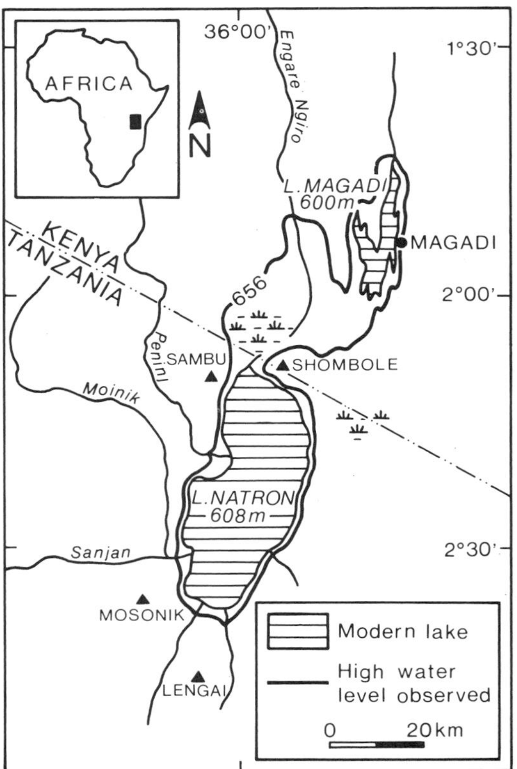

Figure 6-1. Location of Lake Natron and maximum extension of the Late Pleistocene-Holocene paleolakes. (Solid triangles indicate volcanoes.)

GEOLOGICAL SETTING

At the southern tip of the Gregory Rift Valley, northeast of the Ngorongoro volcanic complex, Lake Natron occupies a depression at an altitude of some 600 m. Along its western limit, a fault scarp shows successive layers of basaltic lava flows with zeolites, occasionally interrupted by Plio-Pleistocene lacustrine deposits. Several volcanoes surround the lake. To the south, the Lengai shows frequent activity, with the latest eruption occurring in 1983. Its lava has a strong carbonatitic component, whereas most of the other volcanoes (Gelai, Shombole, Sambu, and Mosonik) range from trachytic to phonolitic types (Dawson, 1962).

Today, Lake Natron is a very shallow (<1 m) water body, some 20 km wide and 60 km long, with extensive trona crusts. During the dry season, it is fed mainly by hydrothermal waters rich in sodium carbonate, pouring in from all directions or collected by a few perennial streams. Two of the rivers (Peninj and Enwase Ngiro) drain larger basins and have reasonably fresh waters (dissolved solids <200 ppm).

PALEOLAKE DEPOSITS

The depression was filled by a larger water body on several occasions during the Late Pliocene and Pleistocene times. Two well-developed volcano-sedimentary formations, intensively faulted (the Humbu and Moinik formations), are observed on the west bank (Isaac, 1965). Preliminary paleomagnetic measurements indicate that the Olduvai (1.87-1.67 m.y.) and possibly the Jaramillo (0.97-0.90 m.y.) events are recorded in these deposits. The Humbu and Moinik formations represent the oldest lacustrine episodes known in the area. High lake levels also existed in more recent times, as shown by the occurrence of an almost continuous belt of biogenic calcareous concretions, ranging from 40 m to 50 m above present lake level. They show a specific morphology, according to their elevation, which is interpreted as a consequence of the ecological requirements of the algal mats at their origin. A few beach-type deposits and flat surfaces make it possible to survey the paleoshoreline accurately. The stromatolitic structures indiscriminantly cover most of the bedrock surfaces and lithologies.

Through detailed morphologic and petrographic examinations of the stromatolites, we were able to establish a precise littoral zonation and to identify several phases of growth. Quite often, the more recent one covers pebbles and boulders eroded from the older ones. The first generations of stromatolites yielded ages beyond the limits of the ^{14}C method (UQ-571: >40,000 B.P.; UQ-638: >31,000 B.P.).

Following the corresponding high lacustrine stands an erosional episode took place. The Moinik and Humbu formations have been significantly eroded by running waters and possibly by wind action as shown, for instance, in the Peninj and Moinik deltas, where fluvial-type paleosurfaces are covered by a layer of gravel and pebbles partly of exotic origin (quartzite from the Precambrian basement cropping out some 30 km farther west). This paleosurface has been fossilized in more recent times, by the recent generation of stromatolites, which are exposed all around the lake. Twelve samples have been radiocarbon dated. The ages (Fig. 6-2) range from 11,600 ± 100 B.P. (UQ-587) to 9,900 ± 150 B.P. (UQ-673). However, most samples fall into a narrow bracket—that is, 10,700-10,300 B.P., which no doubt constitutes the period when the lake was at its high stand.

The surge in algal development that corresponds to this episode also suggests a stable lake level to allow for stromatolite construction and a favorable chemistry of water, drastically different from today's conditions—namely, well-oxygenated and fresh hypohaline waters (Casanova, 1981) without significant turbidity. Two samples yielded slightly older ages (UQ-618: 11,370 ± 140 B.P.; UQ-587: 11,640 ± 100 B.P.), suggesting either that the lake level rise started earlier or that some anomalies did exist in the ^{14}C activity of the lake's total inorganic dissolved carbon (TIDC) in relation to the atmospheric $^{14}CO_2$. However, ^{13}C data, provided in the following discussion, instead indicate an equilibrium between TIDC and atmospheric CO_2.

About 200 measurements of the ^{13}C and ^{18}O contents of the stromatolites have been made. Detailed results will be published later, but it is worth mentioning here the broad paleoenvironmental characteristics suggested by the stable isotope results. $\delta_{PDB}{}^{18}O$ values ranges from about 0‰ to 7‰ (Fig. 6-3). Granting a reasonable temperature range of 25 ± 5°C for the paleolake water during this episode and precipitation of calcite in equilibrium conditions, one may extrapolate $\delta_{SMOW}{}^{18}O$ values between about −2 ± 1 ‰ and +5 ±1‰ (Clayton, 1961) for the lake water. Such isotopic changes reflect variable residence times of the lake water in relation to the evaporation/precipitation balance, inasmuch as the hydrological setting corresponds mainly to runoff inputs. High ^{18}O contents strongly suggest long residence times (Fontes and Hillaire-Marcel, 1982). Taking into account the stabilization of the lake

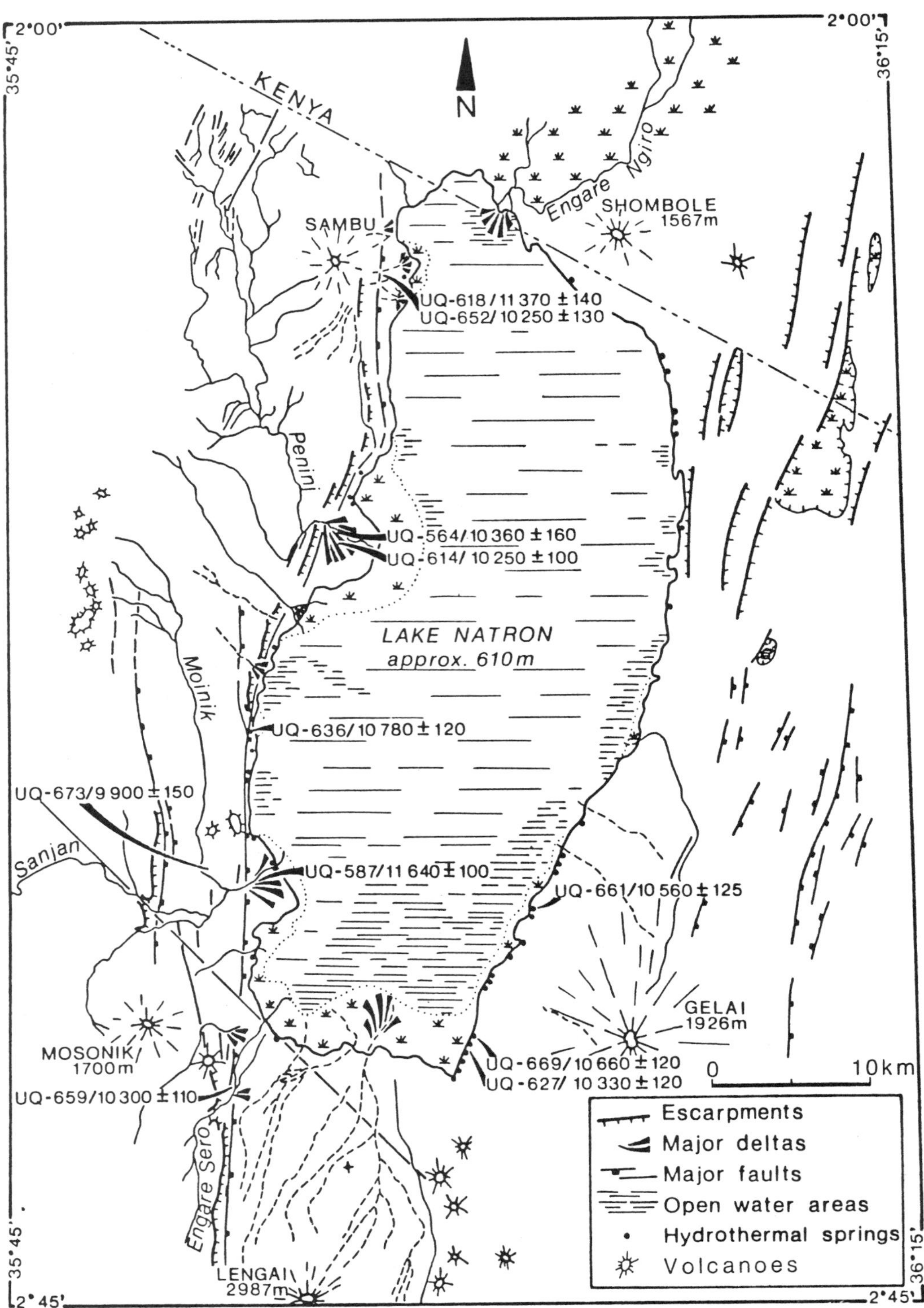

Figure 6-2. ^{14}C ages and location of the sampled stromatolites from recent lacustrine episodes.

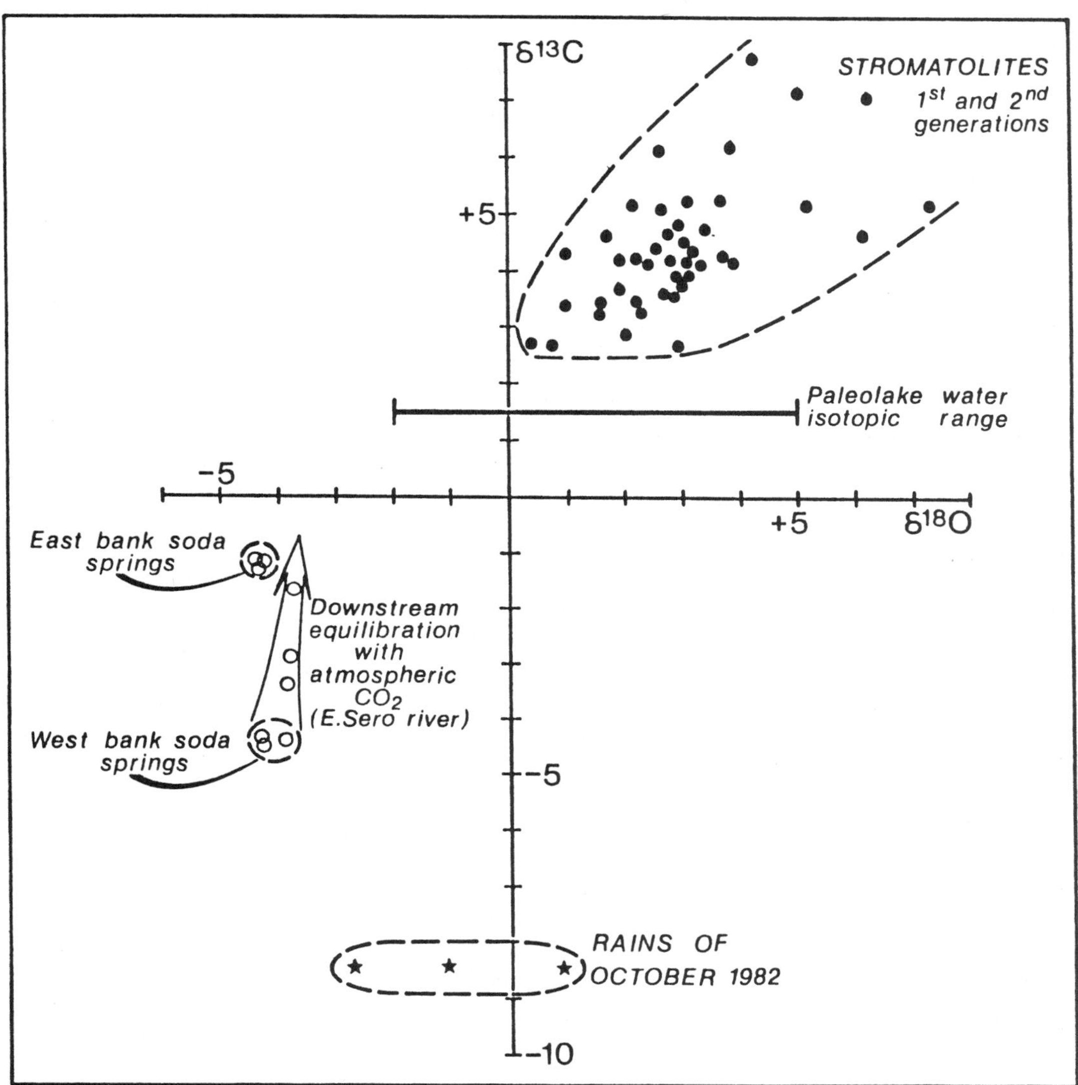

Figure 6-3. Carbon and oxygen isotope compositions of fossil stromatolites, modern spring waters, and rains of October 1982. $\delta^{18}O$ values are reported versus SMOW for waters and versus PDB for carbonates. $\delta^{18}O$ and $\delta^{13}C$ measurements on individual seasonal layers of stromatolites are not reported here.

level to allow for stromatolite growth, one can conclude that during such periods the turnover rate was moderate (i.e., reasonably abundant precipitation and reduced evaporation). Isotopic shifts were usually slow, judging from $\delta_{PDB}{}^{18}O$ determinations of individual laminae of single specimens. The lowest values may be, to some extent, indicative of the isotopic composition of input waters. Therefore, the $^{18}O/^{16}O$ ratio of paleorainfall may have been close to that measured during several rains in October 1982 at the site of Lake Natron (see Fig. 6-3).

The ^{13}C values obtained on fossil stromatolites range from 2.6‰ to 8‰ (see Fig. 6-3). The enrichment in heavy carbon is interpreted as the result of the metabolic activity of algae. During high photosynthetic periods, preferential consumption of $^{12}CO_2$ by the algae induces a significant rise in the $^{13}C/^{12}C$ ratio of TIDC in the immediate surroundings. As a consequence, ^{13}C contents in calcite are correlated to algae development and photosynthetic activity. The lowest $\delta_{PDB}{}^{13}C$ values (about 2.6‰) suggest equilibrium conditions (Bottinga, 1968) between the lake TIDC and the atmospheric CO_2, when algal activity was at its minimum. This finding lends further credibility to the ^{14}C ages obtained on stromatolites: there is no indication that the TIDC in the paleolake had an apparent ^{14}C age in relation to major inputs of deep carbon from spring waters. In fact, $\delta_{PDB}{}^{13}C$ measurements of TIDC in the modern soda springs (pH >9) range from about −4.3‰ on the east bank to about −1.15‰ on the west bank. TIDC seems to equilibrate quite rapidly with atmospheric CO_2 downstream: at the mouths of the main rivers that today feed Lake Natron, the $\delta_{PDB}{}^{13}C$ values are much higher and approach equilibrium conditions with atmospheric CO_2. The postulated equilibrium between TIDC in the paleolake water and atmospheric CO_2 has certainly been favored by the long residence time of the water already suggested by ^{18}O measurements.

CONCLUSION

The recent high-level episodes of Lake Natron, especially the ~10,500 B.P. episode, indicate paleohydrological conditions drastically different from those that currently prevail. The +50 m shoreline (above present lake level) corresponds to a large lake covering some 1,770 km^2 (compared to 1,010 km^2 today) and extending as far north as Lake Magadi (Kenya; see Fig. 6-1).

Further field studies are planned to verify the extent of the stromalitic belt. The ^{14}C age of the recent high stand may be supported in terms of ^{14}C chronology inasmuch as ^{13}C data suggest equilibrium conditions with atmospheric CO_2 or a slight enrichment in heavy isotopes for calcite, in relation to the algae's preferential uptake of light carbon. Precise correlations with other rift lakes would be largely premature, however, considering the different hydrological settings and the lack of information about the probable ^{14}C activity of TIDC in paleolake waters. Therefore, the apparent ^{14}C lag times between some of the reported lacustrine episodes might not be significant. Let us simply mention that Lakes Victoria (Bottinga, 1968) and Bogoria (Tiercelin, 1982), for instance, are known to have had high stands during the same Late Pleistocene-early Holocene period.

Finally, we stress that studies like the present one, which establish the broad lines of paleohydrological situations, do not give precise information about paleoclimate per se. In the present case and in comparison to modern hydrological conditions, the high Lake Natron episode may be equally interpreted as the consequence of more abundant precipitation or as the effect of a change in seasonality; that is, the same amount of precipitation during a longer period would correspond to an increase in the average annual relative humidity and therefore to a decreasing evaporation rate. The drainage basin of Lake Natron is currently characterized by precipitation ranging from 400 mm/yr to 2,000 mm/yr and evaporation ranging from 2,200 mm/yr to 1,400 mm/yr (East African Meteorological Department, 1975). A positive balance (precipitation/evaporation) without any real increase in the water input would be sufficient to fill most depressions of the rift. Subsequently, the recent paleoclimatic changes in this area of eastern equatorial Africa could have simply resulted from lesser seasonally contrasted hydrological regimes, as has already been suggested for some other African areas (Richardson and Richardson, 1972).

ACKNOWLEDGMENTS

We thank the Tanzanian authorities for their friendly cooperation during the field expeditions, which were coordinated by Dr. A. Mturi (Ministry of Culture). This study was also supported by the French Ministry of External Affairs (Decision No. 98/STDF3) and the CNRS-France and CRSNG-Canada (Grant A-9156).

REFERENCES

Bottinga, Y., 1968, Calculation of fractionation factors for carbon and oxygen-isotope exchange in the system calcite-CO_2-water, *Geochim. et Cosmochim. Acta* **33:**49-64.

Casanova, J., 1981, Etude d'un milieu stromatolitique continental: Les travertins plio-pléistocènes du Var (France), Thesis, France: University of Marseille.

Clayton, R. N., 1961, Oxygen isotope fractionation between $CaCO_3$ and water, *Jour. Chemistry and Physics* **34:**724-726.

Dawson, J. B., 1962, The geology of Oldoinyo Lengai, Bull. Volcanol. **24:**350-396.

East African Meteorological Department, 1975, *Climatological Statistics for East Africa,* Nairobi: E. A. Community.

Eugster, H. P., 1980, Lake Magadi, Kenya, and its precursors, in A. Nissenbaum (ed.), *Hypersaline Lakes and Evaporitic Environments,* Amsterdam: Elsevier, pp. 195-232.

Fontes, J. C., and Hillaire-Marcel, C., 1982, Isotopic paleohydrology of African tropical lakes, *Proceedings, 11th INQUA Congress,* Moscow: Nauka Publications.

Fontes, J. C., and Pouchan, P., 1975, Les cheminées du Lac Abbe T.C.: T.F.A.I. Stations hydroclimatiques de l'Holocène, *Acad. Sci. Comptes Rendus* **280:**383-386.

Gasse, F., 1975, L'évolution des lacs de l'afar central (Ethiopie et T.F.A.I.) du Plio-Pléistocène à l'actuel, Thesis, France: University of Paris.

Gasse, F.; Rognon, R.; and Street, F. A., 1980, Quaternary history of the Afar and Ethiopian Rift Lakes, in M. A. Williams and W. Faure (ed.), *The Sahara and the Nile,* Rotterdam: Balkema, pp. 361-400.

Gaven, C.; Hillaire-Marcel, C.; and Petit-Maire, N., 1981, A Pleistocene lacustrine episode in southeastern Libya, *Nature* **290:**131-133.

Hay, R. L., 1967, Chert and its sodium-silicate precursors in sodium-carbonate lakes of east Africa, *Contr. Mineralogy and Petrology* **17:**225-274.

Holdship, A., 1976, The paleolimnology of Lake Manyara. Tanzania: a diatom analysis of a 56-meter sediment core, Thesis, Durham, N.C.: Duke University.

Isaac, G. L., 1965, The stratigraphy of the Peninj beds and the provenance of the Natron australopithecine mandible, *Quaternaria* **7:**101-129.

Johnson, G. D., 1974, Cainozoic lacustrine stromatolites from Hominid-bearing sediments east of Lake Ruddolf—Kenya, *Nature* **247:**520-523.

Kendall, R. L., 1969, An ecological history of the Lake Victoria Basin, *Ecological Monographs* **39:**121-176.

Le Dain, A. Y.; Robineau, B.; and Tapponnier, P., 1980, Les effets tectoniques de l'évènement sismique et volcanique de novembre 1978 dans le rift d'Asal, *Soc. Géol. France Bull.* **22:**817-822.

Lezine, A. M., and Bonnefille, R., 1982, Diagramme pollinique Holocène d'un sondage du Lac Abiyata (Ethiopie, T 42′ Nord), *Pollen et Spores* **24:**463-480.

Nicholson, S. E., and Flohn, H., 1980, African environmental and climatic changes and the general atmospheric circulation in late Pleistocene and Holocene, *Climatic Change* **2:**313-348.

O'Neil, J. R., and Hay, R. L., 1973, $^{18}O/^{16}O$ ratios in cherts associated with the saline lake deposits of east Africa, *Earth and Planetary Sci. Letters* **19:**257-266.

Richardson, J. L., and Richardson, A. E., 1972, History of an African rift lake and its climatic implications, *Ecological Monographs* **42:**499-534.

Street, F. A., and Grove, A. T., 1976, Environmental and climatic implications of late Quaternary lake-level fluctuations in Africa, *Nature* **281:**380-385.

Tiercelin, J. J., 1982, Rifts continentaux: tectonique, climats, sediments, Thesis, France: University of Marseille.

III

Sea-Level Change and Climate

7: The Spectra of Sea Level in a Holocene Time Frame

Rhodes W. Fairbridge
Columbia University

Abstract: On a 10^8-to-10^9-yr time scale, the ocean volume appears to be nearly in a steady state, but fluctuations in volume of up to 3% appear in spectra on a 10^4-to-10^5-yr scale. In addition to short-term disturbances to the ocean surface (wind, tides, currents), relative sea level rises and falls, on a medium- to long-term time scale, due to four variables: (1) transient solid Earth density heterogeneities (Earth's mantle/core dynamics), (2) change of basin shape (e.g., plate tectonics, geoid geometry, hydroisostasy), (3) change in water volume (e.g., water temperature, glacio-eustasy), and (4) change in local or regional crustal motion (e.g., glacio-isostasy, neotectonics, compaction, and sediment loading). These variables are incorporated in the expression of mean sea level (MSL).

The twentieth-century tide-gauge averages of MSL change are gross distortions of global eustasy because of northern mid-latitude clustering and because of station sites on subsiding continental shores. Mid-Pacific atoll stations show little change. Levels of Holocene shorelines disclose regions of relative stability, and the fluctuations correlate with climate proxies. A personal retrospect reviews the history of Holocene studies. Future work calls for better chronology and closer identification of paleoclimatic proxies.

INTRODUCTION

In the 4.6-b.y. evolution of the Earth, three geodynamic-historical principles may be observed to be operative. First, there are the long-term trends of secular change, marked by cumulative effects, thresholds, and punctuated growth. Second, there are cyclic or repetitive changes on all sorts of time scales, from the order of 1 sec to 10^8 yr. Third, there are the more or less sudden events, ranging from a meteorite impact to an intercontinental collision. The first and second categories are theoretically predictable; the third can be expected, but it is not really predictable. In the same class (the third) are stochastic peaks within a steady-state time series. Examples of the very long-term changes are the gradual accretion of denser material in the core and the concomitant thickening of the crust, with a progressive loss of volatiles to the atmosphere and outer space. While there is continuous recycling as a result of plate tectonics, within the planet as a whole there is a secular gravitational sorting in a generalized sense. Continents therefore tend to thicken with time. Probably the core slowly accretes.

We do not know if the mean ocean volume has increased with time. According to Wise (1974) continental "freeboard," and therefore sea level, has remained constant on the average through Phanerozoic time. This finding seems to be borne out by the geomorphic chronology (Fairbridge and Finkl, 1980) and by the eustatic record (Vail, Mitchum, and Thompson, 1977). In terms of 10^8-to-10^9-yr time scales oceanic volume seems to be close to a steady-state situation. On a scale of 10^4-10^5 yr, the volume may fluctuate by more than 3%. On a 10^7-yr scale it is 6%. (Volume of ocean $= 1{,}350 \times 10^6$ km^3; ice volume growth at glacial maxima $= 40 \times 10^6$ km^3.)

Sea level is the surface that represents the plane of contact between the atmosphere and the oceanic hydrosphere. Short-term variations in this surface level are expressed in terms of spectra of tides and wave trains that can be measured by tide gauges (marigraphs, or mareographs) and wave gauges. Long-term variations are identified from evidence of raised beaches, emerged coral reefs, facies boundaries, and stratigraphic unconformities. This chapter reviews those spectra and follows with a personal retrospect of Holocene sea-level studies.

FACTORS THAT MAY CHANGE THE RELATIVE HEIGHT OF MSL

Changes in the height of the sea surface may be brought about in many different ways. There is a fundamental distinction between changes that affect the total volume of the ocean (approximately 1,350 $\times$ 10^6 km^3 at the present time) and those that do not. In addition, there are changes in the elevation of the measuring station. Let us summarize the principal factors.

Sea-level Change Without Variation in Water Volume

Dynamic fluctuations, short-term (steady-state variables such as waves, tides, currents, steric effects, atmospheric pressure anomalies, tsunamis): *time scale* is 1 sec to 100 yr.

Hydro-isostatic changes in basin shape (due to loading or unloading of water that is converted into continental ice sheets and melted again): *time scale* is 10^2-10^5 yr.

Archimedean displacement of water by basin filling (one-way effect of sedimento-eustasy, basin filling by accumulation of sediments and volcanic piles): *time scale* is 10^2-10^8 yr.

Tectono-eustatic changes in basin shape (affecting the capacity of the basin, especially as it relates to variable rates of seafloor spreading and plate subduction): *time scale* is 10^5-10^9 yr.

Geoidal changes (endogenetic effects caused by regional drift of gravitational anomalies perhaps due to migration of core perturbations; gravity effects of changes in surface mass distribution such as volcanic eruptions, delta buildup, ice masses; exogenetic effects due to secular polar migration, in addition to so-called pole tide; exogenetic effects due to change in the planetary spin rate): *time scale* is 10^2-10^9 yr.

Sea-level Change Due to Variation in Water Volume

Glacio-eustasy (fluctuations in the volume of continental ice masses; rapid events related to ice surges and mountain-glacier dynamics; slow, large-amplitude events related to Milankovitch-type climate cycles): *time scale* is 10^1-10^6 yr.

Steric change (variation in the molecular volume of water due to temperature change): *time scale* is 1-10^3 yr. In addition, salinity changes have a notable effect (to 10^9 yr).

Hydrologic change (variation in the rate of overturn of the hydrologic cycle involving longer or shorter residence times in lakes, groundwater, vegetation, etc.): *time scale* is 1-10^3 yr.

Geochemical changes (variations in the water density due to climate and dynamic overturn and to rates of mineralization, diagenesis, and metamorphism involving chemically bound water within mineral lattices): *time scale* is 1-10^7 yr.

Sea-level Change Due to Local Variation in the Height of Land

Glacio-isostasy (response of the Earth's crust and upper mantle to glacial ice loading and unloading, complicated by lithospheric and asthenospheric heterogeneities, variation in elastic and viscosity characteristics with depth and mineral phase changes): *time scale* is 10^1-10^3 (50%, exponential decay to about 3 $\times$ 10^4 yr).

Marginal bulge (peripheral crustal flexure consequent on glacio-isostatic effect on the lithosphere, generating a positive flexure or upwarp of 1-10% of the vertical loading-displacement amplitude, or corresponding downwarp following unloading; during ice buildup or retreat, the marginal bulge moves horizontally in a wavelike displacement): *time scale* is 10^1-10^4 yr.

Extraglacial neotectonics (secular revival of existing structural features, orogenic belts, and lineaments in response to denudational unloading and to horizontal intraplate stress or to plate marginal thermal regimes usually at rates 1-5 mm/yr): *time scale* is 1-10^6 yr.

Sedimentary loading and compaction (concentration of fluvial deltaic loads in specific depocenters creates a secular downwarp that shifts progressively as the delta grows and at rates corresponding to mean discharge. Also localized compaction relating to distribution of peat swamps and muddy oozes): *time scale* is 1-10^6 yr.

Seismic displacement (earthquake disruption of crust relating to extraglacial neotectonics and to volcanism; motion may be up or down or horizontal—i.e., strike-slip): *time scale* is 1 sec to 10 yr.

MSL, THE GEOID, AND CHRONOLOGY

MSL

The plane of sea level is constantly disturbed on the short term by winds, tides, and currents, but these variables can be measured and mathematically smoothed out to establish an MSL. The first tide staff was installed in 1682 in Amsterdam (Fairbridge, 1961*a*), no doubt as a dipstick to help navigation, but it also serves to help determine the rate of subsidence of the western Netherlands. One might imagine that such a vital question for a people living for the most part beneath present MSL would have long since been a matter of public concern, but it was only in the present century that van Veen (1954) had to appeal for more tide gauges in the Netherlands.

At first, tidal observations were merely a matter of convenience to shippers and fishermen and could be achieved by eye on an appropriate marker or a specially marked tide staff. When predictions became more important, a recording tide gauge was developed in Great Britain in 1830, and the first tide tables were published by the Royal Navy's Hydrographic Office in 1833. A portable recording tide gauge was developed in 1881, which was suitable for the worldwide hydrographic mapping then in progress by the British and other navies in areas of their global interests. The first calculations, involving both lunar and solar components, were long and tedious, but it was not until the 1870s that Lord Kelvin, in conjunction with the British Association for the Advancement of Science, developed a giant mechanical calculator. For many years this machine serviced the tidal prediction needs for the world. It was nearly a hundred years later that the modern miniaturized electronic calculators allowed high-speed tidal prediction to be undertaken worldwide in countries with major maritime interests.

The reason for operating tide gauges is primarily to furnish an observational data base for calculating the diurnal and fortnightly tides and, for the use of shipping, to predict tides well into the future. Analysis of the long tide series (preferably at least 20 yr because of the lunar nodal cycle) permits a fairly precise statement of MSL. A knowledge of MSL is required for two very practical reasons. First, it is the ultimate datum plane established by geodesy for determining elevations on the Earth's surface. Thus, Mt. Everest is said to be 8,848 m above MSL. Second, it is employed in establishing datum for submarine sounding, except that the depths indicated on nearshore charts are usually calculated in terms of a datum set at mean lower low water spring tides (MLLWST) because their depths would indicate the minimum safe clearance for shipping regardless of tidal state.

A third use of MSL is for various geophysical purposes and for satellite geodesy. The fact that MSL is established as a rule around the world only at or near the major harbors has an unfortunate consequence of generating a very one-sided coverage, heavily weighted toward northwestern Europe and eastern North America (I return to this problem later). Many Third World nations possess no tide gauges, and the same is true for many oceanic islands. Each country that collects tide-gauge data usually operates through a national survey [in the United States, the National Ocean Survey of the National Oceanographic and Atmospheric Administration (NOAA)], which prepares monthly and annual MSL averages for each of its stations (Hicks et al., 1983).

World data on MSL are collected by the Permanent Service for Mean Sea Level (see also the section on "Organization"), the data being collated and computerized at the Bidwell Observatory of the British Institute of Oceanography, situated near Liverpool, England. Printouts can be obtained for any cooperating station.

The Geoid

The Earth's rotation generates a centrifugal force opposed to gravity that creates an oblate spheroid with an equatorial bulge, the equatorial radius being 11 km greater than the polar radius. If this spheroid were entirely covered by a shallow ocean, subject to both gravitation and rotation, the MSL would constitute an equipotential surface, which is called the *geoid,* or "figure of the Earth." On a mountainous Earth, one can visualize the geoid as the MSL that would be found in a series of water-level canals cut through the continents. The mass of those land areas, however, in spite of isostatic compensation, attracts the mass of the water, so that the geoid and MSL rise toward continental shores, but much less so on isolated islands. Thus, with rock variations in addition, the mean gravitational acceleration measured at the Earth's surface varies considerably. Local inequalities can be largely eliminated by using the perturbations of satellite orbits. Their analysis provides a figure of 1/298.25 for the ratio of the ellipticity of the geoid. The trouble is that this figure is greater than it should be on theoretical grounds, and the explanation seems to lie in the core-mantle density contrast and to heterogeneities in both (see, e.g., the discussion by Stacey, 1977).

It is not surprising, therefore, that even when tidal and other variables are removed, MSL from point to point around the world does not correspond to an ideal spheroid. Indeed, the new satellite geoid discloses remarkable departures, in amounts of as much as 200 m. What is particularly significant is that these major geoid anomalies relate in no way to surface topography, crustal composition, or tectonic structure. Accordingly, Mörner (1976*b*) has proposed that secular changes in MSL may be related to dynamic motions in the core. Evidence for the latter comes from the secular western drift of the nondipole magnetic field, which migrates at an average velocity of 0.2°/yr. From the evidence of numerous paleomagnetic analyses of lake sediment cores, Creer (1981) has demonstrated a corresponding east-west time lag between the profiles obtained in Europe and North America, both with respect to declination and inclination angles. The rate obtained for the past few centuries is not constant, however, and around 2,000-3,000 B.P. it rose to about 0.6°/yr, while earlier in the Holocene it was rather steady at 0.1-0.2°/yr.

From the sea-level point of view, two assumptions would have to be made, and these would require careful and imaginative testing. The first working hypothesis is Mörner's idea that the core motions are sufficiently rapid and effective to create the observed and repeated rise and fall of sea level that is claimed for the Holocene (Fairbridge, 1961*a;* Mörner, 1969, 1976*a;* Schofield, 1973; Ters, 1973; Tooley, 1978). The amplitude of these fluctuations is in the range of 1 m to 6 m. The second hypothesis is that the geomagnetic drift that Creer (1981) has observed is related to the same core convection that Mörner invokes for his geoid flexures. Rather than core motions, I prefer to attribute at least some of the local variances to the changing geometric form of embayments during sea-level change, a process I call the "Grant Effect," from the Bay of Fundy example described by Grant (1970). I shall return to these points toward the end of the chapter.

Organization of MSL Studies

Because of the wide range of disciplines required to study sea level in all its aspects, numerous international commissions have been set up within the various scientific unions represented by the International Council of Scientific Unions (ICSU), with the support of the United Nations Educational, Scientific, and Cultural Organization (UNESCO). For example, the short-term sea-level fluctuations, Category 1(A) are documented and analyzed by the Permanent Service for Mean Sea Level of the British Institute of Oceanographic Sciences, under the aegis of the International Association for the Physical Sciences of the Ocean (IAPSO) and partly supported by UNESCO (through ICSU and the Federation of Astronomical and Geophysical Sciences, FAGS).

In the same category, but with longer time scales (10-100 yr), the study of sea level becomes more the concern of the general public, coastal engineering authorities, and specialists in physical history; this aspect is studied by the Coastal Commission of the International Geographical Union (IGU) and at regular conferences of the Coastal Engineering group.

Categories 1(B) through 1(D) are studied by a variety of geophysicists and marine geologists, usually united under the common banner of International Union of Geodesy and Geophysics (IUGG) and regional associations such as the American Geophysical Union (AGU) and the European Geophysical Union (EGU). One special area concerns polar migration, which falls into the purview of the International Council for the Study of Latitude that belongs to the International Astronomical Union (under the umbrella of FAGS).

Category 2 changes of sea level are dominated, as to importance and attention, by 2(A), the glacio-eustatic cycles, that are studied by geologists and united under the International Union for Quaternary Research (INQUA) and the International Union of Geological Sciences (IUGS). The other subcategories, 2(B) (steric), 2(C) (hydrologic), and 2(D) (geochemical), collectively represent less than 1% of the sea-level change developed in a single, 100,000-yr glacial cycle but are important on an annual basis. The glacio-eustatic data are collected and studied on a standing (i.e., continuous) basis worldwide by the Shorelines Commission of INQUA and its five regional subcommissions; comprehensive annotated bibliographies have been published (Richards, Gallagher, and Colquhoun, 1986). Active since the 1960s its labors appear likely to continue for many more years, in the light of an increasing appreciation for the complexity of the problem and thanks to progressive improvements in geological chronometry.

The work of the International Geological Correlation Project (IGCP) 61, a specific project-oriented attack on the question of sea level during the past 15,000 yr, has succeeded in bringing a number of the key problems into focus and has issued a collected set of curves (Bloom, 1977). Not the least of its results has been the preparation of a handbook of field methods (Van de Plassche, 1986). A successor to that body has been IGCP 200 (Pirazzoli, organizer). Perhaps their most important role has been from time to time to assemble 50 to 100 of the leading experts, regional and international, at key sites (in Europe, North America, west Africa, Japan, Australia, Tahiti, Argentina, and Brazil), where the specific field problems have been faced in a practical manner. Pirazzoli (1986) has prepared a comprehensive global series of MSL curves from the data of the Permanent Service for Mean Sea Level (PSMSL).

Category 3 changes are studied by a wide variety of specialists, largely dependent on the time scales and geographic situations, as well as the processes involved and the observational techniques required to establish and quantify them. The geographic distribution and rates of change of motions in coastal regions can be established by tide gauge observations and analyses, as collected by the PSMSL. These can then be integrated into the overall picture of postglacial isostatic rebound by the regional geodetic networks, operated by national geodetic surveys, although this is partly wishful thinking because those surveys are sparse.

Crustal motions identifiable on a 1-to-100-yr scale are studied internationally by geodesists, seismologists, and other geophysicists coordinated by the Commission on Recent Crustal Movements (CRCM) of the IAG. Its meetings are usually held concurrently with those of IUGG, and a series of valuable symposia volumes summarize their results, the last ones being printed as special numbers of the journal *Tectonophysics*. In addition to the international CRCM, there are technical working groups and regional subcommissions. A regular newsletter is issued by the secretary, Dr. Pavel Viskocil (Prague).

Crustal motions on the 10^2-to-10^6-yr scale are studied by geomorphologists, physical geographers, and structural geologists, who are united under the Commission for Neotectonics of

INQUA, which works closely with the INQUA commissions on shorelines and on the Holocene. A quarterly journal, *Neotectonics*, is now appearing (since 1986), edited by Dr. Margaret Winslow (City College, New York). Because of common chronometric problems and concern with past climates, the INQUA specialists work closely with the paleomagnetists and specialists in deep-sea isotopic analysis, pollen analysis (palynologists), ice-core studies, tephrochronology, varve analysis, and dendrochronology.

Chronology of MSL

The first exercise would be to compare MSL fluctuations with the chronology of the various time series in relation to their field distribution. There is an immediate objection here, however, because all the radiocarbon-dated series, including geomagnetic records and sea level, suffer from error bars up to a range of ±250 yr throughout the Holocene prior to the historical era. Thus, the first task is to sharpen this chronology.

A useful start has been made in the form of IGCP No. 158, dedicated to Holocene correlation, which held a critical meeting in London, May 1981 (reported by Fairbridge, 1981*a*). An epoch-making breakthrough has now taken place in the Scandinavian varve chronology that should enable researchers to get a year-by-year climate record for that region (Cato, 1985). It is hoped that this breakthrough will eventually take us back to over 13,000 B.P. From the varves a link can be forged with the various European dendrochronological series, thanks to an ingenious "vernier" system that employs the quasi-biennial climate cycle (Schove, 1978, 1983). Teleconnections between varves of one region and another can then be proposed—for example, Scandinavia, Crimea, Anatolia, Canada, eastern United States. Then, the western North American dendrochronology (and dendroclimatology) can be locked in to the record, again on a year-by-year chronology.

A key factor in chronology is the detailed radiocarbon analyses of the California bristlecone-pine series (Suess, 1978), as a result of which it is now possible to present a ^{14}C flux rate and thus a cosmic ray intensity variability record for about 8,000 yr. Stuiver and Quay (1980) furnish more details for the last millennium. Eddy (1977) and others have proposed that this record can be interpreted as a signal of solar activity and that it is closely matched by the climatic variability of the Holocene.

A theoretical basis is thus provided for assuming climatically induced fluctuations of MSL during the Holocene. However, the chronology and precision of field studies of sea-level data must both be sharpened up in many ways.

SEA LEVEL AND CLIMATE CHANGE (<10^4 yr)

The use of sea-level data as proxies for climate-change time series requires in-depth studies of each locality and a review of all the variables listed in the introduction. The handling of so many variables might appear at first overwhelming and beyond the technical competence of many specialists, but some simplifying steps are feasible by beginning with a selection of limitations in time and space (areal setting):

Time scale: If the climatic framework chosen for consideration is less than, say, 100 yr, then the various slow-acting processes would have minuscule effects and can be ignored; in the first category, we are left with only one subunit [1(A)], although in the second category we have 2(A), 2(B), 2(C), and 2(D), and in the third we have 3(A), 3(B), 3(C), 3(D), and 3(E), ten variables. If the climate changes are on, say, a 10^3-yr frame, then comparable but distinct selections can be made.

Areal setting: A regional simplification will usually assist the study dramatically. On the <100-yr scale it will not affect 1(A) or 2(A) through 2(D), but in regions remote from Late Pleistocene glaciation, 3(A) and 3(B) can be dropped, leaving 3(C) through 3(E), which are as a rule fairly readily resolved.

Regional patterns can be established only by geological fieldwork, which can be supplemented, in the stabler regions, by remote-sensing data; satellite imagery helps us to generalize the regions, and detailed vertical air photography (on a better than 1:50,000 scale) fills in with more precise evidence. Once the regional patterns are established (e.g., historical trends, stratigraphy, lithology, tidal characteristics, oceanographic behavior), fairly extensive regions on the globe can be blocked out. Great care, however, should be taken in certain tectonic and oceanic regions. For example, the collisional plate marginal belts are notoriously subject to alternate blocks of uplift, tilting, and subsidence, even short-term reversals [category 3(E)]. Or, for oceanic examples, one may consider areas affected by El Niño steric warming, tsunamis, and variable upwelling [rapid changes in categories 1(A), 2(B), 2(C), and 2(D)].

RETROSPECT OVER HOLOCENE EUSTASY

A Personal Indoctrination

I became personally concerned and interested in Holocene eustasy during World War II when in the southwest Pacific, where many landings had to be made on islands rimmed by coral reefs. Whereas the contemporary living reefs grew only to mean low-water level, older parts of the islands commonly displayed an undercut platform of emerged reef limestone, typically at about 3 m above low tide, in some places with a series of intermediate steps, often at about 60 cm and 150 cm. On dune coasts there were commonly three sets of vegetated dunes, in places with swamps, lagoons, or belts of beach ridges in between. Beach ridges were often observed from the air in sheaves of nearly identical height and separation but in sets, often three or more, suggesting a shift in the wind-wave systems.

In the literature I found that Daly (1920) and Keunen (1933) had found similar emerged shore platforms in many parts of the tropical and subtropical oceans. On the northeast coast of New Guinea and parts of Indonesia (Wetar, Timor, Sumba), I observed uplifted steps and tilted reef platforms; older, evidently Pleistocene, reefs in places reached over 1,200 m above sea level. From the seismicity and gravity anomalies (the work of Vening Meinesz), it was obvious that these emerged reef platforms were in youthful orogenic uplift belts. We knew nothing of plate tectonics in those days (1944/1945), but Wegener's continental drift seemed to be beautifully demonstrated by the breakup of Gondwanaland and Australia's impinging on the Indonesian-New Guinea island arcs. Outside the actively mobile belts, the reef platforms were observed to be at the same levels, island after island, over thousands of kilometers. We could thus predict the levels at our landing sites with almost centimeter precision.

Swedish Signposts

Both Daly (1920) and Kuenen (1933) regarded the main 3 m platform as mid-Holocene, on geomorphological grounds. Older platforms existed (even at the same elevation), but these were deeply dissected by erosion channels that had been cut through during lowered (glacial) stands of sea level; these older reef limestones were deeply indurated and affected by karst weathering. One day in 1946 Curt Teichert, who had been a great inspiration and helped enormously with the reef studies, spotted a reference to a paper in Swedish by Sten Florin (1944), describing pollen analysis and the dating of the main beach ridges in southern Sweden by correlation with the varve chronology of Baron de Geer. They occur in three principal groups. I reasoned that to build up a massive beach ridge on an isostatically rising coast, ΔE (rate of eustatic rise) must equal ΔI (rate of isostatic rise) over a fairly protracted period, probably several hundred years. And then, where our emerged beach ridge is separated from the next one by up to some tens of kilometers, the isostatic emergence must have been counteracted by a very rapid and protracted sea-level fall ($\Delta I > \Delta E$). Therefore, the Swedish isostatic rise, which could not be fluctuating, must be modulated by a eustatic oscillation that could only be global in its expression.

I took a long shot and suggested the Pacific-Indian Ocean shore platforms should be correlated with the mid- to late Holocene of southern Sweden (Fairbridge, 1948, 1981*b*). This suggestion had the effect of infuriating many of the coastal pundits of the day, who were impressed by the absence of such data elsewhere—notably, on the eastern and Gulf coasts of the United States. But that absence is hardly surprising when one considers that almost the whole of that coast, from Long Island to Brownsville, Texas, is either in the subsiding marginal bulge area or in a zone of sediment-loading subsidence. Only the Florida segment is more or less stable and shows emerged reefs and beach ridges (Fairbridge, 1987; Stapor and Tanner, 1977; Stapor, 1975).

A few years later, Libby had developed the radiocarbon-dating technique, and Ed Deevey at Yale was kind enough to date shell samples I had collected from the associated raised beaches in Western Australia. I felt reluctant to date the coral because we did not know then how much of it was from older reefs, but whole, well-preserved shells from protected embayments could not be carried up there by storms or tsunamis. This Western Australia coast neither is marked by upwelling nor suffers from "old water" contamination due to glacial melting. When the radiocarbon dates came back, they closely matched those of the principal *Littorina* beaches of Sweden (Fairbridge, 1961*a*).

Climatic Correlation

The next question was about climate. In the low latitudes the amplitude of temperature fluctuations during the mid- to late Holocene had been less than 0.5°C over more than decadal intervals, and such changes are difficult to detect in paleoecological indicators like shell species. In the high latitudes, such as northern Scandinavia and northern Canada, temperature fluctuations over a few decades or centuries could exceed 5°C, and this situation is reflected in the pollen diagrams. Palynology as a science was developed in Scandinavia, and already by the 1950s there was a large store of data, confirming the strong climate fluctuations that had been deduced from nineteenth-century botanical studies. The early dating (in sidereal years) was based on the de Geer varve chronology, but when the radiocarbon dates began to come in, considerable confusion ensued, especially for the time around the mid-Holocene. In the 1950s we knew nothing about archeomagnetism, the secular drift of the mean ^{14}C flux rate, or the periodic "Suess wiggles." It was not until 1981, through the efforts of the late Derek Schove, that a small conference was called in London, under the aegis of IGCP No. 158, to direct attention to the wonderful potential presented by the absolute dating (in sidereal—i.e., calendar—years) of the Suess bristlecone-pine curve, with its ^{14}C flux rate values, and the many other time series now available such as ice cores, varves, pollen, and stalactites (Fairbridge, 1981*a*).

In spite of the shortcomings that existed in the 1950s, it seemed to me that the evidence for climatic fluctuations went hand-in-hand with the evidence of sea-level fluctuations. There were scattered data on glacier advances and retreats in the responsive mid-latitudes (I called them the sensitive latitudes, Fairbridge, 1961*b*). Glaciological studies by H. W. Ahlmann (1953) and his students from Stockholm provided persuasive actualistic arguments, and I received much encouragement from the kindness of Ahlmann.

A Eustatic Curve

My first attempts at drawing a Holocene eustatic curve in the 1940s were on the blackboard and quickly erased. The first printed version (Fairbridge, 1958) depicted a nicely generalized sine curve, gradually diminishing in wavelength and amplitude to the present time. The negative phases were less easily demonstrable, but in 1954, while a visiting professor at Scripps Institution of Oceanography at La Jolla, the late Francis Shepard taught me the basics of scuba diving, and then I discovered submerged undercuts and beach rocks down to about −3 m. Echo sounding disclosed deeper platforms and sampling produced more beach rocks, littoral shells, and corals from various places around the world. I was always careful to try to avoid the unstable orogenic belts, so that on a 5,000-

to-10,000-yr time scale, the neotectonic disturbances would tend to be minor.

By 1960 approximately 100 ^{14}C dates had accumulated worldwide, collected in all sorts of places by many different scientists. It was really a hodgepodge of data, but I screened the dates carefully and rejected all those from suspect areas, including, according to some of my friends, "all those that did not agree with my curve." I was encouraged by journal editors to publish, and thus the 100-sample Fairbridge Curve appeared in *Scientific American* (1960), *Physics and Chemistry of the Earth* (1961*a*), and *Quaternaria* (1962).

Shepard, who had been so kind to me in La Jolla, regarded this curve as outrageous, and Russell from Louisiana could find no evidence for it on the Gulf Coast. Nor could Saskia Jelgersma and her colleagues in the Netherlands. The explanation is fairly easy to find. In rapidly subsiding areas, in contrast to uplift regions like Scandinavia, the ΔI value is always negative, and when ΔE is also negative, the effect is commonly like a brief stillstand or a slowdown in the local sea-level rise. Furthermore, if one is strictly honest about the error bars of the ^{14}C dates and of the elevation interpretation, one would be completely justified in saying that the Holocene (Flandrian) sea-level rise was a simple, mathematically smooth curve, exponentially flattening out as it approaches the present day.

This idealized, mathematically averaged Shepard-type sea-level curve suffers from three flaws: (1) it ignores the punctuated records of sea level in the Scandinavian and Canadian uplift areas (according to Mörner, 1969, of all the Holocene curves available, Shepard's curve is the only one that is completely impossible); (2) it ignores the punctuated geomorphology and stratigraphy of low-latitude reef areas, eolianite regions, and beach ridge belts (see, e.g., Schofield, 1973, 1980); and (3) it ignores the climatological data, the glaciological, palynological, and other indications, that would appear to demand a glacio-eustatic and steric response.

It should be appreciated that the Fairbridge-type sea-level curves also have shortcomings. First, as noted earlier, in regions of strong subsidence the record tends to be smoothed out when $\Delta I = \Delta E$ and both are negative. The fact that almost all the field evidence is subsurface and cannot be observed without comprehensive drilling programs also greatly limits the data base (but see Van de Plassche, 1982; Fairbridge and Lowrie, 1987). Second, on open coasts, subject to heavy swell and periods of intense storminess, especially in regions subject to surges, the amplitude of the surge, wave set-up, and swash so greatly exceeds that of the claimed eustatic fluctuation that evidence of the latter is most commonly erased (cf. Heyworth and Kidson, 1982), even in areas of abundant sea-level fluctuation data from embayments, lagoons, and estuaries—e.g., the coasts of Brazil (Fairbridge, 1976) and Argentina. On headlands and open coasts, there are generally no traces of Holocene fluctuations.

The third shortcoming is that on many coasts the regressive facies tend to be buried (as on beach ridge strandplains) or overgrown by biological action (as on coral reefs). Thus, Chappell (1983) reported finding no trace of negative fluctuations on the Great Barrier Reef (although the adjacent mainland possesses numerous punctuated strandplains). Finally, the role of pure glacio-eustasy in modulating Holocene sea-level curves (every region is somewhat different) was overstated in the early papers (Fairbridge, 1960, 1961*a*, 1962). The existence of steric change and tidal range potential was recognized at that time but was not adequately appreciated or quantified. Nearly half the observed sea-level rise of recent decades has been ascribed to a steric component (Gornitz, Lebedeff, and Hansen, 1982). Another part is attributed to mid-latitude glacio-eustasy (Meier, 1984), but the number of glaciers actually monitored is regrettably small, and the mass balance of major ice sheets has proved to be very difficult to estimate. Thus, the eustatic factor is still elusive in actualistic terms.

Change in Tidal Range and the Grant Effect

Change in tidal range, the Grant Effect mentioned earlier, seems to be a very important factor on the 10^2-to-10^3-yr time scale. Grant (1970) demonstrated the extraordinary increase in tidal range on the Bay of Fundy during the progressive rise of Holocene sea level (combined eustatic rise plus isostatic adjustments, dominated in the late Holocene by marginal bulge subsidence). In northwestern France numerous rias present a topographic form that favors the development of the Grant Effect. They are on the right-hand side (for maximum Coriolis effect) with respect to the rising tide entering the funnel-shaped English Channel. As a result, extremely high tides develop in the rias, and one, the Rance, near St. Malo, has been very efficiently exploited for hydroelectric power. The fluctuations on the Ters (1973; Chapter 12) Brittany curve have a larger amplitude than any other, but they are very accurately dated with radiocarbon (mostly peat samples) and tied in to the climate record with hundreds of pollen diagrams. It thus represents the world's most detailed Holocene sea-level curve. Tidal range effects are only just beginning to be appreciated—for example, in the Netherlands (Beets et al., 1986).

The amplitudes of sea-level fluctuation on the Ters curve exceed those indicated on the Fairbridge (1961*a*) curve (essentially an open-coast subtropical synthesis) by factors of 2 to 6. Its amplitudes exceed by a factor of 5 to 10 those of low tidal-range coasts like northwestern England (Tooley, 1978) or southwestern Sweden (Mörner, 1969, 1976*a*). The Ters field evidence was checked and confirmed by successive inspections by members of the appropriate commissions and IGCP groups. Two facts emerge conclusively from her wealth of data: (1) Although the amplitudes vary, the timing of the Holocene sea-level fluctuations seems to follow the curves of Fairbridge (subtropics), Mörner (southwestern Sweden), Tooley (northwestern England), and Schofield (1973, New Zealand); and (2) the regressions closely match the cooling trends in the Scandinavian-Canadian pollen curves and the principal neoglacial readvances in the mountain glaciers of both hemispheres (Denton and Karlén, 1973). Simple histograms of coastal peat samples, reflecting transgressive versus regressive trends, display corresponding and astonishingly prominent peaks and valleys (Geyh, 1980; Shennan et al., 1983). The case for major climatic fluctuations in the mid- to high latitudes during the Holocene and their correlation with sea-level fluctuations now seems to be persuasively demonstrated.

Beach Ridges and Storminess Cycles

The next decisive turning point in this retrospect was introduced by the thesis survey of the glacio-isostatically uplifted beach ridges of the Hudson Bay carried out by Hillaire-Marcel (1976). By a happy chance I was invited by the Université de Québec à Montréal to assist in his planning and thesis defense (at the University of Paris). The field evidence is dramatic: long staircases of up to 185 emerged beach ridges, reaching in places to 315 m above sea level, and dating back to the glacial ice retreat in Hudson Bay at about 8,300 B.P. The mean frequency of the beach cycle is 45 yr (Fairbridge and Hillaire-Marcel, 1977), attributed to peak crescendos of storminess and tentatively correlated with solar periodicities; probably lunar cycles are also present (Guiot, 1985). The curve is further modulated by 350-to-1,000-yr trends that seem to correlate with the global eustatic/climatic record. From the old Hudson Bay Co. fur trading records it became possible to relate the beach-building and higher sea-level episodes to warm climate spells of a few years (5-10 yr in every 45-yr cycle). Reference to the past three centuries of either meteorological or these proxy-weather data suggest that the storminess cycles would tend to return whenever the Arctic high-pressure cell, usually over northeastern Canada in summer, would shift to the northwest, permitting polar frontal storms to enter the region from the southwest. With an extended summer (up to 5 mo, in place of 3 mo of open, ice-free water) storms could build up massive beach ridges (Fairbridge, 1983). The warmest summers in the past three centuries were 1869-1871, coinciding with a peak in the Gleissberg 80-yr auroral cycle; a sunspot, Wolf number RM = 141; and the building of Hudson Bay beach no. 3 (Fairbridge and Hillaire-Marcel, 1977).

Hillaire-Marcel's work triggered an interest in beach ridges in general. They represent datable time series that are not restricted to the high Arctic but occur in all latitudes—anywhere, in fact, where there is a plentiful supply of sand or other loose material associated with a gentle offshore gradient that permits beach ridge progradation (Bird, 1984). Using air photos, I had counted about 100 such ridges in the Point Peron area of Western Australia that had accumulated since the Older Peron high sea-level stage of around 5,500 B.P. (Fairbridge, 1950). On revisiting Western Australia in 1984, I had the good fortune to meet Peter Woods, who had just completed a soil thesis (1983) on the beach ridge terrain of Point Peron and other coastal sites to the north (Jurien Bay) and south (Australind). With extensive drilling and radiocarbon dating, he had established a beach ridge chronology with a 45-yr cyclicity and longer cycles in the 350-to-1,000-yr range. This was exactly what we had discovered on the Hudson Bay, but now we see a case in the subtropics of the Southern Hemisphere that is subject to the quite different storminess cycles of the circum-Antarctic/Southern-Ocean westerlies.

I prepared a review of beach ridge chronologies for the S.E.P.M. Armstrong Price Festschrift volume (for 1987). In certain favored latitudes, near the margins of prevailing wind belts, the Holocene beach ridges disclose complete reversals in the ridge-building directions. The earliest observations of this sort were from Cape Krusenstern on the northern Bering Sea coast of Alaska at 67°N. Six wind reversals have been recorded over the past 5,000 yr (Moore and Giddings, 1962); in the colder times the beach accretion is from northwesterly winds, but in warmer cycles the polar front moves north and the winds are from the south. As in the Hudson Bay, these weather systems apply only to the short summer season when the sea is not frozen in.

A second group of 280 beach ridges was mapped and dated by Curray and Moore (1964) from Nayarit on the east side of the Gulf of California in Mexico. Again, there were five prevailing wind reversals in about 4,500 yr. In the cooler phases the longshore drift is from the northwest, in the warmer times from the southeast. Along the coast of China (Gulf of Bohai) comparable beach ridges with cheniers are correlated with both sea-level fluctuations and the shifting mouth of Huanghe, the Yellow River (Zhao et al., 1980), which also seems to be a climatically related phenomenon.

Climate Linkage

These Pacific climate systems appear to be interconnected, the best link being provided by a long Holocene core from the Santa Barbara Basin off California analyzed by Pisias (1978). From studies of the microfossils it was possible for him to establish an alteration of warm and cold cycles, with a complex periodicity in the range of 350 yr to 1,000 yr, comparable to other Holocene time series. Cold cycles, representing a strengthening of the south-setting California Current, were able to lower the mean February monthly surface-water temperature by 8-10°C, within two to three centuries, an astonishing departure. During a cold regime, the climate in southern California could only sustain a Sonora-type desert cactus flora (Heusser, 1978). Alternating with these cold (arid) phases were warm intervals comparable in many ways with the 1982-1983 El Niño condition when sea level rose sterically more than 30 cm and heavy rains fell in the normally semiarid areas.

A particularly striking cooling phase may be taken as one example. Around 3,600-3,200 B.P. (^{14}C) wind reversals to northerlies set in both in western Alaska and the Gulf of California. An expansion of the polar high-pressure cells seems to have occurred, accompanied by a southward shift of both the polar and subtropical frontal systems. In the Santa Barbara Basin a dramatic increase in the cold California Current is indicated. In the drought-sensitive White Mountains of eastern California there was an abrupt decrease in tree-ring widths (LaMarche, 1974). This was no local phenomenon but was repeated in the Hudson Bay area of Quebec (Filion, 1983), where the cold phases are marked also by forest fires and dune building. An abrupt sea-level drop (beach no. 75) is ^{14}C dated at 3,300 B.P.

In northwestern Europe this same cooling at 3,600-3,200 B.P. characterized the end of the warm and moist Atlantic (A-2) stage, a pollen-based chronozone, and the beginning of the Subboreal (SB-1). In the Alps there was a major glacier readvance (Löbben Neoglacial), and in Australia the high sea level of the Late Peron phase was terminated by a major drop [Crane Key Emergence; in Sweden, Mörner's (1976*a*) Postglacial Regression PR-7]. This catalog could be continued with evidence of this turning point at the end of the mid-Holocene from around the world. It is emphatically not a local or regional phenomenon. Climatic data from the Nile (Fairbridge, 1984, p. 182), confirmed by ancient Egyptian papyrus documentation, indicates catastrophic droughts and one of the cultural dark ages centered on 4,100 B.P. (sidereal year around 3,600 ^{14}C yr). The same centuries of drought led to the destruction of the Harappan civilization at Mohenjodaro in the Indus Valley (Bryson and Swain, 1981).

Hydro-isostasy and the Geoid

This is an appropriate point to return briefly to two interesting hypotheses, mentioned earlier, that have enjoyed some attention during the 1970s. Both are well reasoned and qualitatively justified, but (in my opinion) both have been grossly overworked and founded on grossly inadequate field data.

First, there is the principle of hydro-isostasy (Higgins, 1969), which rightly argues that postglacial unloading of the high-latitude glaciated regions involves loading of ocean floors by the redistributed water. However, the chronology of Clark and co-workers (1978) is without merit, and practical tests relative to the tilting of continental margins have failed completely (Faure et al., 1980). I can detect no systematic latitudinal difference on the levels of mid-Holocene high sea levels (3-4 m) on a global basis or between mid-oceanic and continental shores. It would appear that the viscosity assumptions of Clark and co-workers are somewhat in error, and that the loading adjustment is much slower than they had deduced. As Walcott (1975) suggested, the glacial marginal bulges are still slowly subsiding.

Second, there is Mörner's "geoidal eustasy" (1976*b*, 1986). He rightly points out that the satellite-derived global geoid displays certain regions of highly anomalous gravity. If these anomalous density sources lie in the core, and if the mantle and core are moving differentially in a fairly rapid manner, then we might reasonably expect that the shifting geoid anomalies would lead to the creation of numerous otherwise inexplicable raised beaches or drowned features. Quite the contrary, however, relatively stable coasts show no such anomalies, and neotectonic, paleotidal, and other mechanisms seem to be quite capable of explaining the observed data. Besides, the initial assumptions are questionable and the chronology weak. Mörner (1986) attempts also to redefine the widely accepted term *eustasy* to suit his convenience. Geoidal changes of sea level (due to change in mass distributions, spin rate, or polar axis) have been clearly recognized (Fairbridge, 1961*a*) but have nothing to do with eustasy.

Modern Tide-Gauge Data

A word or two is necessary concerning the use of modern tide-gauge data as a eustatic guide. In my 1961*a* paper (also in Fairbridge and Krebs, 1962) a tacit assumption was made that actualistic data should indicate global eustatic trends, provided that each tide station was carefully screened to remove neotectonic or accidental noise. Accordingly, in the analyses we removed the data from rapidly subsiding deltas, recent volcanic belts, and glacio-isostatic uplift areas. Curiously, however, our global average of MSL rise (1.2 mm/yr) was very similar to others obtained from nonscreened data. Twenty years later, Gornitz et al. (1982), using a vastly expanded data base, obtained exactly the same value. The agreement was most satisfactory, but unfortunately we both have been misled.

The twentieth-century rise in atmospheric CO_2 seems to call for a greenhouse-effect temperature rise, and a logical next step would be a eustatic sea-level rise, so 1-2 mm/yr seemed just about right. Concern about sea level is reinforced by global reports of beach erosion (Bird, 1984, 1985), but the whole construction is in my opinion quite flawed. This is not the place for in-depth criticism, but recall that CO_2 has large natural fluctuations relating to biological metabolism rates that are largely regulated by the climatically controlled trade winds and upwelling rates. Thus, the nonanthropogenic CO_2 is controlled primarily by climate and biology, not vice versa. Second, natural fluctuations of mean atmospheric temperatures are seen from the Holocene record to have periods of 350-1,000 yr, so a twentieth-century statistical base for global temperature rise is meaningless. Seawater temperature adds a steric component to sea level, but worldwide studies covering the interval 1856-1981, while indicating important fluctuations, suggest no secular change (Folland et al., 1984).

But then, why do most of the tide-gauge records indicate that sea level has been rising throughout the twentieth century? The rise of relative sea level is quite simply explained. More than 90% of the tide stations are located at river mouths, on estuaries and deltas, which are situations that are known geologically to be subsiding, either because of sediment compaction or tectonic downwarping (Pirazzoli, 1986). The only tide gauges that are far removed from plate boundaries, sediment loading, and glacio-isostatic influences are on mid-oceanic islands. Most of the Pacific atolls disclose no secular (eustatic) trends in MSL.

I had been thinking about this problem some years ago while serving as president of the INQUA Shorelines Commission. I devised a global survey that would help to place the tide-gauge data in their true perspective. To our more than 100 members in almost all coastal nations we sent circulars and gathered information that was later collated on a series of world maps (Colquhoun et al., 1981). For each locality (where possible), three numerical values were given: (1) the rate and sign of twentieth-century MSL change; (2) the elevation of the principal mid-Holocene shoreline (ca. 5,000 B.P.); and (3) the elevation of the principal last interglacial shoreline (Eemian, ca. 125,000 B.P.). Little consistency is shown except for points in very stable areas.

What these INQUA maps disclose are the general patterns of world coastal neotectonic displacement. As might be expected, the plate boundary, glacio-isostatic, and deltaic situations show big departures. Unfortunately, the localities with minimal displacement are for the most part also those with few tide gauges. Unfortunately, most of the tide stations are not in positions that can accurately reflect twentieth-century eustatic trends. Thus, it is evident that our whole data base for assuming a twentieth-century eustatic rise is hopelessly flawed.

A totally new methodological basis for tide-gauge analysis is now needed, one that will consider each station by itself, examining geodetic data, fresh-water discharge, salinity, temperature, ocean currents, Holocene levels, and Pleistocene levels. Only then will we be able to speak more confidently about twentieth-century eustatic trends.

CONCLUSION

1. Any change in MSL and/or local or relative sea level (RSL) on a more than 12-mo time scale involves a broad spectrum of variables; fourteen are specified here, in three categories:

(a) Sea-level change due to change in form and dimensions of basins, without variation in water volume (dynamic, hydro-isostatic, tectono-eustatic, geoidal, Archimedean—mainly factors affecting the capacity and shape of the basin and the displacement of water in it);
(b) Sea-level change due to variation in water volume (glacio-eustasy, hydrologic exchange rate of cycle, change in steric water quality, and salinity);
(c) Sea-level change due to local variation in the height of the land (glacio-isostasy, marginal bulge adjustment, extraglacial neotectonics, sedimentary loading and compaction, and seismic displacement).

2. The specialists involved in sea-level studies are from so many disciplines and adhere to so many unions and commissions that special efforts are needed by the scientific community to unite and integrate their studies.

3. Global information about MSL in the present century is grossly distorted by the numerical weighting toward northwestern Europe and eastern North America caused by the uneven distribution of tide gauges.

4. Averaging of tide-gauge results is further distorted because of the frequent location of gauges at river mouths, estuaries, and deltas where crustal loading and sediment compaction lead to constant subsidence. Thus, the global sea-level rise of the present century can be largely equated with subsidence, except in places such as mid-Pacific atolls.

5. Sea-level curves for the past 10,000 yr are now available for many parts of the world. Each is a regional curve and must be subjected to review and corrections in terms of neotectonics, compaction, paleotidal changes, oceanographic factors, and so on before a eustatic approximation can be established.

6. There is sufficient similarity between all the major curves and syntheses that a high degree of confidence can be placed on the timing of fluctuations. The amplitude of those fluctuations increases in areas of large tidal range (the Grant Effect) and decreases in subsidence areas (and disappears when $\Delta I = \Delta E$).

7. Palynological and other paleoecologic and stratigraphic studies prove conclusively that the major negative oscillations of Holocene sea level correspond to cooling climate trends. In higher latitudes, mean annual temperatures may be depressed by up to 5°C over intervals of several decades or even centuries.

8. The cold cycles are exceptional departures within the generally warmer-than-present envelope of Holocene interglacial temperature mode, occurring at intervals of around 350-1,000 yr.

9. Improved chronology in Holocene sea-level and climatic time series is greatly needed, where at present a radiocarbon-dating error of up to ±250 yr is sometimes present. New chronologies on a year-by-year basis are now being provided by tree rings, varves, and ice cores. A complicated task now lies ahead: an expansion and refining of the annual chronologic bases and their integration with the various MSL and climatic time series.

10. The provision of a ^{14}C flux rate for nearly the entire Holocene through the dendrochronological series provides the opportunity of using this value as an inverse measure of solar activity. The predictive potential of planetary motions in helping to explain the latter is treated in Chapter 26.

REFERENCES

Ahlmann, H. W., 1953, *Glacier Variations and Climatic Fluctuations*, Bowman Memorial Lecture Series, Washington, D.C.: American Geographical Society, 51p.

Beets, D. J., et al., 1986, The impact of changes in tidal range on barrier systems at rising sea level: An example from the western Netherlands, *I.G.C.P. 200 Sea-Level Conference, Cork.* (abstract).

Bird, E. C. F., 1984, *Coasts: An Introduction to Coastal Geomorphology*, 3rd ed., Oxford: Basil Blackwood, 320p.

Bird, E. C. F., 1985, *Coastline Changes—A Global Review*, Chichester, N.Y.: Wiley-Interscience, 219p.

Bloom, A. L., 1977, Atlas of sea-level curves, *International Geological Correlation Project 61*, Ithaca, N.Y.: Cornell University, Department of Geological Sciences, 114p.

Bryson, R., and Swain, A. M., 1981, Holocene variations of monsoon rainfall in Rajasthan, *Quat. Research* **16:**135-145.

Cato, I., 1985, The definitive connection of the Swedish geochronological time scale with the present, and the new date of the zero year in Doviken, northern Sweden, *Boreas* **14:**117-122.

Chappell, J., 1983, A revised sea level record for the last 300,000 yrs from Papua New Guinea, *Search* **4:**99-101.

Clark, J. A.; Farrell, W. E.; and Peltier, W. R., 1978, Global changes in postglacial sea level: A numerical calculation, *Quat. Research* **9:**265-287.

Colquhoun, D. J., et al., 1981, *World Shorelines Map*, 3 parts, Columbia, S.C.: INQUA Commission on Quaternary Shorelines.

Creer, K. M., 1981, Long period geomagnetic secular variations since 12,000 yr B.P., *Nature* **292:**208-212.

Curray, J. R., and Moore, D. G., 1964, Holocene regressive littoral sand, Costa de Nayarit, Mexico, in L. M. J. U. Van Straaten (ed.), *Deltaic and Shallow Marine Deposits*, Amsterdam: Elsevier, pp. 76-82.

Daly, R. A., 1920, A recent worldwide sinking of the ocean level, *Geol. Mag.* **57:**246-261.

Denton, G. H., and Karlén, W., 1973, Holocene climatic variations, their pattern and possible courses, *Quat. Research* **3:**155-205.

Eddy, J. A., 1977, Climate and the changing sun, *Climate Change* **1:**173-190.

Fairbridge, R. W., 1948, Notes on the geomorphology of the Pelsart group of the Houtmans Abrolhos Islands, *Royal Soc. Western Australia Jour.* **33** (for 1946/1947):1-43.

Fairbridge, R. W., 1950, The geology and geomorphology of Point Peron, Western Australia, *Royal Soc. Western Australia Jour.* **34**(for 1947/1948):35-72.

Fairbridge, R. W., 1958, Dating the latest movements of the Quaternary sea level, *New York Acad. Sci. Trans.* **20**(Ser. 2):471-482.

Fairbridge, R. W., 1960, The changing level of the sea, *Sci. American* **202**(5):70-79.

Fairbridge, R. W., 1961*a*, Eustatic changes in sea-level, in L. H. Ahrens, F. Press, K. Rankama, and S. K. Runcorn (eds.), *Physics and Chemistry of the Earth,* vol. 4, London: Pergamon Press, pp. 99-185.

Fairbridge, R. W., 1961*b*, Convergence of evidence on climatic change and ice ages, *New York Acad. Sci. Annals* **95**(1):542-579.

Fairbridge, R. W., 1962, World sea-level and climatic changes, *Quaternaria* (Rome) **6:**111-134.

Fairbridge, R. W., 1976, Shellfish-eating preceramic indians in coastal Brazil, *Science* **191:**353-359.

Fairbridge, R. W., 1981*a*, Holocene wiggles, *Nature* **292:**670.

Fairbridge, R. W., 1981*b*, Holocene sea-level oscillations, in L. K. Königsson and K. Paabo (eds.), *Striae, Florilegium Florinis Dedicatum,* vol. 14, pp. 131-141.

Fairbridge, R. W., 1983, Isostasy and eustasy, in D. Smith and A. Dawson (eds.), *Shorelines and Isostasy,* London: Academic Press, pp. 3-25.

Fairbridge, R. W., 1984, The Nile floods as a global climatic/solar proxy, in N. A. Mörner and W. Karlén (eds.), *Climatic Changes on a Yearly to Millennial Basis,* Dordrecht: Reidel, pp. 181-190.

Fairbridge, R. W., 1987, Spectrum of lunar-solar tides in relative sea-level fluctuations, *Soc. Econ. Paleontologists and Mineralogists Spec. Publ.,* in press.

Fairbridge, R. W., and Finkl, C. W., Jr., 1980, Cratonic erosional unconformities and peneplains, *Jour. Geology* **88:**69-86.

Fairbridge, R. W., and Hillaire-Marcel, C., 1977, An 8,000-yr paleoclimatic record of the 'Double-Hale' 45-yr solar cycle, *Nature* **268:**413-416.

Fairbridge, R. W., and Krebs, O. A., 1962, Sea level and the southern oscillation, *Geophys Jour.* **6:**532-545.

Fairbridge, R. W., and Lowrie, A., 1987, Mississippi delta-lobe switching during Holocene eustatic fluctuations, *Coastal Sediments '87* (submitted).

Faure, H.; Fontes, J. C.; Herbrard, L.; Monteillet, J.; and Pirazzoli, P. A., 1980, Geoidal change and shore-level tilt along Holocene estuaries: Senegal River area, West Africa, *Science* **210:**421-423.

Filion, L., 1983, Thèse. Université Laval, Department of Géographie, Québec. (summary in *Nature* **309:**543-546, 1984.)

Florin, S., 1944, Havsstrandens förskjutningar och bebyggelsseutvecklingen i östra Mellansverige under senkvartär tid, *Geol. Fören. Stockholm Förh.* **66:**551-624.

Folland, C. K.; Parker, D. E.; and Kates, F. E., 1984, Worldwide marine temperature fluctuations, *Nature* **310:**670-673.

Geyh, M. A., 1980, Holocene sea-level history: Case study of the statistical evaluation of ^{14}C dates, *Radiocarbon* **22**(3):695-704.

Gornitz, V.; Lebedeff, S.; and Hansen, J., 1982, Global sea level trend in the last century, *Science* **215:**1611-1614.

Grant, D. R., 1970, Recent coastal submergence of the Maritime Provinces, *Canadian Jour. Earth Sci.* **7:**679-689.

Guiot, J., 1985, *Réconstructions des champs thermiques et barométriques de la région de la Baie d'Hudson depuis 1700,* Donnesview, Ontario: Environement Canada SEA.

Heusser, L. E., 1978, Marine pollen in Santa Barbara Basin, California, *Geol. Soc. America Bull.* **89:**673-678.

Heyworth, A., and Kidson, C., 1982, Sea-level changes in southwest England and Wales, *Geologists' Assoc. Proc.* **93:**91-111.

Hicks, S., et al., 1983, *Sea Level Variations for the United States 1855-1980,* Rockville, Md.: NOAA/NOS, 170p.

Higgins, C. G., 1969, Isostatic effects of sea-level changes, in *Quaternary Geology and Climate,* Washington, D.C.: National Academy of Science, pp. 141-145.

Hillaire-Marcel, C., 1976, Déglaciation et relèvement isostatique à l'est de la Bai d'Hudson, *Cahiers de Géographie Québec* **20:**185-220.

Kuenen, P. H., 1933, *Geology of Coral Reefs, The Snellius Expedition,* vol. 5, *Geological Results, Part 2,* Utrecht: Kemink en Zoon, N.V.

LaMarche, V. C., Jr., 1974, Paleoclimate inferences from long tree-ring records, *Science* **183:**1043-1048.

Meier, M., 1984, Contribution of small glaciers to global sea levels, *Science* **226:**1418-1421.

Moore, G. W., and Giddings, J. L., 1962, Record of 5,000 years of Arctic wind direction recorded by Alaskan beach ridges, *Geol. Soc. America Spec. Paper 68,* 232p.

Mörner, N. A., 1969, The late Quaternary history of the Kattegatt Sea and the Swedish West Coast; Deglaciation, shore level displacement isostasy and eustasy, *Sveriges Geol. Undersökning* Års. **63**(3):487.

Mörner, N. A., 1976*a*, Eustatic changes during the last 8000 years in view of radiocarbon calibration and new information from the Kattegatt region and other northwestern European areas, *Palaeogeography, Palaeoclimatology, Palaeoecology* **19:**63-85.

Mörner, N. A., 1976*b*, Eustasy and geoid changes, *Jour. Geology* **84:**123-151.

Mörner, N. A., 1986, The concept of eustasy, *Jour. Coastal Research,* **1**(spec. issue):49-51.

Newman, W. S.; Fairbridge, R. W.; and March, S., 1971, Marginal subsidence of glaciated areas: United States, Baltic and North Seas, in M. Ters (ed.), *Etudes sur le Quaternaire dans le Monde*, Paris: VIII INQUA (1969), pp. 795-801.

Pirazzoli, P., 1986, Secular trends of relative sea-level (RSL) changes indicated by tide-gauge records, *Jour. Coastal Research* **1**(spec. issue):1-26.

Pisias, N. G., 1978, Paleoceanography of the Santa Barbara Basin during the last 8,000 years, *Quat. Research* **10**(3):366-384.

Richards, H. G.; Gallagher, W. B.; and Colquhoun, D. J., 1986, Annotated Bibliography of Quaternary Shorelines, Fourth Supplement 1978-1983, *Jour. Coastal Research* (Spec. Issue No. 2).

Schofield, J. C., 1973, Post-glacial sea levels of Northland Auckland, *New Zealand Jour. Geology and Geophysics* **16**(3):359-366.

Schofield, J. C., 1980, Postglacial transgressive maximum and second-order transgressions of the Southwest Pacific Ocean, in N. A. Mörner (ed.), *Earth Rheology, Isostasy and Eustasy*, Chichester, N.Y.: Wiley-Interscience, pp. 517-521.

Schove, D. J., 1978, Tree-ring and varve scales combined, c. 13,500 B.C to A.D. 1977, *Palaeogeography, Palaeoclimatology, Palaeoecology* **25:**209-233.

Schove, D. J. (ed.), 1983, *Sunspot Cycles*, Benchmark Papers in Geology Series, vol. 68, Stroudsburg, Pa.: Hutchinson Ross Publ. Co., 393p.

Shennan, I.; Tooley, M. J.; Davis, M. J.; and Haggart, B. A., 1983, Analysis and interpretation of Holocene sea-level data, *Nature* **302:**404-406.

Stacey, F. D., 1977, *Physics of the Earth*, 2nd ed., New York: Wiley, 324p.

Stapor, F. W., 1975, Holocene beach ridge plain development, north-west Florida, *Zeitschr. Geomorphologie* (Suppl. 22):116-144.

Stapor, F. W., and Tanner, W. F., 1977, Late-Holocene near sea-level data from St. Vincent Island and the shape of the late Holocene near sea level curve, in W. F. Tanner (ed.), *Coastal Sedimentology*, Tallahassee, Fla.: Florida State University, pp. 35-68.

Stuiver, M., and Quay, P. D., 1980, Changes in atmospheric carbon-14 attributed to a variable Sun, *Science* **207:**11-19.

Suess, H. E., 1978, La Jolla measurements of radiocarbon in tree-ring dated wood, *Radiocarbon* **20:**1-18.

Suess, H. E., 1980, The radiocarbon record in tree rings of the last 8,000 years, *Radiocarbon* **22:**200-209.

Ters, M., 1973, Les variations du niveau marin depuis 10,000 ans, le long du littoral Atlantique Français, *Bull. Assoc. Française du Quaternaire* **36**(INQUA Suppl.):114-135.

Tooley, M. J., 1978, *Sea-level Changes in Northwest England during the Flandrian Stage*, Oxford: Oxford University Press, 232p.

Vail, P. R.; Mitchum, R. M.; and Thompson, S., 1977, Global cycles of relative change of sea level, *Am. Assoc. Petroleum Geologists Mem.* **26:**83-97.

Van de Plassche, O., 1982, Sea-level change and water-level movements in the Netherlands during the Holocene, *Med. Rijks Geol. Dienst*, **36**(1), 93p.

Van de Plassche, O., 1986, *Sea-level Research: Manual for Collection and Evaluation of Data*, Norwich (U.K.): GeoBooks, 600p.

van Veen, J., 1954, Tide-gauges, subsidence-gauges and flood-stones in the Netherlands, *Geologie en Mijnbouw* (n.s.) **16:**214-219.

Walcott, R. I., 1975, Recent and late Quaternary changes in sea level, *EOS* **56**(2):62-71.

Wise, D. U., 1974, Continental margins, freeboard and the volumes of continents and oceans through time, in C. A. Burke and C. L. Drake (eds.), *The Geology of the Continental Margins*, Berlin: Springer-Verlag, pp. 45-68.

Woods, P. J., 1983, Evolution of, and soil development on, Holocene beach ridge sequences, West Coast, Western Australia, Ph.D. dissertation, University of Western Australia.

Zhao, X. T.; Zhang, J. W.; Jiao, W. J.; and Li, G. Y., 1980, Chenier ridges on the west coast of the Bohai Bay, *Kexue Tongbao* (Beijing) **25**(2):243-247.

8: New Evidence for Eustatic Components in Late Holocene Sea Levels

Donald J. Colquhoun and Mark J. Brooks
University of South Carolina

Abstract: Changes in Holocene sea level through time have been assigned to eustatic, isostatic, and neotectonic processes. Eustatic changes demand global expression through linkage in world climate or changes in ocean water or basin volume, while isostatic adjustment and neotectonic distortion involve regional or local geophysical parameters. A Holocene sea-level curve is being developed for the southeastern United States through stratigraphic study in existing marshes and archeologic study in both marsh and interriverine areas. The curve is evaluated both regionally and through local intensive investigations in light of regional and local tectonic elements. Both major and some minor trends in sea-level change are shown to include significant eustatic components.

INTRODUCTION

The interrelationship between sea level and the global climatic systems has been accepted for many years and is particularly obvious in comparing glacial and interglacial stages. Almost all the several Plio-Pleistocene high sea levels that are recorded on the middle and lower coastal plains of the southeastern United States record warm-water conditions and are thought to represent interglacial time. Their colder counterparts are missing in the coastal plain stratigraphic record, except in lakes. Such periods are represented by terrestrial erosional unconformities formed during times of low sea-level stand in the glacial stages and are preserved by burial associated with the subsequent interglacial submergence.

Fairbridge (1961), in presenting global evidence for smaller Holocene sea-level fluctuations, suggested that these minor changes in sea level could also have been the result of changes in the global climatic system. Such changes could have resulted from variation in the ratio of glacial-ice to ocean-water mass or from change in ocean-water volume dependent on temperature fluctuation (Gornitz, Lebedeff, and Hansen, 1982). These small-scale changes would be of eustatic origin and should be reflected on detailed sea-level change curves constructed in many areas of the world. In fact, they are not (see, e.g., Bloom, 1977), being only rarely noted, and often in discordance in time of development. As a result of these observations, an assignation of isostatic adjustment (see, e.g., Clark, Farrell, and Peltier, 1978) or neotectonics (see, e.g., Mörner, 1976) has been suggested. Such changes would not be expected to be time coincident on the world's coasts or to be of similar magnitude.

Small-scale sea-level changes have been shown to have occurred in middle and late Holocene times by the geological and archeological records from the southeastern United States and herein are shown to have significant though unmeasured eustatic components. In-

This paper has been adapted from D. J. Colquhoun and M. J. Brooks, 1986, New evidence from the southeastern United States for eustatic components in the late Holocene sea levels, *Geoarchaeology* **1**:275-291, with permission from the publisher, John Wiley & Sons, Inc.

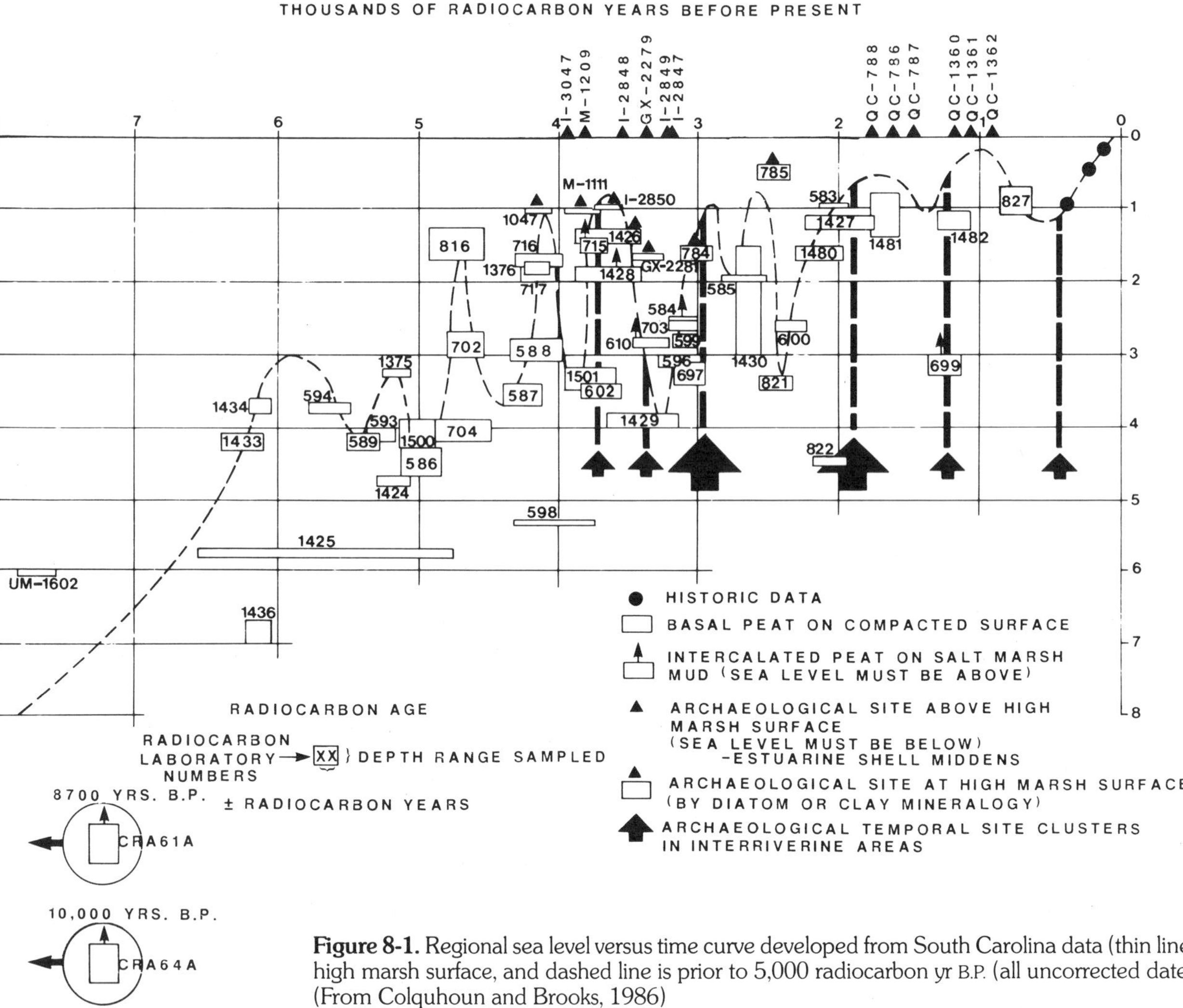

Figure 8-1. Regional sea level versus time curve developed from South Carolina data (thin line is high marsh surface, and dashed line is prior to 5,000 radiocarbon yr B.P. (all uncorrected dates). (From Colquhoun and Brooks, 1986)

vestigations of Holocene-age changes in sea level in the southeastern United States have been made in Georgia (Howard and de Pratter, 1980; de Pratter and Howard, 1977) and South Carolina (Colquhoun et al., 1980).

The sea level versus time curve developed for South Carolina (Fig. 8-1) is based on a variety of sea-level indicators derived from coastal marsh stratigraphy and from archeology of both interriverine and coastal estuarine environments. These indicators and the linkages to sea level are reported in detail later in the chapter. The curve is based on data obtained from coastal localities between about 32° and 34°N latitude and has been updated periodically (Brooks et al., 1979; Colquhoun et al., 1980, 1981; Colquhoun, 1982). The regional tectonic framework of Cretaceous and Tertiary lithostratigraphic units is reasonably well known (Colquhoun et al., 1983). Microseismic analysis by Talwani (1982) delineates activity in the area during the interval 1970-1980. Poley (1984) has summarized releveling data during the same period. Two summary volumes, *U.S. Geological Survey Professional Papers 1028* (Rankin, 1977) and *1313* (Gohn et al., 1983), summarize deep drilling, stratigraphic, biostratigraphic, seismic, gravity, and magnetic observations in the region. Tectonic elements lie athwart and in the vicinity of localities used in regional curve construction, and thus neotectonics, in addition to both eustatic changes and isostatic adjustments, may be present.

To distinguish neotectonic components possibly present in the regional curve, ongoing intensive investigations were initiated in 1981. Three areas were selected for study north of, as well as within, all known major tectonic features (Fig. 8-2). These areas included the Pee Dee River Valley and Winyah Bay in the northern coastal zone (lat. 33°30′N); the Cooper, Wando, Ashley, and Stono valleys, which lie central (lat. 32°45′N); and the North and South Edisto valleys to the south (lat. 32°30′N). Data reported for the statewide study are supported in part by unpublished intensive investigations in individual estuaries.

SEA-LEVEL CHANGE CURVE

The present assessment of sea-level change for the region (see Fig. 8-1) includes data collected from South Carolina through summer 1983. The curve possesses the following characteristics:

1. rapid general rise from prior to 10,000 radiocarbon yr to approximately 6,000 radiocarbon yr B.P. (uncorrected for secular changes in initial activity)
2. slow general rise since about 6,000 B.P.
3. small oscillations (positive and negative, on the order of ±1 m or less, and of irregular frequency, 300-500 yr) superimposed on the general rise in sea level.

HOLOCENE STRATIGRAPHY

Late Pleistocene through middle Holocene rivers cut valleys through Plio-Pleistocene, Tertiary, and Cretaceous sediments to maximum depths greater than 20 m within the three study areas. Intertidal (by associated lithologies), freshwater (by diatom analysis) peats that were deposited on compacted older sediments are common within 5 m of the present high marsh surface but are rare at deeper levels. These peats have yielded ages between about 6,300 B.P. and modern (for all ^{14}C dates and sources, see Table 1 of Colquhoun and Brooks, 1986).

Few data exist for older or deeper occurrences of peat or wood. Ruby (1981) reported one date of 10,000 B.P. at a depth of about −9.5 m. The material was fresh water. The depth is only a maximum for the high marsh surface. He reported a second date of 8,700 B.P. at a depth of approximately −8.5 m. Moslow and Colquhoun (1981) report a date of about 7,000 from −4 m. These data, when coupled with the associated upper and lower delta plain sediments in which these peats were deposited indicate a rapid rise in sea level from prior to about 10,000 B.P. until about 6,000 B.P.

A change from rapid sea-level rise to slow rise occurred near 6,000 B.P. In Figure 8-1 a high stand (but below present sea level) is

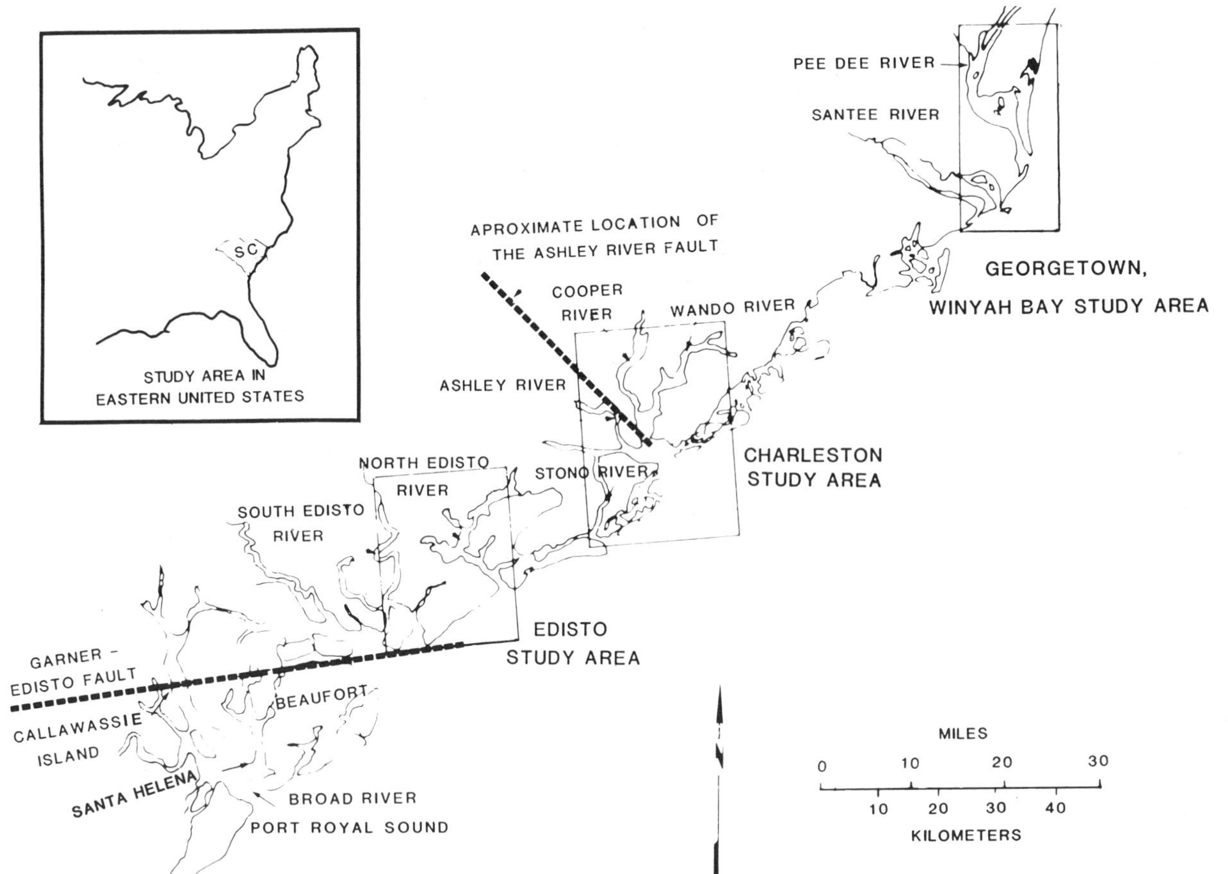

Figure 8-2. Location of three intensive study areas (marsh stratigraphy and archeology) in relation to active tectonic elements discussed in the text. (From Colquhoun and Brooks, 1986)

recorded at 4,800 B.P. based on one date from the Pee Dee River Valley. A second high stand is recorded about 4,200 B.P. by lag shell deposits, (Ruby, 1981) near Cape Romaine in the north (lat. 33°N), by Moslow and Colquhoun (1981) for Kiawah and Seabrook islands in the south-central area, and by McCants (1982) for the South Edisto Estuary. The date has also been reported from intertidal fresh-water peat data from several areas (Brooks et al., 1979; Colquhoun et al., 1981). These two high stands occur within 1-2 m of the present high marsh surface and indicate stability or, at the most, slow general rise in sea level, occurring through to the present.

The locations for the 4,200 B.P. observations occur in sediments either side of known active tectonic elements in the region (see Fig. 8-2). Cape Romaine lies north of and near the margin of Southeast Georgia Embayment downwarping and north of the Charleston Fault Area, active in historic time, while Kiawah and Seabrook islands and the North and South Edisto valleys lie south of that faulting, within the area of Southeast Georgia Embayment downwarping and on the northern (upward or stable) block of the Garner-Edisto Fault. Shelf and coastal plain widths being comparable for the entire area (32°-34°N lat.), the break in rate of rise is therefore eustatic in nature, with isostatic adjustment.

The general rise in sea level (see Fig. 8-1) from regional data appears to be characterized by small (±1 m) fluctuations in sea level at irregular intervals of 300-500 yr since at least 5,300 B.P. (i.e., that time interval where sufficient data are present to express them). Fluctuations probably occurred prior to 5,300 as well, though data are insufficient to show them. Some of these regionally determined small fluctuations coincide with similar conditions reported from the lowlands of northwestern Europe (Colquhoun et al., 1981), strongly suggesting that these changes were also eustatic. However, the data used to construct the South Carolina curve are drawn from areas that may contain neotectonic components associated with both downwarping and faulting. Data based on age and elevation of intertidal fresh-water peats are not sufficient to demonstrate coincidence in sea-level curves in each of the estuarine systems being intensively studied, and thus the magnitude of the eustatic components is unknown. Archeological data (following section) have indicated a coincidence between time of sea-level rise (submergence), as indicated by marsh stratigraphy, and increases in the number of interriverine archeological sites in the Cooper River area (Brooks et al., 1979). Intensive study of interriverine areas associated with the Pee Dee River and Winyah Bay (lat. 33°30′N); the Cooper, Wando, and Stono rivers (lat. 32°45′N); and the North Edisto Estuary (lat. 32°30′N) shows similiar temporal clusterings of archeological sites (Colquhoun, 1982, 1984). Thus, though coincident small-scale fluctuations in sea level cannot be differentiated in each estuary through marsh stratigraphy, changes in sea level are evidenced by the archeology of interriverine areas along 160+ km of the coastal zone. Coincident changes from sets of sites lying athwart known tectonic elements indicate that a eustatic component in sea-level change is present, though the small magnitude of that component has not been measured in each estuary.

ARCHEOLOGY AND SEA LEVEL

The prehistoric subsistence settlement variability recorded for the lower Coastal Plain of South Carolina, and generally for coastal areas of the southeastern United States, is linked causally to cultural, biological, and physical environmental processes and their interactions. The noncultural (physical and ecological) environmental factors provide the linkage to Holocene sea-level changes and, in part, appear to be regional manifestations of global climatic patterns.

Estuarine Sites

Identified archeological sites located within existing estuarine areas and dating from around 12,000 B.P. to 5,000 B.P., early in human occupance of the region, represent riverine or interriverine sites lacking shell middens. These sites were established prior to local estuarine development. The relatively few known sites of this period that are located within estuarine areas are located in the more erosion-resistant areas, usually at considerable distance up river valleys (Fig. 8-3). Other such sites presumably have been destroyed and/or drowned by marine transgression. Firm archeologic evidence exists for the initial development of the estuarine systems by 4,200 B.P. The earliest known marine shell midden deposits on the South Carolina coast date to this time (e.g., Calmes, 1967; Combes, 1975; Hemmings, 1970; Michie, 1973, 1974, 1976; Sutherland, 1974; Trinkley, 1975). Large shell middens dating from between 4,200 B.P. and 3,300 B.P. are typically located in the seaward parts of estuaries, usually adjacent to major channels (see, e.g., Fig. 8-3). Many of these deposits have been extensively eroded by subsequent sea-level transgression. Within present salt marshes the bases of early shell middens extend to 0.80-1.20 m below the existing high marsh surface (Colquhoun et al., 1981; unpublished data in files of the South Carolina Institute of Archeology). Many of these early sites contained a broad range of estuarine and terrestrial subsistence remains that, in conjunction with the considerable diversity of artifact assemblages, may indicate rather intensive multiseasonal habitation. Between 3,300 B.P. and 1,000 B.P. there was a general trend for new shell middens to occur farther inland and upslope within the estuaries, correlating with the slow sea-level rise (see Fig. 8-1) and associated estuarine expansion. By extension from de Pratter and Howard's (1977) work on the northern Georgia coast, many sites established on the South Carolina coast during interpreted sea-level regressions and low stands in the interval around 3,300-2,500 B.P. may also be submerged beneath present marshes or buried under more recent deposits seaward of the present shoreline, if not actually destroyed by subsequent transgressions. By about 2,000 B.P., shell midden sites tended to become noticeably

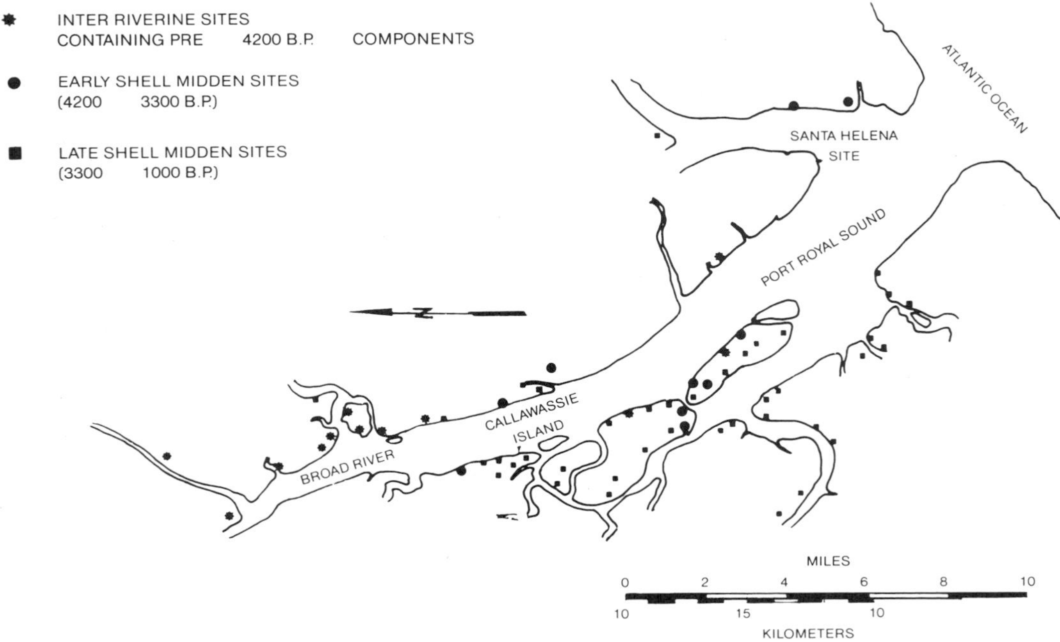

Figure 8-3. Locations of early and late shell middens and interriverine archeologic sites in relation to Broad River and Port Royal Sound, South Carolina. Callawassie Island and Santa Helena are discussed in the text in relation to Little Ice Age sea-level change. (From Colquhoun and Brooks, 1986; after Michie, 1980)

smaller, more numerous, and more dispersed. Shell middens in the 2,000-1,000 B.P. age range are usually located adjacent to existing small tidal creeks (the midden bases are above or just slightly below the present high-marsh surface) and/or on slightly higher ground along existing estuarine margins (Scurry and Brooks, 1980; Michie, 1980; Colquhoun et al., 1981; unpublished data, files of D. Colquhoun and M. Brooks, and the South Carolina Institute of Archeology).

Upland Interriverine Sites

Upland interriverine sites probably represent primarily the exploitation of acorns, hickory nuts, and deer. These sites are numerous, typically small, widely dispersed, and contain monotonously similar artifact assemblages of low diversity, suggesting that the sites represent short-term, seasonal utilization involving a narrow range of procurement activities. Specific archeological, ecological, and ethnohistorical supportive data for the inferred functions of these sites may be found in Brooks and Scurry (1978).

Archeological survey data from interriverine locales adjacent to estuaries indicate a similar temporal pattern to that of the estuarine areas in terms of relative frequency of sites before and after about 4,200 B.P. Sites younger than 4,200 B.P. consistently account for 75% to, in some instances, 100% of the sites discovered in given study areas, and most date younger than 2,000 B.P. Data and arguments have been presented elsewhere relating the relative abundance and geographical distribution of these sites through time to, in part, wetter conditions indirectly related to sea-level changes and accompanying estuarine development and expansion (Brooks, 1980; Colquhoun et al., 1981; Brooks and Canouts, 1981). Thus, similar to the estuarine areas in mid- and late Holocene times, the trend in upland interriverine areas was toward smaller, more numerous, and more dispersed sites, which is attributed to hydrologic changes that resulted in increased dispersion of habitats containing high-density, relatively lower-cost subsistence resources.

TECTONIC ELEMENTS

Major tectonic elements in the region are shown in Figure 8-4, a structure contour map drawn on a middle/late Eocene unconformity that is usually easily distinguished on electric and gamma-ray logs. The map is part of a series of 19 developed to illustrate the surface and subsurface stratigraphy and structure of the South Carolina Coastal Plain Province (Colquhoun et al., 1983). The Cape Fear Arch is present north of latitude 33°30′ as a mildly upwarping to relatively stable tectonic shelf. South of latitude 33°N the Southeast Georgia Embayment is encountered, downwarping from at least the Thanetian to the present. Near latitude 32°45′N the Charleston Fault Area occurs (Lennon, 1985). Initiated in at least the Lutetian, it has remained active to the present.

The three intensive study areas (see Fig. 8-2) were chosen to include and to bracket these tectonic elements in which movement could be determined since middle Eocene, late Eocene, late Oligocene, early Pleistocene, late Pleistocene, middle and late Holocene, A.D. 1886 (the Charleston earthquake), and the period from 1970 to 1985. The field program of coring and drilling for stratigraphic data and excavation and mapping for archeologic data was selected to compare changes of sea level in magnitude and timing of upwarping, downwarping, faulting, and neotectonically distorting blocks. It is assumed that coincident minor changes in sea level, as indicated by indicators in all three areas, and of the same movement sense are manifestations of eustatism.

EUSTATIC CHANGES IN GENERAL SEA LEVEL

The rapid rise in sea level from prior to 10,000 B.P. to about 6,000 B.P., changing to slow rise since about 6000, is eustatic with isostatic adjustment. The change is supported by stratigraphic observations on basal lag shell deposits both at Cape Romaine in the north (lat. 33°N), under Kiawah and Seabrook islands, and in the North

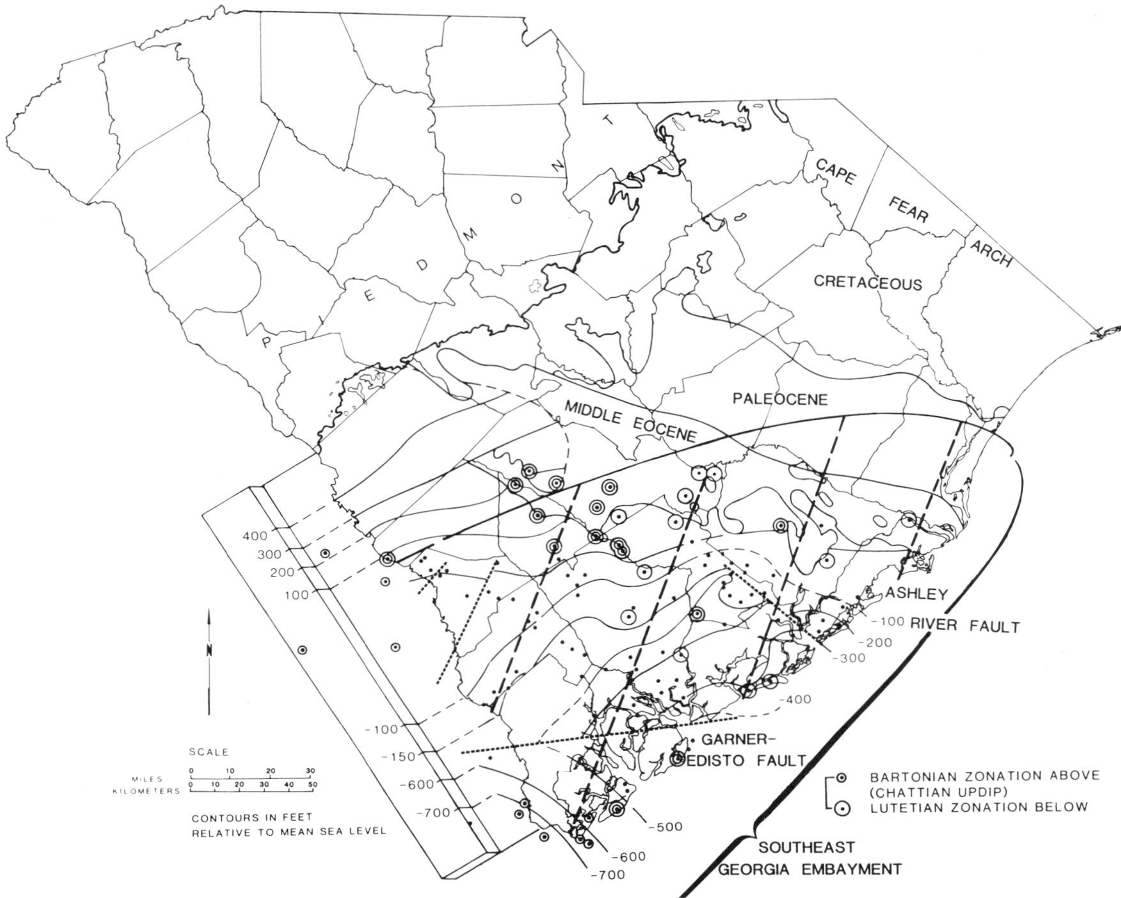

Figure 8-4. Contours drawn on the middle/late Eocene (Lutetian-Priabonian/Bartonian?) unconformity in South Carolina, showing Cape Fear Arch, Southeast Georgia Embayment, and Ashley River and Garner-Edisto Fault trends. Small dots indicate locations of geophysically logged water wells (gamma and electric logging). (From Colquhoun and Brooks, 1986)

and South Edisto estuaries (lat. 32°30′N) and is indicated by basal intertidal fresh-water peat dating in all areas.

The slow observed rise in sea level since 6,000 B.P., indicated by peaks of submergence and emergence illustrated in Figure 8-1, supports the prediction of Clark and co-workers (1978) in postulating a mildly negative coastal zone in the southeastern United States. The isostatic adjustment is observed to occur in all study areas based on peat dating. In addition to these marsh stratigraphic data, similar conclusions with respect to general rise in sea level are supported by estuarine midden data: the earlier (4,200 B.P. to ca. 3,300 B.P.) middens being more submerged (1.2-0.8 m) than the later (2,000 B.P. to ca. 1,000 B.P.) middens, which lie slightly above to slightly submerged with respect to the high marsh.

EVALUATION OF MINOR CHANGES IN HOLOCENE SEA LEVEL

The minor oscillations in middle and late Holocene sea level illustrated in Figure 8-1 have not been observed in detailed marsh stratigraphic studies within the three areas, with the exception of the 4,200 B.P. and 300 B.P. rises in sea level. Sufficient sampling to yield reliable dates has not been accomplished in spite of over 600 cores currently logged.

Brooks et al. (1979), in examining the marsh stratigraphy and interriverine site archeology of the AMOCO property along the Cooper River Valley, near Charleston, South Carolina, noted an interrelationship. Interriverine sites tended to cluster in time, rather than being random, and the clustering seemed to occur during times of sea-level rise, as indicated by marsh stratigraphy (see Fig. 8-1).

To test the clustering, detailed surveys were conducted within four additional areas, positioned with respect to tectonic elements: (1) north of Winyah Bay (lat. 33°15′N) on the flank of the Cape Fear Arch, (2) near the confluence of the Cooper and Wando rivers (lat. 32°45′N) on the northern side of the Ashley River Fault, (3) along the Stono River (lat. 32°45′N) on the south side of the fault and less than 5 km from area 2, (4) south of the North Edisto River (lat. 32°35′N) on the landward block of the Garner-Edisto Fault (see Fig. 8-2).

All areas were located on well-drained former barrier islands of late Pleistocene age. All areas were rural and, because of soil quality, experienced little or no agriculture since the 1700s.

The clustering of dates obtained from interriverine sites surveyed in the four areas is apparent (Fig. 8-4). If the interrelationship between clusters and times of sea-level rise noted at the AMOCO site is regionally valid, then the variation in number of sites discovered in these areas may be explained in light of both eustatism and neotectonism. The results of that comparison conform with known tectonic history of the structural elements.

TECTONIC HISTORY OF STRUCTURAL ELEMENTS

The Cape Fear Arch (see Fig. 8-4) first appeared in Danian time as a mildly positive tectonic element characterized by slightly shallower water deposition (by lithofacies) in Danian, Thanetian, and Chattian time. At other Tertiary time, it is not expressed in regional lithofacies maps. Tertiary sediments, with the exception of Pliocene, are eroded from its crest and lie banded along its southern flank.

The Southeast Georgia Embayment occurs southwest of the southern flank of the Cape Fear Arch. It appears first as a downwarping region during the Thanetian and seems to have been downwarping at all Tertiary stages save the Lutetian, when general faulting occurred along the Ashley River and Garner-Edisto trends. The Ashley River tectonic trend was reported by Talwani (1982) based on microseismic evidence as a deep (4-5 km), steeply dipping feature extending along the Ashley River northwest from Charleston for 10-15 km. Colquhoun (1982) noted a fault cutting Cretaceous and Tertiary strata a few kilometers to the northeast extending from Charleston for 40 km or more to the northwest.

Measurable activity associated with the fault occurs when regional unconformities are examined. These include the offset between the Lutetian-Paleocene boundary of up to 35 m with the southwest side of the fault down and the offset between the Priabonian (Bartonian?)-Lutetian boundary of up to 20 m with the southwest side of the fault down.

Younger pre-Pliocene unconformities have too high regional relief locally to allow estimation of fault displacement, although stratigraphically, the introduction of additional Chattian and Miocene units suggests continued downthrow to the southwest. Stratigraphic evidence indicates sporadic movement in the early Pleistocene. The Wicomico Formation, which exists as a single submergence-emergence unit to the north of the Santee River, is subdivided into two broad barrier island and marsh terraces (Wicomico and Penholoway of Cooke, 1936), the latter as a spitlike feature in the north, which rapidly thins to lie against the Wicomico Terrace immediately south of the projected Ashley River Fault, near St. George, South Carolina (Colquhoun, 1967). Such stratigraphy can be explained by upthrusting north of the fault during marine submergence, while

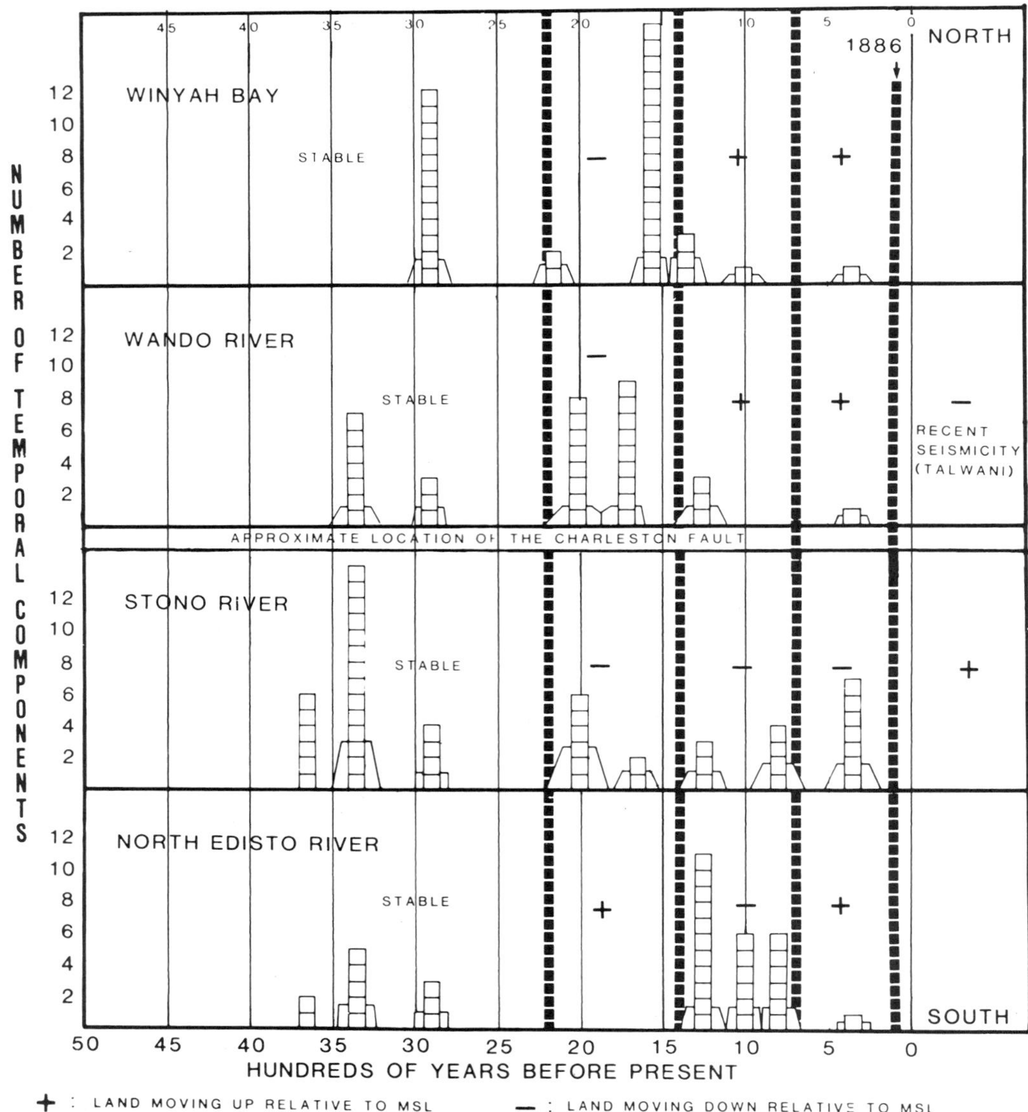

Figure 8-5. Interriverine site data indicating neotectonic and eustatic changes in sea level. (From Colquhoun and Brooks, 1986)

Winyah Bay is on the southern flank of the Cape Fear Arch. Wando River is on the northeastern block (upward moving or stable) of the Ashley River Fault. Stono River is on the southwestern block (downward moving) of the Ashley River Fault. North Edisto River is on the northern block (upward moving or stable) of Garner-Edisto Fault, within the Southeast Georgia Embayment.

the shoreline rested at the Dorchester Bar. The resulting shallowing then allowed reworking of shelf sediment, which migrated landward to form the Penholoway Terrace. The feature is unique when one examines geomorphic maps of South Carolina (Colquhoun, 1969) and Georgia (Hoyt and Hails, 1970).

Movement on the Ashley River Fault has not been demonstrated during the late Pleistocene deposition of the Talbot or Wando formations (McCarten, Lennon, and Weems, 1984). The topographic maps of the region are currently mapped at a 5-ft (1.52 m) contour interval (U.S. Geological Survey topographic maps for Berkeley, Dorchester, Charleston, and Colleton counties, S.C.), which is not sufficient. Seismically induced features such as sand blows have been noted in late Pleistocene sediment (Cox, 1984, pers. comm.) in the region. These features were commonly associated with A.D. 1866 activity (Dutton, 1888). It is assumed that activity continued during late Pleistocene time, though the amount of displacement cannot be measured at a 5-ft (1.52 m) contour interval.

During the late Holocene (Fig. 8-5) little variation in the number of interriverine archaeological sites is noted prior to 1,400 B.P. when comparing the Wando area (on the north) with the Stono area (on the south side of the Ashley River Fault), though five relatively coincident clusters are apparent. Little significant neotectonic movement is assumed. From 1,400-700 B.P., the Winyah Bay and Wando sites indicate similar development of clusters, while significantly more are encountered at the Stono site and many more at the North Edisto. Sea-level rise is enhanced on the southern block of the Ashley River block, with neotectonic addition to the normal eustatic rises, causing concomitant reduction in food supply of primary nature and resulting in exploitation of Tertiary sources. Similar observations occur in the 1,400-700 B.P. time block (see Fig. 8-4). The A.D. 1,600-1,700 rise is reflected in all areas, but most strongly at the Stono site immediately south of the fault. The Charleston Earthquake of 1886 is indicated separating neotectonic movement, positive on the northern block and negative on the southern since 1,400 B.P., from the reverse condition since. Positive neotectonic movement on the southern block compared to the northern is summarized by Talwani (1982) based on microseismic analyses during the 1970s. Recently completed analyses of leveling data obtained since the 1940s indicate similar movement according to Poley (1984).

The Garner-Edisto tectonic trend was reported by Colquhoun et al. (1983). The fault strikes east-west from a few kilometers northeast of Savannah, Georgia, to offshore of Edisto Island, South Carolina. It is probably the landward extension of the Saint Helena Banks Fault reported by Behrendt et al. (1983) based on high-resolution shallow seismic interpretation. Movement on the Garner-Edisto Fault is apparent in the middle Eocene, late Eocene, and late Oligocene. Possible movement is present in Miocene stratigraphic displacement through the Seravallian. In all cases the sense of movement is down to the south and up to the north, with largest displacement at the western end of the fault, near Savannah. During late Holocene time (see Fig. 8-5), examination of the North Edisto site (which lies on the north block) reveals stability until 2,200 B.P. when positive (upward) movement is interpreted. From 1,400 B.P. through 700 B.P., downward movement in conformity with the Ashley River Fault is apparent, and perhaps greater. The cluster developed near 1,700-1,600 B.P. is similar to those found at the Wando and Winyah Bay sites, indicating either stability or possible further upward movement, certainly in contrast with the downward movement of the south block of the Ashley River Fault.

In summary, between Winyah Bay and the Savannah River, tectonic movement indicated through Tertiary stratigraphic examination of the Cape Fear Arch, Southeast Georgia Embayment, and the Ashley River and Garner-Edisto faults is observed to be in conformity with geomorphic and stratigraphic early Pleistocene mapping and neotectonic movements interpreted through archeologic studies in the late Holocene through 1886. Data collected since the 1950s both microseismic and releveling, indicate the reverse. All studies, however, support the existence of tectonic elements, as shown in Figures 8-2 and 8-4.

AMOUNT OF EUSTATIC CHANGE

The amount of eustatic sea-level change associated with the smaller fluctuations is small. The general curve (see Fig. 8-1) indicates variation on the order of 1 m, a figure that should be regarded as maximum considering the nature of the data, the fact that maximum submergence is unknown, and the fact that the basal peats may have floated to a higher elevation than that at which they originally formed. These minor sea-level changes probably include both eustatic and neotectonic components. The neotectonic components include both tectonic downwarping associated with the Southeast

Georgia Embayment, as well as episodic faulting associated with the Ashley River and Garner-Edisto faults. In the case of archeologic control, the eustatic movements, as indicated by the clusters of interriverine sites, are effected by neotectonic change. Such change is indicated by number of sites within clusters, which indicates movement that either enforces or subtracts from the eustatic rise.

Reference to Colquhoun et al. (1981) indicates that many of the control points used in curve construction in Figure 8-1 come from localities lying within the negative (downward) warping or faulting areas associated with the Garner-Edisto or Ashley River faults. These data are so indicated on Figure 8-1 by small arrows. In these cases the elevation indicated may actually be somewhat too low, having been displaced from its original position by tectonics.

CLIMATIC CHANGE ASSOCIATED WITH SMALL SEA-LEVEL OSCILLATIONS

Pollen spectra from the southeastern United States, and principally from Florida (Watts, 1980; Brown, 1981), indicate major change in climax forests near 5,000 B.P., the change being toward wetter conditions. Minor changes on the scale of the smaller sea-level oscillations are not noted.

The fall in sea level shown in Figure 8-1 near A.D. 1300, reaching a minimum near A.D. 1550 and rising since, has been noted in all areas intensively studied in South Carolina. In the Winyah Bay area this fall is noted in a change of subsistence-settlement patterns and an increase in interriverine sites. In the Charleston area it is noted similarly, while in the Edisto area similar observations are supplemented with a basal peat date. Farther south, in the Port Royal Sound area, Michie (1980) has noted many shell midden sites occurring near A.D. 1200 and after A.D. 1700 with intervening sites missing. The area, Callawassee Island, lies well inland, near the present fresh-water/salt-water boundary. Therefore, a slight drop in sea level, such as occurred during historic time at Santa Helena (1577-1587) (Colquhoun et al., 1981), which lies seaward, could explain the apparent lack of sites at Callawassee. *Crassostrea virginica* beds, the main staple in the middens, moved down the estuary too far from the island for local exploitation.

While regional climatic change has not been recorded in the southeastern United States to explain this most recent small fluctuation, the observations are well known in the Nordic literature where they form part of that body of knowledge known as the Little Ice Age.

CONCLUSION

The southeastern United States sea-level change curve, therefore:

1. indicates eustatic change from rapid rise to slow rise in sea level about 6,000 B.P.,
2. indicates small isostatic negative adjustment since about 6,000 B.P.,
3. indicates small (±1 m) oscillations in sea level over the entire period in which data are sufficient to demonstrate,
4. suggests that these changes cannot be explained by neotectonism because they are much too large,
5. suggests that these changes cannot be explained by isostatism because they reverse and are too short,
6. suggests that these changes are eustatic since they are developed in both neotectonically positive and negative areas and in a region that should show constant isostatic response,
7. reinforces Fairbridge's (1961) suggestion that these changes are tied into the global climatic cycle,
8. suggests that such changes are on the order of climatic change involved in the Little Ice Age.

ACKNOWLEDGMENTS

Financial support is acknowledged from the United States Geological Survey, National Science Foundation, South Carolina Department of Health and Environmental Control, Water Resources Commission, the Water Resources Research Institute at Clemson University, and E. I. du Pont de Nemours for aid in data aquisition, and preparation of aspects of these investigations.

REFERENCES

Behrendt, J. C.; Hamilton, R. M.; Ackerman, H. D.; Henry, V. J.; and Bayer, K. C., 1983, Marine multichannel seismic-reflection evidence for Cenozoic faulting and deep crustal structure near Charleston, S.C., in G. S. Gohn (ed.), *Studies Related to Charleston, S.C. Earthquake of 1886—Tectonics and Seismicity,* Washington, D.C.: U.S. Geological Survey Professional Paper 1313, pp. J1-J29.

Bloom, A. L., 1977, *Atlas of Sea-level Curves,* International Geological Correlation Program Project 61, Ithaca, N.Y.: Cornell University, 122p.

Brooks, M. J., 1980, Late Holocene sea level variability and prehistoric adaptations in the lower coastal plain of South Carolina, Master's thesis, Arizona State University, Tempe, 143p.

Brooks, M. J., and Canouts, V., 1981, Environmental and subsistence change during the late prehistoric period in the lower coastal plain of South Carolina, in D. J. Colquhoun (ed.), *Variations in Sea Level on the South Carolina Coastal Plain,* Columbia: Department of Geology, University of South Carolina, pp. 45-72.

Brooks, M. J.; Colquhoun, D. J.; Pardi, R. R.; Newman, W. S.; and Abbott, W. H., 1979, Preliminary archaeological and geological evidence for Holocene sea level fluctuations in the lower Cooper River Valley, South Carolina, *Florida Anthropologist* **32**(3):85-103.

Brooks, M. J., and Scurry, J. D., 1978, An intensive archaeological survey of Amoco Realty Property in Berkeley County, South Carolina, with a test of two subsistence-settlements hypotheses for the prehistoric period, in *Research Manuscript Series 47,* Columbia, S.C.: University of South Carolina, Institute of Archaeology and Anthropology, pp. 43-63.

Brown, J., 1981, Palynologic and petrographic analyses of Bayhead Hammock and marsh peats at Little Salt Spring Archaeological Site (8-So-18), Florida, Master's thesis, Department of Geology, University of South Carolina, Columbia, 52p.

Calmes, A., 1967, Test excavations at three Late-Archaic shell mounds on Hilton Head Island, Beaufort County, South Carolina, *Southeastern Archaeological Conf. Bull.* **8:**1-22.

Clark, J. A.; Farrell, W. E.; and Peltier, W. R., 1978, Global changes in postglacial sea level: A numerical calculation, *Quat. Research* **9**:265-287.

Colquhoun, D. J., 1967, Coastal plain terraces in the Carolinas and Georgia, in H. E. Wright (ed.), *Proceedings of the International Association of Quaternary Research,* vol. 7, Washington, D.C.: National Academy of Sciences, pp. 105-120.

Colquhoun, D. J., 1969, *Geomorphology of the Lower Coastal Plain of South Carolina,* Columbia: South Carolina Division of Geology, State Development Board, Map Series 15, 36p.

Colquhoun, D. J., 1982, An investigation of Holocene neotectonic deformation in the Charleston, South Carolina region, compared to areas to the north and to the south, *U.S. Geological Survey, Summaries of Technical Reports* **15:**78-82.

Colquhoun, D. J., 1984, An investigation of Holocene neotectonic deformation in the Charleston, South Carolina region, compared to areas to the north and to the south, *U.S. Geological Survey, Final Technical Report.*

Colquhoun, D. J., and Brooks, M. J., 1986, New evidence from the southeastern United States for eustatic components in the late Holocene sea levels, *Geoarchaeology* **1:**275-291.

Colquhoun, D. J.; Brooks, M. J.; Abbott, W. H.; Stapor, F. W.; Newman, W.; and Pardi, R. R., 1980, Principles and problems in establishing a Holocene sea level curve for South Carolina, in J. D. Howard (ed.), *Excursions in Southeastern Geology,* Guidebook 20, Boulder, Colo.: Geological Society of America, pp. 143-159.

Colquhoun, D. J.; Brooks, M. J., Michie, J.; Abbott, W. B.; Stapor, F. W.; Newman, W.; and Pardi, R. R., 1981, Locations of archaeological sites with respect to sea level in Southeastern United States, *Striae* **14:**144-150.

Colquhoun, D. J.; Woolen, I. D.; Van Nieuwenhuise, D. S.; Padgett, G. G.; Oldham, R. W.; Boylan, D. C.; Bishop, J. W.; and Howell, P. D., 1983, *Surface and Subsurface Stratigraphy Structure and Aquifers of the South Carolina Coastal Plain,* Report to the Department of Health and Environmental Control, Ground Water Protection Division, Columbia: Office of the Governor, State of South Carolina, 78p.

Combes, J. D., 1975, The archaeology of Kiawah Island, in *Environmental Inventory of Kiawah Island,* Columbia, S.C.: Environmental Research Center, Inc., pp. A-2-A-32.

Cooke, C. W., 1936, Geology of the Coastal Plain of South Carolina, *U.S. Geol. Survey Bull. 867,* 188p.

dePratter, C. B., and Howard, J. D., 1977, Environmental changes on the Georgia coast during the prehistoric period: Early Georgia, in J. D. Howard and C. B. dePratter (eds.), *Excursions in Southeastern Geology,* Guidebook 20, Boulder, Colo.: Geological Society of America, pp. 1-65.

dePratter, C. B., and Howard, J. D., 1977, History of shoreline changes determined by archaeological dating: Georgia coast, U.S.A., *Gulf Coast Assoc. Geol. Socs. Trans.* **27:**251-258.

Dutton, C. E., 1888, The Charleston Earthquake of August 31, 1886, *U.S. Geol. Survey Ann. Rept. 9,* pp. 203-528.

Fairbridge, R. W., 1961, Eustatic changes in sea level, in L. H. Ahrens, K. Rankama, F. Press, and S. K. Runcorn (eds.), *Physics and Chemistry of the Earth,* vol. 4, New York: Pergamon Press, pp. 99-185.

Gohn, G. (ed.), 1983, Studies related to the Charleston,

South Carolina earthquake of 1886—Tectonics and seismicity, *U.S. Geol. Survey Prof. Paper 1313,* 375p.

Gornitz, V.; Lebedeff, S.; and Hansen, J., 1982, Global sea level trend in the past century, *Science* **215:**1611-1614.

Hemmings, E. E., 1970, Preliminary report of excavations at Fig Island, in *The Notebook,* vol. 2, Columbia: University of South Carolina, Institute of Archaeology and Anthropology, pp. 9-15.

Howard, J. D., and dePratter, C. B. (eds.), 1980, *Excursions in Southeastern Geology,* Guidebook 20, Boulder, Colo.: Geological Society of America, 258p.

Hoyt, J. H., and Hails, J. R., 1970, Pleistocene shoreline sediments in coastal Georgia: Deposition and modification, *Science* **155:**1541-1543.

Imperato, D. P., 1984, Holocene Depositional History of the North Edisto Tidal Basin, Master's thesis, Geology Department, University of South Carolina, Columbia.

Lennon, G. W., 1985, Identification of a Northwest Trending Seismogenic Graben near Charleston, South Carolina, Master's thesis, Geology Department, University of South Carolina, Columbia.

McCants, C. Y., 1982, *Evolution and Stratigraphy of a Sandy Tidal Flat Complex within a Mesotidal Estuary,* Columbia: Department of Geology, University of South Carolina, 108p.

McCarten, E. M.; Lemon, J.; and Weems, R. E.; 1984, Geologic map of the area between Charleston and Orangeburg, South Carolina, *U.S. Geol. Survey Map I-1472.*

Michie, J. L., 1973, Archaeological indications for sea level 3,500 years ago, *South Carolina Antiquities* **5:**1-12.

Michie, J. L., 1974, A second burial from Daw's Island, shell midden 38bu9, Beaufort County, South Carolina, *South Carolina Antiquities* **6:**37-47.

Michie, J. L., 1976, The Daw's Island shell midden and its significance during the formative period, in *Program of the Third Annual Conference of South Carolina Archaeology,* Columbia: University of South Carolina, Institute of Archaeology and Anthropology, pp. 8-15.

Michie, J. L., 1980, An intensive shoreline survey of archaeological sites in Port Royal Sound and the Borad River estuary, Beaufort, South Carolina, in *Research Manuscript Series 167,* Columbia: University of South Carolina, Institute of Archaeology and Anthropology, 140p.

Mörner, N. A., 1976, Eustasy and geoid changes, *Jour. Geology* **85:**123-151.

Moslow, T. F., and Colquhoun, D. J., 1981, Influence of sea level changes on barrier island evolution, *OCEANIS* **7**(4):432-454.

Poley, C. M., 1984, Recent vertical crustal movements in the South Carolina plain, Master's thesis, Department of Geology, University of South Carolina, Columbia, 95p.

Rankin, D. W., 1977, Studies related to the Charleston, S.C., Earthquake 1886—A preliminary report, *U.S. Geol. Survey Prof. Paper 1028,* pp. 1-15.

Ruby, C. H., 1981, Clastic facies and stratigraphy of a rapidly retreating cuspate foreland, Cape Romain, S.C., Ph.D. dissertation, Department of Geology, University of South Carolina, Columbia, 207p.

Scurry, J. D., and Brooks, M. J., 1980, An intensive archaeological survey of the South Carolina Port Authority's Bellview Plantation, Charleston, South Carolina, *Research Manuscript 158,* Columbia: University of South Carolina, Institute of Archaeology and Anthropology, 115p.

Sutherland, D. R., 1974, Excavations at the Spanish mound shell midden, Edisto Island, S.C., *South Carolina Antiquities* **5:**25-36.

Talwani, P., 1982, Internally consistent pattern of seismicity near Charleston, S.C., *Geology* **10:**654-658.

Trinkley, M. B., 1975, A typology of Thom's Creek pottery for the South Carolina Coast, Master's thesis, Department of Archaeology, University of North Carolina, Raleigh, 528p.

Watts, W. A., 1980, The late-Quaternary vegetation history of the south-eastern United States, *Ann. Rev. Ecology and Systematics* **11:**387-409.

9: Aminostratigraphy of Coastal U.S. Quaternary Marine Deposits–Deciphering the Timing of Interglacial Sea Levels and the Thermal Histories of Coastal Regions

John F. Wehmiller
University of Delaware

Daniel F. Belknap
University of Maine, Orono

Abstract: Amino acid enantiomeric (D/L) ratios in mollusks have been used by several workers to establish local or regional aminostratigraphic sequences in mid- to high-latitude coastal deposits that lack fossil types suitable for radiometric dating methods. Many examples exist in which local aminostratigraphic sequences appear consistent with mappable stratigraphic or geomorphic relative age sequences. Local aminostratigraphies can be calibrated if independently derived dates are available for one or more of the sites. Correlation among sites requires knowledge of regional temperature differences and kinetic models of the racemization phenomenon.

On the Pacific coast of North America between 48°N and 25°N, marine terrace or basin deposits with aminostratigraphic age estimates correlative to Stages 3, 5, 7, 9, 11, 13, and 15 of the marine isotopic record (Shackleton and Opdyke, 1973) appear to have been recognized in at least one location. The majority of these sites represents early or late Stage 5; a few Stage 3 sites have been recognized in regions of high uplift rates, and most of the late Stage 5 or Stage 3 sites have molluskan faunas indicating waters cooler than present (Kennedy, Lajoie, and Wehmiller, 1982). Most of the pre-Stage 7 sites are represented by stratigraphically superposed basin sequences (Ventura and Los Angeles basins) that have been exposed by uplift and erosion. Latitudinal correlation of Pacific coast aminostratigraphic sequences is facilitated by the availability of four independently dated calibration sites and by the relatively smooth and gradual latitudinal gradient of effective temperature (ca. 0.6°C/°lat.).

On the Atlantic coast between Nova Scotia and Florida, as many as four aminostratigraphic age groups are recognized in different regions, particularly in Florida, the Carolinas, and the Chesapeake Bay region. Latitudinal correlation of these local sequences is more complicated than in the case of the Pacific coast sites because of the steeper and more irregular gradient of effective temperature that has probably affected the region during the late Pleistocene. Amino acid D/L data can be used to infer the nature of these temperature gradients using results from latitudinally spaced calibration localities.

Taken as a whole, the Pacific coast terrace sequence appears to have at least one representative of each of the interglacial high sea levels inferred from the oxygen-isotope record for the past 0.5 m.y. Fewer interglacial high sea-level records, representing approximately 1 m.y., are preserved on the Atlantic coastal plain.

INTRODUCTION

There has been a long-standing interest in Quaternary coastal sea-level records that relate to ice volume histories and local or regional tectonic movements. These records can also be used to understand regional paleoecological zonation, and in selected higher-latitude situations, marine and glacial sequences can be synthesized into a multicomponent climatic record. These studies usually require one or more chemical dating methods for their chronostratigraphic control, and a substantial discipline of geochemical research now exists that is focused on the development and application of dating methods that are appropriate for coastal and other types of Quaternary sequences. Radiocarbon and uranium-series methods have dominated this field, but each has limitations in time range or sample type. As a result, mid- and high-latitude coastal Quaternary records (those without reef-building corals) have been difficult to date by any but relative stratigraphic methods.

Amino acid racemization (AAR) has become an alternative to the application of isotopic methods for dating of Quaternary deposits, especially those containing fossil mollusks. This chapter reviews briefly the progress of this dating method in its applications to Quaternary deposits of the coasts of North America (primarily the United States) between about 25° and 50°N latitude. In so doing, we summarize the apparent number and distribution of mid- to late Quaternary sea-level maxima recorded on both coasts and discuss how these records probably relate to the marine isotopic record (Shackleton and Opdyke, 1973; Mix and Ruddiman, 1984) of ice volume fluctuation. A review of this topic can be found in Wehmiller (1982).

AMINOSTRATIGRAPHY PRINCIPLES

The phenomenon of AAR involves the time-dependent change in the relative proportions of the D (dextro, or right-handed) and L (levo, or left-handed) amino acid forms. In principle, the D/L value increases monotonically from 0 to 1 with increasing fossil age. In addition to fossil age, factors that affect the reaction include temperature, fossil type (shell, bone, etc.), family or genus, and natural diagenetic processes other than simple racemization (i.e., contamination by extraneous amino acids with D/L values different from those indigenous to the fossil). Because the temperature dependence of AAR is a significant variable in the translation of measured D/L values into absolute age estimates, the most common use of AAR is as an aminostratigraphic tool, in which D/L values from closely spaced but discontinuous sites are grouped into statistically significant *aminozones*. This chemical stratigraphic approach then establishes a local relative age sequence and relies on the assumption that the analyzed deposits have all experienced similar effective temperatures (the kinetic average of all temperatures to which the samples have been exposed; see Wehmiller, 1982). Absolute age estimates for one or more of these aminozones are possible if independent age data are available for at least one site within the sequence. Kinetic models of racemization, which incorporate models for temperature histories of a region, can also be used independently or in combination with other dating methods (Wehmiller, 1982).

Correlation between latitudinally spaced aminostratigraphic sequences developed in this manner requires a combination of the following: a knowledge of the past and present differences in temperature for each region, kinetic modeling that incorporates these temperature differences, and/or independent age control for different aminozones at different temperatures. Latitudinally spaced samples with independent age control from central California to southern Baja California have proven to be particularly helpful in constructing such regional aminostratigraphic sequences and in establishing the resolving power of aminostratigraphy for late Pleistocene samples. The following section reviews these Pacific coast studies.

PACIFIC COAST, NORTH AMERICA

Figure 9-1 is a map showing the location of Quaternary basin and terrace deposits from Puget Sound, Washington, to Baja California Sur, Mexico, for which aminostratigraphic data are available. At seven of these sites local

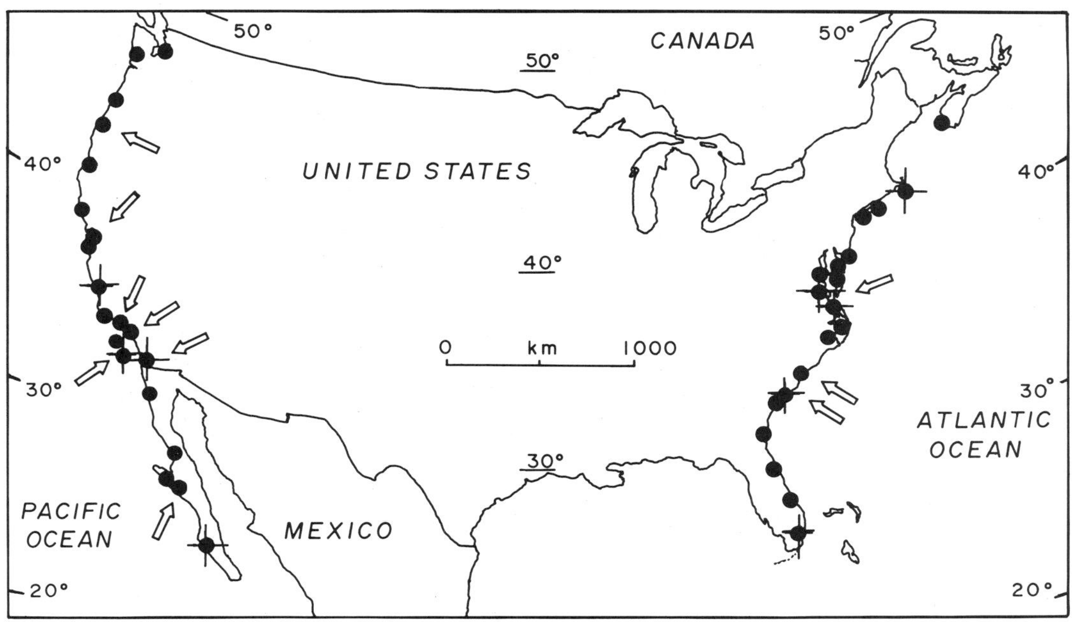

Figure 9-1. Map of United States, northern Mexico, and southern Canada, showing sites for which aminostratigraphic data are available.

Sites marked with ✦ have associated isotopic data. Sites marked with ⇨ have outcrops in which samples with substantially different ages are found in stratigraphic sequence. Sites marked with ● have aminostratigraphic data, but no isotopic data.

aminostratigraphic sequences have been constructed using several genera of mollusks from vertical terrace flights or superposed stratigraphic sequences (Baja Peninsula, San Diego, San Nicolas Island, Los Angeles Basin, Ventura-Sea Cliff, San Francisco Bay, and Cape Blanco, Oregon). In addition, four of these sites have U-Th dates on solitary corals, thereby providing absolute age control for the regional Pacific coast aminostratigraphy. These dated sites all have isotopic ages of 120,000-130,000 yr, indicating correlation with coral terrace deposits found at numerous localities worldwide that collectively indicate a eustatic sea-level maximum at about 125,000 B.P., corresponding to early Stage 5 of the marine isotopic record. Muhs (1983) has recently added calibrated aminostratigraphic data for San Clemente Island, near Los Angeles.

Figure 9-2 shows how aminostratigraphic data from dated samples can be used to estimate the integrated thermal history for these late Pleistocene calibration localities. Shown in Figure 9-2 are model isochrons predicted for 120,000-yr samples that have been exposed to

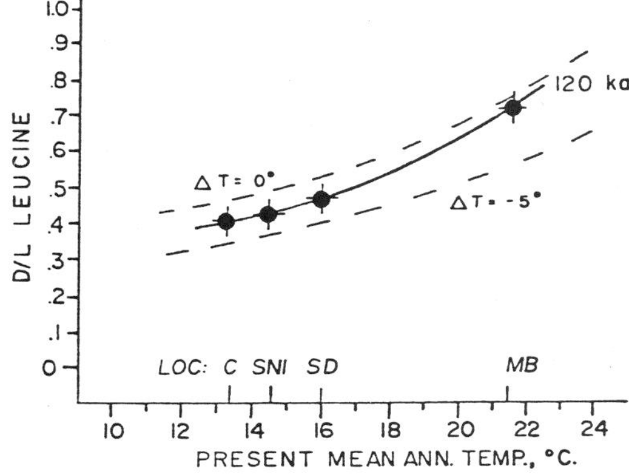

Figure 9-2. Latitudinal presentation of Pacific coast calibrated leucine D/L data compared with kinetic model isochrons.

Isochrons are drawn from the kinetic model of Wehmiller and Belknap (1978) for two temperatures: equal to present at each latitude and 5°C less than present at each latitude. The calibrated data (those fixed to 120,000-U-Th dated sites) fall between the two model isochrons, indicating that effective temperatures have been roughly 2-4°C cooler than present. Locality abbreviation: C, Cayucos, California; SNI, San Nicolas Island, California; SD, San Diego, California; MB, Magdalena Bay, Baja California Sur, Mexico.

effective temperatures equal to, or slightly less than (0° and 5°C), present temperatures at the sites. These isochrons were calculated for one amino acid (leucine) in one molluskan genus *(Protothaca)* using a kinetic model developed by Wehmiller and Belknap (1978, 1982; see Wehmiller, 1982, for further discussion). The calibration isochron, fixed to the U-Th dated samples, represents an effective temperature approximately 2-4°C below modern temperatures. Because these effective temperatures represent the integrated thermal history of samples, they do not uniquely define any one temperature during that history. Nevertheless, when compared with modern temperatures for a site and with the ^{18}O model for the timing of late Pleistocene temperature fluctuations, reasonable estimates of full glacial maximum temperature reductions can be derived from these effective temperature values (Wehmiller, 1982). By this approach, the estimated full glacial (Stage 2, 18,000 B.P.) temperature reduction for coastal California is between 3° and 6°C, with greater reductions at higher latitudes. Full glacial temperatures near southern Baja appear to have been only 2-3° cooler than present. A similar range of sea-surface full glacial temperature reductions has been inferred from micropaleontologic data (Imbrie, McIntyre, and Moore, 1983). The marine-dominated nature of coastal climate from 48°N to 25°N is reflected in the similar slopes of latitudinal temperature gradients for modern air- and sea-surface, full glacial sea-surface, and effective temperatures inferred from the AAR data at dated sites. The smooth latitudinal temperature gradient and the apparent lack of extensive temperature variation during the late Pleistocene of the Pacific coast has simplified the development of kinetic models of molluskan AAR. Such is not the case for the U.S. Atlantic coastal plain, so the models developed with Pacific coast data permit qualitative estimates of more complex Atlantic coastal plain temperature histories, which are discussed in the section on the Atlantic Coast.

Figure 9-3 shows the calibration isochron of Figure 9-2 and additional isochrons derived from this calibration using the leucine-*Protothaca* kinetic model of Wehmiller and Belknap (1978, 1982). These isochrons are drawn to represent the time range of each of the interglacial isotopic stages as estimated by Shackleton and Opdyke (1973). Implicit in the position of these isochrons is the assumption that effective temperatures for pre-Stage 5 sites have been the same as those for nearby Stage 5 calibration sites. In a few cases these model isochrons have been able to be tested with independent age information for pre-Stage 5 sites. Their positions and slopes at different D/L values indicate the sensitivity of any kinetic model age estimate to temperature differences between control and undated sites.

Shown also in Figure 9-3 are data points representing mean values of multiple sample analyses at marine terrace and basin sites along the Pacific coast. Nearly 50% of the 36 sites represented in this figure have D/L values that imply an age within isotope Stage 5. Not surprisingly, terrace/basin sites representing the pre-Stage 5 portion of the record are less frequently observed, but Stage 9 appears to be better represented than Stage 7. Sites from a region between Ventura and Goleta, California, which are undergoing rapid late Quaternary uplift, have D/L values implying an age of approximately 40,000 yr (Stage 3; Lajoie et al., 1979). In many cases the molluskan fauna at the sites with either late Stage 5 or Stage 3 aminostratigraphic age estimates are indicative of cooler than present water conditions during the time of terrace formation (Kennedy, Lajoie, and Wehmiller, 1982). Thus, it appears that these terrace records are consistent with a model of global or regional cooling from late Stage 5 through Stage 2 as would be inferred from various isotopic records.

The Pacific coast terrace record as summarized in Figure 9-3 documents periods of terrace formation corresponding to the seven most recent ice volume minima of the marine isotope record. Though this correlation is not proof of the accuracy of the aminostratigraphic age estimates, it shows the approximate frequency of formation or preservation of late Pleistocene and older coastal terrace deposits. Not revealed in this sequence, however, are several vertical terrace sequences (Palos Verdes Hills, San

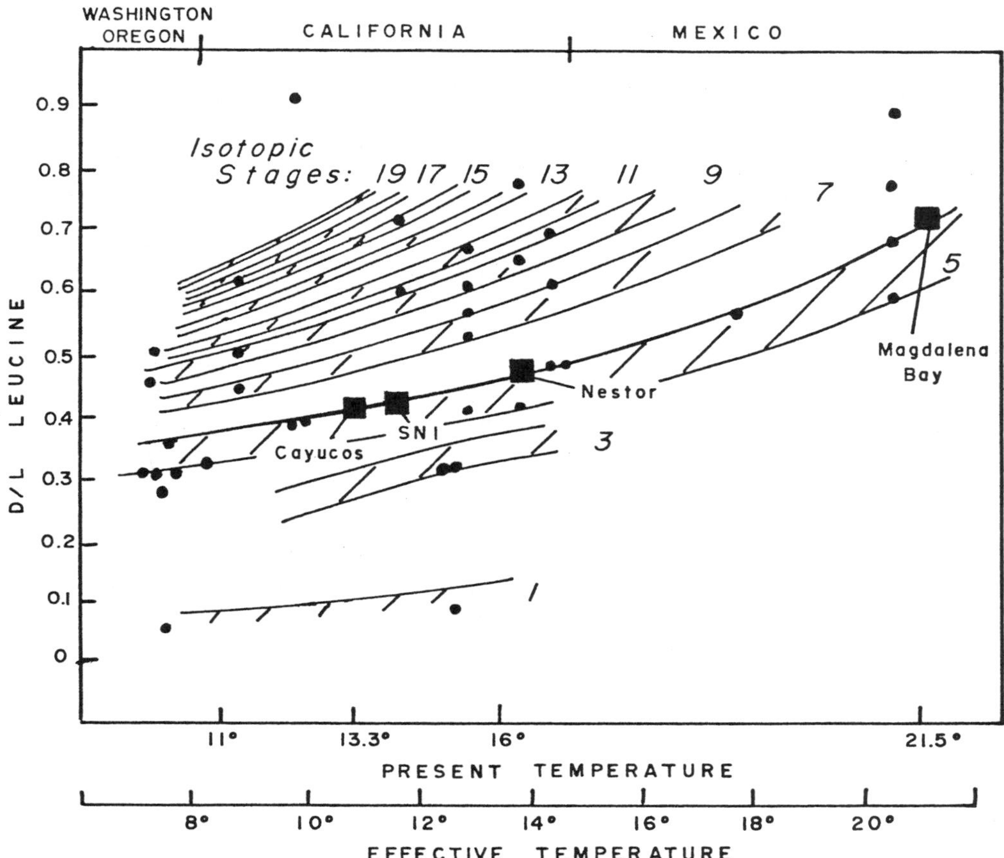

Figure 9-3. Pacific coast aminostratigraphic model.

Isochron bands drawn for pre- and post-Stage 5 age ranges using the calibrated 120,000-yr isochron in Figure 9-2 as the basis for extrapolation of the leucine kinetic model of Wehmiller and Belknap (1978). Data points represent mean values of multiple shell analyses at basin and terrace sites between the Puget Lowland, Washington, and Baja California Sur, Mexico. The four calibration sites (Cayucos; San Nicolas Island; Nestor Terrace, San Diego; and Magdalena Bay) shown in Figure 9-2 are identified here. The present temperature scale shows the current mean annual air temperatures at the latitudes of the calibration sites; the effective temperature scale defines the effective temperatures inferred from the calibration site D/L data and the leucine kinetic model. An effective temperature derived from U-Th calibrated data for one latitude is assumed to be representative of effective temperatures for other Pleistocene samples at the same latitude. It is not the correct effective temperature for Holocene samples because these samples have not been exposed to any glacial-age temperature reductions. (After Wehmiller, 1982, Fig. 6)

Clemente Island, San Nicolas Island) from which few fossils have been available in spite of the large number (approximately 12 in each case) of clearly resolved terraces. The limited data available for these particular terrace sequences (Wehmiller et al., 1977; Muhs, 1983) suggest that all the ice volume minima for the past 1 m.y. may be represented.

ATLANTIC COAST, NORTH AMERICA

Figure 9-1 shows the sites or regions along the Atlantic coast from Nova Scotia to Florida for which aminostratigraphic data have been obtained. The regional aminostratigraphy between New Jersey and South Carolina has been debated in papers by McCartan et al. (1982) and Wehmiller and Belknap (1982). The Florida peninsula has been studied primarily by Mitterer (1974, 1975). Most of the isotopic dating control for the Atlantic coastal plain has been presented by Cronin et al. (1981), and the implications of these dates for aminostratigraphic correlations have been discussed by Wehmiller and Belknap (1982). Even though aminostratigraphic data appear to conform to local biostratigraphic sequences, several fundamental conflicts exist when local aminostratigraphic sequences along the coastal plain are compared with the available radiometric data (Wehmiller and Belknap, 1982), and the construction of smooth latitudinal isochrons (as in the Pacific coast case) does not appear to be possible for the Atlantic coastal plain at the present time. The implications of this situation for local and regional thermal histories are outlined after the local aminostratigraphic sequences along the Atlantic coast are reviewed.

Figure 9-4 presents the Atlantic coast aminostratigraphic data in a manner similar to that shown in Figure 9-3. D/L leucine data for *Mercenaria* (or data converted from other genera or amino acids) are plotted versus latitude. The present temperature gradient through this coastal region is smooth and varies from 7°C at 44°N to 24°C at 26°N. Those data with associated U-Th data or with some stratigraphic control are so indicated. In only a few cases (four) have samples with substantial age differences actually been obtained from stratigraphic sequences in the same outcrop, so many of these local sequences are in fact inferred from multiple lithostratigraphic, biostratigraphic, and geomorphic criteria. Such interpreted sequences provide a bit less rigorous control with which to test amino acid data for stratigraphic reliability.

From Figure 9-4 it can be inferred that the maximum number of aminostratigraphic units recorded so far in any region is between three and five, including the basal units (with D/L

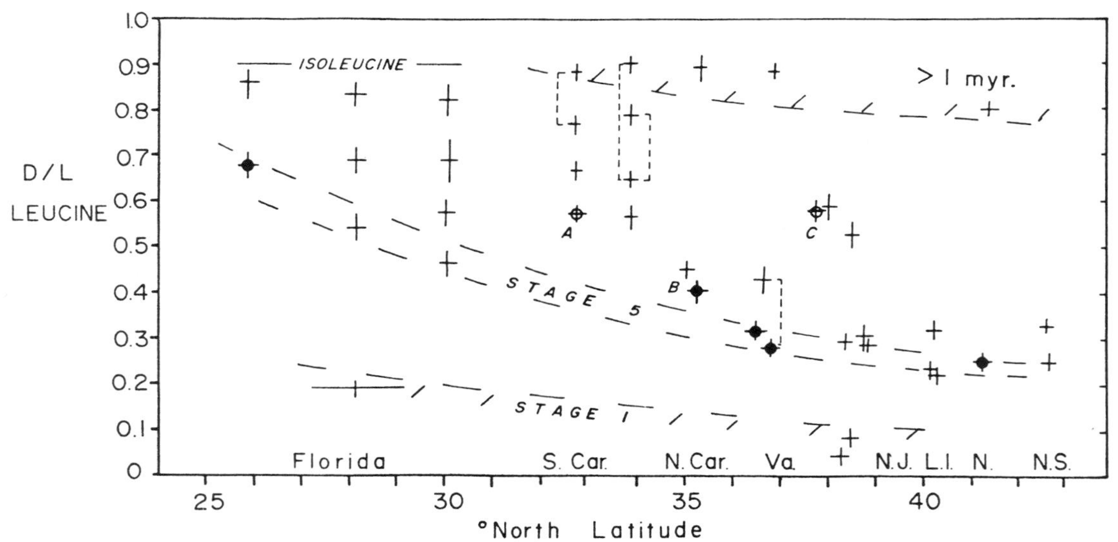

Figure 9-4. Atlantic coast aminostratigraphic model, for leucine in *Mercenaria*.
Data points shown as ◆ and ⊕ have associated U-Th isotopic data. Data points with connecting vertical lines represent samples from local (single outcrop) stratigraphic sequences. Data points for the southern region of the figure, representing the Florida peninsula and labeled "isoleucine," are inferred leucine results derived from the isoleucine data of Mitterer (1975). Isochron bands representing Stage 1, Stage 5, and >1 myr are shown. The Stage 5 isochron is modeled after that seen in Figure 9-3 and is consistent with most, but not all, of the isotopic data for these coastal plain sites, as discussed in the text and elsewhere. (After Wehmiller and Belknap, 1982)

values greater than 0.9) that include the Yorktown, James City, and Waccamaw Formations in Virginia, North Carolina, and South Carolina respectively. As this basement is everywhere greater than 1 m.y. in age, it appears that no more than four mid- to late Quaternary depositional events have been preserved or recorded in the coastal plain. Because even this number of aminostratigraphic units exceeds the number of biostratigraphic units (Cronin, 1980) or depositional events (McCartan et al., 1982) inferred for the region, we are attempting to refine our knowledge of the chronostratigraphic resolving power of amino acid data. Corrado and Hare (1981) and Mitterer (1975) have each recognized the same number (four to five) of aminostratigraphic units in South Carolina and Florida respectively.

In Figure 9-4, isochron bands are labeled Stage 1, Stage 5, and >1 m.y. These regions are constrained by radiometric or biostratigraphic data. The Stage 5 isochron band is drawn to include data from sites with 100,000-to-130,000-yr U-Th coral dates, as well as to include data from many sites long considered as last interglacial (Wehmiller and Belknap, 1982). The Stage 5 isochron band in Figure 9-4 is also modeled after that shown in Figure 9-3 and is consistent with the present Atlantic coast latitudinal temperature gradient.

Point A of Figure 9-4 represents one major conflict with the Stage 5 isochron as it is drawn. Point A includes data from four sites in the Charleston, South Carolina, region that have U-Th coral dates of 80,000-90,000 yr (McCartan et al., 1982). The samples from these sites are so extensively racemized that they have been interpreted to be on the order of 250,000 yr in age based on the latitudinal aminostratigraphic framework shown in Figure 9-4 (Wehmiller and Belknap, 1982). Only by invoking effective temperatures for the Charleston region that are essentially the same as the present temperatures in the area can these D/L values be reconciled with the isotopic dates. If the Charleston area effective temperature remained constant (equal to present) during the late Quaternary, then the aminostratigraphic data from sites to the north (to Virginia) imply that large (ca. 12-15°C) temperature differences must have existed for significant periods (greater than 20,000 yr during the past 100,000) between regions that today have temperature differences of 2-3°C. Though palynologic data can be interpreted as indicating a steeper-than-present full-glacial temperature gradient in the Carolinas (Delcourt, Delcourt, and Webb, 1982), such a large thermal difference does not appear to be reasonable. This conflict is of fundamental importance for the aminostratigraphic method because it emphasizes the role of thermal history options in racemization age estimates: if the South Carolina region has really been warm, then correlation and age estimates for the region between South Carolina and Virginia would shift substantially (by one full isotopic stage). Conversely, if the aminostratigraphic correlations through the South Carolina region (based on smooth latitudinal temperature gradients, similar to the modern temperature gradient) are correct, then the U-Th isotopic dates for the Charleston region are incorrect.

Another fundamental age-temperature conflict is identified by points B and C of Figure 9-4. These points represent groups of results that have been considered (on the basis of both U-Th and biostratigraphic data) to be about 200,000 yr (Stage 7) old (McCartan et al., 1982). The Stage 7 age for point B is consistent with its position relative to the Stage 5 isochron, but the age equivalence of points B and C (present mean annual temperatures of 17°C and 14°C respectively) would require an inversion of effective temperature gradients to explain the more extensively racemized site at a higher latitude. This issue is similar to that outlined earlier for point A in that the aminostratigraphic data identify fundamental questions regarding either temperature histories, U-Th data, or the resolving power of the racemization dating method.

Aminostratigraphic data have proved useful (and controversial) in the issue of mid-Wisconsinan (ca. 30,000-40,000 yr) marine deposits at or above present sea level on the Atlantic coast. Belknap (1982) has studied several sites for which finite ^{14}C dates on shells have been obtained but which yield aminostratigraphic age estimates of 100,000 yr or greater.

One of the more problematic of these sites may be the Port Washington gravel pits on Long Island (see Fig. 9-4). This site has yielded *Crassostrea* D/L data represented by the 0.34 value (converted to an equivalent *Mercenaria* data point) in Figure 9-4. The shells from this site are more racemized than those from a neighboring Stage 5 calibration locality (Sankaty Head, Nantucket; Wehmiller and Soren, 1984) and therefore have a pre-Stage 5 (probable Stage 7) aminostratigraphic age estimate. Finite ^{14}C dates on these shells and on wood and peat samples from the Port Washington gravel pits have been used in support of a mid-Wisconsinan age for the warm climate interval recorded by the pollen at this site (Sirkin and Stuckenrath, 1980).

CONCLUSION

The aminostratigraphic dating method shows great promise for deciphering the chronostratigraphic record of Quaternary coastal deposits, but several ongoing issues require that the method be continually evaluated for its reliability. In most, if not all, cases where rigorous local stratigraphic control exists, aminostratigraphic sequences conform to this control. Pacific coast data suggest that the resolving power of the method is on the order of one interglacial stage for most of the mid- to late Quaternary, providing thermal gradients have been similar to present latitudinal gradients. The tectonic setting of the Pacific coast region appears to have preserved more records of the past million years of sea-level fluctuations, either in thick sequences of basinal sediment or in extensive flights of terraces. Atlantic coast data, interpreted by analogy with the Pacific coast model of a smooth latitudinal gradient of effective temperature, reveal several conflicts when compared with U-Th isotopic data. These conflicts identify some fundamental geochemical and paleoclimatic issues that remain to be resolved. Fewer (three or four) aminostratigraphic age units are recognized on the Atlantic coast than on the Pacific, probably due to the poorer resolution of Quaternary sea-level fluctuations inherent to the tectonics of the Atlantic margin. One of the more interesting challenges for the aminostratigraphic method is the establishment of a highly refined chemical stratigraphy for this complex region.

REFERENCES

Belknap, D. F., 1982, Amino acid racemization data for C-14 dated "mid-Wisconsinan" mollusks of the Atlantic coastal plain, *Geol. Soc. America Abs. with Programs* **14**(1-2):4.

Corrado, J., and Hare, P. E., 1981, Aminostratigraphy of the South Carolina coastal plain by monospecific fossil analyses *(Mulinia lateralis Say), Carnegie Inst. Washington Year Book* **80:**387-389.

Cronin, T. M., 1980, Biostratigraphic correlation of Pleistocene marine deposits and sea levels, Atlantic Coastal Plain of the southeastern United States, *Quat. Research* **13:**213-229.

Cronin, T. M.; Szabo, B. J.; Ager, T. A.; Hazel, J. E.; and Owens, J. P., 1981, Quaternary climates and sea levels of the U.S. Atlantic Coastal Plain, *Science* **211:**233-240.

Delcourt, H. R.; Delcourt, P. A.; and Webb, T., III, 1982, Dynamic plant ecology: The spectrum of vegetational change in space and time, *Quat. Sci. Rev.* **1:**153-176.

Imbrie, J.; McIntyre, A.; and Moore, T. C., Jr., 1983, The ocean around North America at the last glacial maximum, in S. Porter (ed.), *Late Quaternary Environments of the United States,* vol. 1, *The Late Pleistocene,* Minneapolis: University of Minnesota Press, pp. 230-236.

Kennedy, G. L.; Lajoie, K. R.; and Wehmiller, J. F., 1982, Aminostratigraphy and faunal correlations of late Quaternary marine terraces, Pacific coast, U.S.A., *Nature* **299:**545-547.

Lajoie, K. R.; Kern, J. P.; Wehmiller, J. F.; Kennedy, G. L.; Mathieson, S. A.; Sarna-Wojcicki, A. M.; Yerkes, R. F.; and McCrory, P. A., 1979, Quaternary marine shorelines and crustal deformation, San Diego to Santa Barbara, California, in P. L. Abbott (ed.), *Geological Excursions in the Southern California Areas,* San Diego, Calif.: San Diego State University, Department of Geological Sciences, pp. 3-16.

McCartan, L.; Owens, J. P.; Blackwelder, B. W.; Szabo, B. J.; Belknap, D. F.; Kriausakul, N.; Mitterer, R. M.; and Wehmiller, J. F., 1982, Comparison of amino acid racemization geochronometry with lithostratigraphy, biostratigraphy, uranium-series coral dating, and magnetostratigraphy in the Atlantic Coastal Plain of the southeastern United States, *Quat. Research* **18:**337-359.

Mitterer, R. M., 1974, Pleistocene stratigraphy in southern Florida based upon amino acid diagenesis in fossil *Mercenaria*, *Geology* **2:**425-428.

Mitterer, R. M., 1975, Ages and diagenetic temperatures of Pleistocene deposits of Florida based on isoleucine epimerization in *Mercenaria*, *Earth and Planetary Sci. Letters* **28:**275-282.

Mix, A. C., and Ruddiman, W. F., 1984, Oxygen isotope analyses and Pleistocene ice volumes, *Quat. Research* **21:**1-20.

Muhs, D. R., 1983, Quaternary sea level events on northern San Clemente Island, California, *Quat. Research* **20:**322-341.

Oldale, R. N.; Valentine, P. C.; Cronin, T. M.; Spiker, E. C.; Blackwelder, B. W.; Belknap, D. F.; Wehmiller, J. F.; and Szabo, B. J., 1982, The stratigraphy, structure, absolute age, and paleontology of the Upper Pleistocene deposits at Sankaty Head, Nantucket Island, Massachusetts, *Geology* **10:**246-252.

Shackleton, N. J., and Opdyke, N. D., 1973, Oxygen isotope and paleomagnetic stratigraphy of equatorial Pacific core V28-238: Oxygen isotope temperatures and ice volumes on a 10^5 yr. and 10^6 year scale, *Quat. Research* **3:**39-55.

Sirkin, L., and Stuckenrath, R., 1980, The Portwashingtonian warm interval in the northern Atlantic Coastal Plain, *Geol. Soc. America Bull.* **91:**332-336.

Wehmiller, J. F., 1982, A review of amino acid racemization studies in Quaternary mollusks: Stratigraphic and chronologic applications in coastal and interglacial sites, Pacific and Atlantic coasts, United Sates, United Kingdom, Baffin Island, and tropical islands, *Quat. Sci. Rev.* **1:**83-120.

Wehmiller, J. F., and Belknap, D. F., 1978, Alternative kinetic models for the interpretation of amino acid enantiomeric ratios in Pleistocene mollusks: Examples from California, Washington, and Florida, *Quat. Research* **9:**330-348.

Wehmiller, J. F., and Belknap, D. F., 1982, Amino acid age estimates, Quaternary Atlantic coastal plain: Comparison with U-series dates, biostratigraphy, and paleomagnetic control, *Quat. Research* **18:**311-336.

Wehmiller, J. F.; Lajoie, K. R.; Kvenvolden, K. A.; Peterson, E.; Belknap, D. F.; Kennedy, G. L.; Addicott, W. O.; Vedder, J. G.; and Wright, R. W., 1977, Correlation and chronology of Pacific coast marine terraces of continental United States by amino acid stereochemistry—Technique evaluation, relative ages, kinetic model ages, and geologic implications, *U.S. Geol. Survey Open-File Rept. 77-680*, 196p.

Wehmiller, J. F., and Soren, J., 1984, Aminostratigraphy of marine deposits, central and eastern Long Island, New York, *Geol. Soc. America Abs. with Programs* **16**(1):70.

10: Prediction of Effects of Sea-Level Change from Paralic and Inner Shelf Stratigraphic Sequences

J. C. Kraft
University of Delaware

D. F. Belknap
University of Maine, Orono

J. M. Demarest
Exxon Production Research Co., Houston, Texas

Abstract: Holocene Atlantic transgressive barriers and associated paralic lithosomes continue to migrate and alter form as a direct causal effect of sea-level changes. A local relative sea-level curve derived from our research shows a relatively rapid rate of sea-level rise and resultant transgression in the early Holocene and a gradual slowing in the mid-Holocene, to an average of 15 cm/century over the past 2,000 yr. This rate has doubled since the 1880s to approximately 30 cm/century.

Coastal lithosomes have been mapped in three dimensions from drill core evidence, dating back to 11,000 B.P. Construction of cross-sections and paleogeographic maps of Holocene stratigraphic sequences is based on the Doctrine of Uniformitarianism. A corollary to uniformitarianism, "the past is equally a key to the interpretation of the present and to the prediction of the future," allows the derivation of cross-sections and maps for the remainder of the Holocene interglacial in the Delaware coastal zone.

Local evidences of three higher relative sea stands in the past 1.8 m.y. suggest that the Holocene transgression and relative sea-level rise will not exceed that of Pleistocene. Knowing the average rate of sea-level rise for the past 2,000 yr (15 cm/century) and rates of shoreline erosion projected from paleogeographic reconstructions and measured from maps of the past 150 yr, we can predict that the present interglacial will peak in 1,000-4,000 yr and that coastal landforms will transgress landward an average of 2-4 km (with extremes in the Delaware estuary of 5-10 km).

Variants of this approach are presented as related to future paralic-inner shelf stratigraphic sections, present-day stratigraphic sections, and those of mid- to-late Holocene time. Appropriate paleogeographic and predictive coastal maps are related to the geologic sections. Although simplistic in approach, the predictions are based on hundreds of Holocene and Pleistocene data points from the study area. Should preliminary concepts such as the greenhouse effect on sea-level rise be confirmed, rates of change of coastal landforms and sea-level prediction would dramatically change time/rate projections, by one or two orders of magnitude.

INTRODUCTION

A plethora of papers, news releases, comments, and so forth have called our attention to the supposedly forthcoming greenhouse effect. Scenarios of CO_2 increase in the atmosphere and ocean and resultant warming trends from 2° to 5°C are predicted. Accompanying these scenarios are suggestions that an immediate response in the rate of sea-level rise of the world's oceans will ensue. These predicted levels of rise vary from 5-17 cm by A.D. 2000 to 0.52-3.45 m by A.D. 2100 (Hoffman, Keyes, and Titus, 1983; Keyes et al., 1984; Revelle, 1983; Seidel and Keyes, 1983). In the study area tide gauges show that sea level is already rising 33 cm/century relative to land (Hicks, Debaugh, and Hickman, 1983). Similarly, a committee of the National Research Council (Carbon Dioxide Assessment Committee, 1983, p. 2) translates this concept into "a global sea-level rise of about 70 cm (per century) in comparison with the rise of about 15 cm over the last century." They suggest that the public would probably adapt by the building of dikes and/or an orderly retreat from the seaward edge of the coastal zone (Nierenberg, in Carbon Dioxide Assessment Committee, 1983). Ryan (1983) noted that the NRC report included a large number of uncertainties and appeared to be struck by the conservative nature of the report and appeal for calm, not panic. Schelling (1983, p. 61), referring to Boston Harbor, noted "professional guess work suggests that at today's values the cost of defending against a 5-meter rise of sea level is less, perhaps by an order of magnitude, than the value preserved." Precisely the opposite figure of at least one or two orders of magnitude increase in cost could apply to the majority of U.S. coastlines that are undeveloped or developed for tourist or recreation uses.

Geological factors of known and unknown origins also will have a strong impact on the human ability to occupy the coastal zone in the coming centuries. For instance, the long-term average rate of sea-level rise relative to land in the Mid-Atlantic Bight in coastal Delaware on a 2,000-yr basis is approximately 15 cm/century. This rate has doubled to approximately 33 cm/century since about 1920 (Hicks, Debaugh, and Hickman, 1983; Hicks, 1972; Kraft, 1976; Kraft and John, 1976; Belknap and Kraft, 1977; Demarest, 1981). Further reference should be made to the fact that a number of areas in the coastal United States (let alone the world) are in dramatically negative tectonic settings, such as those of the Mississippi Delta and Delaware (Gable and Hatton, 1983; Nummedal, 1983) (Fig. 10-1).

Should the western Antarctic ice sheet surge, an even more massive problem could exist. Such a detachment by melting at its base could result in 2-5 m of global sea-level rise (Ryan, 1983). Most likely, such a catastrophe, if it should happen, would occur over hundreds of years. The matter is discussed in detail by Stuiver et al. (1981) and by Davies (1981) regarding Wilson's theory related to last interglacial evidences for such an event in South Africa. Hollin (1972) has suggested that such a western Antarctic ice sheet surge could happen in a matter of days or weeks.

The average interglacial sea-level rise in coastal Delaware over long-term geologic time has been up to 3-6 m above present. Partial and entire Pleistocene paralic sequences of lagoon, marsh, and transgressive barrier lithosomes have been identified in the Atlantic coastal plain of Delaware. These lateral and vertical sequences are identical to the modern Holocene analog. Quaternary paralic facies of the Omar Formation (Demarest, 1981) form a complex of transgressive lithosomes stranded intact at various high sea stands of interglacials, up to 3-6 m above present (Fig. 10-2). Thus, from a geologist's point of view, we are approaching the norm of 3-to-6-m higher sea stands; the only question is one of time: degree of magnitude of reversal in thousands of years or a human-induced acceleration to 100-200 yr in the future.

Clearly, in spite of a massive literature and much discussion, too little is proven about the geological aspects of sea-level rise, causes, effects, cyclicity, fluctuating or smooth sea-level curves, and global eustatic and/or local relative coastal responses. Some geologists believe that sea level rose to its present position about 6,000-7,000 yr ago during the so-called climatic optimum and has fluctuated above and below that

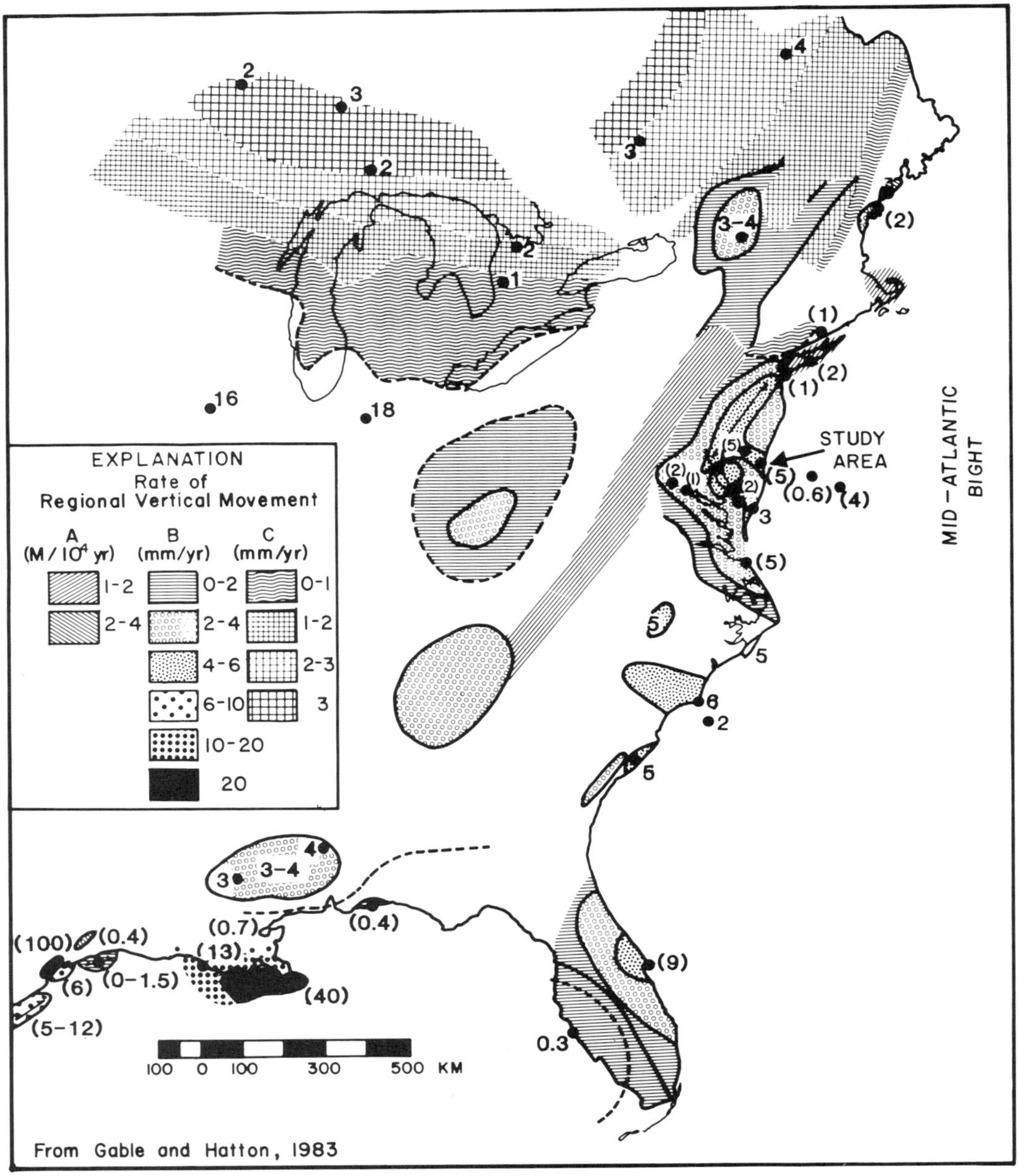

Figure 10-1. A neotectonic map showing areas of coastal subsidence and coastal rise relative to the ocean. Figures in parentheses indicate subsiding areas in millimeters per year. The mid-Atlantic coast is in the center of one of the largest subsiding embayments in the United States other than the massive subsidence ongoing in the Mississippi River delta. It is possible that this subsidence may be caused in part by factors such as withdrawal of groundwater as opposed to eustatic sea-level rise. (From Gable and Hatton, 1983)

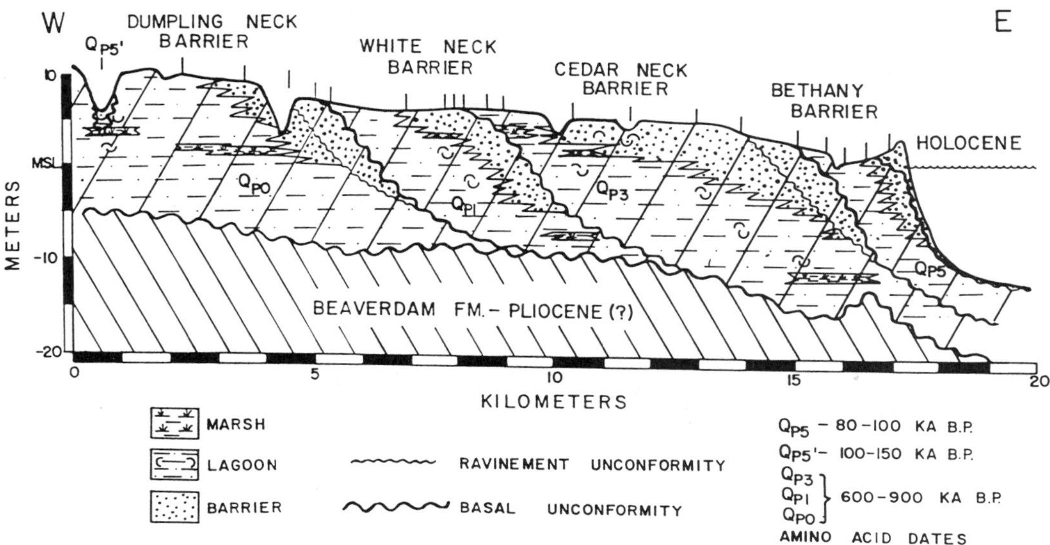

Figure 10-2. Transgressive paralic sedimentary sequences of the Atlantic coastal plain, showing stratigraphic facies relationship and representing at least four Pleistocene epoch interglacials (based on amino acid racemization dating techniques). (After Demarest, Biggs, and Kraft, 1981)

level ever since (Fairbridge, 1960, 1961, 1976; Mörner, 1971), whereas many others believe it has never been above the present during the Holocene (Bloom, 1970; Shepard, 1963). Researchers now commonly agree that no one is able to define a truly global eustatic sea-level curve for the Holocene epoch of geologic time (past 10,000-15,000 yr). In a recent meeting (IGCP, 1983), a consensus was reached that local geological effects are of prime importance in defining both long- and short-term past geological variants and resultant local relative sea-level variation curves (Fig. 10-3). For instance, Gable and Hatton (1983) noted that while much of the nation's shoreline is relatively tectonically stable, large portions include local uplift and subsiding areas. In the Gulf of Mexico, major subsidence is noted at rates of 40 mm/yr at the tip of the Mississippi Delta. The Delaware coastal zone lies in one of the largest subsiding coastal zones in the nation (see Fig. 10-1). A subsiding embayment from the Carolina capes northward and centered in the Delmarva Peninsula and extending onward along the Fall Zone to northern New Jersey occurs on the northwest flank of the Baltimore Canyon Trough (Geosyncline). This has been an area of intensive study of past sea-level changes from the early works of Stuiver and Daddario (1963) and Newman and Rusnak (1965) to workers in Delaware (Kraft, 1976; Belknap and Kraft, 1977; Kraft and John, 1976; Kraft et al., 1975). Geological factors in some areas counter or minimize global or eustatic sea-level fluctuations, whereas in other areas geological factors enhance or exacerbate the situation. The majority of the western coast of the United States is in a tectonically rising setting; nevertheless, coastal erosion is a severe problem.

Pilkey et al. (1981) discussed the relationship of sea-level rise to coastal erosion and warned of dire economic consequences should

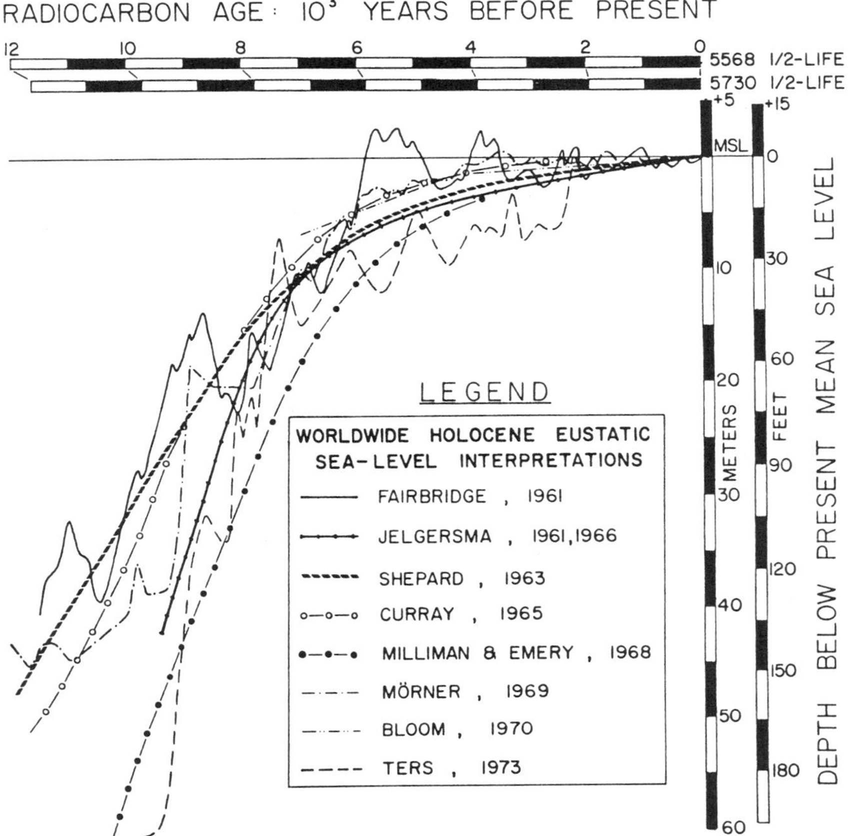

Figure 10-3. Some of the more widely publicized global sea-level curves. Obviously they cannot all be correct indicators of Holocene eustasy. Accordingly, local relative geological factors and climatological factors have affected the evidence from which these curves are derived. In general the longer-term (many millennia) geological fluctuations of the curve cannot be used for short-term prediction. However, the large number of curves showing dramatic fluctuation suggests that these fluctuations might be predictable. If so, an understanding of the geological fluctuations of a local relative curve such as that of the Delaware coastal zone (heavy smooth line) is critical in terms of predicting rates of sea-level rise in the short term.

people continue to occupy the coastal zone as they do at present. These authors suggest careful consideration in national planning of U.S. policy in the coastal zone. Almost universal agreement regarding this problem exists among the geologists. A noted example of dissent, however, is that of O'Brien (1982, 1984), a civil engineer, based on engineering considerations. Criticisms of this kind tend to lack intensive study of the geologic literature, and some are based on outdated geological concepts.

The Bruun Rule (Schwartz, 1967) states that an equilibrium profile exists between rising sea levels and coastal erosion. There are many problems in applying the Bruun concept. Over the years many geologists and engineers have attempted to apply the Bruun Rule in accounting for shoreface landward shifts and deposition in the inner shelf zone. Very few have introduced the third dimension into their analyses. Schwartz and Milicic (1980) noted that the Bruun Rule is not a rule regarding coastal erosion but a description of the response of coastal erosion and inner shelf deposition to sea-level rise, presumably under a profile of equilibrium. Bruun (1980) noted that coastal erosion was, of course, a three-dimensional event and, in 1983, elaborated, stating that his assumptions were based on a "closed material balance system between (1) the beach and near shore and (2) the offshore bottom profile." He noted that the concept was two-dimensional and that care should be taken in expanding it into a three-dimensional system. He also noted that composition of beach and bottom materials and bottom geomorphology were important. Fisher (1980), in an analysis of Rhode Island shoreline retreat, suggested that sediment loss from shoreline erosion would not exclusively move offshore as predicted by the Bruun Rule. Seventy-six percent of the eroded beach material in Rhode Island could be accounted for in washover and inlet (landward) deposits; the remaining 24% "may therefore be deposited offshore as proposed by the Bruun Rule" (Fisher, 1980, pp. 52-53). It seems clear that Bruun never meant to go further than to attempt to describe a two-dimensional equilibrium profile relating eroded sediment to a rising equilibrium profile in the inner shelf. In reality many other factors are involved.

From a global climatologic point of view, Barnett (1983, p. 15) has noted the many complex and interrelated factors involved in studies of recent changes in sea level and concluded, "It is not possible at this time to explain reliably the apparent increase in RSL (relative sea-level rise)." Pilkey et al. (1981) have noted that if we simply apply the Bruun Rule, a 30+ cm/century rise in sea level along much of the Atlantic coast "could be responsible for the serious erosion that confronts many beach communities" (as stated by J. Titus in a 1983 discussion of sea-level rise and barrier islands). Diamante (1982) noted the significance of only 1 mm/yr, a "truly enormous value."

> For example, a value of 3 mm/year would suggest that the British coastline experienced nearly 10 ft (2.4 m) of sea-level rise since the Norman conquest, and the Dutch coast should have experienced more than 3 ft (91 cm) rise since the days of Rembrandt. There is little evidence for rise of this magnitude. Either there is a dramatic recent increase in the rate of sea-level rise or the estimates contain unresolved biases, and evidence exists to support both conclusions. (paraphrased from Diamante, 1982)

We know there was a major warming trend in the several centuries after A.D. 1000 leading to the major Viking expansions westward into Greenland and other areas. Likewise, we know that there was a time referred to as the Little Ice Age from approximately A.D. 1500 to A.D. 1850. Further, temperatures 5,000 yr ago appear to have been higher than at present, possibly accompanied by higher global sea levels during the climatic optimum (Mörner, 1971). It is not clear whether these times of known temperature changes were always accompanied by dramatic sea-level fluctuations. In summary, as Laudan (1983, p. 280) notes in paraphrasing William Conybeare (1822), "The great and fundamental problem . . . of theoretical geology was [is] the problem of changing levels of land and sea." To all occupants of the U.S. coastal zone, this statement has great importance.

RELATIVE SEA-LEVEL CHANGES

Sea level is rising relative to land. The coasts of Delaware are an area in which this rise is above the average for the remainder of the Atlantic and Gulf coasts (with the exception of the Mississippi Delta) (Figs. 10-4, 10-5, 10-6). We know that the rates of relative sea-level rise in our coastal sector since the 1920s are double that of the recent geologic past (2,000 yr). Problems of major coastal erosion related to sea-level rise have been defined along the Delaware coastal zone (Kraft et al., 1979) and approximately 50 other studies in our coastal zone. Should Environmental Protection Agency (EPA) and National Research Council (NRC) warnings have any validity, then geological studies of the nature of the changes in rate of sea-level rise in the recent past and of the amplitudes of the relative geological sea-level rise of the past 1,000 yr become imperative.

The Delaware coastal marsh comprises 13%

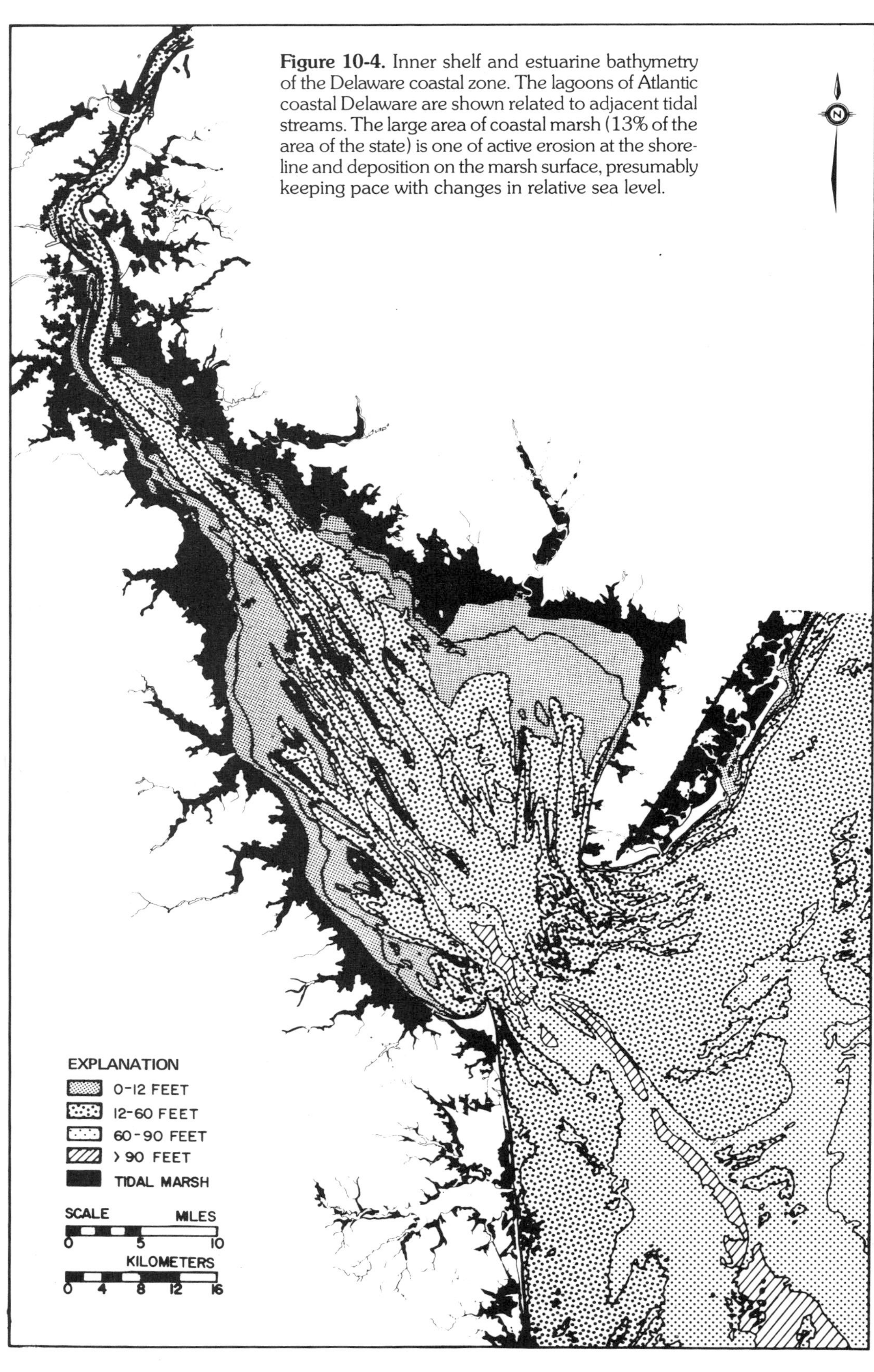

Figure 10-4. Inner shelf and estuarine bathymetry of the Delaware coastal zone. The lagoons of Atlantic coastal Delaware are shown related to adjacent tidal streams. The large area of coastal marsh (13% of the area of the state) is one of active erosion at the shoreline and deposition on the marsh surface, presumably keeping pace with changes in relative sea level.

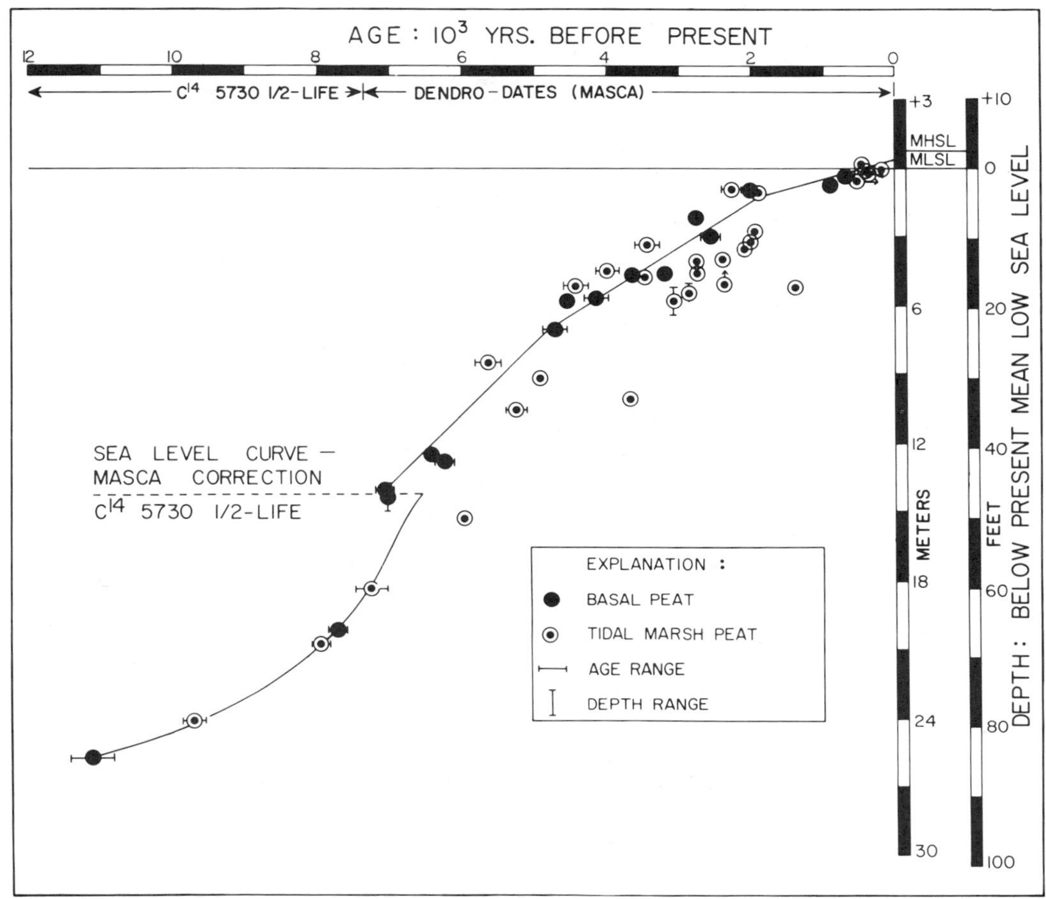

Figure 10-5. Some of the data used to derive a local relative sea-level rise curve for coastal Delaware. The curve shown is a smooth average based only on dates and depths of basal peats. The other data shown include dates on materials that would be expected to be below or above sea level and dates on tidal marsh peats that vary greatly in the area, possibly because of differences in elevation at deposition and/or compaction of the peats and muds.

of Delaware. If we take the EPA's figures (Hoffman, Keyes, and Titus, 1983) as a guide, then most assuredly several events will occur: a landward transgression of several kilometers and/or massive engineering works and projects that probably will cost billions of dollars in the coming century. The Delaware Department of Natural Resources and Environmental Control already monitors coastal erosion, change in landform, and the expenditure of large sums of state (and federal) monies in an attempt to hold the coastal zone in its present position. They do this by techniques such as sand nourishment of the beach and construction of coastal defense structures. These techniques will definitely not be sufficient economically or geologically should the greenhouse effect occur. Therefore, the formation of predictive curves of sea-level rise (ranges from maximum to minimum levels) and the projection of sea-level rise rates from the very recent past and the nature of the longer geological process become critical. Other areas of the nation will of course encounter similar problems. However, other than in Louisiana, the coastal zone of Delaware and adjacent areas will be one of the maximum impact in terms of subsiding shorelines and landward movement of the coastal environments.

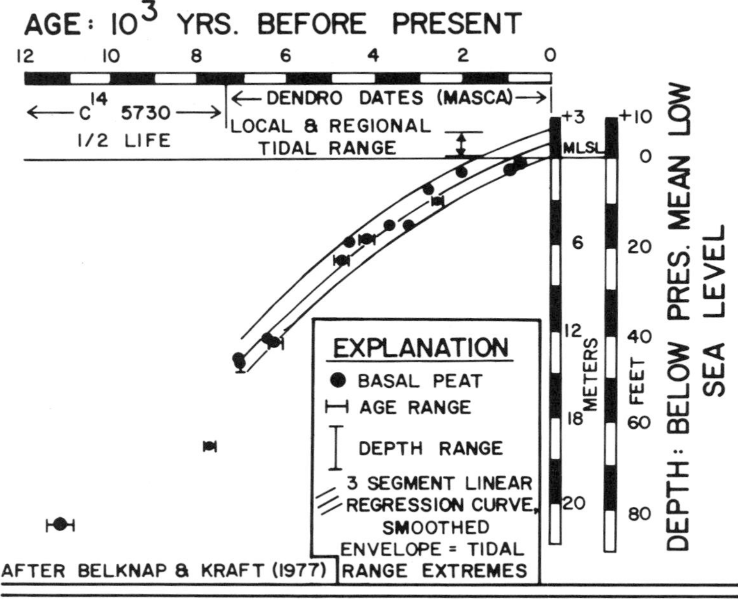

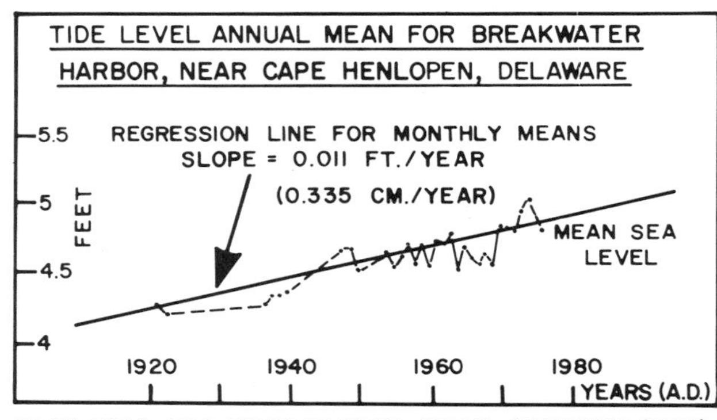

Figure 10-6. A regression analysis through basal peat data showing a sea-level rise curve for the Delaware coastal zone over the past 7,000 yr. Compare this with the lower portion of the figure, which shows a regression line through tidal gauge data that indicates a present relative sea-level rise of 33 cm/century, roughly double the geologic rate of the past 2,000 yr.

We are aware that a large portion of the continental shelf to the east of the Delaware coastal zone included environments very similar to those of the present coastal zone (Edwards and Emery, 1977; Emery et al., 1967). We must assume that our earlier Holocene predecessors, the American Indians, occupied these coastal zones. Over the 15,000 yr since the beginning of the waning of the last major ice sheets across North America, these peoples would have migrated with the coastal zone in a landward direction as the transgression and sea-level rise of the Holocene epoch occurred. Indeed, on land, there is definite evidence that peoples were in the area up to 10,000-12,000 B.P.

Occupancy of the coastal zone since the late 1600s has been significantly different. In particular, since about 1940 it has been deemed desirable to hold the line of the coast against erosion mainly to protect the large financial investments in coastal developments. A critical issue in the near future will be one of philosophy as well as economics. Will we continue to attempt to occupy the present coastal zone regardless of the effects of sea-level rise? Certainly the major port cities will continue to be occupied even with the most extreme sea-level changes; however, the costs of necessary shoreline defense may be enormous.

A very different socioeconomic issue must be addressed regarding the occupancy of the coastal zone in its present location for environmental, wildlife, and recreational needs. On the one hand, it is not evident that the value of the properties (land and buildings thereupon) will in any way be equal to the costs of engineering a structured shoreline in the Dutch or German method. On the other hand, it is possible that decisions will be made to alter completely our rationale for occupancy of the coastal zone. Probably these rationales will slowly evolve over the next several decades if it becomes more evident that the greenhouse effect and its implications are probable or that there are some questions about the EPA (Barth and Titus, 1984) and NRC alerts and their scientific bases. There are some questions based on an analysis of the recent geological past that might suggest scenarios other than that of the greenhouse effect (Edwards and Emery, 1977; Kraft and John, 1976).

DETERMINATION OF SEA LEVELS AND PROJECTIONS INTO THE FUTURE

The broad fringing marshes and other coastal geomorphic variants of coastal Delaware are shown in Figure 10-4. Erosion rates vary from an average of 1 m/yr along the Atlantic Ocean coast to extremes of 6 m/yr along the shoreline of the Delaware estuary. Approximately 13% of the area of the state of Delaware is within the tidal zone or wetland area. Over the short-term geological past, sea level has risen at a rate of approximately 15 cm/century (see Fig. 10-5). Since about 1930, sea level has doubled this rate of local relative sea-level rise to approximately 33 cm/century (see Fig. 10-6). Thus, the potential exists for determining whether sea level has risen over the recent past in a fluctuating manner, a steady straight line, or an accelerating smooth-curve format. Research is now underway in Delaware in an attempt to make comparisons with the work in other coastal areas (Colquhoun, 1981; Fairbridge, 1976; DePratter and Howard, 1977, 1981).

Geologists work under two basic geological dictums: Walther's Law of Correlation of Sedimentary Facies and the Doctrine of Uniformitarianism. The latter, "The present is the key to the past," is most useful in terms of this proposed research. But could we not equally say, "The past is a key to the present and to the prediction of future geological events?" The pioneering works of Fairbridge (1960, 1961, 1976, Mörner (1971), Coleman and Smith (1964), Milliman and Emery (1968), Curray (1964), and Jelgersma (1961) and reviews by Hopley (1983), Kidson (1982), and Bloom (1977) have set the stage for Holocene epoch sea-level studies and the many hundreds of researchers who have worked on this subject. Kraft and John (1976), Belknap and Kraft (1977), and Kraft (1982) have analyzed the Delaware coast for the longer-term sea-level fluctuations of the Holocene epoch. In adjacent states, Newman and Rusnak (1965) and Stuiver and Daddario (1963) constructed curves with small amounts of data but that are of great use in comparison with more recent relative sea-level data from the subsiding study area. In making relative sea-level rise interpretations for coastal Delaware, we have assumed that our data are not sufficient to determine whether sea level has risen in a fluctuating or cyclic fashion as opposed to a smooth-curve format throughout the Holocene epoch. However, many geologists have no doubts that global eustatic sea levels rose above the present sea level 6,000-7,000 yr ago and have fluctuated above and below present sea level ever since.

Colquhoun's study (1981) appears to have demonstrated that sea level has indeed fluctuated on a several-hundred-year basis from 5,000

B.P. to 2,000 B.P., with suggestions that these fluctuations may have continued into the past millennium. Further, Colquhoun (1981) correlates his sea-level peaks with transgressive events as interpreted in the Netherlands. Kidson (1982) showed the dilemma with which one is faced in using the geologic interpretations of many researchers around the world in terms of predicting geologic events into the future. Cinquemani and co-workers (1981) showed the dilemma of attempting to use radiocarbon dates of basal peats (indicators of sea level) from widespread areas along the eastern coast of the United States. Each study of sea-level fluctuation clearly must be done in the relative sea-level sense in a localized area. Data from the East Coast of the United States from Cinquemani and co-workers show that sea-level positions versus recent geologic time from Georgia to the northeastern United States vary widely because of tectonic and many other factors.

Should a major sea-level rise occur, effectively global eustatic, some coastal zones may respond with greater or less local relative sea-level rises due to other geologic factors. It is important to note that a summary of the situation made by Vita-Finzi in 1973 (p. 59) is now almost universally accepted by geologists: "The idea of coastal stability has been described in some quarters as a myth which casts doubt on all attempts to define a truly eustatic curve of sea-level rise, while a eustatic curve of global applicability comes to be dismissed as chimerical."

Role of the Marsh Surface and Marsh Stratigraphy

Coastal geologists specializing in salt-water marsh studies have shown that the surface of the marsh is a direct reflection of present sea level, modified by tidal variants and many other geologic factors. Tide ranges plus geology of the barrier-nonbarrier coastal setting can lead to wide variations in tidal marsh deposition surface levels as related to local mean sea level. However, it is also known that the topography of a marsh surface varies considerably from the leading edge of the transgression at the fringe of the high marsh to those portions of the lower marsh that grow into the salt- and brackish-water interface. It might appear that one can determine the position of sea level and a record of rates of sea-level rise over the recent past by simply taking a core in the marsh and making a series of dating curves of rates via short-term actual sedimentation measurement models or via the ^{210}Pb and ^{137}Cs radionuclide techniques. Church and Somayajulu (1981) noted the usefulness of these techniques and some of the pitfalls in the sense of possible vertical migration of some of the isotopes.

However, many geologists miss the point. As clearly defined by Allen (1977), deposition in the marsh occurs in many fashions. Levels of marsh deposition on the sandy substrate of the back of a barrier vary greatly from the elevation of marsh surfaces comprised of the same species of marsh flora. The center of a marsh that overlies a relatively greater thickness of mud varies in elevation from similar marsh-flora depositional surfaces on the fringes of a lagoon or estuary. Only at the leading edge of the transgression where 1 m or 2 m of marsh sedimentation has occurred can one expect to encounter a continuous depositional record of marsh history and determine with reasonable accuracy the rate of sea-level rise. Marshes are classified into high, medium, and low marsh based on their floras and their topography relative to sea levels and tidal amplitudes. Allen (1977) noted that much of marsh deposition might be lateral accretion in the sense of point bars on marsh creeks and may include slump blocks in a stratigraphic record that might not be recognized in a core. Certainly any dating of rate of sea-level rise by either ^{14}C or ^{210}Pb and ^{137}Cs techniques, at random, across a wide marsh will lead to varying rates and results in determining positions of sea level in the very recent past as well as the earlier Holocene epoch. In wide marshes overlying thick mud sequences, compaction becomes a major factor in leading to similar dates of deposition now at different levels.

Another factor is that of the tidal amplitude. In effect, a local sea-level position will be measured at mean high tide or spring high tide levels. Thus, in a complex estuarine Atlantic coastal lagoon setting like that of Delaware, tidal ranges in the marsh and marsh fringe vary from 20 cm to 152 cm (see Fig. 10-4). This divergence in sea-level positions must be considered an important factor in both our understanding the rates and nature of coastal erosion and migration of coastal sedimentary environmental lithosomes as driven by sea-level rise. Farther north in the estuary of the lower Delaware River, tidal amplitudes reach nearly 200 cm. The bay is surrounded by brackish and medium-salinity tidal marshes.

The worst scenario of any studies that make predictions using the marsh surface as an indicator of sea level would be the indiscriminate use of data subject to the highly variable rates of sedimentation, elevation, and geomorphic positions, as noted by Allen (1977). Stumpf (1981) hypothesized that the great majority of deposition in marshes occurs during the high-intensity storm-flood event. Regardless, it really does not matter since the micromarsh stratigraphic/sedimentologic record does indeed record positions of sea level. The critical factor is consistency in use of the highest sea-level parameters based on their leading-edge sedimentary sections and carefully selected basal marsh peats. Consistent attention to predictive results based on the high variation in marsh surface elevations at peak sea level in the various submarsh environments is imperative.

Colquhoun (1981) and DePratter and Howard (1981, 1977) noted that archeological sites can provide important indicators and limitations regarding the study of marsh sediment techniques in analyzing recent past sea-level fluctuations. The potential use for these techniques in the Delaware coastal zone is being examined. In one area of the Delaware coastal zone a series of recurved spits, covered by dune sands, is being overridden by marsh fringe accommodating sea-level rise over the past millennia. In this particular case archeological dating shows a continuous sequence of occupation of the oldest spit, first occupied approximately 2,000 yr ago, to the last American Indian occupancy sites of about A.D. 1600 on one of the youngest recurved spits. Projections brought forward from such a setting may be useful in interpreting the potential of fluctuations of a geologic nature. If so, projection of the amplitudes of geologic fluctuation into the twenty-first century may provide important limits to the highly variable geologic and/or greenhouse effects in possible sea-level scenarios (in a limiting and/or additive and subtractive sense).

Coastal Erosion and Sea-Level Rise in Atlantic Coastal Lagoons: Rehoboth Bay

Swisher (1982) completed an intensive analysis of rates and processes of shoreline erosion under all the geomorphologically varied coastal zones of Rehoboth Bay (Atlantic coastal lagoon). Based on air photo analysis of six different sets of photographs from 1938 to 1984, coastal erosion rates of an average of 0.4-0.7 m/yr, with extremes of coastal erosion from 1.2 m to 3-4 m/yr were delineated. The average tidal amplitude for Rehoboth Bay as measured by Karpas (1978) was 30 cm. However, this tidal amplitude really varies from approximately 80 cm to the southeast of Rehoboth Bay at Indian River Inlet to 20 cm in the extreme northwest of Rehoboth Bay at Love Creek. We have observed many cases of wave-sapping into marsh sediments of the order of 30-40 cm amplitude (Fig. 10-7). The U.S. Department of Commerce Tide Tables (1987) present an average tidal amplitude for Rehoboth Bay of 20 cm (probably not updated in many decades). In the 1930s Indian River Inlet to Indian River Bay and Rehoboth Bay was widened, deepened, and stabilized by rock jetties. From then to the present, Rehoboth Bay has undergone a catastrophic rise in sea level on the short term (approximately 2×). Thus, it becomes an ideal model in which to observe both coastal erosion rates related to accelerated sea-level rise and, in terms of rates, of marsh rise responding to a possible doubling of average amplitudes of the tides. Normal marsh fringes into coastal lagoons merge gradually into the water, with the lower tidal marsh flora growing into the edge of the lagoon.

Figure 10-7. Wave erosion of low salt marsh (*Spartina alterniflora*) sediment fringing Rehoboth Bay, Delaware.

To the contrary, the tidal marshes of Rehoboth Bay are cliff-faced and up to 40 cm in height.

The small wave-sapped cliff on the fringe of Rehoboth Bay is clearly out of synchronization with the present tidal amplitude and is undergoing severe erosion. Massive erosion is well documented since the 1930s in Rehoboth Bay via small islands covered by a marsh cap (U.S. Geological Survey, 1928). Presently only one island exists, Marsh Island, covered with a low marsh *(Spartina alterniflora)*. Nearby to the south, until 1980, was the last remnant of Big Piney Island. Big Piney Island is recorded on the USGS map as having had an area of 7.5 acre in 1928 (originally 15 acre according to Mr. Charles Horne, owner). The island was covered by a pine forest. In 1967, only one trunk of a pine tree remained. Now the island has disappeared. Little Piney Island disappeared between 1964 and 1968, and Windmill Island is reported in the historic literature but cannot be located. All this evidence indicates a massive change in the tidal amplitude over a very short time (50 yr). The tidal amplitude change doubled from an average 20 cm to an average 40 cm. These types of observations should enable us to envisage many scenarios of catastrophic (short-term) sea-level rise of the coming century.

Holocene Stratigraphy, Paleogeography, and Prediction of Future Geographies

Coastal erosion has been occurring along the Delaware coastline throughout the entire Holocene epoch. From an original shoreline position approximately 75 km seaward on the outer edge of the Atlantic shelf, the coastal environments have migrated landward and upward by the mechanisms of coastal erosion, eustatic sea-level rise, and probably subsidence of the flank of the Baltimore Canyon Trough (Geosyncline). This migration of coastal environments to the northwest occurred across a deeply incised late Wisconsinan topography dominated by the ancestral Delaware River, its tributaries, and the Delaware shelf valley and possible deltas. The rate of coastal erosion along the coasts of Delaware, based on 150 yr of

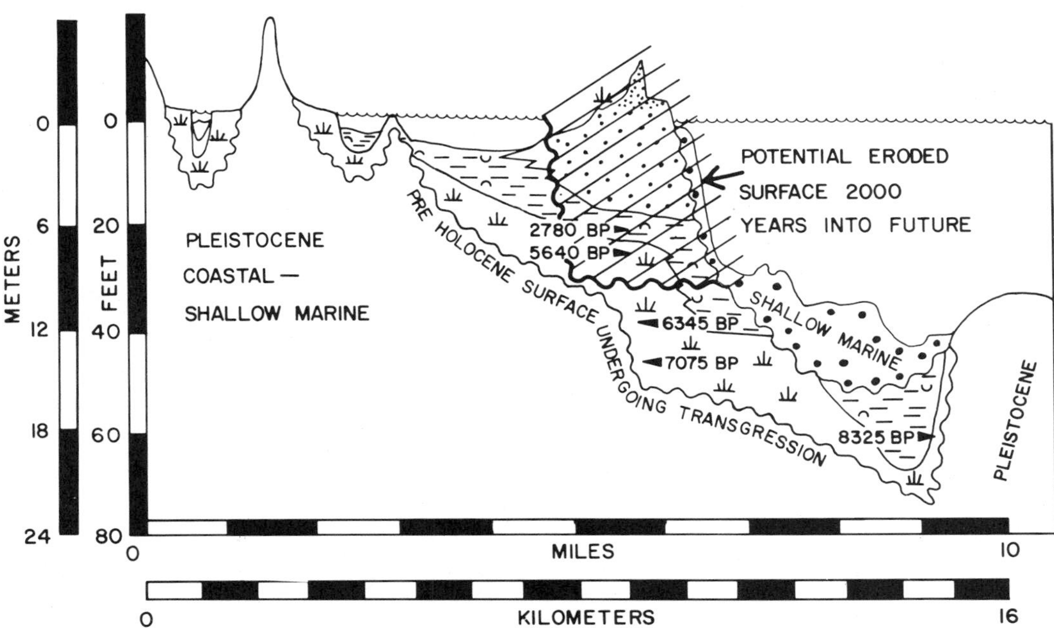

Figure 10-8. A geological cross-section across Rehoboth Bay to the inner shelf of the Atlantic Ocean. The dark striped area is the predicted zone of erosion by retreat of the shoreface or ravinement surface over the next 2,000 yr, based on past geologic conditions, or 100-200 yr into the future under the greenhouse effect. (After Kraft and Belknap, 1975, Fig. 7)

precise mapping by engineers and the predecessors of National Oceanographic and Atmospheric Administration (NOAA), varies from 1 m to 6 m/yr, and the process is pervasive. A number of papers have dealt with the stratigraphy of the Holocene stratigraphic units in the Delaware coastal zone (Kraft, 1971; Kraft et al., 1975, 1979; Kraft, Allen, and Maurmeyer, 1978; Kraft and Belknap, 1975; Kraft and John, 1979; Marx, 1981; Maley, 1981; Weil, 1976; Belknap and Kraft, 1985). Numerous stratigraphic cross-sections of the Holocene stratigraphic units have been constructed. Furthermore, the areal distribution of the various coastal sedimentary environmental lithosomes and their thicknesses have been mapped. Basal remnants of the Holocene stratigraphic sequences have been preserved on the inner shelf and in the Delaware estuary. Presently, erosion in the shoreface is removing the Holocene (and Pleistocene headland) stratigraphic record to depths of approximately 10 m as the transgression continues. Figure 10-8 is a stratigraphic cross-section across the Atlantic coastal lagoon, Rehoboth Bay, its adjacent barrier, and the inner shelf. A projection of the portion of the stratigraphic record that will undergo erosion in the near-term future (ca. 100-200 yr) should the greenhouse effect prove to be as projected or over a period of 2,000 yr based on longer-term stratigraphic projections after Kraft and Belknap (1975) is shown. Figure 10-9 is a cross-section of the wide peripheral marsh (no barrier) adjacent to the Delaware estuary approximately 50 km from the Atlantic Ocean. At Port Mahon along the mid-Delaware Bay western coastline, erosion is currently occurring at rates up to 10 m/yr. Here again, greater than 3 km of erosion might be

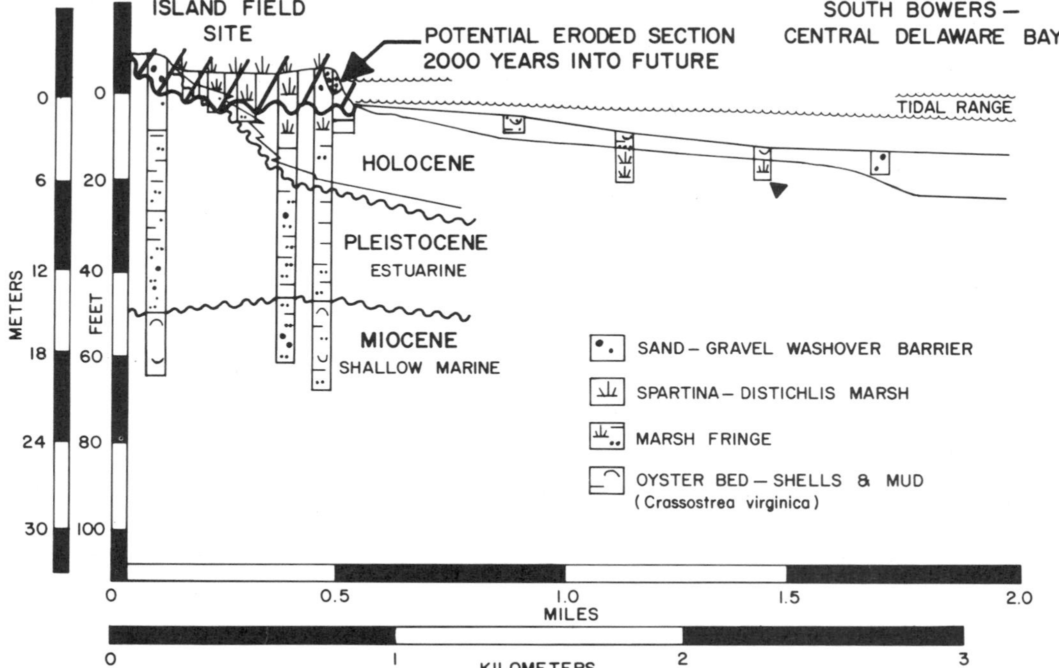

Figure 10-9. A geological cross-section of the Port Mahon coastal zone of the Delaware estuary approximately 50 km from the Atlantic Ocean. The dark striped area is the zone of predicted erosion that will accompany a 2-m sea-level rise. (After Kraft and Belknap, 1975, Fig. 9)

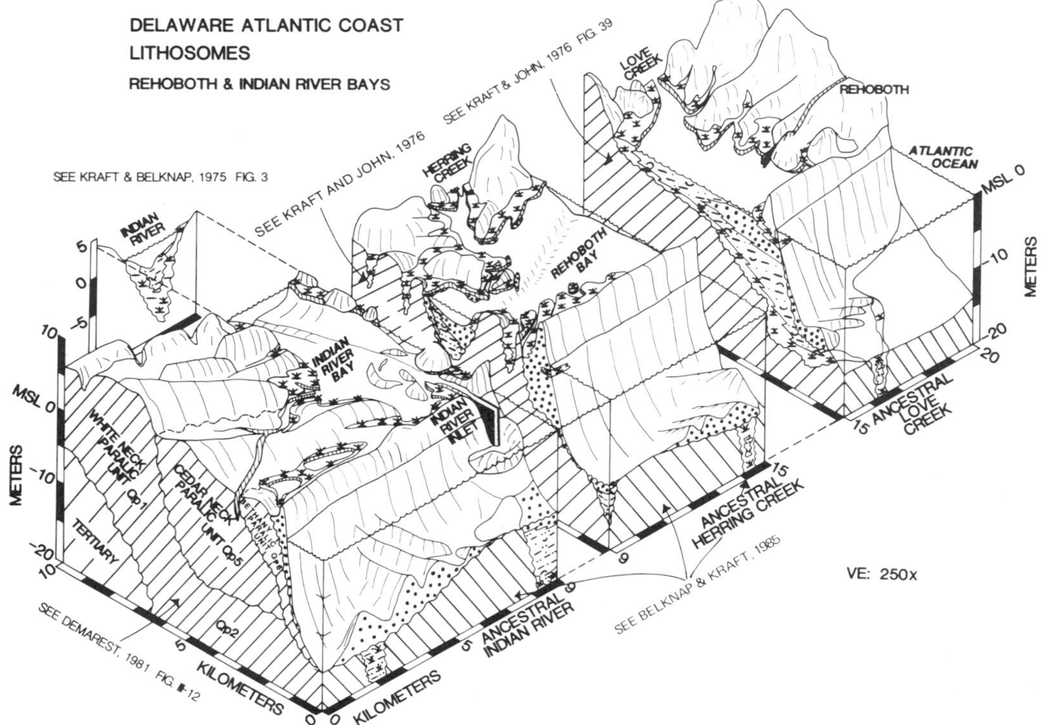

Figure 10-10. An isometric construction of the Delaware Atlantic coastal zone showing the relationship of the present coastal sedimentary depositional environments to shoreface erosion and the pretransgressive Wisconsinan age topography.

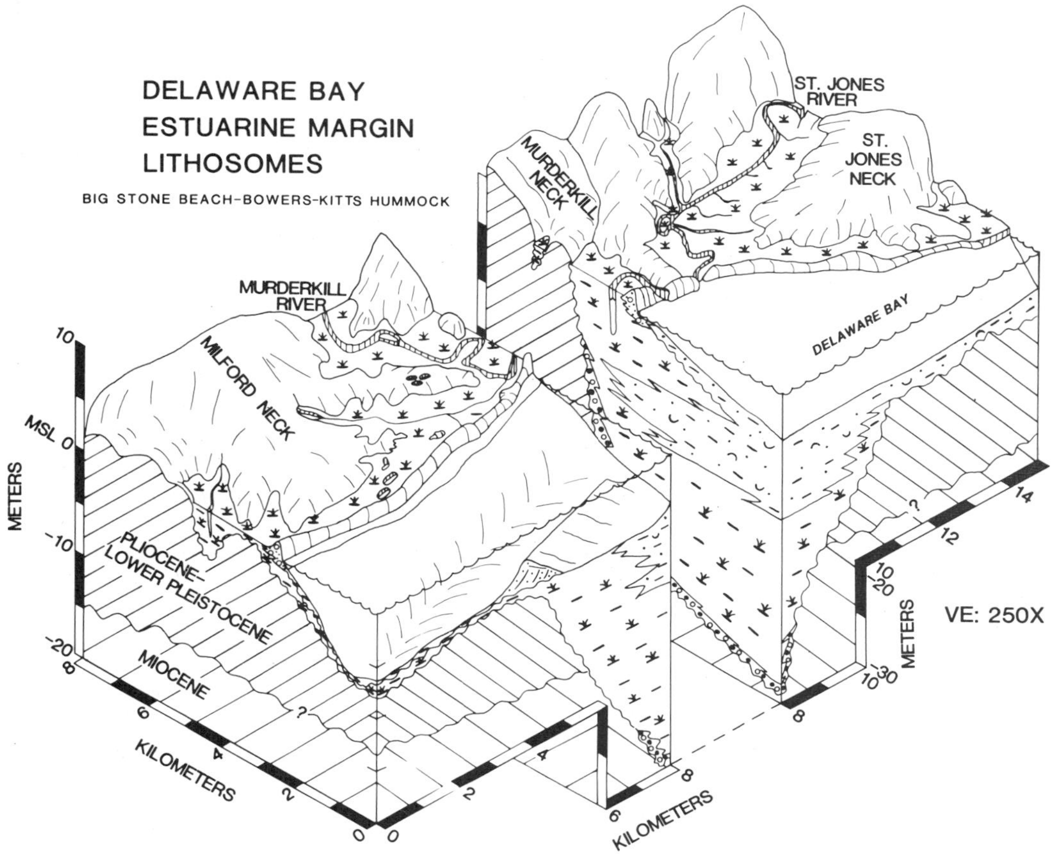

Figure 10-11. An isometric diagram of the coastal zone along the western side of the Delaware estuary showing the relationship between present marsh deposition, the transgressive barrier, and the pretransgression surface. The valley of the Murderkill River encountered in the center of the diagram was first flooded by salt water as evidenced by marsh peat sediment dates approximately 10,000 B.P.

projected in the short term with a catastrophic sea-level rise or over the longer-term geologic transgression (thousands of years).

Based on the dozens of geologic cross-sections of Holocene stratigraphy along the Delaware Atlantic coast, and on isopachous maps and pretransgression surface maps, it is possible to construct three-dimensional diagrams of the present coastal morphology as related to subsurface cross-sections (Figs. 10-10 and 10-11). These isometric diagrams allow us to relate the present surface undergoing inundation with that of the past pretransgression surface, an early Holocene-late Wisconsinan trellislike drainage system and tributary to the ancestral Delaware River. Sediments of the various depositional environments are shown in a three-dimensional analysis of their separate lithosomes such as the coastal marsh, coastal lagoons, estuary, and Atlantic coastal barriers and spits and estuarine barriers.

Figure 10-12 shows many variants of predictive sea-level curves into the short- and long-term future of the next 1-200 yr. Straight-line averaged trend of sea-level rise over the past 2,000 yr (based on basal salt marsh peaks only) is extended to the present (see also Figs. 10-5 and 10-6). Various scenarios of sea-level rise are presented. Those of the EPA (Hoffman, Keyes, and Titus, 1983) are extended from the present several hundred years into the future. Using tangents to the regression analysis sea-level rise curve (Fig. 10-6), lines of projection are also shown based on sea-level rise since the 1920s,

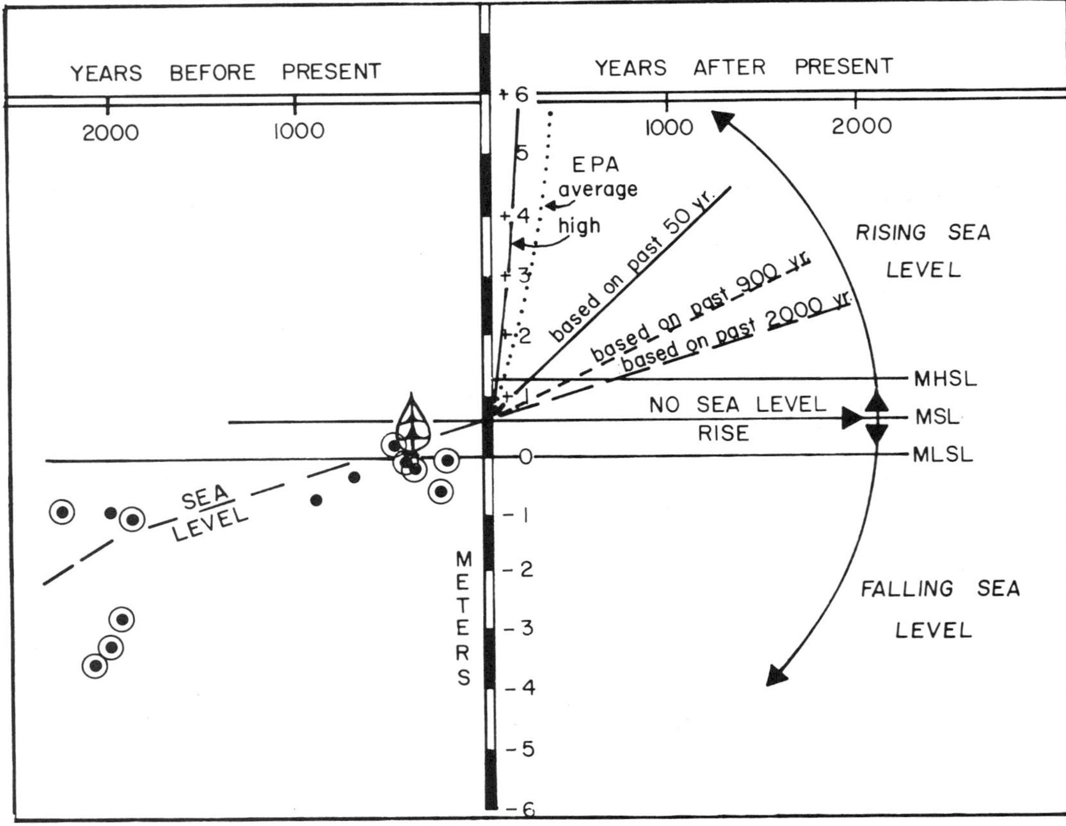

Figure 10-12. Various scenarios of sea-level rise curves as projected by the EPA and local relative sea-level change studies in coastal Delaware.

the past 900 yr, and the past 2,000 yr. It must be recognized of course that none of these scenarios may be correct. There may be no sea-level rise; indeed, there may be a falling sea level into the future. Should the long-term scenario be one of continuing sea-level rise until a peak interglacial such as those in the Pleistocene epoch, then the curves may be projected to approximately 6 m above present (see Fig. 10-2) based on evidence in inland Delaware (Demarest, Biggs, and Kraft, 1981). Should the longer-term Holocene epoch curve be oscillating, as shown by Colquhoun (1981) and DePratter and Howard (1977, 1981), then entirely different scenarios must be projected into the future. These scenarios might include drops in sea level below present or oscillations of sea level within an ever-increasing rise toward approximately 6 m above present. Until the driving mechanisms of sea-level change during the present time and that of at least the latter part of the Holocene epoch are well known, projections into the future cannot be based on firm data; rather, they lie in the realm of geologic hypothesis. Nevertheless, the probability of rise above present level appears most likely, and the possibility of catastrophic change is well supported in some quarters (Hollin, 1972; Barth and Titus, 1984; Revelle, 1983).

Interpretations of three-dimensional forms of Holocene coastal sedimentary environmental lithosomes and an understanding of the morphology of the pretransgression surface provide us with sufficient information to construct

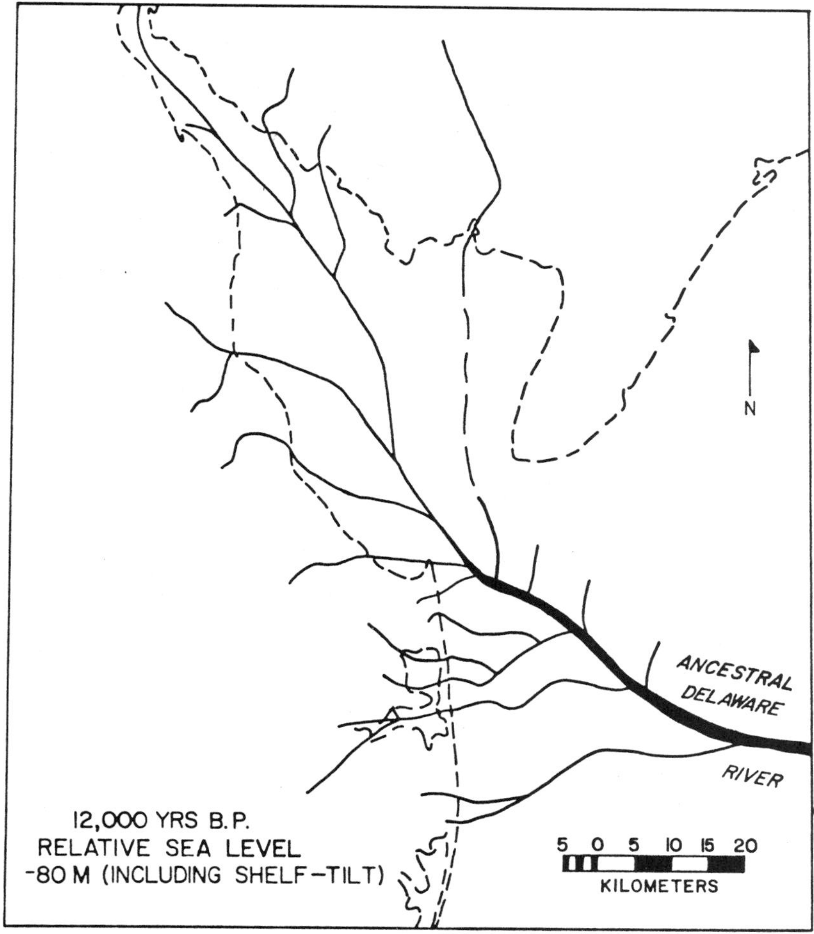

Figure 10-13. A paleogeographic reconstruction of the pre-Holocene epoch-Late Wisconsinan age ancestral Delaware River and its tributaries in the Delaware coastal zone.

paleogeographic maps of the inner shelf and coastal zone of the Delaware estuary and Atlantic coast. Figure 10-13 shows the trellislike drainage system of the ancestral Delaware River and its tributaries in the Delaware coastal zone approximately 12,000 B.P. Eustatic sea-level rise may be projected across this topography. Part of this projection must include the major aspect of coastal erosion and reformation of coastal landforms. By 7,000 B.P. we believe that a lagoon-barrier and estuarine system existed on the inner shelf and in the area of the present Delaware estuary (Fig. 10-14). The interfluves between the tributaries to the ancestral Delaware River were being eroded by Atlantic coastal wave processes, and a lagoon-barrier system developed. The Pleistocene paralic lithosomes provided the sediments for redistribution of sand by littoral transport into the barrier and barrier-spit systems and muds into the estuaries and lagoons. Figures 10-13 and 10-14 may be regarded as fairly accurate because they are based on intensive studies over 15 yr by a number of people, with excellent drill control and a good seismic base for understanding the pretransgression surface on the inner shelf

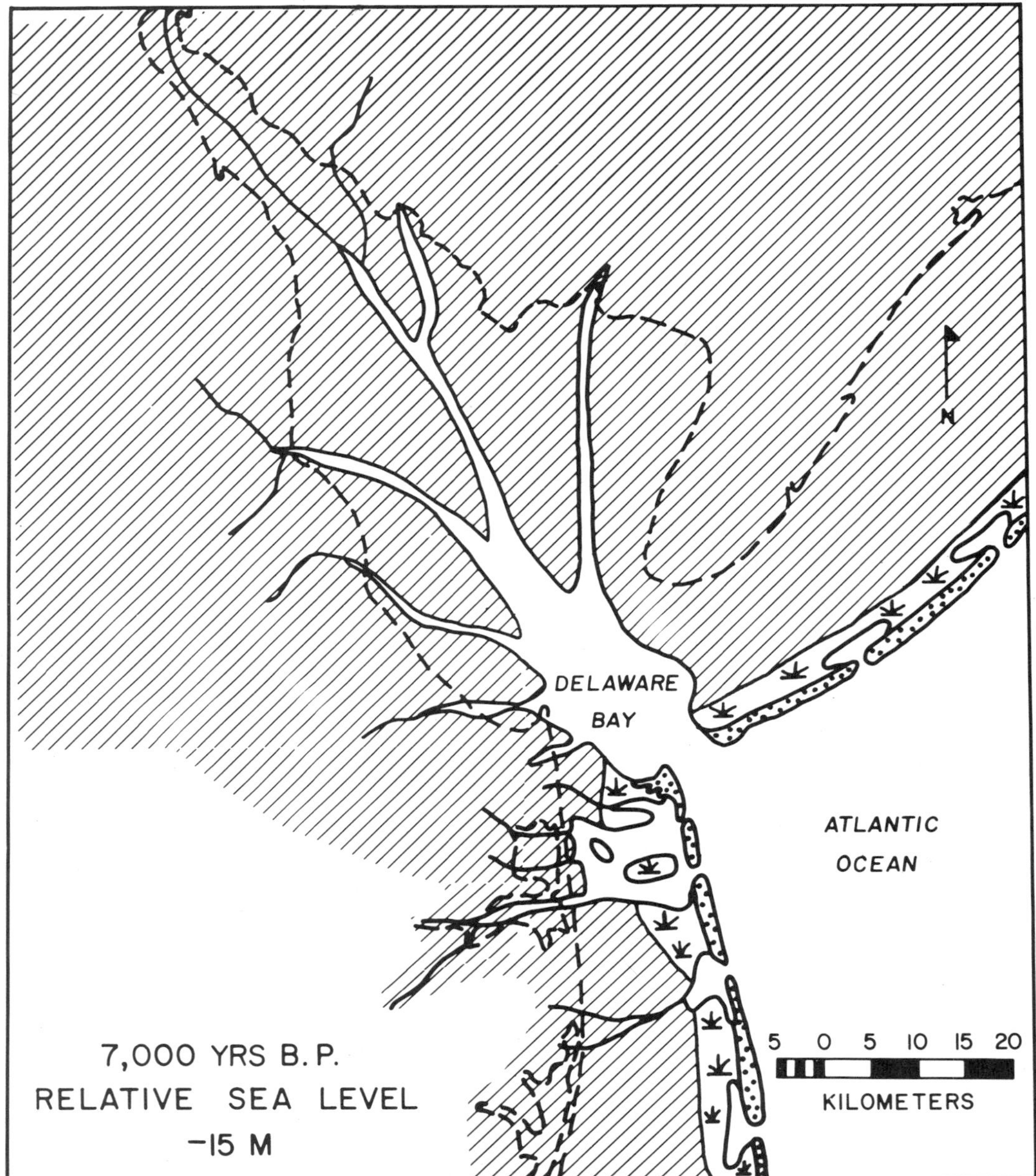

Figure 10-14. A schematic reconstruction of Delaware coastal zone paleogeography of approximately 7,000 B.P. with a relative sea level of −15 m.

(Belknap and Kraft, 1985; Kraft and John, 1976; Weil, 1976). Further, our understanding of sea-level rise based on local studies and those of other areas along the Atlantic coast may be used to establish relative sea levels.

By 2,000 B.P. (Fig. 10-15) the lagoon barrier in Atlantic coastal Delaware was well established and is now well understood, with relative sea level at approximately −2.5 m. Using present morphology; rates of coastal erosion; processes of erosion, transport, and deposition; and the various sea-level rise scenarios as

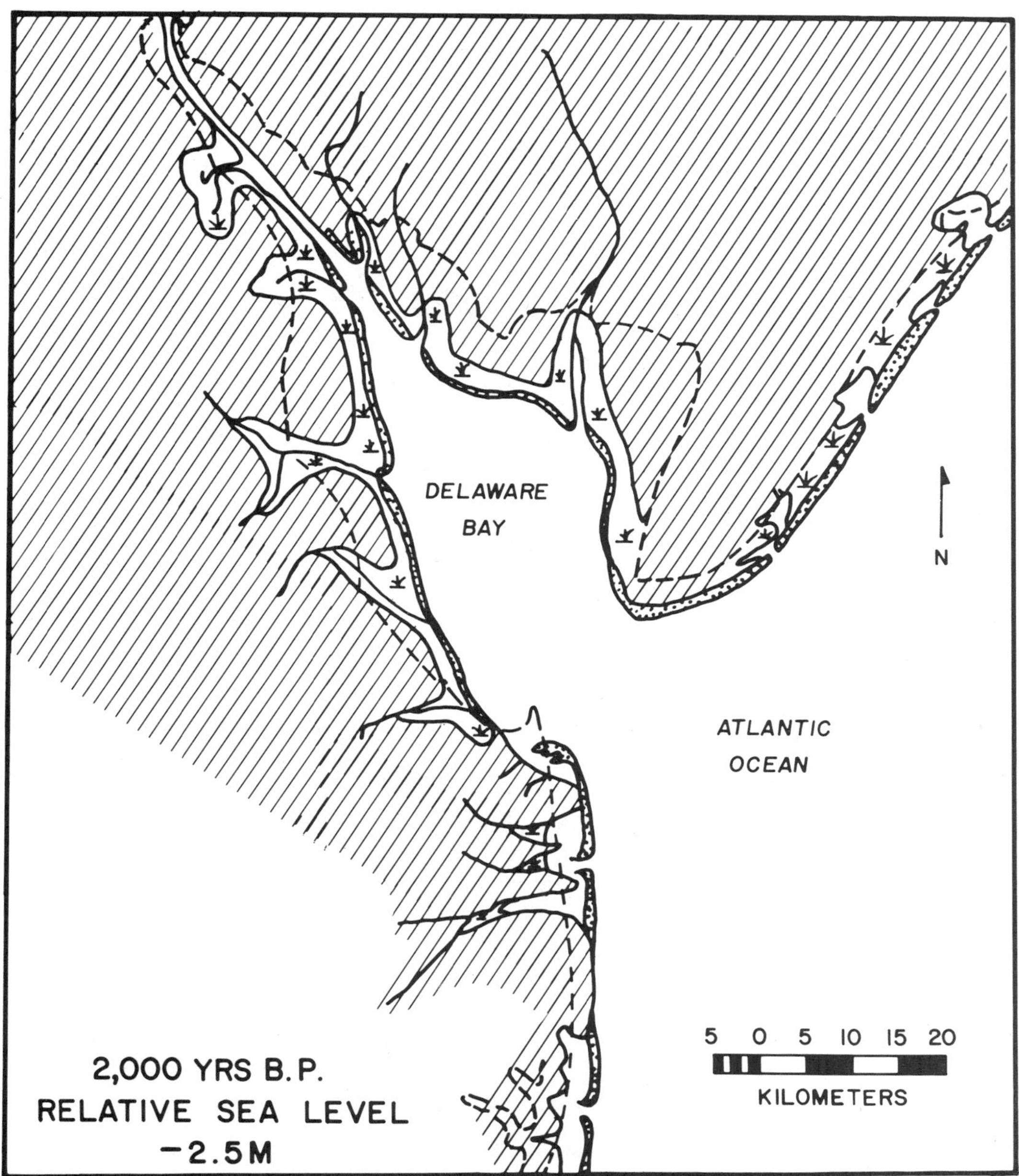

Figure 10-15. A paleogeographic reconstruction of the Delaware coastal zone approximately 2,000 B.P. with a relative sea level of −2.5 m.

shown in Figure 10-12, we have projected in a schematic fashion the morphology and loci of the various elements of the Delaware coastal zone into the future. Figure 10-16 shows four scenarios based on sea-level rise as projected in Figure 10-12. The Delaware estuary will be much larger. Its shoreline will have moved up to 10 km landward. The Atlantic coast of Delaware will have a lagoon-barrier system analogous to the present but somewhat more restricted as higher elevations are encountered by the landward transgressing system. Nevertheless,

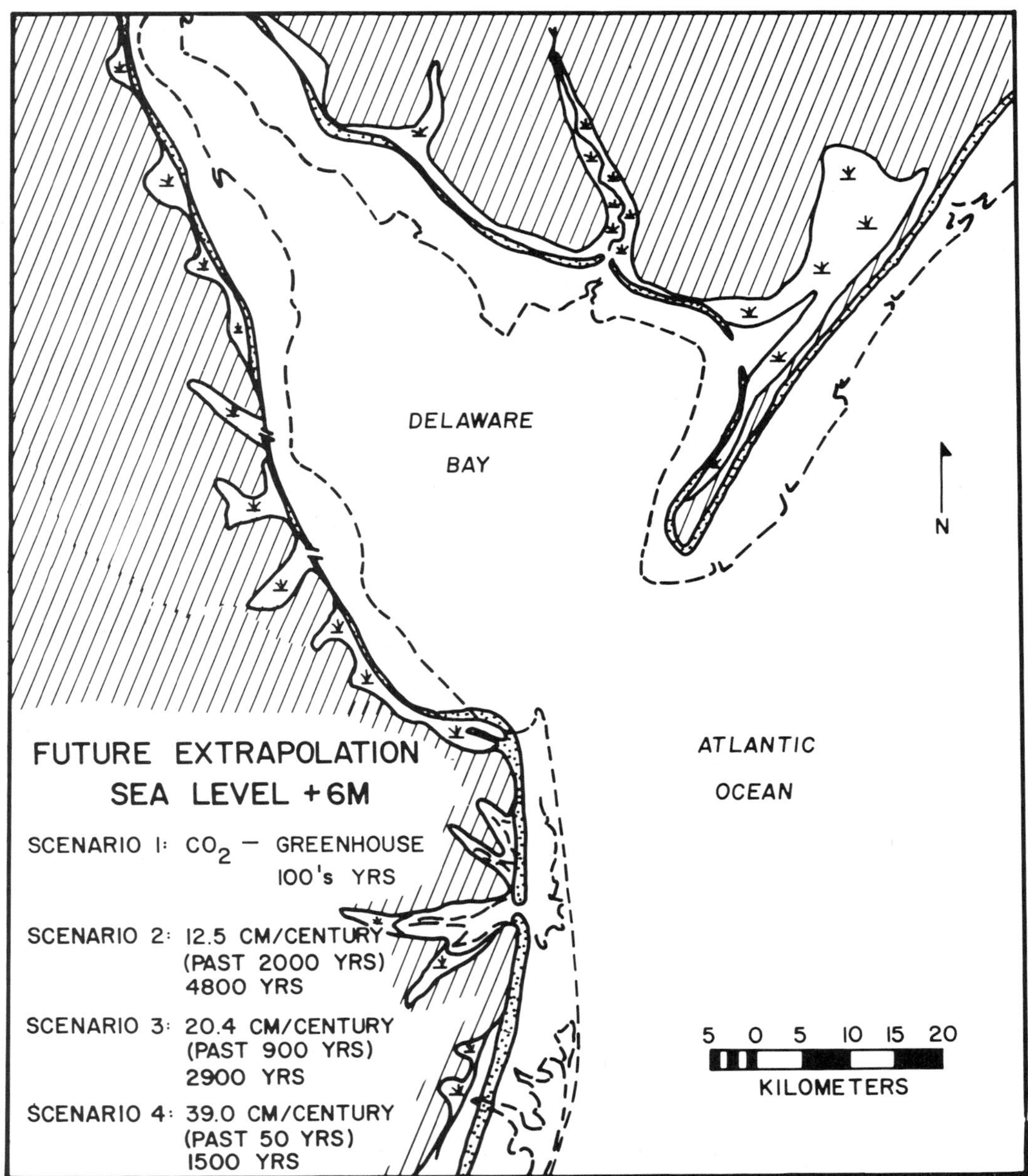

Figure 10-16. A geologic projection of Delaware coastal zone geographies after a landward transgression accompanying a rise of sea level to approximately +6 m.

The dashed line shows the present coast. Various time scenarios and local relative sea-level rise rates are as projected on Figure 10-12. Scenario 1: the extreme rate of sea level rise as predicted in the carbon dioxide greenhouse effect; scenario 2: 4800 years into the future based on the locally determined average rate of sea level rise of 12.5 cm/century over the past 2000 years; scenario 3: 2900 years into the future based on the locally determined average rate of 20.4 cm/century over the past 900 years; and scenario 4: 1500 years into the future based on the average rate of 39.0 cm/century recorded over the past 50 years.

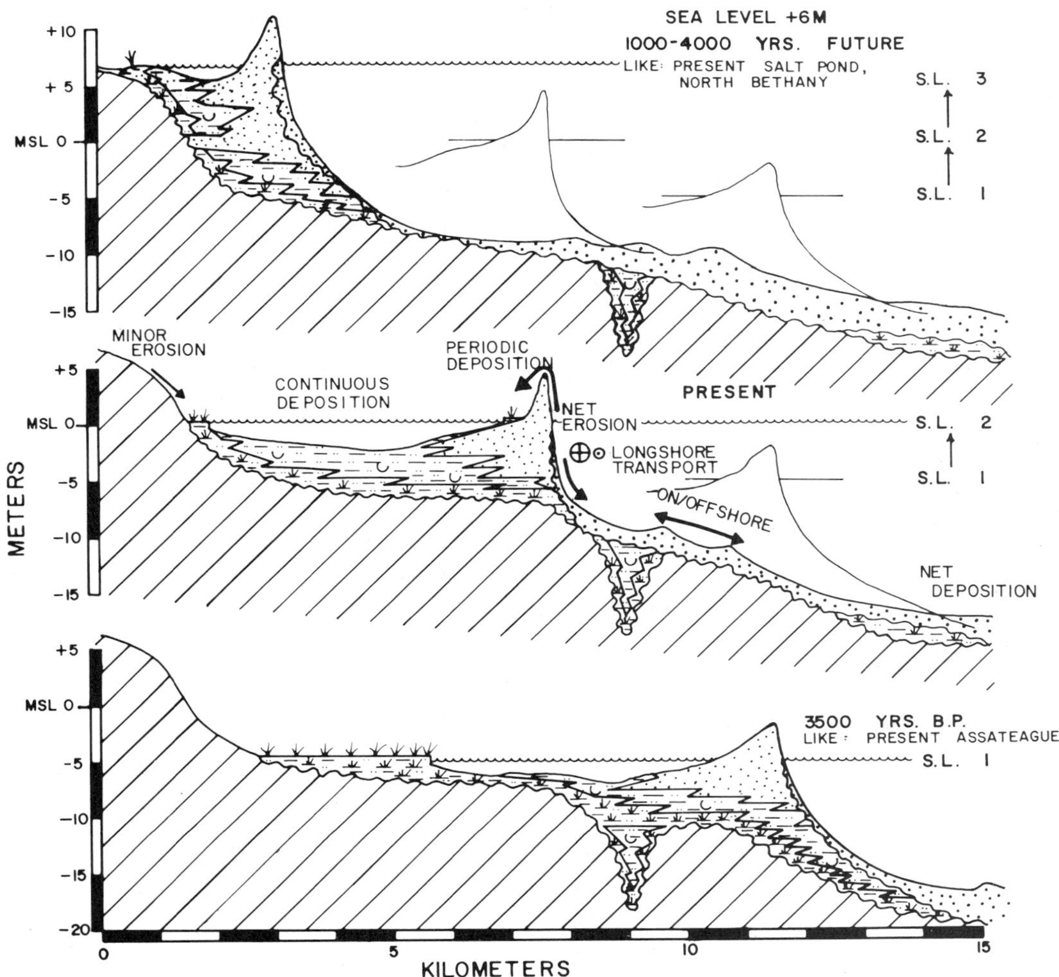

Figure 10-17. A geologic profile across the Atlantic coastal lagoon-barrier-inner shelf area of Delaware showing the present profile and Holocene stratigraphy (middle profile) as related to a paleogeographic reconstruction of 3,500 B.P. (lower profile) and a projected future profile and cross-section with a sea level 6 m above present. (After Belknap and Kraft, 1985, Fig. 12)

in a natural system (with no human attempts to stop the erosion) the Delaware Atlantic coastal zone will have moved landward from 3 km to 10 km and upward to a sea level of approximately +6 m (Figs. 10-16, 10-17, and 10-18). These projections are based on analysis of the Pleistocene epoch paralic sediments of a number of other interglacials in the immediate coastal zone.

CONCLUSION

Future projections are fraught with difficulty. We cannot, of course, know which projection to use, so the best we can do at present is to analyze the situation using known rates of coastal erosion, past rates of environmental shift in the landward transgression, and various scenarios of sea-level rise or the extremes or averages of

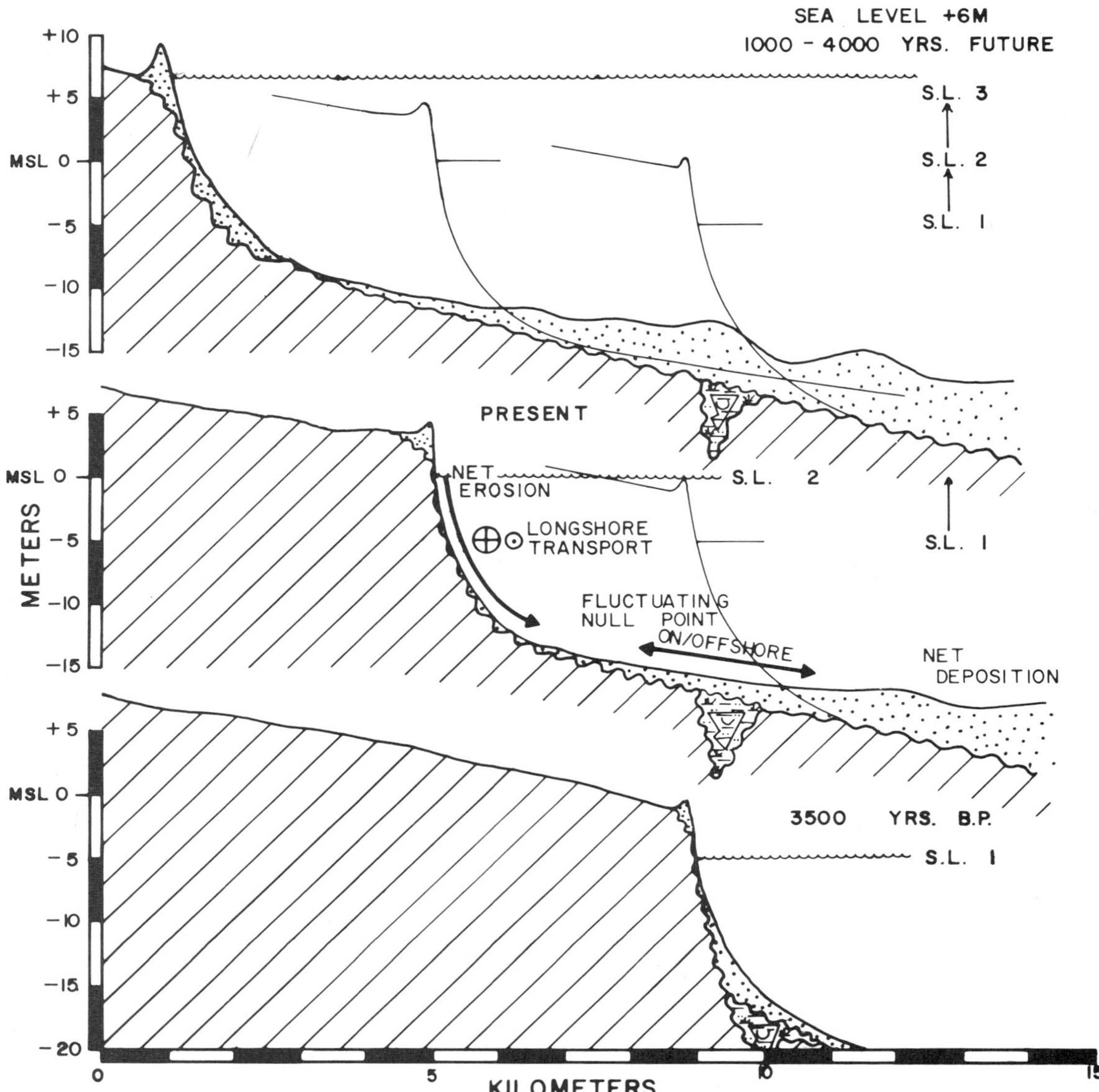

Figure 10-18. A geologic profile across an Atlantic coastal highland comprised of Pleistocene sediments. The present profile and nearshore stratigraphy (middle profile) as related to a paleogeographic reconstruction of 3,500 B.P. (lower profile) and a projected future profile and cross-section with a sea level 6 m above present are shown. (After Belknap and Kraft, 1985, Fig. 12)

the greenhouse effect model. Regardless, should a continually rising sea level relative to land (eustatic plus other causative means) occur, the transgression to a future geography as depicted in Figures 10-16, 10-17, and 10-18 must be considered as inevitable, the only question being one of rate. A major caveat is that of the actions of people. Large amounts of money are presently being spent to hold the line of Atlantic coastal Delaware and a few small barrier areas along the Delaware estuary. However, should the cost of holding the line rise by orders of magnitude involving billions of dollars within hundreds of years, then an economic and philosophical dilemma is formed. The public has not had to deal with this matter. It has had to deal only with the spending of millions of dollars every few years in attempts to nourish the beach and to build groins, jetties, and breakwaters. Questions of occupance of the Atlantic coastal zone of Delaware and the high costs and urgency of planning for short- and long-term alternates are not only evident but also urgent.

ACKNOWLEDGMENTS

This research has been supported by a number of grants from the National Oceanic and Atmospheric Agency-Office of Sea Grant, Department of Commerce (Project Numbers, R/G10-NA85AA-D-SG033 and R/G12-NA86AA-D-SG040) and the Department of Natural Resources and Environmental Control of the State of Delaware. The U.S. Government is authorized to produce and distribute reprints for governmental purposes, not withstanding any copyright notation that might appear hereon. The field and laboratory facilities of the Department of Geology and the College of Marine Studies at the University of Delaware were used in this research. A large number of graduate student colleagues' research in previous years has been important in the preparation of this paper as acknowledged in the text.

REFERENCES

Allen, E. A., 1977, *Petrology and Stratigraphy of Holocene Coastal-Marsh Deposits along the Western Shore of Delaware Bay,* DEL-SG-20-77, Delaware Sea Grant College Program, University of Delaware, Newark. 287p.

Barnett, T. P., 1983, Recent changes in sea level and their possible causes, *Clim. Change* **5:**15-38.

Barth, M. C., and Titus, J. G., eds., 1984, *Greenhouse Effect and Sea Level Rise,* New York: Van Nostrand Reinhold Co., 325p.

Belknap, D. F., and Kraft, J. C., 1977, Holocene relative sea-level changes and coastal stratigraphic units on the northwest flank of the Baltimore Canyon Trough Geosyncline, *Jour. Sed. Petrology* **47:**610-629.

Belknap, D. F., and Kraft, J. C., 1981, Preservation potential of transgressive coastal lithosomes on the U.S. Atlantic shelf, *Marine Geology* **42:**429-442.

Belknap, D. F., and Kraft, J. C., 1985, Influence of antecedent geology on evolution of barrier systems, *Marine Geology* **63:**235-262.

Bloom, A. L., 1970, Paludal stratigraphy of Truk, Ponape and Kusaie, eastern Caroline Islands, *Geol. Soc. America Bull.* **81:**1895-1904.

Bloom, A. L. (ed.), 1977, *An Atlas of Sea Level Curves,* Final Report of IGCP Project No. 61, Sea Levels, Ithaca, N.Y.: Department of Geology, Cornell University, 114p.

Bruun, P., 1980, The "Bruun Rule," discussion on boundary conditions, in J. J. Fisher and M. L. Schwartz (eds.), *Proceedings of the Per Bruun Symposium.* International Geographic Union, Commission on the Coastal Environment, Bellingham, Wash.: Bureau of Faculty Research, Western Washington University, pp. 79-83.

Bruun, P., 1983, Review of conditions for uses of the Bruun Rule of erosion, *Coastal Eng.* **7:**77-89.

Carbon Dioxide Assessment Committee, National Research Council, 1983, *Changing Climate* (W. A. Nierenberg, chairman), Washington, D.C.: National Academy Press. 496p.

Church, T. M.; Lord, C. J., III; and Somayajulu, B. L. K., 1981, Uranium, thorium, and lead nuclides in a Delaware salt marsh sediment, *Estuarine, Coastal and Shelf Sci.* **13:**267-275.

Cinquemani, L. J.; Newman, W.; and Pardi, R. R., 1982, Holocene sea level changes and vertical movements along the East Coast of the United States: A preliminary report, in D. J. Colquhoun (ed.), *Holocene Sea Level Fluctuations, Magnitudes and Causes,* UNESCO-IGCP No. 61—Sea Levels 1981 Business Meeting, Columbia, S.C.: Department of Geology, University of South Carolina, pp. 13-33.

Coleman, J. M., and Smith, W. G., 1964, Late recent rise of sea level, *Geol. Soc. America Bull.* **75:**833-840.

Colquhoun, D. J., 1981, Variation in sea level on the South Carolina coastal plain, in D. J. Colquhoun (ed.), *Variation in Sea Level on the South Carolina Coastal Plain,* UNESCO-IGCP No. 61—Sea Levels 1981 Business Meeting, Columbia, S.C.: Department of Geology, University of South Carolina, pp. 1-44.

Conybeare, W., 1822, *Outlines of the Geology of England and Wales,* London: W. Phillips, 470p.

Curray, J. R., 1964, Transgressions and regressions, in R. L. Miller (ed.), *Papers in Marine Geology,* Shepard Commemorative Volume, New York: Macmillan, pp. 175-203.

Curray, J. R., 1965, Late Quaternary history, continental shelves of the United States, in H. E. Wright, Jr., and D. G. Frey (eds.), *The Quaternary of the United States,* Princeton: Princeton University Press, pp. 723-735.

Davies, O., 1981, A review of Wilson's theory that the last interglacial ended with an ice-surge, the South African evidence therefor, *Annals of the Natal Museum* **24:**701-720.

Demarest, J. M., II, 1978, The Shoaling of Breakwater Harbor-Cape Henlopen area, Delaware Bay, 1842-1971, M.S. thesis, University of Delaware, Newark, 169p.

Demarest, J. M., II, 1981, Genesis and preservation of Quaternary paralic deposits on Delmarva Peninsula, Ph.D. dissertation, University of Delaware, Newark, 253p.

Demarest, J. M.; Biggs, R. B.; and Kraft, J. C., 1981, Time-stratigraphic aspects of a formation: Interpretation of surficial Pleistocene deposits by analogy with Holocene paralic deposits, southeastern Delaware, *Geology* **9:**360-365.

DePratter, C. B., and Howard, J. D., 1977, History of shoreline changes determined by archaeological dating: Georgia Coast, *Gulf Coast Assoc. Geol. Socs. Trans.* **27:**252-258.

DePratter, C. B., and Howard, J. D., 1981, Evidence for a sea level lowstand between 4500 and 2400 years B.P. on the southeast coast of the United States, *Jour. Sed. Petrology* **51:**1287-1295.

Diamante, J. M., 1982, Proposed NOAA absolute sea level monitoring system (a letter) to T. Pyle (NOAA File # OA/Cx21).

Edwards, R. L., and Emery, K. O., 1977, Man on the continental shelf, *New York Acad. Sci. Annals* **288:**245-256.

Emery, K. O.; Wigley, R. L.; Bartlett, A. S.; Rubin, M.; and Barghoorn, E. S., 1967, Freshwater peat on the continental shelf, *Science* **158:**1301-1307.

Fairbridge, R. W., 1960, The changing level of the sea, *Sci. American* **202**(5):70-79.

Fairbridge, R. W., 1961, Eustatic changes in sea level, in L. H. Ahrens, F. Press, K. Rankama, and S. K. Runcorn (eds.), *Physics and Chemistry of the Earth*, vol. 4, New York: Pergamon Press, pp. 99-185.

Fairbridge, R. W., 1976, Shellfish-eating Preceramic Indians in coastal Brazil, *Science* **191:**353-359.

Fisher, J. J., 1980, Shoreline erosion, Rhode Island and North Carolina coasts—Test of Bruun Rule, in J. J. Fisher and M. L. Schwartz (eds.), *Proceedings of the Per Bruun Symposium*, International Geographical Union, Commission on the Coastal Environment, Bellingham, Wash.: Western Washington University, pp. 32-54.

Gable, D. J., and Hatton, T., 1983, Maps of vertical crustal movements in the conterminous United States over the last 10 million years, *U.S. Geol. Survey Misc. Geol. Inv. Map I-1315.*

Hicks, S. D., 1972, Vertical crustal movements from sea level measurements along the East Coast of the United States, *Jour. Geophys. Research* **77:**5930-5934.

Hicks, S. D.; Debaugh, H. A., Jr.; and Hickman, L. E., Jr., 1983, *Sea Level Variations for the United States 1855-1980*, Washington, D.C.: Tides and Water Levels Branch, U.S. Department of Commerce, National Oceanic and Atmospheric Administration, National Ocean Service, 170p.

Hoffman, J. S.; Keyes, D.; and Titus, J. G., 1983, *Projecting Future Sea Level Rise, Methodology, Estimates to the Year 2100, and Research Needs*, Washington, D.C.: U.S. Environmental Protection Agency, Office of Policy and Resources Management, 121p.

Hollin, J. T., 1972, Interglacial climates and Antarctic ice-surges, *Quat. Research* **2:**401-408.

Hopley, D., 1983, *Australian Sea Levels in the Last 15,000 Years: A Review*, Occasional Paper No. 3, Australian Report for IGCP No. 61, Monograph Series, Townsend: James Cook University of North Queensland, 103p.

International Geological Correlation Program (IGCP), 1983, *International Symposium on Coastal Evolution in the Holocene, Abstracts of Papers*, Project 200, Sea Levels and Resultant Coastal Change, International Union of Geological Scientists and UNESCO, Setagaya, Japan: Komazawa University, 159p.

Jelgersma, S., 1961, Holocene sea-level changes in the Netherlands, *Geol. Stichting Med. C-VI*, 101p.

Jelgersma, S., 1966, Sea-level changes during the last 10,000 years, in *Proceedings, International Symposium on World Climates from 8000 B.C. to 0 B.C.*, London: Royal Meteorological Society, pp. 54-71.

Karpas, R. M., 1978, The hydrography of Indian River and Rehoboth: Delaware's small bays, Master's thesis, College of Marine Studies, University of Delaware, Newark, 179p.

Keyes, D. L.; Hoffman, J. S.; Seidel, S. R.; and Titus, J. G., 1984, EPA's studies of the greenhouse effect, *Science* **223:**538-540.

Kidson, C., 1982, Sea level changes in the Holocene, *Quat. Sci. Rev.* **1:**121-151.

Kraft, J. C., 1971, Sedimentary facies patterns and geologic history of a Holocene marine transgression, *Geol. Soc. America Bull.* **82:**2131-2158.

Kraft, J. C., 1976, *Radiocarbon Dates in the Delaware Coastal Zone*, Delaware Sea Grant Rept. DEL-SG-19-76, Newark: University of Delaware, 20p.

Kraft, J. C., 1982, The uses of RC-14 dates; 5568 ½ life vs. MASCA correction, in D. J. Colquhoun (ed.), *Holocene Sea Level Fluctuations, Magnitudes and Causes*, UNESCO-IGCP No. 61—Sea Levels 1981 Business Meeting, Columbia, S.C.: Department of Geology, University of South Carolina, pp. 104-105.

Kraft, J. C., and Belknap, D. F., 1975, Transgressive

and regressive sedimentary lithosomes at the edge of a late Holocene marine transgression, in *Extraits des Publications du 9e Congrès,* Nice: Congrès International de Sedimentologie, pp. 85-99.

Kraft, J. C., and John, C. J., 1976, *The Geological Structure of the Shorelines of Delaware,* Delaware Sea-Grant Rept., DEL-SG-14-76, Newark: University of Delaware, 107p.

Kraft, J. C., and John, C. J., 1979, Lateral and vertical facies relation of transgressive barrier, *Am. Assoc. Petroleum Geologists Bull.* **63:**2145-2163.

Kraft, J. C.; Allen, E. A.; and Maurmeyer, E. M., 1978, The geological and paleogeomorphological evolution of a spit system and its associated coastal environments: Cape Henlopen spit, Delaware, *Jour. Sed. Petrology* **48:**211-225.

Kraft, J. C.; Allen, E. A.; Belknap, D. F.; John, E. J.; and Maurmeyer, E. M., 1975, *Delaware's Changing Shoreline (Geologic Processes and the Geology of Delaware's Coastal Zone),* Technical Report No. 1, Dover: Delaware Coastal Zone Management Program, Delaware State Planning Office, 319p.

Kraft, J. C.; Allen, E. A.; Belknap, D. F.; John, C. J.; and Maurmeyer, E. M., 1979, Processes and morphologic evolution of an estuarine and coastal barrier system, in S. P. Leatherman (ed.), *Barrier Islands from the Gulf of St. Lawrence to the Gulf of Mexico,* New York: Academic Press, pp. 149-183.

Laudan, R., 1983, Geological thought from Hutton to Suess, *Science* **219:**280.

Maley, K. F., 1981, A transgressive facies model for a shallow estuarine environment, Master's thesis, University of Delaware, Newark, p. 184.

Marx, P. R., 1981, A dynamic model for an estuarine transgression based on facies variants in the nearshore of western Delaware Bay, Master's thesis, University of Delaware, Newark, p. 183.

Milliman, J. D., and Emery, K. O., 1968, Sea levels during the past 35,000 years, *Science* **162:**1121-1123.

Mörner, N.-A., 1969, The Late Quaternary history of the Kattegatt Sea and the Swedish west coast, *Sveriges Geol. Undersökning, Ser. C, No. 640,* 487p.

Mörner, N.-A., 1971, Late Quaternary isostatic, eustatic, and climatic changes, *Quaternaria* **14:**65-83.

Newman, W. S., and Rusnak, G. A., 1965, Holocene submergence of the eastern shore of Virginia, *Science* **148:**1464-1466.

Nummedal, D., 1983, Rates and frequency of sea-level changes: A review with an application to predict future sea levels in Louisiana, *Gulf Coast Assoc. Geol. Soc. Trans.* **33:**361-366.

O'Brien, M. P., 1982, The Shoreline Debate: I. Saving the American Beach: A Position Paper by Concerned Coastal Geologists; II(a). A letter to President Reagan; and II(b). Memorandum covering a Document Entitled "Saving the American Beach: A Position Paper by Concerned Coastal Geologists," *Shore and Beach* **50:**3-8.

O'Brien, M. P., 1984, One Man's Opinion #1, *Shore and Beach* **52:**13-14.

Pilkey, O. H.; Howard, J. D.; Brenninkmeyer, B. M.; Frey, R. W.; Hine, A. C.; Kraft, J. C.; Morton, R.; Nummedal, D.; and Wanless, H., 1981, *Saving the American Beach: A Position Paper by Concerned Coastal Geologists to President Reagan,* Skidaway, Ga.: Institute of Oceanography, 16p.

Revelle, R. R., 1983, Probable future changes in sea level resulting from increased atmospheric carbon dioxide, in Carbon Dioxide Assessment Committee, National Research Council, *Changing Climate,* Washington, D.C.: National Academy Press, pp. 433-448.

Ryan, P. R., 1983, High sea levels and temperatures seen next century, *Oceanus* **26:**63-67.

Schelling, T. C., 1983, Climatic change: Implications for welfare and policy, in W. A. Nierenberg (ed.), *Changing Climate,* Washington, D.C.: National Academy Press, pp. 449-477.

Schwartz, M. L., 1967, The Bruun theory of sea-level rise as a cause of shore erosion, *Jour. Geology* **75:**76-92.

Schwartz, M. L., and Milicic, V., 1980, The Bruun Rule: An historical perspective, in J. J. Fisher and M. L. Schwartz (eds.), *Proceedings of the Per Bruun Symposium,* Bellingham: Bureau of Faculty Research, Western Washington University, pp. 6-12.

Seidel, S., and Keyes, D., 1983, *Can We Delay a Greenhouse Warming?, The Effectiveness and Feasibility of Options to Slow a Build-Up of Carbon Dioxide in the Atmosphere,* Washington, D.C.: U.S. Environmental Protection Agency, Strategic Studies Staff, Office of Policy and Resources Management, 180p.

Shepard, F. P., 1963, Thirty-five thousand years of sea level, in T. Clements (ed.), *Essays in Marine Geology,* Los Angeles: University of Southern California Press, pp. 1-10.

Stuiver, M., and Daddario, J. J., 1963, Submergence of the New Jersey coast, *Science* **142:**941.

Stuiver, M.; Denton, G. H.; Hughes, T. J.; and Fastook, J. L., 1981, History of the marine ice sheet in West Antarctica during the last glaciation; a working hypothesis, in C. H. Denton and T. J. Hughes, (eds.), *The Last Great Ice Sheets,* New York: John Wiley, pp. 319-436.

Stumpf, R. P., 1981, The Process of Sedimentation in a Salt Marsh, M.S. thesis, College of Marine Studies, University of Delaware, Newark, 125p.

Swisher, M. L., 1982, The Rate and Causes of Shore Erosion around a Transgressive Lagoon, Rehoboth

Bay, Delaware, M.S. thesis, College of Marine Studies, University of Delaware, Newark, 210p.

Ters, M., 1973, Les variations du niveau marin depuis 10,000 ans le long du littoral Atlantique Français, Le Quaternaire: Géodynamique, stratigraphie et environment, travaux Français recents, in Proceedings, INQUA, 9th International Congress, Christchurch, New Zealand: INQUA, pp. 114-135.

U.S. Department of Commerce, 1987, *Tide Tables, High and Low Water Prediction, East Coast of North and South America Including Greenland,* Rockville, Md.: National Oceanic and Atmospheric Administration, 289p.

U.S. Geological Survey, 1928, *Rehoboth Quadrangle (15 Min.),* Edition of 1928, Washington, D.C.: Department of the Interior.

Vita-Finzi, C., 1973, *Recent Earth History,* New York: Wiley, 138p.

Weil, C. B., Jr., 1976, A Model for the Distribution Dynamics and Evolution of Holocene Sediments and Morphologic Features of Delaware Bay, Ph.D. dissertation, University of Delaware, Newark, 408p.

11: The Quaternary Coastal Record in the Aegean

C. Vita-Finzi
University College, London

Abstract: Since the turn of the century the Quaternary marine deposits of the Aegean have been investigated within the glacioeustatic framework favored by students of the Mediterranean record. Because the beds in question are manifestly warped and faulted, their dating is based chiefly on stratigraphic criteria and paleontology, and their evaluation is used as a measure of the uplift they have undergone.

In the eastern Gulf of Corinth the Pleistocene marine terraces contain molluskan faunas that reflect environment rather than age, and they commonly overlie Neogene deposits. Their chronology, therefore, has to be derived from radiometric data. Shells from so-called Tyrrhenian beaches at heights of between 0.5 m and 70 m were selected with the help of microscopy and X-ray diffraction and dated by the radiocarbon method. They gave ages of 30,000-42,000 yr.

The corresponding deposits and faunal assemblages indicate intertidal and shallow subtidal conditions, but the ages do not necessarily support the contentious thesis of an interstadial transgression because the amount of tectonic deformation is not known. They bear on paleoclimates in two other ways. First, the ages make possible the time correlation of the coastal sequences with those obtained in lake basins and river valleys inland and dated independently by archeological or radiometric methods. Second, they permit direct comparison of the deep-sea isotopic record with stable isotopic measurements on the inshore and coastal faunas and, hence, a direct assessment of the glacioeustatic factor without prior assumptions about the nature of the shoreline evidence.

INTRODUCTION

One recurring theme in the work of Rhodes Fairbridge is the interaction between climate, eustasy, and tectonics. Another is the need to integrate the available data when trying to understand any part of the record (see, e.g., Fairbridge, 1971). This chapter argues that keeping the various strands of the Quaternary narrative separate, at least until one of them can be traced with confidence, is sometimes more productive. The case is bolstered by reference to the fossil beaches of the Gulf of Corinth where, despite decades of intensive research, the relative importance of eustasy and tectonics, and hence, of climatic fluctuations, in fashioning the beaches remains in dispute. The tendency has been to view the Quaternary of the Aegean coasts as the product of glacioeustatic fluctuations complicated by Earth movements and thus to beg rather than to ask the crucial questions.

It is clear that different areas will experience different patterns of sea-level change in response to a particular glacial oscillation as regards timing, sign, and magnitude. Why, then, is the term *Flandrian* still used and not pensioned off? What does the *Würm regression* precisely signify? Indeed, the trend is, if anything, toward a homogenized global sequence shorn of local anomalies as glacioeustasy is fused with the new paleoclimatic orthodoxy of the deep-sea isotopic record.

The solution proposed here is to date different components of the evidence before rather than after fitting them into the eustatic or climatic sequence under review. It is an unoriginal idea but evidently in need of renewed airing.

Quite apart from the opportunity it offers for

testing chronologies, independent dating improves the prospect of picking out underlying periodicities, first because slight discrepancies in timing are not obliterated in the search for coincidences and second because the evidence may reflect changes in environmental components such as temperature and salinity rather than in the farrago we call climate. Pérès (1967, p. 454) had observed that "the Pliocene and Quaternary history of the Mediterranean cannot be interpreted as a succession of warm and cold faunas proceeding from changes in temperature." In his opinion, the key is the system of currents through the Strait of Gibraltar. In an area as tectonically active as the Mediterranean, it is equally hazardous to equate high sea levels with interglacial conditions.

GULF OF CORINTH

The marine Quaternary deposits and landforms of the Gulf of Corinth (Fig. 11-1) have traditionally been investigated within a eustatic framework rooted in the glacial sequence Penck and Brückner proposed for the Alps in 1909. Shorelines now above sea level are consequently equated with an interglacial or interstadial period, and a sequence is deemed incomplete unless it includes representatives of the Calabrian, Sicilian, Milazzian, Tyrrhenian (usually subdivided into Eutyrrhenian and Neotyrrhenian but in some accounts into three stages), and Versilian beaches. In addition, two sets of lacustrine or brackish-water deposits related to events in the Caspian and Black seas are identified, the older, or Apscheronian, corresponding roughly with the Calabrian and the younger, or Tchaudian, with the Sicilian (Bousquet et al., 1976; Herforth and Richter, 1979). In short, many of the ideas advanced at the turn of the century have survived little change (Depéret, 1913; Dufaure and Zamanis, 1979).

The stability of this chronological scheme may help to explain why in 1982, two decades after Fairbridge (1961) deplored the lack of radiocarbon dates bearing on Quaternary sea levels, none had been obtained for the classic shorelines of the Gulf of Corinth: there was little point in dating features whose age was already known and which in any case lay well beyond the reach of the ^{14}C method. Two uranium-series ages had been determined, one of 49,000 ± 20,000 yr on the New Corinth terrace and one of 235,000 +40,000/−30,000 yr for the Old Corinth terrace, but a high $^{234}U/^{238}U$ ratio was taken to indicate that both ages were too low, and the value obtained for the New Corinth terrace was not acceptable because it conflicted with the Tyrrhenian age ascribed to that feature (Sébrier, 1977).

Whether or not all beaches containing a typically Tyrrhenian fauna are of the same age, and whether U-series ages on coral are invariably acceptable, may be doubted. In any case many of the so-called Tyrrhenian beaches in the area lack the appropriate fossils, and in addition, some of the fossils regarded as diagnostic of the various Mediterranean terrace units are not restricted to a single stratigraphic unit and occur in younger or older marine deposits (Dufaure, Keraudren, and Sébrier, 1975; Schröder, 1975). An interesting example is *Chlamys septemradiata*, which is sometimes used for identifying Calabrian units (Bousquet et al., 1976; Keraudren, 1970) but is still found in the Mediterranean (Dufaure, Keraudren, and Sébrier, 1975). Pérès (1967), who sees *C. septemradiata* as a distinctive member of Würm death assemblages, infers from its survival in the Alboran Sea that, as at the present day, water temperatures during the period of Würmian cooling varied from place to place within the Mediterranean. He also shows the importance of salinity, deep-sea temperatures, and currents for the distribution of different species and concludes that remnants of the Tyrrhenian faunas as well as the cold-loving faunas of the Würmian period survive locally where the environment is suitable.

The scope for confusion does not end there. Some of the so-called beaches are ambiguously marine, and there is consequently disagreement over how far they have been uplifted. The extent to which their faunas have undergone redeposition is also of concern since it colors their dating as well as the ecological interpretation.

RADIOCARBON DATING

It was accordingly decided to submit mollusks from some of the lower terraces for ^{14}C dating

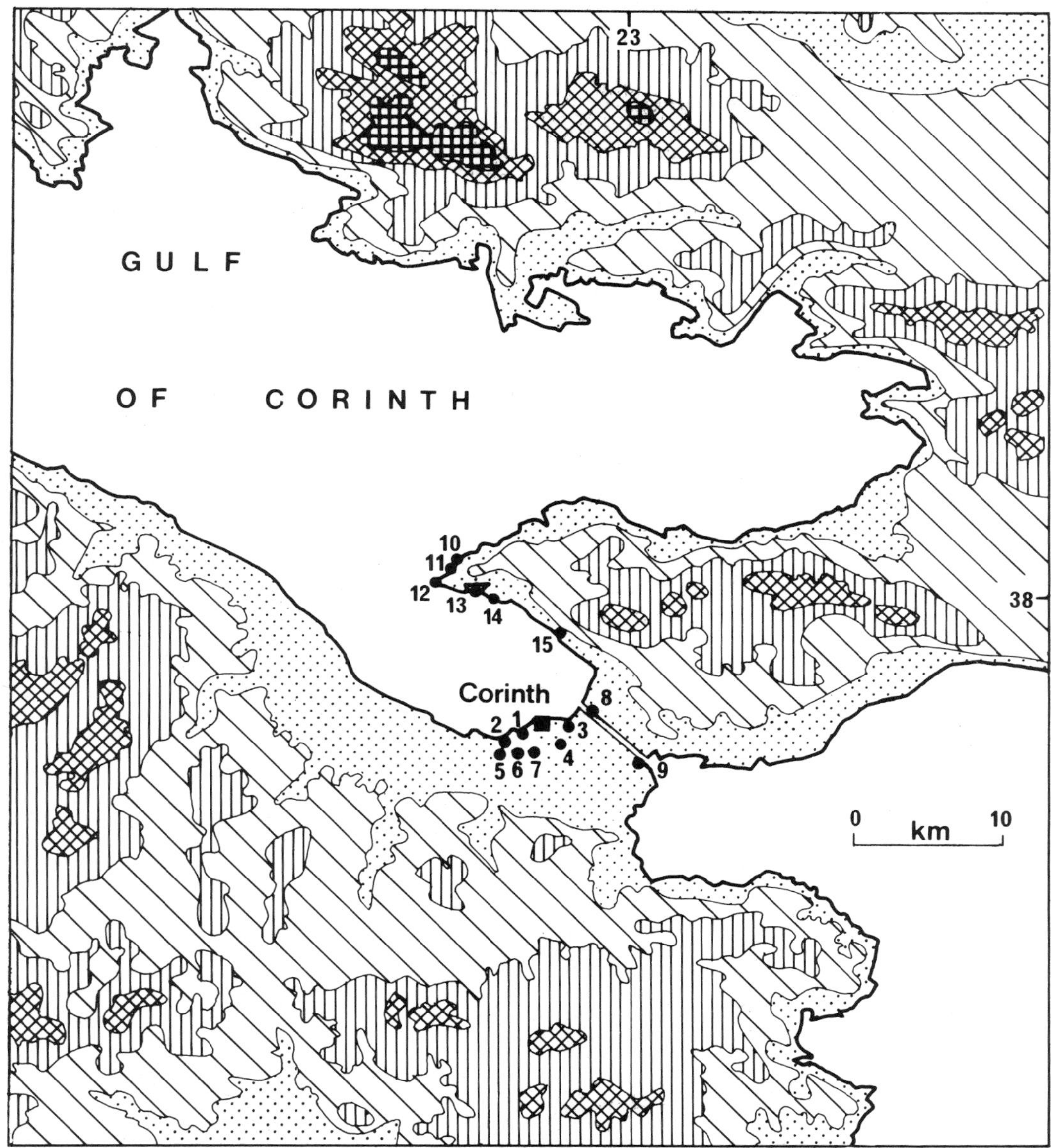

Figure 11-1. Map showing dated sections mentioned in text.

without any prior assumption about their age from paleontological or altimetric considerations. The key criteria adopted for selection were an unambiguous marine origin for the feature and the absence of any sign of contamination of the shell sample. Richards (1982) has shown that death assemblages of mollusks in the Mediterranean can be used to identify the corresponding sea level to within a few centimeters and, hence, that there are times when redeposited material is preferable to shells in growth position for dating former shorelines. For example, of the species listed in Table 11-1, *Pinna fragilis*, as its name implies, would not withstand prolonged transport and the dated specimens had hardly been displaced, but the corresponding waterline is difficult to specify.

The validity of ^{14}C ages on shell carbonate is

Table 11-1. ^{14}C Ages for Gulf of Corinth

Location	Sample Number	Height above High Water (m)	Species	X-Ray Diffusion[a]	^{14}C age (yr B.P.)[b]	Laboratory Number	$\delta^{13}C$ ‰ (A)	$\delta^{13}C$ ‰ (B)	$\delta^{18}O$ ‰
1	3c	26	*Cerastoderma glaucum*	A	$30{,}830^{+970}_{-860}$	SRR-2342	1.1[c]	1.3	1.6[d]
	2a	25	*C. glaucum*	A	$31{,}530^{+670}_{-610}$	SRR-2243	1.3	1.3	1.3
	2g	25	*Cerithium vulgatum*	A	$31{,}960^{+990}_{-880}$	SRR-2462	2.3	2.77	2.50
	1o	19	*Crassostrea* sp.	C	Inner: $30{,}260^{+980}_{-860}$	SRR-2463	2.3	1.23	1.39
					Outer: $32{,}000^{+1{,}000}_{-890}$		1.7		
	4	19	*Pecten jacobeus*	AC	>42,000[e]	SRR-2461	−1.2	−1.15	2.60
2	912/1	10.5	*Venus verrucosa*	A	$30{,}840^{+590}_{-560}$	SRR-2248	−0.7	−1	0.4
3	913/1	≈24	*Glycymeris glycymeris*	A	$35{,}680^{+1{,}170}_{-1{,}020}$	SRR-2341	1.3	1.6	0.1
4	914B	70	*Crassostrea* sp.	C	$34{,}870^{+990}_{-880}$	SRR-2333	−2.4	−1.3	−1.9
5	701	≈25	*Mytilus galloprovincialis*	C +2% A	$41{,}770^{+2{,}750}_{-2{,}060}$[f]	SRR-2343	−2	−1.5	−1.4

6	B6/103	≈90	*Crassostrea* sp.	C	$31{,}800^{+670}_{-620}$	SRR-2340	−3.6	−2.8	−2.9
7	B2/101	≈90	*Crassostrea* sp.	C	$30{,}670^{+570}_{-530}$	SRR-2337	−3.5	−5.2	−4.3
8	302a	12	*Pinna fragilis*	A +9% C	$38{,}320^{+1{,}590}_{-1{,}340}$	SRR-2240	1.7	1.4	1.6
9	5	0.5	*Crassostrea* sp.	C	$35{,}680^{+1{,}450}_{-1{,}210}$	SRR-2339	−2.5	−1.7	−0.3
10	103a	23	*G. glycymeris*	A	$33{,}580^{+850}_{-760}$	SRR-2241	1.6	1.7	1.1
11	103b	20	*Arca noae*	A	$31{,}820^{+660}_{-630}$	SRR-2245	1.2	1	−0.5
12	915a	8	*G. glycymeris*	A	$32{,}440^{+670}_{-610}$	SRR-2242	1.8	1.6	1
13	601	1.7	*Notirus irus*	A	6,890±90	SRR-2244	−1.3	−1	1.5
14	502a2	23.5	*A. noae*	A	$32{,}060^{+740}_{-690}$	SRR-2247	0.4	0	0.3
15	501a	7.5	*Spondylus gaederopus*	A +3% C	$36{,}180^{+1{,}200}_{-1{,}030}$	SRR-2246	0.2	1.1	0.7

[a] Calculated by peak intensity method. A = aragonite with no detectable calcite; C = vice versa.
[b] Normalized to $\delta^{13}C = 0‰$.
[c] A was determined on an aliquot from the bulk CO_2 prepared for dating. B was determined on CO_2 from a separate subsample of shell by hydrolysis with H_3PO_4 at 25°C.
[d] Same as note b.
[e] Age reported as >38,600 by dating laboratory ($\delta^{14}C = -1{,}000 \pm 1.9‰$).
[f] Age reported as >40,420 by dating laboratory ($\delta^{14}C = -994.5 \pm 1.6‰$).

not widely accepted, especially when the results lie close to the limits of the method. The dates listed in Table 11-1, with the exception of 601, are suspect on both counts, yet there is no reason why unaltered shell material that has been carefully pretreated should not give reliable results to the very limits of instrumental and statistical resolution. Nevertheless cross-checks of various sorts were applied to test the ages for stratigraphic as well as chemical validity.

All the samples were screened according to the procedures outlined in Vita-Finzi (1980). After mechanical cleaning with dental burrs, the shells were sectioned and acetate peels taken for inspection by light microscopy and comparison with modern specimens to see whether the original calcite or aragonite structures were unaltered. The specimens that passed this test were sampled across the entire width of the shell for X-ray diffraction. With aragonitic species even a trace of calcite called for renewed cleaning; if this did not eliminate the problem, the sample was rejected unless the scanning electron microscope (SEM) showed that leaching was likely to eliminate the calcite. With calcitic shells (notably, *Crassostrea*) X-ray analysis is of course less decisive, and greater reliance was placed on the SEM for sample selection.

All these techniques employ subsamples. X-ray diffraction, which has a resolution of about 1%, uses a few milligrams. SEM specimens are even less substantial (Fig. 11-2), and an absence of evidence of contamination is not evidence of purity. The $\delta^{13}C$ values obtained by the dating laboratory for age normalization (Column A in Table 11-1), though probably distorted by fractionation induced during chemical conversion of the sample for radiometric counting (D. D. Harkness, pers. comm.), provide a useful additional test of sample quality because a value lower than that characteristic of the species could reflect the influence of meteoric waters

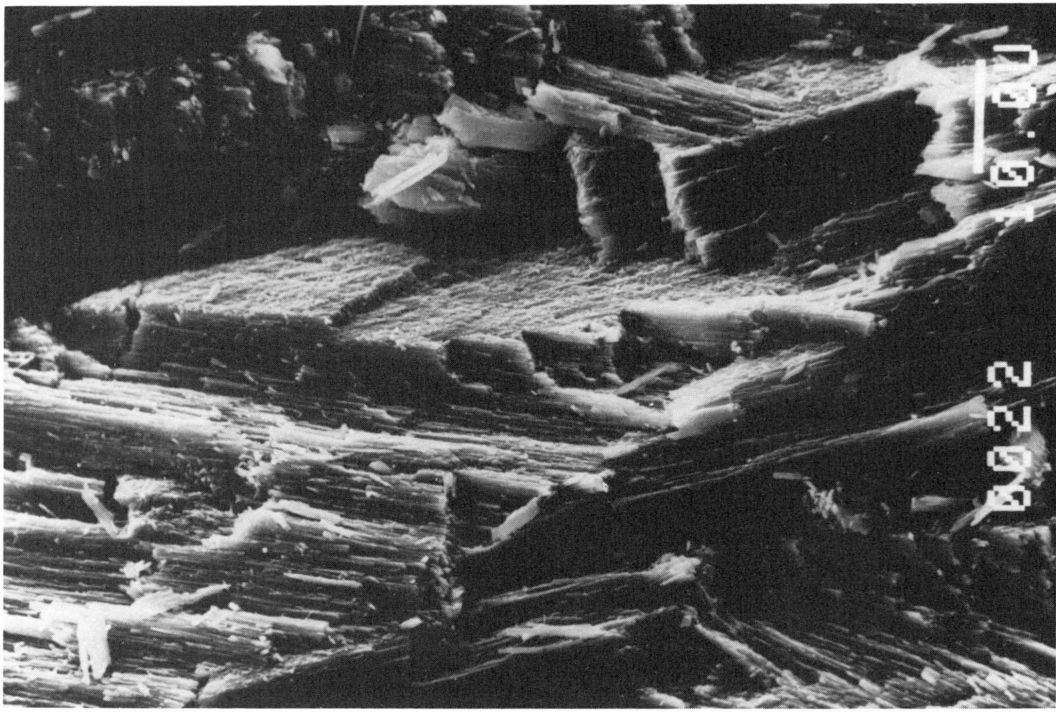

Figure 11-2. Scanning electron micrograph of *Glycymeris glycymeris* from sample 501 showing unaltered aragonitic laths. Scale bar = 1μm.

and thus of modern carbon. There is a hint of this effect in some of the *Crassostrea* samples listed in Table 11-1. As we have seen, it is with these species that other tests of contamination are less satisfactory than for aragonitic mollusks. The $\delta^{13}C$ values given in Column B are geochemically more dependable but they too apply to a small part of the sample, though admittedly to a portion selected at random. The agreement between the two sets of figures is fair, and none of the specimens emerges as devoid of merit. The $\delta^{18}O$ readings are not intended as guides to specimen purity, but it is reasonable to suppose that values departing markedly from the range normally associated with inshore waters would deserve scrutiny.

To test the validity of the screening and the success of any cleaning applied before submission, sample 1o was subjected to sequential dating whereby an age is determined on material obtained from the outer part of a sample by acid leaching and a second age is then determined on the inner part of the sample (Olsson and Blake, 1961/1962). The greater age obtained for the outer fraction may reflect superficial contamination by dead carbon, perhaps from waters percolating from limestone bedrock, but as the inner and outer fractions overlap at 2 standard deviations, any such effect appears to be slight. Had the shells given an age far lower than the true one because of superficial contamination by young carbon, the discrepancy would of course have been in the opposite sense. It is possible that at such levels of activity all the numbers are entirely spurious and any agreement purely the result of chance—or of undetected contamination of very old samples—but at present there is no evidence in support of this view.

One item of evidence in fact suggests that all the ages may prove to be, if anything, too great. Two samples of water were taken in the Gulf of Corinth by Heezen and co-workers (1966), one at the surface and the other at a depth of 800 m. The former gave a $^{14}C/^{12}C$ reading 4.5% lower than average north Atlantic water and the latter one 4.1% lower, perhaps because dead carbon is derived from the local limestones and there is limited water exchange between the Ionian Sea and the gulf. Similarly low values were recorded at a depth of 2,340 m in the Levant Basin.

Sample 601, the only fossil shoreline to give a Holocene age, was checked by running a duplicate by a first-order technique that relies on the absorption of CO_2 in a base (Vita-Finzi, 1983) rather than the benzene method used for the samples listed in Table 11-1. The result was 7,100 ± 1,300 yr (UCL-10), in excellent agreement with the conventional date as well as with the stratigraphic evidence for the section (Fink and Schröder, 1971). The Pleistocene dates are outside the reach of the first-order method, and the one remaining test is contextual.

At Location 1 (Fig. 11-3), a section in the New Corinth terrace about 0.5 km west of modern Corinth investigated, among others, by Keraudren (1970) and Freyberg (1973), marine sands and marls rest unconformably on a Tertiary gravel and are in turn covered by dune sands interbedded with colluvium. The stratigraphy and fauna suggest gradual deepening followed by rapid emergence and the deposition of regressive dunes. The unconformity between the dunes and the marine sands is weak enough to indicate a very brief erosional phase. Samples 1o and 4, being on redeposited shell, set a limiting age to the overlying beds; note that X-ray diffraction analysis of the *Pecten* material is inconclusive because the species is of mixed mineralogy and contamination is thus difficult to assess. The agreement between the ages for samples 2a and 2g, and the slightly lower age for 3c higher in the section, carry conviction in view of the good quality of the samples. The *Cerastoderma glaucum* shells of 2a were in growth position, and those of 3c, though displaced, were still articulated.

Locations 2 and 3 are western and eastern extensions respectively of the New Corinth terrace, and their ages are consistent with this view. Of the remainder, only locations 10 and 11 are of value for testing, rather than being tested by, their ^{14}C ages, and here again the older age is stratigraphically lower. Elsewhere, as at location 1, the dubious samples—namely, those with a wholly or partly calcitic composition or with appreciably negative $\delta^{13}C$ values—

Figure 11-3. Section at location 1 showing gravel-capped Tertiary marls overlain by marine beds and aeolianites.

give the greatest ages where one might expect contamination to tend toward spuriously low ages. Taken in conjunction with the inner/outer results for sample 1o, this result carries some reassurance.

PALEOCLIMATIC APPLICATIONS

During the 1981 Corinth earthquakes, parts of the coast underwent uplift whereas others subsided (Jackson et al., 1982). In some places there is evidence of earlier movements in the opposite sense (Vita-Finzi and King, 1984). Coupled with the spasmodic nature of the activity, such reversals make it difficult to extract the eustatic component from the shoreline record by postulating uniform rates of deformation. It is also evident that correlation by height is unwise even within the restricted confines of the gulf, a warning already voiced by Hey (1978) for different parts of the Mediterranean. Nevertheless, the dated terraces can be used to link climatic changes derived from fluviatile and lacustrine evidence to the local eustatic record.

The uppermost unit at location 1 has previously been equated with the Kokkinopilos beds of northwestern Greece, simply because both deposits consist of red silts and sands interdigitating with eolianites and overlie fossil beach deposits (Higgs and co-workers, 1967). Confirmation (or rejection) of this proposal would have implications for various facets of the climatic history of Greece. The Kokkinopilos beds have been interpreted as the products of flash floods and increased mechanical weathering at a time of predominantly arid conditions because in Epirus they consist of coalescing alluvial fans apparently laid down by ephemeral floods at a time when the headwaters were well supplied with angular debris and the valley floors experienced iron and manganese enrichment between flows (Higgs et al., 1967). Evidence that these conditions prevailed as far south as the Gulf of Corinth would clearly color attempts to model the corresponding climatic patterns, especially if it could be shown that emergence was then prevalent.

As we have seen, the dunes and colluvium at location 1 were laid down during a phase of emergence during the past 31,000 yr. In Epirus the valley fills have been dated in two ways. First, they have yielded redeposited Mousterian artifacts and at one site contain an advanced Paleolithic chipping floor. At the nearby shelter of Asprochaliko a date of 26,100 ± 900 yr B.P. (I-1956) has been obtained for an advanced Paleolithic horizon and presumably also dates the latter part of the Kokkinopilos phase of deposition. Second, Kokkinopilos beds pass into lacustrine deposits in the Ioannina Basin. Fossil beaches preserved in the lakeside shelter of Kastritsa date from between 20,000 ± 480 and 19,900 ± 370 yr B.P. (sample nos. I-2468 and 2465), and it has been surmised that the lake level was above the present for an unknown length of time before this time interval and fell after it in fluctuating manner. The local pollen evidence indicates steppe conditions, which favors the view that—granted lake level was controlled solely by climate—its rise stemmed more from a fall in evaporation than an increase in precipitation (Bailey et al., 1983).

The resolution in our data is not adequate to pin down the onset and end of any climatic oscillation or any lag between the coastal and continental response to such oscillations. But there is now enough information to show that high lake level, flash flood aggradation, and the deposition of eolianites were taking place between 30,000 yr and 20,000 yr B.P. in Epirus and the coasts of the Gulf of Corinth. Yet, in a tectonically active area regressions cannot always be identified with certainty in the shoreline evidence, and even if allowance is made for depth and aspect, the fauna may reflect shifts in the circulation pattern or offshore habitats that have no simple climatic interpretation. Where the parochial nature of the fauna may score is in the provision of carbonate samples suitable for stable-isotope analysis.

Mollusks from beach deposits have tended to be neglected in this connection on the grounds that they are likely to include specimens from disparate environments that range widely in age (Keith et al., 1964), but data from such assemblages have been found valuable in the paleoclimatic analysis of elevated coral terraces in Barbados (Shackleton and Matthews, 1977) and New Guinea (Aharon, Chappell, and Compston, 1980) as well as of Holocene deposits on the Israeli coast (Kaufman and Magaritz, 1980). Preliminary analysis of stable isotopes from fossil beaches and middens in southern Iran (P. A. R. Ireland, pers. comm.) suggests that the vagaries of redeposition on a beach do not necessarily increase the scatter of isotopic values displayed by mollusk assemblages, although shells collected for food can on occasion be linked to a well-defined source area (Emiliani et al., 1964).

Shackleton and Matthews (1977) used the isotopic signature of the beach faunas to determine their relationship to the oceanic record. Independent dating of such sequences permits discrepancies between the two lines of evidence to emerge. Coupled with tectonic studies designed to isolate the eustatic component, such data could in due course reveal how closely sea level is associated with oceanic paleotemperature and perhaps also the extent to which nonclimatic factors, such as diagenetic recrystallization of foraminifera (Killingley, 1983) or the release of hydrothermal waters enriched in ^{18}O at times of high ridge activity, have influenced the deep-sea isotope history. Only then will it be possible to uncover "the manner and rates of sea-level oscillation in relation to climatic changes" (Fairbridge, 1961, p. 138).

ACKNOWLEDGMENTS

I thank the Natural Environment Research Council for field support and ^{14}C ages and D. D. Harkness, G. C. P. King, and S. J. Phethean for their help.

REFERENCES

Aharon, P.; Chappell, J.; and Compston, W., 1980, Stable isotope and sea-level data from New Guinea supports Antarctic ice-surge theory of ice ages, *Nature* **283:**649-651.

Bailey, G. N.; Carter, P. L.; Gamble, C. S.; and Higgs, H. P., 1983, Asprochaliko and Kastritsa: Further investigations of Palaeolithic settlement and economy in Epirus (North-West Greece), *Prehistoric Soc. Proc.* **49:**15-42.

Bousquet, B.; Dufaure, J.-J.; Keraudren, B.; Péchoux, P.-Y.; Philip, H.; and Sauvage, J., 1976, Les corrélations stratigraphiques entre les faciès marins, lacustres et continentaux du Pleistocène en Grèce, *Soc. Géol. France Bull.* **18:**413-418.

Depéret, C., 1913, Observations sur l'histoire géologique pliocène et quaternaire du Golfe et de l'Isthme de Corinthe, *Acad. Sci. Comptes Rendus* **156:**427-431, 659-663, 1048-1052.

Dufaure, J.-J.; Keraudren, B.; and Sébrier, M., 1975, Les terrasses de Corinthie (Grèce): Chronologie et déformations, *Acad. Sci. Comptes Rendus* **281D:**1943-1945.

Dufaure, J.-J., and Zamanis, A., 1979, Un vieux problème géomorphologique: les niveaux bordiers au sud du Golfe de Corinthe, *Assoc. Géographes Français Bull.* **57:**341-350.

Emiliani, C.; Cardini, L.; Mayeda, T.; McBurney, C. B. M.; and Tongiorgi, E., 1964, Paleotemperature analysis of fossil shells of marine mollusks (food refuse) from the Arene Candide cave, Italy and the Haua Fteah Cave, Cyrenaica, in H. Craig, S. L. Miller, and G. J. Wasserburg (eds.), *Isotopic and Cosmic Chemistry,* Amsterdam: North-Holland, pp. 133-156.

Fairbridge, R. W., 1961, Eustatic changes in sea level, in L. H. Ahrens, F. Press, and R. K. Runcorn (eds.), *Physics and Chemistry of the Earth,* vol. 4, London: Pergamon Press, pp. 99-185.

Fairbridge, R. W., 1971, Quaternary sedimentation in the Mediterranean region controlled by tectonics, paleoclimates and sea level, in D. J. Stanley (ed.), *The Mediterranean Sea,* Stroudsburg, Pa.: Dowden, Hutchinson & Ross, pp. 99-113.

Fink, R., and Schröder, B., 1971, Anzeichen eines holozänen Meereschochstandes an der Landenge von Korinth, *Neues Jahrb. Geologie u. Paläontologie Monatsh.* **5:**265-270.

Freyberg, B. V., 1973, Geologie des Isthmus von Korinth, *Erlanger Geologische Abh.* **95:**5-160.

Heezen, B. C.; Ewing, M.; and Johnson, G. L., 1966, The Gulf of Corinth floor, *Deep-Sea Research* **13:**381-411.

Herforth, A., and Richter, D. K., 1979, Eine pleistozäne tektonische Treppe mit marinen Terrassensedimenten auf der Perachorahalbinsel bei Korinth (Griechenland), *Neues Jahrb. Geologie u. Paläontologie Abh.* **159:**1-13.

Hey, R. W., 1978, Horizontal Quaternary shorelines of the Mediterranean, *Quat. Research* **10:**197-203.

Higgs, E. S.; Vita-Finzi, C.; Harris, D. R.; and Fagg, A. E., 1967, The climate, environment and industries of Stone Age Greece: Part III, *Proc. Prehistoric Soc.* **33:**1-29.

Jackson, J. A.; Gagnepain, J.; Houseman, G.; King, G. C. P.; Papadimitriou, P.; Soufleris, C.; and Virieux, J., 1982, Seismicity, normal faulting, and the geomorphological development of the Gulf of Corinth (Greece): The Corinth earthquakes of February and March 1981, *Earth and Planetary Sci. Letters* **57:**377-397.

Kaufman, A., and Magaritz, M., 1980, The climatic history of the eastern Mediterranean as recorded in mollusk shells, *Radiocarbon* **22:**778-781.

Keith, M. L.; Anderson, G. M.; and Eichler, R., 1964, Carbon and oxygen isotopic composition of mollusk shells from marine and fresh-water environments, *Geochimica Cosmochimica Acta* **28:**1757-1786.

Keraudren, B., 1970, Les formations quaternaires marines de la Grèce, *Mus. d'Anthropologie et de Préhistoire de Monaco Bull.* **16:**5-153.

Killingley, J. S., 1983, Effects of diagenetic recrystallisation on $^{18}O/^{16}O$ values of deep-sea sediments, *Nature* **301:**594-597.

Olsson, I. V., and Blake, W., Jr., 1961/1962, Problems of radiocarbon dating of raised beaches based on experience in Spitzbergen, *Norsk Geog. Tidsskr.* **18:**47-64.

Pérès, J. M., 1967, The Mediterranean benthos, in H. Barnes (ed.), *Oceanography and Marine Biology,* vol. 5, London: Allen and Unwin, pp. 449-533.

Richards, G. W., 1982, Intertidal molluscs as sea-level indicators, Ph.D. dissertation, University of London, 357p.

Sébrier, M., 1977, Tectonique récente d'une transversale à l'Arc Egéen: Le Golfe de Corinthe et ses régions périphériques, Thesis, 3 ème cycle, University of Paris XI, 76p.

Shackleton, N. J., and Matthews, R. K., 1977, Oxygen isotope stratigraphy of late Pleistocene coral terraces in Barbados, *Nature* **268:**618-620.

Vita-Finzi, C., 1980, C-14 dating of Recent crustal movements in the Persian Gulf and the Iranian Makran, *Radiocarbon* **22:**763-773.

Vita-Finzi, C., 1983, First-order ^{14}C dating of Holocene molluscs, *Earth and Planetary Sci. Letters* **65:**389-392.

Vita-Finzi, C., and King, G. C. P., 1984, The seismicity, geomorphology and structural evolution of the Corinth area of Greece, *Philos. Trans. R. Soc. London A* **314:**379-407.

12: Variations in Holocene Sea Level on the French Atlantic Coast and Their Climatic Significance

Mireille Ters
Sorbonne University, Paris

Abstract: Following the maximum Würm emergence (ca. −120 m, between ca. 17,000 B.P. and 20,000 B.P.), the glacio-eustatic rise of sea level occurred discontinuously through a succession of submergences and emergences. Between 8,250 B.P. (C-14) and today, seven major submergences and six smaller emergences took place. On the French Atlantic coast no evidence of a Holocene marine level higher than at present has been observed.

The fluctuations of the generally rising sea are inferred to reflect climatic changes. This view is supported by correlations between sea levels and neoglacial advances in Europe and in North America, with palynological data, and the $^{18}O/^{16}O$ ratios measured in the Camp Century, Greenland, ice core. Prehistoric and historic climate data likewise are in good agreement with the concept of climate control of sea level.

INTRODUCTION

Exact data for reconstructing eustatic sea levels require an area of tectonic stability. This requirement rules out regions of known recent neotectonic movements and those affected by glacio-isostasy. To avoid such difficulties, I have concentrated my research (1) in a relatively stable area (the American Massif) of France and its borders; (2) in an area large enough for results to be of general significance (600 km north to south, 480 km east to west); (3) on deposits whose sedimentologic characteristics, fauna, and flora provide information on contemporary sea level (such as brackish-water deposits; schorre, i.e., salt marsh sediments; and coastal peats); and (4) on stratigraphic successions of nonmarine, brackish-water, and marine deposits that contain proof of submergences and emergences.

In my synthesis, I have included the considerable previous research on Holocene sea levels in the same area by Bourdier, Castaing, Elhai, Feral, Gabet, Giresse, L'Homer, Larsonneur, Lefèvre, Mariette, Verger, Houault, Morzadec-Kerfourn, and Giot. My principal conclusions, however, are based on the results of boreholes I have drilled on the strand plains and beaches at Camiers, Saint Marc, Le Palus, Bréhec, Brétignolles, and other sites and also on my previous studies of the estuaries of the Loire (Ters, Planchais, and Azema, 1968), the Seine (Ters et al., 1971), and the Somme (Ters et al., 1980) (Fig. 12-1).

The curve showing the submergences and emergences during the Flandrian transgression (Fig. 12-2) is based on more than 200 radiocarbon dates. Each point on the curve represents mean sea level (MSL) (in meters) above or below today's MSL (NGF, i.e., French national MSL datum). The points take into account the local tidal conditions as well as the sample data. Along the coast studied, the tide ranges between 5 m and 15 m. Other indicators designate that MSL was equal to or lower than that shown. For example, if the base of a coastal peat bed is found at −2 m MSL and the local tidal range is 12 m, then the corresponding MSL at time of peat growth was at or below −8 m. The interpretation of the stratigraphic successions is based on analysis of the sediment, the macrofauna, the microfauna, and the pollen.

OSCILLATIONS IN THE HOLOCENE RISE OF SEA LEVEL

Sea Level Between 10,050 B.P. and 8,000 B.P. (^{14}C)

Not enough data are available for drawing a continuous sea-level curve for the entire study area, but some episodes may be traced. Near the mouth of the Gironde estuary, a brackish-water sediment (silt with ostracods) has been dated at 10,050 ±200 B.P. (135 in Table 12-1; hereafter the numbers in parentheses will refer to the list in that table); the corresponding MSL can be estimated at about −60 m (Castaing, Feral, and Klingebiel, 1971). The local tidal range is nearly 7 m.

In the northern part of the English Channel coast of France, peat dated at 9,700 ±200 B.P. (129) was collected from a depth of −55 m (NGF). This finding implies a maximum MSL at about −60 m. On the eastern channel coast, Preboreal fresh-water peat has been recovered from boreholes (Larsonneur, 1971). This peat indicates that the contemporary high-water mark stood lower than the level from which the samples came (128, 130, 132, and 133).

During the Preboreal and the Boreal, the sea rose very rapidly as the larger ice caps melted; between about 9,700 B.P. and 8,200 B.P., the rise was more than 25 m (>5 cm/yr). Between 8,200 B.P. and 7,900 B.P., the sea seems to have risen 15 m in 300 yr (averaging 5 cm/yr). This happened during the xerothermophilous phase when *Quercus ilex* attained its maximum and was recorded on pollen charts of Normandy (Elhai, 1963) and west-central France (Paquereau, 1964; Planchais, 1971).

Sea Levels near Le Havre

Details of the Boreal rise have been provided in a complete succession 36 m thick based on boreholes for the new harbor at Le Havre. At depths between −23 m and −17 m, marine and brackish-water sediments are interstratified with peat dated at about 8,200-7,900 B.P.

The first half of the Boreal and up to 8,200 ±190 (122). MSL stood below −32 m (peat and clay lacking foraminifera or marine ostracods in the Le Havre boring). However, in a peat dated at 8,470 ±170 B.P. (124), *Chenopodiaceae* are fairly abundant.

The first submergence (at about 8,150 B.P.; 121). *Bucella frigida* (a brackish-water ostracod) first appears in clay deposits that contain a Boreal flora and *Chenopodiaceae*, indicating an MSL at about −22 m.

The first emergence (below −25 m, ca. 8,050 ±170 B.P.; 119). A slight emergence of at least 3 m is indicated. The brackish-water deposits are overlain by a fresh-water peat and then by a solidified lacustrine limestone, composed of ostracod shells, that underlies large parts of the Seine estuary at depths between −21 m and −22 m.

The second submergence (about −21 m, ca. 7,900 B.P.; 117). Between about 8,000 B.P. and 7,900 B.P., the sea rose again and deposited laminated intertidal silts at depths between −20.4 m and −17.5 m. These silts contain abundant foraminifera and marine ostracods. The great abundance of *Chenopodiaceae* indicates a deposit very similar to schorre (salt marsh deposits such as those of the Wadden Zee made only at high spring tides), but they diminish near the top of the silts, indicating another emergence.

The second emergence (about −24 m, ca. 7,820 ±170 B.P.; 116). Fresh-water peat overlies the laminated silt at −17.25 m. A total absence of *Chenopodiaceae* indicates that the coast line was substantially modified by an emergence of several meters. The flora is Boreal (M. Denefle, pers. comm.). Insect remains studied by P. J. Osborne and F. W. Shotton indicate that the climate was close to that of today (Sanquer, 1968).

(text continues on page 218)

2°
0°
2°
4°
50°
48°
Calais
Wissant
Wimereux
Hardelot
CAMIERS
Merlimont
Marquenterre
St-FIRMIN
PARIS
LE HAVRE
Bernières
Asnelles
Carentan
Cherbourg
Nacqueville
Montmartin
Lingreville
St-Pair
St Jean-le-Thomas
Avranches
Pontorson
Mt-St-Michel
Cancale
RENNES
St-MARC
BREHEC
Lannion
Locquirec
Plouescat
Goulven, Kerlouan, Guissény
ARGENTON
TRÉOMPAN
Porspoder
Audierne
Beg-an-Dorchen
Er Lannic
Penerf
Redon
St-Gildas-de-Rhuys
Hoëdic
BELLE-ILE
Grande-Brière
Petit-Mars
St-Brévin
La Plaine-sur-Mer
Préfailles
La Vendette
FROMENTINE

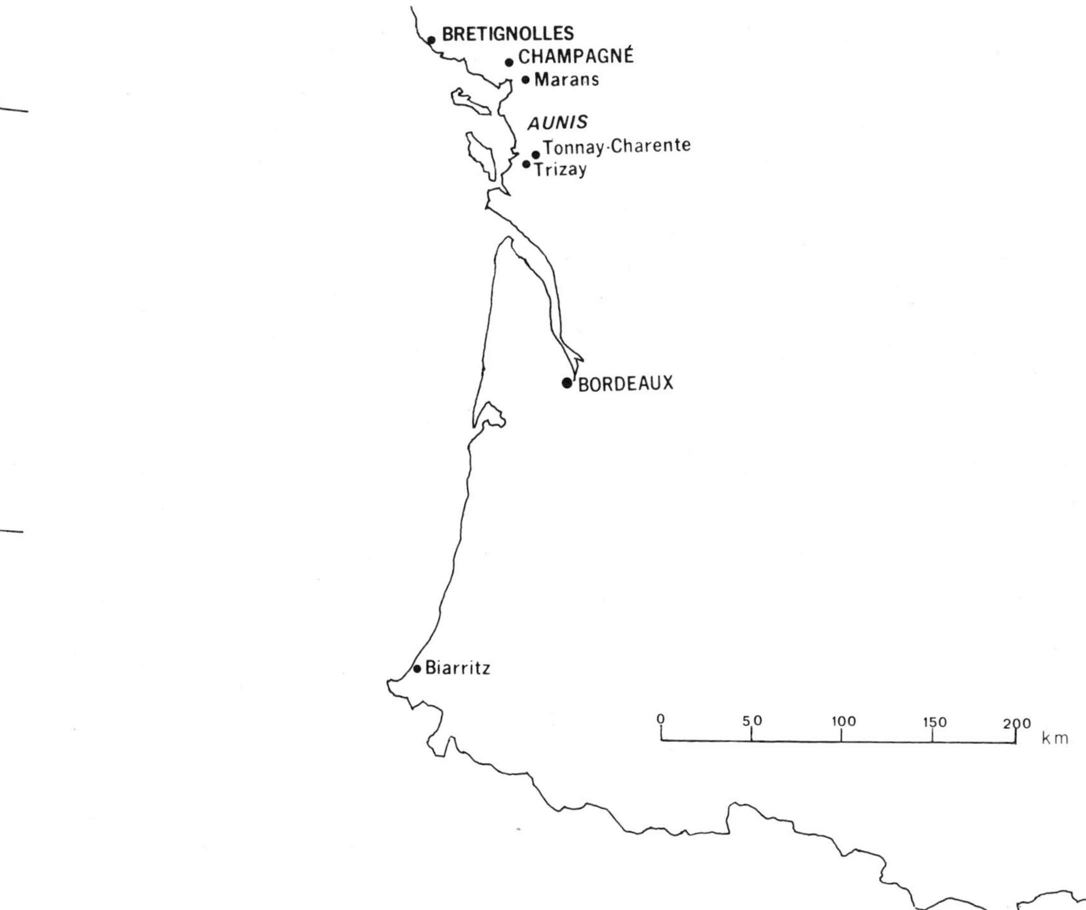

Figure 12-1. Index map of the French Atlantic coast showing localities studied.

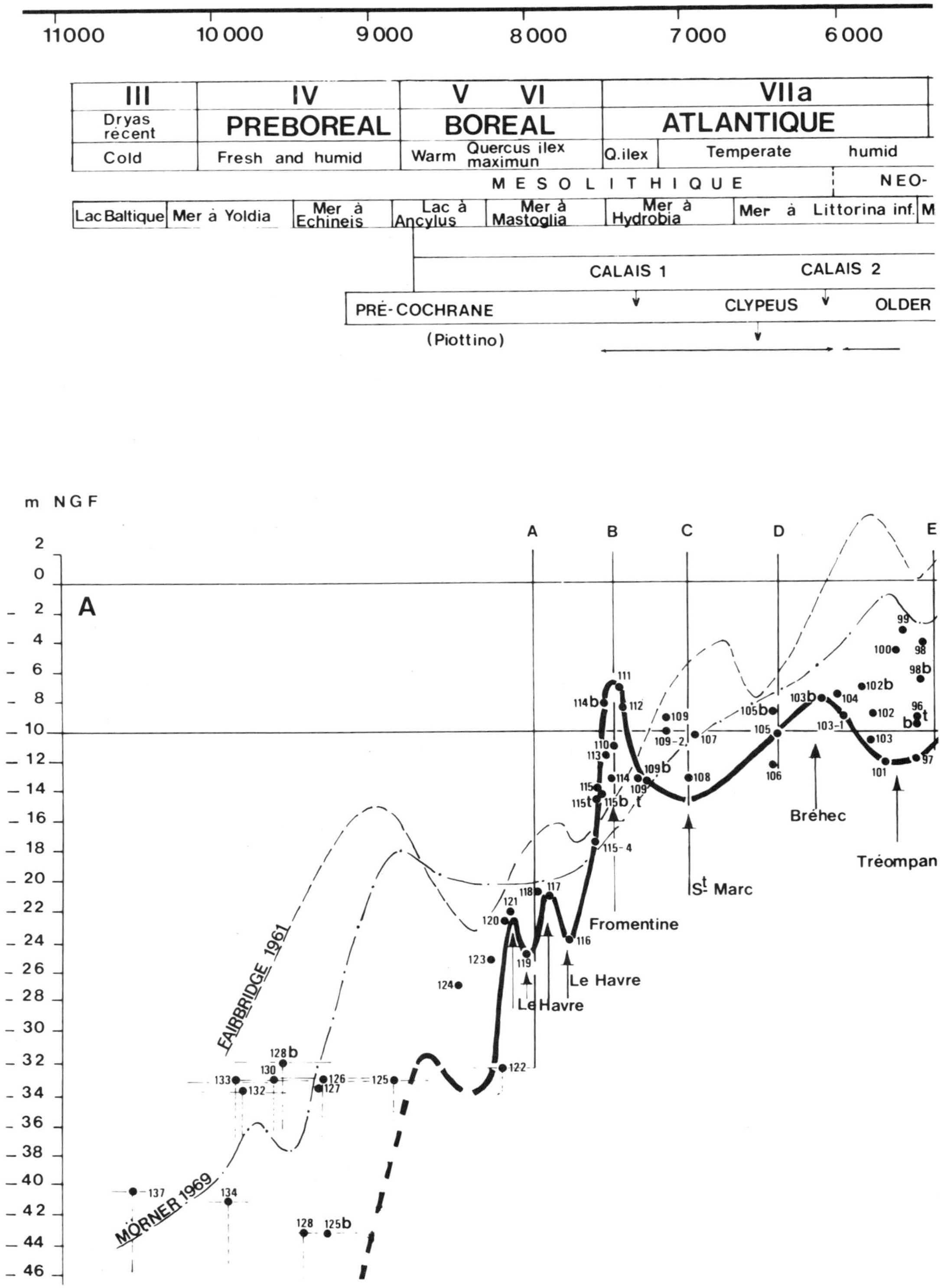

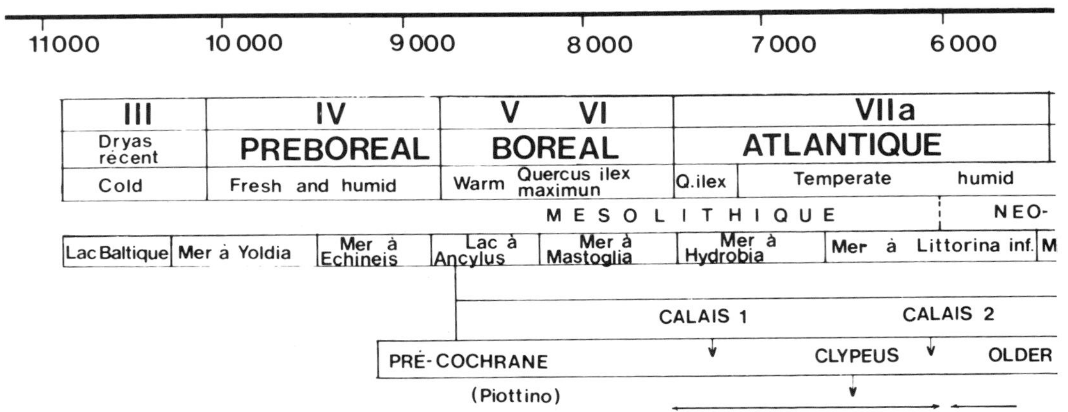

m NGF

A

A B C D E

FAIRBRIDGE 1961

MÖRNER 1969

Fromentine

Le Havre

Le Havre

St Marc

Bréhec

Tréompan

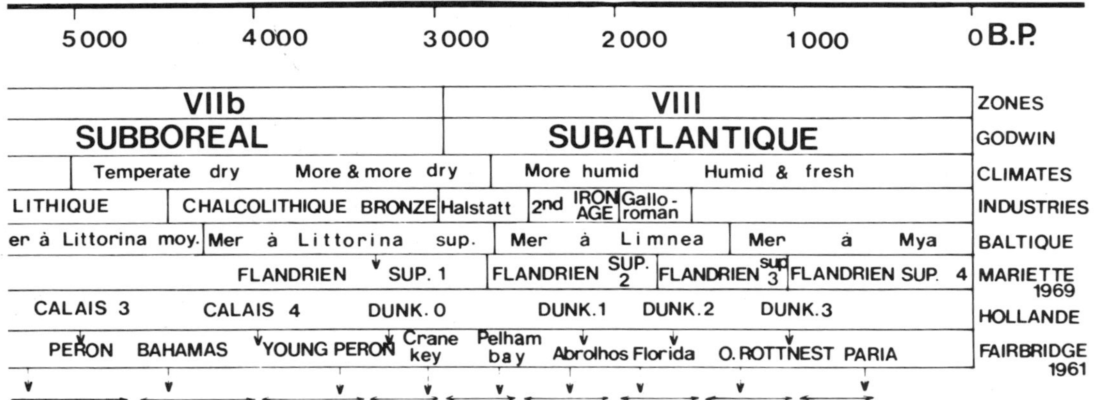

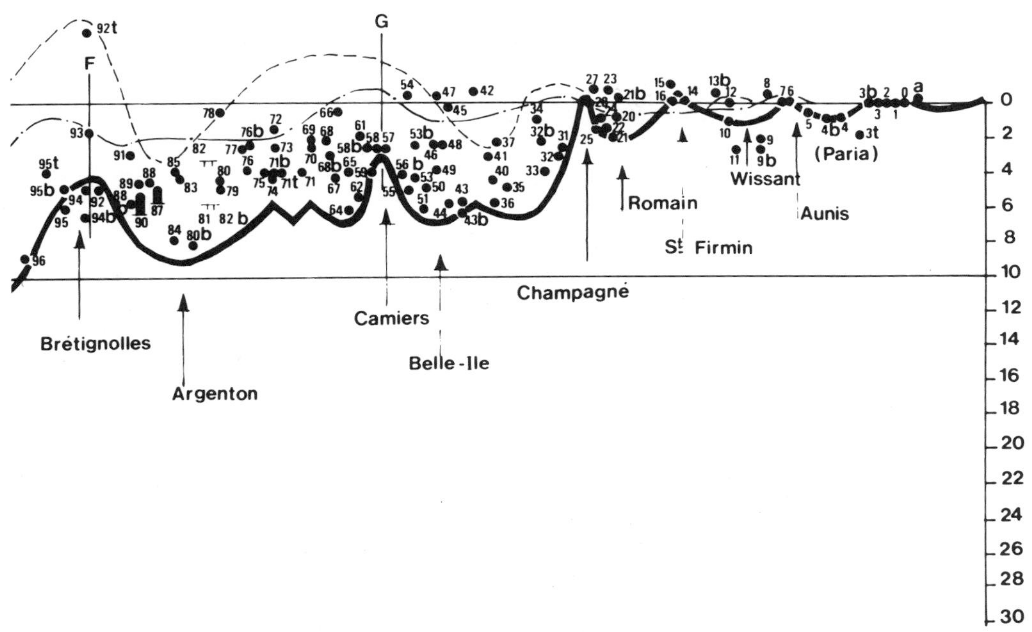

Figure 12-2. Curve showing oscillations in the Flandrian transgression along the French Atlantic coast, from 10,000 B.P. to the present, compared with various indicators of Holocene climate and sea level (see text).

Unbroken line: mean sea level. Points shown on the curve are based on determinations having small vertical uncertainty. Points above the curve indicate only that mean sea level stood below the indicated level. Boreholes are shown by letters as follows: A, Le Havre; B, Fromentine; C, Saint Marc; D, Bréhec; E, Treompan; F, Bretignolles; and G, Camiers.

Table 12-1. Radiocarbon Dates

Sample Number	^{14}C Date (B.P.)	Laboratory Number (Gif-sur-Yvette and others)	Locality	Formation or Prehistoric Site	Elevation (m NGF)	Mean Sea Level during Deposition (m)	References
a	180 ± 90	2,872	Salenelles (Somme)	Coastal shingle bank (shells)	+4	= 0	Ters (unpubl.)
0	450 ± 90	1,355	Bréhec (Côtes-du-Nord)	Coastal shingle ridge	+6.50	= 0	Ters and Pinot (1969)
1	472 ± 95	1,799	Anse de la Houle Cancale (Ille-et-Vilaine)	Coastal shingle ridge	+8	= 0	Verger (1972)
2	560 ± 95	762	Berck-Plage (Pas-de-Calais)	Sand with humus	+2 HM	= ±0	Etienne and Robert (in *Radiocarbon* 1969)
3	580 ± 120	Gsy 238	Beg-an-Dorchen (Finistère Sud)	Coastal shingle ridge	+4.50	= 0	Giot (in *Radiocarbon* 1969)
3 bis	610 ± 90	2,129	Broúage (Charente-Maritime)	Wood (into salt marsh)	+2.20	= 0	Regrain (unpubl.)
4	780 ± 100	Sa 224	Pénerf (Morbihan)	Shells in clay	+4 ?	+1 ?	Rivière (in *Radiocarbon* 1965)
4 bis	810 ± 90		Montmartin (Manche)	Brackish clay	±6	±1	Giresse (1969)
5	980 ± 100	841	Marquenterre (Somme)	Fresh-water peat	+2.5	= −1	Lefèvre (1972)
5 bis	1,066 ± 100	(Pi 19)	Saint-Michel-en-l'Herm (Vendée)	Oyster shells over schorre	+2	= 0	Patte, Ters, and Verger (1961)
6	1,100 ± 90	891	Beg-an-Dorchen (Finistère Sud)	Coastal shelly sand ridge	+3	= 0	Giot and Guilcher (1968)
7	± 1,100		Littoral Aunis (Charente-Maritime)	Salt marsh	+3	= 0	Gabet (1970)
8	1,220 ± 100		Pénerf (Morbihan)	Shells in marine clay	+3	+0.5	Rivière (1965)
9	1,250 ± 120	388	Pontorson (Ille-et-Vilaine)	Coastal silt	±7	−2	Giresse (in *Radiocarbon* 1969)
9 bis	1,260 ± 100	1,144	Avranches (Manche)	Coastal silt	?	−3.15	Giresse (1969)
10	1,380		Crabec (Manche)	Shelly sand	+2	−1 to −1.70	Giresse, Dangeard, and Hommeril, 1967
11	1,390 ± 100	1,300	Lividie en Plounéour Trez (Finistère Nord)	Clay with Hydrobia	+2	< −2.70	Morzadec-Kerfourn (1969)
11 bis	1,400 ± 100		Pont de la Roque (Manche)	Marine clay		−1.8 to −2.2	Giresse (1969)
12	1,430 ± 120	389	Saint-Jean-le-Thomas (Manche)	Coastal silt	+7.5	= 0	Giresse (in *Radiocarbon* 1969)

12 bis	1,500 ± 150	Gsy 59	Lingreville (Manche)	Silt	+1 to +2	?	Elhaï (1963)
12 ter	1,500 ± 100	1,307	Pénerf (Morbihan)	Silt	+3.5	?	Rivière (in *Radiocarbon* 1965)
13	1,600 ± 105		Le Yaudet, Lannion (Côtes-du-Nord)	Peat	?	?	Giot (1968)
13 bis	1,521 ± 95		Briqueville (Manche)	Peat	+7.4	+0.4	Elhaï (1963)
14	1,680 ± 120	391	Hauteville (Montmartin)	Coastal silt	+7	= 0	Giresse (in *Radiocarbon* 1969)
15	1,730 ± 100	1,356	Saint-Firmin (Somme)	Shelly coastal ridge	+6	0 to +1	Ters and Pinot (1969)
16	±1,730	—	Le Curnic en Guissény (Finistère Nord)	Coins issued under Tetricus in fishtank	+4.5	= 0	Sanquer, 1968
18	1,968 ± 110	(Pise)	Marans (Charente-Maritime)	Oysters	−2	= 0	Ters (unpubl.)
20	2,050 ± 110	1,558	La Petite Aiguille (Charente-Maritime)	Gallic salt pan	+3	< −1	Gabet (1971)
21	2,060 ± 100	1,271	Trizay (Charente-Maritime)	Fresh-water peat	+1.3	< −1.7	Gabet (1971)
21 bis	2,070 ± 100	2,871	Erondelles (Somme)	Shell bed	+6	+0.5	Ters (unpubl.)
22	2,100 ± 110	1,294	Camiers (Pas-de-Calais)	Brackish clay	+2.5	= −1.5	Mariette (1969)
23	2,100 ± 100	1,185	Sainte-Radegonde (Vendée)	Coastal ridge (shells)	+6	+0.5 ?	Ters and Pinot (1969)
24	2,100 ± 110	1,294	Camiers (Pas-de-Calais)	Gallic salt pan	+3	< −1	Mariette (1969)
25	2,130 ± 150	172	Nacqueville (Manche)	Fresh-water peat	+2.5	< −1.5	Elhaï (1963, and in *Radiocarbon* 1966, 8)
26	2,130 ± 150	171	Le Becquet (Manche)	Coastal peat	+2.5	≤ −1.5	Delibrias and Elhaï (in *Radiocarbon* 1966, 8)
27	2,180 ± 105	817	Argenton (Finistère Nord)	Brackish peat (A)	+8.4	< +0.60	Morzadec-Kerfourn (1969)
28	2,200 ± 100	1,800	Champagné (Vendée)	Coastal ridge (shells)	+3.4	0 to +0.5	Verger (1972)
29	±2,220	—	Moulin de la Rive (Locquirec) (Finistère Nord)	Shells	−3	±0	Delibrias and Giot (1971)
31	2,330 ± 105	818	Santec (Finistère)	Brackish peat	+2	< −2.8	Morzadec-Kerfourn (1969)
32	2,350 ± 150	(Sa 190)	Saint-Jacques-de-Rhuys (Morbihan)	Oak tree in peat	−1.20	< −4	Labeyrie and Delibrias (in *Radiocarbon* 1964)
32 bis	± 2,450	—	Lingreville (Manche)	Wooded peat	−2.10 below HM	< −2.10	Elhaï (1963)
33	2,450 ± 110	794	Chalézac (Charente-Maritime)	Halstatt dwelling peat	0	< −4	Gabet (1969)

(continued)

Table 12-1. *(continued)*

Sample Number	*^{14}C Date (B.P.)*	*Laboratory Number (Gif-sur-Yvette and others)*	*Locality*	*Formation or Prehistoric Site*	*Elevation (m NGF)*	*Mean Sea Level during Deposition (m)*	*References*
34	2,480 ± 100	2,607	Saint-Marc en Tréveneuc (Côtes-du-Nord)	Peat	+6.35	< −1	Ters (unpubl.)
35	2,670 ± 110	1,100	Audierne (Finistère Sud)	Shell in beach sandstone	−1.5	< −4.5	Giot and Guilcher (in *Radiocarbon* 1971)
36	2,680 ± 200	348	Ster-Vraz, Belle-Ile-en-Mer (Morbihan)	Fresh-water peat	−2.80	< −5.3	Planchais and Corillion (1968)
37	2,700 ± 100	1,178	Asnelles (Calvados)	Brackish peat	+2	< −2.5	Larsonneur (1971)
40	2,750 ± 150	160	Le Curnic en Guissény (Finistère Nord)	Brickwork	0 to −0.5	< −4.5	Giot (1968)
41	2,770 ± 300	(Sa 43)	Grande Brière (Loire-Atlantique)	Upper black peat	−0.7	< −3.15	Vince (1958)
42	±2,850		Argenton (Finistère Nord)	Peat	+4.5	< +0.6	Morzadec-Kerfourn (1969)
43	2,890 ± 230	(Ny 53)	Ster-Vraz, Belle-Ile (Morbihan)	Fresh-water peat	−2.80	< −5.3	Coppens (in *Radiocarbon* 1968, 2)
43 bis	2,910 ± 90		Marquenterre	Fresh-water peat	0	< −6	Ters et al. (1980)
44	2,970 ± 230		Ster-Vraz, Belle-Ile (Morbihan)	Fresh-water peat	−2.80	< −5.3	Coppens (1967)
45	3,020 ± 110	712	Lampaul-Plouarzel (Finistère Nord)	Brackish peat	+3.5	< 0	Morzadec-Kerfourn (1969)
46	3,050 ± 110		Carentan (Coquebourg) (Manche)	Brackish peat	+1.5	< −2.50	Larsonneur (1971)
47	±3,050		Porspoder (Finistère Nord)	Brackish peat	+4.2	< +0.3	Morzadec-Kerfourn (1969)
48	3,060 ± 160	173	Saint-Pair (Manche)	Brackish peat	+4	< −2.5	Elhaï (in *Radiocarbon* 1966, 8)
49	3,060 ± 110	842	Marquenterre (Somme)	Fresh-water peat	0.5	< −3.5	Lefèvre (1972)
50	3,075 ± 110		Le Yaudet en Lannion (Côtes-du-Nord)	Peat	−2	< −6	Giot (1969)
51	3,130 ± 110	2,953	Le Croesty en Arzon, Morbihan	Black clay	0	= −6	Ters (unpubl.)
52	3,170 ± 130	1,640	Brétignolles (Vendée)	Peat bed within dune	+4	< +1	Ters (unpubl.)
53	3,175 ± 160	159	Le Curnic en Guissény (Finistère Nord)	Charcoal	0 to −0.5	−4.5 ?	Giot (in *Radiocarbon* 1966)

53 bis	3,180 ± 140	1,295	Camiers (Pas-de-Calais)	Brackish water habitation	+1.5	< −2.5	Mariette (1971)
54	±3,200		Porspoder (Le Gratz) (Finistère Nord)	Silt	4.5	+0.5	Morzadec-Kerfourn (1969)
55	3,220 ± 110	(47 C)	Le Curnic en Guissény (Finistère Nord)	Peat			Giot (in *Radiocarbon* 1966)
56	3,310 ± 200	Gsy 278	Le Curnic en Guissény (Finistère Nord)	Roots in fresh-water peat	0 to −1	< −4.5	Morzadec-Kerfourn (1969)
56 bis	3,320 ± 200	(Sa 223)	Saint-Aubin-sur-Mer (Calvados)	Peat	−2	< −4	Roth (unpubl.)
57	3,340 ± 110	1,241	Montmartin (Manche)	Shells in sand	+4.5	< −2.3	Giresse (1969)
57 bis	3,380 ± 110	2,244	Estuaire Charente (Charente-Maritime)	Oyster shells in coastal shingle ridge	−4.3	?	Gabet (in *Radiocarbon* 1972)
58	3,390 ± 120	710	Porsguen en Plouescat (Finistère Nord)	Brackish peat	+2.5	< −2.2	Morzadec-Kerfourn (1969)
58 bis	3,390 ± 100	1,300	Plounéour-Trez (Finistère Nord)	Brackish peat	+2	< −2.7	Morzadec-Kerfourn (1969)
59	3,400 ± 130	1,639	Camiers (Pas-de-Calais)	*Scrobicularia* in brackish clay	0	= −4	Ters (unpubl.)
60	3,400 ± 130	1,179	Asnelles (Calvados)	Peat with *Chenopodiaceae*	+1.9	< −2.6	Larsonneur (1971)
61	3,430 ± 110	3,421	Brétignolles	Wood in peat	+1	< −2	Ters (unpubl.)
62	3,460 ± 200	(Sa 202)	Bernières-sur-Mer (Calvados)	Clayey peat	−1.4	< −5.4	Hommeril (1964)
63	±3,500	—	Plounéour-Trez (Finistère Nord)	Peat	+2	< −2.7	Morzadec-Kerfourn (1969)
64	±3,550	—	Marquenterre (Somme)	Fresh-water peat	±−2	±−6	Lefèvre (1972)
65	3,550 ± 130	1,638	Camiers (Pas-de-Calais)	Upper peat	0	< −4	Ters (unpubl.)
66	3,600 ± 110	1,992	Brétignolles (Vendée)	Fresh-water peat	+2.5	< −0.5	Ters (unpubl.)
67	3,620 ± 125	815	Argenton en Landunvez (Finistère Nord)	Brackish peat	0	< −3.9	Morzadec-Kerfourn (1969)
68	3,660 ± 115	714	Trézien (Finistère Nord)	Wood in fresh-water peaty clay	+1.5	< −2.2	Morzadec-Kerfourn (1969)
68 bis	3,675 ± 60	Fontes	La Mailleraye (Seine)	Peat	−0.60	< −2.60	Huault and Lefèvre (1983)
69	3,750 ± 100	713	Brignogan (Finistère Nord)	Fresh-water peat	+2.5	< −2.2	Morzadec-Kerfourn (1969)
70	3,720 ± 200	397	Merlimont (Pas-de-Calais)	Brackish clay	+1	< −2.5	Etienne and Robert (1969)
71	±3,750	—	Saint-Hippolyte (Charente-Maritime)	Old Bronze Age site	0	< −4	Gabet (1971)
71 bis	3,900 ± 200	Fontes	Vernier marsh	Peat with *Sphagnum*	+0.5	< −4	Huault (1983)

(continued)

Table 12-1. *(continued)*

Sample Number	^{14}C Date (B.P.)	Laboratory Number (Gif-sur-Yvette and others)	Locality	Formation or Prehistoric Site	Elevation (m NGF)	Mean Sea Level during Deposition (m)	References
72	3,950 ± 200	1,293	Hardelot (Pas-de-Calais)	Habitation in peat	+2.5	< −1.5	Mariette (1971)
73	3,950 ± 140	1,013	Asnelles (Calvados)	Fresh-water peat	+1.8	< −2.7	Elhaï and Larsonneur (1969)
74	3,970 ± 135	816	Argenton (Finistère Nord)	Brackish peat (C)	−0.5	< −4.4	Morzadec-Kerfourn (1969)
75	±4,000		Hardelot (Pas-de-Calais)	Peat and dwelling site (end of Neolithic)	0 to +4	< −4 to 0	Mariette (1972)
76	4,100 ± 300	(Sa 42)	Grande Brière (Morbihan)	Remains of boat in peat	+2.1	< −3.9	Vince (1958)
76 bis	4,110 ± 300	(Sa 53)	Grande Brière (Morbihan)	Fresh-water peat	+0.7	< −2.30	Vince (1958)
77	4,120 ± 140	711	Porsguen en Plouescat (Finistère Nord)	Charcoal at base of peat	+2	< −2.7	Morzadec-Kerfourn (1969)
78	4,250 ± 250	282	Plouguerneau (Le Corréjou) (Finistère Nord)	Charcoal in peat	+4	< 0.5	Morzadec-Kerfourn (1969)
79	±4,250		Argenton (Finistère Nord)	Fresh-water peat	−1 ?	< −5	Morzadec-Kerfourn (1969)
80	4,260 ± 300	(Sa 46)	Grande Brière (Morbihan)	Stalk in peat	−1.30	< −4.5	Vince (1958)
80 bis	4,280 ± 250	Fontes	Vernier marsh	Black peat	−4.50	< −8	Huault (1980)
81	±4,300		Lerret en Kerlouan (Finistère Nord)	Base of passage grave	−0.85	< −6.05	Giot (1968)
82	±4,300		Roc'h ou Braz (Le Kernic) (Finistère Nord)	Base of passage grave	+1.70	< −3.10	Giot (1968)
82 bis	±4,300		La Vendette (Vendée)	Dolmen	−2.5	< −6	Mounès (unpubl.)
83	4,480 ± 300	(Sa 41)	Grande Brière (Morbihan)	Fresh-water peat	−1.25	< −4.25	Vince (1958)
84	±4,500		Argenton (Finistère Nord)	Brackish peat	−3.90	< −7.80	Morzadec-Kerfourn (1969)
85	±4,500		Wimereux (Pointe aux Oies) (Pas-de-Calais)	Neolithic site in brackish clay	−1	< −4	Mariette (1969)

86	4,520 ± 140	793	Tonnay-Charente (Charente-Maritime)	Wood in coastal shingle ridge	−9	?	Gabet (1971)
87	±4,600		Er Lannic (Morbihan)	Base of a Cromlec'h	+0	< −5	Giot (1968)
88	4,630 ± 300	(Sa 39)	Grande Brière (Morbihan)	Roots in peat	−1.30	< −4.70	Vince (1958)
88 bis	4,650 ± 100	—	Marquenterre	Peat	−0.5	< −6	Ters (1983)
89	4,700 ± 130	1,243	Montmartin (Manche)	Shelly sand	−2	< −4.70 to −5.20	Giresse (1969)
90	±4,700		Men Ozac'h (Finistère Nord)	Base of a menhir	−2	< −6.40	Giot (1968)
91	±4,750		Goulven (Finistère Nord)	Fresh-water peat	+1.7	< −3	Morzadec-Kerfourn (1969)
92	4,910 ± 120	2,108	Brétignolles (Vendée)	Brackish clay, with *Scrobicularia* and *Cardium*	−2	= −5	Ters (unpubl.)
93	4,980 ± 120	Gsy 75	Brignogan (Finistère Nord)	Fresh-water peat	+3	< −1.7	Morzadec-Kerfourn (in *Radiocarbon* 1966)
94	4,990 ± 120	2,109	Brétignolles (Vendée)	Wood in brackish clay	−2	= −5	Ters (unpubl.)
94 bis	5,000 ± 100	(M.C. 463)	Saint-Pair (Manche)	Wood in peat	+1	−6	Giresse (1969)
95	5,080 ± 140	843	Marquenterre (Somme)	Fresh-water peat	−2	< −6	Lefèvre (1972)
95 bis	5,080 ± 140	2,606	Saint-Marc en Tréveneuc Côtes-du-Nord)	Fresh-water peat	+1.15	< −5	Ters (unpubl.)
96	5,300 ± 140	1,358	Bréhec (Côtes-du-Nord)	Tree trunk in peat	−2	< −9 to −12	Ters (unpubl.)
96 bis	5,500 ± 120	Fontes	Vernier marsh	Wooded peat	−5.20	< −9	Huault (1983)
96 ter	5,445 ± 120	Fontes	Vernier marsh	Brown peat	−5.30	< −9	Huault (1980)
97	±5,500		Bréhec (Côtes-du-Nord)	Fresh-water peat	−6	< −12	Ters (unpubl.)
98	5,510 ± 250	345	Le Curnic en Guissény (Finistère Nord)	Habitation site in peat (charcoal)	±0 ?	< −4 ?	Giot (in *Radiocarbon* 1970)
98 ter	5,520 ± 110		Marquenterre	Fresh-water peat	−1.50	< −6	Ters (1980)
98 bis	5,520 ± 150	844	Marquenterre (Somme)	Fresh-water peat	−2.50	< −6.50	Lefèvre (1972)
99	5,650 ± 150	1,014	Asnelles (Calvados)	Fresh-water peat	+1.6	< −2.9	Larsonneur (1971)
100	5,680 ± 250	371	Asnelles (Calvados)	Fresh-water peat	+0.1	< −4.3	Larsonneur (1971)
101	±5,700		Tréompan en Ploudalmézeau (Finistère Nord)	Peat	−6.90	−12 ?	Morzadec-Kerfourn (1969)
102	5,770 ± 150	766	Tréompan (Finistère Nord)	Brackish peat	−4.4	< −8.80	Morzadec-Kerfourn (1969)
102 bis	5,900 ± 100	2,525	Brétignolles (Vendée)	Brackish ooze	−4	< −7	Ters (unpubl.)
103	5,900 ± 100	—	Estuaire Gironde (Gironde)	*Ostrea edulis* shell	−9	−11	Castaing et al. (1971)
103.1	6,090 ± 120	Fontes	Vernier marsh	Brackish clay	−5.90	−9	Huault (1980)
103 bis	±6,200	—	Bréhec (Côtes-du-Nord)	Shell beneath fresh-water peat	−2	< −8	Ters (unpubl.)

(continued)

Table 12-1. *(continued)*

Sample Number	^{14}C *Date (B.P.)*	*Laboratory Number (Gif-sur-Yvette and others)*	*Locality*	*Formation or Prehistoric Site*	*Elevation (m NGF)*	*Mean Sea Level during Deposition (m)*	*References*
104	6,070 ± 130	2,978	Saint-Marc (Côtes-du-Nord)	Peat	−2	< −8	Ters (unpubl.)
105	6,400 ± 130	2,151	Saint-Marc-en-Tréveneuc (Côtes-du-Nord)	Tree trunk on surface of fresh-water peat	−3	= 10	Ters (unpubl.)
105 bis	6,450 ± 160	845	Marquenterre (Somme)	Fresh-water peat	−2.70	−8.70	Lefèvre (1972)
106	6,555 ± 100	Fontes	Vernier marsh	Fresh-water peat	−9.70	−13	Huault (1983)
107	6,960 ± 105	—	Carentan (Appeville) (Manche)	Peat on schorre clay	−3	< −10	Elhaï (1963)
108	7,050 ± 160	2,527	Saint-Marc (Côtes-du-Nord)	Base of fresh-water peat	−6	< −13	Ters (unpubl.)
109	7,150 ± 300	396	Merlimont (Pas-de-Calais)	Peat on brackish clay	−4	< −9	Etienne and Robert (in *Radiocarbon* 1969)
109.2	7,220 ± 175	Fontes	La Meilleraye (Seine)	Diatomite	−7	< −10	Huault and Lefèvre (1983)
109 bis	7,370 ± 140	5,097	Saint-Marc	Fresh-water peat	−7	< −13	Ters (unpubl.)
109 ter	7,400 ± 80	Fontes	La Mailleraye (low Seine)	Peat between two deposits of marine silts	−10.50	< −13.50	Huault and Lefèvre (1974)
110	7,420 ± 110	(Gro 2043) (rectifié)	Fromentine (Vendée)	Brackish peat	−7	−11	Verger and Florschutz (1960)
111	±7,450	—	Carentan (Manche)	Schorre clay	−5 to −3	−8 to −6	Elhaï (1963)
112	±7,500	—	Marais de Petit Mars (Loire-Atlantique)	Brackish clay with *Chenopodiaceae*	−3.84 to −4.34	−7 to −8	Planchais (1971)
113	±7,500	—	Asnelles (Calvados)	Clay with organic remains	−7	< −11.5	Larsonneur (1971)
114	±7,500	—	Marais de Redon (Ille-et-Vilaine)	Brackish clay	−11.50	±−13	Morzadec-Kerfourn (1965)
114 bis	7,550 ± 170	1,218	Coquebourg, Carentan (Manche)	Silt with organic remains	−4.60	< −8	Larsonneur (1971)
115	7,580 ± 250	(W 705)	Anse Duguesclin, Cancale (Ille-et-Vilaine)	Peat	−7	< −14	Bourcart and Boillot (1960)
115 ter	7,620 ± 140	4,886	Marquenterre	Fresh-water peat	−8	< −14	Ters et al. (1980)
115 bis	7,630 ± 100	Fontes	Vernier marsh	Fresh-water peat	−12.30	< −15	Huault (1983)
115.4	7,640 ± 60	Fontes	La Mailleraye (Seine-Maritime)	Peat	−14.80	−17.80	Huault and Lefèvre (1974)

116	7,820 ± 170	1,406	Le Havre (Seine-Maritime)	Fresh-water peat	−17.50	< −24	Ters and Pinot (1969)
117	±7,900		Le Havre (Seine-Maritime)	Laminated silt	−18	= −21	Ters and Pinot (1969)
118	7,980 ± 190	763 bis	Fort-Mahon (Somme)	Fresh-water peat	−17	< −22	Etienne and Robert (in *Radiocarbon* 1969)
119	8,050 ± 170	1,403	Le Havre (Seine-Maritime)	Fresh-water peat	−21.70	< −25	Ters and Pinot (1969)
120	8,130 ± 190	1,019	12 km W du Havre, Bay of the Seine	Peat	−26.7	< −31	Larsonneur (1971)
121	±8,150		Le Havre (Seine-Maritime)	Fresh-water clay	−20.30	±−22	Ters and Pinot (1969)
122	8,200 ± 190	1,022	Manche au large de Cherbourg	Peat	−29	< −32.5	Larsonneur (1971)
123	8,250 ± 220	1,401	Le Havre	Peat	−21.5 to −22.7	< −25	Guyader (unpubl.)
124	8,470 ± 170	1,402	Le Havre (Seine-Maritime)	Fresh-water peat	−22	< −27	Ters and Pinot (1969)
125	8,850 ± 300	1,238	Submarine valley of the Seine	Fresh-water peat	−29	< −33	Guyader (unpubl.)
125 bis	9,300	—	Dogger Bank	Peat	−39	< −43	Veenstra (1965)
126	9,340 ± 300	746	Manche au large du Havre	Fresh-water peat	−27.70	< −33	Larsonneur (1971)
127	9,400 ± 200	1,990	Manche septentrionale	Pine trunk	±−54	< −55	Ters (unpubl.)
128	9,470 ± 130	1,021	Manche Cherbourg (Bay of Becquet)	Fresh-water peat	−34.40	< −43	Larsonneur (1971)
128 bis	9,600 ± 300	—	Le Havre (Seine-Maritime)	Peat	±−27	< −32	Michel (1968)
129	9,700 ± 200	1,995	Manche septentrionale	Fresh-water peat	−55	< −60	Ters (unpubl.)
130	9,730 ± 300	745	Au large du Havre	Peat	−27.40	< −32	Guyader (unpubl.)
131	9,750 ± 200	764	Fort-Mahon (Somme)	Peat	−16	< −20	Etienne and Robert (in *Radiocarbon* 1969)
132	9,880 ± 230	1,023	Bay of Cherbourg	Fresh-water peat	−22.90	< −34	Larsonneur (1971)
133	9,900 ± 300	744	45 km au large du Havre	Fresh-water peat	−26.75	< −33	Guyader (unpubl.)
134	9,955 ± 100	—	Manche	Fresh-water peat	−36	< −41	Larsonneur (1971)
135	10,050 ± 200	—	Estuary of Gironde	Silty, brackish sediment	−5	±−60	Castaing et al. (1971)
136	10,200 ± 230	850	Grande Vasière, Pointe Penmarc'h	*Cyprina islandica*	−110 to −120		Clémarec (1969)
137	10,565 ± 100	—	Manche	Fresh-water peat	−36	±40	Larsonneur (1971)

The Fromentine Submergence (Vendée) (−7 m, ca. 7,500 B.P., ^{14}C)

The upward trend of the curve (Fig. 12-3) is based on several dates. From the second emergence at Le Havre (about −24 m, around 7,820 B.P.) to the crest of the Fromentine submergence, the sea rose very rapidly. A borehole shows 20 m of alternating marine and continental sediments. Thin layers of fresh-water peat are intercalated from +2.50 m to −17.50 m NGF (Huault and Lefèvre, 1983*a*, 1983*b*).

In the Seine Valley (Vernier marsh). At −12.3 m, a woody peat of late Boreal age is dated at 7,630 ±100 B.P. (115 bis). It contains many diatoms showing a fresh-water environment, but no *Chenopodiaceae.* It is covered by brackish clay and, above that, by marine sands containing foraminifera and ostracods (deposits of the Fromentine submergence). Judging from the local tide range, MSL contemporaneous with the peat probably stood as low as −15 m.

In the Marquenterre marsh (north of France, Rue 3 borehole; Ters et al., 1980). In this marsh, a nearly contemporaneous peat (7,620 ±140 B.P.; 115 ter) rich in *Chenopodiaceae* was formed very near the shore. It is covered by rolled pebbles and marine sands from the Fromentine submergence. The corresponding MSL was around −14 m.

Near Cancale (Brittany), a peat site at −7 m (115) has been dated at 7,580 ±250 B.P. It indicates an MSL at or below −14 m.

At Fromentine (Vendée), pollen analysis and micropaleontologic studies of borehole sediments (Verger and Florschutz, 1960) have provided a detailed succession composed of a peat, a schorre (salt marsh) deposit containing *Chenopodiaceae,* and a clay rich in foraminifera, overlain by strata indicating brackish conditions. According to Florschutz (pers. comm.), the basal peat, at −7.05 m to −7.15 m, dated at 7,420 ±110 B. P. (110), contains an early Atlantic flora. Therefore, at that time, MSL stood below −10 m and probably around −11 m. Between −6.9 m and −6.6 m, the peat gives way to schorre clay. The overlying marine deposit ends at −4.4 m and is overlain by schorre sediments between −4.4 m and −4.2 m. Taking into account the local tide range of about 6 m, MSL was at −7 m at the maximum of this submergence.

Near Carentan (borehole 1, Appeville), Elhai (1963) studied a brackish-water deposit at between −3 m and −5 m depth whose age lies near the Boreal/Atlantic transition (111). The great abundance of *Chenopodiaceae* denotes a schorre deposit, and sea level may be estimated at −6 m to −8 m. This deposit is covered by Atlantic fresh-water peat dated at 6,960 ±105 B.P. (107).

At Coquebourg (Veys Bay), a borehole described by Huault and Larsonneur (1972) has revealed alternating continental, brackish-water, and marine deposits between −12 m and −4 m (NFG); a silt containing plant debris was dated at 7,550 ±170 (114 bis). This silt is covered by a transgressive marine shelly deposit with *Cardium edule* (between −4 m and −3 m), which corresponds to an MSL at −8 m to −7 m. Next above it are brackish-water deposits and, above them, fresh-water silt, indicating emergence. The maximum submergence at −8 m to −7 m corresponds exactly with the two values previously mentioned.

In the marshes of Petit-Mars (north of Nantes), Planchais (1971) found a blue brackish-water clay containing *Chenopodiaceae, Armeria,* and *Hystricosphera* between −3.84 m and −4.3 m, which implies an MSL at the Boreal/Altantic transition of −7 m to −8 m (112).

In the Redon marshes, brackish silt containing *Hystricosphera* (40% of all tree pollen) covers continental deposits between −11.5 m and −8 m. This deposit represents the last 500 yr of the Boreal period (zones VI b and c) and the Boreal/Atlantic transition at about 8 m below modern sea level (114; Morzadec-Kerfourn, 1969).

At Merlimont (Pas de Calais), peat at −4 m, dated at 7,150 ±300 B.P., overlies brackish-water clay. This finding indicates a maximum sea level at the time at about −8 m (109).

The ups and downs of sea level between 7,700 B.P. and 7,400 B.P. may be traced in the successions that include, from base upward, continental sediments passing upward into schorre clay, then to marine sands, above which are brackish-water deposits, and above them, nonmarine strata. In areas of northwestern

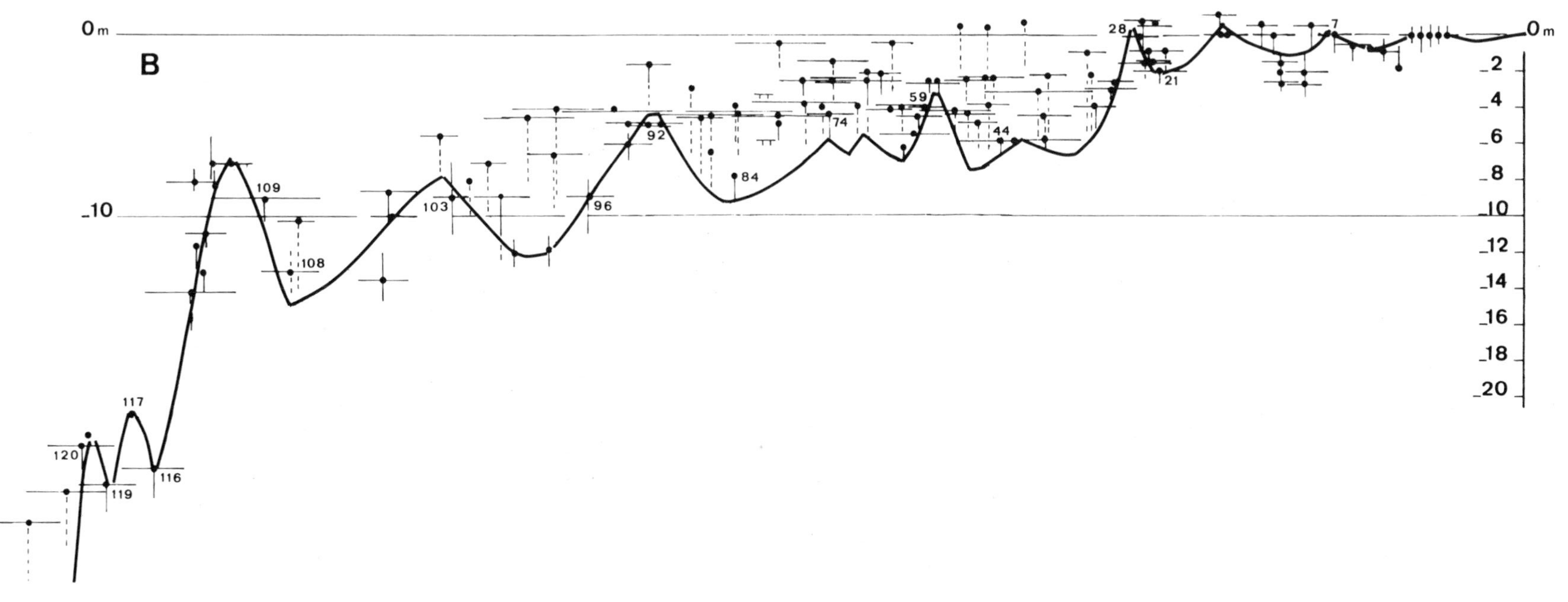

Figure 12-3. The French sea-level curve showing ranges of errors in radiocarbon dates and in elevations.

Europe that have not been affected by neotectonic movements, researchers believe that at the time of the Boreal/Atlantic transition (around 7,500 B.P. or thereabouts; Godwin, 1956), sea level stood at about −7 m.

During the Fromentine submergence (between 7,660 B.P. and 7,500 B.P.), the climate seems to have been milder than today's. In the Vernier marsh borehole, *Quercus ilex* is found only in sands deposited during this maximum (Huault and Lefèvre, 1983*b*).

The Saint Marc Emergence (Tréveneuc, Côtes du Nord) (About −14 m, ca. 7,300 B.P.)

The existence of this emergence is proved by a study of peat deposits in the cove of Saint Marc for which we have boreholes penetrating 14 m of sediment. At −7 m NGF, a peat (without *Chenopodiaceae*) is dated at 7,370 ±140 B.P. (109 bis). This finding implies an MSL equal to or lower than −14 m. At −6 m, a peat has been dated at 7,050 ±160 B.P. (108). At this level, the flora is of Atlantic type, as at −7 m, and very few *Chenopodiaceae* are present, which suggests that the sea was not close by. Because the local tide range is 12.85 m, the −6-m peat could have formed only when the sea stood lower than −13 m. This peat now extends beneath the sea (at 8.40 m below MSL, which suggests that the general level was even lower than −13 m).

At Carentan, Elhai (1963) noted a brackish deposit of late Boreal age (probably from the Fromentine submergence), covered at −3 m by fresh-water peat containing an Atlantic flora dated at 6,960 ±105 B.P. (107). Considering the tide range of 9 m and a total absence of *Chenopodiaceae*, this deposit implies a sea level of well below −10 m.

Between −9.7 m and −9.5 m in the Vernier marsh borehole (in the Seine Estuary), a fresh-water peat dated at 6,555 ±100 B.P. (106) overlies the sands of the Fromentine submergence. This implies an MSL at or lower than −13 m. The lack of *Chenopodiaceae* implies that the shore was not very close.

The Bréhec Submergence (Plouha, Côtes du Nord) (About −8 m, ca. 6,100 B.P.)

At Saint Marc, a woody peat accumulated up to −4 m; above this level, a rapid rise by the sea left the tree trunks lying in the mud. One of these trees (a willow), has been dated at 6,400 ±130 B.P. (105); therefore, at about 6,400 B.P., MSL was around −10 m.

At Bréhec, a deposit of shelly sand several meters thick has accumulated in the western part of the bay. It is rich in brackish foraminifera and ostracods; also, it contains *Cardium edule* and *Scrobicularia plana*. Accretion would have been up to the high-water mark. Today, however, the top of the sand lies at −2 m (MSL), implying an MSL during deposition of −7 m to −8 m (the local tidal range is 12 m). Near the present land surface, this deposit is overlain by gray marsh clay and that, in turn, by peat dated at 5,300 ±140 B.P. (96; Ters and Pinot, 1969).

At Le Palus (Côtes du Nord), a borehole by the author reached shelly beach sand situated at almost the same level as at Bréhec (−1.2 m) beneath peat containing an Atlantic flora.

In the Bay of Veys (Manche), coastal silt was deposited around 6,200 B.P. (103 bis), with a relative MSL of about −7 m (Giresse, 1969).

In the Vernier marsh borehole (Huault and Lefèvre, 1983*b*), marine sands intercalated between fresh-water peats dated respectively 6,555 ±100 B.P. (106) and 6,090 ±120 B.P. (103.1) were deposited between −8.50 m and −6.20 m (NGF). They correspond to the Bréhec submergence (ca. 6,100 B.P., at −8 m). They contain large amounts of dinoflagellates, with foraminifera and ostracods (especially between −8.30 m and −6.50 m). At this part of the Atlantic stage, the high values of arboreal pollen indicate the maximum forest cover of the landscape during the entire Holocene.

In the Gironde estuary, Castaing, Feral, and Klingebiel (1971) observed *Ostrea edulis* at −9 m and dated at 5,900 ±100 B.P. (103).

The Tréompan Emergence (Ploudalmézeau, Finistère) (At −12 m, ca. 5,700 B.P.)

Another fall in sea level at this time left several clear indications. In the Vernier marsh borehole (Fig. 12-4), a woody peat, rich in dinoflagellates (Huault and Lefèvre, 1983*b*) and dated at 6,090 ±120 B.P. (103.1), covers marine sands of the Bréhec submergence. This succession implies a drop of the sea to below −10 m NGF.

At Tréompan, peat deposits exposed on the beach at low tide have been studied by Morzadec-Kerfourn (1969). A sample was collected at −4.4 m (local sea level below −8.8 m). The peat is also visible by divers to −2.5 m beneath low-tide level. Taking into consideration local sedimentation factors, the fall of sea level at about 5,700 B.P. (101) was to −12 m at least.

In the Vernier marsh borehole, a peat, separated by 1 m of clay and silt from the one dated at 6,090 ±120 B.P., has been dated at 5,445 ±120 B.P. (96 ter; Huault and Lefèvre, 1983*b*). The position of the peat, lying between −5.2 m and −5 m, indicates that MSL was then at or lower than −9 m (96 bis) (Fig. 12-4).

In the Marquenterre (Rue 3 borehole; Ters et al., 1980), an almost contemporaneous peat (dated 5,520 ±110 B.P.; 98 ter) is intercalated between the marine sands of the Grehec and Brétignolles submergences.

At Le Curnic (en Guissény, Finistère), Neolithic dwellings studied by Giot and dated at 5,510 ±250 B.P. (98) have been found in a peat containing an Atlantic flora. The elevation suggests that contemporary sea level stood below −4 m (Giot et al., 1960).

At Bréhec, the peat overlying the sands of the Bréhec submergence and containing a brackish fauna (103 bis) spreads over the modern lower beach of −2 m from MSL down to the level of the lowest tides (−6 m) and beyond. This implies a minimum MSL of −12 m. Pollen analysis from the peat by Denèfle (pers. comm.) confirms a late Atlantic formation. Because *Chenopodiaceae* are not present, the deposit must have been formed at some distance from the sea. A tree trunk in situ at the surface of the peat has been dated at 5,300 ±140 B.P. (96).

At Abbéville, a comparison may be made between this drop in sea level and the deepening of channels in the Somme valley (Sursomme and Portelette channels) cut in tufa and peat of Boreal and early Atlantic ages. These channels ultimately were infilled by marine sand ("lower blue sands") containing Chassean pottery (Bourdier, 1969; dated at 5,200-4,800 B.P.).

The Brétignolles Submergence (Maximum about −5 m, ca. 5,000 B.P.)

After the abrupt Tréompan retreat, the sea returned to a new high level, averaging −8 m, some time after 5,300 ±140 B.P. (96). At Bréhec, trees of this date growing on the surface of the fen were laid flat by the transgression.

At Brétignolles, the author's boreholes on the modern beach between Les Osselins and La Parée penetrated a brackish-water deposit between −2 m and −5 m MSL. Near the top of this deposit, a clay containing *Cardium edule* and *Scrobicularia plana* has been dated at 4,910 ±120 B.P. (92), and 20 cm lower in the same deposit, a piece of wood was dated at 4,990 ±120 B.P. (94). The microfauna of foraminifera and ostracods within this clay is entirely brackish; it indicates an enclosed environment. This schorre deposit is very rich in *Chenopodiaceae* and, according to Denèfle (pers. comm.), is situated at the Atlantic/Subboreal limit. Local thickness is 6 m, so this rise would have been around −5 m. Near the top, the marine influence diminishes and the facies becomes progressively a fen peat with a Subboreal flora.

In the Marquenterre region, fresh-water peat deposits of the Subboreal have been analyzed

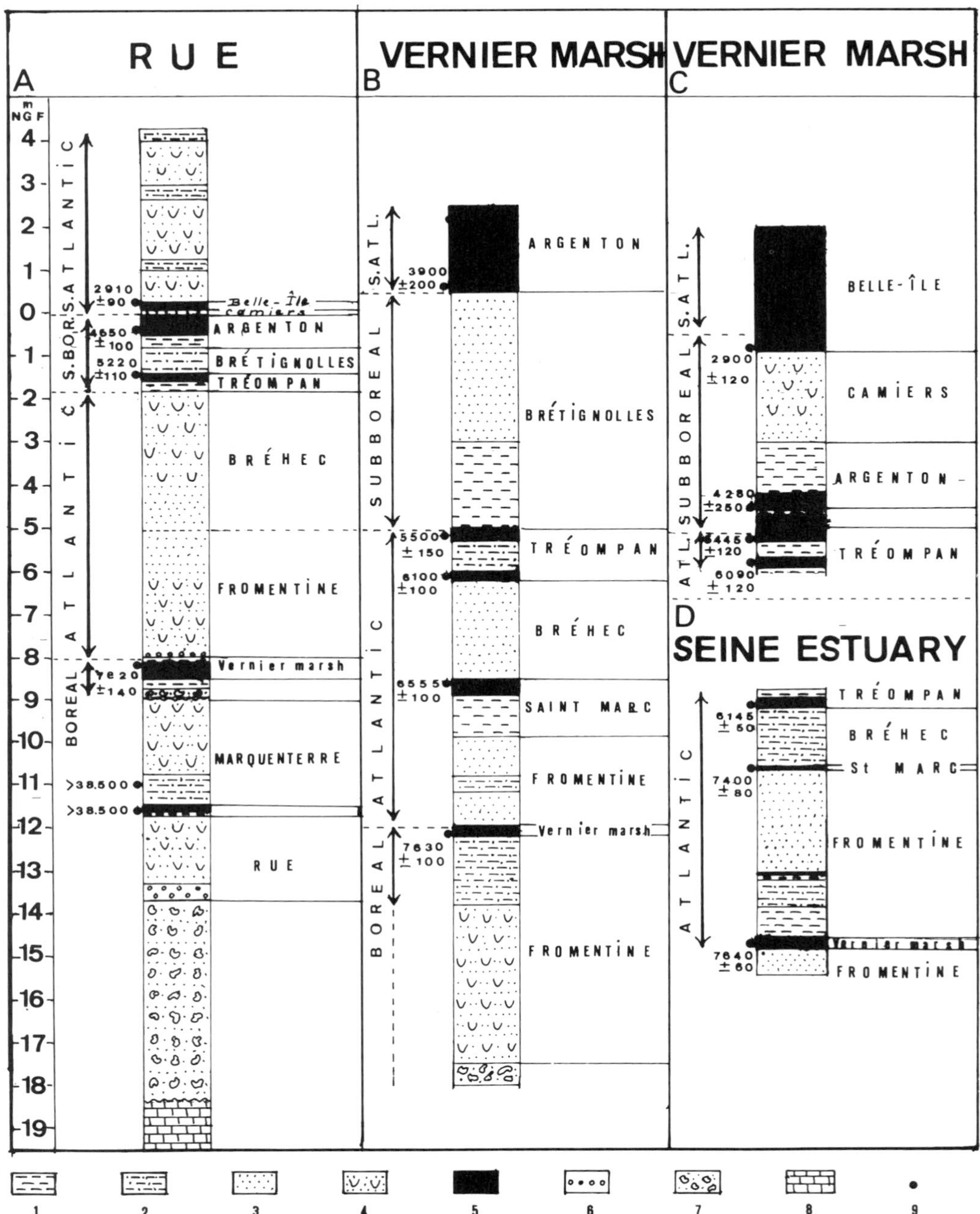

Figure 12-4. Boreholes into the Holocene record of alternating marine and nonmarine sediments. A, Rue 3 bore (Marguenterre); B, Vernier marsh, bore 2 (Seine estuary); C, Vernier marsh, bore 1 (Seine estuary): D, Seine estuary, bore 31. *Lithology:* 1, clay; 2, silty clay; 3, silt; 4, shelly sand; 5, peat; 6, marine pebbles; 7, sandy shingles; 8, Cretaceous limestone; 9, ^{14}C date.

by Planchais (1971); one deposit at −2 m, dated at 5,080 ±140 B.P. (95), indicates an MSL below −5 m.

In the Vernier marsh borehole (Huault and Lefèvre, 1983*b*), a layer of fine-grained marine sands, 4 m thick and rich in *Chenopodiaceae* and dinoflagellates, is intercalated between freshwater peats dated at 5,500 ±120 B.P. (96 bis) and 3,900 ±200 B.P. (71 bis). This deposit is very likely contemporaneous with those of the Brétignolles submergence.

In the Brière (Loire Atlantique), peats dated at 4,630 ±300 B.P. (88) and 4,480 ±300 B.P. (83) overlie marine or brackish ooze culminating at −1.3 m (Giot, 1968). From this finding we may infer that after a brief rise, the sea retreated to a mean level of approximately −4.5 m. Underlying the peat is another marine layer, probably of the same age as the Brétignolles deposit.

The Argenton Emergence (Below −8 m; ca. 4,500 B.P.)

Based on evidence ranging from palynology and prehistory to climatic and sedimentological data, it is now known that between 4,800 B.P. and 3,500 B.P., an emergence of considerable importance took place. Moreover, this emergence lasted longer than previous Holocene emergences. Following the peak of the Brétignolles submergence, the best evidence available on the subsequent retreat of the sea is provided by now submerged peat beds that overlie marine deposits containing an Atlantic flora. In order of age, the following may be cited:

At Montmartin, Calvados, a shelly sand dated at 4,700 ±130 B.P. (89) is placed at −2 m from MSL. This would indicate that MSL at the time of deposition stood at or about −4.70 m (Giresse, 1969).

In the Brière region, Giot (1968) found peat deposits dated at 4,630 ±130 B.P. (88), 4,480 ±300 B.P. (83), and 4,260 ±300 B.P. (80) overlying brackish-water clays at −1.30 m. Taking into account local sedimentation, these peat beds imply an MSL below −4.50 m.

In the Somme Valley, at between −2 m and 0 m in the La Portelette borehole, Abbéville, peat is intercalated within two marine deposits. These are the lower blue sands containing human artifacts from the middle and latest Bronze Age. This alternation constitutes the proof of an emergence between the end of the Neolithic and the beginning of the Bronze Age. In other boreholes at Abbéville, peat 1-4.5 m thick overlies the lower blue sands (Fig. 12-4; Commont, 1910).

On the beach at Argenton (Finistère), a series of superimposed peat beds, each separated by marine deposits, has been described by Morzadec-Kerfourn (1969), who has taken care to show the climatic significance of these alternating layers. From top downward the succession is as follows:

- Coastal marsh deposit (layer B), with transitional Subboreal/Subatlantic flora: a low sea level about −3.90 m dated at about 3,610 ±125 B.P. (67).
- Peat (C), transition at the base from a lacustrine formation to another one rich in *Chenopodiaceae* and *Hystricosphera* toward the top; implies a low sea level of −4.40 m at about 3,970 ±135 B.P. (74).
- Peat (D), without *Chenopodiaceae* or *Hystricosphera*, suggesting an MSL well below −5 m (79).
- Peat (E), at −3.90 m, with *Hystricosphera* and *Chenopodiaceae*, contains Subboreal flora; inferred MSL (taking into account the local tidal range) at below −7.80 m (84).

Therefore, between about 4,600 B.P. and 3,500 B.P., an emergence, albeit a brief one, took place to an MSL that reached −8 m (Morzadec-Kerfourn, 1969).

In the Marquenterre (Rue 3 borehole, Fig. 12-4), a peaty deposit, which overlies a brackish-water clay, began to form around 4,650 ±100 B.P. (88 bis). The inferred MSL was below −6 m NGF. In this Subboreal peat, the remains of thermophilous trees are less abundant than they are within the deposits formed during the Fromentine and Bréhec submergences.

The Vernier marsh borehole cut a peat whose base marks the passage from the Subboreal to the Subatlantic (at +0.50 m NGF and dated at 3,900 ±200 B.P.; 71 bis). The pollen content records a progressive drying of the marsh (disappearance of *Chenopodiaceae*, dinoflagellates, and marine diatoms) (Huault and Lefèvre, 1983). MSL is inferred to have been below −5 m. The elevation and pollen content of this peat are very similar to those of peat (D) at Argenton (described in a following paragraph).

During the same period, fen peat formed at various levels. All suggest an MSL between −0.50 m and −6 m or lower (78, 73, 68, 64, 65, and 62).

The Prehistoric Record. Menhirs (or standing stones), passage graves, and Old Bronze Age dwellings confirm that between about 4,800 B.P. and 3,500 B.P., sea level was low. A list of megalithic monuments, now partly submerged, has been prepared by Giot (1968). For the Brittany coast, these include

1. The menhir at Men Ozac'h (Plouguerneau), the base of which lies at −6.40 m referenced to high-water level (90) and thus suggests that the sea stood below this level at about 4,700 B.P.
2. The menhir circles of Er Lannic, constructed between about 5,000 B.P. and 4,500 B.P. (87) when the sea stood below −5 m.
3. The covered alley at Larret, Kerlouan, built about 4,300 B.P., in which the base of the lower supports lies about 6.85 m below the high-water mark (tidal range, 12 m; 81).
4. The passage grave at Roc'h ou Braz (Le Kernic, Plouescat), built about 4,300 B.P. (82), now lies within the intertidal zone (77), but the base of the lower supports, now at +1.70 m, implies that the building could have been built only if sea level stood at least 3.10 m lower than today.
5. Menhirs in south Brittany and in the Bay of Bourgneuf are partly covered by Flandrian marine clays (at Luzeronde, Ile de Noirmoutier), La Pierre Folle, Bourgneuf en Retz (J. Mounes, pers. comm.).
6. The siliceous sandstone dolmen of La Vendette rests on Lutetian limestone, at about −2.50 m below MSL, and thus must have been erected when the sea stood below −6 m.

Today, several Neolithic sites lie under water. At the Pointe aux Oies, near Wimereux, Mariette (1971) discovered a habitation site in brackish-water clay at −1 m from MSL. This site dates from about 4,500 B.P. (85) and suggests that the sea formerly stood below −4 m. At Hardelot, Pas de Calais, Mariette noted two other such sites in the peat. One dates from about 4,000 B.P.; it is situated between 0 m and +4 m and therefore implies that the Neolithic sea stood at −4 m (75). The other lies at +2 m and has been dated at 3,950 ±200 (72); its implied sea level is below −1.50 m. At Saint Hippolyte, Charente, Gabet (1971) described an Old Bronze Age dwelling from about 3,750 B.P. (71) at modern sea level. It has been covered by 4 m of marine sediment, thus implying that at time of construction, MSL was below −4 m.

Late Neolithic or Bronze Age caskets have been found on the strand line in Brittany (Giot, 1968, 1969). They indicate a sea level lower than today's. In a borehole made by the author at Camiers, several peats and brackish-water sediments are interlayered. These imply that a few slight emergences interrupted the general rise of sea level at about 3,800 B.P. and 3,550 B.P.

The Camiers Submergence (Pas de Calais) (About −4 m, ca. 3,400–3,100 B.P.)

The site at Camiers, on the Canche estuary, contains a sequence of alternating marine, brackish, and peaty deposits (Mariette, 1971; Ters and Pinot, 1969). From bottom to top, these deposits include marine sands; peat containing a Subboreal flora, dated at 3,550 ±130 B.P. (65) and by the absence of *Chenopodiaceae*, suggesting a temporary emergence between 0 m and +1.50 m; blue clay containing *Hydrobia* and *Scrobicularia*, with shells at the base dated at 3,400 ±130 B.P. (59). From this schorre (salt marsh) deposit, MSL at the time can be fixed at −4 m (tidal range is 8 m). In the same region (Rue 3 borehole), at about 0 m NGF, a silty deposit is intercalated between two peats dated at 4,650 ±100 B.P. and 2,910 ±90 B.P. respectively. The foraminifera (*Jadammina polystoma*, and species of *Nonion* and *Elphidium*) are typical of the schorre. These silts were deposited during the Camiers submergence.

At Asnelles, Calvados, a brackish peat at +1.90 m and containing abundant *Chenopodiaceae* has been dated at 3,400 ±130 B.P.

(60). The corresponding MSL is estimated to have stood somewhat below −2.60 m. At Porsgeuen, Plouescat, Finistère, a Subboreal peat containing abundant *Chenopodiaceae* has been dated at 3,390 ±120 B.P. near the top (58); (Morzadec-Kerfourn, 1969). The inferred MSL thus lay below −2.20 m. The base of the underlying peat is later than 4,120 ±140 B.P. (77).

In the Charente estuary, a shell bed dated at 3,380 ±110 B.P. (57 bis) at +4.30 m lies between layers of estuarine sediments (Gabet, 1971).

At Montmartin, Calvados, a shelly sand dated at 3,340 ±110 B.P. (57) appears at 4.50 m. Its implied MSL is about −2.30 m.

At Porspoder, Le Gratz, Finistère, a silt containing *Plantago* maritime overlies schorre with *Obione* and *Salicornia.* It is inferred to be a storm-beach deposit, lying about 0.50 m above today's high-water mark. It has been dated at 3,200 B.P. or thereabouts (54) and indicates that the sea was close (Morzadec-Kerfourn, 1969).

Archeological evidence from the Bronze Age supports the probability that around 3,350 B.P. a submergence crested at about −2.50 m. At Camiers, a Late Bronze Age dwelling lying at 1.50 m above MSL has been built on a brackish-water clay containing *Scrobicularia.* Within this dwelling are flint artifacts, bones of domestic animals, Urnfield-type pottery, and shells dated at 3,180 ±140 B.P. (53 bis; Mariette, 1971). At Abbéville, the marine upper blue sands lying between 0 m and +3 m contain Chalcolithic remains. These sands resulted from an upward oscillation of the sea. The tide range here is 11 m, so the implied minimum sea level is −2.50 m. At between +3 m and +5 m, the sands are overlain by calcareous tufa containing objects from the middle and late Bronze age (Erondelle, LaChaussee-Tirancourt) (Bourdier, 1969; Commont, 1910).

The schorre formation at Camiers (at 0 m) has been dated at 3,400 ±130 B.P. (59), implying a former MSL of about −4 m. This deposit accumulated up to −1.50 m, after which a new emergence enabled another Late Bronze Age dwelling to be built around 3,180 ±140 B.P. (53 bis). The maximum MSL of about −2.50 m was thus reached between these two dates, or about 3,250 B.P.

The Belle Isle-en-Mer Emergence (Morbihan) (About −8 m, ca. 3,000 B.P. and 2,600 B.P.)

This low level is contemporaneous with both Halstatt and the second Iron Age. It contains two retreat phases, about 3,000 B.P. (to −8 m) and 2,600 B.P., separated by an upward oscillation at around 2,750 B.P. During the retreats, several peat beds formed near the coastline.

At LeYaudet, near Lannion, Côtes-du-Nord, the peat dated at 3,075 ±110 B.P. (50) indicates an MSL of from −6 m to −8 m (Giot, 1969).

At Saint-Pair, Manche, a peat dated at 3,060 ±160 B.P. (48) suggests that MSL stood below −2.50 m.

At Belle Isle-en-Mer, the peat of Ster Vraz has been dated in three places, giving 2,970 ±230 B.P. (44), 2,890 ±230 B.P. (43), and 2,680 ±200 B.P. (36). The fresh-water floral succession is Subboreal/Subatlantic transitional (Planchais and Corrilon, 1968). The altitude of −5 m MSL indicates that MSL at the time stood below −8 m (local tidal range is 6 m).

In the Marquenterre (Rue 3 borehole), a layer of peat whose top has been dated at 2,910 ±90 B.P. (43 bis) is intercalated between silts deposited in the Camiers submergence (below) and marine shelly sands (above). Sea level stood at or lower than −6 m (Ters et al., 1980). A climatic deterioration is indicated by the low percentages of *Quercus* (6.4%), *Tilia* (1.7%), and *Ulmus* (0.2%). *Chenopodiaceae* are virtually absent.

In the Grande Brière area, peat formed at around 2,770 ±300 B.P. (41), implying that the sea stood below −3.15 m. At Asnelles, peat containing *Chenopodiaceae* formed around 2,710 ±100 B.P. (37). Contemporary MSL lay below −2.50 m. At Lingreville, woody peat formed at 2.10 m below the high-water mark (32 bis). The flora places this deposit on the climatic limit between Subboreal and Subatlantic (Elhai, 1963). At Saint-Jacques de Rhuys, Morbihan, the trunk of an oak tree in peat implies an MSL of −4 m (23). At Santec, Finistère, brackish-water peat formed at 2.80 m below the high-water mark. It has been dated at 2,330 ±105 B.P. (31), and it formed as the sea rose to the crest of the Champagné submergence. In the Redon marshes, a peat bed at −1.60 m MSL has been found between two silts lacking *Hystricosphera* (Morzadec-Kerfourn, 1969). An MSL below −4.60 m is implied.

The prehistoric record has provided ample evidence of an emergence at this time. Bricked salt pans 4.50 m below the high-water mark at Le Curnic, Guissény, have been dated at 2,750 ±150 B.P. (40). At Chalezac, Charente, a Halstatt dwelling site at modern sea level and dated at 2,450 ±110 B.P. (33) has been buried beneath 2 m of marine sediments (Gabet, 1971). At time of construction, the sea stood at least 4 m below today's MSL. On the islet of Gaignoc (Landeda, Finistère), an Iron Age fortification, probably accessible from land at low tide, ceased to be used toward the end of this period, probably because, at about 2,300 B.P., the sea rose abruptly.

The Champagné Submergence (Vendée) (Maximum Level about +0.50 m, ca. 2,200 B.P.)

In the Marais Poitevin (Poitevin marsh), the shelly shingle banks of Champagné and Sainte Radegonde, several kilometers long and more than 5 m thick, have been dated at 2,200 ±100 B.P. (28) and 2,100 ±100 B.P. (23). Locally the tops of these banks exceed the high-water mark by 1 or 2 m. They are probably storm-beach or spring-tide deposits.

At Argenton (Guen Trez), Subatlantic peat (A) formed between 0 m and 1 m above the high-water level. The initial environment was one of fresh water. Later, the environment turned brackish. A sample from a depth of 15 cm has been dated at 2,180 ±105 B.P. (27; Morzadec-Kerfourn, 1969).

In the Marquenterre (Rue 3 borehole), a peat dated at 2,910 ±90 B.P. (43 bis) is overlain by shelly marine sands rich in foraminifera and ostracods. Pollen of *Quercus*, *Ulmus*, and *Tilia* are more abundant than in the peat. These deposits are probably products of the Champagné submergence.

The Roman Emergence (Sea Level at about −1.50 m, ca. 2,000 B.P.)

The Champagné submergence was short lived. During the La Tène and Roman periods and the early Romano-Gallic period, between about 2,100 B.P. and 1,900 B.P., MSL fell.

At Camiers, brackish-water clay dated at 2,100 ±110 B.P. (24) accumulated up to +3 m above MSL. A Gallic salt pan found on the surface of the clay was constructed when sea level stood at about −1 m.

At Nacqueville, Calvados, fresh-water peat dated at 2,130 ±150 B.P. (25) at 1.50 m below high-water level implies a lower-than-present sea level. In the Charente region, at Trizay, a peat dated at 2,060 ±100 B.P. (21) is situated at 1.70 m below the high-water mark. At La Petite Aiguille, a salt pan dated at 2,050 ±110 B.P. (20) denotes an MSL of below −1 m (Gabet, 1971).

Historical sites dating from the time of the Roman emergence are to be found in many places. For example, in Ardres, Pas de Calais, the Noires-Terres site excavated by Sommé and Cabal (1972) contains Roman and Gallic remains. The optimum occupation time was during the second half of the second century A.D. The site was abandoned about A.D. 268 (probably at the time of the Saint Firmin submergence). At Wissant, Pas de Calais, the Roman-Gallic site observed by Mariette (1971) was built between 3.50 m and 2.20 m below present high-water level, indicating a sea level appreciably lower than at present. On both the Channel and the Atlantic coasts, several sites are known where Roman remains have been found below present high-water marks (Lebesconte, 1898). At Pontaubault, near Mont Saint Michel, the upper peat, lying between marine clays, contains Roman-Gallic objects such as coins, bricks, and amphoras. In the Lys marshes, Nord, Aurelian (A.D. 161-180) and Posthuman (A.D. 258-268) coins have been found between a peat layer and the underlying coastal silt.

The Saint Firmin Submergence (Near the Crotoy, Somme) (Sea Level between 0 m and +0.50 m, dated ca. 1,700 B.P. [= A.D. 250])

From the second century A.D. onward, sea level rose noticeably. In the following century, the rate of rise increased. Sometime after A.D. 300, the maximum, somewhere approaching present MSL, was reached. Submergence appears to have been extremely fast. At Saint Firmin, a shell bank, containing chiefly *Cardium edule* and dated at 1,730 ±100 B.P. (15), lies against the western flank of the older shingle ridge at Mayoc. The younger ridge extends to 2 m above the present high-water mark. However, as at Sainte Radegonde (23), the shell bank could be a storm beach (Ters and Pinot, 1969).

At Hauteville, Manche, coastal silt dated at 1,680 ±120 B.P. (14) could have been deposited by a sea having the same mean level as today's (Giresse, 1969).

At Lingreville, Manche, a peat bed dated 1,500 ±150 B.P. (12 bis) is situated at 0.40 m above today's high-water mark (Elhai, 1963). This peat implies a sea level somewhat higher than at present.

In the Bresle estuary, Somme, the sea invaded a land surface covered by Roman-Gallic pottery and forests (Bourdier, 1969).

In the Bay of Mont Saint Michel, peat containing Roman-Gallic remains has been covered by marine deposits containing Frankish objects from the third century A.D. Underneath the Couesnon viaduct, a gray coastal silt containing Frankish remains separates a Roman-Gallic peat from another peat dating from the Middle Ages. The same sequence is observable at La Vilde-Bidon and in the Lys marshes (Nord).

The Wissant Emergence (Sea-Level Minimum about −1 m, 1,300 B.P., or Sixth to Eighth Century A.D.)

At Wissant, the location of a landing stage on the shore 500 m out from the present coastline has been interpreted by Mariette (1971) as the result of a drop in sea level during the Middle Ages. At Les Noires-Terres, Ardres, reoccupation of an older site during the high Middle Ages was made possible by an emergence (Sommé and Cabel, 1972). At Lividic, Finistère, clay containing *Hydrobia* and dated at 1,390 ±100 B.P. (11) terminates upward at +2 m (Morzadec-Kerfourn, 1969). The corresponding sea level must have been somewhat below today's.

At Crabec, Manche, a shell-bearing sand dated at about 1,380 B.P. (10) terminates upward at +2 m. Taking into account the tidal range, this sand suggests an MSL at or below −1 m. At Pontorson, coastal silt dated at 1,250 ±120 B.P. (9) suggests an MSL of −1 m to −2 m. In the Couesnon estuary, peat from the high Middle Ages overlies a coastal silt containing Frankish remains (Lebesconte, 1898). This implies an emergence.

From all these data, I conclude that around the seventh century A.D., sea level dropped slightly.

The Aunis Submergence (At about Present Level, ca. 1,100 B.P. [= A.D. 850])

In the Charente region, remains of salt marshes at about present sea level have been dated from about 1,100 B.P. (7; Gabet, 1970). At Beg an Dorchen (Finistère), a shell-and-shingle ridge, the accretion of which has been dated at 1,100 ±90 B.P. (3 bis), lies at the present-day high-water mark (Giot, 1969). At Les Noires-Terres, Ardres, a dwelling site was abandoned in the tenth century A.D.

Fluctuations of Sea Level during the Past Thousand Years

During this time, the sea approximated its present level. Shelly deposits on storm beaches 0.50-1 m above today's high-water mark are known from several localities. At Bréhec, they have been dated at 450 ±90 B.P. (0); in the LaHoule cove near Cancale, at 472 ±95 B.P. (1); and at Beg an Dorchen, at 580 ±120 B.P. (3). These local deposits do not necessarily indicate important movements of sea level. Such movements (a rise on the order of centimeters since 1850), however, are implied by marsh data.

CORRELATIONS WITH PREVIOUS WORK

Fairbridge Curve

The generalized sea-level curve I have prepared (see Fig. 12-2) strikingly resembles Fairbridge's (1961) curve (dashed line, Fig. 12-2). Both show seven submergences since 8,200 B.P. The two curves are not identical, however. I have not found any indications of levels higher than at present around 5,800-5,000 B.P. and 4,000-3,000 B.P. as on the Fairbridge curve. Because the data shown in Fairbridge's curve have been collected worldwide, they may include points from certain regions affected by crustal movements unlike those that have affected the French coast. Sea level at 10,000 B.P. is almost certain to have been decidedly lower than Fairbridge's curve suggests.

Mörner's Curve

Mörner's (1969) curve (dashed and dotted line, Fig. 12-2), based on his study of many marine and nonmarine sedimentary sequences in Scandinavia, displays oscillations that are less pronounced than those of Fairbridge's or my curve. Specific examples are the relatively small-scale variations of Mörner's curve at about 5,700 B.P., 5,100 B.P., and 3,200 B.P. The postglacial isostasy of Scandinavia makes it difficult to extract purely eustatic changes in a region of continually varying relative sea level.

Sea-Level Curves Based on the North Sea Coast (Holland and Belgium)

Careful studies from this area provided the first detailed indications of the complexity of late Glacial and Holocene coastal changes. Data have come from many boreholes. The fossils, pollen, and sediments have been comprehensively analyzed; numerous radiocarbon dates provide time control (Hageman, 1969; Jelgersma, 1966; van der Heide and Zagwijn, 1967). The interbedding of peats, brackish-water deposits, coastal sediments, and marine layers implies numerous oscillations of sea level. The ages, but not the ranges, of these oscillations are in accord with the results from the French Atlantic coast. The Calaisian, Dunkirkian, and Upper Flandrian submergences correlate with those of my graph (Fig. 12-2). Among the recent low levels on my graph, a so-called Roman retreat has been shown in Belgium (Paepe, 1972), and the so-called Carolingian retreat corresponds to my Wissant emergence. I have also indicated suggested correlations between the Baltic area and the French Atlantic coast on Graph A (see Fig. 12-2).

Shepard's Curve

Shepard (1963) fitted 77 dated points taken from relatively stable areas of North America onto a linear curve. The values for points dated at 5,000 B.P., 6,000 B.P., 7,000 B.P., and 8000 B.P. do not differ much from the points of these ages on my curve. Shepard's curve, however, reflects *transgressive maxima*. It does not take into account emergences proved by the alternation of marine and nonmarine sediments found all along the same coasts at approximately the same dates.

Suggate's Curve

The New Zealand sea-level curve (Suggate, 1968) matches my curve in its display of six submergences between 7,200 B.P. and 2,000 B.P. In general, the dates of both submergences and emergences match between New Zealand and France. The chief differences are the absence on the French coast of the indications on the New Zealand coast at 4,200 B.P. and 2,000 B.P. of a sea level higher than at present. It is not likely that sea-level oscillations between two regions as different geophysically and as far apart geographically as northwestern France and New Zealand would correspond perfectly. Therefore, the similarities are all the more remarkable. The differences of MSL values between the two curves increase slightly and progressively with decreasing age starting at 8,000 B.P. For example, the differences are 0 m at about 8,000 B.P., 4 m at about 7,000 B.P., 5 m at about 6,800 B.P., and 6.5 m at about 6,500 B.P.

CORRELATIONS WITH HOLOCENE GLACIAL ADVANCES, COLD PERIODS, AND WARM PERIODS

In Canada, Alaska, Scandinavia, the Alps, the Patagonian Andes, and Africa, advances of glaciers have coincided with colder intervals during the Holocene epoch. The principal known facts about such glacial advances are summarized in Figure 12-5. The following sections correlate glacial advances and sea levels.

The Preboreal Cold Phase

The deglaciation phase of Norway, between 20,000 B.P. and 8,500 B.P., is well known. The Lofoten glacial advance took place between 10,200 B.P. and 9,500 B.P. (Møller and Sollid, 1972). It could match the Loch Lomond advance in Scotland and the Piottino phase in the Ticino Alps (10,000-9,500 B.P.). In the French areas here described, however, only a few scattered points indicate oscillations of sea level during this interval (128, 129, and 135).

The Le Havre Emergences (ca. 8,000 B.P. and 7,800 B.P.)

Pollen analyses of samples in the Saint Gotthard Massif show a large and rapid increase in *Abies, Picea,* and *Betula* percentages. Such changes indicate a cold cycle. Several abundance peaks of these species took place between 8,000 B.P. and 7,500 B.P. (Zoller, Schindler, and Röthlisberger, 1966). The pollen changes reflect complex climatic variation during this time interval. In the Canadian Arctic, a glacial readvance between 8,200 B.P. and 8,000 B.P. is indicated by the Cockburn moraines (Bryson et al., 1969). This glacial readvance appears to correspond with a crest in the Boreal submergence between 8,200 B.P. and 7,800 B.P. and also with the time when a lake formed in the Seine estuary.

The Fromentine Submergence (ca. 7,500 B.P.)

Pollen analysis of samples from the Vernier marsh borehole near the estuary of the river Seine indicates a warm interval (Huault and Lefèvre, 1983*b*). The thermophilous *Quercus ilex* is present in samples indicating the beginning of the Atlantic stage (7,600-7,500 B.P.) and is not present in any younger layers. This indication of maximum warmth coincides with the Fromentine submergence.

The Saint Marc Emergence (ca. 7,300 B.P. to 6,600 B.P.)

In the Stubai Alps, pollen zone VII is dated from 7,100 B.P. to about 6,500 B.P. (Mayr, 1964). In the Saint Gotthard Massif, an abundance peak of *Abies* has been noted at about 6,950 B.P. (Zoller, Schindler, and Röthlisberger, 1966). In the Swiss Alps generally, important cold phases (Misox oscillations) are known to have occurred between 7,500 B.P. and 6,500 B.P. In Baffin Land, glaciers advanced at about 6,700 B.P. (Bryson et al., 1969).

The Tréompan Emergence (ca. 5,800–5,300 B.P.)

In the Stubai Alps, glacial advance VIId has been dated at 5,300 B.P. or thereabouts (Rootmos I) (Mayr, 1964). In the Valtelline region of the Alps, the Daun kataglacial period (having three phases) has been bracketed between the dates of 6,000 B.P. and 5,000 B.P. (Venzo, 1971). In the Cascade Ranges (United States), in Baffin Land, and in the Patagonian Andes, glaciers readvanced between 5,800 B.P. and 5,000 B.P. The probable maximum around 5,300 B.P. has been named the Flint phase (Denton and Karlén, 1973).

The Brétignolles Submergence (5,000–4,900 B.P.)

In the Tuktoyaktuk Peninsula (on the coastal plain of the Alaskan North Slope), changes in the plant cover reflect climatic fluctuations during the Holocene (Ritchie and Hare, 1971). From 8,500 B.P. to 5,550 B.P., the forest cover of fir in association with birch was unbroken. From 5,500 B.P. to about 4,000 B.P., a more temperate and wetter climate is implied by considerable thickets of alder and the more dispersed colonies of fir. Within the area now taken up by tundra, an in situ trunk of *Pica glauca* dated at 4,940 ±140 B.P. has been found. The size and width of the tree rings show that radial growth at the time was greater than in fir trees of the present day in the Inuvik taiga (Ritchie and Hare, 1971). This tree's age coincides with the time of the Brétignolles submergence. After 4,500 B.P., the climate became cooler; trees were replaced by moorland-type tundra with dwarf birches (this matches the Argenton emergence). In Patagonia, major glacial retreats have been dated between about 5,050 B.P. and 4,490 B.P. (Denton and Karlén, 1973).

The Argenton Emergence (ca. 4,800–3,500 B.P.)

The climatic cooling during the first part of the Subboreal period is a well-known phenomenon. In Switzerland, the Piora cold oscillation started at about this time (Zoller, Schindler, and Röthlisberger, 1966). At about 4,600 B.P., the Oberaar glacier was advancing. In the Stubai Alps, stage VII of glacial advance (Rootmos 2) has been dated at about 4,500 B.P. (Mayr, 1968). In the Swiss Jura, *Abies* attained a maximum during the Neolithic age (between about 4,750 B.P. and 3,750 B.P.). Simultaneously, *Picea* and *Betula* advanced (Leroi-Gourhan and Girard, 1971).

According to Frenzel (1966, 1973), during the first half of the Subboreal in the Northern Hemisphere, continental climate tendencies developed. In Denmark, during the Subboreal, the quantity of ivy that was present throughout the Boreal and the Atlantic declined (Iversen, 1954). Also, elm, oak, and lime decreased, a typical result of climatic cooling. In Sweden, a cold, dry episode at about 4,250 B.P. (sidereal time, by varve dating) has been inferred by Granlund (1932). In Finland, at about 4,800 B.P., the climate became colder and damper (Vasari, 1972). In Siberia, broad leaved warm-weather forest species started to die out around 4,700-4,500 B.P., and towards 3,000 B.P. the forest associations capable of resisting cold were expanding (Serebryanny, 1969). This second cool phase (3,000 B.P.) is contemporaneous with the Belle Isle emergence.

In the Canadian Arctic, glaciers readvanced somewhat after 5,000 B.P. Toward 4,800 B.P., another readvance has also been inferred (Porter and Denton, 1967). On Baffin Island and in the Glacier Bay region, glaciers doubled in length at about 4,700 B.P. Notable advances are also invoked at about 4,350 B.P. and 3,700 B.P. (Karlstrom, 1956; Goldthwait, 1966). In the United States, a cold period has been inferred for about 4,300 B.P. (Deevey and Flint, 1957).

In the Loire basin, *Vitis* pollen, which had been present in samples dated from the first half of the Boreal (and again from those dated as Atlantic), was not found in samples from the first half of the Subboreal (Planchais, 1971). In the Gironde region at this time, water plants decreased and xerothermophilous plants developed (Paquereau, 1964). This situation probably resulted from a temporary withdrawal of the coastline.

The Camiers Submergence (ca. 3,400–3,100 B.P.)

A climatic warming melted glaciers in British Columbia between 3,280 B.P. and 3,170 B.P. (Denton and Karlén, 1973).

The Belle Isle Emergence (ca. 3,000–2,200 B.P.)

In the Stubai Alps, intensive glacial advance between 2,800 B.P. and 2,500 B.P. has been described (Heuberger, 1954). Mayr (1968) has inferred two stages of glacial advance, one between 2,900 B.P. and 2,600 B.P. (Stage IXa) and the other between 2,500 B.P. and 2,100 B.P. (Stage IXb). In the Saint Gotthard Massif, just at the beginning of the Subatlantic, *Abies*, *Picea*, and *Betula* increased; somewhat before 2,830 B.P., the quantities of *Hedra* and *Ilex* decreased (Godwin, 1956). In Scandinavia, at 3,150 B.P. and 2,250 B.P., trees were replaced by *Sphagnum* in bogs, implying colder and drier episodes (Granlund, 1932).

In arctic Canada, a glacial readvance has been dated at around 2,800 B.P. (Ritchie and Hare, 1971). Comparable readvances have been inferred for the La Sal Mountains, Utah, and in the Sierra Nevada between 3,300 B.P. and 2,400 B.P. The maximum glacial advance has been dated at about 3,000 B.P. In Swedish Lapland, glaciers readvanced between 2,700 B.P. and 2,370 B.P. (Denton and Karlén, 1973). At Mount Northcote, New South Wales, Australia, a cold phase is known to have started about 3,000 B.P. (Costin, 1972).

The Roman Emergence (ca. 2,000 B.P.)

A glacial advance at this time has been noted on Baffin Island and on Mount Rainier, United States.

The Saint Firmin Submergence (ca. 1,700 B.P.)

This sea-level rise is contemporaneous with the period of intense deglaciation in North America between A.D. 0 and A.D. 400.

The Wissant Emergence (ca. 1,300 B.P.)

In the Stubai Alps, Stage Xb is placed between 1,600 B.P. and 1,200 B.P. At about 1,200 B.P., Iceland was entirely surrounded by sea ice. In Alaska, between about 1,325 B.P. and 1,075 B.P., glaciers advanced (Denton and Karlén, 1973).

The Past Thousand Years

The glacial advance of 900-700 B.P. (stage Xd of the Stubai Alps) partly coincides with the cold, dry episodes noted in Sweden at about 750 B.P. (Granlund, 1932) and with the Paria emergence of 700-750 B.P. (Fairbridge, 1961). In Switzerland between 665 B.P. and 740 B.P., the Aletsch and Grindelwald glaciers advanced (Denton and Karlén, 1973). During the the first half of the twelfth century, grape vines were grown commercially in England, but cold was recorded during the second half of the thirteenth and early in the fourteenth centuries (LeRoy Ladurie, 1967). During the Little Ice Age—notably, in the middle of the sixteenth century (maximum at about A.D. 1530)—cold periods were experienced in France, the Alps, Norway, and Alaska. In the second half of the seventeenth century—notably, in A.D. 1690—pack ice entirely surrounded Iceland. Cold peri-

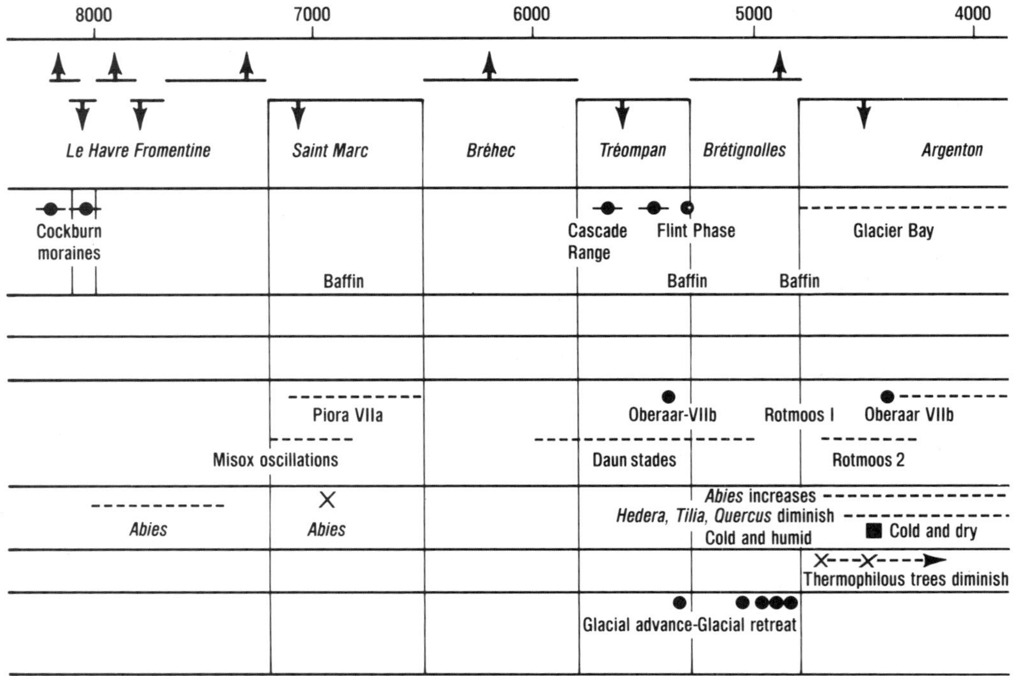

Figure 12-5. Correlations between low sea levels and episodes of glacial readvances during the Holocene.

ods such as the Fernau stage (A.D. 1600-1616) and the Napoleonic stage (A.D. 1800-1825) in the Alps (Venzo, 1971) were so short that they surely affected sea levels very little.

During the fourteenth and fifteenth centuries, shelly shingle banks developed in Brittany just above the present-day high-water level. Collectively, the radiocarbon dates (0, 1, and 3) indicate a submergence between the Paria and Little Ice Age emergences.

To sum up, the data provided in Figure 12-5 support the close correlation and synchroneity between low sea levels and dated glacial advances. The oscillations appear to be worldwide phenomena of chiefly climatic and eustatic origin.

CORRELATIONS WITH ICE-CORE DATA

Climatic fluctuations have been recorded by fluctuations in the ratios of the stable isotopes ^{18}O and ^{16}O measured in samples of ice cored at Camp Century, Greenland (Dansgaard et al., 1971). The oscillations on the resulting curve relate to the temperature at which North Atlantic seawater evaporated and in the precipitation that built up the ice cap. Thermal maxima (in sidereal years) are indicated at 8,000 B.P. (7,700 C-14), 7,200 B.P. (6,500 C-14), 5,900 B.P. (5,100 C-14), 5,100 B.P. (4,500 C-14), 4,700 B.P. (4,100 C-14), 3,100 B.P. (2,800 C-14), 2,200 B.P. (2,000 C-14), and 1,200 B.P. (1,100 C-14). These dates coincide almost perfectly with my sequence of submergence maxima. Only the double warm peak of 5,100 B.P. (4,500 C-14) and 4,700 B.P. (4,100 C-14) is not represented by submergent French Atlantic coastal deposits. Despite slight time lags connected with radiocarbon dates, the emergent sea levels are also indicated by the results from the ice cores.

Despite the somewhat imprecise basis for dating on the curve of Dansgaard et al. (1971) from the Greenland ice core, the correlations among the ice-core results, Fairbridge's (1961) curve, and my results are remarkable. The lengths and intensities of the seven climatic optima during the past 8,000 yr shown by the ice-core

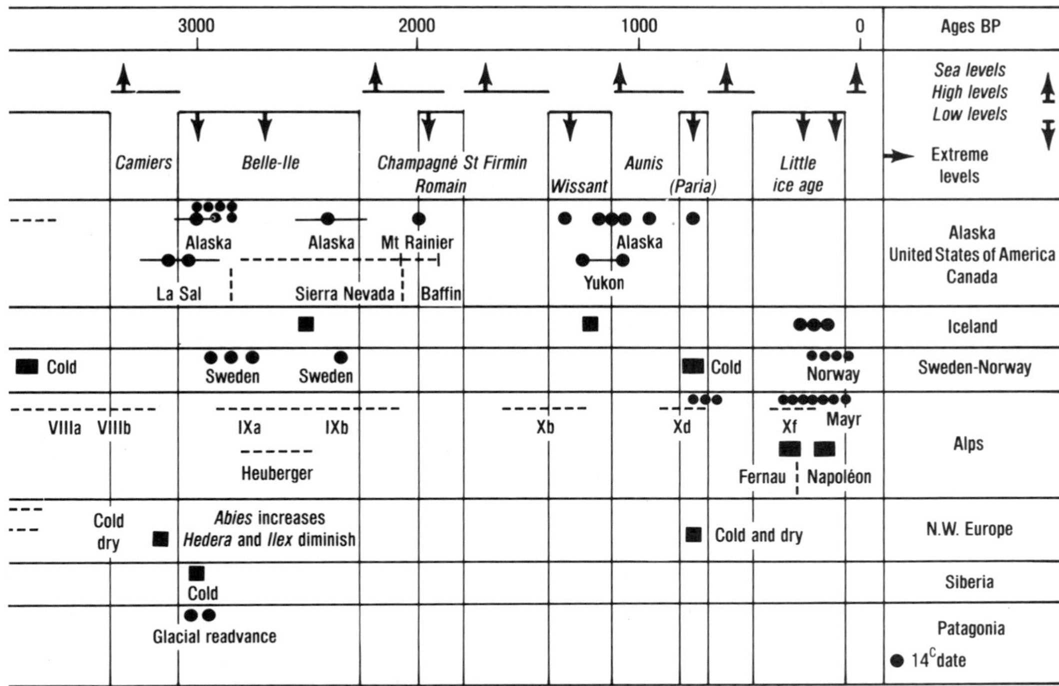

results correspond closely with submergent mean world sea levels.

CAUSES OF SEA-LEVEL CHANGES DURING THE HOLOCENE

Without prejudice to other possible causes, such as changes in the structure of the ocean floors, local tectonism, isostatic loading by the water of the transgression, and amounts of groundwater influencing local conditions, I conclude that the major cause of the Holocene rise in sea level has been glacial melt. The upward and downward oscillations match closely all kinds of paleoclimatic data. In particular, these include:

oscillations of the ratios of the isotopes $^{18}O/^{16}O$ in ice cores from Camp Century, Greenland;
advances and retreats of glaciers, as in arctic Scandinavia, the Alps, the Andes, Patagonia, Africa, and various localities in North America;
variations in plant cover, such as have been recorded on pollen charts for western Europe for a long period of time.

Despite the importance of local conditions, advances and retreats of glaciers are good worldwide climatic indicators. Nevertheless, because each glacial advance and retreat may be fairly short, data from glaciers are not as precise or as continuous as pollen charts. Variations in sea level depend directly on the amount of ice locked up in glaciers. The sea rises when glaciers melt and drops when they advance. These changes may be caused by solar activity. Worldwide climatic data are now beginning to be compared and correlated as I have done. Further confirmation of my results should be sought by numerous comparable studies based on rigorous control by worldwide dating methods.

CONCLUSION

The study area here summarized has been deliberately limited to a short and relatively stable area along the French Atlantic and French Channel coasts. Despite certain shortcomings of the data base, my results should serve as a guideline to the principal stages of the Flandrian transgression from 10,000 B.P. to the present. For example, I consider the total number of radiocarbon-dated samples used, 174, to be inadequate. Moreover, the radiocarbon method gives relatively imprecise results. However, the results, which are based on a great diversity of evidence—geomorphologic, sedimentologic, paleontologic, palynologic, and prehistoric—are internally consistent.

On my curve, the broader oscillations correspond to significant global phenomena. Nevertheless, more complete documentation and more precise dates are desirable. The groupings seen in the coastal data clearly indicate the maxima and minima. Data from widely dispersed localities, used by some other investigators, are obviously more difficult to coordinate than data taken from the stratigraphic relationships at individual localities. The relative sea-level changes commonly are indicated by subsurface data from one place (e.g., at Le Havre, Saint Marc, Argenton, Brétignolles, and Abbéville).

Far from being regular, the Flandrian transgression took place in oscillatory steps. Seven principal maxima appeared during 8,200 yr. Perhaps 5-10 negative fluctuations were associated with the maxima (see Fig. 12-2 and 12-3). The most complete single succession is at Abbéville (Commont, 1910). Here are the records of six marine submergences and emergences. From 7,400 B.P. onward, MSL had reached the position of −7 m (with respect to MSL of today). Subsequent oscillations were confined to the range between −15 m and +0.50 m. During the past 8,000 yr, the average duration of each major oscillation has been on the order of 1,000 yr (see Le Havre, Fromentine, Brehec, Brétignolle, Camiers, Champagné, and Aunis submergences). Other, shorter oscillations are also indicated. The amplitudes of each minor fluctuation diminished progressively from 8,000 B.P. to 2,000 B.P. During the past 1,700 yr, their ranges were <1 m.

This 1,000-yr cyclicity of sea level is not dissimilar to the finer climatic variations noted within the Riss and Würm glacial periods as seen in the caves in Perigord and in the French Midi.

Seen in general, the mean level of the oceans has risen 60 m during the past 10,000 yr. This figure does not reflect a constant eustatic rise but one repeatedly interrupted by brief reversals. The rate of sea-level rise seems to have been particularly fast during the Preboreal, toward the middle and at the end of the Boreal (9 cm/yr between 8,250 B.P. and 8,150 B.P.), at the beginning and the end of the Subboreal (1.5 cm/yr between 5,400 B.P. and 5,000 B.P.), and at the beginning of the Subatlantic (1.3 cm/yr between 3,600 B.P. and 3,360 B.P.). The rates of the falls appear to have been as fast as those of the rises; the falling levels were accompanied by periodic glacial readvances. After 7,400 B.P., the rate of rise was much slower. This episode may have coincided with the final melting of the Scandinavian and North American ice caps.

Along the French coast studied, no indications have been found of a sea level noticeably higher than at present. The only exceptions are dated about 2,100 B.P. and 1,600 B.P. (Dunkirkian I and II, nearly 0.5 m above today's MSL). The sea did not remain long at these two levels, however. During the Holocene, the French coast may have subsided slowly; it lies on the margin of the glacio-isostatic uplifted area. About 5,000 B.P., the Brétignolles submergence crested at −5 m; evidence for the sea at this level then can be seen all the way from Picardy to the Vendée. All along its 600-km length from north to south, the elevations of markers for each high level or low level maintain similar values. The French Atlantic coast reacted as a single tectonic entity. The fact that all the points on my curve lie beneath those of Fairbridge's and Mörner's curves indicates that the Earth's crust did not behave uniformly all over. Moreover, today's ocean level does not correspond to a regular, smooth geodetic surface pattern. Many more well-documented studies with supporting geophysical evidence from the stabler sectors are clearly necessary before we can thoroughly understand the mechanisms of marine submergences and emergences and correctly interpret them in time and space.

Along the French Atlantic and French Channel coasts, intensive studies have revealed a record of subsurface sediments extending from the Boreal to the present day. Marine and nonmarine sediments are interbedded. Peats were deposited when the area was emergent and marine sands or silts when it was submerged. Samples recovered from boreholes have been dated by the radiocarbon method and analyzed for their pollen, sedimentology, and invertebrate paleontology. The same general relationships have been found throughout the area. Thus, the results obtained all along the 600-km length of this passive margin are of general eustatic significance.

REFERENCES

Bourdier, F., 1969, Etude comparée des depots quaternaires des bassins de la Seine et de la Somme, *Assoc. Géol. Bassin Paris Bull. No. 21*, pp. 169-220.

Bryson, R. A.; Wendland, W. M.; Ives, J. D.; and Andrews, J. T., 1969, Radiocarbon isochrones on the disintegration of the Laurentide ice sheet, *Arctic and Alpine Research* **1**(1):1-14.

Castaing, P.; Feral, A.; and Klingebiel, A., 1971, Paléogéographie de l'Holocène sur le plateau continental au large de l'embouchure de la Gironde, *Soc. Géol. France Comptes Rendus No. 6*, pp. 325-327.

Commont, V., 1910, Terrasses fluviatiles de la vallée de la Somme, *Soc. Géol. du Nord* **39:**185-210.

Commont, V., 1910, Note sur les tufs et les tourbes de la vallée de la Somme, *Soc. Géol. du Nord* **39:**210-248.

Costin, A. B., 1972, Carbon-14 dates from the Snowy Mountains area, southeastern Australia, and their interpretation, *Quat. Research* **2**(4):579-590.

Dansgaard, W.; Johnsen, S. J.; Clausen, H. B.; and Langway, C. C., 1971, Climatic record revealed by the Camp Century ice core, in K. K. Turekian (ed.), *Late Cenozoic-Ice Age Climates*, New Haven, Conn.: Yale University Press, pp. 37-56.

Deevey, E. S., and Flint, R. F., 1957, Postglacial hypsithermal interval, *Science* **125:**182-184.

Denton, G. H., and Karlén, W., 1973, Holocene climatic variations. Their pattern and possible cause, *Quat. Research* **3**(2):155-205.

Elhai, H., 1963, La Normandie occidentale entre la Seine et le Golfe normand-breton. Etude morphologique. Thèse, University Bière, Bordeaux.

Fairbridge, R. W., 1961, Eustatic changes in sea level, in L. H. Ahrens, F. Press, K. Rankama, and S. K. Runcorn (eds.), *Physics and Chemistry of the Earth*, vol. 4, New York: Pergamon Press, pp. 99-185.

Frenzel, B., 1966, Climatic change in the Atlantic/Sub-Boreal transition in the Northern Hemisphere: Botanical evidence, in J. S. Sawyer (ed.), *World Climate from 8000 to 0* B.C., London: Royal Meteorological Society, 229p.

Frenzel, B., 1973, *Climatic Fluctuations of the Ice Age*, A. E. M. Nairn (trans.), Cleveland and London: Case-Western Reserve University Press, 306p.

Gabet, C., 1970, La transgression flandrienne en Aunis et Saintonge. Nouvelles observations sur les marais, *Cong. Soc. Savants, Tours (1968)*, pp. 35-40.

Gabet, C., 1971, La phase terminale de la transgression flandrienne sur la littoral charentais, *Quaternaria* **14:**181-188.

Giot, P. R., 1965, Information archéologiques, circonscription de Rennes, *Gallia Préhistoire* **8:**33-40.

Giot, P. R., 1968, La Bretagne au peril des mers holocènes, in *La Préhistoire: Problèmes et Tendances*, Paris: Éditions du Centre National de Recherche Scientifique.

Giot, P. R., 1969, Chronique des datations radiocarbone armoricaines, *Annales Bretagne* **76:**153-162.

Giot, P. R.; L'Helgouach, J.; Briard, J.; Waterbolk, H. T.; van Zeist, W.; and Muller-Wille, M., 1960, Une station du Néolithique primaire armoricain: le Curnic en Guissény (Finistère), *Soc. Préhistorique France Bull.* **57**(1-2):38-50.

Giresse, P., 1969, Essai de sédimentologie comparée des milieux fluvio-marins du Gabon, de la Catalogne et du Sud-Cotentin, thèse, University Caen.

Godwin, H., 1956, *The History of the British Flora. A Factual Basis for Phytogeography*, Cambridge: Cambridge University Press (2nd ed., 1975).

Goldthwait, R. P., 1966, Evidence from Alaskan glaciers of major climatic changes, in J. S. Sawyer (ed.), *Climate from 8000 to 0* B.C., London: Royal Meteorological Society, pp. 40-53.

Granlund, E., 1932, De Svenska hogmossarnas geologi, *Sveriges Geol. Undersökning Arsb. 26*, 193p.

Guilcher, A., and Giot, P. R., 1969, Bretagne-Anjou, *Livret-Guide de l'Excursion C-16, Congrès 8*, Paris: INQUA, 79p.

Hageman, B. P., 1969, Development of the western part of the Netherlands during the Holocene, *Geologie en Mijnbouw* **48**(4):373-388.

Heuberger, H., 1954, Gletschervorstösse zwischen Daun und Fernau-Stadium in den nordlichen Stubaier Alpen (Tirol), *Zeitschr. Gletscherkunde u. Glazialgeologie* **3:**91-98.

Huault, M. F., and Larsonneur, C., 1972, La baie des Veys et les marais de Carentan: Histoire postglaciaire, *Assoc. Géol. Bassin Paris Bull.*, no. 33, pp. B6-B10.

Huault, M. F., and Lefèvre, D., 1983*a*, Un depôt holocène exceptionnel dans la basse vallée de la Seine: la diatomite calcifiée de la Meilleraye sur Seine (France), *Assoc. Française Etude Quaternaire Bull.* **4:**171-181.

Huault, M. F., and Lefèvre, D., 1983*b*, A mire environment during the Holocene: The Marais Vernier (France), *Quat. Studies in Poland* **4:**229-236.

Iversen, J., 1954, The late-glacial flora of Denmark and its relation to climate and soil, *Denmarks Geol. Undersökning Ser. 2, Na 80,* pp. 87-119.

Jelgersma, S., 1966, Sea-level changes during the last 10,000 years, in J. S. Sawyer (ed.), *World Climate from 8000 to 0 B.C.*, International Symposium Proceedings, London: Royal Meteorological Society, pp. 54-71.

Karlstrom, T., 1956, Tentative correlation of Alaskan glacial sequences, *Science* **125:**73-74.

Kerfourn, M.-Th., 1965, Analyse pollinique des sédiments flandriens de la vallée de la Vilaine aux environs de Redon (Ille-et-Vilaine), *Soc. Géol. et Minéral. Bretagne Bull.* (1962-1963), pp. 147-157.

Larsonneur, Cl., 1971, Manche centrale et Baie de Seine: géologie du substratum et des depôts meubles, thèse, University Caen.

Lebesconte, P., 1898, Periodes géologiques gallo-romaine et franque. Leurs relations avec le Quaternaire, le Pliocene et l'époque moderne, *Bull. Soc. Sci. Med. Ouest* **7:**354-408.

Lefèvre, P., 1972, Sur quelques particularites granulométriques et chimiques des sédiments post-flandriens et actuels de la plaine maritime picarde, 8th Cong. INQUA (Paris), 1969, *Assoc. Francais Etude Quaternaire Bull. Supplement No. 4,* pp. 715-721.

Leroi-Gourhan, A., and Girard, M., 1971, L'abri de la Cure à Baulmes (Suisse). Analyse pollinique, *Soc. Suisse Préhistoire et Archéologie Annuaire* **5:**7-15.

LeRoy Ladurie, E., 1967, *Histoire du climat depuis l'an mil,* Paris: Flammarion. (1971 translation by Barbara Bray, *Times of Feast, Times of Famine,* New York: Doubleday, 426p.)

Mariette, H., 1971, L'archéologie des depôts flandriens du Boulonnais, *Quaternaria* **14:**137-150.

Mayr, F., 1964, Untersuchungen über Ausmass und Folgen der Klima- und Gletscherschwankungen seit dem Beginn der Postglacialen Wärmezeit, *Zeitschr. Geomagnet.* (new series) **8**(3):257-285.

Mayr, F., 1968, Postglacial glacier fluctuations and correlative phenomena in the Stubai Mountains, eastern Alps, Tyrol, in *Proceedings of the 7th INQUA Congress, 1965,* vol. 14, Boulder, Colo.: University of Colorado, pp. 167-177.

Møller, J. J., and Sollid, J. L., 1972, Deglaciation chronology of Lofoten-Vesteralen-Ofoten, North Norway, *Norsk Geog. Tidsskr.* **26**(3):101-133.

Mörner, N.-A., 1969, The Late Quaternary history of the Kattegatt Sea and the Swedish west coast, *Sveriges Geol. Under., Ser. C, No. 640,* 487p.

Morzadec-Kerfourn, M.-Th., 1969, Variations de la ligne de rivage au cours du Post-Glaciaire, le long de la côte nord du Finistère. Analyses polliniques de tourbes et de depôts organiques littoraux, *Assoc. Française Etude Quaternaire Bull.* **4:**285-318.

Paepe, R., 1972, La plaine maritime entre Dunkerque et la frontière belge, *Soc. Études Géog. Belgique Bull.* **29:**47-66.

Paquereau, M. M., 1964, Flores et climats post-glaciaires en Gironde, *Soc. Linnéen Bordeaux Actes* **101:**1.

Patte, E.; Ters, M.; and Verger, F., 1961, Sur l'origine humaine des buttes coquillières de Saint-Michel en l'Herm (Vendeé), *Bull. Inst. Oceanogr. Monaco No. 1211,* 7p.

Planchais, N., 1971, Histoire de la végétation post-wurmienne des plaines du Bassin de la Loire, d'après l'analyse pollinique, thèse, Univ. Montpelier.

Planchais, N., and Corillon, R., 1968, Recherche sur l'évolution recente de la flore et de la vegetation de Belle-Ile-en-Mer (Morbihan), d'après l'analyse pollinique de la tourbière submergée de Ster-Vras, *Soc. Bot. France Bull.* **115:**441-458.

Porter, S. C., and Denton, G. H., 1967, Chronology of neoglaciation in the North American Cordillera, *Am. Jour. Sci.* **265:**177-210.

Ritchie, J. C., and Hare, F. K., 1971, Late Quaternary vegetation and climate near the Arctic tree line of northwestern North America, *Quat. Research* **1**(3):331-342.

Rivière, A.; Vernhet, S.; Arbey, F.; and others, 1966, Sur les terrains récents des côtes atlantiques, *Acad. Sci. Comptes Rendus,* Series D **262:**5-8.

Sanquer, R., 1968, Decouvertes récentes aux environs de Brest (période romaine): le reservoir à poissons du Curnic, en Guissèny, *Annales Bretagne* **75:**246-265.

Serebryanny, L. R., 1969, L'apport de la radio-chronometrie à l'étude de l'histoire tardi-quaternaire des regions de glaciation ancienne de la Plaine russe, *Revue Géographie Phys. et Géologie Dynam.*, **11:**293-302.

Shepard, F. P., 1963, Thirty five thousand years of sea level, in T. Clements (ed.), *Essays in Marine Geology in Honor of K. O. Emery,* Los Angeles: University of Southern California Press, pp. 1-10.

Sommé, J., and Cabal, M., 1972, La plaine maritime dans la région d'Ardres (Pas-de-Calais) et le site archéologique des Noires-Terres: *Université de Lille, Cahiers Géographie Physique No. 1,* pp. 29-43.

Suggate, R. P., 1968, Post-glacial sea-level rise in the Christchurch metropolitan area, New Zealand, *Geologie en Mijnbouw* **47**(4):291-297.

Ters, M.; Delibrias, G.; Denèfle, M.; Rouvillois, A.; and Fleury, A., 1980, Sur l'évolution géodynamique du Marquenterre (Basse Somme) à l'Holocène et durant le Weichselien ancien, *Assoc. Française Étude Quaternaire Bull.* **17:**11-23.

Ters, M., and others, 1971, Sur le remblaiement Holocène dans l'estuaire de la Seine, au Havre (Seine-Maritime), France, *Quaternaria* **14:**151-174.

Ters, M., and Pinot, J.-P., 1969, *Livret-Guide, Excursion A10*, Congrès 8, Paris: INQUA, 110p.

Ters, M.; Planchais, N.; and Azema, C., 1968, L'évolution de la basse vallée de la Loire, à l'aval de Nantes, à la fin du Würm et pendant la transgression flandrienne, *Assoc. Française Étude Quaternaire Bull.* **5:**217-246.

van der Heide, S., and Zagwijn, W. H., 1967, Stratigraphical nomenclature of the Quaternary deposits in The Netherlands, *Geol. Stichting Med. No. 18*, pp. 23-29.

Vasari, Y., 1972, Climatic changes in Arctic areas during the last ten thousand years, in Y. Vasari, H. Hyvaerinen, and S. Hicks (eds.), *Climatic Changes in Arctic Areas During the Last Ten-Thousand Years*, Oulu, Finland: Universitat Ouluensis, Acta, ser. A, Sci. Rerum. Nat., No. 3, Geol. No. 1, pp. 239-252.

Venzo, S., 1971, Gli stadi tardo-wurmiani e post-wurmiani nelle Alpi insubriche valtillinesi, *Soc. Italiana Sci. Nat. e Museo Civico Storia Nat. Milano Atti* **112/2:**161-276.

Verger, F., 1972, Le gisement coquillier de l'anse de la Houle, près de Cancale, est d'age historique, *Acad. Sci. Comptes Rendus*, Series D **275:**649-650.

Verger, F., and Florschutz, F., 1960, Sur l'existence à Fromentine (Vendée) d'une couche de tourbe du debut de l'Atlantique, *Acad. Sci. Comptes Rendus* **251:**891-893.

Zoller, H.; Schindler, C.; and Röthlisberger, H., 1966, Postglaziale Gletscherverstände und Klimaschwankungen in Gotthard-massiv und Vorderrheingebeit, *Naturf. Gesell. Basel Verh.* **77**(2):97-164.

IV

Short-Term Climate (10–10^2 yr) and Periodicity

13: Abrupt Geomagnetic Variations–Predictive Signals for Temperature Changes 3–7 yr in Advance

Goesta Wollin, John E. Sanders, and David B. Ericson
Columbia University

Abstract: A forecasting method for seasonal and annual surface-air temperatures based on the study of the variations in the rate of change of geomagnetic horizontal intensity is suggested. Preliminary results of a study on a time scale of years and decades indicate that temperatures for the Northern Hemisphere during the next 20 years generally will trend downward.

When geomagnetic effects are represented by the Geomagnetic Activity Recurrence Index, an 11-yr cycle of opposite phase to the sunspot cycle and the solar-radio flux cycle is apparent. Other parameters that vary cyclically on periods of about 11 yr and that are closely lag correlated with geomagnetism, sunspot activity, and solar-radio flux include cosmic-ray flux, solar-proton events, nitric-oxide molecules, ozone trend, Earth's rotation rate, sea-surface temperatures, the thickness of the 100-500-mb pressure field, sea-level pressure, and surface-air temperatures.

INTRODUCTION

Bigelow (1898) wrote:

> That there is a causal connection between the observed variations in the forces of the Sun, the terrestrial magnetic field, and the meteorological elements has been the conclusion of every research into this subject for the past 50 years. The elucidation of exactly what the connection is and the scientific proof of it is to be classed among the most difficult problems presented in terrestrial physics. . . . The bibliography is large—covers a century—and embraces such names as . . . Gauss, Sabine, . . . Faraday, Wolf, . . . Steward, Schuster, . . . Airy, . . . Kelvin, and many others.

Since Bigelow wrote this, more than a thousand papers have been published on the subject of geomagnetic activity and the weather. Notable is a paper by Egyed (1961), which reported a relationship between a 27-day recurrence in the temperature distribution and the 27-day recurrence of geomagnetic activity. It was one of the many outstanding papers resulting from the important 1961 Conference on Solar Variations, Climatic Change, and Related Geophysical Problems of which Professor Rhodes W. Fairbridge was conference editor. Since 1961, comprehensive reviews of geomagnetism-weather relationships have been published by King (1975), Wilcox (1975), Roberts (1979), Herman and Goldberg (1979), Prohaska and Willett (1981), Frazier (1982), and Spruit and Roberts (1983). The literature on the subject tends to be contradictory, and the reported results have sometimes been confused and disjointed. Particularly useful and critical approaches to the literature have been published by Pittock (1978, 1980). An appreciable influence of geomagnetic activity on the weather is not widely accepted. For example, when Wollin et al. (1973) published their first results of a study of short-term geomagnetic and temperature variations, the results were considered by Sternberg and Damon (1979) not to be valid. The influence of geomagnetic activity on the weather is not used for forecasting purposes, and a widely accepted physical mechanism for any connection between geomagnetism and weather has not emerged.

Nevertheless, a few common features appear so widely in the otherwise disparate literature that they suggest the existence of a valid basis for the conclusion that geomagnetic variations and weather changes are related. First, meteorological adjustments occur after geomagnetic fluctuations (King, 1975; Wilcox, 1975). Second, the largest meteorological adjustments that appear to be related to geomagnetic variations occur during winter (Wilcox, 1975).

In this chapter, we use a new method in this field of research for determining the values of geomagnetic components: the rate-of-change method. We use this method because plotted absolute values of geomagnetic components at a given observatory typically show long-continued secular variation in the same sense but not necessarily or usually at a constant rate. We report magnetic-observatory records that show that changes in the rate of secular variation are often sharp and that these abrupt changes are predictive signals for temperature fluctuations 3-7 yr in advance.

We present annual and seasonal variations, make some predictions for the Northern Hemisphere, and end by showing that nine geophysical factors are highly lag correlated with geomagnetism, sunspot activity, and solar-radio flux. Likewise, these nine factors display cyclic variation of about 11 yr.

ANNUAL VARIATIONS

Wollin, Ryan, and Ericson (1981) reported the results of a study covering the period from 1935 to 1970 in which they compared annual geomagnetic and temperature data from 29 pairs of magnetic observatories and weather stations in the Northern Hemisphere. Of these, 27 pairs lag correlated. Annual surface-air temperatures were used, and both annual absolute values and rate-of-change values of geomagnetic horizontal intensity *(H)* and geomagnetic total intensity *(F)* were plotted. These authors used the horizontal component because Nelson, Hurwitz, and Knapp (1962); Akasofu and Chapman (1972); and Alldredge (1975) had concluded that geomagnetic activity is especially marked in *H*. The plotted absolute values of *H* and *F* do not correlate with the temperature curves. By contrast, 25 curves of variations in the rate of change of *H* lag correlated with paired temperature curves. The 25 curves of variations in the rate of change of *F* based on data from the same observatories did not correlate with the temperature changes at the paired weather stations. Two curves of the variations in the rate of change of *F* based on data from two magnetic observatories lag correlated with the temperature records from nearby weather stations. No correlation was found between the records of any of the geomagnetic components from two magnetic observatories and temperatures from paired weather stations. Temperature trends lagged variations in the rate of change of *H* and *F* by 1-3 yr. The correlation was inverse; that is, increase in the rate-of-change values of *H* and *F* was followed by decrease in temperatures and vice versa. The lagged correlation coefficients ranged from −0.52 to −0.86.

In Figure 13-1 annual variations in the rate of change of *H* from the Kakioka magnetic observatory (36°13′N and 140°11′E) are compared with the annual mean surface-air temperature changes from the Tokyo weather station (35°41′N and 139°45′E). The time scales are displaced to illustrate more clearly the lag correlation between the two curves (see examples in Brown, 1974; Wollin, Ryan, and Ericson, 1981). The figure shows that the lag in temperature trends against geomagnetic variation is 3 yr. The curves were matched visually by shifting the time scale for the geomagnetic curve with respect to the temperature curve. Then the correlation coefficient was computed, and it is −0.75. The correlation coefficients were sub-

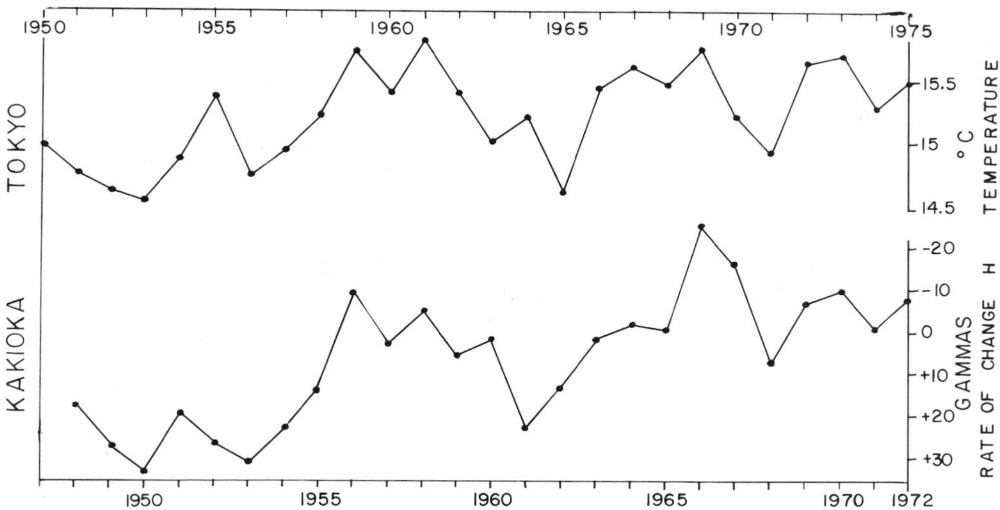

Figure 13-1. Annual geomagnetic intensity *(H)* rate-of-change values from Kakioka compared with annually averaged surface-air temperatures from Tokyo. Note the 3-yr displacement of the time scales.

sequently computed for lags of −5 yr to +5 yr at yearly intervals from the chosen lag, and all coefficients were less than −0.75 or +0.75.

SEASONAL VARIATIONS

We consider the months for each season as follows: December, January, and February for winter; March, April, and May for spring; June, July, and August for summer; and September, October, and November for fall. Seasonal variations in the rate of change of *H* from Kakioka are compared with seasonally averaged surface-air temperature changes from Tokyo in Figure 13-2. The curves were plotted so that the value for the season of one year is followed by the value of the same season of the next year. The time scales are displaced so that temperature changes lag geomagnetic variations by 3 yr. The correlation coefficients (Table 13-1) were obtained by the same statistical treatment described previously. For the winter, the 3-yr lag coefficient is −0.64; for the spring, −0.31; for the summer, −0.52; and for the fall, −0.61.

In Figure 13-3 seasonal variations in the rate of change of *H* from the Tucson Magnetic Observatory (32°14′N and 110°50′W) are compared with time-lagged seasonally averaged surface-air temperature changes from the Phoenix weather station (33°26′N and 112°01′W). The anticorrelation between the winter curves seems to be best.

Seasonal variations in the rate of change of *H* from the San Juan Magnetic Observatory (18°07′N and 66°09′W) are compared with time-lagged seasonally averaged surface-air temperature changes from the San Juan weather station (18°28′N and 66°07′W) in Figure 13-4. The visual best fit lag between temperature trends and geomagnetic variations for all seasons is 7 yr. It can be observed visually that the anticorrelation between the winter curves is best.

As has been shown in Figures 13-2, 13-3, and 13-4 and in Table 13-1, the best correlated surface-air temperature adjustments to variations in the rate of change of geomagnetic horizontal intensity *H* occur during the winter. This result is of interest because Wilcox (1975) has suggested that the finding of any geomagnetic effects that show this feature should be considered in the search for physical mechanisms.

Regarding seasonal and annual weather/climate forecasting, Lamb (1982) concluded that how far it may ultimately be possible to

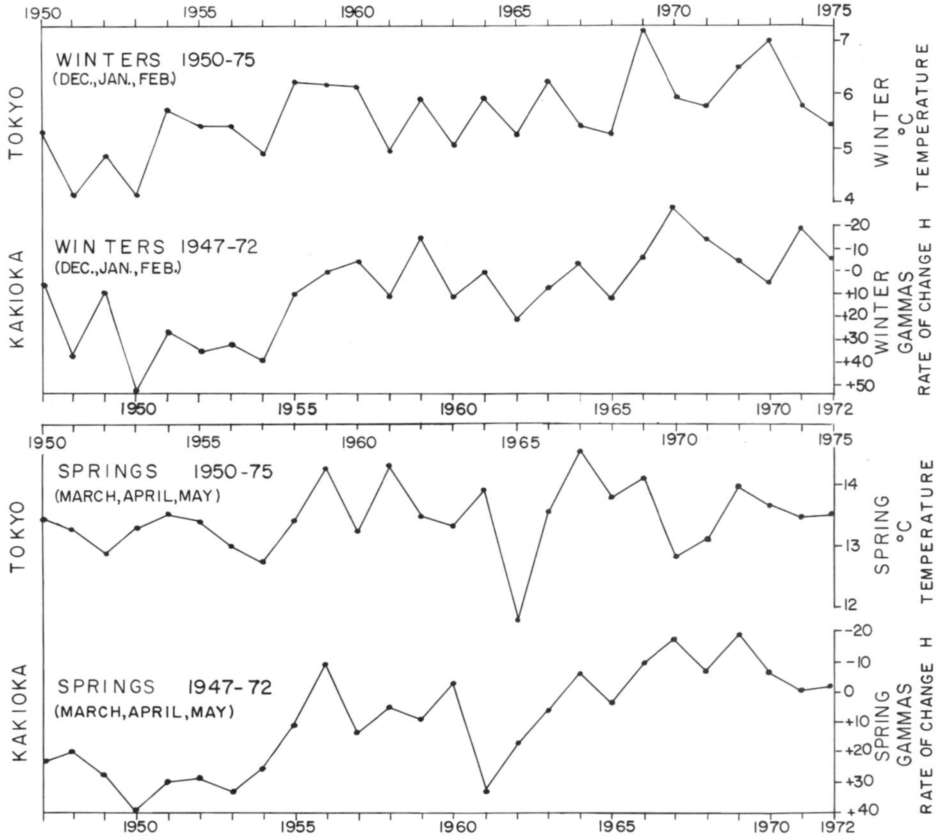

Figure 13-2. Seasonal geomagnetic intensity *(H)* rate-of-change values from Kakioka compared with seasonally averaged surface-air temperatures from Tokyo. The curves were plotted so that the season value for one year is followed by the value for the same season of the next year. Note the 3-yr displacement of the time scales.

forecast one season or one year ahead with the methods now in use cannot be adequately judged in the present state of knowledge. The method we have reported here may be considered an encouraging development in the forecasting of seasonal and annual temperature trends.

FUTURE TEMPERATURE TREND IN THE NORTHERN HEMISPHERE

Lamb (1977) identified four distinct climatic periods within the present interglacial, which began roughly 11,000 yr ago. The first was a warm period that followed the latest ice age, causing its end. This warm period peaked between about 7,000 yr and 5,000 yr ago. It was followed by a colder period that was at its coldest between about 2,900 yr and 2,300 yr ago. After this cold period came a warm interval that reached a peak about 1,000-800 yr ago. The fourth main climatic period marked a return to colder conditions, which because it was colder than anytime since the last ice age proper, has been dubbed the Little Ice Age. The climax of the Little Ice Age came between 550 yr and 125 yr ago. During this last distinct climatic period within the present interglacial, it is interesting to note that the interval from about 1910 to 1960 was the mildest 50-yr period of any comparable interval throughout the past

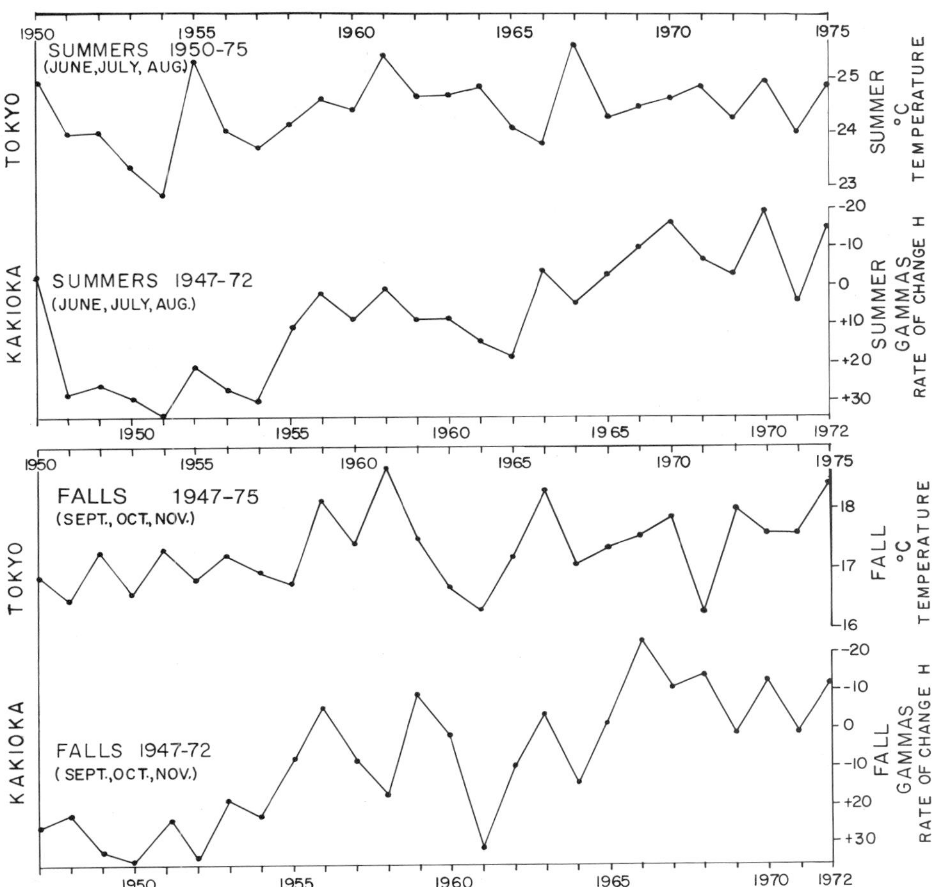

millennium. Was this little warm interval a passing phase within the Little Ice Age, or does it mark the onset of a climatic optimum?

The instability apparent in the climatic situation since 1960 has led to a position in which the leaders of meteorological and climatic research have given conflicting advice about probable future trends. One of the sources of errors for the forecasters is the fact that the warming attributable to the increase of CO_2, although clear in the laboratory and in theory, has not proven as applicable in the global environment context where feedback (i.e., consequential) effects operating through oceans and water vapor in the atmosphere may greatly alter the outcome.

Computer simulations of future climatic trends —that is, numerical modeling—predict a global warming trend for the next hundred years. However, it is interesting to note that, in his comprehensive review of numerical weather/climate forecasting, Thompson (1983) concluded:

> There are now, and always will be, three major sources of errors in numerical predictions. These are: 1) not totally removable errors in the specification of initial conditions; 2) defects of the physical model and its mathematical formulation; and 3) approximations of numerical representation.

The preliminary results of our study (to be published in *Ski Area Management*, May/June 1987) of forecasting the temperature trend for the Northern Hemisphere indicate that the temperature trend for the years to AD 2010 or so will decrease rather than increase as is generally predicted. However, we found an anomaly

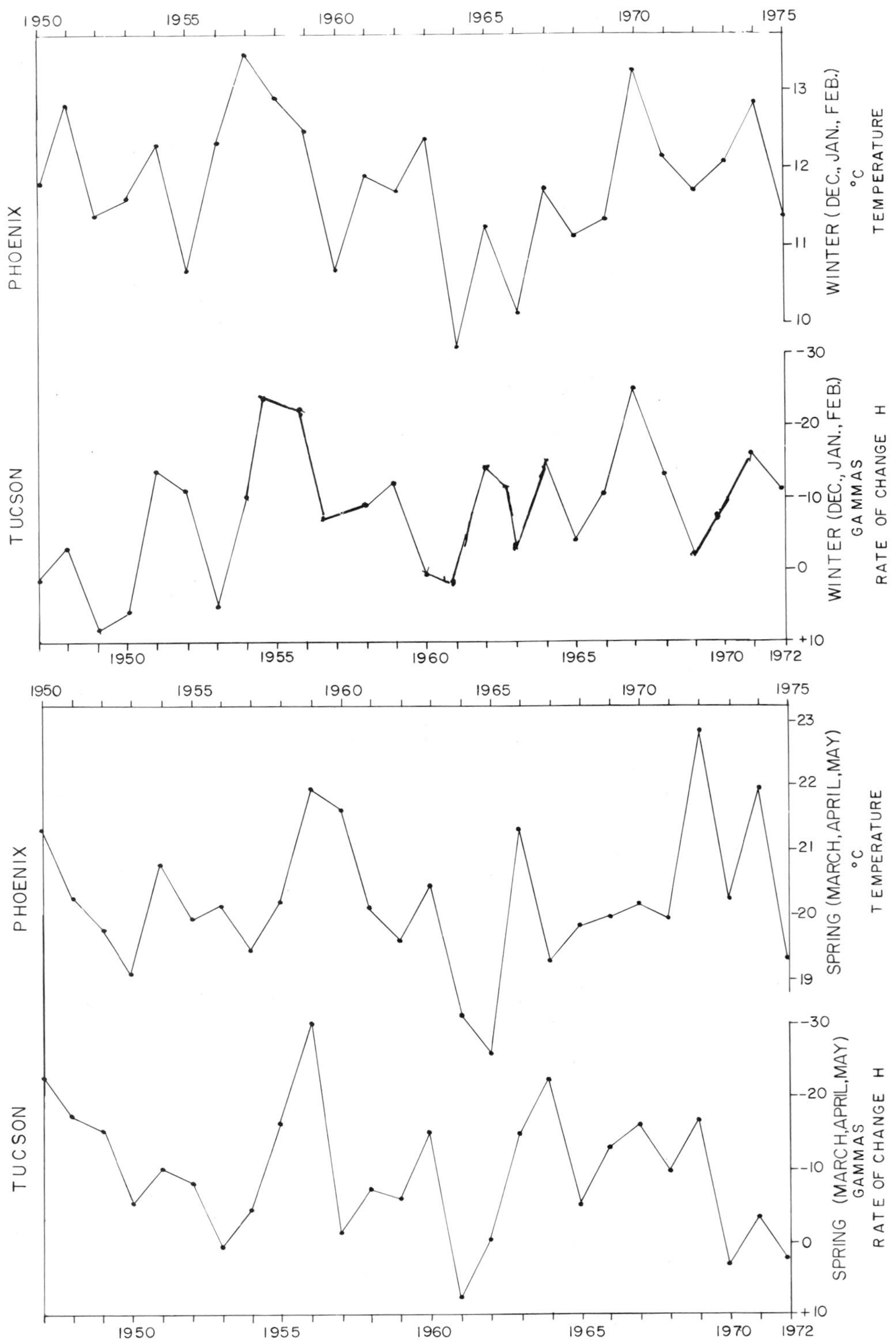
PHOENIX
WINTER (DEC., JAN., FEB.)
°C
TEMPERATURE
TUCSON
WINTER (DEC., JAN., FEB.)
GAMMAS
RATE OF CHANGE H
PHOENIX
SPRING (MARCH, APRIL, MAY)
°C
TEMPERATURE
TUCSON
SPRING (MARCH, APRIL, MAY)
GAMMAS
RATE OF CHANGE H

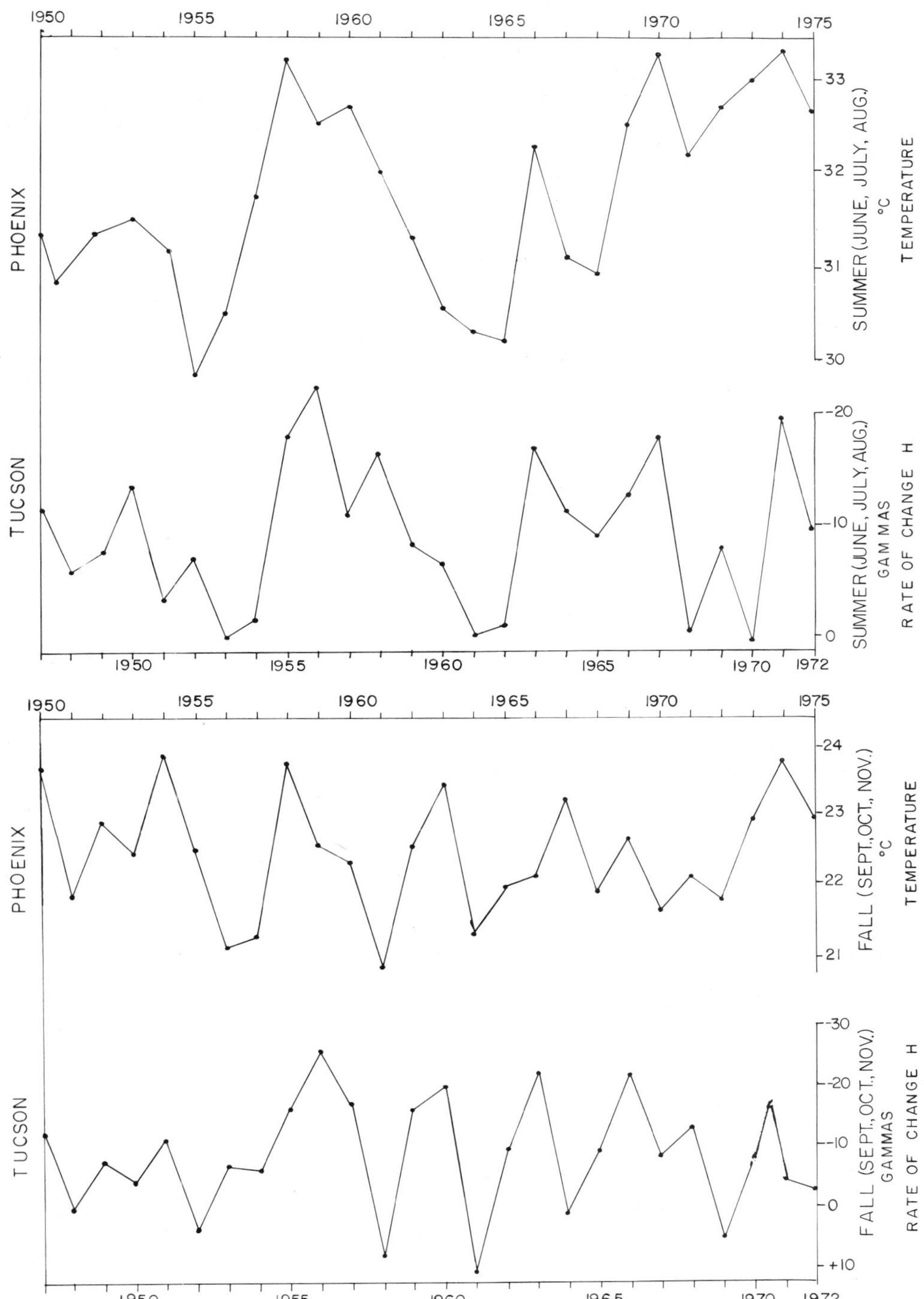

Figure 13-3. Seasonal geomagnetic intensity *(H)* rate-of-change values from Tucson compared with seasonally averaged surface-air temperatures from Phoenix. The curves were plotted so that the season value for one year is followed by the value for the same season of the next year. Note the 3-yr displacement of the time scales.

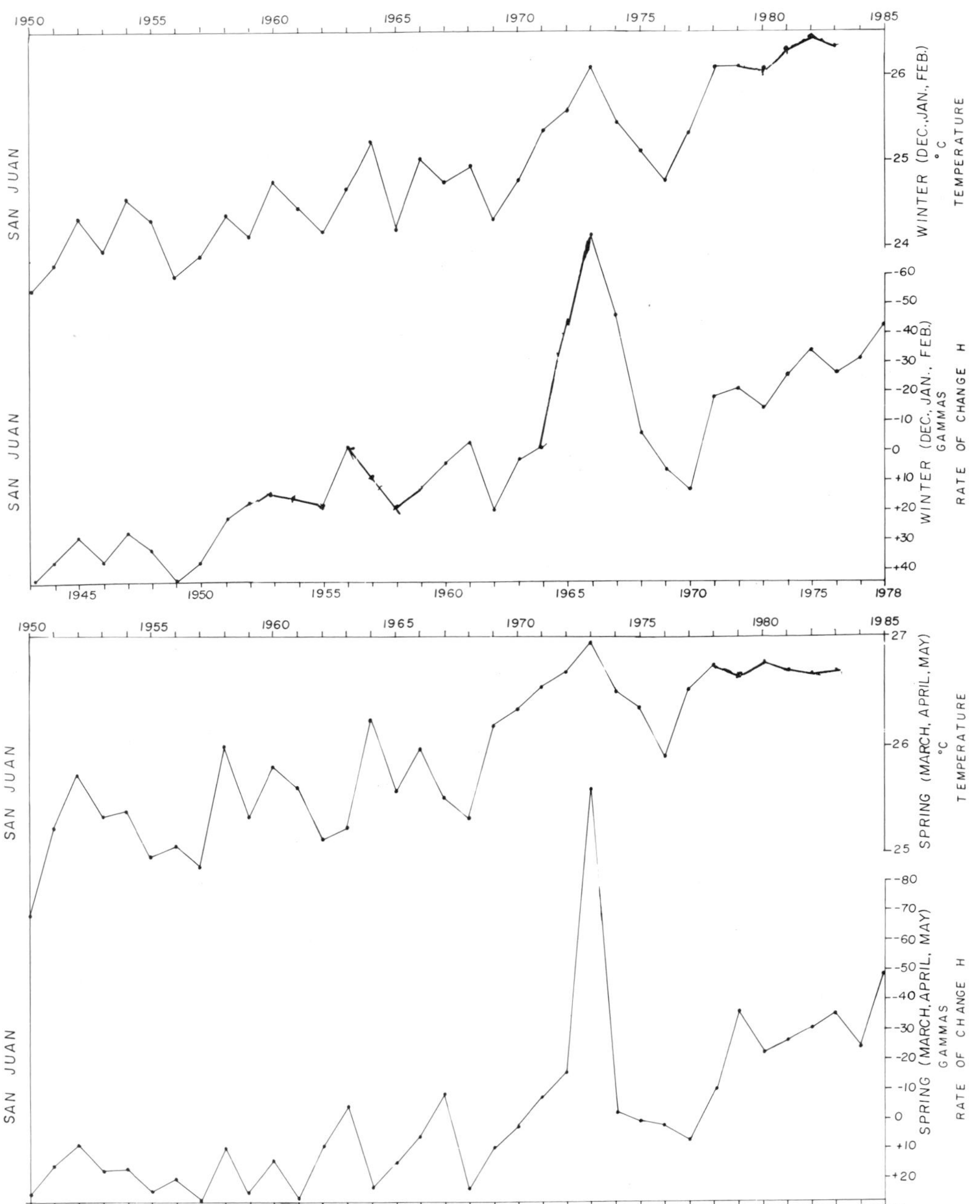

Figure 13-4. Seasonal geomagnetic intensity *(H)* rate-of-change values from San Juan Observatory compared with seasonally averaged surface-air temperatures from a San Juan weather station. The curves were plotted so that the season value for one year is followed by the value for the same season of the next year. Note the 7-yr displacement of the time scales.

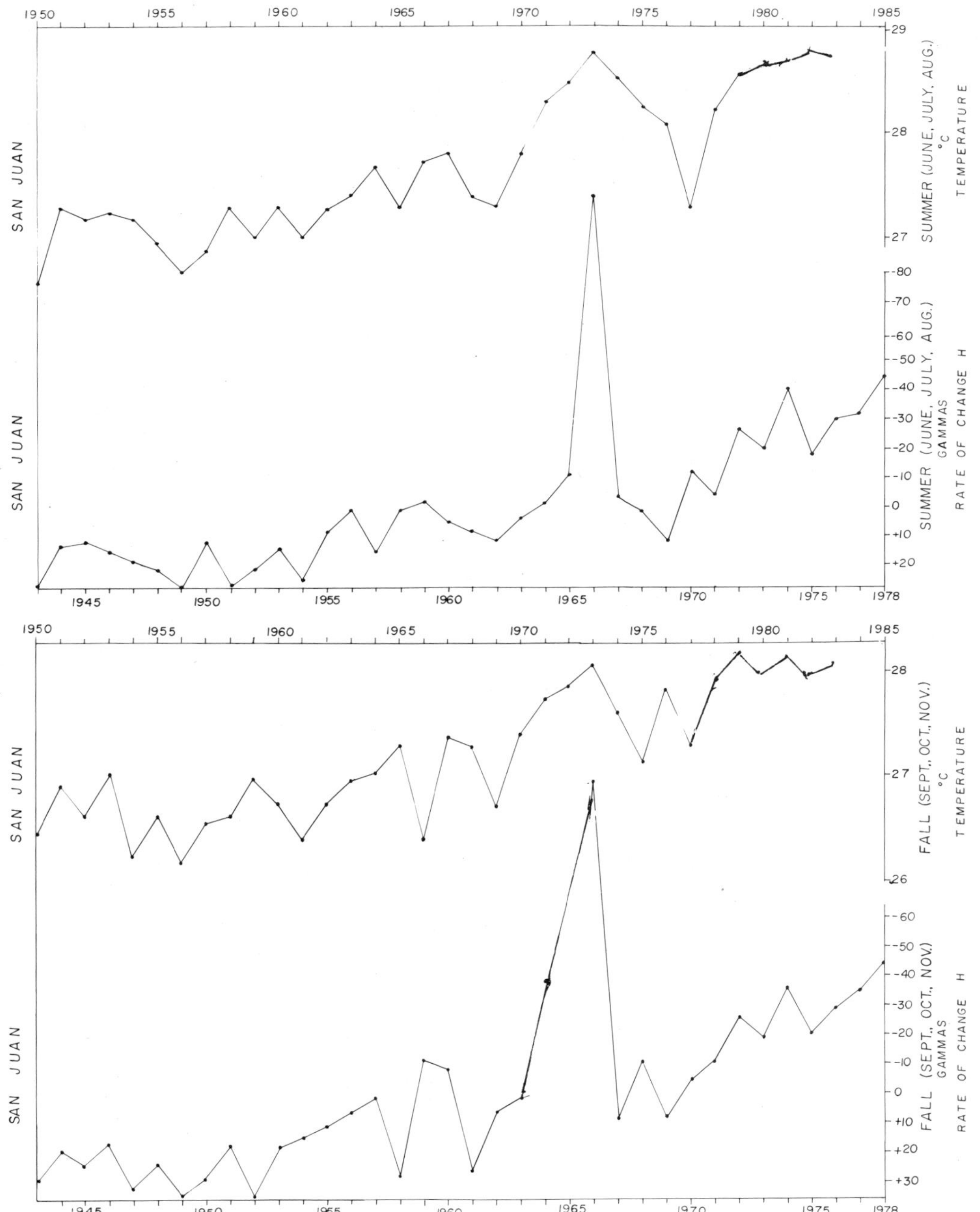
SAN JUAN
SUMMER (JUNE, JULY, AUG.)
°C
TEMPERATURE
SAN JUAN
SUMMER (JUNE, JULY, AUG.)
GAMMAS
RATE OF CHANGE H
SAN JUAN
FALL (SEPT., OCT., NOV.)
°C
TEMPERATURE
SAN JUAN
FALL (SEPT., OCT., NOV.)
GAMMAS
RATE OF CHANGE H
1950 1955 1960 1965 1970 1975 1980 1985
1945 1950 1955 1960 1965 1970 1975 1978
29 28 27
-80 -70 -60 -50 -40 -30 -20 -10 0 +10 +20
28 27 26
-60 -50 -40 -30 -20 -10 0 +10 +20 +30

Table 13-1. Correlation Coefficients for Various Shifts of Time Scales

		Lag (yr)										
Kakioka	*Tokyo*	*−2*	*−1*	*0*	*1*	*2*	*3*	*4*	*5*	*6*	*7*	*8*
Winter	Winter	0.04	−0.38	0.16	−0.19	−0.26	−0.64	−0.12	−0.04	−0.03	0.09	0.09
Spring	Spring	−0.02	−0.06	−0.07	0.04	−0.11	−0.31	−0.24	−0.02	0.19	−0.20	−0.03
Summer	Summer	0.06	0.13	0	0.32	0.17	−0.52	−0.19	−0.20	−0.19	−0.01	−0.31
Fall	Fall	0.42	0.04	0.01	−0.16	−0.34	−0.61	−0.27	−0.02	−0.12	0.09	0.05

Note: The correlation coefficients of the lags chosen by visually matching each pair of geomagnetic and temperature curves shown in Figure 13-2 are underlined.

in the area around Puerto Rico and Bermuda where we predict a trend toward warmer temperatures. We predict that the cooling trend in the Northern Hemisphere since 1950 will continue, with shorter-term warm fluctuations superposed, as the one between 1979 and 1984, for some decades further and that the next ice age will come within 1,000-2,000 yr.

CYCLIC GEOPHYSICAL PARAMETERS CORRELATED WITH GEOMAGNETIC VARIATIONS

Although we have shown that surface-air temperature changes follow variations in the rate of change of geomagnetic horizontal intensity, we do not assume that geomagnetism is the basic cause of weather. The basic cause may be changes in planetary configurations as have been suggested, for example, by Johnson (1946), Gribbin (1973), Zhenqiu and Zhisen (1980), and Fairbridge (1984). An important cause of the changes in weather, we assume, is variations of the energy received by the Earth from the Sun in the form of light emission by the hot photosphere. We suggest that the geomagnetic effects may be connected with important modulations of the main solar energy output. Some of the parameters that may move cyclically with the geomagnetic effects are presented here.

To represent the geomagnetic effects we chose the Geomagnetic Activity Recurrence Index developed by Sargent (1979). The data used in developing the index were the aa-index values of Mayaud (1973), the rotation charts of geomagnetic activity developed by Bartels (1932), solar-flare data, coronal-hole maps, and interplanetary scintillation data. The recurrence index is an indicator of geomagnetic stability, and it has an 11-yr cycle of opposite phase to that of sunspot numbers. A high recurrence index indicates that the geomagnetic activity is being caused by recurrent solar-wind streams and gives a crude estimate of how dominant the solar-wind-stream structure is at the time of interest. The recurrence index is a good measure of the level of geomagnetic activity that would be induced in a planetary magnetosphere in the outer solar system (i.e., beyond about 3 AU). The top smoothed curve in Figure 13-5 is from the Geomagnetic Activity Recurrence Index.

To illustrate more clearly the correlations among the twelve parameters in the figure, we plotted increases downward in the geomagnetic recurrence index, cosmic-ray flux, NO molecules, and ozone trend and displaced the time scales so that the figure shows a 6-mo lag of cosmic rays behind sunspots, a 1-yr lag of ozone behind NO molecules, and a 1.5-yr lag of the 100-500-mb pressure field behind sea-surface temperatures.

As can be seen in Figure 13-5, all parameters have cyclic variations of about 11 yr. It is of interest to note that in the figure, a decrease in the geomagnetic recurrence index is followed by an increase in surface-air temperatures, the same anticorrelation that is shown in Figures 13-1 through 13-4.

The deposition rates of cosmic rays can be related to variations in geomagnetism and sunspot activity. High sunspot activity results in solar wind's deflecting away cosmic rays. Surface-air temperature changes are directly or indirectly associated with variations in geomagnetism and sunspot activity. Thus, as we found (see Fig. 13-5), cosmic-ray variations display direct correlation with the geomagnetic recurrence index and anticorrelation with the surface-air temperatures.

Reid et al. (1976) have suggested that a reduced geomagnetic field correlates with an increase in solar-proton events and a decrease in ozone; Alldredge (1976) has reported that a decrease in annual means of geomagnetic horizontal intensity correlates with an increase in sunspot activity; Suess and Lockwood (1980) have shown that a decrease in the geomagnetic recurrence index correlates with an increase in sunspot activity and an increase in solar-radio flux at 10.7 cm; and Namias (1973) found that an increase in sea-surface temperatures correlates with an increase in the 100-500-mb pressure field—the same correlations that we found in Figure 13-5.

As can be seen in Figure 13-5, the solar-

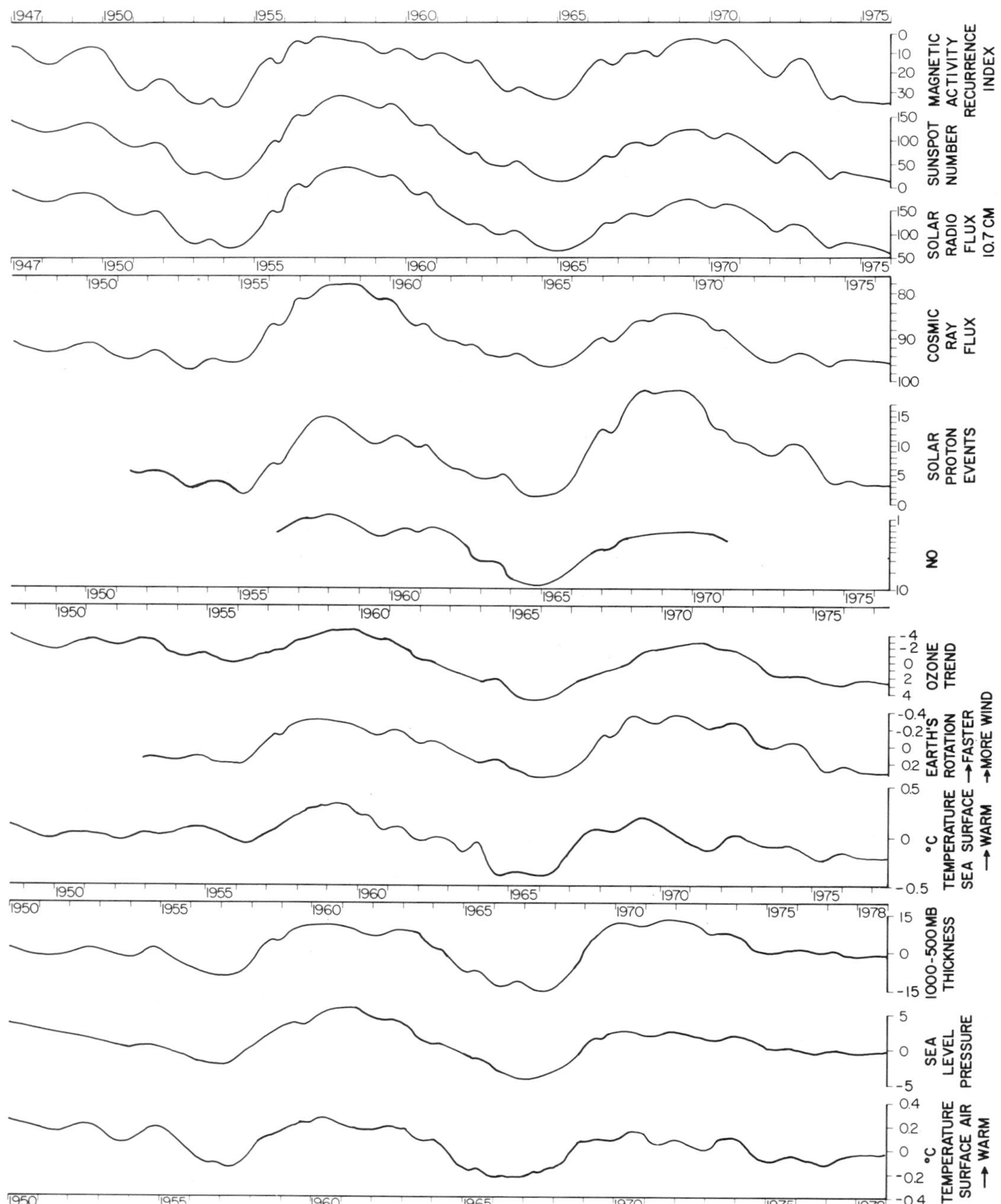

Figure 13-5. Cyclic solar and geophysical parameters correlated with geomagnetic variations.

Shown from the top are Geomagnetic Recurrence Index (after Sargent, 1979), sunspot numbers by Waldmeier (1975, 1979), solar-radio flux at 10.7 cm (after Covington, 1977), cosmic-ray flux (after Lockwood and Webber, 1981), solar-proton events (Shea and Smart, 1977; Kahler, 1982), NO molecules (after Ruderman and Chamberlain, 1975), ozone trend for the Northern Hemisphere (Christie, 1973; DeLuisi, 1982), the Earth's rotation rate (after Lambeck, 1980), sea-surface temperatures in the north Pacific (after Namias, 1978), variation of the thickness of the 1,000-to-500-mb pressure field for the Northern Hemisphere region from 25°N to the pole (after Boer and Higuchi, 1981), principal-component analysis of mean sea-level pressure for the North Atlantic-European sector of the Northern Hemisphere (after Kelly, 1977), and surface-air temperatures for the Northern Hemisphere (from Angell and Korshover, 1977; Jones, Wigley, and Kelly, 1982). As can be seen by the vertical scales, increases in some of the parameters are plotted downward. Note the three displacements of the time scales.

proton events curve and the NO curve display negative correlation. According to Crutzen, Isaksen, and Reid (1975), the correlation between those parameters should be positive because solar-proton events are the main stratospheric sources of NO. The NO curve in Figure 13-5 is from Ruderman and Chamberlain (1975), and its variations are assumed to be caused by the presence of nuclear test explosion products in the upper atmosphere. A product of nuclear test explosions is fractions of N atoms in the excited states, and according to Crutzen et al., these fractions react with NO

$$N + NO \rightarrow N_2 + O,$$

leading to a net loss of NO. Thus, it is the presence of fractions of N atoms in the excited states, produced by nuclear test explosions, that results in the negative correlation between the solar-proton events and the NO curve in Figure 13-5.

TIME LAGS

The 6-mo lag of cosmic rays behind sunspots has been illustrated by Quenby (1967). Rein Silberberg (pers. comm.) has suggested that the reason for the lag may be as follows:

The region of solar modulation of cosmic rays extends to about 50-70 AU (about 9×10^9 km), and the solar disturbances propagate outward at the speed of the solar wind (about 470 km/sec. Thus:

$$\begin{aligned}\text{Time lag} &= \frac{\text{distance to modulation boundary}}{\text{speed of solar wind}}\\ &= 9 \times 10^9/470\\ &= 1.9 \times 10^7 \text{ sec}\\ &\approx 7 \pm 2 \text{ mo}\end{aligned}$$

The time lag of 1 yr of ozone behind NO molecules is thought to be caused by the fact that once the destructive NO molecules have been introduced into the ozone layer, it takes at least 1 yr before they can be removed by mixing down to the ground (Prinn, 1980).

We suggest that the time lag of 1.5 yr of the 1,000-to-500-mb pressure field behind sea-surface temperatures may result from the relatively slow and sluggish motions of ocean circulation and the ocean's great heat capacity and areal extension. For example, Namias (1970) has reported that the upper layer of the north Pacific Ocean "remembers" its initial thermal state for up to 2 yr.

CONCLUSION

The evidence presented in this chapter suggests that the correlation between geomagnetism and weather may be real and not simply a coincidence and that, if further investigated, may be exploited in long-range weather forecasting. The conclusions are as follows:

1. Numerous solar and geophysical factors display cyclic variations.
2. One or more causal mechanisms may underlie these striking variations.
3. Even in the absence of an understanding of the causes, the measurable variations in the rate of change of geomagnetic horizontal intensity seem to offer a powerful tool for predicting temperature trends 3-7 yr in advance.

At this stage, however, the most important conclusion that emerged from our findings is that more research in this field needs to be done.

REFERENCES

Akasofu, S.-I., and Chapman, S., 1972, *Solar-Terrestrial Physics*, Oxford: Clarendon Press, 901p.

Alldredge, L. R., 1975, A hypothesis for the source of impulses in geomagnetic secular variation, *Jour. Geophys. Research* **80:**1571-1578.

Alldredge, L. R., 1976, Effects of solar activity on annual means of geomagnetic components, *Jour. Geophys. Research* **81:**2990-2996.

Angell, J. K., and Korshover, J., 1977, Estimate of the global change in temperature, surface to 100 mb, between 1958 and 1975, *Monthly Weather Rev.* **105:**375-385.

Bartels, J., 1932, Terrestrial-magnetic activity and its relations to solar phenomena, *Terr. Magnetism and Atmos. Electricity* **37:**1-52.

Bigelow, F. H., 1898, *U.S. Dept. Agriculture Weather Bur. Bull. 21.*

Boer, G. J., and Higuchi, K., 1981, Seasonal climatic variability, *Atmosphere-Ocean* **19:**90-102.

Brown, G. M., 1974, A new solar-terrestrial relationship, *Nature* **251:**592-594.

Christie, A., 1973, Secular or cyclic change in ozone, *Pure and Appl. Geophysics* **106–108:**1000-1009.

Covington, A. E., 1977, A working collection of daily 2800 MHz solar flux values 1946-1976, Rept. No. ARO-5, Ottawa: National Research Council of Canada, 86p.

Crutzen, P. J.; Isaksen, I. S. A.; and Reid, G. C., 1975, Solar proton events: Stratospheric sources of nitric oxide, *Science* **189:**457-459.

DeLuisi, J. J., ed., 1982, *Geophysical Monitoring for Climatic Change No. 9,* Washington, D.C.: U.S. Government Printing Office, p. 28.

Egyed, L., 1961, Temperature and magnetic field: *New York Acad. Sci. Annals* **95:**72-77.

Fairbridge, R. W., 1984, Planetary periodicities and terrestrial climate stress, in N.-A. Mörner and W. Karlén (eds.), *Climatic Changes on a Yearly to Millennial Basis,* Dordrecht, The Netherlands: D. Reidel Publishing Co., pp. 509-520.

Frazier, K., 1982, *Our Turbulent Sun,* Englewood Cliffs, N.J.: Prentice-Hall, 198p.

Gribbin, J., 1973, Planetary alignments, solar activity and climatic change, *Nature* **246:**453-454.

Herman, J. R., and Goldberg, R. A., 1979, *Sun, Weather, and Climate,* Washington, D.C.: National Aeronautics and Space Administration, 360p.

Johnson, M. O., 1946, *Correlation of Cycles in Weather, Solar Activity, Geomagnetic Values, and Planetary Configurations,* San Francisco: Phillips and Van Orden Co., 149p.

Jones, P. D.; Wigley, T. M. L.; and Kelly, P. M., 1982, Variations in surface air temperatures: Part 1. Northern Hemisphere, 1881-1980, *Monthly Weather Rev.* **110:**59-70.

Kahler, S. W., 1982, The role of the big flare syndrome in correlations of solar energetic proton fluxes and associated microwave burst parameters, *Jour. Geophys. Research* **87:**3439-3448.

Kelly, P. M., 1977, Solar influence on North Atlantic mean sea level pressure, *Nature* **269:**320-322.

King, J. W., 1975, Sun-weather relationships, *Aeronautics and Astronautics* **13:**10-19.

Lamb, H. H., 1977, *Climate: Present, Past and Future,* vol. 2: *Climatic History and the Future,* London: Methuen, 835p.

Lamb, H. H., 1982, *Climate, History and the Modern World,* London: Methuen, 387p.

Lambeck, K., 1980, *The Earth's Variable Rotation,* London: Cambridge University Press, 79p.

Lockwood, J. A., and Webber, W. R., 1981, A study of the long-term variations and radial gradient of cosmic rays out to 22AU, in *Conf. Papers, 17th Internat. Cosmic Ray Conf.* **3:**259-262.

Mayaud, P. N., 1973, A hundred year series of geomagnetic data 1868-1967, *IAGA Bull. No. 33,* The Netherlands: Moppel.

Namias, J., 1970, Macroscale variations in the sea surface temperatures in the North Pacific, *Jour. Geophys. Research* **75:**565-582.

Namias, J., 1973, Thermal communication between the sea surface and the lower troposphere, *Jour. Phys. Oceanography* **3:**373-378.

Namias, J., 1978, Multiple causes of the North American abnormal winter 1976-77, *Monthly Weather Rev.* **106:**279-295.

Nelson, J. H.; Hurwitz, L.; and Knapp, D. G., 1962, *Magnetism of the Earth,* Washington, D.C.: U.S. Government Printing Office, 79p.

Pittock, A. B., 1978, A critical look at long-term sun-weather relationships, *Rev. Geophysics and Space Physics* **16:**400-420.

Pittock, A. B., 1980, Enigmatic variations, *Nature,* **283:**605-606.

Prinn, R. G., 1980, Chemical cycles and circulation of stratospheric ozone, in S. P. Parker (ed.), *McGraw-Hill Encyclopedia of Ocean and Atmospheric Sciences,* New York: McGraw-Hill, pp. 43, 50.

Prohaska, J. T., and Willett, H. C., 1981, The application of eigenanalysis to the study of solar-climatic relationships, Cambridge, Mass.: Solar Climatic Research Institute, 54p.

Quenby, J. J., 1967, Relationship between cosmic rays and sunspots, in S. Flügge (ed.), *Handbuch der Pysik,* vol. XVI/2, Berlin: Springer Verlag, pp. 310-371.

Reid, G. C.; Isaksen, I. S. A.; Holzer, T. E.; and Crutzen, P. J., 1976, Influence of ancient solar-proton events on the evolution of life, *Nature* **259:**177-179.

Roberts, W. O., 1979, Introductory review of solar-terrestrial weather and climate relationships, in B. M. McCormac and T. Seliga (eds.), *Solar-Terrestrial Influences on Weather and Climate,* Dordrecht, The Netherlands: D. Reidel Publishing Co., pp. 29-40.

Ruderman, M. A., and Chamberlain, J. W., 1975, Origin of the sunspot modulation of ozone: Its implications for stratospheric NO injection, *Planetary and Space Sci.* **23:**247-268.

Sargent, H. H., III, 1979, A geomagnetic activity recurrence index, in B. M. McCormac and T. Seliga (eds.), *Solar-Terrestrial Influences on Weather and Climate,* Dordrecht, The Netherlands: D. Reidel Publishing Co., pp. 101-104.

Shea, M. A., and Smart, D. F., 1977, Significant solar proton events 1955-1969, in A. H. Shapley and H. W. Kroel (compilers), *Solar-Terrestrial Physics and Meteorology,* SCOSTEP Working Document 2, Washington, D.C.: National Academy of Sciences, pp. 119-134.

Spruit, H. C., and Roberts, B., 1983, Magnetic flux tubes on the sun, *Nature* **304:**401-406.

Sternberg, R. S., and Damon, P. E., 1979, Re-evaluation of possible historical relationship between magnetic intensity and climate, *Nature* **278:**36-38.

Suess, S. T., and Lockwood, G. W., 1980, Correlated variations of planetary albedos and coincident solar-interplanetary variations, *Solar Physics* **68:**393-409.

Thompson, P. H., 1983, A history of numerical weather prediction in the United States, *Am. Meteorol. Soc. Bull.* **64:**755-769.

Waldemeier, M., 1975, Annual mean sunspot numbers 1700-1974, in A. H. Shapley, H. W. Kroel, and J. H. Allen (compilers), *Solar-Terrestrial Physics and Meteorology,* SCOSTEP Working Document, Washington, D.C.: National Academy of Sciences, pp. 106-107.

Waldmeier, M., 1979, Monthly sunspot numbers 1964-1978, in A. H. Shapley, C. D. Ellyett, and H. W. Kroel (compilers), *Solar-Terrestrial Physics and Meteorology,* SCOSTEP Working Document 3, Washington, D.C.: National Academy of Sciences, p. 92.

Wilcox, J. M., 1975, Solar activity and the weather, *Jour. Atmos. and Terrest. Physics* **37:**237-256.

Wollin, G.; Kukla, G. J.; Ericson, D. B.; Ryan, W. B. F.; and Wollin, J., 1973, Magnetic intensity and climatic changes, 1925-1970, *Nature* **242:**34-37.

Wollin, G; Ryan, W. B. F.; and Ericson, D. B., 1981, Relationship between annual variations in the rate of change of magnetic intensity and those of surface air temperature, *Jour. Geomagnetism and Geoelectricity* **33:**545-567.

Zhenqiu, R., and Zhisen, L., 1980, Effects of motions of planets on climatic changes in China, *Kexue Tongbao* **25:**417-422.

14: Short-Term Paleoclimatic Changes: Observational Data and a Novel Causation Model

Nils-Axel Mörner
Stockholm University and
Grönby Independent
Research Center, Sweden

Abstract: Major short-term climatic changes during the past 20,000 yr are found to have had a duration of about 50-150 yr and to have had a geographical extent that is regional or even hemispherical, and sometimes only local, but never global (as previously assumed). This finding opens new possibilities of discriminating between different possible causation mechanisms. A rotational-gravitational-oceanographic model is proposed, where the ultimate driving force is a planetary beat. This model reveals a basic simplicity behind the general complexity and multiple interaction.

INTRODUCTION

Climate changes and shifts in a broad spectrum of cycles and time ranges and the effects of these changes may be locally, regionally, hemispherically, or globally induced (see Mörner, 1984*a*, Fig. 1). Since the early 1970s, paleoclimatology has concentrated largely on the so-called Milankovitch variables (e.g., Berger et al., 1984), while contemporary climatology has been focused on the so-called weather machine. The climatic and paleoclimatic changes of intermediate length or cyclicity have, because of this, largely been ignored. At the Second Nordic Symposium on Climatic Changes and Related Problems, held in Stockholm in 1983, emphasis was placed on climatic changes and shifts that take place on a yearly to millennial basis (Mörner and Karlén, 1984). We especially analyzed the amplitude, duration, and global significance of the major short-term climatic changes during the past 20,000-30,000 yr (i.e., the period for which there is a reasonably firm chronology allowing regional and global comparisons) and their relation to other geophysical variables. This led to the formulation of a new theory for the origin and character of those changes (Mörner, 1984*b*, 1984*c*), which is discussed further in this chapter.

THE LATE CENOZOIC

The Late Cenozoic is characterized by a steep temperature gradient between the equatorial and polar regions (in contrast to the Late Cretaceous and the Early Cenozoic, when the temperature gradient was small and the polar regions had a boreal type of climate). This contrast is a combined function of the Earth's distribution of land and sea, its rate of rotation, and the circulations in the atmosphere and oceans. Consequently, global and regional changes in weather and climate are strongly dependent on these variables (in the long-term as well as the short-term ranges). This temperature gradient is also the background to the glacial/interglacial alternations that have set the character of the major climatic changes during the past 2.5 m.y. (cf. Mörner, 1978*a*).

MAJOR SHORT-TERM CLIMATIC CHANGES

During the past 20,000 years or so, there have been several major climatic changes of short duration. Southern Scandinavia has been a classic area for the identification and recording of these changes (Mörner, 1984*a;* Berglund and Mörner, 1984). We have, therefore, critically analyzed both the available records and their geographical significance (Mörner and Karlén, 1984).

Regionality

Mörner (1984*c*) found the following regarding the geographical extent and validity of major short-term climatic changes:

> No events of strictly global significance;
> Several events of opposite, compensational, character in the two hemispheres (A.D. 1840-1940, A.D. 1940-1980, and 10,000 B.P.) or in different regions (13,000 B.P., 11,000 B.P., and 2,500 B.P.;
> Events that were only locally induced (Older Dryas, Little Ice Ages).

This means that these major short-term climatic changes represent redistributions of heat over the globe and not general rises and falls as previously assumed (e.g., Flohn, 1979). Consequently, we must talk about regionality instead of globality. This revision implies a drastic change of older concepts. It agrees well, however, with the recent change of the old concept of eustasy from having global validity to regional validity (Mörner, 1976*a*, 1980*a*, 1983*a*). Mörner (1984*c*) claimed that the novel idea of a nonglobality of major short-term changes was based on a very careful analysis. Regionality instead of globality is of fundamental importance because it calls for quite different causative mechanisms (Mörner, 1984*c*, Table 1).

Time Range and Amplitudes

The short-term climatic oscillations analyzed represent changes covering time periods of 50-150 yr. The amplitudes, however, range from fractions of 1°C (the 1940-1980 pattern) to some 10°C. Special attention was drawn to (cf. Mörner, 1984*c*): (1) the 0.5°C event 1840-1940; (2) the about 1°C events of the Little Ice Ages; (3) the 2.5°C event at about 2,500 B.P. (4) the 5°C (or greater) event at about 11,000 B.P. (the onset of the Younger Dryas Stadial); and (5) the 10°C (or greater) event of isotopic substage 5d (and 5b), which seems to have been a very short (but severe) cooling event.

Mörner (1984*c*) stressed that these changes were very similar, representing some sort of family of curves; that the differences merely lay in the amplitude; and that they, therefore, might be expected to have had a similar origin.

Characteristics of the Changes

Mörner (1984*c*) found that none of the rapid climatic changes and shifts had a global extension, but only a hemispherical or regional extension of a type where the change is counter-balanced by compensational changes of opposite sign in other regions. The implication of this observation can be summarized in the following five points:

1. The climatic changes analyzed cannot originate from global rises and falls in temperature but must represent redistributions of heat over the globe—regionally or hemispherically and sometimes even locally.
2. The duration of these redistributions of heat over the globe is in all cases about 50-150 yr.
3. Redistributions of heat, including storing and transformation of heat, which last for 50-150 yr, seem to be possible only via ocean circulation changes.
4. Ocean circulation changes can be driven by several different mechanisms. The duration seems to preclude a wind-driven mechanism (cf. Lamb, 1979) and suggests a combined rotational-gravitational-oceanographic origin such as proposed by Mörner (1984*b*, 1985, 1987*a*).
5. The time (i.e., the duration) and space (i.e., the

geographical extent factors offer the possibility of discriminating among different possible causative mechanisms (see Mörner, 1984*c*, Table 1).

Correlations and Combination with Other Geophysical Variables

Climatic changes cannot be understood in isolation from other related geophysical changes. I have tried to demonstrate in several papers that a complex interaction exists between different geophysical variables (Mörner, 1978*a*, 1980*a*, 1981*a*, 1984*b*, 1984*d*, 1987*a*). There is, for example, a correlation between paleoclimate, paleomagnetism, and paleogeoid changes that may be taken to indicate that internal processes play a significant role, even in paleoclimatology (e.g., Mörner, 1984*b*, Figs. 2-4, 1986*a*).

Long-Term Phenomena. There is an interesting correlation between the main archeomagnetic cycle of the past 9,000 yr, the atmospheric ^{14}C production cycle, and the paleogeoid cycle as determined from global (eustatic) sea-level differences (Mörner, 1978*a*, 1980*a*, 1984*b*, Fig. 4). The weakening of the dipole moment and corresponding deformation of the toroidal field (Yukutake, 1972) at around 8,000 B.P. and 2,500 B.P. must have caused a corresponding speeding up of the mantle (and crust) motion with respect to the core. This change would create various forms of stresses in the mantle and lithosphere and might generate deformations of the phase-transitional boundaries, giving rise to both geoid deformations and geodynamic effects (Mörner, 1984*b*, Figs. 8-10, 1986*a*). It seems more likely, however, that it was rotational changes caused by the rapid redistribution of water masses after the last continental glaciation that triggered the other changes. Any change in the rate of rotation may significantly affect the Earth's weather and climate because of its effects on the atmospheric angular momentum and the ocean circulation (i.e., angular momentum).

Short-Term Phenomena. In southern Scandinavia, it has been possible to analyze in detail and by integration the records of sea-level changes, temperature, and paleomagnetism. In total, 16 low-amplitude oscillations over the past 10,000 yr were identified in all three records (Mörner, 1978*a*, 1980*a*, 1984*b*, Fig. 6). The sea-level changes represent a regional eustasy that has been found to apply for the whole of northwestern Europe and the northeastern Atlantic (Mörner, 1979*a*). This record had earlier (Mörner, 1973) been shown to exhibit a remarkably detailed correlation with climatic warm/cold changes in high-sedimentation-rate cores in the Atlantic (Wollin, Ericson, and Ewing, 1971). New paleoceanographic records from the Denmark Strait west of Iceland (Kellogg, 1984) include clear peaks and troughs that seem to be more or less identical to the southern Scandinavian records (Mörner, 1984*b*, Fig. 6). Furthermore, Colquhoun et al. (1983) have claimed that the South Carolina sea-level and human habitation records show a correlation between sea-level and climate and that these fluctuations correlate with those in Europe (a statement questioned, however, by Mörner, 1981*b*). Because the recorded changes are apparently not global in extent, there can be only one explanation to the correlations established between the records from southern Scandinavia, the Atlantic, the sea west of Iceland, and (not so clearly) South Carolina: namely, changes in the Gulf Stream activity as proposed by Mörner (1984*d*, Figs. 11 and 12, 1987*a*).

Lamb (1979) was able to demonstrate that a delicate balance exists between cold Arctic water and warm Atlantic water in the areas around Iceland and that sea surface temperature is liable to fall by 3 to 5°C (or even more) in times as short as years and decades. Lamb showed that this was indeed the case during the Little Ice Ages of Europe.

Taira (1981) found a high degree of agreement between the sea-level changes in Japan and those in northwestern Europe. He also found a correlation between sea-level changes and seismotectonic events. Wollin and co-workers (1980) showed that there is a high correlation (with a 3-yr lag) between changes in the north Pacific sea-surface temperature near Japan and seasonal magnetic intensity changes in Japan, Alaska, and Arizona.

The regional eustatic changes in Brazil are regarded by the writer as quite different from those in Europe (Mörner, 1981*a*, 1983*a*). During the past 5,000 yr, there were two distinct high-amplitude regressions (an earlier low-amplitude regression may be seen at around 6,900 B.P.). The climatic changes recorded in the Amazonas (Servant et al., 1981) show a similar sequence of three humidity/aridity cycles. The coastal regressions seem to correspond to drastic aridity phases with heavy deforestation inland. The drastic regressions in the regional eustatic curve of Brazil may represent major geoid changes (with corresponding inland effects via lowering of the groundwater as proposed by Mörner, 1978*b*) or possibly anomalously large deformations of the dynamic sea level (with corresponding meteorological effects farther inland). It is of great interest that the Brazilian regressions appear to be directly counterbalanced by transgressions in Europe (Mörner, 1981*a*, Fig. 2).

The repeated appearances of the El Niño/Southern Oscillation anomalies in the central and eastern Pacific have had drastic meteorological, oceanographic, and biological effects (Cane, 1983; Barber and Chavez, 1983; Philander, 1983; Newell and Hsiung, 1984; Rasmussen and Wallace, 1983). The El Niño events have a strong effect on upwelling along the Peruvian coast and, therefore, affect the biological productivity and consequently the CO_2 content of the atmosphere (Newell and Hsiung, 1984). During the past 2,000 yr, Stuiver et al. (in Bojkov, 1983, Fig. 5) have recorded several rapid high-amplitude changes in atmospheric CO_2. By analogy with the present-day El Niño events, one may expect that these CO_2 peaks represent periods with drastically decreased upwelling (and increased downwelling)—that is, some sort of super El Niño lasting for some decades up to a century (i.e., some 50-150 yr as is the case with the major short-term climatic changes). Rosen et al. (1984) and Eubanks, Dickey, and Steppe (1984) showed that the 1982-1983 El Niño event is visible in the atmospheric angular momentum (a speeding up) and in the slowing down of rotation of the solid Earth (and by solid Earth we include the liquid outer core). Some of the interchange of momentum must have gone to the oceans and may have been responsible for the circulation changes recorded (Mörner, 1987*a*). With more momentum transferred to the hydrosphere, larger and longer-lasting effects of the El Niño type can easily be achieved (Mörner, 1985, 1987*a*).

A NEW ROTATIONAL-GRAVITATIONAL-OCEANOGRAPHIC MODEL

The discovery of a nonglobality and a 50-to-150-yr duration of the major short-term climatic changes (Mörner, 1984*b*) made it possible to discriminate among different possible causative mechanisms and to propose a new model for those changes (Mörner, 1984*b*, Figs. 8-13; 1984*c*, Fig. 7).

The apparent nonglobality speaks against solar sources, interplanetary shielding, and atmospheric shielding (gases, vapor, dust). The relatively long duration speaks against atmospheric changes.

Redistribution and storing of heat in the oceans have the right "memory" to concur with a 50-to-150-yr duration and the right nonglobal distribution to concur with the regional, hemispherical, and sometimes local character of the major short-term climatic changes.

Short-term changes in the ocean circulation can be brought about by rotational changes (i.e., angular momentum), geoidal deformations of the sea level and the internal ocean stratification (rapid glacio-eustatic changes could have some effect), and changes in the atmospheric circulation (i.e., the wind pattern).

Rotation

Changes in the Earth's radius (e.g., sea level), mass distribution (glaciations, plate tectonics, etc.), electromagnetic coupling between the core and the mantle, hydrosphere/lithosphere coupling (friction, topography, land/sea distribution, atmosphere/hydrosphere interaction (surface friction, mountain pressure), and planetary orbits including the Earth/Sun orbital relations and the Earth/Moon distance will all affect the Earth's rate of rotation. Because the Earth is not a solid body but includes the atmosphere, the hydrosphere, the low-viscosity asthenosphere, and the fluid outer core, there can be significant differential changes in rotation (Fig. 14-1), though the total momentum remains constant (except for the case of radius changes and possible interplanetary exchange of momentum).

Changes in the angular momentum of the atmosphere show a high degree of correlation with changes in the rate of rotation of the solid Earth for annual and shorter frequency signals (Hide et al., 1980; Barnes et al., 1983; Rosen and Salstein, 1983; Eubanks et al., 1983), and indicate that the atmosphere plays a dominant role for the changes in the length of the day (LOD) on the order of a year and shorter periods. Rosen and Salstein (1983, Fig. 6) showed that the seasonal changes were caused predominantly by the variability of the major jet streams. They found that the nonseasonal signal originated from regional imbalances in angular momentum in the regions 10-25°S and 20-35°N.

The 1982-1983 El Niño event (Rosen et al., 1984; Eubanks, Dickey, and Steppe, 1984) represents an interannual change in the Earth's rate of rotation (slowing) and in the atmospheric angular momentum (speeding). Eubanks et al. (1983, Figs. 8 and 9) showed that there is a secular trend in the LOD record for 1976-1981 that is not compensated by atmospheric angular momentum changes. They suggested a core/mantle origin. A hydrospheric compensation is, however, equally possible (Mörner, 1985, 1987*a*).

Hide et al. (1980) as well as Lambeck (e.g., 1980) concluded that the decade variations in the LOD were compensated and caused by core/mantle dislocations. Lambeck (1980) estimated the maximum contribution to the observed variance from the atmosphere to be about 10-20%. Courtillot and co-workers (1978) found that there were significant geomagnetic changes at around 1910 and 1967 and that these events represented periods of major rotational changes (deceleration/acceleration shifts).

Lambeck and Casenave (1976) observed that, since 1820, there was a fairly good correlation between changes in the Earth's rate of

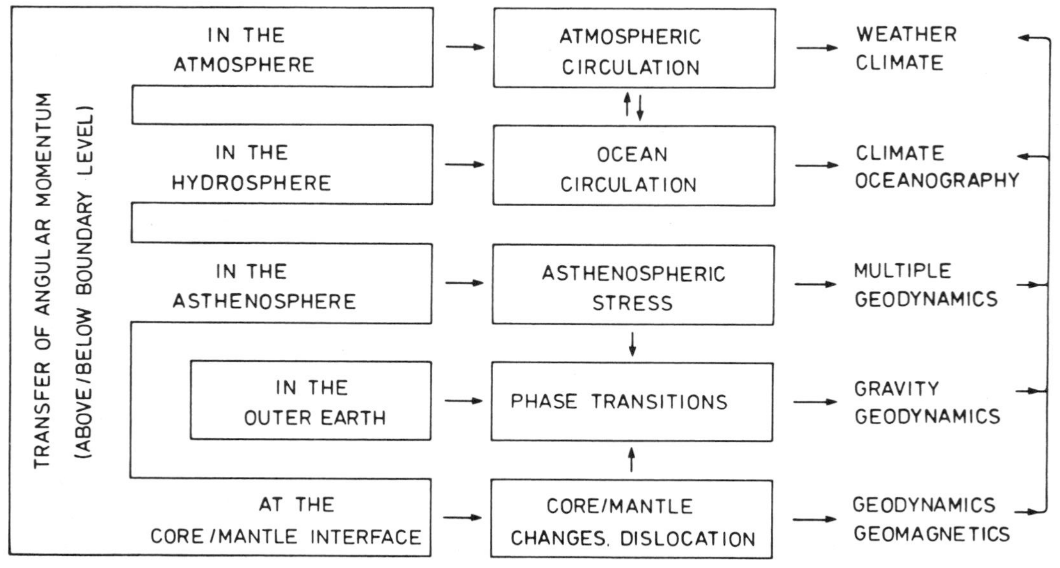

Figure 14-1. Differential rotation and multiple geophysical effects.

rotation and various climatic records, suggesting that acceleration periods correlated with periods of warming and increased zonal winds and that periods of deceleration correlated with periods of cooling and decreased zonal winds, despite the fact that the atmospheric compensation amounted to only some 10% of the total interchange of momentum.

The multiple effects of differential rotation—that is, the interchange of momentum between the Earth's solid, viscous, fluid, and gaseous layers—are illustrated in Figure 14-1 (cf. Mörner, 1984*b*, Fig. 10). The annual and intra-annual changes in rotation seem to be almost completely compensated and caused by the changes in atmospheric angular momentum. The interannual and decade changes in rotation may be dominated by core/mantle dislocations (and changes in the electromagnetic coupling), though effects at the other boundaries (see Fig. 14-1) seem also to have occurred.

The deceleration/acceleration changes at 1840, 1900, and 1967 were all possibly related to geomagnetic changes, indicating core/mantle dislocations. The 1900-1910 event was also associated with distinct geodynamic effects (Mörner, 1984*b*, p. 497), indicating asthenospheric and lithospheric stress changes. The 1875 and 1930 acceleration/deceleration events do not correlate with any major geomagnetic changes and, therefore, seem to have been compensated and caused by changes in the atmosphere and ocean circulation. The climatic and geodynamic effects related to the last century's LOD variations are further discussed elsewhere (Mörner, 1986*b*, 1987*a*).

Gravity (Geoid)

Through sea-level analyses, the writer has concluded that the Earth's geoidal surface has changed considerably and rapidly with time, in the short-term as well as the long-term range (Mörner, 1976*a*, 1980*a*, 1981*c*, 1983*a*, 1984*e*, 1987*b*; Newman et al., 1980; Newman, Marcus, and Pardi, 1981). The origin of geoid deformations is explained by various types of mass redistributions within the Earth, that is, for example, glacial volume changes, mass redistribution, phase transitions, partial melting changes, and core/mantle dislocations. Some of these changes (phase transitions and core/mantle dislocations) are found to occur very rapidly (Mörner, 1984*e*), which reveals the activity of a highly dynamic Earth. Changes in the Earth's rate of rotation and in the tilt of the rotational axis could affect the main geoidal rotational ellipsoid (Fairbridge, 1961; Mörner, 1983*a*, Table 5.1).

Deformation of the oceanic equipotential surface—namely, the geoid—will by necessity be accomplished by similar deformations of the continental geoid (i.e., the "geoidal groundwater level"; Mörner, 1982, 1987*c*), the interoceanic stratification, and various equipotential surfaces in the atmosphere (e.g., the tropopause and the 500-mb level) as shown by Mörner (1984*b*, Figs. 8 and 9). The atmospheric and hydrospheric changes will induce changes in weather and climate. The stability and altitude of the tropopause (cf. Gage and Reid, 1981; Paine, 1983) are of fundamental importance for the weather and climate experienced on the Earth's surface. The 500-mb level is another important level for the tropospheric circulation (e.g., King, 1974; Mörner, 1983*b*; Bucha, 1984). The importance of gravity-induced fluctuations in the mesosphere has been stressed by Ebel (1983). The potential vorticity (Ertel, 1942) in the lower stratosphere is of fundamental importance for the dynamics of the troposphere (Paine, 1983). Finally, geoid deformations and rotational changes are related in a reversible cause-effect relationship.

Oceanography

Lamb (1979) showed that the sea-surface temperature was able to change very rapidly and very drastically and that this ability had significant effects on weather and climate in the surrounding areas. He also showed (1979, Fig. 11) that the sea-surface temperature variations in the Atlantic during the past 200 yr exhibited significant latitudinal differences (including counterbalancing).

Circulation in the oceans takes much longer than circulation in the atmosphere. Consequently, the ocean changes last longer and have a longer memory. Therefore, we must seek the origin of the major short-term climatic changes (i.e., decades to century-long) in ocean circulation changes (Mörner, 1984*b*, 1984*c*, 1987*a*). Also, ocean circulation changes will give rise to a differential redistribution of heat just as was found in the case of the major short-term climatic changes (Mörner, 1984*c*).

Figure 14-2 gives a generalized picture of the Earth's main oceanic circulation system and the five areas of major coastal upwelling. Eight main flows are indicated: (1) the southward flow of cold Arctic water, (2) the northeastward flow of warm equatorial water in the north Atlantic (the Gulf Stream) and in the Pacific (the Kuroshio Drift), (3) the southward flow along the Californian and Euro-African coasts (with related coastal upwelling), (4) the North and South Equatorial Currents, (5) the splitting branches of these currents with one branch from the South Equatorial Current being lost to the Northern Hemisphere (both in the Atlantic and in the Pacific), (6) the southern branch of the South Equatorial Current, (7) the three major northeastward currents in the Southern Hemisphere (the Peruvian, Benguela, and Western Australian cold currents) with corresponding coastal upwelling, and (8) the cold circum-Antarctic West Drift.

As shown by Mörner (1984*b*, Fig. 12), there is a feedback coupling between the oceanic flow of water and the rate of rotation. Decreased rotation will bring water from the equator to the poles, a process which will increase the rate of rotation again (because the radius shortens and mass goes poleward) and reverse the flow and finally once again reverse the rate of rotation. This is in full agreement with the vibrational tendency and frequency-changing cycles in the south Scandinavian and north Atlantic records (Mörner, 1984*b*, Figs. 6, 11, and 12; 1973). This process explains both the correlations established between Japan (Kuroshio dependent) and northwestern Europe (Gulf Stream dependent) and the negative correlations often found between records from the Northern and Southern Hemispheres.

As shown previously, the annual and intra-annual changes in the Earth's rotation seem to be completely compensated (or caused) by changes in the angular momentum in the atmosphere. This means that the hydrosphere remained closely coupled to the solid Earth and that the oceanographic effects are caused by changes in the wind pattern and mountain pressure. This seems also to be the dominant process during the 1982-1983 El Niño/Southern Oscillation event, although an additional loss

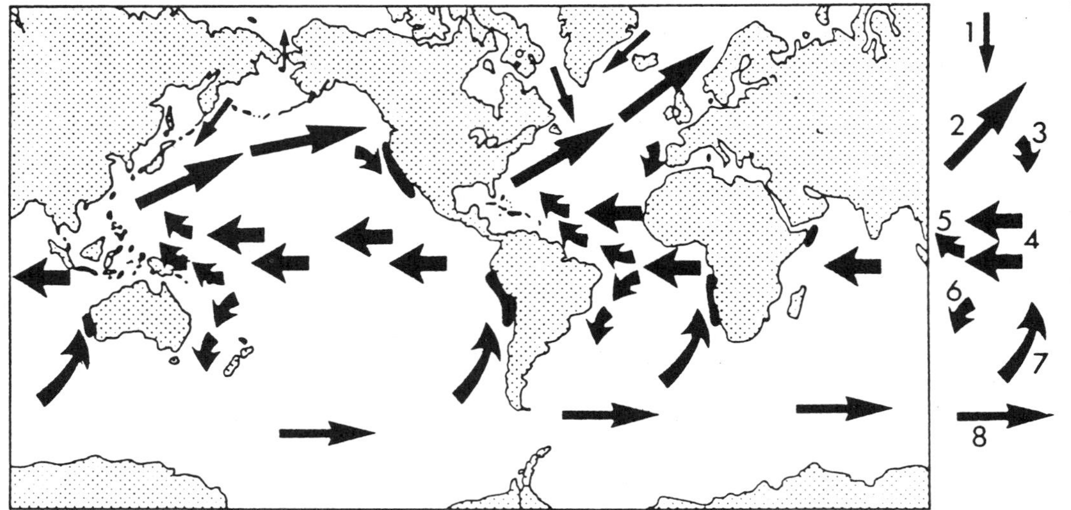

Figure 14-2. The Earth's main ocean circulation system (arrows 1-8) and five areas of coastal upwelling.

of angular momentum from the solid Earth is recorded in 1982-1983, and is likely to have been transferred to the hydrosphere (Mörner, 1987*a*).

The interannual changes in rotation are not compensated by angular momentum changes in the atmosphere (e.g., Eubanks, Dickey, and Steppe, 1984, Figs. 8 and 9). This fact suggests a considerable, probably dominant, contribution from angular momentum changes in the hydrosphere.

The decade changes in rotation (LOD) have little effect in the atmospheric angular momentum (at the most, some 10-20%) and in some cases may be directly linked to differential rotation at the core/mantle boundary and must sometimes be more or less predominantly caused/compensated by angular momentum changes in the hydrosphere. This means that hydrospheric decelerations are likely to correlate with increased rate of rotation of the mantle-and-crust (decreased LOD) and decreased geomagnetic dipole moment, while hydrospheric accelerations are likely to correlate with decreased rate of rotation (increased LOD) and increased geomagnetic dipole moment.

It therefore seems quite significant that a correlation seems to exist between warm periods, high sea levels, and low geomagnetic intensity on the one hand and cold periods, low sea levels, and high geomagnetic intensity on the other hand (Mörner, 1978*a*, 1980*a*, 1984*b*).

Earthquakes and Volcanism

The partly melted, low-viscosity (some 10^{18}-10^{21} poise) asthenosphere is a layer where accelerations and decelerations of the mantle should tend to produce differential motions but, because of the viscosity, can produce only stress (and flow) variations. These stress variations, however, may be extremely important for various geodynamic processes (Mörner, 1984*b*, Figs. 8 and 9). They are likely to affect earthquakes, volcanism, plate tectonics, and phase-transitional boundaries (Mörner, 1986*a*, 1986*b*, 1986*c*).

The 1900-1910 rotational and geomagnetic changes were linked to significant changes in seismic activity as seen in an active plate boundary area such as Venezuela and in an old craton such as Fennoscandia (Mörner, 1984*b*, p. 497). This indicates that the Earth's entire lithospheric plate system was affected by the rotational changes.

Plotting the geographical distribution of earthquakes from 1920 to 1978, Mogi (1979) was able to show that, from 1940 to 1960, there was a general shift from a global spreading to a concentration at high latitudes. Since the 1930-1960 period represents a deceleration period, mass is expected to have tended to move from the equatorial region toward the polar regions with the earthquakes following the expanding areas (Mörner, 1986*b*, 1986*c*).

Rotational changes may also set up propagating waves in the asthenosphere, explaining certain migrational patterns (including a deformation front) in the earthquake occurrence (e.g., Alexander, 1982). The Holocene crustal undulations suggested by Taira (1976) for Japan may be expressions of the same process.

Planetary Beat

The rotational-gravitational-oceanographic changes found to control the short-term climatic changes and related geophysical terrestrial variables need an ultimate, pulsating, driving force. Mörner (1984*b*, Fig. 13) proposed that this force was a planetary beat that independently affects the Sun and the Earth (Fig. 14-3). He cited evidence that an 11-yr rhythm is independently set up in the Sun and the Earth and argued that the long-term Milankovitch variables also, and maybe even predominantly, had an effect on the Earth's climate via internal processes (and not just insolation).

The gravitational interaction in our multibody planetary system will continually affect orbital distances and planetary rates of rotation, causing independent rotational and gravitational changes in the Sun and the Earth (Mörner, 1984*b*, 1984*d*). The effects on the Sun could lead to fluctuations in luminosity and solar wind that would affect the Earth's atmosphere (Bucha, 1984; Lean, 1984; Paine, 1983) and hence the

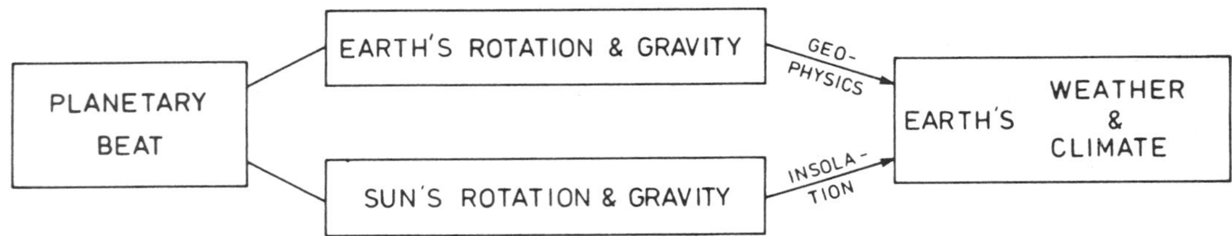

Figure 14-3. Independent terrestrial and solar responses to the planetary beat and their effect on the Earth's weather and climate.

weather and climate. The effects on the Earth will lead to a complex interaction of geophysical changes (rotation, gravity, heat flow, geomagnetism) that, via differential rotation and deformation of equipotential surfaces, will affect weather and climate. This situation is schematically illustrated in Figure 14-3 (and further discussed in Mörner, 1984*b*). This mechanism means that the complex climate and weather "machines" are, in fact, a simple (in physical-mathematical terms) matter of mass, momentum, and energy (Mörner, 1985, 1987*a*).

Applications

The model presented was formulated to explain the short-term—that is, decade-to-century-long—climatic changes. With minor changes, especially in dominant factors, the model also applies to shorter and longer time period changes.

The annual and intra-annual changes in weather and climate are predominantly forced by atmospheric circulation changes—namely, angular momentum. The hydrosphere closely follows the solid Earth's changes in rotation. Tidal forces (also a form of planetary beat) affect the Earth. Changes in luminosity and solar wind affect the Earth, as well.

The decade-to-century-long changes in climate are controlled by differential rotation where the oceanic circulation changes (angular momentum) play a dominant role and by geoidal changes that affect the ocean and the atmosphere. The ultimate driving force is the planetary beat.

The 10^4-to-10^5-yr-long changes in climate are mainly forced by the so-called Milankovitch variables and the rotational and gravitational response by the Earth, with the hydrospheric changes in a central position. Simultaneous insolation changes are thought (e.g., Berger et al., 1984) to play the dominant role for climatic changes. Their role may be small, and even insignificant, however, provided the major forcing comes via the Earth's geophysical response.

The very long-term climatic changes must predominantly be caused by the oceanic circulation changes (both lateral and vertical) controlled by rotation changes, tectonic land/sea redistribution, eustatic changes in sea level, and the heat transfer from the mantle to the oceans.

OBSERVATIONAL DATA IN VIEW OF THE MODEL

There are numerous observational data on various types of climatic changes during the past 20,000 yr (cf. Mörner, 1984*b*, 1984*d*). This section is intended as a test of the hypothesis (Mörner, 1987*a*).

At around 13,000-12,700 B.P., there was a sudden and drastic change in marine biota around southern Scandinavia (Mörner, 1984*a*, p. 5). This change represents a general increase in the Gulf Stream transport and a northward migration of the Polar Front (Ruddiman and McIntyre, 1981). This redistribution of water masses would imply a decreasing angular momentum of the hydrosphere and an increasing rate of rotation of the solid Earth. An increase of the rate of rotation of the mantle and crust occurs when the dipole moment decreases (Yukutake, 1972). It is, therefore, of great interest that the Gothenburg Geomagnetic Excursion (e.g., Mörner, 1977) is found in direct association with this paleoclimatic event. From the point of possible asthenospheric stress, one may note that the uplift of central Fennoscandia began at the same time (Mörner, 1979*b*).

The well-known Older Dryas Stadial was a local Scandinavian event (Mörner, 1984*c*). Despite this, similar short cooling events occurred in the Great Lakes region and in Mexico, which suggests simultaneous disturbances elsewhere, too.

The onset of the Younger Dryas Stadial at around 11,000 B.P. marks one of the most prominent short-term coolings that has been recorded. It represents a drastic southward migration of the Polar Front. This migration would imply an increase in the hydrospheric angular momentum and a corresponding deceleration of the solid Earth. A similar cooling seems to have occurred in the Southern Hemisphere (increase of the northeastward oceanic flows with related increase in upwelling and stronger penetration of cold Antarctic winds into low latitudes) but not, for example, in Mexico.

The Pleistocene/Holocene boundary (and the end of the Younger Dryas Stadial) at around 10,000 B.P. marks an abrupt change in climate: a distinct warming in most of the Northern Hemisphere but coolings and glacial readvances in South America and Mexico (cf. Mörner, 1984*c*). The sudden appearance of a new, more temperate marine fauna in southern Scandinavia and the North Sea is evidence of a suddenly intensified Gulf Stream flow, which in turn suggests a decrease in angular momentum of the hydrosphere and an acceleration of the solid Earth. The negative correlation with the South American records may be caused by an intensified loss of warm water from the Southern to the Northern Hemispheres (arrow 5 in Fig. 14-2). The asthenospheric inflow of mass reached its peak just at 10,000 B.P. in Fennoscandia (Mörner, 1979*b*, 1978*b*, p. 492, 1980*c*). Numerous neotectonic events occurred at around this time (Mörner, 1980*b*). Paleomagnetic records from Scandinavia may be taken to suggest a shift from a strong to a weak geomagnetic field, which is likely to have caused an increase in the rate of rotation of the mantle and crust (i.e., just as the observational data suggest).

At around 8,000-7,700 sidereal years B.P., there was a considerable decrease of the dipole moment (which would imply an accelerating mantle and crust), a sudden collapse of the remaining ice cap over the Canadian Arctic, a very large eustatic rise in sea level (that should decrease the rate of rotation), an increase in Gulf Stream transport (which would imply a decreasing angular momentum of the hydrosphere), and a general warming tendency. Asthenospheric and lithospheric stress increases are evidenced by the initiation of a new type of uplift (linear with time) and the creation of a major hinge zone in the uplift (Mörner, 1979*b*, 1980*c*). Furthermore, there are a few records of a possible short geomagnetic flip in association with the onset of the marine transgression at about 8,000 B.P. The 16 short-term changes in temperature, sea level, magnetism, and Gulf Stream activity (Mörner, 1984*b*, Fig. 6) represent pulsating decreases and increases in the Gulf Stream flow due to differential rotation between the hydrosphere and the solid Earth.

Decreases of angular momentum in the hydrosphere increase the Gulf Stream transport and suppress the southward penetration of Arctic water. This leads to a warming in Europe and the north Atlantic and a somewhat higher sea level. The corresponding acceleration of the solid Earth is linked to decreases in the dipole moment (and deceleration of the core). This explains the recorded correlations between high temperature, high sea level, stronger Gulf Stream flow, and low geomagnetic intensity.

Increases of angular momentum in the hydrosphere decrease the Gulf Stream transport and increase the southward penetration of cold Arctic water. This leads to a cooling in Europe and the north Atlantic and a somewhat lower sea level. The corresponding deceleration of the solid Earth is linked to an increase in dipole moment with decreased tendency for core/mantle dislocation. This explains the correlations between low temperature, low sea level, weaker Gulf Stream flow, and strong geomagnetic intensity.

Furthermore, it is significant that there seems to be a general correlation between the behavior of the Gulf Stream and the Kuroshio Drift.

The correlation between sea-level changes and tectonic activity suggested by Taira (1981) for Japan implies corresponding fluctuations in the asthenospheric stress—that is, just what differential rotation is likely to produce (see Fig. 14-1; Mörner, 1984*b*). The absence of correlations with the Southern Hemisphere could be due to the quite different circulation systems in the two hemispheres. It may be significant, however, that there seems to be a remarkable correlation in eustasy between Europe and New Zealand (Mörner, 1976*b*, Fig. 11) and that the main oceanic current affecting New Zealand is the southward branch of the equatorial current (arrow 6 in Fig. 14-2), which should experience similar increases and decreases as the major northeastward oceanic flows in the north Atlantic and north Pacific.

Wollin, Ryan, and Ericson (1980) found a good correlation (with a 3-yr lag) between the magnetic intensity and the sea-surface temperature in the north Pacific for the period 1950-1970; strong magnetic intensity correlates with low temperature, and weak magnetic intensity correlates with high temperature. This is exactly the same relationship found in this analysis. It should, therefore, represent dipole moment and rate of hydrospheric rotation (the Kuroshio flow). This is the same result as for the 16 Holocene fluctuations in southern Scandinavia and the north Atlantic.

The two high-amplitude regressions in Brazil and aridity phases farther inland are likely to represent large-scale geoid deformations. Significant variations in the South Equatorial Current and its southern branch along the Brazilian coast may, however, also be used to explain at least a part of the recorded changes.

The significant cooling event at around 2,500 B.P. in northwestern Europe represents the end of the Holocene climatic optimum. In North America and Greenland, there was a cooling trend at the same level but also an earlier one at around 3,500 B.P. In Mexico (Heine, 1984), there was a cooling and glacial readvance. In South America (Servant, 1984), there was a drastic warming (and glacial retreat) at the same time (Mörner, 1984*c*), indicating compensational changes (probably a matter of reduced transfer of warm water to the Northern Hemisphere; see arrow 5 in Fig. 14-2). At about the same time there was a shift from increasing to decreasing dipole moment, which should correspond to a rotational shift from deceleration to acceleration (of the mantle and crust). The acceleration of the solid Earth may be compensated (partly) by decreased angular momentum of the hydrosphere. The picture, however, is not so clear in this case.

The high-amplitude oscillations in atmospheric CO_2 content (Bojkov, 1983, Fig. 5; Mörner, 1987*a*, Fig. 10) are likely to represent fluctuations between decreased upwelling (high CO_2 content) and increased upwelling (low CO_2 content) due to a decrease and an increase of angular momentum of the hydrosphere (arrows 3 and 7 respectively in Fig. 14-2).

The Little Ice Ages of Europe represent expansions of cold Arctic water and dramatic southward migrations of the Polar Front (Lamb, 1979). This migration would correspond with increased angular momentum in the hydrosphere. Significant atmospheric, geomagnetic, and solar variations are linked to these periods. The causative problem remains complicated.

The deceleration/acceleration changes in 1840, 1900, and 1960, together with the acceleration/deceleration changes of 1875 and 1930, in the rate of rotation of the solid Earth have already been partly discussed. Major geomagnetic changes occurred at 1840, 1910, and 1967 (Courtillot, Ducruix, and Le Mouel, 1978). The 1900-1910 event was linked to drastic stress changes affecting earthquake frequency both in Venezuela and Fennoscandia (e.g., Mörner, 1986*b*, 1986*c*). It also changed the Chandler wobble amplitude (Anderson, 1974). If the rotational changes were (totally or partly) compensated by angular momentum in the hydrosphere, one would expect to see a cool period up to about 1840, a warming period up to about 1875, a cooling period up to about 1900, a warming period up to about 1930, a cooling period up to about 1970, and probably a warming period after that. This finding is in good agreement with the general climatic changes in the Northern Hemisphere (Mörner, 1986*b*, 1987*a*). In the Southern Hemisphere, however, the trends have been rather the opposite (cf. Mörner, 1984*c*, Fig. 5).

The 1982-1983 El Niño event indicates that there was a significant interchange of angular momentum between the solid Earth (decrease) and the atmosphere (increase). A 0.4 ms transfer of angular momentum from the solid Earth to the hydrosphere is also recorded (Mörner, 1987*a*).

CONCLUSION

Major short-term climatic changes have been found to be on the order of 50-150 yr duration and to have a regional (and sometimes hemispherical) but never global extent. These facts, together with their correlations with other geophysical variables, led to the formulation of a new causative model.

The model proposed is a rotational-gravitational-oceanographic model, where the oceanographic circulation redistributes heat in response to rotational changes and gravitational deformations of equipotential surfaces. The ultimate driving force is a planetary beat caused by interaction within our revolving planetary multibody system.

This model brings climatic changes down to a simple matter of mass, momentum, and energy (Mörner, 1985, 1987*a*). A substantial amount of observational data appears to support the theory.

The model proposed explains the planetary-solar-terrestrial interaction and the multiple interaction of various terrestrial geophysical variables (cf. Mörner, 1984*b*). It shows complexity and interaction (cf. Mörner, 1978*a*) at the same time as it implies a basic simplicity.

REFERENCES

Alexander, G., 1982, Quakewatch, *Science* **30:**38-45.

Anderson, D. L., 1974, Earthquakes and the rotation of the Earth, *Science* **186:**49-50.

Barber, R. T., and Chavez, F. P., 1983, Biological consequences of El Niño, *Science* **222:**1203-1210.

Barnes, R. T. H.; Hide, R.; White, A. A.; and Wilson, C. A., 1983, Atmospheric angular momentum fluctuations correlated with length of the day changes and polar motion, *Royal Soc. [London] Proc.,* Series A, **387:**31-73.

Berger, A. L.; Imbrie, J.; Hays, J.; Kukla, G.; and Saltzman, B., eds., 1984, *Milankovitch and Climate,* Dordrecht, The Netherlands: Reidel.

Berglund, B., and Mörner, N.-A., 1984, Late Weichselian deglaciation and chronostratigraphy of Southern Scandinavia: Problems and present "state of the art," in N.-A. Mörner, and W. Karlén (eds.), *Climatic Changes on a Yearly to Millennial Basis,* Dordrecht, The Netherlands: Reidel, pp. 17-24.

Bojkov, R. D., 1983, Report of the WMO (CAS) meeting of experts on the CO_2 concentration from pre-industrial times to I.G.Y., *World Meteorological Organization Proj. Research Monitoring Atmospheric CO_2, Report 10,* 34p.

Bucha, V., 1984, Mechanism for linking solar activity to weather-scale effects, climatic changes and glaciations in the northern hemisphere, in N.-A. Mörner and W. Karlén (eds.), *Climatic Changes on a Yearly to Millennial Basis,* Dordrecht, The Netherlands: Reidel, pp. 415-448.

Cane, M. A., 1983, Oceanographic events during El Niño, *Science* **222:**1189-1195.

Colquhoun, D. J.; Brooks, M. J.; Brown, J. G.; and Stone, P. A., 1983, Correlation between sea level and climatic change in the middle Holocene of the southeastern United States, IUGG, 18th General Assembly, Hamburg 1983, *Abstracts ICL,* p. 89.

Courtillot, V.; Ducruix, J.; and Le Mouel, J.-L., 1978, Sur une accélération récente de la variation séculaire du champ magnétique terrestre, *Acad. Sci. Comptes Rendus* **287D:**1095-1098.

Ebel, A., 1983, Gravity waves—Sources of momentum, heat and turbulent kinetic energy in the middle atmosphere, IUGG, 18th General Assembly Hamburg 1983, *Abstracts IAMAP,* p. 247.

Ertel, H., 1942, Ein neuer hydrodynamischer Wirbersatz, *Meterol. Zeit.* **59:**277-281.

Eubanks, T. M.; Dickey, J. O.; and Steppe, J. A., 1984, The 1982-83 El Niño, the Southern Oscillation and changes in the length of the day, *Jet Propulsion Laboratory (Pasadena), Geodesy Geophysics Preprint No. 111,* 9p.

Eubanks, T. M.; Steppe, J. A.; Dickey, J. O.; and Callahan, P. S., 1983, A spectral analysis of the Earth's angular momentum budget, *Jet Propulsion Laboratory (Pasadena), Geodesy Geophysics Preprint No. 102,* 50p.

Fairbridge, R. W., 1961, Eustatic changes in sea level, *Phys. Chemistry of the Earth* **4:**99-185.

Flohn, H., 1979, On time scales and causes of abrupt paleoclimatic events, *Quat. Research* **12:**135-149.

Gage, K. S., and Reid, G. C., 1981, Solar variability and the secular variation in the tropical tropopause, *Geophys. Research Letters* **8:**187-190.

Heine, K., 1984, The classical Late Weichselian climatic fluctuations in Mexico, in N.-A. Mörner and W. Karlén (eds.), *Climatic Changes on a Yearly to Millennial Basis*, Dordrecht, The Netherlands: Reidel, pp. 95-115.

Hide, R.; Birch, N. T.; Morrison, L. V.; Shea, D. J.; and White, A. A., 1980, Atmospheric angular momentum fluctuations and the changes in the length of the day, *Nature* **286:**114-117.

Kellogg, T. B., 1984, Late-glacial-Holocene high-frequency changes in deep-sea cores from the Denmark Strait, in N.-A. Mörner and W. Karlén (eds.), *Climatic Changes on a Yearly to Millennial Basis*, Dordrecht, The Netherlands: Reidel, pp. 123-133.

King, J. W., 1974, Weather and the Earth's magnetic field, *Nature* **247:**131-134.

Lamb, H. H., 1979, Climatic variations and changes in the wind and ocean circulation: The Little Ice Age in the northeast Atlantic, *Quat. Research* **11:**1-20.

Lambeck, K., 1980, Changes in the length-of-day and atmospheric circulation, *Nature* **286:**104-105.

Lambeck, K., and Casenave, A., 1976, Long term variations in the length of day and climatic change, *Royal Astron. Soc. Geophys. Jour.* **46:**555-573.

Lean, J. L., 1984, Solar ultraviolet irradiance variations and the Earth's atmosphere, in N.-A. Mörner and W. Karlén (eds.), *Climatic Changes on a Yearly to Millennial Basis*, Dordrecht, The Netherlands: Reidel, pp. 449-471.

Mogi, K., 1979, Global variation of seismic activity, *Tectonophysics* **57:**T43-T50.

Mörner, N.-A., 1973, Climatic changes during the last 35,000 years as indicated by land, sea and air data, *Boreas* **2:**33-53.

Mörner, N.-A., 1976*a*, Eustasy and geoid changes, *Jour. Geology* **84:**123-151.

Mörner, N.-A., 1976*b*, Eustatic changes during the last 8,000 years in view of radiocarbon calibration and new information from the Kattegatt region and other northwestern European coastal areas, *Palaeogeography, Palaeoecology, Palaeoclimatology* **19:**63-85.

Mörner, N.-A., 1977, The Gothenburg Magnetic Excursion, *Quat. Research* **7:**413-427.

Mörner, N.-A., 1978*a*, Paleoclimatic, paleomagnetic and paleogeoidal changes: Interaction and complexity, in *Evolution of Planetary Atmosphere and Climatology of the Earth*, Toulouse: CNES, pp. 221-232.

Mörner, N.-A., 1978*b*, Paleogeoid changes and paleoecological changes in Africa with respect to real and apparent paleoclimatic changes, *Palaeoecol. Africa* **10:**1-12.

Mörner, N.-A., 1979*a*, South Scandinavian sea level records: A test of regional eustasy, regional paleoenvironmental changes and global paleogeoid changes, in K. Suguio, T. R. Fairchild, L. Martin, and J.-M. Flexor (eds.), *Proceedings of the 1978 International Symposium on Coastal Evolution in the Quaternary*, Sao Paulo, pp. 77-103.

Mörner, N.-A., 1979*b*, The Fennoscandian uplift and Late Cenozoic geodynamics: Geological evidence, *GeoJournal* **3:**287-318.

Mörner, N.-A., 1980*a*, Eustasy and geoid changes as a function of core/mantle changes, in N.-A. Mörner (ed.), *Earth Rheology, Isostasy and Eustasy*, New York: Wiley & Sons, pp. 535-553.

Mörner, N.-A., 1980*b*, A 10,700 years' paleotemperature record from Gotland and Pleistocene/Holocene boundary events, *Boreas* **9:**283-287.

Mörner, N.-A., 1980*c*, The Fennoscandian uplift: Geological data and their geodynamical implication, in N.-A. Mörner (ed.), *Earth Rheology, Isostasy and Eustasy*, New York: Wiley & Sons, pp. 251-284.

Mörner, N.-A., 1981*a*, Eustasy, paleoglaciation and paleoclimatology, *Geol. Rundschau* **70:**691-702.

Mörner, N.-A., 1981*b*, Holocene sea level fluctuations on a global scale: Evidence for geoidal eustasy, *INQUA Neotectonics Comm. Bull.* **4:**100.

Mörner, N.-A., 1981*c*, Space geodesy, paleogeodesy and paleogeophysics, *Annales Géophysique* **37:**69-76.

Mörner, N.-A., 1982, Water and water, *INQUA Neotectonics Comm. Bull.* **6:**72-76.

Mörner, N.-A., 1983*a*, Sea levels, in R. Gardner and H. Scoging (eds.), *Mega-Morphology*, Oxford: Oxford University Press, pp. 73-91.

Mörner, N.-A., 1983*b*, Illusions and problems in water budget synthesis, in A. Street-Perrot, M. Beran, and R. Ratcliffe (eds.), *Variations in the Global Water Budget*, Dordrecht, The Netherlands: Reidel, pp. 419-423.

Mörner, N.-A., 1984*a*, Introduction, in N.-A. Mörner and W. Karlén (eds.), *Climatic Changes on a Yearly to Millennial Basis*, Dordrecht, The Netherlands: Reidel, pp. 1-13.

Mörner, N.-A., 1984*b*, Planetary, solar, atmospheric, hydrospheric and endogene processes as origin of climatic changes on the Earth, in N.-A. Mörner and W. Karlén (eds.), *Climatic Changes on a Yearly to Millennial Basis*, Dordrecht, The Netherlands: Reidel, pp. 483-507.

Mörner, N.-A., 1984*c*, Concluding remarks, in N.-A. Mörner and W. Karlén (eds.), *Climatic Changes on a Yearly to Millennial Basis*, Dordrecht, The Netherlands: Reidel, pp. 637-651.

Mörner, N.-A., 1984*d*, Terrestrial, solar and galactic origin of Earth's geophysical variables, *Geog. Annaler* **66A:**1-9.

Mörner, N.-A., 1984*e*, Geoidal topography: Origin and time consistency, *Marine Geophys. Research*, **7:**205-208.

Mörner, N.-A., and Karlén, W., eds., 1984, *Climatic Changes on a Yearly to Millennial Basis*, Dordrecht, The Netherlands: Reidel, 667p.

Mörner, N.-A., 1985, The Earth/ocean interchange of momentum and major climatic changes on a decadal basis, *Abstracts IAMAP/IAPSO Joint Assembly, Honolulu, Hawaii*, pp. 31, 44, 68, 94, 118.

Mörner, N.-A., 1986*a*, Internal processes and lithospheric changes, *INQUA Neotectonics Comm., Bull.* **9:**8-13.

Mörner, N.-A., 1986*b*, Global neotectonics, arcs and geoid configuration, in F. C. Wezel (ed.), *The Origin of Arcs*, Amsterdam, The Netherlands: Elsevier, pp. 79-91.

Mörner, N.-A., 1986*c*, Variations in seismic, neotectonic and volcanic activity in response to differential rotation of the Earth's layered system, *Revista CIAF* **11:**205-209.

Mörner, N.-A., 1987*a*, Ocean circulation changes and redistribution of energy and mass on a yearly to century time-scale, in T. Wyatt (ed.), *Long Term Changes in Marine Fish Populations*, in press.

Mörner, N.-A., 1987*b*, Models of global sea level changes, in M. J. Tooley and I. Shennan (eds.), *Sea Level Changes*, London: Blackwell, pp. 332-355.

Mörner, N.-A., 1987*c*, Dynamic and gravitational groundwater levels: A two-layer groundwater model. *Jour. Geol. Soc. India* **29:**128-134.

Newell, R. E., and Hsiung, J., 1984, Sea surface temperature, atmospheric CO_2 and the global energy budget: Some comparisons between the past and the present, in N.-A. Mörner and W. Karlén (eds.), *Climatic Changes on a Yearly to Millennial Basis*, Dordrecht, The Netherlands: Reidel, pp. 533-561.

Newman, W. S.; Marcus, L.; Pardi, R.; Paccione, J.; and Tomacek, S., 1980, Eustasy and deformation of the geoid: 1000-6000 radiocarbon years B.P., in N.-A. Mörner (ed.), *Earth Rheology, Isostasy and Eustasy*, New York: Wiley & Sons, pp. 555-567.

Newman, W. S.; Marcus, L. F.; and Pardi, R. R., 1981, Paleogeodesy: Late Quaternary geoidal configurations as determined by ancient sea levels, *IAHS Publ. 131*, pp. 263-275.

Paine, D. A., 1983, A climate hypothesis describing the solar-terrestrial system as a frequency domain with specific response characteristics, *EOS* **64:**425-428.

Philander, S. G. H., 1983, El Niño Southern Oscillation phenomena, *Nature* **302:**295-301.

Rasmusson, E. M., and Wallace, J. M., 1983, Meteorological aspects of the El Niño/Southern Oscillation, *Science* **222:**1195-1202.

Rosen, R. D., and Salstein, D. A., 1983, Variations in atmospheric angular momentum on global and regional scales and the length of day, *Jour. Geophys. Research* **88:**5451-5470.

Rosen, R. D.; Salstein, D. A.; Eubanks, T. M.; Dickey, J. O.; and Steppe, J. A., 1984, An El Niño signal in atmospheric angular momentum and Earth rotation, *Jet Propulsion Laboratory (Pasadena, Geodesy Geophysics Preprint No. 109*, 9p.

Ruddiman, W. F., and McIntyre, A., 1981, The North Atlantic Ocean during the last glaciation, *Palaeogeography, Palaeoecology, Palaeoclimatology* **35:**145-214.

Servant, M., 1984, Climatic variations in the low continental latitudes during the last 30,000 years, in N.-A. Mörner and W. Karlén (eds.), *Climatic Changes on a Yearly to Millennial Basis*, Dordrecht, The Netherlands: Reidel, pp. 117-210.

Servant, M.; Fontes, J. C.; Rieu, M.; and Saliège, J.-F., 1981, Phases climatiques arides Holocènes dans le sud-ouest de l'Amazonie (Bolivia), *Acad. Sci. Comptes Rendus* **292**(2):1295-1297.

Taira, K., 1976, A wave-like pattern of Holocene crustal warping in eastern Asia, *Palaeogeography, Palaeoecology, Palaeoclimatology* **19:**249-254.

Taira, K., 1981, Holocene tectonism in eastern Asia and geoidal changes, *Palaeogeography, Palaeoecology, Palaeoclimatology* **36:**75-85.

Wollin, G.; Ericson, D.; and Ewing, M., 1971, Late Pleistocene climates recorded in Atlantic and Pacific deep-sea sediments, in K. K. Turekian (ed.), *The Late Cenozoic Glacial Ages*, New York: Wiley & Sons, pp. 199-214.

Wollin, G.; Ryan, W. B. F.; and Ericson, D. B., 1980, Short period climatic variations as a geomagnetism-ocean-atmosphere problem, manuscript.

Yukutake, T., 1972, The effects of change in the geomagnetic dipole moment on the rate of the Earth's rotation, *Jour. Geomagnetism and Geoelectricity* **24:**19-47.

15: Earth's Quantum Climate

Douglas Alan Paine
Cornell University

Abstract: In 1948, Jules Charney provided the impetus for numerical weather prediction by scaling the primitive equation set to reduce the complexity of the short-range forecast problem. Newtonian physics was thus proven capable of defining the structural evolution of cyclone families and fronts. Because such meteorological phenomena are scale dependent, it made considerable conceptual sense to isolate their appropriate forcing functions in space and time.

However, to choose a specific space and time scale when we are intent on addressing the physics of climate seems counterproductive. For instance, it would appear that the coupled atmospheric-oceanic-biospheric system functions as an efficient absorber and emitter of solar-terrestrial radiation by maintaining a high degree of scale interactivity. Evidently, a formal description of this function requires that we view the atmosphere as a veritable space/time continuum—that is, an infinite array of space and time events that happen simultaneously at many scale levels, all of which merge into a purposeful whole while sustaining Earth's life support function.

In this chapter, we seek to build a bridge between the scale-specific Newtonian worldview and the scale-interdependent studies of contemporary physics. By scaling Planck's quantum theory to accommodate a waveform that is both absorbed and emitted by the atmosphere, a unified theory of climate emerges. Climate change becomes synonymous with angular momentum adjustments at specific atmospheric levels, while the resultant macroquantum domain is found to undergo abrupt transitions between preferred nonequilibrium states. Because these jumps are achieved through the onset of diabatic physics associated with either a changing albedo (e.g., volcanic aerosol, snow cover, deforestation, etc.) or a changing concentration of trace gases (e.g., H_2O, O_3, CO_2), the theory appears ideally suited to studies of climate modification.

By redeploying quantum principles to solve complex problems of atmospheric, oceanographic, biospheric, and solar physics, we stand to gain the ability to study the interactivity among domains that were previously isolated by their highly disparate spatial and temporal identities.

INTRODUCTION

This chapter proposes that principles of quantum physics apply to the exchange of angular momentum between the solar system and the terrestrial environment. In effect, we say that a proven conceptual tool, which was first formulated within the atomic and molecular domains, can be restated within the space/time scales commonly associated with the atmospheric, ocean-lithospheric, and biospheric domains.

My purpose in pursuing this discussion bears directly on the diagram that served as a kind of Logos for the Barnard College Climate Symposium, held in honor of Professor Fairbridge. Participants at this gathering will undoubtedly recall the drawing by his colleague, John Sanders of Columbia University. It depicted the trajectory of the sun's orbit relative to the solar system's center of mass. Beginning around 1988, the angle swept out by the sun's center in relation to this barycenter will change from its characteristic counterclockwise direction of unfoldment, into a clockwise mode that is expected to last into the middle of the 1990s. This negative angular momentum anomaly, as pointed out by Jose (1965), last occurred in 1811. Because the event has a 178-yr return frequency, its previous appearances in A.D. 1633 and 1455 may have coincided with the so-called Maunder and Spörer Minima or extended episodes of greatly diminished sunspot activity (Robbins, 1983). The event thus raises a number of important questions about how the complex geophysical system, not only on the solar system level but also within the individual planetary atmospheres, might respond to such an anomaly.

A portion of the scientific community-at-large has coincidently proposed that a so-called geobiosphere program of observation should commence in 1988, 30 yr after the highly successful 1957-1958 International Geophysical Year. To quote from a news article appearing in *EOS* [Feb. 21, 1984]:

> An international, interdisciplinary program to study the closely coupled system of the terrestrial environment and the life that inhabits it has been proposed for later this decade. As currently outlined, the International Geosphere-Biosphere Program (IGBP) would encompass at least a decade of research and would involve a host of nations. IGBP would embrace studies of physical, biological, and ecological processses. A major challenge [of the program] will be that of understanding the causes and effects of climate change.

The article went on to cite the program chairman, Herbert Friedman:

> Progress in understanding global change will require extensive and well-organized observations made over much of the earth and over long periods of time. The scope of such an effort requires international cooperation and interdisciplinary emphasis. Coordinated efforts between adjacent scientific disciplines and programs of synoptic observations focused on common, interrelated problems that affect the earth as a whole [are needed].

Recent experience suggests that we are already performing some of the observational techniques envisioned by Friedman. For instance, exchanges of angular momentum are currently used to gauge the response characteristics among coupled geophysical systems. Satellite remote sensing and balloon-borne radiosondes are able to detect changes in atmospheric angular momentum and Earth's length of day (LOD), the latter to an accuracy of 0.01 ms. Thus, when the record-setting, equatorial Pacific Ocean warming event of 1982-1983 led to major storminess along the western U.S. coastline, Rosen et al. (1984) were among the first to report an apparent El Niño signal registered in late January 1983. The global atmospheric angular momentum peaked at nearly 25% beyond its annual cyclic range (from 75×10^{24} to 175×10^{24} kg m^2 s^{-1}), while the LOD anomaly was 3.5 times the root-mean-square scatter.

What we would appear to lack to enter into the proposed IGBP is a common theoretical tool for examining such data, one which would aid us in constructing mutually understandable problem-solving techniques. I believe the lack of a unifying conceptual framework is a legacy from our continuing involvement with the scale-dependent Newtonian paradigm: that is, each of us in our respective disciplines is trained in methodologies that emphasize differences of scale, and these differences erect formidable barriers when it comes to sharing a common understanding of the global system seen as a whole.

Quite a different picture emerges, however, if we were to adopt the foundations of quantum mechanics that were designed to accommodate differences of scale. Then, rather than centering our attention on the differences that serve to separate disciplines, we could conceivably gain access to a more general (less restrictive) syntax by which systems of many disparate scales process information to adjust to their constantly changing environment.

AN ATMOSPHERIC HARMONIC OSCILLATOR

The principle involving the conservation of angular momentum has persisted as one of the mainstays in scientific investigation ever since the time of Johannes Kepler. At the beginning of the twentieth century, Planck found that the exchange of angular momentum at the atomic length scale occurred in indivisible packets, or quanta. In 1905, Einstein went on to describe how photons, the quanta of light, gave rise to stimulated radiative emissions (photoelectric effect). The question I examine here is whether we might entertain the notion that the Earth's atmosphere also absorbs and emits quanta of angular momentum in the scale range associated with the meteorological spectrum. Unlike the radiative frequencies measured at the atomic length scale, where wavelengths are commonly registered in nanometers (1 nm = 10^{-9} m), meteorological wavelengths typically fall within the range of 1 km to 10^4 km, the latter being the scale distance from the equator to the poles. Presumably, *meteorological quanta* would involve atmospheric processes defined at fractions of a kilometer (<10^3 m).

To describe theoretically quantum processes of angular momentum exchange within an atmospheric domain, this section adopts the hierarchical methodology proposed by Bohm and Hiley (1975). To provide the foundation for a quantum approach to the atmosphere, we first envision a supersystem-system-subsystem composite as consisting of:

1. the total number of air parcels comprising a given limited atmospheric domain structure (supersystem),
2. an individual air parcel (system, or localized domain structure),
3. the subsystem of gaseous molecules comprising the air parcel.

In this microscale approach to the problem, the air parcel is hypothesized to assume the role of an elemental oscillator whose function is to communicate the quantum potential within the limited atmospheric domain. The quantum potential is defined by a temperature difference,

$$\tau \equiv T_o - T_c,$$

where T_o is the reference level temperature maintained by parcels that are nonaccelerative, and T_c is the critical temperature attained when a limited atmospheric domain (and its parcels) enter into a fully accelerative condition. Both temperatures within the Earth's lower atmosphere often lie within the range of 270 K to 330 K, or well above absolute zero. If the angular momentum exchange were initiated by the complete condensation of the system's water vapor content, T_o and T_c would be defined by the potential (θ) and equivalent potential (θ_e) temperatures respectively.

Consider the nonaccelerative state of the totality of air parcels comprising a limited atmospheric domain (supersystem) to be the stable state. Consider the transition toward the fully accelerative state to be a transition toward an unstable state:

	Stable Condition→→	*Unstable Condition*
Physics:	Geostrophic	Ageostrophic
	Adiabatic	Diabatic
	Hydrostatic	Nonhydrostatic

What effect does radiation impinging on a limited atmospheric domain have with regard to the hypothesized transition between the stable and unstable states?

Let us first define the stable environment of the total number of air parcels, which constitutes the supersystem associated with a given limited atmospheric domain, to be representative of hydrostatic equilibrium. The resultant buoyant force at any level is zero,

$$0 = -g - \frac{1}{\rho_e}\frac{\partial P}{\partial z}, \qquad \textbf{(15-1)}$$

where the inverse distribution of density in the environment (ρ_e) multiplied by the pressure gradient is exactly counterbalanced by g, the gravitational acceleration. Let us further assume that a given parcel always adjusts its pressure to the pressure of the environment. The following equation describes the time rate of change of parcel motion:

$$\ddot{z} = -g - \frac{1}{\rho_p}\frac{\partial P}{\partial z}, \qquad \textbf{(15-2)}$$

where ρ_p designates the parcel's perturbation density.

Eliminating $\delta P/\delta z$ from (15-1) and (15-2),

$$\ddot{z} = g\frac{(\rho_p^{-1} - \rho_e^{-1})}{\rho_e^{-1}}. \qquad \textbf{(15-3)}$$

Utilizing the equation of state ($P = \rho RT$) for the parcel and the environment respectively, (15-3) is rewritten

$$\ddot{z} = -g\frac{(T_e - T_p)}{T_e}, \qquad \textbf{(15-4)}$$

where T_e and T_p refer to the environmental and parcel temperatures respectively. Later, we will find it instructive to substitute $T_e = T_o$ as descriptive of an environmental reference temperature, where $T_o \leq T_p \leq T_c$.

At this point we wish to consider how a net radiation exchange between the parcel (system) and its environment (supersystem) influences the vertical motion ($\dot{z}$) of a given elemental parcel. In general, the radiative flux for a blackbody (perfect absorber and emitter of energy) is given by

$$F = \sigma T^4. \qquad \textbf{(15-5)}$$

The reader will note that we are taking the unusual step of defining a thermodynamical concept (Stefan-Boltzmann law) at the system level: the air parcel, composed of a subsystem of gaseous molecules, is being assigned a task traditionally reserved for the atomic or molecular level.

Once we have incorporated (15-5) into the final definition of an elemental oscillator functioning within the scale level of atmospheric dynamics, the importance of this step will be clear. First, the scale factor(s) needed to rewrite Planck's law for meteorological processes comes from replacing Boltzmann's constant (k, given in erg K^{-1}) with the atmospheric constant C_p (also given in erg K^{-1}) used in Poisson's definition of potential temperature, θ:

$$T \equiv \theta\left(\frac{P}{P_o}\right)^{R/C_p},$$

where P_o is an arbitrary pressure reference level of 100 kPa (1,000 mb) found near mean sea level, and R and C_p are the dry gas constant and specific heat of a gas at constant pressure. Because Planck's law builds on the definition of entropy change being equated to a system's change of energy divided by temperature, where for an atmospheric process

$$dS \equiv \frac{dU}{T} = C_p d(\ln \theta),$$

the possibility of scaling this radiation law to accommodate changes of atmospheric angular momentum and entropy (involving changes of θ, weighted by C_p) may be evident (see Hess, 1979).

However, there is a second, perhaps unforeseen, result of invoking (15-5) at this stage in the theory. When we link the thermal exchange of energy achieved by a parcel to changes observed within a limited atmospheric domain, we are in effect saying that limited volumes of the atmosphere will be found to exhibit macroquantum properties. By hypothesizing the relevance of blackbody physics applied to these scale levels, we are formally suggesting that superconductant physics—at a temperature reference level well above absolute zero— attends atmospheric exchanges of angular momentum with both of its bounding environments—namely, the solar wind and solid body Earth.

Evidently, blackbody physics and quantum processes must be an integral part of all superconductant phenomena. We therefore seek to describe mathematically the manner and means by which the quantized signal, τ, defined by the temperature difference involving $T_o(\theta)$ and $T_c(\theta + \Delta\theta)$, is communicated from subsystem to system to supersystem. To capture the essence of superconductance, we must assume that the localization or manifestation of this signal at any one of these levels is tied to the simultaneous (to use David Bohm's descriptive phrase) spontaneous fusion of the same signal into the next higher level. We shall find that our hypothesis eventually comes to center on the function of time ordering in the interplay or exchange of the quantum signal among the subsystem (molecular)-system (air parcel)-and supersystem.

For an idealized spherical parcel, the net radiation flux is

$$F_{net} = 4\pi r_p^2 \sigma (T_e^4 - T_p^4), \quad \textbf{(15-6)}$$

and this expression is equivalent to the total change in energy per unit time for the parcel due to radiational processes:

$$\frac{4}{3}\pi r_p^3 C_p \rho \left.\frac{dT_p}{dt}\right|_R = 4\pi r_p^2 \sigma (T_e^4 - T_p^4). \quad \textbf{(15-7)}$$

Solving for $\left.\frac{dT_p}{dt}\right|_R$,

$$\left.\frac{dT_p}{dt}\right|_R = \frac{3\sigma(T_e^4 - T_p^4)}{C_p \rho r_p}. \quad \textbf{(15-8)}$$

The density in (15-8) is to be considered a kind of mean density, and σ is the Stefan-Boltzmann constant.

Since $r_p = f(P_e, T_p)$ and density may be replaced by mass *(M)* divided by volume *(V)*

$$P_e = \rho R T_p = \frac{M}{V} R T_p = \frac{M}{4/3\,\pi r_p^3} R T_p,$$

we solve for r_p, the radius of the spherical parcel,

$$r_p = \left(\frac{M}{4/3\pi}\frac{RT_p}{P_e}\right)^{1/3}. \quad \textbf{(15-9)}$$

Substituting (15-9) into (15-8),

$$\left.\frac{dT_p}{dt}\right|_R = \frac{3\sigma(T_e^4 - T_p^4)T_p^{-1/3}}{C_p \rho \left(\dfrac{MR}{\frac{4}{3}\pi P_e}\right)^{1/3}}. \quad \textbf{(15-10)}$$

After the use of a Taylor series expansion and an order of magnitude argument, as outlined in the Appendix, we obtain

$$\left.\frac{dT_p}{dt}\right|_R = \frac{12\sigma T_e^3 (T_e - T_p)}{C_p \rho \left(\dfrac{3MRT_e}{4\pi P_e}\right)}. \quad \textbf{(15-11)}$$

Despite the apparent complexity of expression (15-11) for the time rate of temperature change of the oscillating parcel due to radiation impingement, it is well to recall that the right-hand side of (15-11) is a function of *z*, the position of the parcel relative to its initial temperature T_o, which was originally equal to T_e.

In that regard, we must allow the parcel to change its temperature when moving back and forth from the reference state ($T_p = T_o$). This temperature change due to vertical motion involves the adiabatic lapse rate γ_d:

$$\left.\frac{dT_p}{dt}\right|_{motion} = -\gamma_d \dot{z}. \quad \textbf{(15-12)}$$

Thus, the total temperature change of the parcel is

$$\left.\frac{dT_p}{dt}\right|_{total} = -\gamma_d \dot{z} + \frac{f(z)}{C_p \rho}(T_e - T_p), \quad \textbf{(15-13)}$$

where

$$f(z) = \frac{12\sigma T_e^3}{\left(\dfrac{3}{4}\dfrac{MR}{\pi P_e}\overline{T_e}\right)^{1/3}}.$$

Now we have two differential equations, (15-4) and (15-13),

$$\ddot{z} = -g\frac{(T_e - T_p)}{T_e} \quad \textbf{(15-14a)}$$

and

$$\frac{dT_p}{dt} = -\gamma_d \dot{z} + \frac{f(z)}{C_p \rho}(T_e - T_p). \quad \textbf{(15-14b)}$$

Observing that

$$\frac{d}{dt}(T_p - T_e) = \frac{dT_p}{dt} + \gamma \dot{z}, \quad \textbf{(15-15)}$$

where γ is the actual temperature lapse rate through the environment, we obtain from

(15-14*a*) and (15-14*b*) a combined hydrothermodynamic description for the oscillating parcel:

$$\frac{d}{dt}(T_e - T_p) = \dot{z}(\gamma_d - \gamma) - \frac{f(z)}{C_P\rho}(T_e - T_p). \qquad \textbf{(15-16)}$$

Let $\tau = (T_e - T_p) = (T_o - T_c)$, and after differentiating (15-16) once with respect to time, we have

$$\ddot{\tau} + \frac{f(z)}{C_P\rho}\dot{\tau} + \frac{g}{T_e}(\gamma_d - \gamma)\,\tau = 0. \qquad \textbf{(15-17)}$$

This equation has the form

$$P(x)\frac{d^2y}{dx^2} + Q(x)\frac{dy}{dx} + R(x)y = S(x),$$

descriptive of a harmonic oscillator, where the functions P, Q, and R reduce to coefficients. If the coefficient of τ in (15-17) is positive, the solution for τ is a sinusoidal function of time; that is, the parcel will oscillate about its original position with a fundamental frequency ν_o given by

$$\nu_o = \frac{\sqrt{\frac{g}{T_e}(\gamma_d - \gamma)}}{2\pi}. \qquad \textbf{(15-18)}$$

In the stable case, this may be verified by substitution for τ into (15-17),

$$\tau = A\cos 2\pi\nu t + B\sin 2\pi\nu t,$$

where A and B are constants of integration.

PLANCK'S LAW: FROM ATOM TO ATMOSPHERE

Equations (15-17) and (15-18) provide us with a mathematical structure sufficiently similar to Planck's starting point that we may now turn to the key steps taken in his original (1901) derivation. In doing so, we will want to be alert for the point at which we can exchange the fundamental constants of electromagnetic radiation theory with those of hydrothermodynamics—specifically, $k \twoheadrightarrow C_P$. We must do so while still retaining the essential map of information exchange in both domains: that is, changes of entropy divided by changes in energy are equated to the inverse of temperature.

In the previous section we saw that a quantum signal in the atmospheric domain has been defined in terms of a temperature difference. The zero-, first-, and second-order derivatives taken with respect to time of this temperature signature (originating at the subsystem level) has given mathematical substance to an approach to quantum atmospheric dynamics. To reach this viewpoint, we began with a system, or oscillating air parcel, responding to a field of radiation. The parcel was envisioned to transit between two constant entropy states as the supersystem (limited atmospheric domain) underwent a change of energy.

Planck likewise imagined a linear oscillator with a natural frequency ω_o inserted into a field of radiation. Its dimensions are supposed small compared to the surrounding wavelengths. The oscillator is both damped and excited by the radiation, depending on the relation between the frequency of radiation (ν) and ω_o. The relevant equation is

$$\ddot{x} + 2\rho\dot{x} + \omega_o^2 x = \frac{e}{m}E_x, \qquad \textbf{(15-19)}$$

where m and e are the mass and charge of the oscillating electron, E_x is the x component of the electrical field associated with the radiation, and $2\rho\dot{x}$ is the force of the damping. By analyzing (15-19), Planck found that the energy of the oscillator (U_ν) at frequency ν was related to the energy density of the radiation (U) by

$$U = \frac{c^3}{8\pi\nu^2}U_\nu. \qquad \textbf{(15-20)}$$

where c is the speed of light, a limiting velocity in the atomic realm. Thus, he could consider the energy of the oscillator rather than the radiant field. This turns out to be a crucial simplification.

Two radiation laws were current at Planck's time:

Wien's Law (high ν, small λ)

$$U(\nu, T) = \frac{\alpha k A}{c^3} \nu^3 e^{-\alpha\nu/T} \qquad \textbf{(15-21)}$$

Rayleigh-Jeans Law (small ν, long λ)

$$U(\nu, T) = \frac{8\pi\nu^2}{c^3} kT \qquad \textbf{(15-22)}$$

Let the oscillator have "temperature" T and entropy S, and as before,

$$dS = \frac{1}{T} dU. \qquad \textbf{(15-23)}$$

Thus, for Wien's Law,

$$U = \frac{\alpha k A \nu}{8\pi} e^{-\alpha\nu/T} = A_1 e^{-\alpha\nu/T}, \qquad \textbf{(15-24)}$$

so that

$$T^{-1} = -(\alpha\nu)^{-1} \ln(U/A_1) = \frac{dS}{dU}, \qquad \textbf{(15-25)}$$

and hence,

$$\frac{d^2S}{dU^2} = -\frac{1}{\alpha\nu U}\bigg|_{\textit{from Wien's Law}}. \qquad \textbf{(15-26)}$$

From Rayleigh-Jeans Law we find

$$U = kT, \qquad \textbf{(15-27)}$$

so

$$\frac{1}{T} = \frac{k}{U} = \frac{dS}{dU}, \qquad \textbf{(15-28)}$$

and hence,

$$\frac{d^2S}{dU^2} = -\frac{k}{U^2}\bigg|_{\textit{from Rayleigh-Jeans Law}} \qquad \textbf{(15-29)}$$

Equations (15-26) and (15-29) would appear to be an ideal point for substituting for the constants (α, k) relative to the atomic and molecular scale lengths, versus those identified with atmospheric scale lengths. This step will eventually be confirmed by noting that the next step in the derivation—seen in hindsight from Planck's Law—amounts to allowing quanta (packets of angular momentum) emitted at the shortest meteorological wavelength to be reabsorbed among the long-wave class. This complex process of scale interactivity periodically closes, or bridges, the spectral gap found in the intermediate or mesoscale range of meteorological events. We shall return to this point for amplification after completing the derivation.

Planck now chose to interpolate between these two laws with an expression that would give the correct law at either end of the spectrum. He put

$$\frac{d^2S}{dU^2} = -\frac{1}{\alpha\nu U + U^2/k} = -\frac{k}{U}\left(\frac{1}{\alpha\nu k + U}\right). \qquad \textbf{(15-30)}$$

Now we take

$$\frac{d^2S}{dU^2} = -\frac{1}{\alpha\nu}\left[\frac{1}{U(1 + U/\alpha\nu k)}\right] \qquad \textbf{(15-31)}$$

by the method of partial fractions

$$= -\frac{1}{\alpha\nu}\left[\frac{1}{U} - \frac{(\alpha\nu k)^{-1}}{(1 + U/\alpha\nu k)}\right].$$

Thus

$$\frac{dS}{dU} = -\frac{1}{\alpha\nu}\{\ln[(\alpha\nu k)^{-1}U]$$

$$- \ln[1 + (\alpha\nu k)^{-1}U]\} + \text{const}$$

$$= -\frac{1}{\alpha\nu} \ln\left[\frac{(\alpha\nu k)^{-1}U}{1 + (\alpha\nu k)^{-1}U}\right] + \text{const}. \qquad \textbf{(15-32)}$$

Now the condition that, as $U \to \infty$, $T \to \infty$ $\to \to dS/dU = 1/T \to 0$ so the constant of integration is zero. Thus,

$$\frac{1}{T} = -\frac{1}{a\nu} \ln \left[\frac{(a\nu k)^{-1}U}{1 + (a\nu k)^{-1}U} \right]$$

$$e^{a\nu/T} = \left[\frac{(a\nu k)^{-1}U}{1 + (a\nu k)^{-1}U} \right]^{-1}$$

$$(a\nu k)^{-1}Ue^{a\nu/T} = 1 + (a\nu k)^{-1}U \qquad \textbf{(15-33)}$$

and

$$U(e^{a\nu/T} - 1) = a\nu k, \qquad \textbf{(15-34)}$$

so with the definition of the new constant $h \equiv ak$, the quantized nature of the radiation emerges:

$$U = \frac{h\nu}{e^{h\nu/kT} - 1}, \qquad \textbf{(15-35)}$$

or, from (15-20),

$$U_\nu = \frac{8\pi\nu^2}{c^3} \frac{h\nu}{e^{h\nu/kT} - 1}, \qquad \textbf{(15-36)}$$

and the intensity of the radiation is

$$I_\nu = \frac{2h\nu^3}{c^2(e^{h\nu/kT} - 1)}. \qquad \textbf{(15-37)}$$

While Planck's derivation is complete, we return to the aforementioned expressions (15-26) and (15-29) where the second-order derivative of entropy was used to partition energy between the long and short wavelengths of the electromagnetic spectrum. Substituting C_P for k and h^*/C_P for a, the respective equations derived from the atmospheric analogue to (15-35), (15-36), and (15-37) become:

$$U = \frac{h^*\nu}{e^{h^*\nu/C_PT} - 1}, \qquad \textbf{(15-38)}$$

Table 15-1. Magnitude of Fundamental Constants Appearing Within Quantum Dynamics of Atomic and Atmospheric Processes

Atomic Spectrum	*Atmospheric Spectrum*	*Units*
$k = 1.380 \cdot 10^{-16}$	$1 \cdot 10^7 = C_P$	erg K^{-1}
$c = 2.998 \cdot 10^{10}$	$3 \cdot 10^3 = c^*$	cm s^{-1}
$h = 6.625 \cdot 10^{-27}$	$2.1 \cdot 10^{10} = h^*$	erg s^1

$$U_\nu = \frac{8\pi\nu^2}{c^{*3}} \frac{h^*\nu}{e^{h^*\nu/C_PT} - 1}, \qquad \textbf{(15-39)}$$

$$I_\nu = \frac{2h^*\nu^3}{c^{*2}(e^{h^*\nu/C_PT} - 1)}. \qquad \textbf{(15-40)}$$

The asterisks attached to h^* and c^* designate constants involved in describing the most efficient information exchange to and from the meteorological spectrum, where the atmosphere's quanta have energy given by

$$E = h^*\nu \qquad \textbf{(15-41)}$$

and a variable wavelength set by c^* and the frequency of the incoming (outgoing) radiation:

$$c^* = \lambda\nu. \qquad \textbf{(15-42)}$$

Empirical as well as theoretical considerations (Paine, 1983) indicate that the constants we have been dealing with undergo a considerable shift, by many orders of magnitude, when we compare their respective numerical values as established within the atomic and atmospheric domains (see Table 15-1).

MAXWELLIAN ANALOGUE FOR ATMOSPHERIC DYNAMICS

Our final formulation focuses on the limiting velocities c and c^*, which govern the fastest admissible signal within the electromagnetic and hydrothermodynamic (atmospheric) domains. We employ c^* to formulate a description of the wave phenomena accomplishing the most efficient angular momentum transfer through vertical levels of the atmosphere. This presents us with a field intensity, analogous to the role given to E_x in (15-19), which alternately damps and excites the parcel described by expression (15-17). A by-product of this endeavor is that the discrete or discontinuous nature of the quanta we set out to describe will now find its complement in a relativistic setting that emphasizes the continuum approach of field theory.

In Paine et al. (1975), a theory for the conservation of three-dimensional spin (or atmospheric angular momentum) led to the presentation of two coupled wave equations:

$$\nabla \times \vec{V}_{\mathrm{geo}} = + \frac{1}{c^*} \frac{\partial \vec{V}_{\mathrm{ageo}}}{\partial t} \quad \textbf{(15-43a)}$$

and

$$\nabla \times \vec{V}_{\mathrm{ageo}} = - \frac{1}{c^*} \frac{\partial \vec{V}_{\mathrm{geo}}}{\partial t}, \quad \textbf{(15-43b)}$$

where the two vector quantities $\vec{V}_{\mathrm{geo}}$ and $\vec{V}_{\mathrm{ageo}}$ refer to the geostrophic and ageostrophic wind components. The geostrophic wind represents a balance between the atmospheric mass and momentum fields (stable state, nonacceleratiive flow), while the ageostrophic wind refers to an imbalanced condition (unstable state, accelerative flow).

The paired equations (15-43*a*) and (15-43*b*) are analogous to James Clerk Maxwell's formulation for an electromagnetic wave, and its accompanying fields, propagating in a vacuum (absence of free space charges):

$$\nabla \times \vec{H} = + \frac{1}{c} \frac{\partial \vec{E}}{\partial t} \quad \textbf{(15-44a)}$$

and

$$\nabla \times \vec{E} = - \frac{1}{c} \frac{\partial \vec{H}}{\partial t}, \quad \textbf{(15-44b)}$$

where $\vec{H}$ and $\vec{E}$ refer to the magnetic and electric field intensities, and c is the speed of light.

The coupled wave equations (15-43*a*) and (15-43*b*) offer a macroscopic view of the problem we first confronted on the microscale: that is, the supersystem-system-subsystem methodology has now been shifted upward, in a manner of speaking, to encompass:

the environment (supersystem) surrounding the limited atmospheric domain undergoing information exchange,
the limited atmospheric domain as system,
the parcel described in (15-17) as subsystem.

To "read" (15-43*a*) and (15-43*b*), we would say the complexity of angular momentum exchange to and from the atmosphere reduces to two simultaneous, dynamical processes: first, the absorption of geostrophic rotation leads to a local time rate of change in the ageostrophic field, weighted by the inverse of c^*; second, the absorption of ageostrophic rotation leads to a local time rate of change in the geostrophic field intensity, weighted by the negative inverse of c^*.

DISCUSSION AND INTERPRETATION

We now have the mathematical basis for approaching the atmosphere as a macroquantum domain. To briefly review the steps taken: First, we derived an equation descriptive of an elemental oscillator. Excitation of the oscillator maximizes the flow of free energy into and out of a limited atmospheric volume. This energy is tied to the volume's increase and decrease of diabatic potential, as measured in terms of its sensible or latent heat flux. To achieve the desired equation, we had to envision a nonhydrostatic process, or fully accelerative state within atmospheric dynamics.

We are led to conclude that intense convective processes are synonymous with a quantum approach to atmospheric physics. Our inability to fix these microscale events at specific moments of time and spatial locale within a Cartesian grid network thus appears not so much a fault of Newtonian determinism, as it is symptomatic of a more fundamental indeterminancy embedded within the physics of the atmosphere. To state this in another way, the inability to pinpoint a convective event as one that happens at a particular (isolated) scale level is inherent in the quantum view that these events are actually spread across scale levels in a fashion already envisioned by contemporary physics, beginning at the atomic and subatomic lengthscales.

Next, we sought to express the energy of the oscillator in terms of its temperature and change of entropy. If we view the increase or decrease of diabatic potential in terms of this formalism, we find that indivisible packets of energy, or quanta, are responsible for raising or lowering the quantum potential of the system. We may also view this change of potential in terms of a system's redistribution of angular momentum. This redistribution is accomplished by a wave process whose fundamental identity is given by the wave-trace velocity that acts as a limiting velocity for the system. If this system is the lower atmosphere, then energy and momentum originally confined within the meteorological spectrum of waves are free to escape this domain and radiate to upper atmospheric levels once this limit has been reached.

There is an effective metaphor for understanding why we may approach a quantum process from either a wave or a particle point of view. Consider an external gravity wave within the fluid medium of the ocean as it approaches a sloping beach. The phase velocity of the wave, *c*, will eventually be matched by fluid parcels accelerating to equal this critical velocity. (The parcels are forced to accelerate when fewer and fewer of them take on the task of distributing the wave's energy over an ever diminishing fluid volume.)

When a fluid parcel comprising the wave begins to exceed the speed of the wave's forward movement, the wave breaks. Absorption of kinetic energy in the zone of wave breaking action could just as easily be attributed to the wave or to some idealized particle giving definition to this energy exchange. In fact, from the mathematical description of the process, the quanta of energy absorbed or emitted wherever waves break is predicted by the inability to distinguish between the energy elements making up the wave and the wave itself treated as a distinct entity. So long as the wave remains unbroken, it is a part of an indivisible energy continuum; once it breaks, its portion of the energy continuum becomes discrete.

Just as we earlier saw that the quest for a mathematical description of an elemental oscillator (Planck's electron) brought us to the view that convective processes in the atmosphere are quantum events, the wave-particle duality centers our attention on the ensemble effect of convection. Namely, satellite pictures reveal that convective activity erupts wherever meteorological waves rapidly amplify and break. Apparently, wave-breaking activity, which is as ubiquitous as the convective event itself within the atmosphere, could be referred to as a kind of "quantum foam." Although the froth of wave-breaking activity is seen as disorder erupting within the context of classical determinism, there is a new kind of order attached to the way in which systems of diverse scale identity (lithosphere, atmosphere, biosphere, and solar wind) exchange energy and information across their boundaries.

Let us look at a specific example of an atmospheric quantum event. A trough of low pressure advancing at the forward edge of cold air crossing the relatively warm Great Lakes in late fall or early winter often gives rise to vigorous convective activity. In the most intense instances, the largescale wave exciting the efficient flux of sensible and latent heat into the tropospheric column can actually be observed to form higher frequency waves appearing as a series of intense convective squall-lines. The convection leaves its signature by altering the low-level temperature field, a process that can be mapped by the changing pressure topography of isentropic (constant entropy) surfaces. A traditional meteorologist would anticipate that the distribution of geostrophic momentum would undergo appreciable change in such turbulent flow.

The quantum approach that we have been pursuing would also anticipate that the intense flux of diabatic potential into the troposphere (often equaling or exceeding the solar parameter) necessitates a redistribution of angular momentum. This redistribution occurs not only within the tropospheric column, but within the stratosphere and extending upward to, in the most severe instances, ionespheric levels.

The coupled wave equations referred to in (15-43*a*) and (15-43*b*) describe an acoustically modified, internal-type of gravity wave originating wherever intense diabatics are found. This wave form propagates vertically upward and amplifies under the exponential reduction of density with height. Wherever the trace velocity distinguishing a particular wave packet enters into a layer where the background geostrophic flow speed matches this velocity, the ageostrophic momentum making up the wave is reabsorbed. The net result is one of emitting and reabsorbing wave energy in such a way so as to efficiently redistribute angular momentum along the atmosphere's vertical axis. Each zone of wave absorption can be expected to first build a momentum shelf that results in the development of Kelvin-Helmholtz waves and an outbreak of clear-air turbulence. This reestablishes a temperature anomaly similar to the one associated with the initial onset of a quantum atmospheric process.

In the present instance, the gradient of geopotential energy (which is proportional to the geostrophic windspeed at any given level) serves the same function as a sloping beach in dictating where wave energy will be absorbed and accumulated. The ensuing vertical motion resulting from this ageostrophy excites fields of adiabatic compression and expansion to reproduce the temperature signal whose origins were originally diabatic in character. This is why the quantum approach to this scale-interactive process comes to center upon wave packets of a given frequency, ratioed against the resulting change of enthalpy as expressed in each of the exponents found within Eqs. (15-38) through (15-40).

CONCLUSIONS AND SUGGESTIONS FOR FUTURE RESEARCH

In 1979, the National Science Foundation's journal *Mosaic* devoted an entire issue to unifying themes in research. One article (Lansford, 1979) addressed the many competing theories that seek to explain how volcanic aerosols, carbon dioxide buildup, ozone depletion, and solar variability are among fifty or more factors thought to influence climate. Lansford (1979) went on to state that a theory capable of unifying the physics of climate was "still in the wings." The formalism of quantum mechanics applied within the realm of geophysics may lead to this desired unification.

A new theory advanced within one discipline often begins to transform many other disciplines as scientists begin to test the theory's central tenets. As an example, plate tectonics has had a far reaching impact, not only within the geological sciences, but bearing on disciplines as diverse as biology and paleoclimatology. This is an interesting twist of events, because the precursor to plate tectonics was continental drift. The latter theory's acknowledged founder, Wegener, was a meteorologist by training. However, continental drift was far less effective than plate theory in stimulating new thinking because it lacked a rich, dynamical context for testing the validity of Wegener's hypothesis (Glen, 1982).

When the notion arose that Earth's varying magnetic field must be recorded in its ever-changing and moving crust, the essential thesis of plate tectonics could be verified. Recent hypotheses of mass extinctions related to specific impact craters are now benefitting from this same imprint of Earth's magnetism, only here we are no longer speaking of merely a geological theory, but one that has begun to influence fields as broad as astronomy and concepts as far reaching as biological evolution. It is relevant to the theory advanced in this chapter that all of these topics are couched in the language of angular momentum exchange.

For instance, any torque applied to Earth's crust varying in proportion to the changing distribution of ice volume or sea level could disrupt the submantle convective eddies believed to be responsible for the magnetic polarity of Earth. Or, as another example, the discovery in the last decade of a pulsating sun with discernible acoustic waves adds a new dimension to studies involving the radiation of energy through the convective mantle of the solar sphere. Both the terrestrial and solar phenomena just cited bear the footprint of the quantum process at work when it comes to establishing how energy is released from either the deep interior of the sun, the Earth, or the Earth's atmosphere. Until we understand these processes in sufficient detail we are unlikely to make much headway in advancing the physics of Earth's changing climate.

In the quantum approach to the Earth sciences, we find that climate change coincides with changes in the angular momentum budget of Earth's atmosphere. The atmosphere, in turn, may be expected to respond to both changes in the lithosphere as well as to changes that may have their roots deeply connected to the angular momentum budget of the solar system itself. I suspect we will soon have a means to not only detect these ongoing changes (as mentioned in the introduction in conjunction with the 1982-83 El Niño event), but of someday reading these changes written in the distant past of the geological record. In a sense, we may already have this means at our disposal since polarity shifts in Earth's magnetic field could be viewed as following from a chain reaction of quantum adjustments within the geophysical sphere. If the paleoclimatologist, the geologist, and biologist all eventually require the same essential data source to interpret the history of their respective realms (and quantum mechanics may be the conceptual vehicle by which we begin to make sense of this common data pool), we will have advanced along the road to having a common language for sharing information between the physical and biological sciences.

ACKNOWLEDGMENTS

Anton Chapin and William Pensinger provided essential portions of the derivation and interpretation, respectively, that led to this atmospheric version of quantum mechanics. I express my appreciation also to Dr. John Dutton of the Pennsylvania State University: notes taken during his course dealing with radiation theory first acquainted me with Planck's derivation. This work is currently supported through U.S. Department of Agriculture Hatch Project NY(C) 125442, as well as funds from the Associanziones Specola Solare Ticinese (Switzerland).

APPENDIX

Using Taylor series

$$f(x+a) = f(x) + af'(x) + \frac{a^2}{2!}f''(x) + \ldots,$$

where we let $a = (T_p - T_e)$ and

$$T_p = T_e + (T_p - T_e) = T_e + a,$$

we obtain, neglecting higher-order terms,

$$T_p^4 - T_e^4 \simeq 4T_e^3(T_p - T_e).$$

Multiplying by -1 and substituting into (15-10) gives

$$\left.\frac{dT_p}{dt}\right|_R = \frac{12\sigma T_e^3(T_e - T_p)T_p^{-1/3}}{C_p\rho\left(\dfrac{MR}{4/3\pi P_e}\right)^{1/3}}.$$

In the same way as above we may write, after an order of magnitude argument,

$$T_p^{-1/3} \simeq \left[1 - .33\left(\frac{T_p - T_e}{T_e}\right)\right] T_e^{-1/3} \simeq T_e^{-1/3}.$$

REFERENCES

Bohm, D. J., and Hiley, B. J., 1975, On the intuitive understanding of nonlocality as implied by quantum theory, *Foundations of Physics* **5:**93-109.

Glen, W., 1982, *The Road to Jaramillo,* Stanford: Stanford University Press.

Hess, S. L., 1979, *Introduction to Theoretical Meteorology* (reprint ed.), Huntington, N.Y.: Krieger Publishing Company.

Jose, P., 1965, Sun's motion and sunspots, *Astron. Jour.* **70:**193-200.

Lansford, H., 1979, Modeling the climate system, *Mosaic* **10:**2-8.

Paine, D. A., 1983, A climate hypothesis describing the solar-terrestrial system as a frequency domain with specific response characteristics, *Eos* **64:**425-428.

Paine, D. A.; Zack, J. W.; Moore, J. T.; and Posner, R. J., 1975, A theory for the conservation of 3-D relative vorticity which describes the cascade of energy-momentum leading to tornadic vortices, in *Ninth Conference on Severe Local Storms,* Preprint Volume, Boston: American Meteorological Society, pp. 131-138.

Robbins, R. W., 1983, An analysis of cycles in sunspot data, *Cycles* **34:**187-192.

Rosen, R. D.; Salstein, D. A.; Eubanks, T. M.; Dickey, J. O.; and Steepe, J. A., 1984, An El Niño signal in atmospheric angular momentum and earth rotation, *Science* **225:**411-414.

V

Long-Term Climate (10^3–10^7 yr) and Periodicity

16: Paleoclimatic Variability at Frequencies Ranging from 10^{-4} Cycle per Year to 10^{-3} Cycle per Year–Evidence for Nonlinear Behavior of the Climate System

P. Pestiaux
Université Catholique de Louvain-la-Neuve

J. C. Duplessy
Laboratoire Mixte CNRS-CEA, Gif-sur-Yvette, France

A. Berger
Université Catholique de Louvain-la-Neuve

Abstract: The paleoclimatic variability at frequencies ranging from 10^{-4} cycle per year (cpy) to 10^{-3} cpy is investigated using a set of four deep-sea cores from the Indian Ocean. After determination of the upper and lower frequency limits of the significant spectral peaks, three frequency bands of high paleoclimatic variability are defined: they are centered around the spectral maxima located at 10.2 kyr, 4.6 kyr, and 2.3 kyr. The frequency localization of the spectral peaks is then refined by high-resolution spectral analysis.

Some of the resulting peaks have frequencies that are close to those previously detected in other paleoclimatic records. Additional peaks are also in good correspondence with the response of a nonlinear climatic oscillator forced by insolation variations. This correspondence suggests that the climatic system is characterized by a highly nonlinear response to orbital forcing and that the precessional contribution of the forcing interacts strongly with the precipitation-temperature feedback used in the model. This interaction is particularly efficient in the Indian Ocean, which is influenced by the monsoon circulation.

INTRODUCTION

Long-term climatic changes during the Pleistocene are of primary importance in understanding the behavior of the climatic system. Past climates are reconstructed from indirect evidence concerning the volume of former ice sheets and the changes of flora and fauna over the continents or within the oceans. Such geological proxy data can be found in marine or continental records. But the deep-sea paleoclimatic indicators are the most complete in the sense that they are distributed over the world ocean, they cover different time scales, and they can be dated by radiometric methods (Kukla, 1978).

Most efforts in the interpretation of paleoclimatic indicators recorded in deep-sea cores have been focused in two directions. The first one has been the reconstruction of the sea-surface temperatures during the last ice age by isotopic and faunal analyses on a global spatial scale and in a restricted time range (Imbrie and Kipp, 1971; Hecht, 1973; CLIMAP, 1976). The second one has been the study of climatic sensitivity to orbital forcing. To test quantitatively the relationship between the fluctuations of Pleistocene ice sheets and the astronomical parameters of the Earth's orbit, different data analysis methods have been applied to deep-sea records covering a sufficiently large time interval to allow the identification of spectral maxima close to the astronomical ones—namely, those around 100 kyr, 41 kyr, 23 kyr, and 19 kyr (Hays, Imbrie, and Shackleton, 1976; Pisias and Moore, 1981; Pestiaux and Berger, 1984*b*).

The complexity of the climate system, including at least the ocean and the cryosphere as long-term components, implies that not only orbital frequencies (10^{-5}-10^{-4} cpy) have to be considered but also higher frequencies in the range of 10^{-4} cpy to 10^{-2} cpy. A complex system composed of interconnected subsystems with different characteristic times can produce a response with quasiperiodic behavior at linear combinations of the forcing frequencies. The choice between the possible combinations depends on the type of nonlinearities and on the values of the internal parameters (Minorsky, 1962). The detection of these combinations in proxy records can be used to study theoretically the proper mathematical equations modeling the system and giving rise to these combinations or to calibrate the parameterization of existing nonlinear climatic models. Indeed, the interest in detecting such high frequencies and differentiating among them has been expressed recently to verify the results found by nonlinear climatic oscillators forced by the variations of the Earth's orbital parameters (Le Treut and Ghil, 1983).

A well-known difficulty in this program stems from the fact that the benthic infauna burrow the uppermost 5-20 cm of sediment (Berger and Heath, 1968; Guinasso and Schink, 1975), blurring and smoothing the paleoclimatic signal. This bioturbation leads to a complete suppression of any variability at time scales shorter than 10-6 kyr in deep-sea cores that have a sedimentation rate of a few centimeters per thousand years. Hence, deep-sea cores with a sedimentation rate lower than 10 cm/kyr are unsuitable for the investigation of paleoclimatic variability at frequencies higher than 10^{-4} cpy (Goreau, 1980; Pisias, 1983; Pestiaux and Berger, 1984*a*).

This chapter presents a set of deep-sea cores characterized by high sedimentation rates and studies their climatic variability in the frequency range of 10^{-4} cpy to 10^{-3} cpy. The next section describes the data base used, stressing those characteristics that are important for the final interpretation of the results. The following section discusses the basic chronologies and constructs the time scales of the planktonic and/or benthic oxygen-isotopic records. After briefly outlining our approach to time-series analysis, we present the results, starting with the most significant spectral peaks. Next, we use the method of maximum entropy spectral analysis to increase the resolution of these spectral peaks and hence improve their climatic interpretation. Then we define three frequency bands of increase in paleoclimatic variability and compare them to other experimental and theoretical evidence of such paleoclimatic variability. The final section is a discussion of climatic implications.

DATA BASE

Two necessary conditions for reliable identification of high-frequency paleoclimatic variability are those of high sedimentation rate and refined sampling. Following these two criteria, four deep-sea cores collected during cruises OSIRIS I, II, and III of the French ship M/S *Marion Dufresne* in the Indian Ocean (Table 16-1) have been selected, each of them being sampled every 10 cm. The oxygen-isotopic composition ($\delta^{18}O$) of planktonic and/or benthic foraminifera has been measured for each sample following the method described in Duplessy (1978).

The $\delta^{18}O$ variations of planktonic foraminifera that live in surface waters are controlled in a complex way by sea-surface $\delta^{18}O$ variations and by sea-surface temperatures, while the $\delta^{18}O$

Table 16-1. Location and Depth of Cores

Core	*Latitude*	*Longitude*	*Depth (m)*
MD77191	07°30′N	76°43′E	1,254
MD76131	15°32′N	72°34′E	1,230
MD76135	12°26′N	50°31′E	1,895
MD73025	43°49′S	51°16′E	3,284

variations of benthic foraminifera are mostly sensitive to the variations of deep-water characteristics. The main controlling factor is sea-water $\delta^{18}O$, although minor deep-water temperature changes have been evidenced during the last climatic cycle (Duplessy, Moyes, and Pujol, 1980; Duplessy, 1983). We describe each of the records now to ensure a meaningful interpretation of the results. The oxygen-isotope records are presented as a function of the depth in Figure 16-1, the record on the left being from the species living closest to the surface. As can be seen from Figure 16-2, three cores are located in the northern Indian Ocean particularly close to the coast, leading to a large sediment contribution of terrigeneous detrital input from the continent. The analyzed planktonic species is *Globigerinoides ruber* for the three cores taken from the northern Indian Ocean (Duplessy, 1982). Core MD77191 (Fig. 16-1*a*) is located at the tip of India and is influenced by the water circulation in the Arabian Sea as well as in the Bay of Bengal and, therefore, by the monsoon circulation. The planktonic foraminifera are submitted to surface conditions depending on salinity. Since the depth of the core is 1,250 m, the benthic species *(Uvigerina)* lived in intermediate water masses whose physical characteristics are dependent on the mixing between the Indian Ocean deep waters and those coming from the Red Sea or sinking in the northern Arabian Sea (Wirtky, 1973). Core MD76131 (Fig. 16-1*d*), located off Bombay in the eastern Arabian Sea, is also influenced by the monsoon variability, especially through the effect on salinity of the coastal run-off. The second analyzed species is *Globorotalia menardii*, a planktonic foraminifer that lives up to a depth of 200 m or even deeper (Duplessy, Blanc, and Be, 1981) and is therefore less sensitive to surface effects.

Along the eastern coast of Arabia, intense upwelling takes place during the summer monsoon when the wind blows from the southwest parallel to the coast. The planktonic $\delta^{18}O$ record of core MD76135 (Fig. 16-1*c*) located in that area reflects surface conditions that depend on the wind-controlled upwelling intensity (van Campo, Duplessy, and Rossignol-Strick, 1982). Since the core depth is 1,895 m, the benthic isotopic record mainly reflects the $\delta^{18}O$ variations of the Indian Ocean deep water originating from the southern ocean.

The last core analyzed (MD73025, Fig. 16-1*b*) is different from the preceding ones since it is located in the southern Indian Ocean, in the belt of siliceous high productivity. Its high sedimentation rate comes, therefore, from the high proportion of diatoms contained in the sediment. The planktonic isotopic record (*Neogloboquadrina pachyderma* left coiling) is influenced by the surface hydrology, the input of Antarctic meltwater, and the local ratio of precipitation to evaporation. These factors explain the strong difference between the planktonic and the benthic isotopic records as it is illustrated by the difference in the relative amplitudes of the substages within oxygen-isotopic stage 5 (Emiliani, 1955).

CHRONOLOGY

Core MD77191 exhibits the highest sedimentation rate and is therefore limited in time to the past 18,000 yr. Eleven age controls (Table 16-2) have been estimated by measuring the ^{14}C content of organic matter in the sediment. Considering that the first two dates may be within the bioturbation zone, a constant sedimentation rate of 52.6 cm/kyr has been estimated to a depth of 630 cm. At that level, around 12 kyr, the sedimentation rate drops drastically to a value of 13.9 cm/kyr, suggesting a dramatic change in the sedimentation rate associated with the modification of the climatic regime.

The chronology of the three other cores has been established on the basis of unambiguous

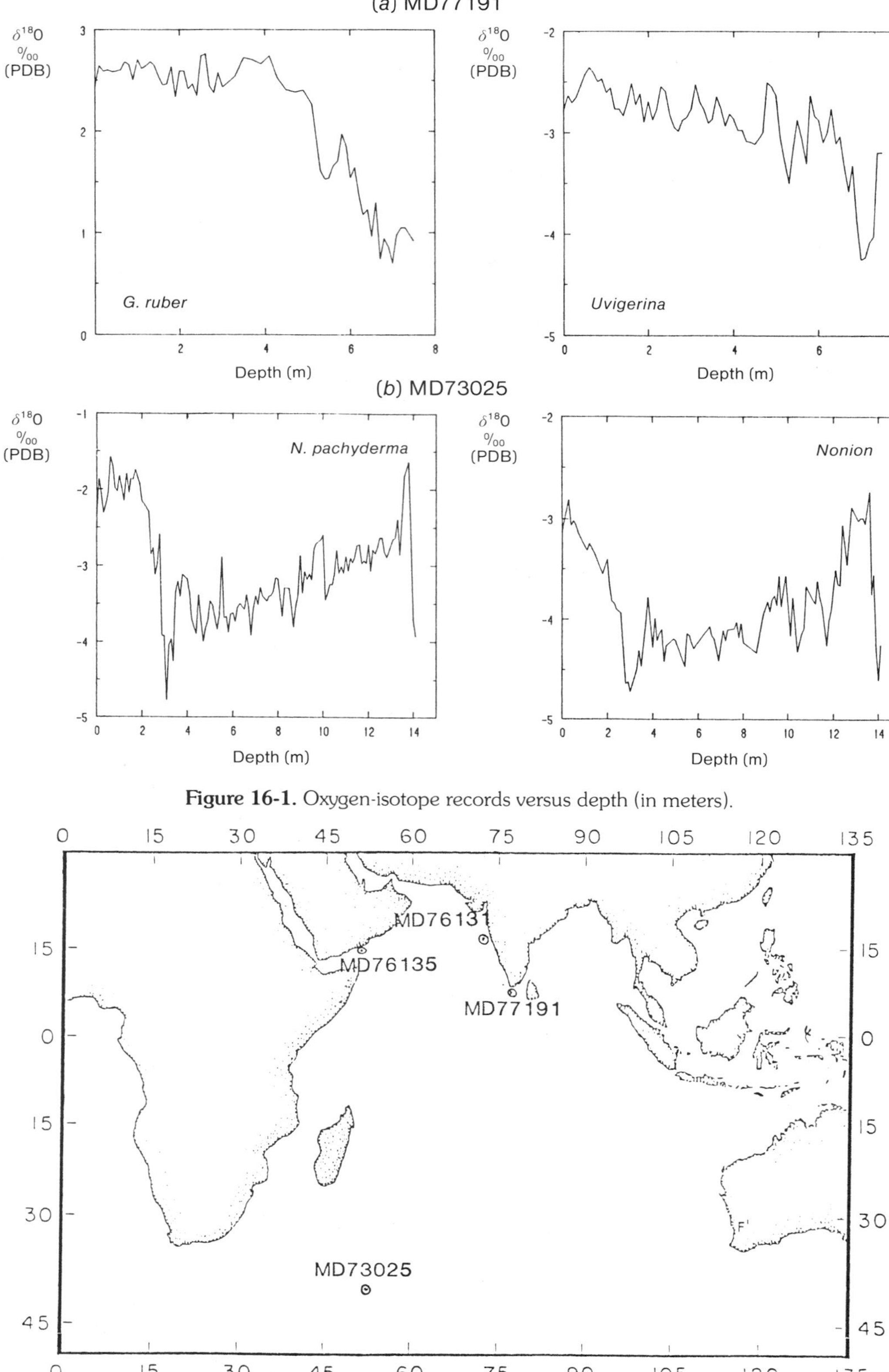

Figure 16-1. Oxygen-isotope records versus depth (in meters).

Figure 16-2. Location of the cores.

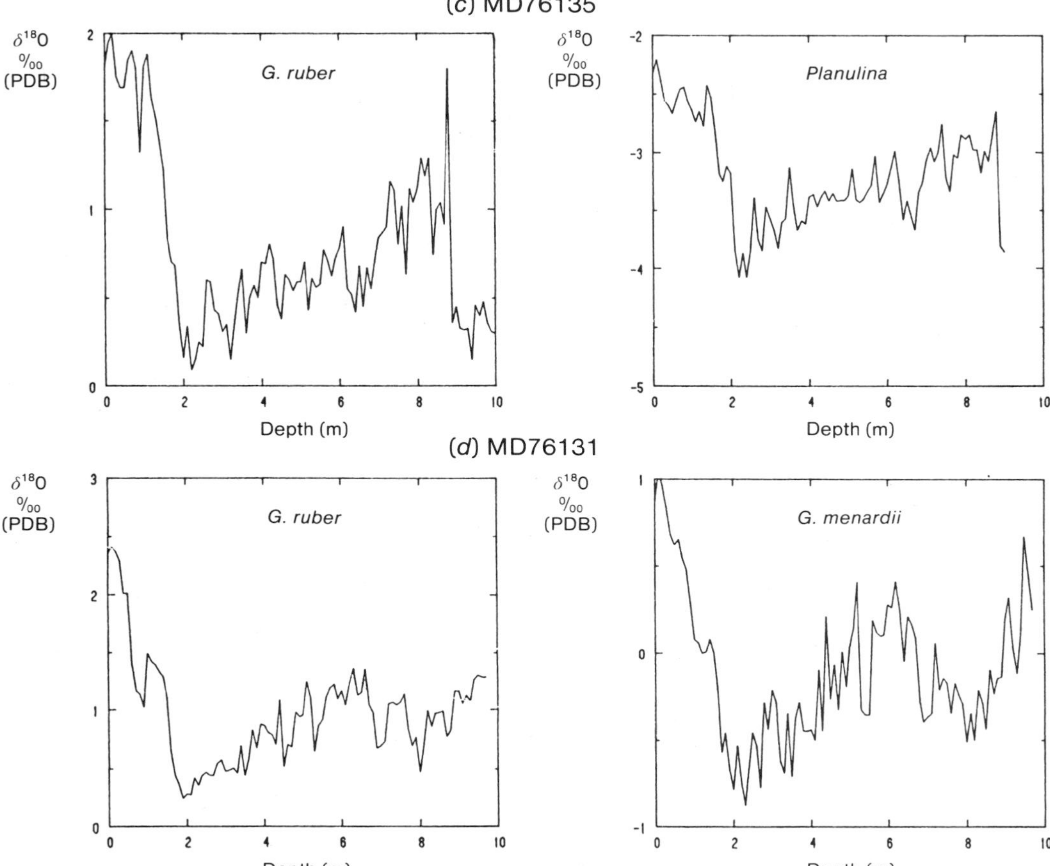

Table 16-2. Age Control of Core MD77191

Depth (cm)	*^{14}C Ages (yr B.P.)*
0	2,250 ± 230
60	2,900 ± 240
150	3,100 ± 250
250	5,000 ± 290
400	7,760 ± 350
480	9,160 ± 440
500	10,180 ± 480
540	10,970 ± 520
600	11,730 ± 570
640	13,400 ± 680
700	17,700 ± 500

Note: ^{14}C ages are corrected for the age of the surface water.

isotopic signatures in each of them (Table 16-3). The top of the core is assumed to reflect modern conditions, and the isotopic stage transition 1/2, corresponding to the abrupt climatic change from the last glacial maximum to the last interglacial, has been dated by ^{14}C (Sancetta, Imbrie, and Kipp, 1973; Duplessy, Blanc, and Be, 1981); an age of 11.1 kyr B.P. has been estimated for the $\delta^{18}O$ increase found between the Allerod and the Younger Dryas events (Duplessy, Blanc, and Be, 1981). The last glacial maximum has been shown to be synchronous to within 1.5 kyr also in the Indian Ocean (Prell et al., 1980), suggesting an approximate age of 18 kyr B.P.

The 4/5 isotopic transition has been dated on the basis of the extrapolation of sedimentation rate found by classical absolute dating methods, the age of 70 ± 5 kyr fitting reasonably well all the proposed age estimations (McIntyre, Ruddiman, and Jantzen, 1972). The 5/6 isotopic transition has been dated using ^{230}Th and ^{231}Pa ratios (Ku, Bischoff, and Boersma, 1972), indicating an age of 127 kyr B.P. with an uncertainty of 6 kyr. The limitation of our study to the last full glacial-interglacial cycle has the advantage that the age models based on absolute dating methods match reasonably well with the more restrictive age models based on the a priori cause-to-effect relationship between insolation and paleoclimatic indicators. Our age model does not introduce any a priori assumption concerning the physical relationship to insolation, and the results of our analyses have to be interpreted therefore as independent verification of the climatic model with which they are compared.

SPECTRAL CHARACTERISTICS

Using the age models described earlier, the sample depths are first transformed into nonequidistant, time-dependent values before interpolating them by cubic splines at constant time intervals corresponding to the average sampling resolution (Pestiaux, 1982). In all the

Table 16-3. Age Models Based on Isotopic Stratigraphy

Isotopic Signature: *Adopted Age:*	*Transition* *1/2* *11 kyr*	*Glacial* *Maximum* *18 kyr*	*Transition* *4/5* *70 kyr*	*Transition* *5/6* *127 kyr*
MD76131	X		X	
MD76135	X		X	
MD73025		X		X

Note: The unambiguous signatures adopted are marked by a cross.

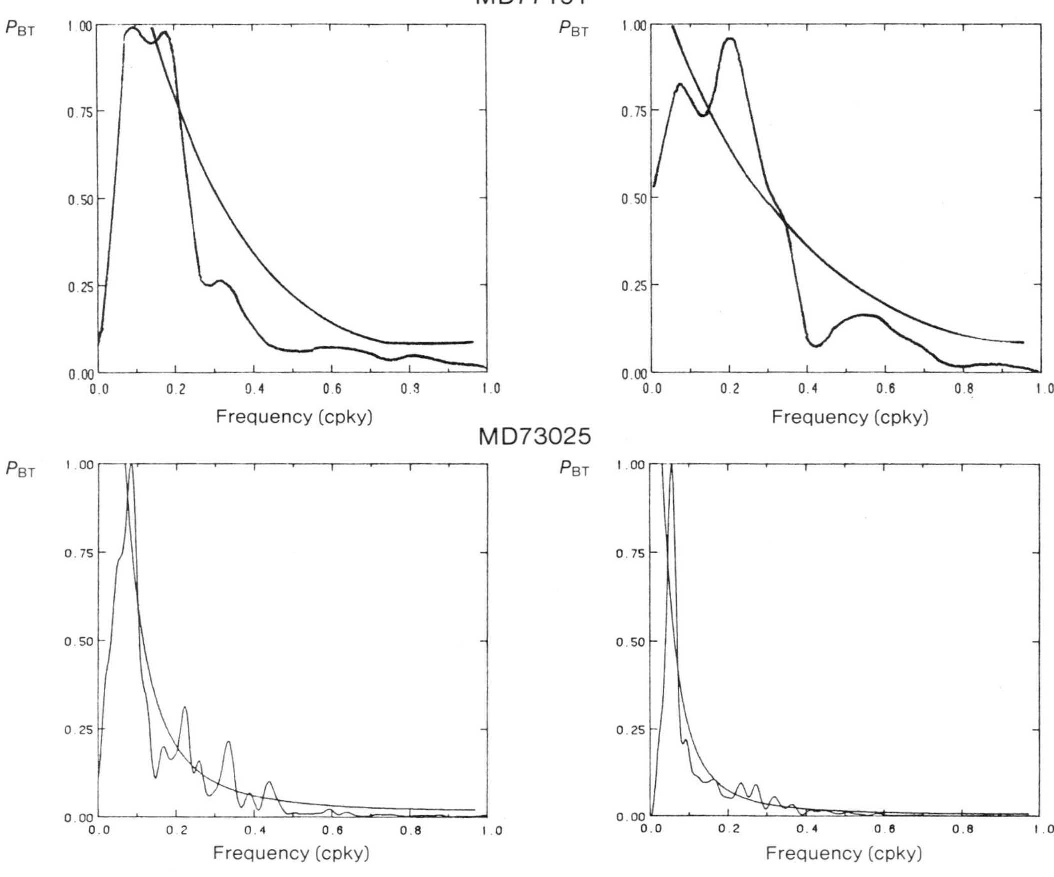

Figure 16-3. Normalized Blackman-Tukey spectra together with the 80% confidence limit of the equivalent red noise process.

records, the high-frequency variability is superimposed on the glacial-interglacial oscillation, implying that the mean is nonstationary. If spectral analysis were applied directly to these original nonstationary records, the shape of the power spectrum would be reddened artificially, leading to an amplification of the low frequencies and therefore masking variability in high frequencies (Hannan, 1958). To remove this effect, it has been necessary to detrend each of the records by fitting an appropriate piecewise polynomial and subtracting it from the original record. This polynomial detrending has been preferred to high-pass filtering since the building of an appropriate filter would have implied the loss of too many values with respect to the length of the records.

A strategy for the spectral analysis of paleoclimatic records is described by Pestiaux and Berger (1984*a*), who suggested the use of a combination of different methods, each of them having unique advantages and drawbacks. To avoid duplication of results, two techniques of spectral analysis have been selected here for their complementarity. The classical lagged correlation spectral analysis (nonparametric approach) is first used for its accuracy in the spectral amplitude estimation and its good statistical properties—that is, for producing reliable bandwidths and confidence intervals of the spectral estimates. This first method is the indirect autocovariance approach commonly used under the name of the Blackman-Tukey (BT) technique. The estimated autocovariance function is computed from the initial time series and smoothly truncated by multiplying it with a set of weights, called a lag window, the Tukey (Hanning) lag window in this case. The power

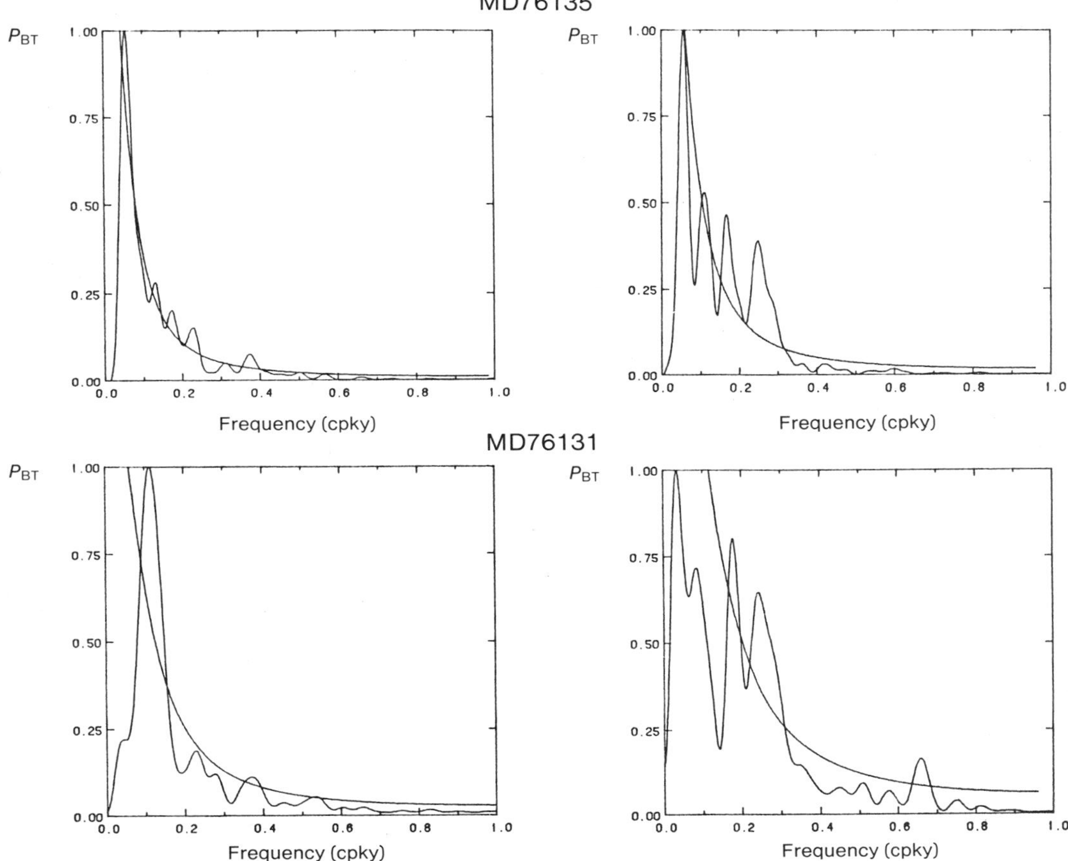

spectrum is then obtained by taking the real discrete Fourier transform of this modified autocovariance function (Hannan, 1958; Jenkins, 1969; Parzen, 1961). The statistical significance of the spectral peaks is tested with regard to a null hypothesis of continuum red noise background, the first autocorrelation coefficient being significantly different from zero at a 5% significance level in all the cases. The power spectra of the eight records analyzed are presented in Figure 16-3 together with the 80% upper confidence interval of the associated red noise power spectrum.

The second technique used is the maximum entropy spectral analysis by least squares (MELSA), which is useful especially when the number of available data is limited or when high-frequency resolution is needed to split neighboring spectral peaks and control the regularity of the detected spectral peaks (Marple, 1980). When the analyzed record presents a spectral peak that varies in time, as is often the case in deep-sea paleoclimatic records, MELSA indicates that a detected broad spectral peak is to be considered as a superposition of two quasiperiodicities, each characterizing some part of the record, which can change smoothly into each other with time. For example, the planktonic record of the core MD73025 exhibits a broad spectral peak around the 12.2 kyr quasiperiodicity when using the BT technique. MELSA shows this to be in fact a trimodal peak with quasiperiodicities at 19 kyr, 13 kyr, and 10 kyr; the two last are the result of a shift of the frequencies of the peaks starting around 70 kyr B.P. The maximum entropy spectra (Fig. 16-4) clearly suggest superposition of nonstationary spectral components that are grouped in bimodal

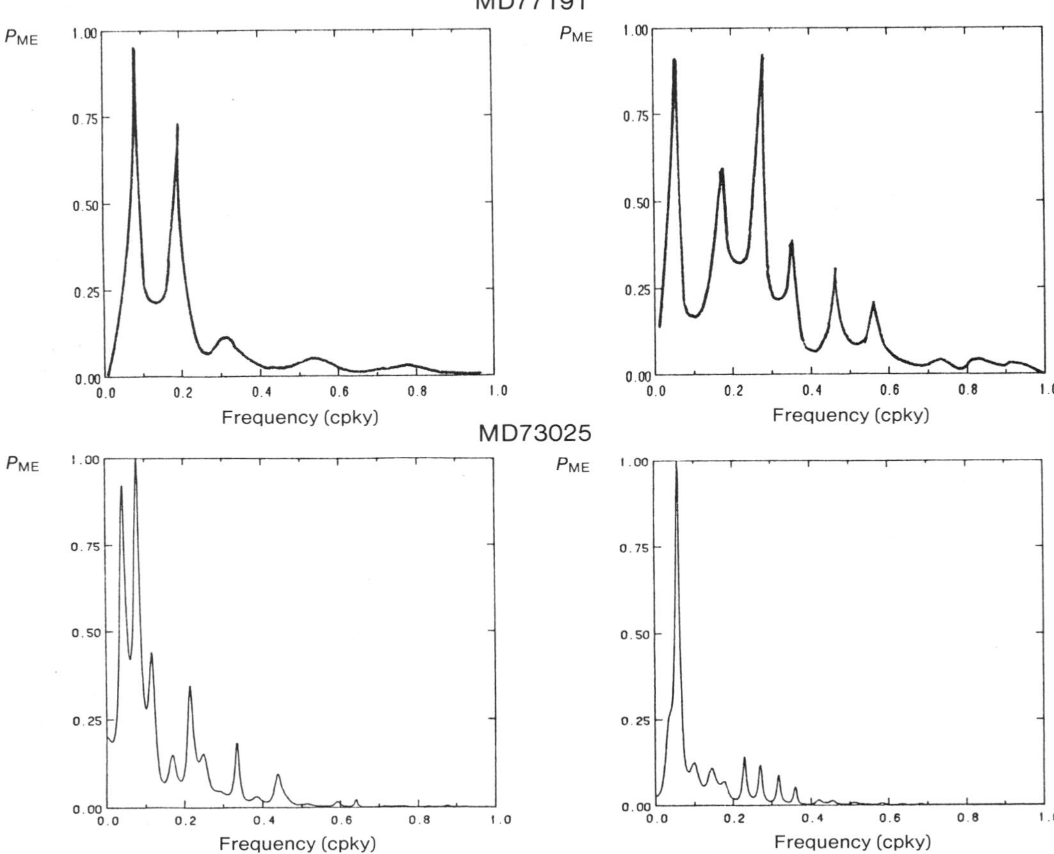

Figure 16-4. Normalized maximum entropy spectra.

or trimodal peaks in the case of a temporary irregularity. Indeed, if one has a T_1 quasiperiodicity in one-half of a record and another T_2 quasiperiodicity in the other half of the same record, the two corresponding frequencies being sufficiently different, the resulting MELSA spectrum will have two distinct spectral peaks at $1/T_1$ and $1/T_2$. However, if the quasiperiodicity changes smoothly rather than abruptly with time and the mean quasiperiodicity is at T_3, a broad bimodal spectral peak will appear at $1/T_3$.

DISCUSSION

Each of the spectral peaks found to be statistically significant (S), by the classical lagged correlation method, have been put between two upper and two lower limits as shown in Table 16-4. The first upper and lower limits (U1 and L1) correspond to the statistical bandwidth of the Tukey window and represent the frequency resolution of the spectral analysis method. To these limits we add an uncertainty of chronology of roughly one-tenth of the estimated quasiperiodicity. The final upper and lower limits (U2 and L2) thus obtained are used as a first indication for the lack of resolution of the lagged correlation method and for the possible differentiation of the peaks, in the same core or from core to core.

Starting with the most confidently dated record (MD77191), the quasiperiodicities around 2.3 kyr and 1.9 kyr, together with their associated upper and lower limits, suggest the definition of a first frequency band (B1) of paleoclimatic variability including the quasiperiodicities rang-

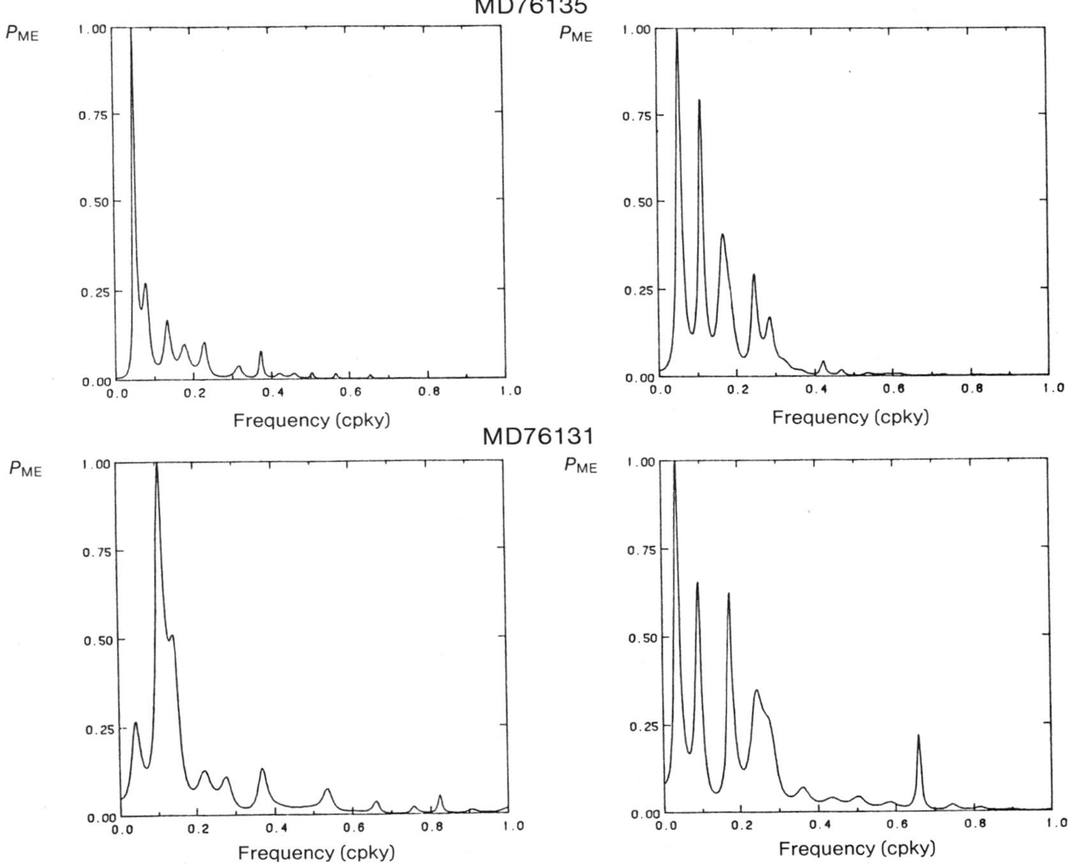

Table 16-4. Statistically Significant Spectral Peaks Together with Their Statistical and Experimental Uncertainty

Core	U2	U1	S	L1	L2
MD77191					
Planktonics	3.4	3.2	2.3*	1.8	1.6
Benthics	2.7	2.5	1.9***	1.5	1.3
MD73025					
Planktonics	16.2	15.2	12.2***	10.2	9.2
	5.4	4.9	4.5***	4.2	3
	3.5	3.2	3***	2.9	2.6
	2.6	2.4	2.3***	2.2	2
Benthics	26.4	24.6	17.6***	13.7	11.9
	5.1	4.7	4.4***	4.1	3.7
	4.2	3.9	3.7***	3.5	3.1
	3.7	3.4	3.2***	3	2.7
MD76135					
Planktonics	26.5	25.8	17.4*	13.1	11.4
	10	9.2	7.8*	6.8	6
	7.1	6.5	5.8**	5.2	4.4
	5.3	4.9	4.5***	4.1	3.6
	3.1	2.8	2.7***	2.6	2.3
Benthics	28.2	26.3	17.6*	13.2	11.4
	11.4	10.5	8.8*	7.6	6.7
	7.4	6.8	6***	5.3	4.7
	4.8	4.4	4.1***	3.4	—
MD76131					
Upper planktonics	11.3	10.5	8.3***	6.9	6.1
	3.1	2.8	2.6*	2.4	2.1
Lower planktonics	6.8	6.3	5.4**	4.8	4.3
	5	4.6	4.1**	3.7	3.3
	1.8	1.6	1.5***	1.4	1.2

Note: Significance levels are as follows: *, .2; **, .1; ***, .05.

ing from 1.3 kyr to 3.4 kyr (B1). The spectral peaks of the other records that have a large part of their extreme possible range in this first frequency band are retained as belonging to this band. Two other frequency bands (B2 and B3) have been defined in a similar way, one including quasiperiodicities ranging from 3.5 kyr to 6.5 kyr (B2) and the last one from 7 kyr to 14 kyr (B3). The maximum entropy spectra (see Fig. 16-4) are then used to increase the frequency resolution and eventually split neighboring spectral peaks (Table 16-5). All the records covering a sufficiently long time interval present a superposition of three different variabilities, with increasing amplitudes when the quasiperiodicity increases. In each of the frequency bands defined, the regularity of the oscillations has been controlled to verify that two spectral peaks located in the same frequency band were the

Table 16-5. Maximum Entropy Quasiperiodicities (kyr) in Each Frequency Band

Core	B3	B2	B1
MD77191			
Planktonics			2.3
Benthics			2.5, 1.8
MD73025			
Planktonics	13	4.5	2.9, 2.3
Benthics		4.3, 3.6	3, 2.8
MD76135			
Planktonics	12.3, 7.4	5.6, 4.3	2.4
Benthics	8.8	5.5, 4.1	
MD76131			
Upper planktonics	9.5		2.7
Lower planktonics		5.6, 4.1	1.5
Mean	10.2	4.6	2.3
Standard deviation	1.2	0.3	0.2

result of a shift in the spectral peaks in that band. These results are illustrated in Figure 16-5 where the statistically significant quasiperiodicities have been pointed with their maximal upper and lower limits U2 and L2. The maximum entropy refinements of these quasiperiodicities have been indicated by crosses. Finally, the quasiperiodicities refined in frequency resolution and belonging to the same frequency band have been averaged to give mean quasiperiodicities of 10.2 kyr, 4.6 kyr, and 2.3 kyr, with corresponding standard deviations of 1.2 kyr, 0.3 kyr, and 0.2 kyr (see Table 16-5).

The increase of paleoclimatic variability about the red noise level in these three frequency bands is confirmed by the detection of spectral peaks in other paleoclimatic indicators from the land and the ocean. For example, *deep-sea isotopic records* with a sedimentation rate around 2 cm/kyr present quasiperiodicities around 11 kyr (Morley and Hays, 1981). *Pollen assemblages* from the Grande Pile record (Molfino, Heusser, and Woillard, 1984) suggest evidence for quasiperiodicities around 9 kyr and 6 kyr. Pisias and co-workers (1973) and Schnitker (1982) detected quasiperiodicities around 5 kyr and 2.5 kyr respectively in a *paleotemperature record derived from foraminiferal assemblages* and in the *faunal composition of deep-sea benthic foraminifera.* Denton and Karlén (1973) found a quasiperiodicity of 2.5 kyr in the postglacial waxing and waning of North American and European *mountain glaciers.* Dansgaard et al. (1984) and Benoist et al. (1982) found quasiperiodicities around 2.5 kyr in oxygen-isotopic records from Greenland and Antarctic ice cores.

Although far from obvious, a possible interpretation of these preferential frequency bands of climatic variability may be given in terms of a nonlinear response of the climatic system forced by insolation variations at the top of the atmosphere. The first feature of the 10.5 kyr, 5 kyr, and 2.5 kyr quasiperiodicities is that they are roughly harmonics of themselves and of the 41 kyr, 23 kyr, and 19 kyr quasiperiodicities found in the insolation (Berger and Pestiaux, 1984). In the case of a linear system submitted to a periodic external forcing with frequency w_f, the response of that system can be expressed as the sum of a sinusoidal oscillation at that frequency, w_f, and of additional sinusoidal oscillations at the characteristic frequencies, w_i, of the system (Minorsky, 1962).

In the case of a nonlinear system the problem is more complicated. The response often contains a component at the forcing frequency, w_f, but this oscillation is no longer necessarily a sinusoid. It can be distorted, more or less asymmetric following the case. The amplitude of the response, A_r, depends not only on the forcing frequency, w_f, but also on the forcing amplitude, A_f (Minorsky, 1962). Furthermore, the amplitude of the response, A_r (w_f, A_f), can be more complex with several possible values and jumps occurring between them. The response can also present harmonic, kw_f, or subharmonic, w_f/k, components, with k being an integer. The choice between them (synchronization) depends in general on the forcing amplitude. The complexity of nonlinear systems is increased further when internal characteristic frequencies, w_i, are introduced. The response of such a system can oscillate with an internal frequency w_i, with a modified internal frequency w_i due to the presence of the external frequency w_f, or even hesitate between these two types of behaviors to give combined oscillations. To summarize, one can say that nonlinear oscillators are characterized on the one hand by harmonics, subharmonics, and combination tones of the forcing frequencies and on the other hand by a frequency-amplitude dependence of the response. Notice that the use of the term *combination tones* for linear combinations with integer coefficients of given frequencies comes from the first study of such tones, in musical acoustics (Helmholtz, 1885).

Le Treut and Ghil (1983) have studied the sensitivity of a simple nonlinear climatic oscillator, developed by Källen and co-workers (1979) and by Ghil and Le Treut (1981), to orbital forcing. This oscillator includes a large part of the components of the climatic system acting on the time scale of glaciation cycles—namely, the global temperature, the global ice volume, and the deformation of the Earth's crust under the ice. The internal frequency (w_i) of the oscillator has been found to be roughly 6 kyr. Taking the insolation forcing frequencies w_{f1}, w_{f2}, and w_{f3} respectively at 1/19, 1/23, and 1/41 kyr^{-1}, the response exhibits two increases of paleoclimatic variability around 10.5 kyr and 6 kyr, with

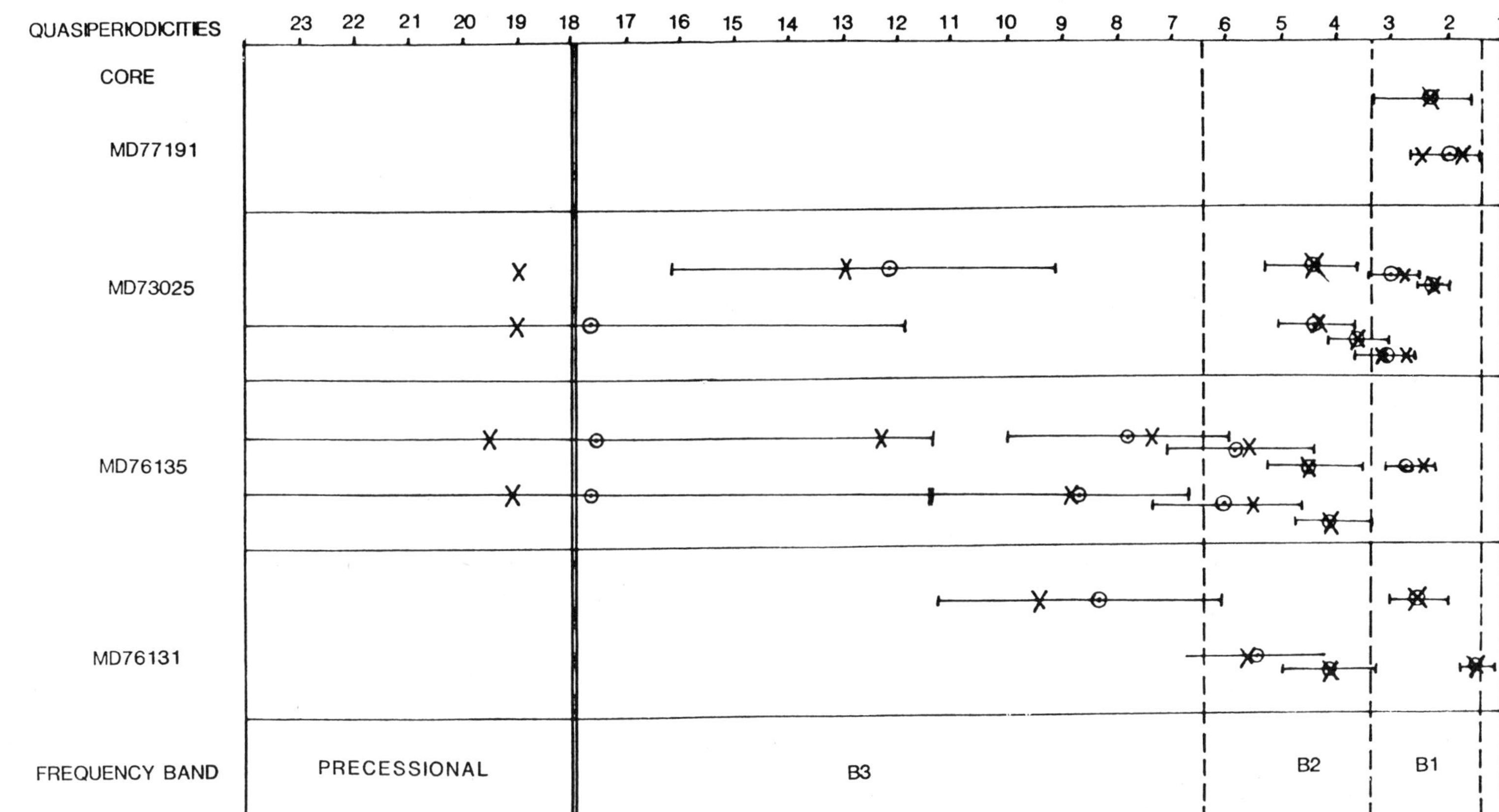

Figure 16-5. Definition of the frequency bands corresponding to an increase of paleoclimatic variability. The statistically significant quasiperiodicities (⊙) are put within their extreme possible ranges L2 and U2 (├──⊙──┤). The maximum entropy refinement of the quasiperiodicities is indicated by crosses (X).

clusters of peaks around these values expressed as combination tones of the forcing frequencies, w_{fi}.

Each of the quasiperiodicities detected in the present study has been decomposed as combination tones of the forcing frequencies, so that each combination $n_i w_{f1} + p_i w_{f2} + q_i w_{f3}$ approximates a quasiperiodicity d_i detected in our records, with n_i, p_i, q_i being positive or negative integers. Defining the order of the combinations O_i by

$$O_i = |n_i| + |p_i| + |q_i|,$$

those integers n_i, p_i, and q_i, which correspond to the minimal order, have been chosen to approximate each detected frequency d_i. As shown in Table 16-6, there exists always at least one possibility of such combination of tones (Kinchin, 1964; Koksma, 1936), and the number of possibilities increases as the order O_i increases. Four combination tones detected in the data and having quasiperiodicities above 5 kyr (the limit of spectral resolution explicitly considered by Le Treut and Ghil [1983]) coincide with their model prediction.

A monsoon amplification of the insolation forcing has been found independently by sensitivity experiments with general circulation models: an increase in insolation leading to an increase in temperature, precipitation, and cloudiness in the monsoon countries (Kutzbach, 1981; Royer, Deque, and Pestiaux, 1983). Insolation variations in low latitudes are governed mainly by precessional effects. The order of the precessional coefficients (n_i, p_i) in Table 16-6 is always higher than the order of the obliquity coefficient (q_i). This finding confirms the importance of the precipitation-temperature feedback, particularly strong in the Indian Ocean due to the monsoon circulations.

CONCLUSION

The detection of combination tones of the astronomical frequencies both in the paleoclimatic data and in the results of the paleoclimatic models is an efficient criterion for the calibration and the validation of paleoclimatic models. In the case of the Indian Ocean, the analysis of combination tones has demonstrated the importance of the precessional components in relation with the monsoon effect. It seems, therefore, that the role of the hydrological cycle is fundamental in these countries. Systematic use of high-frequency resolution spectral techniques

Table 16-6. Adjusted Combination Tones

d_i	n_i	p_i	q_i	O_i	$n_i w_{f1}+p_i w_{f2}+q_i w_{f3}$	*Le Treut and Ghil, 1983*
13.0	1	0	1	2	12.98	X
12.3	2	0	−1	3	12.37	
10.2	1	1	0	2	10.41	X
9.5	2	0	0	2	9.51	X
8.8	0	2	1	3	8.99	
7.4	1	2	0	3	7.17	X
5.6	0	4	0	4	5.76	
5.5	1	3	0	4	5.47	
4.6	0	5	0	5	4.61	
4.3	4	0	1	5	4.26	
4.1	3	2	0	5	4.02	
3.0	2	4	2	8	3.05	
2.9	2	5	1	8	2.88	
2.8	3	4	1	8	2.81	
2.7	2	5	2	9	2.69	
2.5	5	2	2	9	2.51	
2.3	5	4	0	9	2.29	

The linear combinations $n_i w_{f1}+p_i w_{f2}+q_i w_{f3}$ of the astronomical frequencies w_{fi} are chosen to approximate the detected frequency $1/di$.

and of cores with high sedimentation rate would allow us to extend these results obtained for the Indian Ocean to other climatic regions.

ACKNOWLEDGMENTS

We much appreciate the suggestions and comments of M. Ghil, J. C. Jouzel, W. Karlén, J. Labeyrie, H. Le Treut, J. L. Lorius, and C. Nicolis. Thanks are due to J. Antignac and B. Le Coat for their help in the isotopic analyses and to G. Delibrias for providing the ^{14}C ages of core MD77191. Cruises OSIRIS I, II, and III of the French M/S *Marion Dufresne* are supported by TAAF (Terres Australes et Antarctiques Françaises).

This research has been funded by the French CNRS and CEA and by the European Economic Community under contracts CLI-026-B and STI-004-J-C (CD).

REFERENCES

Benoist, J. P.; Glangeaud, F.; Martin, N.; Lacoume, J. L.; Lorius, C.; and Oulahman, A., 1982, Study of climatic series by time-frequency analysis, in *Proceedings of the ICASSP82*, New York: IEEE Press, pp. 1902-1905.

Berger, A., and Pestiaux, P., 1984, Accuracy and stability of the Quaternary terrestrial insolation, in A. Berger, J. Imbrie, J. Hays, G. Kukla, B. Saltzman (eds.), *Milankovitch and Climate*, Dordrecht, Holland: Reidel Publ. Company, pp. 83-111.

Berger, W. H., and Heath, G. R., 1968, Vertical mixing in pelagic sediments, *Jour. Marine Research* **26:**135-143.

Bhattacharya, K.; Ghil, M.; and Vulis, I. L., 1982, Internal variability of an energy-balance model with delayed albedo effects, *Jour. Atmos. Sci.* **39:**1747-1773.

Broecker, W. S.; Gerard, R.; Ewing, M.; and Heezen, B. C., 1960, Natural radiocarbon in the Atlantic Ocean, *Jour. Geophys. Research* **65:**2903-2931.

CLIMAP, 1976, The surface of the ice-age Earth, *Science* **191:**1131-1137.

Dansgaard, W.; Johnsen, S. J.; Clausen, H. B.; Dahl-Jensen, D.; Gundestrup, N.; Hammer, C. H.; and Oescheger, H., 1984, North Atlantic oscillations revealed by deep Greenland ice cores, *Geophysical Monogr.* **29:**288-298.

Denton, G. H., and Karlén, W., 1973, Holocene climatic variations—Their pattern and possible cause, *Quat. Research* **3:**155-205.

Duplessy, J. C., 1978, Isotope studies, in J. Gribbin (ed.), *Climatic Change*, Cambridge: Cambridge University Press, pp. 46-67.

Duplessy, J. C., 1982, Glacial to interglacial contrast in the northern Indian Ocean, *Nature* **295:**494-498.

Duplessy, J. C., 1983, Oxygen and carbon isotopes in benthic foraminifera: Deep water circulation changes during the last Interglacial and the effect of temperature on the oxygen isotope record, IAMAP Symposium ACGP-1 IUGG XVIII General Assembly, Hamburg.

Duplessy, J. C.; Blanc, P. L.; and Be, A. W. H., 1981, Oxygen-18 enrichment of planktonic foraminifera due to gametogenic calcification below the euphotic zone, *Science* **213:**1247-1250.

Duplessy, J. C.; Moyes, J.; and Pujol, C., 1980, Deep water formation in the North Atlantic Ocean during the last ice age, *Nature* **286:**479-482.

Ghil, M., and Le Treut, H., 1981, A climate model with cryodynamics and geodynamics, *Jour. Geophys. Research* **86:**5262-5270.

Goreau, T. J., 1980, Frequency sensitivity of the deep-sea climatic record, *Nature* **287:**620-622.

Guinasso, N. L., Jr., and Schink, D. R., 1975, Quantitative estimates of biological mixing rates of abyssal sediments, *Jour. Geophys. Research* **80:**3032-3043.

Hannan, E. J., 1958, *Time Series Analysis*, New York: John Wiley & Sons.

Hays, J. D.; Imbrie, J.; and Shackleton, N. J., 1976, Variations in the Earth's orbit: Pacemaker of the Ice Ages, *Science* **194:**1121-1132.

Hecht, A. D., 1973, A model for determining Pleistocene paleotemperatures from planktonic foraminiferal assemblages, *Micropaleontology* **19:**68-77.

Helmholtz, H. L. F., 1885, *On the Sensations of Tone as a Physiological Basis for the Theory of Music*, 2nd English ed. (reprinted New York: Dover, 1954).

Imbrie, J., and Kipp, N. G., 1971, A new micropaleontological method for quantitative paleoclimatology: Application to a late Pleistocene Caribbean core, in K. K. Turekian (ed.), *The Late Cenozoic Glacial Ages*, New Haven: Yale University Press, pp. 71-147.

Jenkins, G. M., 1969, General considerations in the analysis of spectra, *Technometrics* **3:**133-166.

Källen, E.; Crafoord, C.; and Ghil, M., 1979, Free oscillations in a climate model with ice-sheet dynamics, *Jour. Atmos. Sci.* **36:**2292-2303.

Kinchin, A. Y., 1964, *Continued fractions*, Chicago: University of Chicago Press.

Koksma, J. F., 1936, *Diophantische Approximationen*, Berlin: Verlag.

Ku, T. L.; Bischoff, J. L.; and Boersma, A., 1972, Age studies of Mid-Atlantic Ridge sediments near 42°N and 20°N, *Deep-Sea Research* **19:**233-247.

Kukla, G., 1978, The classical European glacial stages: Correlation with deep-sea sediments, *Nebraska Acad. Sci. Trans.* **6:**57-93.

Kutzbach, J. E., 1981, Monsoon climate of the early Holocene: Climatic experiment with the Earth's orbital parameters for 9,000 years ago, *Science* **214:**59-61.

Le Treut, H., and Ghil, M., 1983, Orbital forcing, climatic interactions and glaciation cycles, *Jour. Geophys. Research* **88:**5167-5190.

Marple, L., 1980, A new autoregressive spectrum analysis algorithm, *IEEE Trans. on Acoustics, Speech and Signal Processing* **28:**441-454.

McIntyre, A.; Ruddiman, W. F.; and Jantzen, R., 1972, Southward penetrations of the North Atlantic Polar Front: Faunal and floral evidence of large-scale surface water mass movements over the last 225,000 years, *Deep-Sea Research* **19:**61-77.

Minorsky, N., 1962, *Nonlinear Oscillations*, Princeton, N.J.: D. Van Nostrand.

Molfino, B.; Heusser, L. H.; and Woillard, G. M., 1984, Frequency components of a Grande Pile pollen record: Evidence of precessional orbital forcing, in A. Berger, J. Imbrie, J. Hays, G. Kukla, and B. Saltzman (eds.), *Milankovitch and Climate*, Dordrecht, Holland: Reidel Publ. Company, pp. 391-404.

Morley, J. M., and Hays, J. D., 1981, Towards a high resolution, global, deep-sea chronology for the last 750,000 years, *Earth and Planetary Sci. Letters* **53:**279-995.

Nicolis, C., 1984, A plausible model for the synchroneity or the phase shift between climatic transitions, *Geophys. Research Letters* **11:**587-590.

Parzen, E., 1961, Mathematical considerations in the estimation of spectra, *Techtonometrics* **3:**167-190.

Pestiaux, P., 1982, *The Basic Geological Chronology*, Scientific Report 1982/1, Louvain-la-Neuve, Belgium: Institute of Astronomy and Geophysics, Catholic University of Louvain-la-Neuve.

Pestiaux, P., and Berger, A., 1984*a*, Impacts of deep-sea processes on paleoclimatic spectra, in A. Berger, J. Imbrie, J. Hays, G. Kukla, and B. Saltzman (eds.), *Milankovitch and Climate*, Dordrecht, Holland: Reidel Publ. Company, pp. 493-510.

Pestiaux, P., and Berger, A., 1984*b*, An optimal approach to the spectral characteristics of deep-sea climatic records, in A. Berger, J. Imbrie, J. Hays, G. Kukla, and B. Saltzman (eds.), *Milankovitch and Climate*, Dordrecht, Holland: Reidel Publ. Company, pp. 417-445.

Pisias, N. G., 1983, Geologic time series from deep-sea sediments: Time scales and distortion by bioturbation, *Marine Geology* **51:**99-113.

Pisias, N. G.; Dauphin, J. P.; and Sancetta, C., 1973, Spectral analysis of Late Pleistocene-Holocene sediments, *Quat. Research* **3:**3-9.

Pisias, N. G., and Moore, T. C., Jr., 1981, The evolution of Pleistocene climate: A time series approach, *Earth and Planetary Sci. Letters* **52:**450-458.

Prell, W. L.; Hutson, W. H.; Williams, D. F.; Bé A. W. H.; Geitzenauer, K., and Molfino, B., 1980, Surface circulation of the Indian Ocean during the Last Glacial Maximum, approximately 18,000 yr B.P., *Quat. Research* **14:**309-336.

Royer, J. F.; Deque, M.; and Pestiaux, P., 1983, Orbital forcing of the inception of the Laurentide ice sheet? *Nature* **304:**43-46.

Sancetta, C.; Imbrie, J.; and Kipp, N. G., 1973, Climatic record of the past 130,000 years in North Atlantic deep-sea core V23-82: Correlation with terrestrial record, *Quat. Research* **3:**110-116.

Schnitker, D., 1982, Climatic variability and deep ocean circulation: Evidence for the North Atlantic, *Palaeogeography, Palaeoclimatology, Palaeoecology* **40:**213-234.

Van Campo, E.; Duplessy, J. C.; and Rossignol-Strick, M., 1982, Climatic conditions deduced from a 150 kyr oxygen isotope—Pollen record from the Arabian Sea, *Nature* **296:**56-59.

Wirtky, K., 1973, Physical oceanography of the Indian Ocean, in B. Zeitschel (ed.), *The Biology of the Indian Ocean*, New York: Springer, pp. 18-36.

17: Geomorphic Activity on Escarpments and Associated Fluvial Systems in Hot Deserts

Ran Gerson and Sari Grossman
The Hebrew University of Jerusalem

Abstract: Hillslopes having relatively small catchment areas may well reflect major climatic changes. Many escarpments in hot desert terrains possess debris-flow-controlled taluses that formed during prolonged mildly arid to semiarid climatic regimes, or pluvial modes. Under arid to extremely arid climates (interpluvial climatic modes), gullying strips away talus mantles and rapid scarp recession leaves behind talus flatiron relicts. Such relicts, representing the latest stages of pluvial modes, may be used to reconstruct geomorphic development under the changing climates of the Quaternary. The model presented is supported by stratigraphic chemical analysis of desert varnish on gravel exposed over talus relicts. Although absolute dating of such sequences is still wanting, we suggest estimates of the rate of scarp retreat.

Taluses of escarpments in different lithologic settings have different thresholds of reaction to climatic regimes and changes. Among the most dissimilar are taluses derived from limestones overlying sandstones and those derived from densely jointed porphyritic rocks. Relationships between floodplains and adjacent talus slopes may serve as indicators of climatically induced geomorphic stages otherwise undetectable. Extreme aridity and moderate to intense seismicity of the late Holocene are inferred from talus activity such as rockfall and collapse and resulting hillslope deposits.

INTRODUCTION

Most hot desert terrains of the world have experienced arid climates during long periods of time including much of the Late Tertiary and the Quaternary (Hsu et al., 1977; Lamb, 1977; Van Zinderen Bakker, 1978).

Talus mantles on escarpments and hillslopes in arid regions appear to be fairly widespread. Even a casual inspection of their distribution indicates that talus mantles exist or have existed under climates ranging from semiarid through extremely arid, with different available relief conditions and a gamut of lithostratigraphic settings (Figs. 17-1, 17-2, 17-3).

Taluses in terrains that appear to be quite proper for their existence show different degrees of activity: (1) active; (2) nonactive; (3) eroding, gullied, or largely stripped away; and (4) nonexistent. Examination of talus mantles and talus relicts indicates a dependence of their processes, and even their existence, on type of terrain, lithostratigraphy, and climatic regime.

Several climatic regimes may be defined within the hot arid zone (partly following Dan et al., 1981): extremely arid regime, 0-80 mm of mean annual precipitation; arid, 80-150 mm/yr; moderately arid, 150-250 mm/yr; and semiarid, 250-400 mm/yr.

This chapter deals with escarpments and discrete hillslopes carved at least partly in gravel-producing brittle rocks. Most such hillslopes, under hot arid climates, incorporate or have included at some period in the past debris-mantled taluses and a basal concavity passing into a pediment or a braided floodplainlike

Figure 17-1. An erosional escarpment under present-day extremely arid climate, Timna Valley, the Elat Mountains, southern Israel. Mean annual precipitation is about 30 mm. Note the much eroded talus remnants.

basal plain. Characteristic regional cases in point include (1) much of the hot arid-mildly arid portions of the Basin and Range Province and Colorado Plateau in the southwestern United States; (2) plateau margins and other escarpments in arid north Africa, the southern Levant, and the Arabian Peninsula; and (3) escarpments in the arid environments of South Africa.

The widespread distribution of escarpments including active taluses or talus relicts may render them invaluable for paleogeomorphic and paleoenvironmental investigations. The main objective of this chapter is to explore the use of talus slopes, their relicts, and the associated floodplains as indicators of environmental change in areas presently characterized by hot arid climates. Excluded are hillslopes undercut along their base; the discussion involves only protected hillslopes, along which there is a pediment, a bajada, or an alluvial terrace.

Figure 17-2. Debris flows typical to present-day environmental conditions in deserts. Flows seldom reach base of slope or adjacent floodplains (eastern Sinai).

ESCARPMENTS IN ARID ENVIRONMENTS

Escarpments may be defined as steep slopes of appreciable relief (several tens to more than 1,000 m) and considerable length. Escarpments, initiated either by faulting or by intensive fluvial incision, tend to be well preserved in form through gradual recession under arid climatic regimes (King, 1963, 1967). In gravel-producing terrains, under hot arid climates, one often encounters taluses or talus relicts at the foot of a feeding cliff, as major components of an escarpment (Fig. 17-4).

As idealized by Wood (1942) and King (1967), the standard hillslope and many escarpments are composed of four elements: (1) crest, (2) scarp (cliff, free face), (3) debris slope, and (4) pediment. One finds these elements or their relicts along most escarpments. The terminology used here, crest, cliff, talus slope, and floodplain, is somewhat different. The talus slope is formed not only by detritus fallen from the scarp above but mostly by debris-flow deposits. In a steady state of a dynamic equilibrium condition (Chorley and Kennedy, 1971), the talus slope usually occupies some 50-80% of the escarpment height and area. The pediment, in most cases, is an element not separable from the talus but is a continuous concave facet composed of debris-flow deposits grading into fluvial deposits farther from the escarpment (see Figs. 17-3, 17-5). Both the talus proper and the gentle concave facet below it are cut in bedrock and are thinly veneered by debris-flow and fluvial deposits (1-5 m thick; see Figs. 17-5, 17-6). Recession of the escarpment, as well as lowering of the plains at its base, is conducted through bedrock consumption by debris flows, gullying, and fluvial erosion.

Such escarpments are best developed and maintained under the following environmental conditions:

1. A rather rapid creation of available relief of at least several tens of meters, either by faulting (as a fault scarp) or by rapid stream incision (as an erosional scarp).
2. Exposure or partial exposure of hard, brittle rocks in the upper portion or all over the initial escarpment (King, 1967). Where the rocks are layered, horizontal or subhorizontal attitude is essential for escarpment maintenance for long periods of time.
3. Existence of extremely arid to semiarid climates for prolonged periods of time, on the order of 10^5-10^6 yr. Maintenance of relatively steep slopes and escarpments coincides with such climates.
4. Efficient removal of debris from the hillslopes. This allows continuous recession without a complete, irreversible burial. Wash, gullying, and debris-flow activity are pertinent mechanisms. Rockfall and uninterrupted accumulation is infrequently encountered in the desert environment.

As observed by Carson and Kirkby (1972), there is a conspicuous persistence of steep-slope elements in semiarid areas as opposed to a tendency toward obliteration of such elements in periglacial areas. It should be emphasized here that such a persistence is even greater in extremely arid terrains. Complete burial of an escarpment should not be included in a general model of talus activity or destruction in arid environments. There are exceptions to this generalization; some escarpments, such as those composed of pervious and friable sandstones from crest to base, may be almost completely

Figure 17-3. Talus relicts and continuing alluvial terrace built of gravel derived from the upper cliffs composed of limestone and dolomite. Present-day floodplains carry mostly sand, derived from sandstones (underlying the upper cliff carbonate formations) exposed by stripping of taluses, western margins of the southern Arava Rift Valley.

Figure 17-4. Talus relicts remain as triangular, flatironlike facets, between upstream-widening gully basins, Makhtesh Ramon, central Negev. Bedrock lithology is limestones and dolomites overlying sandstones.

Figure 17-5. A detail of a cut across a characteristic talus slope. Debris-flow deposits, some 3 m thick, overlying bedrock composed of sandstone (Makhtesh Ramon, central Negev).

Figure 17-6. Talus relicts, central Arava Valley, Dead Sea Rift, southern Israel. Bedrock lithology is flints overlying chalks.

buried under debris and change their mode of development through cliff obliteration (Schumm and Chorley, 1966; Young, 1972).

TALUS AND CLIFF GEOMORPHIC PROCESSES

Examination of many dozens of talus surfaces, including relict, fossil, and presently active, indicates that four processes are active on talus slopes: (1) debris flows, (2) slope wash, (3) gullying, and (4) collapse and rolling of blocks.

Debris Flows

Debris-flow deposits are the most widely encountered sediments in desert talus cross-sections (see Figs. 17-5, 17-6). These are mostly gravel that range from boulder to grit sizes, mixed with fines; the gravel is matrix supported. The material is unsorted, having a bimodal or trimodal distribution, modes being in the fines, fine gravel, and coarse gravel size ranges. One occasionally finds several debris-flow lobes and levees cutting across each other. Taluses exposing debris-flow structures and textures are encountered in many lithostratigraphic settings including limestone overlying sandstones (north Africa, Sinai and the Negev, southern Jordan, and Saudi Arabia); flint overlying chalk (same general areas as previous); granitic rocks (southern Mojave and northern Sonoran deserts; central Sahara, Sinai, and southern Negev; southern Jordan; and western Saudi Arabia); basalts overlying quartz monzonites (southwestern Arizona); metadiorites and schists (eastern Sinai and southern Negev); and sandstones overlying acid porphyritic rocks (southern Negev).

Usually, the suitable environmental conditions for debris-flow activity on taluses under arid climates are well-developed Reg soil (desert gravelly soil), including 20-35% of fines and salts (Dan et al., 1982) on a talus; an unstable rocky cliff above the talus, prone to releasing large blocks onto the talus below; and a high-intensity rainfall event. When collapse blocks from the cliff travel down over a wet talus, fulfilling the first condition, a debris flow may be triggered. Thus, unlike most debris flows in other environmental settings, debris flows on gravelly talus slopes under arid climates are always triggered by an upper cliff collapse. In this respect they are similar to alpine *sturzstroms*—catastrophic debris streams generated by rockfalls (Hsu, 1975).

Examination of different hot arid areas in the southwestern United States and the Sinai-Negev region leads to definite conclusions. Under most lithological and relief settings in mildly arid to semiarid climatic conditions, one finds prevalence of debris-flow activity on taluses all the way from the upper cliff into adjacent floodplains (exceptions are dealt with later; Gerson, 1981). Under arid to extremely arid conditions, debris-flow activity is very restricted, and in most cases debris flows do not reach even talus footslopes (see Fig. 17-2).

It is possible that many of the so-called screes in hot deserts described in the literature as rockfall accumulations are in fact debris-flow controlled. The author has examined dozens of artificial and natural cuts through taluses down to bedrock in various lithological settings; most of them are composed of debris-flow deposits mantled by weathered gravel on top of Reg soils. The more frequently encountered thickness of talus mantles ranges between 1 m and 5 m.

As observed by Carson and Kirkby (1972), most talus slopes (termed *debris slopes of accumulation*), while operating in a talus mode, are "equilibrium transportation slopes, on which equal quantities of material are being supplied . . . and removed" (p. 342). The process is, in most cases, debris flow (Gerson, 1981). Less frequently, under conditions of densely shattered hard and brittle crystalline rocks, one may find mudless debris flows, composed mostly of fine-to-medium-sized gravel without supporting fines, known as *sieve deposits* (Hooke, 1967).

Slope Wash

Slope wash operates under most lithological and climatic conditions. Only terrains composed of certain rock types such as pervious sandstones or certain shales are immune from slope wash (Lavee, 1973; Yair and Lavee, 1974). It may well be that under the less arid climates, during frontal rainfalls of relatively long duration, integrated flow all the way to the slope foot is quite frequent. In the more arid climates, such as in the central and southern Negev or in southwestern Arizona, partial area activity is the rule (Yair and Lavee, 1974).

Gullying

Gullying and stripping of taluses are dominant at present in the more arid environments, where average rainfall is usually less than 100 mm/yr (see Figs. 17-1, 17-4; Bull and Schick, 1979; exceptions to be noted later). Gullying is much less effective in the less arid climates where debris flow and slope wash together maintain receding talus slopes. A clear indication of the differential timing of colluvial stripping was observed in the Negev Highlands: in the lower elevations, of 400-500 m above sea level, where precipitation currently ranges between 60 mm and 90 mm/yr, colluvial mantles cover the lower 20-30% of the hillslopes. In the higher central Negev Highlands, where elevation is 800-1,000 m above sea level and precipitation is some 90-150 mm/yr, a still continuous colluvial cover of 50-80% is frequently encountered; gullying and stripping have started only very recently.

Rock Block Collapse

Collapse and rolling of rock blocks is the least frequently encountered talus-forming process. It is enhanced by two phenomena: (1) earthquakes, triggering rock block collapse from metastable/unstable cliffs, and (2) extremely arid climates, where continued stripping and gullying, with no appreciable debris-flow activity, dominate the escarpment. When large blocks collapse from the upper cliffs during intensive rainstorms, they tend to accumulate on stable facets of hillslopes and in gullies. Notch formation at cliff-hillslope and cliff-gully contact zones, mostly at contacts between hard rocks overlying erodible ones, is a frequent trigger of rock block collapse.

Formation of Flatironlike Talus Facets

Gullies that widen upstream, into small broad basins, leave between them triangular, flatironlike talus facets termed triangular slope facets by Büdel, 1970 (see Figs. 17-3, 17-4, 17-6). When detached from the receding feeding cliffs, these facets turn into talus relicts, pointing in their form and sediments at a different environmental setting—debris-flow-controlled hillslopes under moderately arid to semiarid climates.

The possibility of detachment of flatironlike or cuestalike debris-mantled facets by intensive weathering and erosion at the scarp foot, as suggested by Twidale (1967) for the Flinders Ranges area (at present in a semiarid climate), was carefully examined in the field. There is no hint of such a sequence of events in the arid areas studied and described herein. It seems that only areas well within the hot subhumid environment in the past, that underwent deep basal weathering, subsequent erosion, and the formation of a definite piedmont angle, are suitable for these processes.

Under the most frequently encountered conditions (debris-flow control and wash/gully control), removal of material along an escarpment is efficiently conducted without piling up accumulated material. Impeded removal of debris (Savigear, 1956) is an exceptional situation along most escarpments in arid environments. The cases where debris is concentrated at the base of a slope is fairly unusual. Rather, a continuous transport by surface wash, gully flow, or debris flow is the rule; events with inadequate transport power during low magnitude rainfall alternate with medium- and high-magnitude events that cause material stalled on the slope or at its base to resume transport to adjacent floodplains. Talus accumulation by rockfall is infrequently encountered. A major limiting factor is weathering: (1) When erosion prevails, weathering yields less debris than erosion can remove, and corrosion keeps bedrock exposed and continuously eroded. (2) When debris flows are a major transporting agent, weathering of the upper cliff and degree of soil evolution over a talus will determine its susceptibility to the debris-flow process.

TALUS RELICTS SEQUENCES—A KEY TO MAJOR CLIMATIC FLUCTUATIONS

Figures 17-3, 17-4, and 17-6 demonstrate a situation frequently found along escarpments in deserts: a series of flatironlike talus relicts aligned along an escarpment or skirting a mesa. The taluses and talus relicts are composed mostly of debris-flow deposits in a relatively thin mantle. They imply debris-flow-controlled phases of scarp recession. The interflatiron erosional gaps were formed by first-order streams incised into taluses and bedrock of a once-continuous talus apron.

Figure 17-7 presents a map of a portion of the northern erosional escarpment of Makhtesh (=erosion cirque) Ramon, central Negev, Israel. The different talus-bajada surfaces are marked according to their relative age. Figure 17-8 illustrates a reconstructed stagewise evolution of such an escarpment. According to the ensuing model, each talus relict is related to an end of a long, relatively wet (moderately arid or semiarid) climatic phase. Each intertalus gap coincides with a whole interpluvial-pluvial cycle (phases D+W, in Fig. 17-8).

Both talus relicts and intertalus gaps are continuously aligned in parallel along lengthy

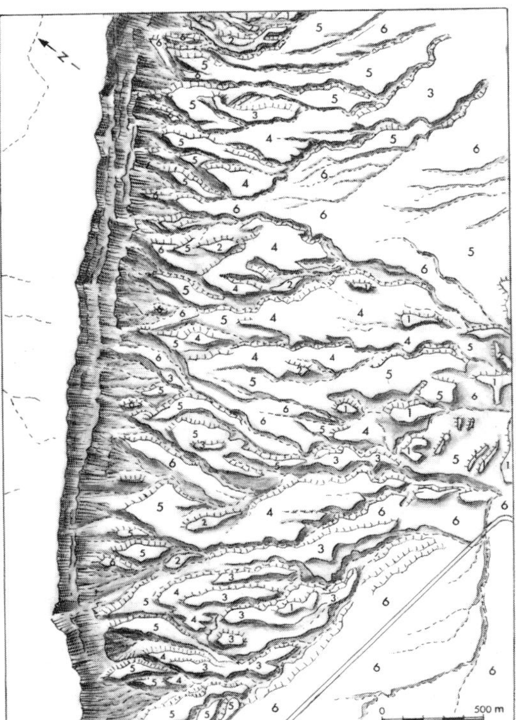

Figure 17-7. A map of a portion of the northern escarpment of Makhtesh Ramon, central Negev. Talus flatiron relicts and related pediment/bajada plains are numbered chronologically, with 1 being the oldest surface.

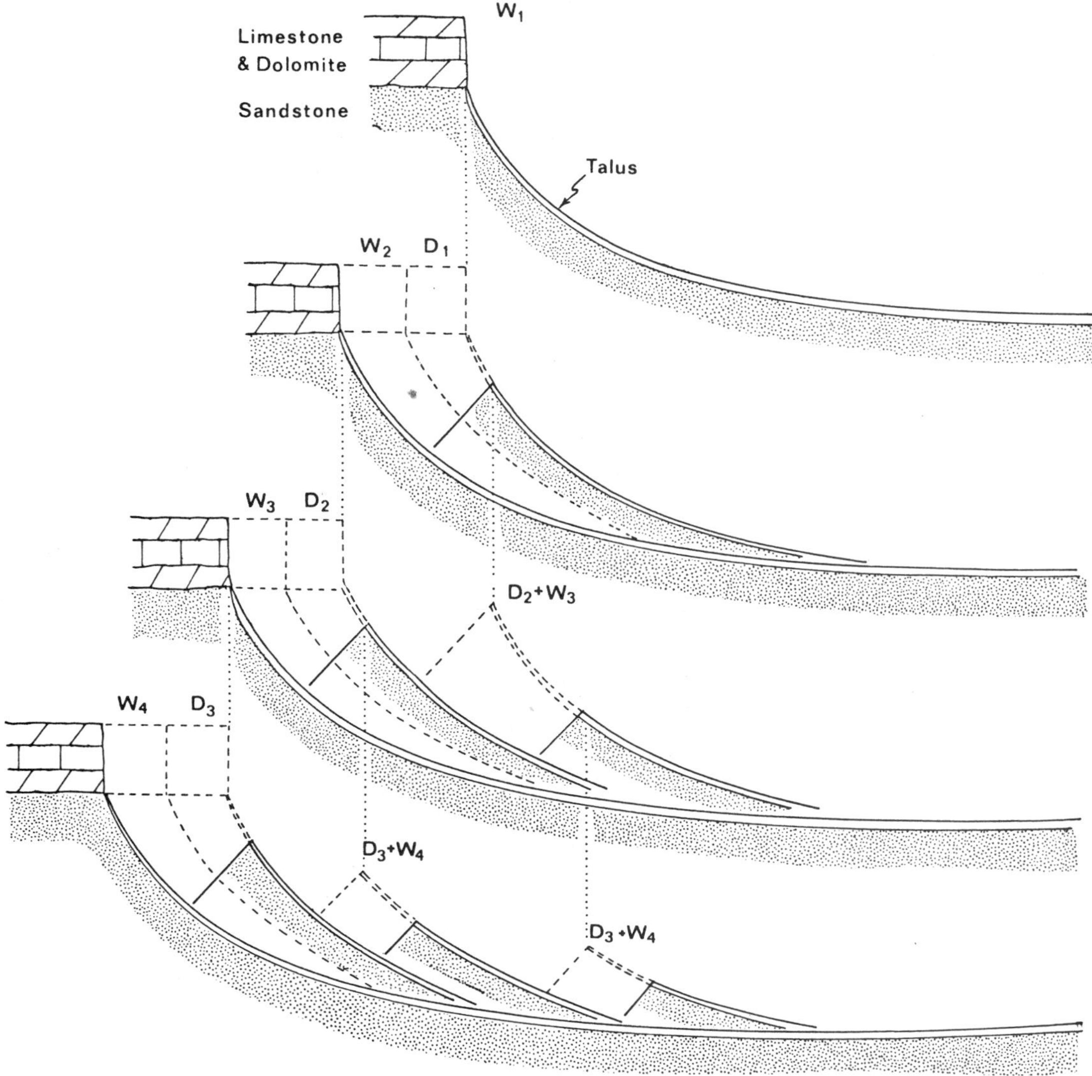

Figure 17-8. A generalized model of the effects of major environmental changes on escarpment retreat. D = dry, arid to extremely arid, climate; W = relatively wet, moderately arid to semiarid, climate.

escarpments (as in Makhtesh Ramon, some 30 km long; see Fig. 17-7). The dominance of debris flows in most taluses, the long time interval (tens of thousands of years) required to convert an eroded escarpment (such as in Fig. 17-1) into a talus-embanked one, the maintenance of such talus through parallel retreat, and the parallel alignment of talus relicts, all point to lengthy wetter climates controlling widespread terrains.

Taluses composed of dense, cohesive debris-flow deposits, including very coarse gravel, are apparently insensitive to relatively slight and brief climatic fluctuations. Such fluctuations probably affect process rate more than type. Only a change from a moderately arid (or semiarid) to an arid or extremely arid climate for several thousands of years may cause a talus to cross a process threshold into a mode of wash- and gullying-controlled activity. Holocene gullying and stripping of Pleistocene veneers is an example. Only one of the most recent talus-bajada relicts has been reasonably well dated, by Mousterian artifacts embedded in them (Goldberg, 1981), to have formed during the Early to Middle Würm, which was a relatively wet period in the eastern Mediterranean region.

Having no absolute age control for most talus relicts and associated alluvial surfaces, one may resort to certain reasoning for supporting relative chronologies. We may assume that tens of thousands of years will elapse while a rugged, eroded escarpment undergoes talus buildup and talus apron maintenance while the escarpment is receding (a wet-mode geomorphic operation, W in Fig. 17-8). Such a stage entails a long pluvial mode. Judging from the Late Quaternary, a favorable environment in the subtropical arid zone coincides with the early periods of a glacial cycle (such as the first half of the Würm; Fairbridge, 1972; Horowitz, 1979). The more arid climate prevails during later glacial times and in the following interglacial period (such as the Holocene); during such a later period erosion and talus stripping is common.

Pluvial/interpluvial changes, then, may roughly be correlated with glacial/interglacial cyclicity. Relatively small fluctuations of short-lived environmental regimes may not cause a gravelly debris-flow-controlled talus to pass through the threshold of gullying and erosion.

We may, then, try to correlate major climatic fluctuations with glacial cycles, roughly alternating every 10^5 to 1.25×10^5 yr (National Academy of Sciences, 1975; Hays, Imbrie, and Shackleton, 1976; Kukla, 1977; see section on comments, problems, etc. for reservations regarding this point). Synchronous with some part of a glacial stage, there prevailed a pluvial climate (effectively wetter environment) in many of the subtropical hot arid zones. For example, long periods of the Late Pleistocene, especially during its earlier half, were certainly wetter than the present (Fairbridge, 1972; Begin, Ehrlich, and Nathan, 1974; Rognon, 1976; Lamb, 1977; Messerli, Winiger, and Rognon, 1980; Goldberg, 1980; Gerson, 1982). The other end of the climatic spectrum is the dry interpluvial, like that of much of the Holocene, synchronous with an interglacial phase.

This general pluvial/interpluvial model is substantiated by the chemical composition of desert varnish on talus flatiron sequences in flint-chalk (see Fig. 17-6) and syenite terrains, which were analyzed for Mn:Fe ratios, from the varnish surface down to some 100 mm below the surface. Dorn and Oberlander (1982) have found that with a more arid climate, the Mn:Fe ratio tends to be lower. A chemical stratigraphic cross-section through a desert varnish that has developed through a sequence of different climatic phases will show high Mn:Fe ratios for wet modes and low ratios for dry modes. In studies of talus sequences, desert varnish on a talus of W_3 age shows only one Mn:Fe peak developed during the W_4 wet mode; samples collected on W_2 talus flatirons show two Mn:Fe peaks developed during the W_3 and W_4 wet modes (Dorn, 1983). These data certainly support the deduced climatic model derived from the field observation of the talus relicts.

TALUS ACTIVITY UNDER DIFFERENT LITHOSTRATIGRAPHIC SETTINGS

There are differences between modes and rates of talus activity in dissimilar lithologic settings. It appears that relict triangular flatirons are best developed along two-formation escarpments, where a hard, weathering-resistant formation overlies an erodible or weatherable one—for example, flint overlying chalk (see Fig. 17-6), limestone overlying sandstone (see Fig. 17-4), or basalt overlying quartz monzonite. Notch development at the formation contact zone is characteristic of the talus erosion, gullying, and stripping mode. It is with a two-formation scarp that talus flatirons and interrelict gaps best represent the effect of different climatic regimes.

A different setting is found in monolithic terrains, such as quartz porphyry, syenite porphyry, and metadiorite. Here, the change of climatic regimes is reflected principally in the type of talus activity and the rate of talus recession.

The variation in lithostratigraphic settings clearly manifests itself in the Elat region (south-

Figure 17-9. Active debris flows in eastern Sinai. The terrain is built of densely shattered porphyritic and fine crystalline rocks.

ern Negev) where there are various rock types within the same watersheds. An example is the Naḫal (=wadi, wash) Avrona watershed in Mount Amram. Here, in a relatively small area of some 4 km^2, one finds marked differences in mode, rate, and timing of talus activity among different lithologic terrains. Relict talus flatirons, built of debris-flow deposits, are typical along escarpments built of limestones overlying sandstones (see Figs. 17-3, 17-4). The erosional gap between different talus relicts is wide and clear. Latest phase talus mantles are being actively stripped away at present, and most taluses are already detached from their feeding cliffs. Whereas taluses and derived fluvial deposits of related alluvial terraces are composed mostly of limestone gravel, present-day floodplains carry mostly sand, derived from sandstones exposed by talus stripping (see Figs. 17-1, 17-3). No present-day debris-flow activity is evident.

The same general pattern is found over wide areas in the Negev and eastern Sinai: where limestones overlie sandstones there is a distinct erosional gap between talus phases, and the most recent talus is being eroded away under the present extremely arid climatic conditions.

Conversely, vast talus generation continues over many hillslopes underlain by quartz porphyry, and gullying has started to affect some of these taluses most recently. In some porphyritic terrains of mostly shattered rocks and fine-to-medium clasts, talus activity continues, debris flows are active, and gully initiation is not visible (Fig. 17-9).

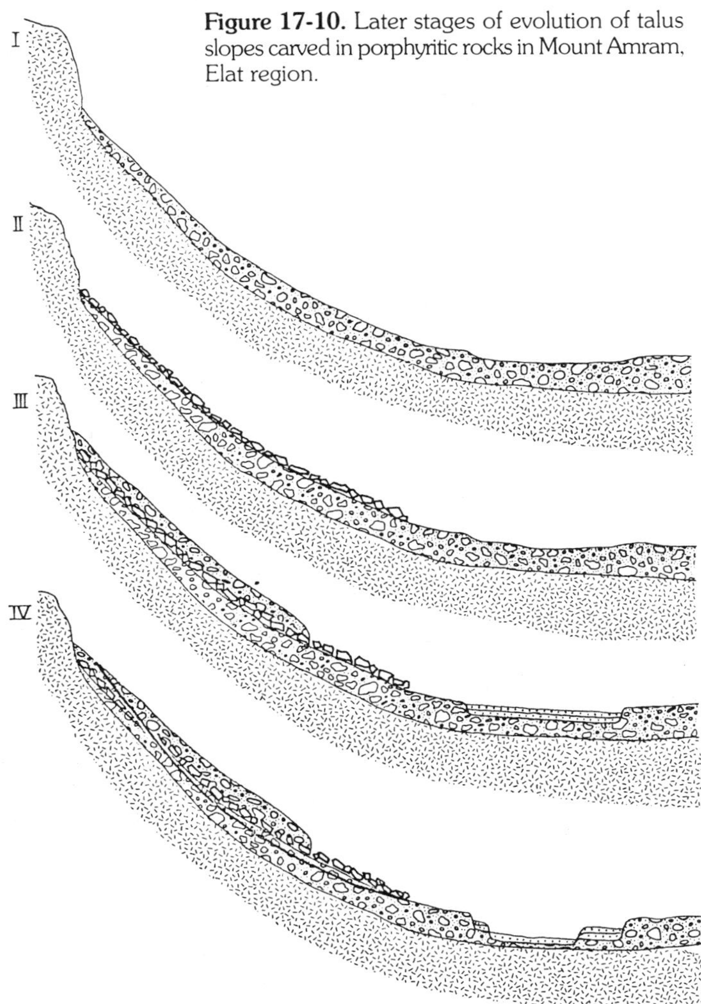

Figure 17-10. Later stages of evolution of talus slopes carved in porphyritic rocks in Mount Amram, Elat region.

Rates and modes of hillslope recession also appear to differ in the two settings, a situation that is demonstrated by the different sequences of evolution. The Mount Amram (southern Negev) area may serve as an example. The stages identified and interpreted in the evolution of the taluses in the igneous terrains there are represented in Figure 17-10:

I. Debris-flow- and wash-controlled talus. Talus coalesces smoothly both topographically and sedimentologically, with the adjacent floodplain. Inferred climatic regime is moderately arid to semiarid.
II. Rock block collapse and wash-controlled talus. Large (0.5-1 m) blocks are transported from the upper cliff and cover much of the talus surface. Inferred environment is extremely arid and/or intensive seismicity.
III. Debris-flow- and wash-controlled talus. Debris flows do not reach footslope. Lower part is controlled only by wash but no gullying. Inferred climate is arid.
IV. Present-day processes. Gullying of talus mantle and some collapse of rock blocks. Climate is extremely arid (mean annual precipitation is about 30 mm).

This stratigraphy is exposed vertically within 2-3 m of the surface in talus mantles. The rate of cliff retreat is relatively slow; net talus accumulation was occurring until very recently.

The behavior of different terrains is exempli-

Table 17-1. Estimates of Scarp Retreat between Successive (Reconstructed) Final Stages of Talus Activity

Lithostratigraphy	*Examined Localities*	*Available Relief*	*Horizontal Distance between Successive Cliff Positions*
Flint overlying chalk	Central Negev, Central Arava Valley	50-100 m	10-20 m
Limestone, dolomite, overlying sandstone	Makhtesh Ramon,	80-150 m	10-25 m
	Timna Valley,	150-300 m	20-40 m
	El Tih escarpment	300-500 m	30-60 m
Igneous and metamorphic rocks	Eastern Sinai, Mount Timna, Growler Mts. (Arizona)	150-300 m	15-30 m

fied in Table 17-1. The less the available relief, the slower the rate of scarp retreat. Also, there is a marked dissimilarity between different lithostratigraphic settings: in two-formation terrains, such as limestone overlying sandstones or flint overlying chalk, the rate of retreat appears to be faster (especially during dry modes) than in terrains composed of massive and brittle igneous and metamorphic rocks.

SOME REQUIREMENTS FOR PALEOCLIMATIC MANIFESTATION OF ESCARPMENTS

Certain conditions are required for the expression of past climates on escarpments:

Relief: It was found that escarpments carved in gravel-producing rocks, but having relief of less than 40 m, do not usually develop flatironlike relicts. This finding is due to the fact that such escarpments are too low to permit the development of properly spaced triangular gully basins.

Hillslope gradient: Hillslopes of low gradients are dominated by slope wash during both wet and dry modes, as is the case of most of the Cretaceous limestone terrains eroded by slowly developing drainage nets in the Negev Highlands and similar landscapes in northern Africa. Here, relatively gentle slopes of 10-20°, veneered by loess, have developed during the wetter

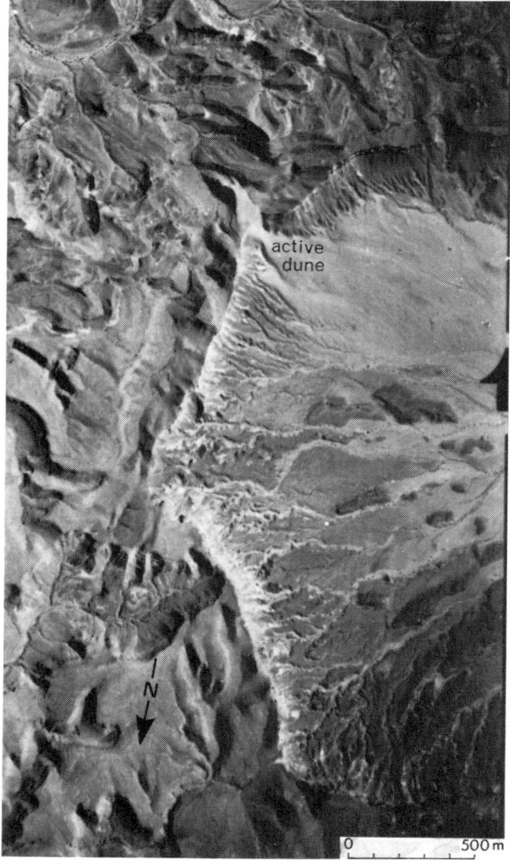

Figure 17-11. A climbing dune in the southeastern end of the Mahmal Valley, Makhtesh Ramon.

climatic periods. Wash of silts and fine gravel is dominant, not debris flows.

Rock type: As mentioned in the introduction, escarpments carved in rocks that do not weather to gravel do not have colluvial protection favoring preservation of talus relicts. Wash is the prevailing process both during wet and dry climatic modes. Also, escarpments developed in monolithologic massive igneous rocks sometimes do not tend to develop talus flatirons since there is no distinct cliff-forming unit at their top.

Streams: Undercutting streams at the base of escarpments may cause accelerated erosion and thus steepening of the escarpment, leading to a morphology that does not preserve the past colluvia.

In terrains where sand and sandstones are effectively exposed to eolian transport during a dry mode, accumulation of sand on hillslopes may serve as an indicator of extreme aridity. Figure 17-11 is an illustration of such a situation; in eastern Makhtesh Ramon (central Negev), where westerly winds prevail, sand is constantly being wind driven toward a west-facing escarpment. The sequence of events in that area is as follows:

1. A wet mode, with debris flows and wash controlled talus and floodplain activity.
2. A drier, arid, mode, when gravelly colluvium derived from the escarpment interdigitates with climbing dunes (Fig. 17-12).
3. Extreme aridity, during which intensive erosion and vigorous action of climbing dunes prevail.

The following evolution for this sequence is proposed, although no means for absolute dat-

Figure 17-12. A cross-section through a talus in the southeastern end of the Mahmal Valley, Makhtesh Ramon. Alternating layers of climbing dunes and colluvium.

ing are available: Stage 1, early Late Pleistocene, long wet mode; Stage 2, either latest Late Pleistocene or Mid-Holocene, both arid but not extremely so; Stage 3, extreme aridity, effective either at the beginning of the Holocene or at the last few millennia (the Late Holocene), as indicated by prolonged destabilization of sand dunes and vast exposure of sandstones from previous talus cover.

TALUS-FLOODPLAIN RELATIONSHIPS

A distinct stage in the evolution of many watersheds in the mountains of Elat and eastern Sinai, is represented by a steady-state equilibrium of alluvial floodplains (no appreciable aggradation or degradation), while adjacent taluses were in a state of growth, encroaching onto the floodplains. This process is best demonstrated along escarpments composed of densely jointed and tectonically shattered porphyritic rocks and small alluvial fans that fringe larger floodplains. Here, both talus deposits and alluvial fans emanating from small side tributaries cover the floodplains of the trunk channels of the area (Fig. 17-13). Stages of landscape development in these watersheds are:

1. Intensive talus activity in both sedimentary and porphyritic terrains. Debris-flow deposits dominate taluses and floodplains alike. Four widespread phases of talus activity and three intertalus erosional phases took place during a relatively long period of time—namely, Middle through Late Pleistocene. Associated high terraces, including debris-flow deposits, are left flanking the major channels.
2. A clear phase of latest stage talus destruction in the sedimentary terrains. Fluvial deposits derived from sedimentary terrains dominate the floodplains. Accumulation of 6-8 m of those sediments is clearly evident.
3. Steady-state to static equilibrium conditions along the floodplains. Neither erosion nor deposition occurs along valley bottoms. Wash and very thin debris flows from igneous terrains lead to partial coverage of floodplains by a thick (up to 15 m)

Figure 17-13. Talus deposits derived from porphyritic rocks cover alluvial sediments derived from a limestone-overlying-sandstone terrain of headwater reaches of Naẖal Amram, Elat region.

talus cover (see Fig. 17-13). Soils and surficial weathering features help date this phase to Early to Middle Holocene (Gerson, 1981, 1982; Amit and Gerson, 1986). Thus, while taluses in terrains composed of sedimentary rocks in the watershed are being eroded away (stage 2), taluses in igneous terrains in the same watersheds and under the same climate are highly active.

4. Incision of narrow channels into formerly wide floodplains and the beginning of gullying in talus slopes that aggraded in stage 3. Latest Holocene extremely arid climatic mode prevails.

The same sequence has been observed in Makhtesh Ramon (central Negev): Taluses derived from densely jointed columnar basalt overlie floodplains of adjacent streams, later to be entrenched to the present alluvial channels, leaving talus-blanketed fluvial terraces.

We are dealing with a region where no absolute dating method appears applicable. Only archeological remains such as surficial structures and old copper mines cut by the latest incision (Conrad and Rothenberg, 1980) help place the last phase in the Late Holocene (past 4,000-3,000 yr). Still, one may use a general time-correlated climatic change model to explain this sequence: First, moderately arid to semiarid climates prevailed for long periods of time, with drier episodes in between. These climates brought about talus buildup and maintenance. Sediment transport was mainly through debris-flow activity. Taluses and floodplains alike are dominated by debris-flow deposits. Much of the Pleistocene is represented by such activity. Periods of talus destruction are relatively dry and are represented by incision into former floodplains and bedrock. The scarcity of fluvial (layered, lenticular, sorted) deposits associated with these terraces stresses the point that the transition from one talus phase to the next was through a slow destruction of taluses by wash and gullying in intervals less arid than the Holocene (when mostly waterlaid fluvial sediments dominate the floodplains). The incision (through debris-flow deposits and bedrock) of an early floodplain (now a terrace) to a lower level (the modern floodplain) is a clear clue to a complete change in climate from moderately arid/semiarid into arid/extremely arid.

Next, arid to extremely arid climatic regimes prevail throughout most of the Holocene. During the Early to Middle Holocene, talus stripping in the sedimentary terrains was almost completed. Fluvial deposits accumulated in the floodplains during the talus gullying episode. While aggradation slowed, talus activity in porphyritic terrains adjacent to floodplains continued. Only the extreme aridity of the Late Holocene brought about the prevalence of wash and gullying of these taluses, concomitant with trunk channel incision (Figs. 17-13, 17-14).

RESPONSE OF FLOODPLAINS TO MAJOR CLIMATIC FLUCTUATIONS

The evidence of floodplain sedimentology and geometry correlates fairly well with that of the hillslopes and escarpments in the headwater reaches. One may usually observe three types of sediments in floodplains associated with the types of escarpments analyzed in the preceding sections:

1. Debris-flow deposits, widespread in alluvial terraces directly associated with talus aprons and found as the predominant sediments less than 1.5 km downstream from the escarpments.
2. Coarse to fine gravel with some sand, usually associated with eroding taluses and an advanced state of eroding desert soils, where regolith is being stripped away and densely shattered bedrock exposed on the hillslopes.
3. Sand and silt, associated with eroding soils on the hillslopes and bedrock exposures that are denuded of available gravel.

There are, of course, many cases in which the types of sediments in the floodplains are not readily associated with a clearly defined hillslope condition. This situation may be related to both transition periods from one environmental regime to another and the derivation of sediments from old alluvial deposits.

The geometry of the floodplains may well reflect the hydrologic and sedimentologic regimes of their feeding escarpments: (1) Close to the mountain front, a predominance of debris-flow deposits is associated with steep alluvial fans. (2) Elongated, narrow, and gently sloping flood-

Figure 17-14. Recent gullying has started to affect talus slopes built of porphyritic rocks, Mount Katherina, southern Sinai.

plains are characteristic where fine gravel and sand prevail and relatively large areas of bare rock are typical to the hillslopes. The alluvial fans in the latter case are located far into the depositional basin. The gradients of the linear floodplains are usually 2-4° gentler than those of the alluvial fans in the first case.

However, as a flight of terraces of alluvial fans and linear floodplains develops, one can see clearly that the lower surfaces are incised into the upper ones and that prominent banks, usually several meters high, bound each surface. As a result, every environmental change, with its tendency to develop a floodplain suitable in sediment or geometry, has to adjust to preexisting topography—slope, depth, width, and position with respect to the escarpment. Hence, one will not always find the floodplains ideally adjusted to the regime imparted by the feeding escarpment. The equilibrium form of such floodplains is partly determined by previously carved topography. A stream may, then, maintain a reach in equilibrium in various geometric configurations, in which gradient, width, and depth interchange in the adjustment to the hydrologic and sedimentologic regimes imposed by the headwater hillslopes. However, since a stream channel may drain appreciable piedmont areas of older alluvial terraces, active floodplains also have to absorb the effects of such terrains.

One has to bear in mind that floodplains draining large catchment areas may be readily responsive in their geometric adjustments to discrete flood events not usually characteristic of the regime. A rather extreme flow may change a floodplain to a degree irreparable by subsequent lesser flow events; the floodplain will not regain its former geometric configuration. It will adjust within the form imposed by the extreme event. With these considerations in mind, we may examine several sequences of formation and abandonment of floodplains at the foot of

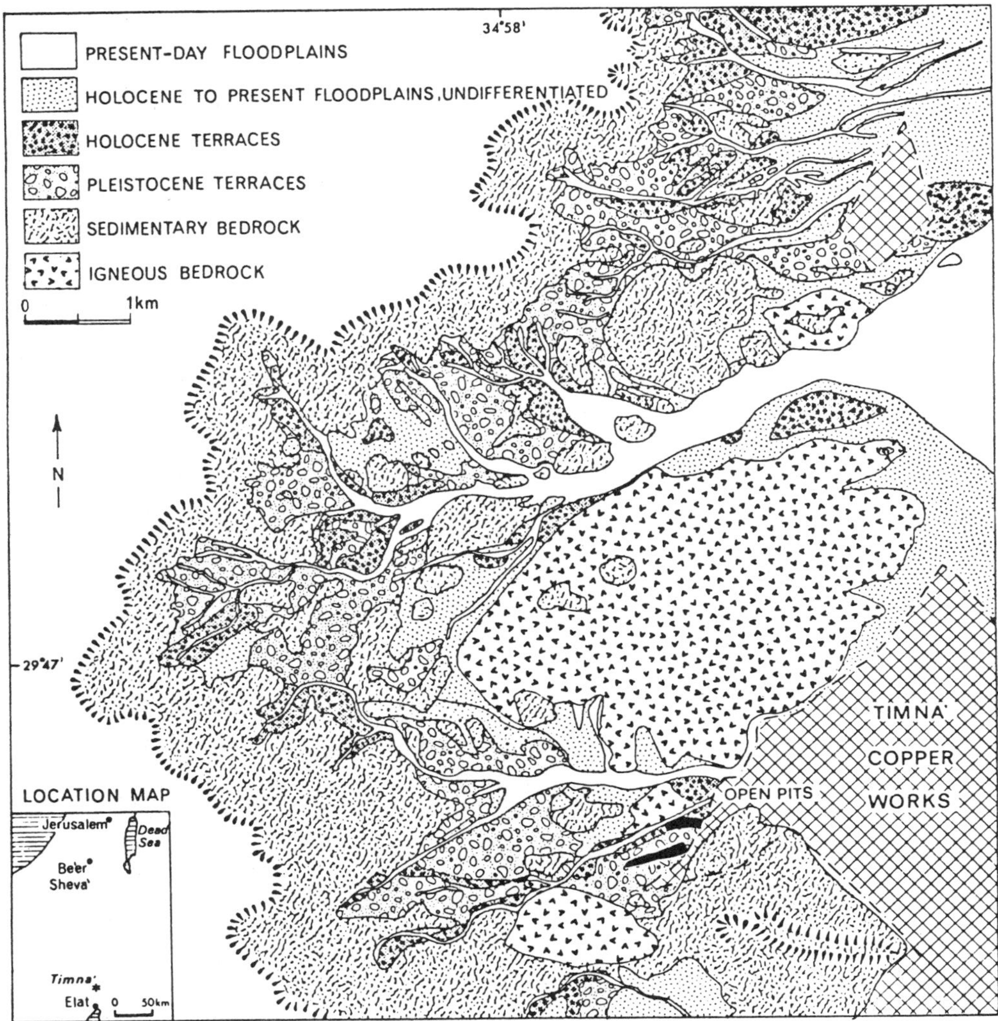

Figure 17-15. Transition from bajada depositional regime to linear floodplain activity in the Timna Valley, Elat Mountains, southern Negev.

escarpments. In the southern Negev there are several sequences of alluvial surfaces that reflect, both in their sedimentary characteristics and their geometric configurations, the effects of major climatic fluctuations. In the Timna area (Fig. 17-15) and in the Avrona watershed, one observes several cycles of floodplain deposition, each consisting of fine alluvial sediments overlain by debris-flow deposits (Figs. 17-16, 17-17). The lower fine-grained, mostly sandy, sediments are associated with escarpments exposing bare rocks with no appreciable talus cover. Sand, contributed by the sandstones exposed at the base of the escarpment, is the prevailing sediment type. Such is also the situation at present: extremely arid climate since the onset of the Holocene with no talus/debris-flow activity. The upper, coarse-grained member is composed mostly of debris-flow deposits, derived from talus-mantled escarpments, in which the limestones, overlying the sandstones, prevail as major sediment contributors. Pleistocene terraces in the Timna Valley, located more than 1.5 km from the escarpment, are built largely of water-

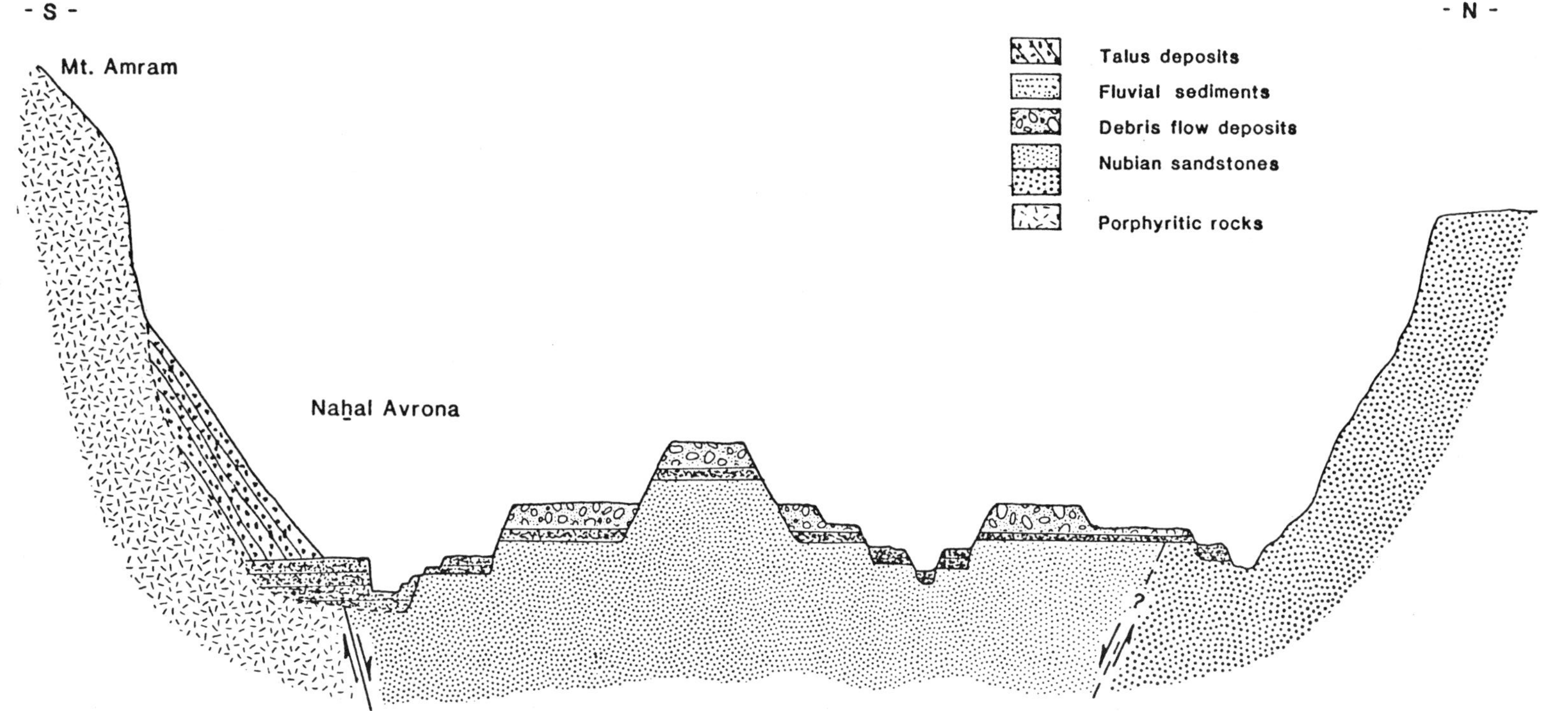

Figure 17-16. A cross-section through Avrona Valley, Elat Mountains. Note debris-flow deposits overlying fluvial sediments and talus deposits overlying Early Holocene fluvial sediments. This cross section represents flight of terraces less than 1.5 km from the feeding escarpment.

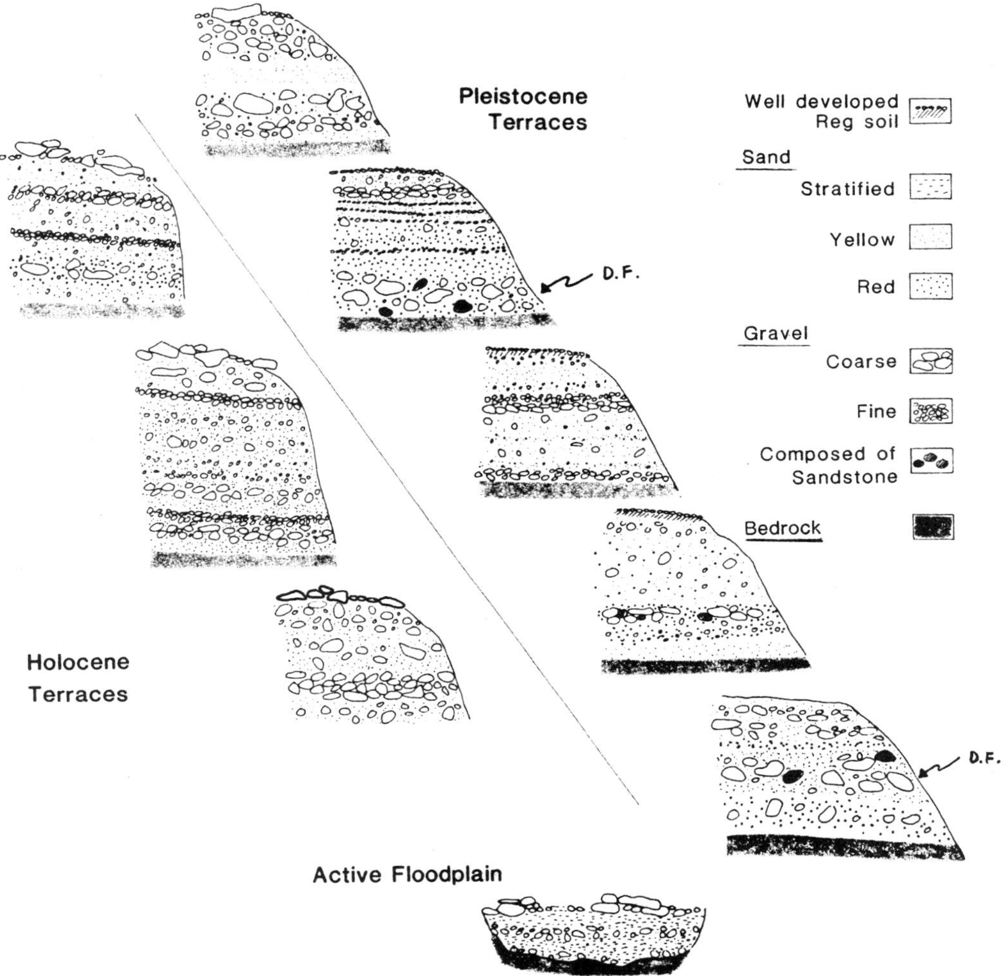

Figure 17-17. Alluvial deposits of high Pleistocene and Holocene terraces in the Timna Valley, Elat Mountains. D.F. = debris-flow deposits.

laid sediments. Only some layers or lenses consist of debris-flow deposits (Fig. 17-17). The direct effects of talus debris-flow activity are only partly observed in terrace fills farther downstream from the escarpment. The Holocene terrace fills are composed mostly of water-laid sediments (Grossman and Gerson, 1986). The same environment was observed also in the central Negev, in areas of similar lithostratigraphy. The general paleoclimatic scheme interpreted from these alluvial terrace sequences fits the talus-sequence model suggested in "Talus Relict Sequences": During a period of extremely arid climatic regime, when talus activity and debris flow processes are at their lowest, bare rock outcrops of the lower parts of the escarpments contribute fluvial sediments—in the present case, sand. During a period of intensive talus activity—a moderately arid to semi-arid climate—debris-flow and coarse water-laid gravelly sediments are derived from the upper parts of the escarpments; in the present case it is mostly limestone clasts with some fines and sand. During a period of talus gullying and stripping—a transition from a relatively wet regime to an extremely dry regime—there is also a deposition of coarse gravelly sediments in the floodplains.

Along many valley bottoms in the southern Negev, eastern Sinai, and the basin margins in southwestern Arizona, one often observes large gravel bars composed of very coarse gravel at the top of the Holocene terrace fills (Fig. 17-17). Such gravel bars were formed under an extremely arid climate during floods of high magnitude that carried cobbles and boulders, which had accumulated for long periods of time in small headwater valleys. Such events occur once in several 10^2 to few 10^3 yr.

Another example is in the northern Sonoran Desert of southwestern Arizona and the southern Mojave Desert of southeastern California, where one often observes the following sequence: (1) well-sorted fluvial gravel with reddened silt and sand, overlain by (2) coarse unsorted fluvial gravel with some sand and silt of yellow-gray color, and (3) these deposits are downcut by the present stream channels or incised with inset low recent terraces. The interpretation suggested for this sequence is: (1) deposition of fine gravel and red silt and sand during a wet-mode climate or the transition into a dry mode. The red sediments are derived from the soils eroded from the hillslopes. Stream flows are still moderate in power. (2) Stripping of the soils exposes coarse gravelly regolith, which is entrained and deposited over stage (1) deposits. Flash floods are common. (3) Exposure of bare rock and depletion of sources of gravel lead to intensive flooding, high stream power, and incision to the present channel.

COMMENTS, PROBLEMS, QUESTIONS, AND RESERVATIONS

1. Correlations between glacial cycles and general wet/dry modes in subtropical regions are still poorly understood (Fairbridge, 1972; Lamb, 1977). However, the Late Quaternary is better known (i.e., Messerli, Winiger, and Rognon, 1980).

2. A more refined definition and characterization of pluvials and interpluvials is sorely needed. Precipitation, temperature, the effects of summer storms in areas of present-day winter storms, reconstruction of soils and vegetation, etc. await characterization for past periods, especially earlier than Late Pleistocene.

3. The lack of chronology for colluvial and fluvial deposits in many desert terrains persists. Few Late Quaternary deposits have been absolutely dated, but most remain on an uncertain time scale.

4. How do extreme events (so-called catastrophic events) fit into the scheme presented here? Extreme rainfall storms activate debris flows, and when these storms occur during a pluvial regime, they only enhance retreat of escarpments including debris-flow-controlled talus facets; such events may not usually cause talus gullying and stripping. The effects of extreme rainfall events during a dry, interpluvial mode are to erode further and expose debris-mantled bedrock, as is the case of the dry Holocene stage.

5. Both in geomorphic and sedimentologic characteristics, models like the one presented here disregard to a certain degree transition regimes from one environmental mode to another and so overlook distinct stages in landscape evolution. Only maturely developed landforms are included. Still, the general framework appears to be valid.

6. Floodplains, being sensitive to discrete flow events and minor climatic fluctuations, may still be correlated in their sediments and geometry with those of their feeding escarpments. However, more phases of evolution are usually recognized in the floodplains than in the hillslopes.

CONCLUSION

Under various environmental settings—climatic, lithostructural, and topographic—talus slopes, bare bedrock, and floodplains are differentially active in the arid zone. Talus relicts may bear evidence to past modes of accumulation or destruction: debris flows, wash, gullying, collapse, and rolling of rock blocks. Relationships of taluses to adjacent floodplains support interpretations of climatic control and climatic change.

The sensitivity of talus mantles to minor climatic fluctuations is less than that of floodplains. This situation is due mainly to the small catchment area of talus slopes, the lower power generated by talus wash, and talus armoring by large gravel. Talus relicts, having smaller catchment than at the times when they were attached to an upper talus and a feeding cliff, are even less sensitive to environmental change. Their stability is assessed by the rather well-developed Reg soil at their surface. Hence explaining the apparent lack of response to minor environmental changes. Taluses, then, are suitable for the assessment and interpretation of long prevailing climate and major climatic fluctuations.

The proposed model—of debris-flow-controlled talus aprons, representing moderately arid to semiarid climatic regimes, and talus apron destruction through stripping and gullying during drier regimes—reflects the response of most escarpments and gravel-producing hillslopes to major climatic changes. However, in some lithologic settings, different nuances may exist: rock types like quartz and syenite porphyries, being hard and brittle, tend to shatter into small gravel and operate in a debris-flow mode even under the influence of the more arid climates. Only under an extremely arid climate will taluses in such a terrain wash and gully away. Only in the most hyperarid environments on Earth, such as the Atacama Desert, are there vast areas of colluvium mantled slopes lacking signs of water erosion.

Relationships between floodplains and bordering taluses help the assessment of different kinds of equilibriums along the former. Encroachment of taluses derived from brittle porphyritic rocks over adjacent floodplains in the southern Negev demonstrates the fact that those floodplains have neither aggraded nor degraded for the better part of the Holocene, until extreme aridity of the Late Holocene produced incision as a general process.

The cycles expressed in the talus sequences are pronounced also in the down-system floodplains. Debris-flow deposits and fluvial deposits may change according to the modes of geomorphic activity on the hillslopes. Deposition and erosion occur following changes in the activity of headwater reaches and according to the ratio between stream power and available sediment load.

REFERENCES

Amit, R., and Gerson, R., 1986, The evolution of Holocene Reg (gravelly) soils in deserts—an example from the Dead Sea region, *Catena* **13:**59-79.

Begin, Z. B.; Ehrlich, A.; and Nathan, Y., 1974, Lake Lisan: The Pleistocene precursor of the Dead Sea, *Israel Geological Survey Bull. 63,* 30p.

Büdel, J., 1970, Pedimente, Rumpfflachen und Rückland-Steilhange, *Zeitschr. Geomorphologie,* n.s., **14:**1-57.

Bull, W. B., and Schick, A. P., 1979, Impact of climatic change on an arid watershed: Nahel Yael, southern Israel, *Quat. Research* **11:**153-171.

Carson, M. A., and Kirkby, M. J., 1972, *Hillslope Form and Process,* London: Cambridge University Press, 475p.

Chorley, R. J., and Kennedy, B. A., 1971, *Physical Geography; A Systems Approach,* London: Prentice-Hall, 370p.

Conrad, H. G., and Rothenberg, B., 1980, *Antikes Kuper im Timna-Tal Der Anschnitt, Zeitschrift für Kunst und Kultur in Bergbau,* vol. 1, Bochum, West Germany: Gerbau Museum, 233p.

Dan, J., 1981, Soil formation in the arid regions of Israel, in J. Dan, R. Gerson, H. Koyumdjisky, and D. H. Yaalon (eds.), *Aridic Soils of Israel: Properties, Genesis and Management,* International Conference on Aridic Soils, Jerusalem, 1981, pp. 17-50.

Dan, J.; Gerson, R.; Koyumdjisky, H.; and Yaalon, D. H. (eds.), 1981, *Aridic Soils of Israel: Properties, Genesis and Management,* Special Publication No. 190, Agricultural Research Organization, Bet Dagan: The Volcani Center.

Dan, J.; Yaalon, D. H.; Moshe, R.; and Nissim, S., 1982, Evolution of Reg soils in southern Israel and Sinai, *Geoderma* **28:**173-202.

Dorn, R. I., 1983, Cation-ratio dating: A new rock varnish age-determination technique, *Quat. Research* **20:**49-73.

Dorn, R. I., and Oberlander, T. M., 1982, Rock varnish, *Progress in Phys. Geography* **6:**317-367.

Fairbridge, R. W., 1972, Climatology of a glacial cycle, *Quat. Research* **2:**283-302.

Gerson, R., 1981, Geomorphic aspects of the Elat Mountains, in J. Dan, R. Gerson, H. Koyumdjisky, and D. H. Yaalon (eds.), *Aridic Soils of Israel: Properties, Genesis and Management,* Special Publication No. 190, Agricultural Research Organization, Bet Dagan: The Volcani Center, pp. 279-293.

Gerson, R., 1982, The Middle East: Landforms of a planetary desert through environmental changes, *Striae* **17:**52-78.

Goldberg, P., 1980, Late Quaternary stratigraphy of Israel: An eclectic view, in P. Sanlaville and J. Chauvin (eds.), *Prehistoire du Levant,* C.N.R.S. Colloque No. 598, Lyon: C.N.R.S., pp. 55-66.

Grossman, S., and Gerson, R., 1987, Fluviatile deposits and morphology of alluvial surfaces as indicators of Quaternary environmental changes in the southern Negev, Israel, in L. E. Frostick and I. Reid (eds.), *Desert Sediments: Ancient and Modern,* Special Publication of the Geological Society of London (in press).

Hays, J. D.; Imbrie, J.; and Shackleton, N. J., 1976, Variations in the earth's orbit: Pacemaker of the ice ages, *Science* **194:**1121-1132.

Hooke, R. LeB., 1967, Processes on arid-region alluvial fans, *Jour. Geology* **75:**438-460.

Horowitz, A., 1979, *The Quaternary of Israel,* New York: Academic Press, 349p.

Hsu, K. J., 1975, Catastrophic debris streams (Sturzstroms) generated by rockfalls, *Geol. Soc. America Bull.* **86:**129-140.

Hsu, K. J.; Montadert, L.; Bernoulli, D.; Cita, M. B.; Erickson, A.; Garrison, R. E.; Kidd, R. B.; Melieres, F.; Muller, C.; and Wright, R., 1977, History of the Mediterranean salinity crisis, *Nature* **267:**399-403.

King, L. C., 1963, *South African Scenery,* 3rd ed., Edinburgh: Oliver and Boyd, 308p.

King, L. C., 1967, *The Morphology of the Earth,* 2nd ed., Edinburgh: Oliver and Boyd, 726p.

Kukla, G. J., 1977, Pleistocene land-sea correlations. I. Europe, *Earth-Sci. Rev.* **13:**307-374.

Lamb, H. H., 1977, *Climate: Present, Past and Future,* vol. 2, *Climate History and the Future,* London: Methuen, 835p.

Lavee, H., 1973, The Relation between Debris Mantle Properties and Runoff Yield in an Extremely Arid Environment, M.Sc. thesis, The Hebrew University of Jerusalem, 71p. (in Hebrew).

Messerli, B.; Winiger, M.; and Rognon, P., 1980, The Saharan and east African uplands during the Quaternary, in M. A. J. Williams and H. Faure (eds.), *The Sahara and the Nile,* Rotterdam: Balkema, pp. 87-118.

National Academy of Sciences, 1975, *Understanding Climatic Change: A Program for Action,* U.S. Committee for the Global Atmospheric Research Program, Washington, D.C.: National Research Council, National Academy of Sciences, 239p.

Rognon, P., 1976, Essai d'interpretation des variations climatiques au Sahara depuis 40,000 ans, *Rev. Géographie Phys. et Géologie Dynam.* **13:**251-282.

Savigear, R. A. G., 1956, Technique and terminology in the investigation of slope forms, *International Geographical Union, Slopes Commission Rept. No. 1,* pp. 66-75.

Schumm, S. A., and Chorley, R. J., 1966, Talus weathering and scarp recession in the Colorado Plateau, *Zeitschr. Geomorphologie* **10:**11-36.

Twidale, C. R., 1967, Hillslopes and pediments in the Flinders Ranges, south Australia, in J. A. Jennings and J. A. Mabbutt (eds.), *Landform Studies from Australia and New Guinea,* Cambridge: Cambridge University Press, pp. 95-117.

Van Zinderen Bakker, E. M., 1978, Late Mesozoic and Tertiary palaeoenvironments of the Sahara region, in E. M. Van Zinderen Bakker (ed.), *Antarctic Glacial History and World Palaeoenvironments,* Symposium Proceedings, Tenth INQUA Congress, August 1977, Birmingham, U.K., Rotterdam: A. A. Balkema, pp. 129-135.

Wood, A., 1942, The development of hillside slopes, *Geol. Assoc. Proc.* **53:**128-140.

Yair, A., and Lavee, H., 1974, Areal contribution to runoff on scree slopes in an extremely arid environment: A simulated rainstorm experiment, *Zeitschr. Geomorphologie* **20**(suppl.):106-121.

Young, A., 1972, *Slopes,* Edinburgh: Oliver and Boyd, 288p.

18: Long-term and Short-term Rhythmicity in Terrestrial Landforms and Deposits

Leszek Starkel
Polish Academy of Sciences

Abstract: Long-term geomorphic changes, taken on the 10^5-yr basis of glacial cycles, are reflected by first-order landforms, particularly in the present-day temperate latitudes. Short-term variations, in contrast, as seen in the interglacial span of the Holocene, are shown by second- and third-order landforms and sedimentary sequences. In the mid-latitudes, the principal cycles are of the order of 2,000-2,500 yr, often separated by short violent interruptions of 100-to-300-yr duration (e.g., as heavy rainfall).

INTRODUCTION

Different time scales of climate cyclicity and other rhythmic phenomena are reflected in terrestrial landforms and deposits. The major cycles of deposition (first-order cycles, after Schumm, 1981) as well as of relief evolution (planation surfaces, cf. Davis, 1912) are both related mainly to *orogenic cycles*. These cycles are disturbed by various shorter rhythmic fluctuations such as tectonics, climatic cycles, and major hydrologic events. Those shorter rhythms are also reflected in the second-order, third-order, and so on sedimentological cycles and in the relief as complexes of landforms in vertical or parallel sequences or as polycyclic forms.

This chapter discusses the nature and diversity of both long-term and short-term rhythmicity reflected in the relief and in sedimentary sequence, as exemplified by the Quaternary features of the temperate zone.

QUATERNARY LONG-TERM CLIMATIC CYCLES

Quaternary cycles of 100 ka, 40 ka, and 20 ka duration are recognized in the various climatic zones (Kukla, 1977), starting with the well-known glacial-interglacial rhythm reflected in the cyclic evolution of soils and ecosystems (Iversen, 1973). Hydrological changes are superimposed on the thermal cyclicity in various ways. In the present-day zone of the temperate forest, a humid phase embraces the second half of the interglacial and anaglacial phase of each cold stage. The glacial cycles explain the eventual growth of ice sheets as well as the late pleniglacial loess deposition and rapid ice retreat (Grichuk, 1973). The alternation of warm and cold stages is reflected both in the change of type and rate of denudation and deposition. The main depositional phases are usually restricted to relatively short time units. Under these conditions, the highest rate of deposition has been proved to have occurred in the cold phases and at a lesser scale in the warm interglacial phases. These two types of depositions are of quite different character (Fig. 18-1). The cold phase processes are related to the existence of permafrost, ice sheets, and cold desert. The warm phase processes are mainly chemical and are responsible for the formation of soils and deep regoliths. Both these groups of landforms, created by different geomorphic process systems, differ in character. They are usually separated in time by erosional forms, created during relatively short transitional phases when two main systems, the

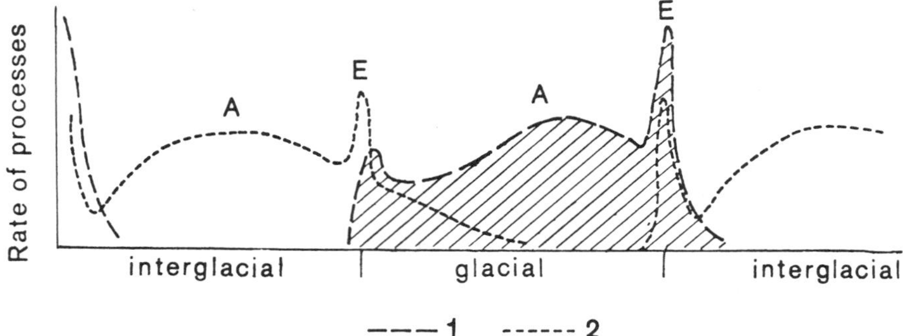

Figure 18-1. The different geomorphic systems during one glacial-interglacial cycle.
1. Rate of processes characteristic of the cold stage (cryogenic, gravitational, eolian, etc.). 2. Rate of processes characteristic of the warm stage (leaching, piping, linear erosion, etc.). A = main phases of deposition (or soil formation), E = transitional phases of erosion related to high intensity of both groups of processes.

slope and the fluvial systems, have reached threshold values and, with abrupt changes, have turned either to more active or to stabler systems (Knox, 1975).

This transformation can best be seen on the longitudinal profiles, on which slope and fluvial systems are united, joined in the transverse section of the valley bottom in the upper reaches of each river (Fig. 18-2). During the cold phase, erosion is a dominant process only in the upper part of a slope. In the lower reaches of the valley, there is a large sediment supply and overloading causing deposition. During the warm phase, the slope processes do not reach a threshold value, and the major parts of the slopes are stable, while soil formation and percolation proceed. The meandering river channels show a tendency to aggradation. The only places of intense fluvial erosion are in the lowest parts of undercut slopes and in headwater areas. Schumm (1977) distinguishes in longitudinal profile a production zone, a transport zone, and a deposition zone. Considering the transformation of the system in time, however, we find some transitional phases, when both the slope and the fluvial systems are unstable. During those phases, an intensive fluvial transport takes place in bedload channels, with substantial oscillatory behavior occurring around the mean values (Thornes, 1983; Fig. 18-3*a*). These phenomena are characteristic of the fluvial system and cause cyclic changes of the channel parameters that reflect climatic cyclicity. A different situation develops in the slope system. The retreat of permafrost and renewed percolation leads to the stabilization of all gentle slopes at the beginning of the warm hemicycle. In contrast, steep slopes can be transformed locally (see Fig. 18-5*b*); during each cold stage, the steep slopes become gentler and more rounded, or they may retreat, which is a one-way rhythmic tendency but not a cyclic change (Fig. 18-3*b*).

Taking into consideration the separate sections of the longitudinal profiles of slope and river valleys, we may distinguish various sequences of facies during Quaternary climatic cycles (see Fig. 18-2*d*). Sequence I is typical of the upper sections of slopes where any remains of older cycles have been effaced (periglacial action). Sequence II is typical of depositional lower sections with a solifluction member as an indicator of the coldest phase (Starkel, 1969). The third one shows an aggradational profile of loess with fossil (paleosol) horizons (Pecsi, 1982; Jersak, 1973). Sequence IV occurs at the edge of the valley side and has a Holocene landslide at its top (Starkel, 1969). Sequences V and VI are typical of upper and middle parts of the valley and are represented by lateral cut and fill. The cold stage strata are mainly bedload sediments of braided rivers, and warm stage deposits are usually suspended-load sediments

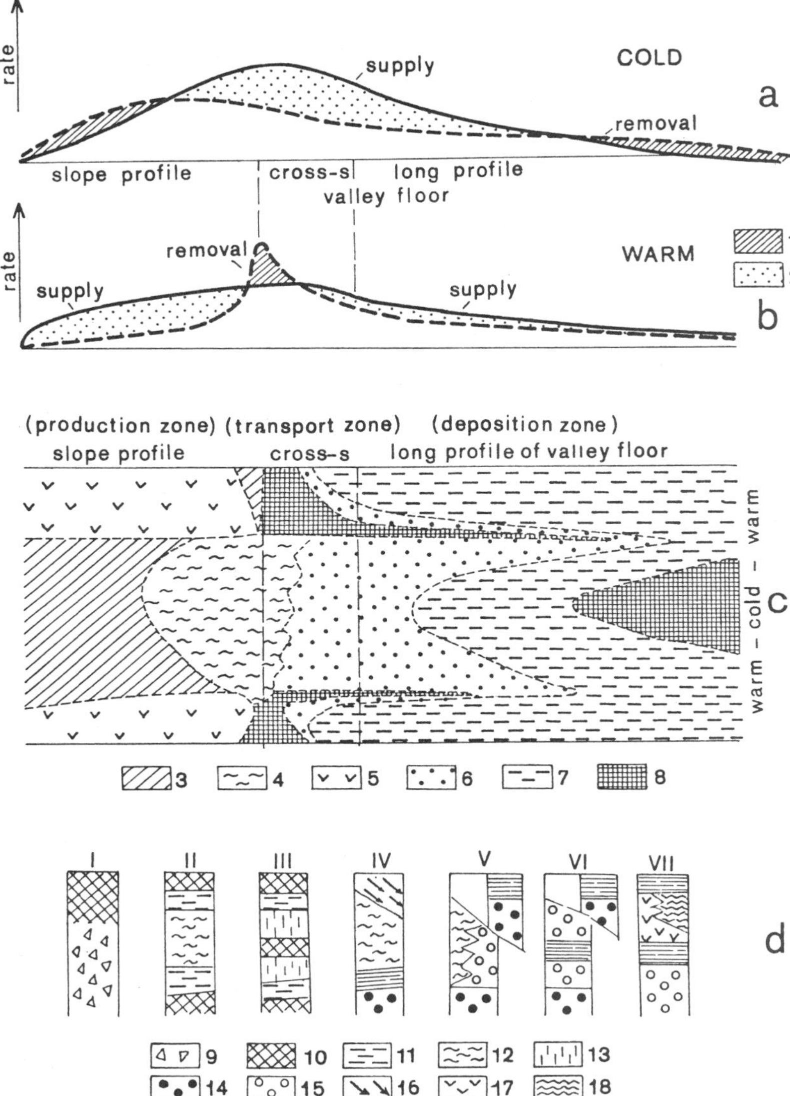

Figure 18-2. Sequence of changes in the longitudinal slope and valley bottom profile during the glacial-interglacial cycle.

(a) The relationship between removal and supply in the longitudinal profile during the cold stage; *(b)* the same relationship during the warm stage; *(c)* glacial-interglacial transect of differentiation of processes and deposition in the longitudinal profile; and *(d)* typical sequences of deposits during one cycle in different reaches of slope and valley profiles. The text explains the profiles labeled with Roman numerals. Symbols: 1, rate of removal (degradation); 2, rate of supply (aggradation); 3, zone of intensive slope degradation; 4, zone of deposition by solifluction, slope wash, and so on; 5, zone of soil formation (leaching, etc); 6, zone of bedload deposition; 7, zone of mixed and suspended load deposition; 8, zone of downcutting.

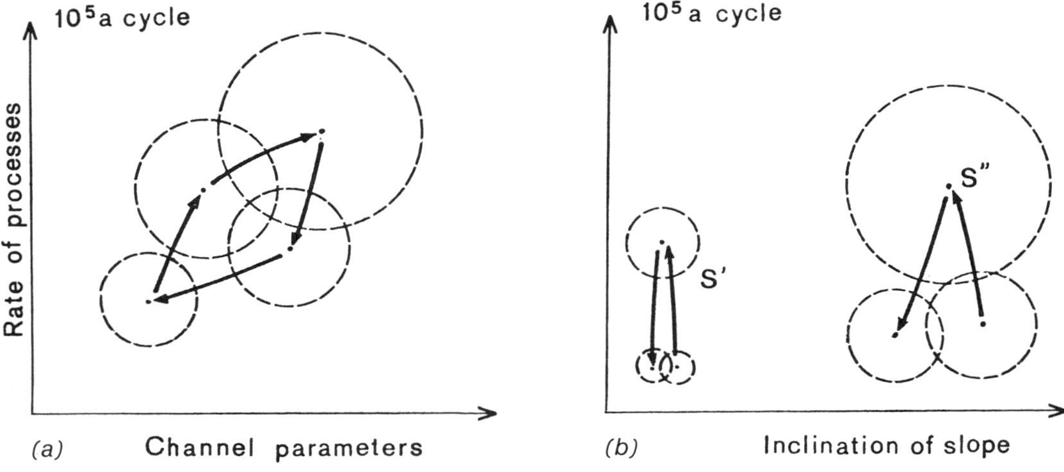

Figure 18-3. Changes of equilibrium during the glacial-interglacial cycle in the channel and slope systems. The channel parameters show sequence of changes after reaching a threshold value combined with variation of the oscillatory behavior around a mean value. The slope system shows rhythmic decline of gradient, very slight on the gentle slopes *(S')*.

of meandering rivers. The last sequence, VII, is typical of the perimarine facies (Hageman, 1969).

Many other depositional sequences reflect the climatic cyclicity. Among them are the sequences in areas periodically invaded by ice sheets that disclose a till in the middle (Woldstedt, 1958). Another case is represented by the fillings of dead-ice hollows with interglacial lacustrine and bog sediments with perhaps a periglacial member of a subsequent glacial stage on top (Dylik, 1967). In the Karst areas, two frost-debris layers are frequently separated by a calcareous (travertine) tufa (Lozek, 1976).

OVERLAPPING CLIMATIC AND TECTONIC CHANGES

Cyclicity of climatic origin is often more evident in areas with distinct uplift or subsidence. In the case of uplift, the parallel alluvial fills are replaced by a sequence of terraces or pediment levels developed on an expanded vertical scale. In a subsiding area, a normal aggradational sequence occurs in which the cyclicity is reflected by change of grain size and facies in the alternating members (Fig. 18-4).

Figure 18-4 shows five basic cases of such overlapping. In relatively stable areas (Case I), parallel fills represent the cold and warm phases of fluvial deposition (Starkel, 1983*a*). Second-order climatic fluctuations are reflected in the internal structure showing secondary fills, as are well known from the last cold stage.

A slight tectonic uplift (Case II) causes the formation of terraces with rocky steps that are usually covered with cold phase gravel bodies. Internal fills related to stadial-interstadial cycles are also visible. Those terraces are typical of the German Mittelgebirge (Brunnacker et al., 1982), Bohemia (Sibrava, 1972), and the flysch Carpathians (Starkel, 1966). Transitional phases are reflected mainly in the lateral shifting of channels and planation. In the Danube gap north of Budapest, those terraces are underlain and dated by the calcareous tufa deposits (Pecsi et al., 1984).

A more intense uplift in the mountain valleys is reflected by numerous, mainly erosional, terraces (Case III, Fig. 18-4). The number of such terraces, their local preservation and distinct differences in height, indicate the rate of tectonic uplift. This rate was very fast in the Rumanian Subcarpathians, with 15 terraces and 500 m downcutting (Starkel, 1969). In such cases even the cycles of shorter duration

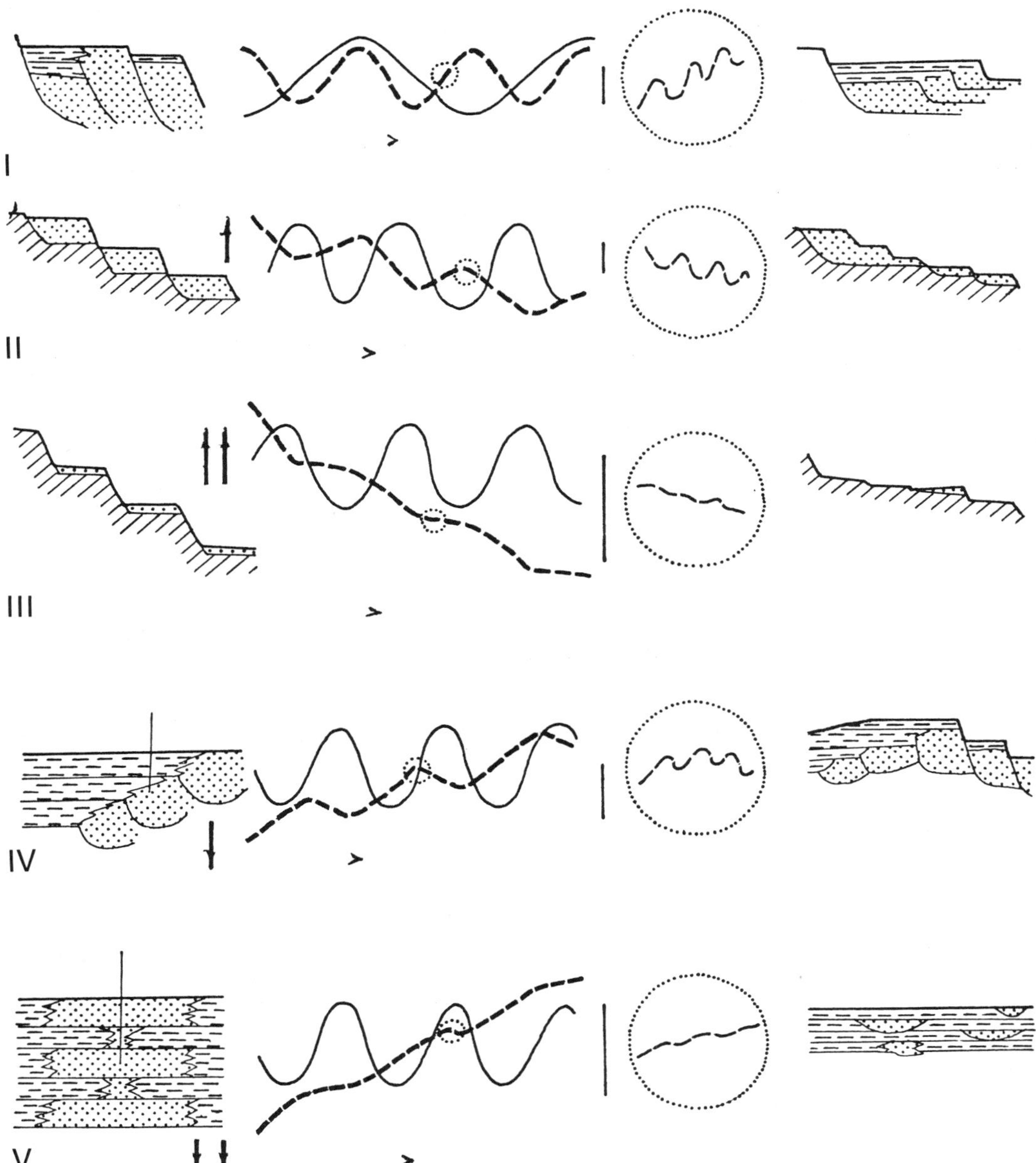

Figure 18-4. Climatic and tectonic factors coincide in the fluvial response.

I, Climatic cycles in tectonically stable areas; II, in conditions of slight tectonic uplift; III, in conditions of intense tectonic uplift; IV, in conditions of slight tectonic subsidence; V, in conditions of intense tectonic subsidence.

The middle column shows the overlapping of climate cycles (down = cold, up = warm) and variations between erosion (down) and aggradation (up). On the left are typical sequences of terrace bodies. On the right are the enlarged fragments of curve with the second-order rhythmic variations and their reflection in alluvial sequences.

Signs: 1 (fine line), curve of temperature; 2 (heavy broken line), curve of sediments yield (down = erosion; up = aggradation); 3 (arrow up), uplift tendency; 4 (arrow down), subsidence; 5 (circle to the right), enlarged fragment of curve; 6 (dots), channel deposits; 7 (dashes), overbank deposits; and 8 (diagonals), bedrock.

(20 kyr) are distinct as separate erosional steps (cf. Sibrava, 1972).

A contrary tendency is visible in the lowlands, which show a slow gradual subsidence during the Quaternary leading to a slight incision in interglacial phases and the fossilization (burial) of interglacial floodplains during cold stages (Case IV). The subsequent fills typically exist side by side, partly due to epigenesis connected with the glacial retreat (Rozycki, 1980).

A continuous subsidence (Case V) is restricted mainly to the basins that are surrounded by highlands. Climatic cycles here are reflected for the most part by the change from coarse to fine-grained sediments and back (Molnar, 1965). Parallel valley fills with systems of paleochannels of variable size exist only in the case of long-distance avulsions (rapid changes) of river channels (Borsy and Felegyhazi, 1983).

SHORT-TERM RHYTHMICITY

Short-term cycles are related mainly to variations in the hydrological regime. The Holocene cycles, each with a duration of about 2-2.5 kyr, consist of two phases. The longer one, with lower precipitation, is stabler; the fluvial or slope system usually does not reach a threshold value, and even if it does (as in highland areas) it only disturbs a metastable equilibrium on a local scale (Fig. 18-5). The second and shorter phase, with high frequency of heavy rainfalls, is usually not more than 100-300 yr long. These extreme events surpass the threshold of the system and lead to a new metastable equilibrium with higher amplitude of the oscillatory behavior around the mean equilibrium value (Starkel, 1983*b*). The result of that change is the formation of the new geometric parameters of the slope (and fluvial) system or the creation of a new sediment fill (or superposition in subsiding areas).

The clearest examples of that rhythmicity are offered by the fluvial system. In many river valleys of the temperate zone, we observe parallel cuts and fills (Becker and Schirmer, 1977), which are more distinct in upland areas or in mountain forelands (Fig. 18-6) than in the lowland river valleys, where the channels change their parameters only slightly (see Fig. 18-5*a*). The shift from a meandering to a braided channel is most common, after which there is a recovery, or adjustment, phase (Selby, 1974; Thornes and Brundsen, 1977). This rhythmicity is reflected in the parallel changes of river discharge *(Qw)* and sediment load *(Qs)*. But

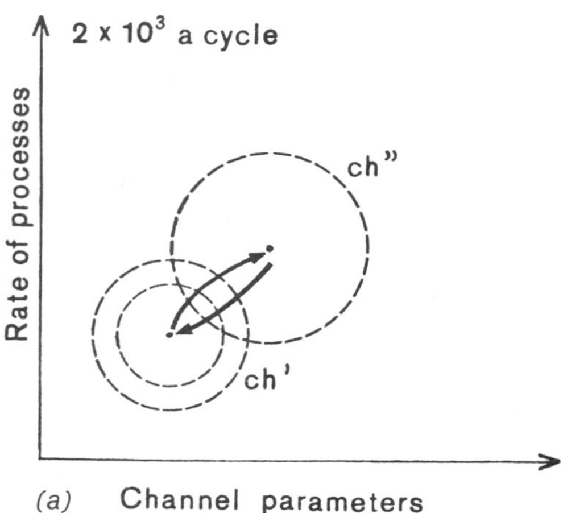

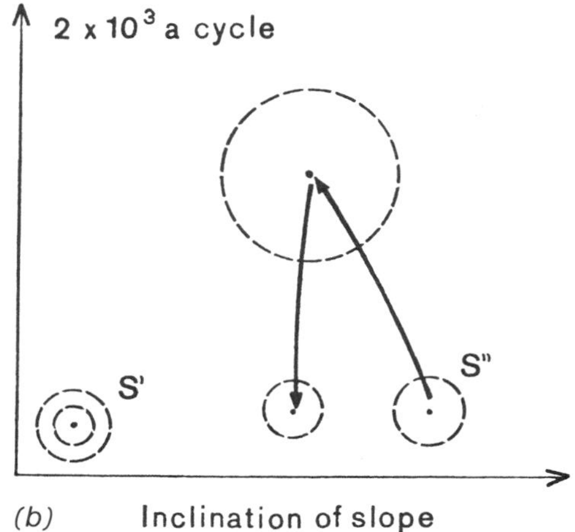

Figure 18-5. Changes of equilibrium during short-term cycles in the channel and slope systems. The channel parameters show very rare change after reaching the threshold value (e.g., from meandering to braided) and usually turn back during the recovery phase. The gentle slopes are stable. Only the steeper ones can be declined during extreme events.

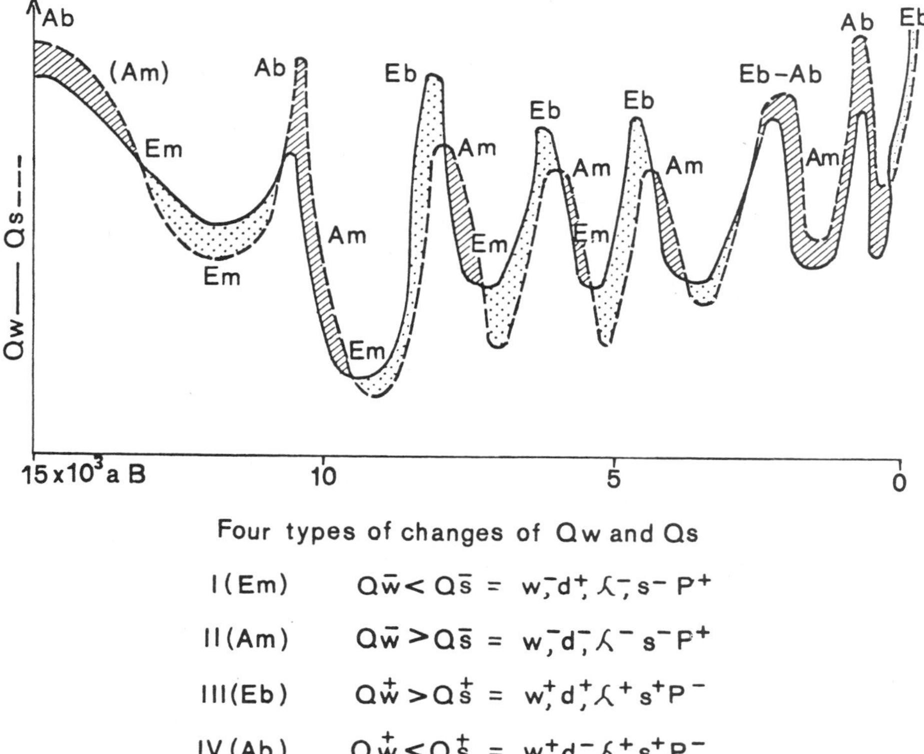

Figure 18-6. Types of sequences of change in the hydrological regime and sediment yield typical of central European valleys during the past 15,000 yr. (After Starkel, 1983*b*)
$\overset{+}{Qw}$ = rise in channel-forming discharge, $\overline{Qw}$ = decrease in discharge, $\overset{+}{Qs}$ = rise in sediment load, $\overline{Qs}$ = decrease in sediment load, *E* = erosion, *A* = aggradation, *m* = tendency to meandering, *b* = tendency to braiding.

depending on the leading factor (rise of *Qw* or *Qs*) during an active phase, there follows not only an increased bedload but also a differentiated tendency to erosion or to aggradation. As a result, the base of each next fill is lower or higher (Starkel, 1983*b*). In this way the last pleniglacial and the Younger Dryas cold phases of activity are reflected by aggradation ($\overset{+}{Qw} < \overset{+}{Qs}$), and in contrast, the mid-Holocene warm phases 8,400-8,000 B.P., 6,500-5,900 B.P., and 5,000-4,500 B.P. are reflected by downcutting (see Fig. 18-6). Only in the youngest Holocene are phases of increased fluvial activity that are reflected by the upbuilding of floodplains due to accelerated (anthropogenic) soil erosion. These cyclic changes are expressed mainly in the relief by new paleochannel systems that vary in size and represent the longer phases of channel stability with lower frequency of extreme events.

The transformation of slopes during these short-term cycles is relatively slight in the temperate forest zone. Gentle slopes show only a small increase in the rate of leaching (see Fig. 18-5*b*). On steeper slopes, however, the equilibrium may be disturbed due to heavy rains, and after a short phase of rapid mass movements, the slope attains a new equilibrium but at a gentler angle (see Fig. 18-5*b*, slopes *S''*). The phases of increased mass movements are characteristic of high mountains and are connected with temporary lowering of the cryonival belt and of the tree line (Fig. 18-7). In the European mountains there is a very good correlation

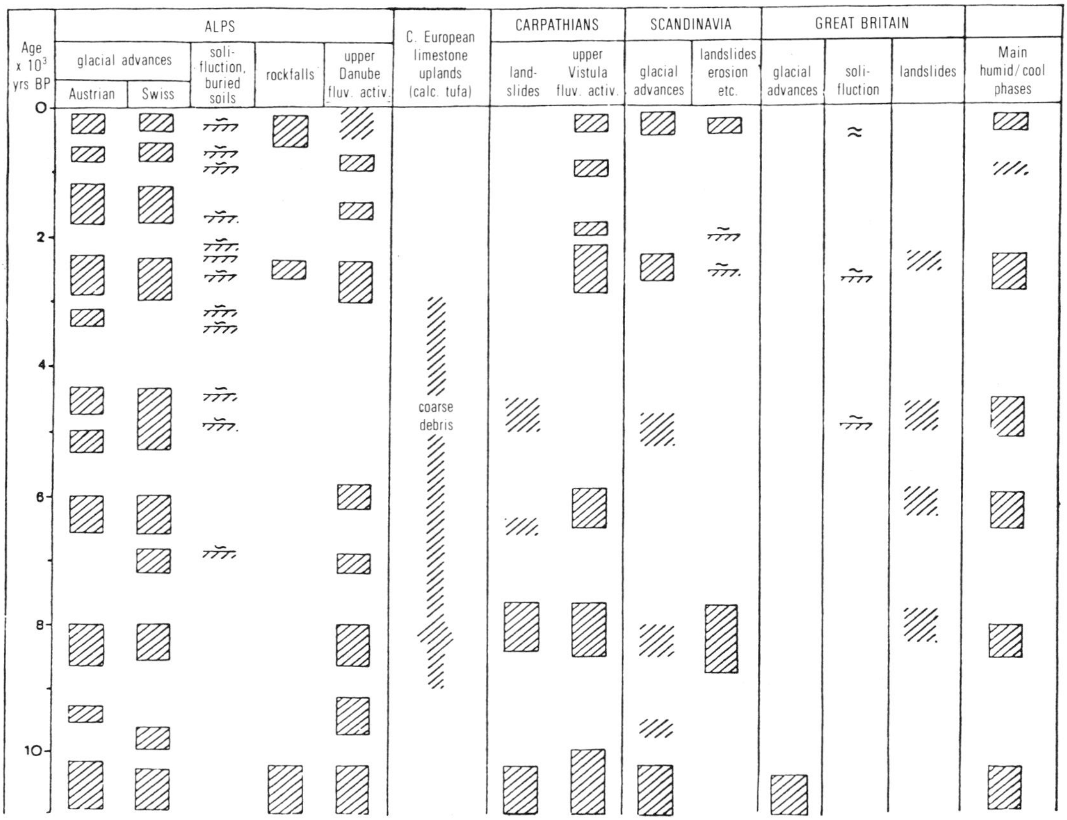

Age x 10^3 yrs BP	ALPS glacial advances Austrian	ALPS glacial advances Swiss	ALPS soli-fluction, buried soils	ALPS rockfalls	ALPS upper Danube fluv. activ.	C. European limestone uplands (calc. tufa)	CARPATHIANS land-slides	CARPATHIANS upper Vistula fluv. activ.	SCANDINAVIA glacial advances	SCANDINAVIA landslides erosion etc.	GREAT BRITAIN glacial advances	GREAT BRITAIN soli-fluction	GREAT BRITAIN landslides	Main humid/cool phases
0														
2														
4						coarse debris								
6														
8														
10														

Figure 18-7. Holocene phases of high geomorphic activity in the European mountains.

between these intensive slope and fluvial processes and the fluctuations of mountain glaciers (Bortenschlager, 1982; Patzelt, 1977), as well as avalanches and congelifluction lobes, especially in the late Holocene (Furrer, Leuzinger, and Amman, 1975; Grove, 1972; see also Fig. 18-7).

Those changes are also reflected in the precipitation of calcareous tufa and in the lake and bog history. Since they are not discussed in Weber's Grenzhorizont and Granlund's RY horizons, it should be emphasized that in many European lakes there are sharp rises of the water level dated at circa 4,500 B.P. that coincide with flooding on many rivers and again circa 2,500-2,000 B.P. (Dolukhanov, 1977; Berglund, 1983; Ralska-Jasiewiczowa and Starkel, 1984). These cyclic fluctuations of lake levels are each less than 2,000 yr long, as was earlier suggested by Schnitnikow (1957).

This short-term cyclicity is superimposed on the long-term cycles. In one complex of landforms it is reflected in the repetition of geomorphic groups, mainly side by side, as in paleomeander systems, frontal moraines, or landslides. In one sedimentological unit of a higher order it is expressed by repetition of lower-range members (units) in the vertical or lateral sequence (cf. Fig. 18-2).

In this discussion on the reflection of rhythmicity in the relief and deposits, the role of separate extreme events must not be forgotten. In the fluvial history, LaMarche (1968) and Baker (1983) gave good examples. In each case the system reaches a threshold value and usually returns to the previous equilibrium during a longer or shorter phase of recovery (Selby, 1974). Such an event can, however, also substitute for the normal phase in the cyclic system. It is common in climatic zones that have a very low frequency of extreme events. One or two events with a recurrence interval of 1,000 yr or so, coinciding with an abrupt change of climate, would then play a positive substantial role in the rhythmic variations of the Quaternary sequence.

Depending on the deviations of the climatic system, the rate of tectonic movements, lithology (sediment supply), and the vertical and areal extent of geomorphic and sedimentological changes, the role of cycles of various duration may be different (Fairbridge, 1980). A single extreme event can be mistaken as a long-term rhythmicity. For example, a rapid rise of the saline Sambhar Lake in Rajasthan (India) by nearly 2 m, which was reflected in the landforms and lacustrine deposits, might be mistaken for a climatic change (Starkel, 1972). Therefore, especially in environments that rarely reach a threshold value, a careful examination of all available data is necessary. Only a wide regional distribution of similar phenomena, correlated by using chronometric methods, may prove that these phenomena reflect long- or short-term climatic rhythmicity.

REFERENCES

Baker, V. R., 1983, Paleoflood hydrological analysis from slackwater deposits, *Quaternary Studies in Poland (Poznań)* **4:**19-26.

Becker, B., and Schirmer, W., 1977, Palaeoecologic study of the Holocene valley development of the River Main, Southern Germany, *Boreas* **6**(4):303-321.

Berglund, B., 1983, Palaeoclimatic changes in Scandinavia and on Greenland—A tentative correlation based on lake- and bog-stratigraphical studies, *Quaternary Studies in Poland (Poznań)* **4:**27-44.

Borsy, A., and Felegyhazi, E., 1983, Evolution of the network of water courses in the North-Eastern part of the Great Hungarian Plain from the end of the Pleistocene to our days, *Quaternary Studies in Poland (Poznań),* **4:**115-124.

Bortenschlager, S., 1982, Chronostratigraphic subdivisions of the Holocene in the Alps: *Striae (Uppsala)* **16:**75-79.

Brunnacker, K.; Löscher, M.; Tillmanns, W.; and Urban, B., 1982, Correlation of the Quaternary terrace sequence in the Lower Rhine Valley and Northern Alpine Foothills of Central Europe, *Quat. Research* **18:**152-173.

Davis, W. M., 1912, *Die erklärende Beschreibung der Landformen,* Leipzig (English translation).

Dolukhanov, P. M., 1977, The Holocene history of the Baltic sea and ecology of prehistoric settlement, *Baltica* **6:**227-244.

Dylik, J., 1967, The main elements of Upper Pleistocene palaeogeography in Central Poland, *Biul. Peryglacjalny* **16:**85-115.

Fairbridge, R. W., 1980, Thresholds and energy transfer in geomorphology, in D. R. Coates and J. D. Vitek (eds.), *Thresholds in Geomorphology,* London: Allen & Unwin, pp. 43-49.

Furrer, G.; Leuzinger, H.; and Amman, K., 1975, Klimaschwankungen während des alpinen Postglazials in Spiegel fossiler Böden. *Naturf. Gesell. Zurich Vier.,* **120**(1):15-31.

Grichuk, V. P., 1973, Vegetation, in *The Paleogeography of Europe during the Late Pleistocene, Reconstruction and Models,* Moscow (in Russian), pp. 182-219.

Grove, J. M., 1972, The incidence of landslides, avalanches and floods in western Norway during the Little Ice Age, *Arctic and Alpine Research* **4:**131-138.

Hageman, B. P., 1969, Development of the western part of the Netherlands during the Holocene, *Geologie en Mijnbouw* **48**(4):373-388.

Iversen, J., 1973, The development of Denmark's nature since the Last Glacial, *Danmarks Geol. Undersögelse [skr.],* V, 7-C, 126p.

Jersak, J., 1973, Litologia i Stratygrafia Lessu Wyzyn Poludniowej Polksi (English summary), *Acta Geog. Lodz 32,* 139p.

Knox, I. C., 1975, Concept of the graded stream, in W. N. Melhorn and R. C. Fremal (eds.), *Theories of Landform Development,* Binghamton, N.Y.: Geology Department, State University of New York, pp. 169-198.

Kukla, G. J., 1977, Pleistocene Land-Sea Correlations, I. Europe, *Earth-Sci. Rev.* **13:**307-374.

LaMarche, V. C., 1968, Rates of slope degradation as determined from botanical evidence, White Mountains, California, *U.S. Geol. Survey Prof. Paper 352-I,* pp. 341-377.

Lozek, V., 1976, Klimatbhängige Zyklen der Sedimentation und Bodenbildung während des Quartärs im Lichte malakozoologischer Untersuchungen, *Ceskoslovenska Akad. Ved (Praha), Rospravy, Rada Matematiko-Prirodovedecka* **86**(8):1-97.

Molnar, B., 1965, Beiträge zur Gliederung und Entstehung der jungstertiären und quartären Schichten des Donau-Theiss Zwischenstromlandes auf Grund der Schwermineralienzusammensetzung (in Hungarian, summary in German), *Földtani Közl (Budapest)* **95:**217-225.

Patzelt, G., 1977, Der zeitliche Ablaut und das Ausmass postglazialer Klimaschwankungen in den Alpen, in Dendrochronologie und postglaziale Klimaschwankungen in Europe, *Erdwiss. Forschung (Wiesbaden)* **13:**249-259.

Pecsi, M., 1982, The most typical loess profiles in Hungary, *Quaternary Studies in Hungary (Budapest)* **3:**145-169.

Pecsi, M.; Schweitzer, F.; and Scheuer, G., 1984, Plio-Pleistocene tectonic movements and the travertine horizons in the Hungarian Mountains, *Studia Geomorph. Carpatho-Balcanica (Kraków)* **17:**19-27.

Ralska-Jasiewiczowa, M., and Starkel, L., 1984, Record of the hydrological changes during the Holocene in the lake, mire and fluvial deposits of Poland, in *Earth Surface Processes and Landforms.*

Różycki, S. Z., 1980, Principles of stratigraphic subdivision of Quaternary of Poland, *Quaternary Studies in Poland (Poznań)* **1:**99-106.

Schnitnikow, A. W., 1957, Changes of humidity on the continent of the Northern Hemisphere (in Russian), *Zap. Gieograf. O-wa SSSR,* **16.**

Schumm, S. A., 1977, *The Fluvial System,* New York: Wiley, 338p.

Schumm, S. A., 1981, Evolution and response of the fluvial system, sedimentologic implications, *SEPM Spec. Publ. No. 31,* pp. 19-29.

Selby, M. I., 1974, Dominant geomorphic events in landform evolution, *Internat. Assoc. Eng. Geology Bull.* **9:**85-89.

Sibrava, V., 1972, Zur Stellung der Tschechoslovakei im Korrelierungssystem des Pleistozäns in Europa, *Sbornik Geol. Ved (Praha)* **A8:**1-218.

Starkel, L., 1966, Evolution of the relief of the Polish East Carpathians in the Quaternary (with the upper San basin as example), *Geographia Polonica* **10:**89-114.

Starkel, L., 1969, Climatic or tectonic adaptation of the relief of young mountains in the Quaternary, *Geographia Polonica* **17:**209-229.

Starkel, L., 1972, The role of catastrophic rainfall in the shaping of the relief of the Lower Himalaya (Darjeeling Hills), *Geographia Polonica* **21:**103-147.

Starkel, L., 1977, The paleogeography of Mid- and East Europe during the Last Cold Stage and West-European comparisons, *Royal Soc. London Philos. Trans.* **A273:**249-270.

Starkel, L., 1983*a*, Climatic change and fluvial response, in R. Gardner and H. Scoging (eds.), *Mega-geomorphology,* Oxford: Clarendon Press, pp. 195-211.

Starkel, L., 1983*b*, The reflection of hydrologic changes in the fluvial environment of the temperate zone during the last 15,000 years, in K. J. Gregory (ed.), *Background to Palaeohydrology: A Perspective,* New York: Wiley, pp. 213-235.

Starkel, L., 1984, The reflection of abrupt climatic changes in the relief and in the sequence of continental deposits, in N.-A. Mörner and W. Karlén (eds.), *Climatic Changes on a Yearly to Millennial Basis,* Dordrecht: Reidel Publ. Co. pp. 135-146.

Thornes, J. B., 1983, Evolutionary geomorphology, *Geography* **68**(3):225-235.

Thornes, J. B., and Brundsen, D., 1977, *Geomorphology and Time,* London: Methuen, 208p.

Woldstedt, P., 1958, *Das Eiszeitalter,* 2nd ed., Stuttgart.

19: Speciation and Cyclic Climatic Change

Thomas M. Cronin
U.S. Geological Survey

Abstract: Paleoclimatology relies heavily on paleontologic proxy data to document patterns of climate change in the geologic record. Two primary reasons for using fossils for paleoclimatology are to reconstruct specific temperature regimes and to identify climatically induced paleoenvironmental changes using temporal sequences of ecologically meaningful faunal or floral assemblages. Most of these studies assume that particular taxa have the same ecological tolerances throughout their stratigraphic ranges and that we can recognize speciation events in fossils. These assumptions are tantamount to the sometimes invalid supposition that all genotypic evolutionary change influencing a species' ecology can be recognized by changes in phenotype and that closely related species can be distinguished from one another by their morphology. Consequently, two key aspects of using fossils for paleoclimatology are often those least studied by paleontologists: the relationship between long-term morphologic stasis (and intraspecific variability) to climatic changes and the paleoclimatic significance of the disruption of within-species stasis as recognized by increased speciation or extinction rates.

Climatic change is a primary mechanism for potentially catalyzing speciation events because it can alter a species' biogeography, isolating populations from one another. Yet examination of species subjected to cyclic climatic changes indicates that speciation usually does not occur during multiple cycles. Cyclic patterns of climate change appear to foster stability in species and sometimes in whole faunas. Conversely, relatively infrequent or one-time climatic transitions that have changed Earth climate from one state to another disrupted within-species stability and caused speciation to occur more frequently than during cyclic change. When evolutionary patterns are modulated by climate, the magnitude and rapidity of the climatic event are less important factors in influencing periods of speciation than are the frequency and duration of the change.

INTRODUCTION

Major advances in understanding natural phenomena often come from melding concepts and empirical data from two or more distinct fields previously considered unrelated. Darwin's observations in fields as disjunct as animal breeding, zoogeography, and paleontology gave him a unique perspective that led to his evolutionary synthesis. Paleoclimatology, clearly an interdisciplinary field, owes much of its progress to the exchange of concepts and data among seemingly unrelated fields. One important area of paleoclimatology—the use of paleontologic proxy data as evidence for climatic reconstruction—applies our knowledge of organisms, their ecology and their evolutionary history, to the detection of patterns of paleoclimatic change. From paleontologic data, we infer causal mechanisms to explain observed patterns. Paleoclimatologists must therefore develop paleontologic tools, apply them to the geologic record, and interpret biological events in light of physical and chemical changes in the atmosphere, lithosphere, oceans, and cryosphere. As we become more sophisticated in our analyses of climatic history, and especially in studies of cyclicity in pre-Quaternary climates, in which a greater proportion of taxa is extinct, the accuracy and validity of our paleontologic proxy data must be carefully scrutinized. In a more general vein, we must ask, Why are fossils useful for paleoclimatology and how good are they? To do this, the biologic process of speciation, and more generally, the process of selection (Vrba, 1984) and the very factors driving evolution, must be examined within the context of environmental change. This chapter briefly explores the nature of paleontologic evidence for climatic change and the theoretical underpinnings found in evolutionary biology.

PALEONTOLOGIC PROXY DATA

Many kinds of paleontologic data are used in paleoclimatology. One category involves morphologic data where a distinct (often functional) morphology tells us about paleoenvironmental conditions, regardless of which taxonomic group possesses it. The correlation of characteristic types of leaf physiognomies with climate zones (Wolfe, 1978; Wolfe and Poore, 1982) is a good example. The development of eyes in species of benthic organisms such as trilobites and ostracodes is a function of the location of species habitat in relation to the photic zone and, as a paleobathymetric tool, is useful for reconstructing eustatic sea-level fluctuations (Benson, 1973). These are both examples of hereditable features. Nonhereditable ecophenotypic variation in morphology also is useful for paleoclimatology. Examples include coiling direction in planktic foraminifers, which is believed to reflect surface-water temperatures (Bandy, 1972), and distinct morphotypes in the benthic foraminifer *Elphidium* (Feyling-Hanssen, 1972).

A second category, which is the focus of this chapter, includes taxonomic paleoclimatic proxy data and, as a corollary, the phenotypic characteristics that bestow on a taxon its individuality and its particular ecology. In this category, fossil occurrences of ecologically sensitive taxa, usually species whose modern ecology is known, are used to infer past conditions. This is the most conventional use of fossils for reconstructing past environments qualitatively and quantitatively. Planktic foraminifers, pollen assemblages, beetles, ostracodes, and mollusks are typical examples of such fossil occurrences. In these instances we rely on our knowledge of the physiological and ecological requirements of species inferred from modern distributions (or less frequently, experimental evidence) to tell us about past conditions. Two basic assumptions are usually made: (1) that we can recognize individual biological species in the fossil record and (2) that no physiological or ecological change has occurred in a species during its entire stratigraphic range.

In most applications of paleontologic data to paleoclimatology, although we may know when and where a particular group is useful (i.e., in what stratigraphic range, geographic area, and sedimentologic setting), we know very little about why a species, subjected to frequent and severe fluctuations in environment, survives with relatively minor intraspecific change and without complete extinction (rather than just local extinction) of the species. To explore the problem of long-term stasis, this chapter unites recent advances in evolutionary theory with paleoclimatology by examining the biological factors that determine why certain taxonomic groups are useful for paleoclimatology.

The alliance between evolutionary theory and climate has a long history. For instance, early in this century, the interpretation of racial variability in vertebrates as a function of climate zones was at the center of the debate between Lamarckian and Darwinian theories (Rensch, 1929; see Rensch, 1983, for a discussion). Indeed, *climatic variation* lies at the heart of the modern evolutionary synthesis in its basic tenet of geographic speciation (see Mayr, 1963, pp. 485-487, for a historical discussion of this topic). More recently, Stanley (1984) has postulated that global refrigeration during the Phanerozoic has been an important factor in causing faunal crises and major periods of extinction. Casey, Wigley, and Perey-Guzman (1983) have found a relationship between six Phanerozoic periods of major cooling and the quantum evolution of radiolarian family groups. On the microevolutionary scale, Lohman and Malmgren (1983) recently found a correlation between migration of ecophenotypes of the planktic foraminifer *Globorotalia truncatulinoides* and climatic change. Although it is not certain whether the observed morphologic trends reflect gradual evolutionary change or ecophenotypic variation due to an unknown environmental factor, the morphologic variability is within that normal for the species and does not involve a speciation event. Vrba (1980) has reviewed several paleontologic studies containing evidence that environmental change plays an important role in driving evolutionary trends at different hierarchical levels. Perhaps more important, she has also pointed out how the integrity of individual species is an important aspect of the relationship between evolution and climate.

These few examples show how the paleontologic record contains evidence that climate has affected organisms at all hierarchical levels from individual populations to the evolution of entire communities and higher taxonomic groups. Yet, with few exceptions, paleoclimatologists have seldom attempted to validate basic assumptions about fossil species by seeking evidence from evolutionary theory and population genetics concerning how and why species survive climatic oscillations. Conversely, evolutionists have virtually ignored the unique and potentially informative evolutionary information available in the literature on climatic change and the fossil species that document it. In the next few sections I suggest that paleoclimatic changes modulate evolution such that certain types of species under specific circumstances are more useful for paleoclimatology than others. Paleontologic evidence of how species respond to cyclic climatic changes is shown to reinforce theories about speciation mechanisms and long-term stability in species and to provide support for the basic assumptions we make when using species for paleoclimatology.

INTEGRITY OF THE SPECIES

Central to this discussion is the concept that species are real entities, consisting of actually or potentially interbreeding populations, each having specific ecological requirements. Individual species of beetles, foraminifers, ostracodes, plants, or other organisms form the essential baseline data for paleoclimatology. The concept that a species is an integrated unit in time and space has had a long and complex history. Van Valen (1982) identifies two related hypotheses that address the concept: (1) punctuated equilibrium (Eldredge and Gould, 1972), which maintains that evolution ordinarily occurs in brief spurts, and long periods of stasis separate the origination of new species, and (2) species integrity (Dobzhansky, 1970), which holds that most species are geographically uniform in most respects. In this chapter I use the term *stasis* to refer to the cohesiveness or integration of a species over its geographic and stratigraphic range. In dealing with the fossil record, I am primarily referring to fossilizable morphologic features as a means to recognize stasis; however, physiological characteristics of a particular species inferred from its present or past distributions must also be considered an aspect of the phenotype that distinguishes a species from related taxa.

Although some paleontologists believe that the median species longevity is between 3 m.y. and 10 m.y., depending on the group of organisms (Stanley, 1979, 1982), the actual mechanisms by which a species maintains its integrity over millions of years is an actively disputed topic in evolutionary science today (Charlesworth, Lande, and Slatkin, 1982; Stanley, 1982; Van Valen, 1982). How can a species tolerate frequent and often rapid environmental change and habitat destruction with seemingly no effect? Among those who accept estimates for the long duration of species, some believe the mechanism of stabilizing selection—that is, the weeding out of extreme phenotypes during successive generations—can account for long-term stasis (Charlesworth, Lande, and Slatkin, 1982). Others believe there are limited ranges of phenotypes constrained by ontogenetic development and morphology (Gould, 1982).

It is significant that some evolutionists now question whether we can really recognize biological species using only fossilized remains, and consequently, they believe that the estimates of 10 m.y. for species' longevities, based on stratigraphic ranges derived from the fossil record, have been grossly overestimated. Charlesworth and co-workers (1982) state, "There seems to be no way in which systematists can arrive at a certain classification of closely related allopatric taxa as separate biological species on the basis of purely morphological criteria" (p. 481). Likewise, Schopf (1982), a paleontologist, concludes that "many morphologic species in the fossil record are simply not distinguishable into biological species given standard paleontological procedures" (p. 1,156). Instead of a median duration of 10 m.y., Schopf believes mean species' durations are only 200,000 yr. These arguments stem from the belief that there is a high percentage of sibling species (perhaps 10-20% of all species) and that speciation is not always accompanied by fossilizable morphologic change. If Schopf's thesis is correct, it constitutes a serious challenge to the use of fossils for paleoclimatology. How can species be used to identify past environments if closely related species, which often inhabited distinct environments, cannot be distinguished from one another? How can we use recurring species or assemblages of species to document successive 100,000-yr cycles if species only live an average of 200,000 yr?

To examine this problem we must discuss how organisms respond to climatic change. First, a species can become extinct, a topic that merits separate treatment and is not discussed here. Second, they can migrate to equitable environments and return to the original area once suitable conditions have returned. Although some populations may become extinct, the species will survive in parts of its geographic range and recolonize its former habitat. Most paleoclimatologic work presumes this sequence of events to occur when successive occurrences of diagnostic species or assemblages are found signifying a cyclic pattern of paleoenvironmental change.

Third, a species can change, either by branching off a new daughter species or by changing its phenotype (either its physiological tolerances or its morphology) to suit the new climatic conditions. If change occurs within the species, the paleontologist must determine if it reflects genetic changes or nonheritable ecophenotypic variation. If speciation occurs, under what conditions does the descendant species form and what is its ecologic preference? Paleontologists must, of course, always be concerned with the possibility that a new species originates with no apparent morphologic divergence from the ancestral species or that a particular species may adapt its physiology to changing environments with no fossilized morphologic change.

In simplest terms, the confident recognition of a fossil species and its distinction from closely related species based on fossilizable parts of the phenotype remain the top priority. In practice, this becomes a question of deciding when speciation occurs and when it does not. The next two sections present evidence that speciation rarely occurs during climatically induced geographic isolation and that many species consistently show morphologic stability during repeated cycles of climatic change, supporting assumptions that species-level data are valid paleoclimatic indicators.

BIOGEOGRAPHY, GEOGRAPHIC ISOLATION, AND SPECIATION

Biogeographic data on fossil and recent species should form the nucleus of any research aimed at unraveling evolutionary relationships of fossil groups and the use of fossils for paleoclimatology. Since the neo-Darwinian "modern synthesis" of evolution was developed in the 1930s and 1940s, geography has been the major factor in the classification of mechanisms of speciation (Mayr, 1942, 1982*a*). Geographic isolation by a barrier is considered a key ingredient for speciation to occur, so most speciation is believed to have occurred allopatrically in small, peripheral populations. Mayr (1982*a*) and Cracraft (1982, 1984) discuss the concepts of allopatric speciation in detail. Lazarus (1983) gives a detailed discussion of speciation in planktic microfossils.

Climatic fluctuations play an important role in effecting short-term, often temporary, alteration of a species' biogeography and therefore in causing geographic isolation. Glacial-interglacial changes and eustatic sea-level fluctuations and their obvious effects on temperature, precipitation, ocean circulation, continental vegetation, and food resources clearly constitute primary mechanisms for isolating populations of a species.

The following two case studies exemplify how climatic change can cause geographic isolation.

The Champlain Sea was the postglacial arm of the Atlantic Ocean that flooded the St. Lawrence and Champlain valleys of eastern Canada and the northeastern United States between 13,000 yr and 10,000 yr ago (Cronin, 1977). It owed its existence to isostatic depression of the crust by the Laurentide Ice Sheet and subsequent inundation by marine water as the ice margin retreated northward during the last deglaciation. The sea was virtually isolated from the Atlantic, connected only by a narrow strait near Quebec City. The fauna of the sea included about 79 species of benthic foraminifera (Cronin, 1979), 34 ostracode species (Cronin, 1981), and a large molluskan fauna (Wagner, 1970). These faunas included eurytopic species tolerant of reduced, fluctuating salinities and water temperatures and stenotopic species that could survive only in water of normal marine salinity. Despite the isolation of the faunas from the main populations along the northeastern North American coasts, there is no evidence for speciation in any of the groups over the brief history of the sea. Yet this must be considered an opportunity for allopatric speciation to occur in an isolated area.

A second example involves a longer period of geographic isolation of formerly contiguous populations of ostracodes in the north Atlantic. Many high-latitude cryophilic ostracode species are amphiatlantic—that is, indigenous to the extensive continental shelves on both sides of the north Atlantic (Hazel, 1970). Most species became dispersed across the many shallow-water habitats in northern Europe, northeastern North America, Iceland, and Greenland, extending their ranges as far south as Cape Cod and the British Isles. During the last glaciation, and presumably earlier glaciations, when the polar front in the central north Atlantic shifted southward and continental glaciers occupied much of the adjacent land masses, these species migrated southward along eastern and western Atlantic coasts (Hazel, 1968; Cronin, 1981; Cronin et al., 1981). Climatic events therefore effectively caused the separation of eastern and western north Atlantic populations for the interval 70,000-10,000 yr ago by the deep waters of the north Atlantic and ice cover in high latitudes. This scenario probably occurred many times during the glacial periods of the past million years. Yet the repeated isolations of populations did not produce different species on opposite sides of the Atlantic for the majority of amphiatlantic taxa.

In these examples, climatically induced geographic isolation did not catalyze speciation, despite the apparent opportunities that existed. These results support Mayr's (1982*b*) view that in the majority of cases, geographic isolation will not result in speciation, although when speciation does occur, it usually occurs in isolated populations. Because these examples represent relatively brief intervals of time within a single Milankovitch climatic cycle of 100,000 yr, let us now turn to other examples in the geologic record to examine the net long-term effect on different groups of organisms of many such climatic cycles occurring over millions of years.

STASIS DURING MULTIPLE CLIMATIC CYCLES

Evidence for cyclic climatic fluctuations of different periodicities is known from many parts of the Phanerozoic, and frequently the sediments reflect a biogenic origin. The following examples are representative of the many documented cases of cyclicity in the geologic record.

Schwarzacher and Fischer (1980) present a convincing case of the causal relationship between Earth orbital perturbations and patterns of limestone-shale bedding in the Cretaceous deposits of Italy. In these sections, limestones consisting of lithified coccolith oozes containing globigerinids, radiolaria, pelagic crinoids, and ammonite aptychi signify cyclic deposition caused by climatic fluctuations affecting surface-water conditions and, in turn, nutrients and the productivity of calcium-precipitating plankton. Spectral analysis of bundles of beds reveals 100,000-yr Milankovitch cycles similar to those known from the Quaternary (Hays, Imbrie, and Shackleton, 1976) occurring for an interval of about 3-8 m.y. of the Cretaceous. If the same species of foraminifers, radiolarians, and other pelagic groups occur throughout most of the section, as is apparently the case for most of the assemblages, we have an example of evolutionary stability within many species during extreme climatic and oceanographic fluctuations.

Another example of cyclic sedimentary deposition can be found in some facies of the Monterey Formation of California (Garrison and Douglas, 1981). This Miocene and Pliocene unit consists of diatomites, cherts, organic mudstones, phosphates, and authigenic carbonates deposited under complex oceanographic, climatic, and tectonic conditions. By comparison with modern analogs, the rhythmic deposits of the Monterey have been associated with climatically induced changes in the position and, perhaps, intensity of the oxygen minimum zone (Pisciotto and Garrison, 1981). Cycle periodicities range from annual to several thousand years, but an external forcing mechanism has not been determined. Nevertheless, numerous species of radiolarians (Weaver, Casey, and Perey, 1981), diatoms (Barron, 1976), and foraminifers (Govean and Garrison, 1981) persist in various facies of the Monterey for several million years. Although long-term climatic changes influenced the overall sequence of microfaunal assemblages between 15 m.y. and 4 m.y. ago, many individual species persisted through long intervals of cyclic climatic and oceanographic change.

Probably the most carefully documented climatic cycles are Quaternary Milankovitch cycles known from fluctuations in percentages of planktic foraminifers and oxygen-isotope ratios in deep-sea core sediments (Imbrie and Kipp, 1971; Hays, Imbrie, and Shackleton, 1976) and sea-level changes on island and continental margins (Bloom et al., 1974; Cronin, 1983). Throughout climatic cycles having periodicities of about 100,000 yr, 41,000 yr, and 23,000 yr, many Northern Hemisphere temperature-sensitive species moved southward during glacial intervals and northward during interglacials with little intraspecific change and no speciation occurring. In morphometric analyses of Quaternary ostracodes from the eastern United States subjected to Milankovitch cycles, morphologic stasis was found to predominate despite presumably severe selective pressures exerted by climatic cooling and sea-level drops that eliminated much of the available continental shelf habitat (Cronin, 1985; in press). Lohman and Malmgren (1983) found gradual change in the morphology of *Globorotalia truncatulinoides* during these Milankovitch cycles, but the variability was not outside that expected for the species. Kellogg (1983) found that a climatic event may have caused the speciation of the radiolarian *Eucyrtidium matuyamai* from *E. calvertense* about 1.9 m.y. but both persisted as discrete species during climatic cycles of the past 1.9 m.y.

In summary, the general pattern frequently but not exclusively encountered in fossil sequences deposited during cyclic climatic changes shows stability within individual species and a lack of appearance of new species. Indeed, the routine documentation of paleoclimatic cycles using species-level data provides one of the strongest arguments for the long-term integrity of species and undermines Schopf's case against the validity of fossil species. If fossil species were not true biological species, they would not work in paleoclimatology. The following summary of speciation mechanisms gives clues on why this might be so.

MECHANISMS OF SPECIATION

The renewed interest in the speciation process is clear from the spate of recent publications on mechanisms of speciations (White, 1978; Eldredge and Cracraft, 1980; Atchley and Woodruff, 1981; Barigozzi, 1982; Milkman, 1982; Lazarus, 1983). Although speciation is a complex and multifaceted process and we really know very little about it (Bush, 1982), a generalized sequence of events can be postulated that incorporates the major points of most models. Ayala (1982) has outlined two stages of the generalized speciation process. During an initial stage an extrinsic event catalyzes the splitting of a daughter species by somehow interrupting gene flow between two populations of the same species (Ayala, 1982). Geographic isolation is probably the most common means of initially interrupting gene flow. Isolated populations become increasingly dissimilar, and reproductive isolating mechanisms may begin to appear. The second stage involves the completion of reproductive isolation when previously isolated populations come into contact with each other again, but little gene flow occurs and additional reproductive isolating mechanisms develop. The first stage of speciation is reversible if previously isolated populations, on elimination of the barriers, are able to form a single gene pool again.

Potential catalysts of speciation as envisioned in this model are ubiquitous in the geologic record of paleoclimatic and paleogeographic changes, but which ones cause speciation and which do not is a perplexing problem for evolutionists today. One possible explanation for the preponderance of climatic changes unaccompanied by speciation is that neither rapidity nor magnitude of the environmental change is the most important factor in determining whether a species evolves to adapt to new conditions. Potential or incipient speciation events frequently occur when climatic changes isolate populations of the same species over geologically short intervals (Cronin, 1983). Nonetheless, the second stage of the speciation process, the establishment of reproductive isolation, is not completed. Speciation is, in essence, aborted. In this scenario, the most plausible explanation is that there simply was not enough time for the thousands, tens of thousands, or even millions of generations needed for genomic reorganization to occur (Carson, 1982) before the climatic cycle restored the original conditions to which the species was adapted. Cycles of differing periodicities will have different effects on organisms with different generation times and population ecologies, but we still have a poor idea of how many generations are necessary to establish reproductive isolation.

Some paleontological data support the idea that the frequency and duration of a climatic event are more important than its amplitude or its suddenness in influencing whether species maintain their integrity over extended periods of time. Support for this explanation comes not only from direct observations of stasis throughout intervals of cyclic climatic change as discussed earlier but also from the positive correlation of periods of increased speciation with times of sustained climatic change corresponding to climatic steps, or transitions in the global climate system (Berger, 1982). As the Earth's climate shifts from one cyclic mode to the next, there seem to be times of biotic crises, and although much of the research on this topic has focused on extinction (e.g., Stanley, 1984), many examples exist of climatic steps catalyzing speciation events or radiations (e.g., radiolaria, Casey, Wigley, and Perey-Guzeman, 1983; ostracodes, Benson, 1983; Cronin, 1985, and in press). Under these circumstances, the lack of cyclicity and the long-term duration of the climatic change appears to afford enough time for many speciation events to be completed. Sustained climatic change disrupts long-term stasis that has persisted for several million years of oscillatory climatic change.

In summary, climatic events in the geologic record can be associated with an observed phase of extinction or diversification, but it should be emphasized that association alone does not necessarily signify a causal relationship. Further, most previous cases of climates affecting organic evolution, whether they concern individual species or entire faunas, have deliberately concentrated on intervals of known faunal turnover, usually at stage boundaries. The far more frequent situations where climate has no appreciable effect except to cause temporary migration of faunas from one area to another have rarely been given adequate attention. These cases, where species persist for several million years of climatic oscillation, provide convincing evidence that the duration of a large percentage of species is on the order of 3-10 m.y. and not the 200,000-yr estimate of Schopf (1982).

If the simplified hypothesis that stasis predominates during climatic cycles and relatively high numbers of evolutionary first appearances of species characterize one-time climatic transitions is corroborated by paleontologists working with different organisms, then a convincing case can be made that climate plays an important role in evolution in influencing when speciation does and does not occur. At present, it appears that climate effectively modulates or adjusts evolutionary trends, but it should be emphasized that the intrinsic biologic properties (ecology, physiology, reproductive strategy) of individual species will determine what types of climatic changes affect them. For many organisms, periodicity is probably more important than magnitude. In any case, the dynamics of evolutionary trends and climatic events can only become clearer by using the fossil record to test various hypotheses about speciation and extinction. Only when these dynamics are understood can we reasonably say when the extrinsic environment is driving evolution and when intrinsic biologic factors predominate.

CONCLUSION

Paleontologic research that uses fossils as paleoclimatic proxy data faces the same problems as studies of fossils within the context of evolutionary theory—when and how speciation occurs and does not occur forms the central problem; biogeography, taxonomy, and morphology provide the tools to solve it. This discussion allows the following testable hypotheses:

1. Most cases of geographic isolation caused by climatic change do not result in the origination of a new species.
2. Many groups of fossils are useful for documenting cyclic climatic change because the very nature of the environmental cycles inhibits speciation. Consequently, biogeographic shifts in species should be true monitors of the environmental change, and an assumption that a species' ecology does not change appreciably will usually be valid.
3. The increasing recognition of climatically induced cycles throughout the geologic record supports estimates of 3-10 m.y. for average species longevity because periods of cyclic change actually foster within-species stasis by preventing genomic reorganization.
4. Major diversifications in some fossil groups represent times of increased rates of species origination and should correspond to gradual one-time transitions in the Earth's climate system. These transitions may not necessarily be times of extinction.
5. The magnitude and rapidity of an environmental climatic change, whether a temperature change, sea-level drop, or other climatically induced event, are not as important for organisms as are the frequency and duration of the event.

ACKNOWLEDGMENTS

I thank Walter Newman and Michael Rampino for inviting me to provide a chapter for this book and Rhodes Fairbridge, whose early papers on sea level incited my interest in studying patterns of environmental change. Thanks go to David Adam, L. E. Edwards, R. Z. Poore, and E. E. Compton-Gooding for helpful reviews of the chapter and insights on the topic discussed.

REFERENCES

Atchley, W. R., and Woodruff, D. S., 1981, *Evolution and Speciation*, Cambridge: Cambridge University Press.

Ayala, F. J., 1982, Gradualism versus punctualism in speciation: Reproductive isolation, morphology and genetics, in C. Barigozzi (ed.), *Mechanisms of Speciation*, New York: Alan R. Liss Inc., pp. 51-66.

Bandy, O. L., 1972, Origin and development of *Globorotalia (Turborotalia) pachyderma* (Ehrenberg), *Micropaleontology* **18:**294-318.

Barigozzi, C., ed., 1982, *Mechanisms of Speciation*, Progress in Clinical and Biological Research, vol. 96, New York: Alan R. Liss Inc.

Barron, J. A., 1976, Revised Miocene and Pliocene diatom biostratigraphy of upper Newport Bay, Newport Beach, California, *Marine Micropaleontology* **1:**27-63.

Benson, R. H., 1973, Ostracodes as indicators of threshold depths in the Mediterranean during the Pliocene, in D. J. Stanley (ed.), *The Mediterranean Sea*, Stroudsburg, Pa.: Dowden, Hutchinson and Ross, pp. 63-73.

Benson, R. H., 1983, Biomechanical stability and sudden change in the evolution of the deep-sea ostracode *Poseidonamicus*, *Paleobiology* **9**(4):398-413.

Berger, W. H., 1982, Climate steps in ocean history—Lessons from the Pleistocene, in W. H. Berger and J. C. Crowell (eds.), *Climate in Earth History*, Washington, D.C.: National Academy of Sciences, Studies in Geophysics, pp. 43-54.

Bloom, A. L.; Broecker, W. S.; Chappell, J. M. A.; Matthews, R. K.; and Mesolella, K. J., 1974, Quaternary sea-level fluctuations on a tectonic coast: New ^{230}Th/^{234}U dates from the Huon Peninsula, New Guinea, *Quaternary Research* **4:**185-205.

Bush, G. L., 1982, What do we really know about speciation?, in R. Milkman (ed.), *Perspectives on*

Evolution, Sunderland, Mass.: Sinauer Associates Inc., pp. 119-128.

Carson, H. L., 1982, Speciation as a major reorganization of polygenic balances, in C. Barigozzi (ed.), *Mechanisms of Speciation*, New York: Alan R. Liss Inc., pp. 411-433.

Casey, R. E.; Wigley, C. R.; and Perey-Guzman, A. M., 1983, Biogeographic and ecologic perspective on polycystine radiolarian evolution, *Paleobiology* **9**(4):313-376.

Charlesworth, B.; Lande, R.; and Slatkin, M., 1982, A neo-Darwinian commentary on macroevolution, *Evolution* **36:**474-498.

Cracraft, J., 1982, Geographic differentiation, cladistics and vicariance biogeography: Reconstructing the tempo and mode of evolution, *American Zoologist* **22:**411-424.

Cracraft, J., 1984, The terminology of allopatric speciation, *Systematic Zoology* **33:**115-116.

Cronin, T. M., 1977, Late-Wisconsin marine environments of the Champlain Valley (New York, Quebec), *Quat. Research* **7:**238-253.

Cronin, T. M., 1979, Late Pleistocene benthic foraminifers from the St. Lawrence Lowlands, *Jour. Paleontology* **53:**781-814.

Cronin, T. M., 1981, Paleoclimatic implications of late Pleistocene marine ostracodes from the St. Lawrence Lowlands, *Micropaleontology* **27:**384-418.

Cronin, T. M., 1983, Rapid sea level and climatic change: Evidence from continental and island margins, *Quaternary Science Reviews* **1**(3):117-214.

Cronin, T. M., 1985, Speciation and stasis in marine Ostracoda: Climate modulation of evolution, *Science* **227:**60-63.

Cronin, T. M., in press, Evolution and paleobiogeography of Neogene and Quaternary marine Ostracoda, U.S. Atlantic Coastal Plain, *U.S. Geol. Survey Prof. Paper 1367.*

Cronin, T. M.; Szabo, B. J.; Ager, T. A.; Hazel, J. E.; and Owens, J. P., 1981, Quaternary climates and sea levels of the U.S. Atlantic Coastal Plain, *Science* **211:**233-240.

Dobzhansky, Th., 1970, *Genetics of the Evolutionary Process*, New York: Columbia University Press.

Eldredge, N., and Cracraft, J., 1980, *Phylogenetic Patterns and the Evolutionary Process*, New York: Columbia University Press.

Eldredge, N., and Gould, S. J., 1972, Punctuated equilibria: An alternative to phyletic gradualism, in T. J. M. Schopf (ed.), *Models in Paleobiology*, San Francisco: Freeman, Cooper and Co., pp. 82-115.

Feyling-Hanssen, R. W., 1972, The foraminifer *Elphidium excavatum* (Terquem) and its variant forms, *Micropaleontology* **18**(3):237-254.

Fischer, A. G., 1981, Climatic oscillations in the biosphere, in M. H. Nitecki (ed.), *Biotic Crisis in Ecological and Evolutionary Time*, New York: Academic Press, pp. 103-132.

Garrison, R. E., and Douglas, R. G., eds., 1981, *The Monterey Formation and Related Siliceous Rocks of California*, Tulsa, Okla.: Society of Economic Paleontologists and Mineralogists, Pacific Section, Special Publication, 327p.

Gould, S. J., 1982, The meaning of punctuated equilibrium and its role in validating a hierarchical approach to evolution, in R. Milkman (ed.), *Perspectives on Evolution*, Sunderland, Mass.: Sinauer Associates, pp. 83-104.

Govean, F. M., and Garrison, R. E., 1981, Significance of laminated and massive diatomites in the upper part of the Monterey Formation, California, in R. E. Garrison and R. G. Douglas (eds.), *The Monterey Formation and Related Siliceous Rocks of California*, Tulsa, Okla.: Society of Economic Paleontologists and Mineralogists, Pacific Section, Special Publication, pp. 181-198.

Hays, J. D.; Imbrie, J.; and Shackleton, N. J., 1976, Variations in the Earth's orbit: Pacemaker of the ice ages, *Science* **194:**1121-1132.

Hazel, J. E., 1968, Pleistocene ostracode zoogeography in Atlantic submarine canyons, *Jour. Paleontology* **42:**1264-1271.

Hazel, J. E., 1970, Ostracode zoogeography in the southern Nova Scotian and northern Virginian faunal provinces, *U.S. Geol. Survey Prof. Paper 529-E*, pp. E1-E20.

Imbrie, J., and Kipp, N. G., 1971, A new micropaleontologic method for quantitative paleoclimatology: Application to a Late Pleistocene Caribbean core, in K. K. Turekian (ed.), *Late Cenozoic Glacial Ages*, New Haven, Conn.: Yale University Press, pp. 71-181.

Kellogg, D. E., 1983, Phenology of morphologic change in radiolarian lineages from deep-sea cores: Implications for macroevolution, *Paleobiology* **9**(4):355-362.

Lazarus, D., 1983, Speciation in pelagic Protista and its study in the planktonic microfossil record: A review, *Paleobiology* **9**(4):327-340.

Lohman, G. P., and Malmgren, B. A., 1983, Equatorward migration of *Globorotalia truncatulinoides* ecophenotypes through the late Pleistocene: Gradual evolution or ocean change, *Paleobiology* **9**(4):414-421.

Mayr, E., 1942, *Systematics and the Origin of Species*, New York: Columbia University Press.

Mayr, E., 1963, *Animal Species and Evolution*, Cambridge: Harvard University Press.

Mayr, E., 1982*a*, Processes of animal speciation, in C. Barigozzi (ed.), *Mechanisms of Speciation*, Progress in Clinical and Biological Research, vol. 96, New York: Alan R. Liss, pp. 1-20.

Mayr, E., 1982*b*, Speciation and macroevolution, *Evolution* **36:**1119-1132.

Milkman, R., 1982, *Perspectives on Evolution*, Sunderland, Mass.: Sinauer Associates Inc.

Pisciotto, K. A., and Garrison, R. E., 1981, Lithofacies and depositional environments of the Monterey Formation, California, in R. E. Garrison and R. G. Douglas (eds.), *The Monterey Formation and Related Siliceous Rocks of California,* Tulsa, Okla.: Society of Economic Paleontologists and Mineralogists, Pacific Section, Special Publication, pp. 97-122.

Rensch, B., 1929, *Das Prinzip geographischer Rassenkreise und das Problem der Artbildung,* Berlin: Borntrager.

Rensch, B., 1983, The abandonment of Lamarckian explanations: The case of climatic parallelism of animal characteristics, in M. Grene (ed.), *Dimensions of Darwinism,* Cambridge: Cambridge University Press, pp. 31-42.

Schopf, T. J. M., 1982, A critical assessment of punctuated equilibrium: I. Duration of taxa, *Evolution* **36**(6):1144-1157.

Schwarzacher, W., and Fischer, A. G., 1980, Limestone-shale bedding and perturbations of the Earth's orbit, in G. Einsele and A. Seilacher (eds.), *Cyclic and Event Stratification,* Berlin: Springer, pp. 72-94.

Stanley, S. M., 1979, *Macroevolution: Pattern and Process,* San Francisco: W. H. Freeman.

Stanley, S. M., 1982, Macroevolution and the fossil record, *Evolution* **36:**460-473.

Stanley, S. M., 1984, Temperature and biotic crises in the marine realm, *Geology* **12:**205-208.

Van Valen, L. M., 1982, Integration of species: Stasis and biogeography, *Evolutionary Theory* **6:**99-112.

Vrba, E. S., 1980, Evolution, species, and fossils: How does life evolve?, *South African Jour. Sci.* **76:**61-84.

Vrba, E. S., 1984, What is species selection?, *Systematic Zoology* **33:**318-328.

Wagner, F. J. E., 1970, Faunas of the Pleistocene Champlain Sea, *Geol. Survey Canada Bull. 181,* 104p.

Weaver, F. M.; Casey, R. E.; and Perey, A. M., 1981, Stratigraphic and paleoceanographic significance of early Pliocene to middle Miocene radiolarian assemblages from northern to Baja, California, in R. E. Garrison and R. G. Douglas (eds.), *The Monterey Formation and Related Siliceous Rocks of California,* Tulsa, Okla.: Society of Economic Paleontologists and Mineralogists, Pacific Section, Special Publication, pp. 71-86.

White, M. J. D., 1978, *Modes of Speciation,* San Francisco: W. H. Freeman.

Wolfe, J. A., 1978, A paleobotanical interpretation of Tertiary climates in the northern hemisphere, *Am. Scientist* **66:**694-703.

Wolfe, J. A., and Poore, R. Z., 1982, Tertiary marine and nonmarine climatic trends, in W. H. Berger and J. C. Crowell (eds.), *Climate in Earth History,* Washington, D.C.: National Academy of Sciences, Studies in Geophysics, pp. 154-158.

20: Terrestrial Climate Change from the Triassic to Recent

William L. Donn
Columbia University

Abstract: During the Cenozoic Era, the meridional temperature gradient in both the Northern and Southern Hemispheres steepened considerably until the present frigid, ice-covered polar conditions developed. Equatorial Pacific temperatures from Cretaceous to Eocene may have been about 10°C lower than at present. From at least the Triassic through Eocene times, polar-latitude climate varied regionally and temporally between warm temperate to subtropical conditions, as demonstrated by a host of fossil fauna and flora including representative forms such as Spitzbergen dinosaurs, Ellesmere Island flying lemurs, Greenland breadfruit trees, Antarctic amphibians and coal measures, among others. Evidence of high oxygen-18 temperatures from the Paleogene Arctic Ocean, although subject to interpretation, certainly does not conflict with these other proxy climate indicators. Various explanations, including the hypothesis of polar wandering, are considered critically.

THE PROBLEM

Terrestrial climate during the late Cenozoic appears to be anomalous when viewed against the geological past. Today both polar regions are frigid and ice covered while the tropical zones are very warm. The meridional temperature gradients in both hemispheres are very strong: about 40°C in the Northern Hemisphere and 60°C or more in the Southern Hemisphere. During much of the past, hemispheric meridional temperature gradients were in the range of 10°C to 20°C. At no time in the past do proxy climate data indicate steep tropical to polar temperature ranges simultaneously in both hemispheres. On the contrary, proxy and oxygen-isotope temperature determinations indicate that frigid ice-age conditions were limited to one hemisphere at a time and that polar regions of both hemispheres were relatively warm during much of Phanerozoic time.

It is noteworthy that the global climate changes referred to here appear to have been controlled by conditions that affected the present high latitudes. Some significant change is indicated for present tropical latitudes. For example, present mean equatorial temperatures are about 26°C. But ^{18}O temperatures for the tropical Pacific Ocean from Cretaceous to Eocene appear to have been about 10°C cooler than at present (Boersma and Shackleton, 1981; Shackleton, 1984). Despite the tropical cooling, the high latitudes of the past show a relatively rich fauna and flora with affinities that vary from warm temperate to subtropical or tropical climates. However, the major terrestrial climate changes were those in which polar regions suffered large changes in temperature, causing these regions to swing between the warm conditions of the Mesozoic and the ice-age conditions of the present and late Cenozoic.

As indicated by Donn and Shaw (1977), the ice caps on Greenland, Antarctica, and related frigid regions are not a relic of former glacial stages. They are rather a natural consequence of the position of these lands in polar latitudes. The ice cover of the Arctic Ocean is related to both its central polar location and its relative thermal isolation from the circulation of the rest of the world ocean (Ewing and Donn, 1956). Donn and Shaw also showed that the change

Lamont-Doherty Geological Observatory of Columbia University LDGO Contribution No. 3777 to late Cenozoic ice-age conditions is explainable by the Earth-atmosphere thermal regime that resulted from continental drift, the principal effects being the increase of high-latitude albedo of land compared to water and the loss of high-latitude ocean heat storage. The consequent radiation balance and related total heat balance led to continental frigidity in high latitudes. Hence, wherever an adequate supply of moisture is present, such as from the oceans around Greenland and Antarctica, ice caps will result.

Paleogeographic reconstructions, such as those of Phillips and Forsyth (1972), Smith and Briden (1977), and Barron et al. (1981), indicate that much of the present arctic continental regions were in high enough latitudes by late Mesozoic and Eocene times to have developed frigid conditions. The record described in the following, however, gives no support to this argument. High-latitude climate supported a wealth of warm temperate fauna and flora through the Eocene, at least. The steepening meridional temperature gradient computed by Donn and Shaw (1977) also did not exist.

SUMMARY OF THE GEOLOGIC RECORD

The following descriptions are a partial summary of fauna and flora of the Arctic and Antarctic record from the Triassic through the Eocene that bear on the climate of this long period of time, more than 150 m.y.

Paleoflora

At least since the Arctic fossil plant study of Heer (1883), it has been known that a high-latitude paleoclimate anomaly exists. He described about 100 species of fossil plants that included cycads and ferns in west Greenland. Also, a rich flora of more than 200 Mesozoic plant species are given by T. M. Harris (in Schwarzbach, 1963). The tropical breadfruit form *Artocarpus* from east Greenland (Cretaceous to Paleocene) is described by Nathorst (1911). Early Tertiary conifers from Spitsbergen are described by Schweitzer (1974) as resembling those of the warm temperate regions of southern Asia to which many of the genera are now restricted as they require a frost-free climate. Explorations in Ellesmere Island beginning in the mid-1970s have revealed a rich flora from Cretaceous to Eocene, including trees reaching 1.5 m in diameter that show pronounced seasonal growth rings (Dawson et al., 1976; West, Dawson, and Hutchinson, 1977; Christie and Rouse, 1976; McKenna, 1980).

Anomalous paleoflora are also reported from the high latitudes of the Southern Hemisphere. King (1961) notes that a warm to tropical environment is indicated by all the Jurassic flora from Gondwanaland. More recently, Jefferson (1982) describes Cretaceous paleoforests on Alexander Island, Antarctica, and concludes that the warmth of the environment of that period is inconsistent with the paleolatitude of 65-75°S currently accepted for the region.

Paleofauna

The present latitudinal ranges of Mesozoic amphibians and reptiles are shown in Figures 20-1 through 20-3 as modified from Colbert (1964), with additional data from Barrett et al. (1968) and Elliot et al. (1970). The high-latitude distribution, with some forms approaching the poles of both hemispheres, is clearly at odds with present conditions. The more limited range in the Southern Hemisphere is probably due to the present lower-latitude continental geography (except Antarctica) as well as more restricted geologic rock ranges. When corrected for changes in continental positions (based on paleomagnetic determinations, A-K on right side of Fig. 20-16), the latitudinal ranges of Triassic forms increase.

The highest latitude ranges in the Northern Hemisphere include Triassic stegosaurs and Cretaceous ornithopod dinosaurs (Colbert, 1964). Colbert (p. 632) notes there "can be little doubt as to the limitation of these huge reptiles to tropical and semitropical climes."

Although the Ellesmere Island expeditions uncovered large numbers and varieties of inver-

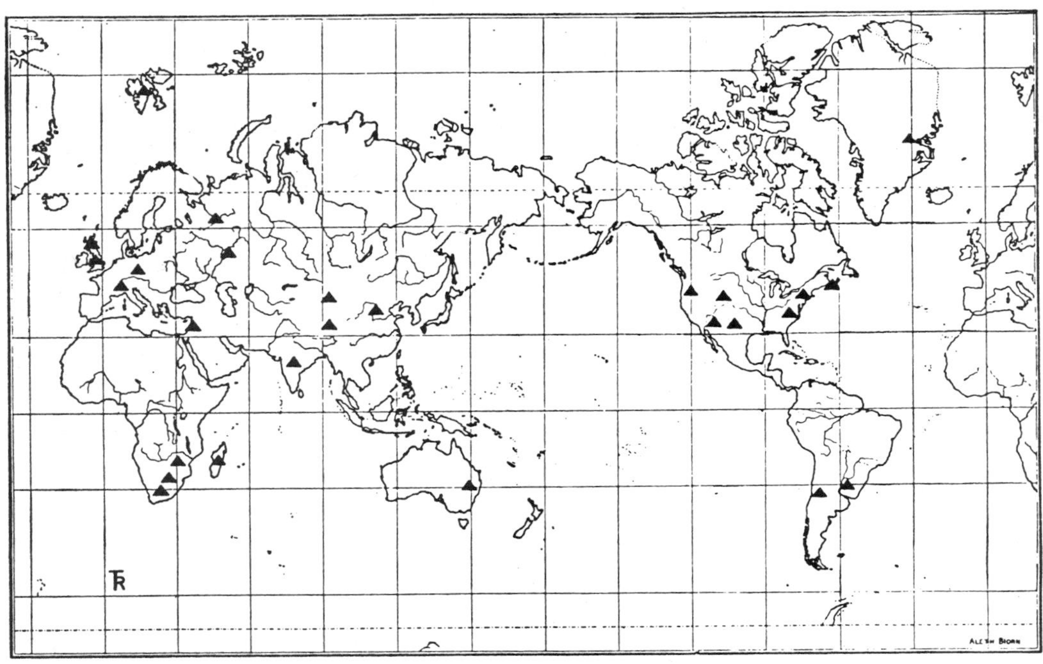

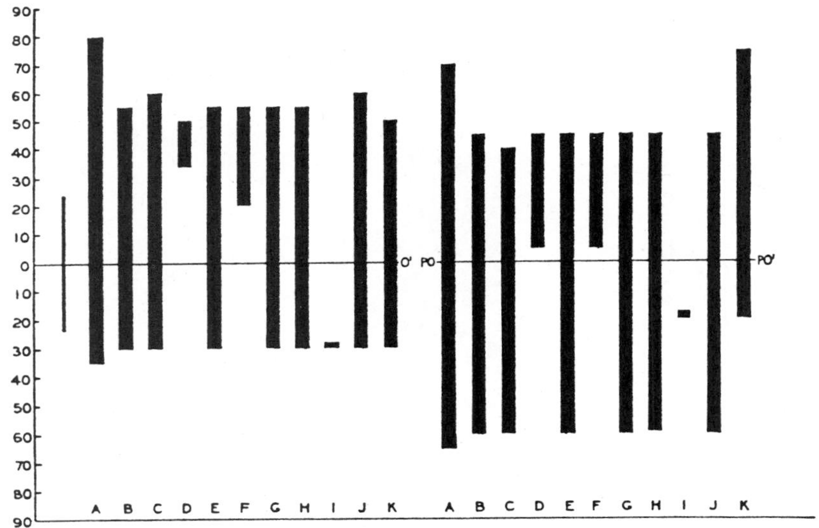

Figure 20-1. Distribution of known Triassic tetrapod faunas. (After Colbert, 1964)

Vertical bars indicate latitudinal spread of known occurrences of certain Triassic terrestrial tetrapods (A-K, left). Latitudinal spread of Triassic terrestrial tetrapods corrected for changes in continental positions based on paleomagnetic data (A-K, right). This diagram indicates a wide Triassic tropical belt, as inferred from the latitudinal spread of the ectotherms. A, Labrynthodont amphibians; B, rynchosaurs; C, procolophonids; D, protosaurs; E, pseudosuchians; F, phytosaurs; G, stagonolepids; H, saurischian dinosaurs; I, ornothischian dinosaurs; J, dicynodonts; and K, ictidosaurs. The narrow bar on the left indicates the limits of the modern tropics.

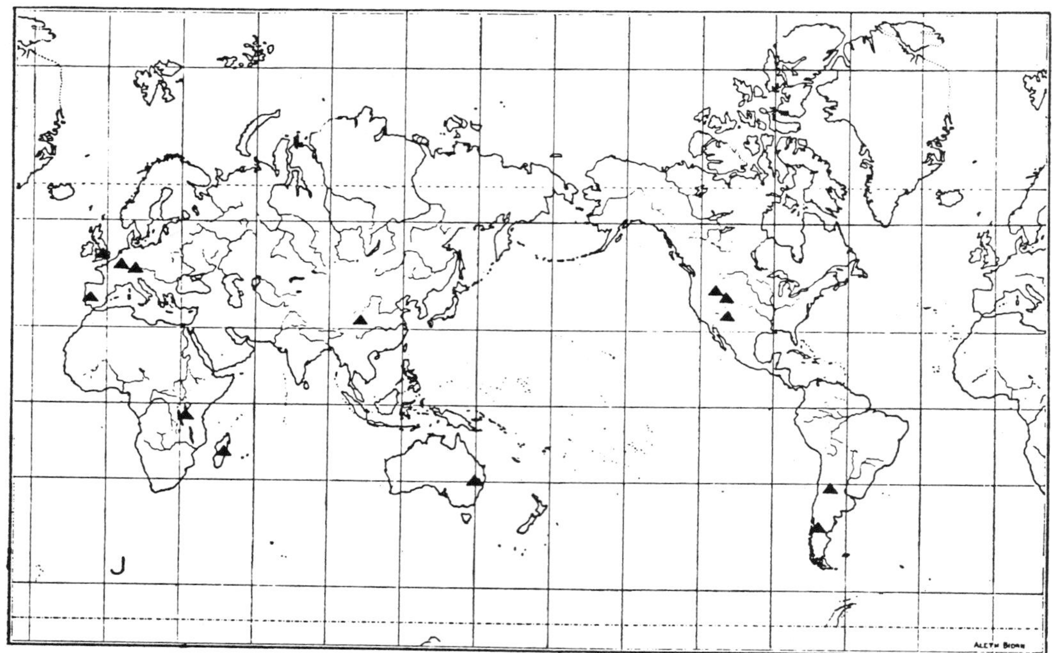

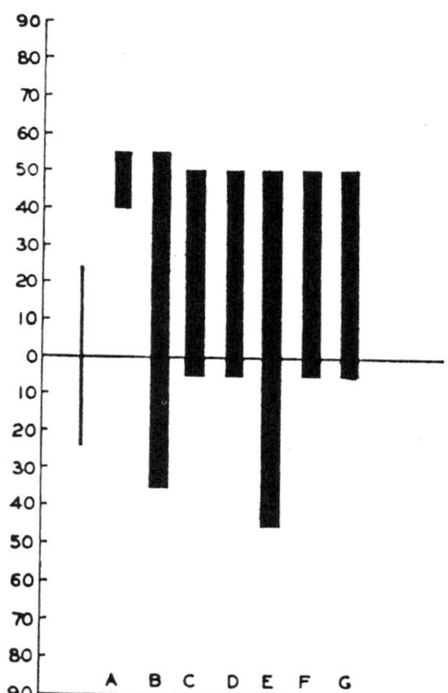

Figure 20-2. Distribution of known Jurassic tetrapod faunas. Vertical bars indicate latitudinal spread of known Jurassic tetrapod faunas. A, turtles; B, crocodilians; C, pterosaurs; D, theropod dinosaurs; E, sauropod dinosaurs; F, ornithopod dinosaurs; and G, stegosaurian dinosaurs.

tebrates and vertebrates, it is probably the latter that give the more obvious climate implications. These include one order of amphibians, three orders of reptiles including crocodiles, and ten orders of mammals that include abundant Dermoptera, or flying lemurs. The latter are currently restricted to southeast Asia, and crocodiles are not found north of Florida and north Africa (Schwarzbach, 1963). From even this limited summary, it is hard to escape the conclusion that high-latitude climate of both hemispheres during the Mesozoic and early Tertiary was quite different from that of the present frigid conditions.

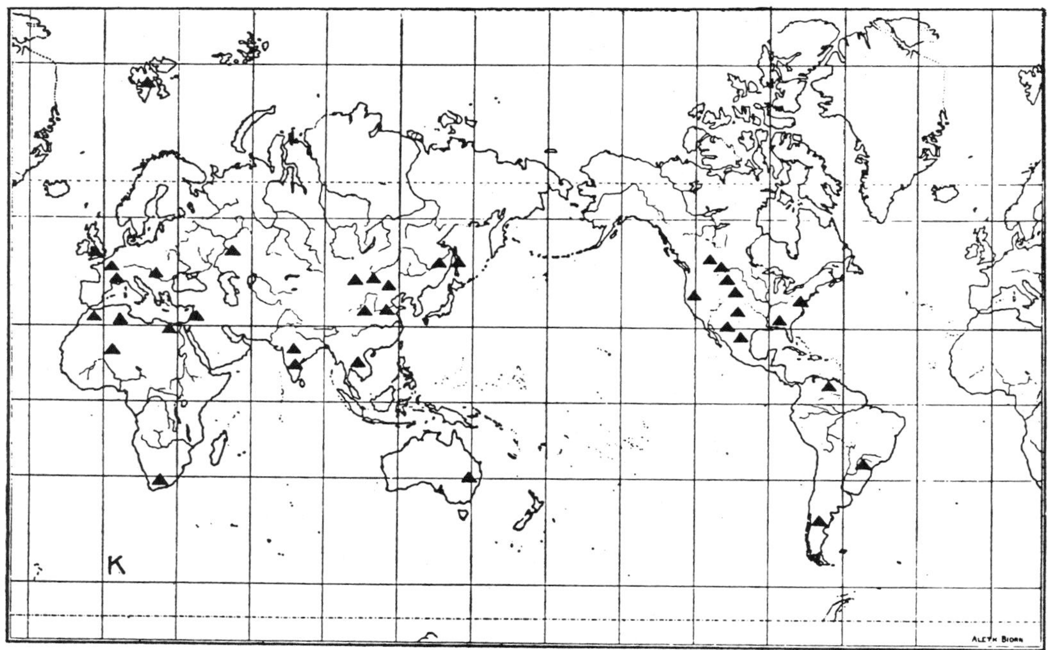

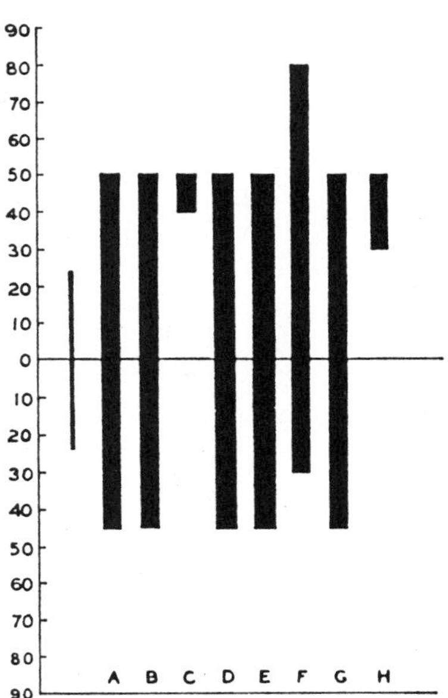

Figure 20-3. Distribution of known Cretaceous tetrapod faunas. Vertical bars indicate latitudinal spread of known Cretaceous tetrapod faunas. A, turtles; B, crocodilians; C, pterosaurs; D, theropod dinosaurs; E, sauropod dinosaurs; F, ornithopod dinosaurs; G, ankylosaurian dinosaurs; and H, ceratopsian dinosaurs.

Oxygen-Isotope Temperatures

In describing ^{18}O temperatures from the Arctic, Bowen (1966) summarized data from the Cretaceous (Santonian); he noted that, compared with the present, the latitude of the 18°C isotherm was shifted from about 32°N to about 60°N, the 20°C isotherm from 30°N to 55°N, and the 24°C isotherm from about 20°N to 48°N. According to Bowen, the average temperature within 20° latitude of the North Pole was 16-17°C at that time. At other times during the Cretaceous as well as the Jurassic, reported polar surface-water temperatures were close to these values and far above the present subfrigid conditions. Furthermore, Lowenstam and Epstein (1959) showed that Atlantic deep-sea temperatures during parts of the Cretaceous must have been 16-17°C compared with about 1.5°C for the present.

In a study of marine paleotemperatures based on deep-sea core studies, Shackleton and Boersma (1981) have determined an Eocene meridional temperature gradient less than half the present value, with the main change being in the higher-latitude temperatures. The review by Savin (1977) shows that temperatures indicated by Cretaceous north Pacific benthic organisms were only a few degrees below temperatures from planktonic organisms. Then, beginning in the Eocene, benthic temperatures dropped almost uniformly to present low values—about 15°C below present planktonic temperatures. This finding also indicates that Cretaceous mid-latitude surface-water temperature was not too different from polar-derived bottom-water temperature and, hence, polar surface-water temperature. The striking Pacific bottom-water temperature decrease during the Cenozoic thus indicates a significant cooling of polar surface waters.

In addition to the ^{18}O temperatures from the more open ocean, a number of newer computations have been made from invertebrates from the Arctic Ocean. An Eocene value of 15°C has been obtained from a scaphopod found in the Eureka Sound Formation of Ellesmere Island (Donn, 1982). Unpublished computations by Donn from several Paleogene *Pecten* sp. forms, collected by Mirancovich of the U.S. Geological Survey from the north slope of Alaska, gave values that averaged 22°C.

Temperature calculations for the Arctic forms are based on the use of ocean salinities like that of present shelf waters. If the Arctic Ocean were a nearly land-locked sea, as proposed by the recent paleogeographic reconstructions of Churkin and Trexler (1981) and Fujita and Newberry (1982), then the problem is introduced as to whether the Arctic Ocean of that time was similar to a large fresh-water lake or more like a Mediterranean Sea. Assuming the former, the ^{18}O temperatures are too high; assuming the latter, they are, if anything, too low. Despite the uncertainty in interpretation of ^{18}O Arctic temperatures, they do at least support the proxy climate data, which indicate unequivocally that the polar worlds of both hemispheres were warm temperate to subtropical for at least the long time interval from the Triassic through the Eocene.

DISCUSSION

The high-latitude fauna, and particularly flora, indicate that not only temperature but also moisture conditions of the Mesozoic and early Tertiary must have been quite different from the present. Much of the present Arctic of North America, as well as Asia, has desert-level precipitation, unlikely to support the vegetation described here or the related herbivores. "The most important indicators of heavy rainfall are coals and other carbonaceous sediments" (Schwarzbach, 1963, p. 160). Mesozoic and Eocene coal beds found in lands that border the present Arctic Ocean, as well as in Antarctica, indicate the presence of both warm and moist paleoconditions in these areas for the time period being discussed.

Both the global uniformity of climate for most of the time involved as well as the direct and proxy indicators of warm, moist climates extending into present barren, frigid, or ice-capped high latitudes pose a perplexing problem to the paleoclimatologist. One proposed explanation is based on a decrease to about zero in the Earth's obliquity, or axial tilt (Allard, 1948; Wolfe, 1977, 1978). This effect would leave the entire globe illuminated, as on present equinoxes. It is proposed that the continuous high-latitude sunlight would provide the warmth indicated by the geologic record. The known periodic variations in obliquity having a maximum range of 2.5° are from well-known gravitational perturbations (e.g., Milankovitch, 1930). There are no known astronomical causes of periodic or irregular long-period changes in the tilt of the Earth's axis, particularly of the large amount required for a change to near zero obliquity. Even if such a change occurred, the tangential rays of the Sun would probably lead to permanent winter conditions in high latitudes rather than warm temperate to semitropical climate. Furthermore, a zero to near zero obliquity would lead to nonvarying annual temperatures with no seasonality; the evidence of prominent tree rings, however, seems to refute the low-obliquity model with its aseasonal characteristics. The exhaustive paleobotanical study, including tree rings, of Creber and Chaloner (1984) seems to give further evidence of significant obliquity. It also seems unlikely that greatly increased obliquity as hypothesized by Williams (1975) to explain Precambrian glacial climates can account for the warmer high latitudes but cooler low latitudes (Croll, 1875; Milankovitch, 1930). The latter is not supported by observation for the Mesozoic-Eocene interval.

Another proposed explanation is a variation in the solar constant resulting from long-period variability in solar luminosity. This problem has been reviewed by Newkirk (1980, p. 293), who considers some possible causes of long-period variability but concludes, "The information on hand is insufficient to clarify the ambiguities that remain to be resolved concerning the outputs of the ancient sun." From a geologic point of view, if solar luminosity increased sufficiently to warm high latitudes to warm-temperate or semitropical levels, we might expect that lower latitudes should have been hotter in the past. There is no indication that such an overheating occurred, and in fact, the ^{18}O data referred to earlier suggest that cooler conditions existed.

A third, possibly obsolete, explanation involves the transport of large volumes of warmer ocean waters from the Pacific into the Arctic oceans during the Mesozoic because of the greater separation between Alaska and Siberia at the time (e.g., Shackleton and Boersma, 1981). However, the dynamics of ocean gyres, which are wind-generated, are controlled primarily by the Coriolis force of the Earth's rotation. The shape and position of the Pacific gyre would not be changed significantly by the size of the passageway into the Arctic Ocean, although an increased interchange would certainly have occurred. Furthermore, with the weaker wind systems that would have existed with the decreased meridional temperature gradients of the time, currents would have probably been weaker, thereby decreasing the ocean heat flux into the Arctic. It is certainly difficult to imagine the increased warmth of the northern Arctic lands from increased heat transport through a larger Alaska-Siberia passageway, and in view of the narrower to nonexistent Atlantic Ocean, northward ocean heat flux would surely have been much less than at present. How, then, can this mechanism be responsible for the greater warmth of Greenland, Spitzbergen, and northwestern Europe? If, in contrast, the Arctic was a nearly closed sea for much of the time, following, for example, Churkin and Trexler (1981), the lack of thermal exchange with the surrounding oceans should have caused cooling equal to or greater than that of the present.

Sandberg (1983) and Berner, Lasaga, and Garrels (1983) propose that times of enhanced global tectonics would also have an increased greenhouse effect from increased CO_2 discharge into the atmosphere. These would be time-limited occurrences that should affect more than polar latitudes. It is also difficult to imagine the rich flora and tetrapod fauna migrating to the lands and islands of perpetual winter darkness following any CO_2 warming effects that can be theorized.

In 1982, Donn suggested that the warm polar paleotemperatures and decreased meridional temperature gradient in Figure 20-4 for the Jurassic could be explained by pole positions very different from those of the present, as indicated by the crosses. This suggestion implies that a strong nondipole magnetic component introduced serious errors in determining paleopole locations on the basis of the axial-dipole theory. The possibility of gross polar wandering from convectional effects and mass redistributions was suggested in the theoretical studies of Gold (1955), as well as of Goldreich and Toomre (1969), who note that a 90° displacement would be related to one convectional cycle (100-150 m.y.). Merrill and McElhinny (1983, p. 173) note, "Testing the axial nature of the field requires the use of paleoclimate information." It may not therefore be proper to use an axial field to interpret paleoclimate data

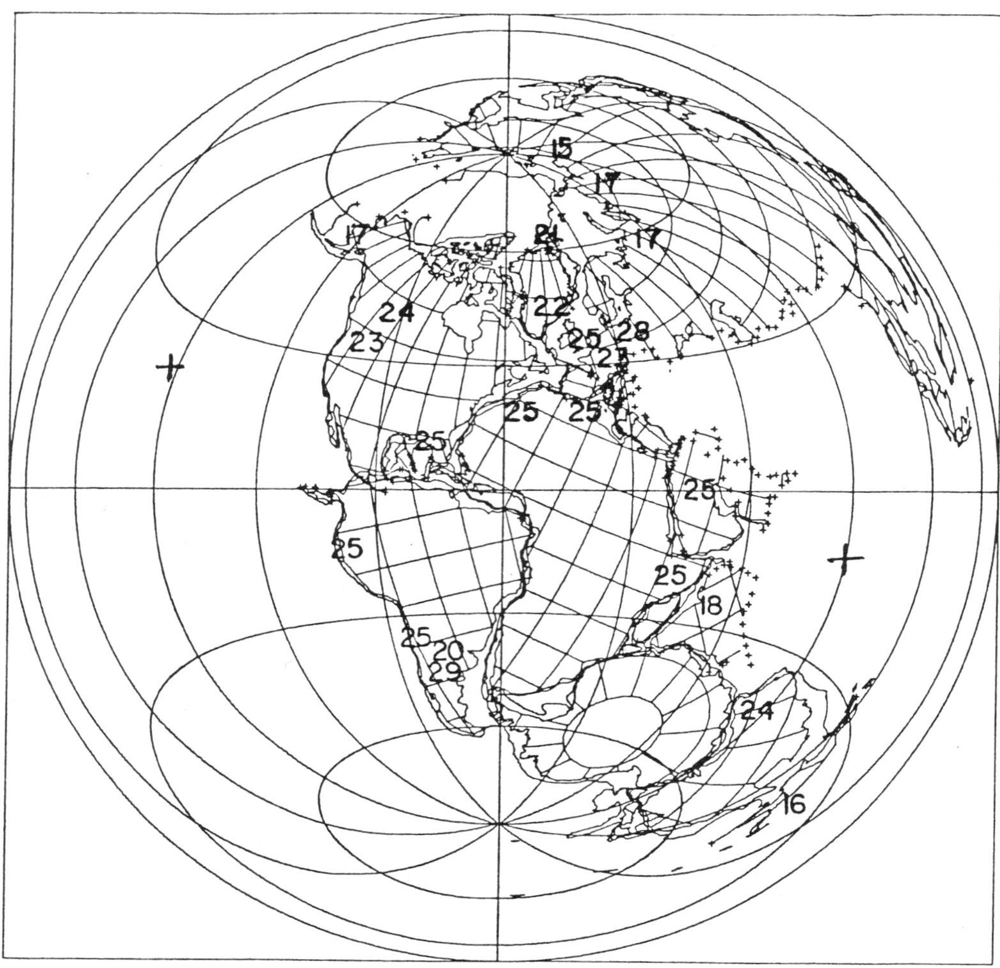

Figure 20-4. Paleogeographic map for early Jurassic showing temperature distribution. The 25° values are taken from carbonate deposits, assuming this to be the lower limit of such deposition. Others are ^{18}O temperatures. Crosses show possible pole positions. (After Smith and Briden, 1977)

until the geography is established independently from the paleomagnetism. The pole position proposed here would have placed most of the continental lands in low latitudes and could thus explain the uniform temperature distribution of Figure 20-4 and the anomalous paleowarmth of the polar regions. Although this model seems capable of explaining the high-latitude paleoclimate enigma, many consequences seem to follow that can be verified.

The proxy fossil data support the view that, from Triassic to Eocene times, polar latitudes experienced warm and often humid climates that were equivalent to present warm temperate to subtropical conditions. Conditions were globally much more equable than at present. Ice sheets were apparently absent. Following the Eocene, the meridional temperature gradient began to steepen because of increased cooling in Arctic and Antarctic latitudes until late Cenozoic glacial conditions developed. Hypotheses such as changing tilt of the Earth's axis, variations in the solar constant, different ocean circulation patterns, and changes in atmospheric CO_2 do not seem tenable on either empirical or theoretical grounds. Polar wandering could provide a terrestrial mechanism for accomplishing the climate conditions and changes described, particularly when accompanied by the effects of continental drift.

REFERENCES

Allard, H., 1948, Length of day in the climates of past geologic eras and its possible effects upon changes in plant life, in A. E. Murneek and R. O. Whyte (eds.), *Vernalization and Photoperiodism,* Chronica Bot., pp. 101-119.

Barrett, P.; Baillie, R.; and Colbert, E., 1968, Triassic amphibians from Antarctica, *Science* **161:**460-462.

Barron, E.; Harrison, E. G.; Sloan, W.; and Hay, W., 1981, Paleogeography, 180 million years ago to present, *Eclogae Geol. Helvetiae* **74:**443-470.

Berner, R.; Lasaga, A.; and Garrels, R., 1983, The carbonate-silicate geochemical cycle and its effect on atmosphere carbon dioxide over the past 100 million years, *Am. Jour. Sci.* **283:**641-683.

Boersma, A., and Shackleton, N., 1981, Oxygen- and carbon-isotope variations and planktonic-foraminifera depth habitats, late Cretaceous to Paleocene, central Pacific Deep Sea Drilling Project Sites 463 and 465, in *Initial Reports of the Deep Sea Drilling Project, 62,* Washington, D.C.: Government Printing Office, pp. 513-517.

Bowen, R., 1966, *Paleotemperature Analysis,* Amsterdam: Elsevier, 265p.

Christie, R., and Rouse, G., 1976, Eocene beds at Lake Hazen, northern Ellesmere Island, *Geol. Surv. Canada Paper 76-1C,* pp. 153-155.

Churkin, M., Jr., and Trexler, J., 1981, Continental plates and accreted oceanic terranes in the arctic, in A. Nairn, M. Churkin, Jr., and F. Stehli (eds.), *The Ocean Basins and Margins,* vol. 5, *The Arctic Ocean,* New York: Plenum Press, pp. 1-20.

Colbert, E., 1964, Climate zonation and terrestrial faunas, in A. Nairn (ed.), *Problems in Paleoclimatology,* New York: John Wiley & Sons, pp. 617-637.

Creber, G., and Chaloner, W., 1984, Climatic indications from growth rings in fossil wood, in P. Benchley (ed.), *Fossils and Climate,* New York: John Wiley & Sons, pp. 49-74.

Croll, J., 1875, *Climate and Time in Their Geologic Relations. A Theory of Secular Changes of the Earth's Climate,* London: Stanford, 557p.

Dawson, M.; West, R.; Langston, W., Jr.; and Hutchinson, J., 1976, Paleogene terrestrial vertebrates: Northernmost occurrence, Ellesmere Island, Canada, *Science* **192:**781-782.

Donn, W., 1982, The enigma of high-latitude paleoclimate, *Palaeogeography, Palaeoclimatology, Palaeoecology* **40:**199-212.

Donn, W., and Shaw, D., 1977, Model of climate evolution based on continental drift and polar wandering, *Geol. Soc. America Bull.* **88:**390-396.

Elliot, D.; Colbert, E.; Breed, W.; Jensen, J.; and Powell, J., 1970, Triassic tetrapods from Antarctica: Evidence for continental drift, *Science* **169:**1197-1202.

Ewing, M., and Donn, W., 1956, A theory of ice ages, *Science* **123:**1061-1066.

Fujita, K., and Newberry, J., 1982, Tectonic evolution of northeastern Siberia and adjacent regions, *Tectonophysics* **89:**337-357.

Gold, T., 1955, Instability at the earth's axis of rotation, *Nature* **175:**526-529.

Goldreich, P., and Toomre, A., 1969, Some remarks on polar wandering, *Jour. Geophys. Research* **74:**2555-2567.

Heer, O., 1883, *Flora Fossilis Arctica,* vols. 1 through 7.

Jefferson, T., 1982, Fossil forests from the lower Cretaceous of Alexander Island, Antarctica, *Paleontology* **25:**681-708.

King, L., 1961, The paleoclimatology of Gondwanaland during the Paleozoic and Mesozoic Eras, in A. E. M. Nairn (ed.), *Descriptive Paleoclimatology,* New York: Wiley-Interscience Publishers, pp. 307-331.

Lowenstam, H., 1964, Paleotemperatures of the Permian and Cretaceous periods, in A. E. M. Nairn (ed.), *Problems in Paleoclimatology,* New York: Wiley-Interscience Publishers, pp. 227-248.

Lowenstam, H., and Epstein, S., 1959, Cretaceous paleotemperatures as determined by the oxygen isotope method, their relations to, and the nature of rudistid reefs, in *International Geological Congress, 20th, Mexico City, 1956, Symposium on Cretaceous,* pp. 65-76.

McKenna, M., 1980, Eocene paleolatitude, climate and mammals of Ellesmere Island, *Palaeogeography, Palaeoclimatology, Palaeoecology* **30:**349-362.

Merrill, R., and McElhinny, M., 1984, *The Earth's Magnetic Field,* New York: Academic Press, 401 p.

Milankovitch, M., 1930, Mathematical climatology and the astronomical theory of climate change, in Koppen-Geiger (ed.), *Handbuch der Klimatology,* vol 1, Part A, Berlin: Borntraeger.

Nathorst, A., 1911, Fossil floras of the arctic regions as evidence of geologic climates, *Geol. Mag.* **5:**8.

Newkirk, G., Jr., 1980, Solar variability on time scales of 10^5 to 10^9 years, in R. Pepin, J. Eddy, and R. Merrill (eds.), *Proceedings of the Conference on the Ancient Sun,* New York: Pergamon Press, pp. 293-320.

Phillips, J., and Forsyth, D., 1972, Plate tectonics, paleomagnetism and the opening of the Atlantic, *Geol. Soc. America Bull.* **83:**1597-1600.

Sandberg, P., 1983, An oscillating trend in Phanerozoic non-skeletal carbonate mineralogy, *Nature* **305:**19-22.

Savin, S., 1977, The history of the earth's surface

temperature during the past 100 million years, *Ann. Rev. Earth and Planetary Sci.* **5:**319-325.

Schwarzbach, M., 1963, *Climates of the Past,* London: D. van Nostrand and Co., 328p.

Schweitzer, H.-J., 1974, Die "Tertiaeren" Koniferen Spitzbergens, *Paleontographica* **149:**1-89.

Shackleton, N., 1984, Oxygen isotope evidence for Cenozoic climate change, in P. Brenchley (ed.), *Fossils and Climate Change,* Chichester: Wiley, pp. 27-47.

Shackleton, N., and Boersma, A., 1981, The climate of the Eocene ocean, *Geol. Soc. London Jour.* **138:**153-157.

Smith, A., and Briden, J., 1977, *Mesozoic and Cenozoic Paleogeographic Maps,* Cambridge: Cambridge University Press, 63p.

West, R.; Dawson, M.; and Hutchinson, J., 1977, Fossils from the Paleogene Eureka Sound Formation, N.W.T., Canada: Occurrence, climate and paleogeographic implications, in R. M. West (ed.), *Paleontology and Plate Tectonics,* vol. 2, Milwaukee Museum Special Publications in Biology and Geology **2,** pp. 77-93.

Williams, G., 1975, Late Precambrian glacial climate and the earth's obliquity, *Geol. Mag.* **112:**441-465.

Wolfe, J., 1977, Paleogene floras from the Gulf of Alaska region, *U.S. Geol. Survey Prof. Paper 997,* 108p.

Wolfe, J., 1978, A paleobotanical interpretation of Tertiary climates in the Northern Hemisphere, *American Scientist* **66:**694-703.

VI

Solar Variations, Cycles, and Possible Causes

21: Sunspot Cycles and Weather History

D. J. Schove
St. David's College, U.K.

Abstract: Solar activity cycles are examined using proxy evidence such as documentary sightings, auroral descriptions, ^{14}C flux rate, and varve periodicities for periods extending back more than 10 millennia. A figure displays the correlation between observed auroral numbers and radiocarbon values based on Stuiver's production index for the period A.D. 300-1900. Other figures also explain methods of varve correlation and chronology in Scandinavia and North America. Using teleconnections based on pattern recognition in solar periodicities, the Quasi-biennial Oscillation, and 3-yr cycles, possible varve correlations are taken back to about 20,000 yr ago and related to radiocarbon-dated glacial advances.

INTRODUCTION

Professor Fairbridge, like myself, has always been fascinated by the mystery of sunspot cycles and again, like myself, has always felt that there must be links with weather, tree rings, and indeed, human history. When we first met in 1960 such links were controversial. Douglass could not find the 11-yr sunspot cycle in tree rings from the seventeenth century and ascribed this to the lack of sunspots in the Maunder Minimum. Indeed, with primitive analytic methods, it was difficult to find the sunspot cycle in tree rings in any century, and real connections came later with the explanation of the wiggles on the radiocarbon curve. However, in 1960 Rhodes Fairbridge was planning a New York symposium on these topics, and he advised me to divide sunspots and tree rings into two separate papers for his 1961 conference. These two papers appeared in the *Annals of the New York Academy of Science* (Vol. 95, Section 1); this volume contained also the first computer-developed spectra on tree rings (in a paper by Bryson and Dutton, 1961) which failed to show the 11-yr cycle, although they did disclose 2-yr and 3-yr cycles. That publication is now out of print, but Professor Fairbridge suggested that I should reprint my papers in the book *Sunspot Cycles* (Schove, 1983*d*), a volume in a series he edited. This I have done (pp. 242-267), adding critical comments. Another paper from that New York Academy of Science symposium, by Anderson (1961), expressed a disbelief in the 11-yr cycle in Paleozoic and Mesozoic varves. Now, after 22 yr—one Hale solar cycle, incidentally—some of the links between sunspots and weather are becoming clear.

SUNSPOT CYCLES SINCE A.D. 300

In the 1930s, I was invited to work on the meteorology in the Chinese dynastic histories by Dr. Needham and his collaborators who supplied translations. We had been collecting dates of what was thought to be lightning, and I noticed that the years in question (e.g., A.D. 300, 311, 322) occurred at 11-yr intervals (Schove, 1983*d*, pp. 109-122). A weak sunspot link with thunderstorms seemed possible, but such intervals were too significant for the so-called lightning to be anything but the aurora, which are known to correlate with the solar cycle. The Far Eastern Records combined with European records of strange visions enabled us to build up, as part of what I call the Spectrum of Time project, the approximate dates of every 11-yr maximum back to before A.D. 800.

Sunspot cycles can thus be determined using as proxies the unusual celestial phenomena that are recorded in what I term *eyewitness history* (Schove, 1983*c*). Originally I simply assumed that there were nine cycles every century, but now it is possible to prove this for specific centuries and to date the cycles. There are nine cycles even in the seventeenth century despite the Maunder Minimum, which some people have taken more literally than was intended in Eddy's (1976) stimulating work. My dates differ little from those established by Wolf (cf. Schove, 1983*d*, p. 88), but I have reduced my magnitudes (p. 14) in light of Eddy's work.

Galileo's students in the sunny Mediterranean lands, where telescopes had already been used by day, probably continued to observe sunspots during the Maunder Minimum, but fear of the Inquisition partly explains why no observations seem to have been recorded there. In northern Europe, in contrast, the needs of navigation encouraged the use of telescopes at night, and cloudy weather often discouraged daytime observing, but even here I have recently found that sunspots were not so infrequent about 1650 as had been assumed (Schove, 1983*d*, p. 145). Hevelius, in a letter to Huygens (October 1658, translated in Schove, 1983*c*, p. 392, thus mentions the spots he saw in August/September 1654 and December 1657 and notes that "from the beginning of 1654, right up to now, whenever he had inspected the sun" he had seen no others, although "previously from my own observations . . . they were much more splendid." Unfortunately, I cannot find his records for the period 1646-1653, but it is clear from the auroral evidence that there was maximum solar activity about 1650.

Another phenomenon is helpful in finding sunspot maxima in weak cycles. For several months in the middle of any cycle (Schove, 1983*d*, p. 2), large sunspots do not develop, as Gnevyshev (1977) and Wittmann (1978) have demonstrated, and in weak cycles there is often a long "Gnevyshev gap" (p. 146) in both sunspots and auroras. In the case of auroras, the double peak helps us to date (Table 21-1) the precise position of the smoothed maximum in the Zurich sense. Minima may be established from evidence of 3-yr minima in auroral frequency (Table 21-2). Clues like this constrain the suggested dates (p. 14) of turning points of the cycles in the seventeenth century.

In New England, auroral displays were very few in the period 1645-1699 (Schove, 1983*d*, p. 146; Rizzo and Schove, 1962), but displays did occur in southern Canada as the "Jesuit Relation" testifies. Further investigations of the

Table 21-1. Auroral Stages and Sunspot Maxima, 1646–1699

Aurorae					*Sunspot Maxima*	
1st Maxima (a)	*2nd Maxima (b)*	*Mid-point (c)*	*Lull (d)*	*Centre (e)*	*Schove (f)*	*Wolf (g)*
1649	53	51.5	50.7	51.1	50.8 ± 1	49
1661	(63)	62.5	(62.5)	62.5	61 ± 1/2	60
1671	76	74	73.8	73.9	73.5 ± 1	75
1681	(85)	83.5	(84.5)	84	85 ± 1	85
1691/2	(96)	94.3	(93.5)	93.9	94.5 ± 1	93

Source: From Schove, 1983*c*, p. 392.
Notes: Numbers in parentheses are uncertain. Dates in *a* and *d* relate to 12-month periods. The Centre (col. *e*) is defined as $(c + d)/2$. The maxima in *f* are from Schove, 1983*a*, table 2. Aurorae are mainly from the British Isles and NW Europe.

Table 21-2. Auroral and Sunspot Minima, 1645–1699

Auroral Minima (*a*)	*Sunspot Minima* (*b*)	*Wolf's Minima* (*c*)
1646.5	45.5	45
1655.5	55.9	55
1668.1	66.7	66
(1678.5)	79.5	79.5
1689.3	89.5	89.5
1698.3	99	98

Source: From Schove, 1983*c*, p. 392.
(*a*) Auroral 3-yr minimum.
(*b*) Sunspot minimum adopted in Schove, 1983, table 2.
(*c*) Sunspot minimum as given originally by Wolf (1858-1968).

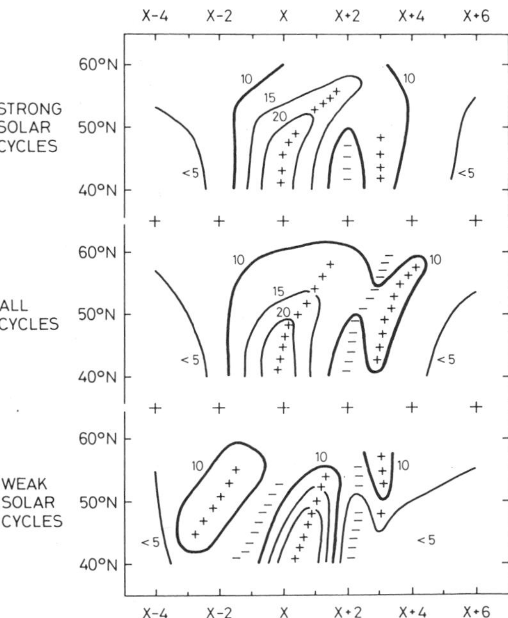

Figure 21-1. Auroral frequency, geomagnetic latitude (actual west European latitude is shown), and the course of the solar cycle (Schove, 1983*a*, cf. Schove, 1983*d*, p. 13). X is the year of official sunspot maximum (1700-1950). Secondary minima are shown by minus signs between two maxima (plus signs). Frequency is expressed as a percentage of all displays in the 11-yr period.

French travel literature listed by Harisse in various bibliographies would enable us to date what both the Jesuits and the North American Indians regarded as miracles.

MEAN LENGTH OF THE 11.1-YR CYCLE SINCE 3000 B.C.

Various attempts have been made to date the 11-yr maxima in the medieval period, but they are usually inconsistent (Schove, 1983*d*, p. 25). The dates of Ding et al. (1983), although based mainly on sunspots rather than aurorae, agree closely with those in the independent revisions of 1983 (Schove, 1983*d*, p. 14 and Appendixes A and B, corrected in the 1984 version). These dates suggest (p. 147) that the mean value since A.D. 300 has been slightly less than 11.11 yr, although the omission of a single maximum would, I admit, suffice to make it 11.12 yr, a value that some so-called planetary enthusiasts have predicted. In the B.C. period, from 3000 B.C. onward, the cycle has been about 11.1 yr or possibly 11.12 yr (Schove, 1983*d*, pp. 25, 300), and this seems about right for the period 649 B.C. to A.D. 302 (pp. 149-167). Since the year A.D. 2087 should be near a sunspot minimum (p. 162), and indeed for other reasons (p. 26), sunspot cycles are expected to become longer and weaker in the next few decades (Fairbridge and Hameed, 1983).

The cycle since A.D. 750 seems to have been less than 11.11 yr because the phase (measured by the remainder when the last two digits of the minimum year are divided not by 11 but by 11.1) has slowly decreased (Schove, 1983*d*, p. 374, col. j). Solar activity has varied considerably (Fig. 21-1). In the even centuries, such as the present, it has usually been strong; in the odd centuries, such as the seventeenth (Maunder Minimum), fifteenth, eleventh, and seventh, it has been weak (pp. 165-166, Fig. 4), because of a 200-yr cycle. The variations are reflected not only in auroral numbers (pp. 138, 373, col. h) but also in fluctuations in radiocarbon production as inferred from the radiocarbon (Suess) wiggles. Professor Fairbridge drew my attention to this radiocarbon effect in 1962, and now, thanks to Stuiver's (1983) decadal tree-ring analysis of production, his figures can be adapted to provide a "calculated auroral index" that

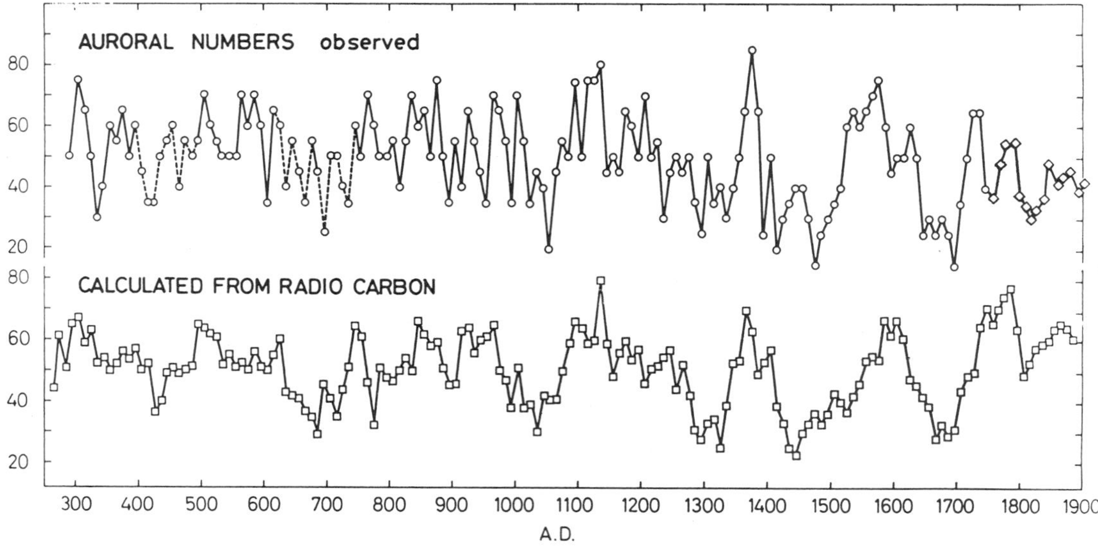

Figure 21-2. Auroral numbers and calculated index for solar fluctuations, A.D. 275-1900. (Top) Auroral frequency (Schove, 1983*d*, p. 138, slightly revised as in Appendix C, col. h). (Bottom) Calculated frequency based on Stuiver's index of radiocarbon production (an inverse relationship) (Schove, 1983*d*, p. 372, Appendix C, col. g).

agrees with the (slightly smoothed) observed data, as shown in Figure 21-2.

The earliest auroral catalogue (unfortunately lost, unless it is waiting to be found in some Han tomb) is that of Liu Xiang, who collected 151 divine lights between 205 B.C. and 32 B.C. These lights were more frequent during certain reigns (179-156 B.C., 140-86 B.C.) than in others (Dai and Chen, 1980; Schove, 1983*d*, p. 26); similar fluctuations were noted in Roman aurorae (p. 166), which were few from 160 B.C. to 135 B.C..

LONG CYCLES

Spectral analyses of sunspots and aurorae have been made separately (e.g., Youji, Baorong, and Yongming, 1983) and in combination (Schove, 1983*d*, pp. 16, 193), and the results (pp. 378, 386) are seldom consistent, partly because the cycles vary with the period investigated, suggesting that the cycles are due to currents beneath the surface of the Sun rather than to permanent astronomical cycles. However, the measurement of tree-ring radiocarbon production as the decadal and annual results of Stuiver (1983) and Pearson Pilcher, and Baillie (1983) are processed will provide a good test over several thousand years; then even the fixed (e.g., tidal) cycles of astronomical geometry may become detectable. Stuiver has meanwhile examined separately and in combination the periods A.D. 1-1000 and A.D. 1000-1850 using decadal values of radiocarbon production.

Cyclicity found by Stuiver (1983) is as follows:

26.5 yr: Found in both periods A.D. 1-1000 and A.D. 1000-1850.

33.5 yr: Found in A.D. 1000-1850 only.

42.5 yr: This had been found from the Schove (1983*d*, pp. 149-168) synthesis (Schove, 1983*d*, pp. 16, 154-158, and 378) and occurs in both periods, although in the second millennium cyclicity occurred also at 45.5 yr.

59 yr: This cycle—the Wolf Cycle (Schove, 1983*d*, p. 193)—had been found in various analyses, including that of Far Eastern sunspots by Ding et al. (1983). In the first millennium Stuiver (1983) found its significance is "just above red noise"; in the second it is not quite 2 sigma. In the Schove amplitudes (1983*d*, p. 154) this cycle appeared as 58.3 yr, but Bain (1976) found 58 yr, using my larger set of sunspots and aurorae, and also the auroral numbers of 1962 (Schove, 1983*d*, pp. 136-141). Bain (1976) found cyclicity also at 51 yr, 53 yr, 62 yr, and 65 yr (Schove, 1983*d*, p. 193).

The so-called Gleissberg 80-yr cycle did not appear, but this is believed to affect only the phase, although (Schove, 1983*d*, p. 373, Appendix C, col. j) the data have not been analyzed (cf. pp. 193, 213). [The cycle of about 93 yr (pp. 193, 378) does not appear, no doubt because of phase changes.]

103/110 yr: Some cyclicity appears in the first millennium only.

123 yr: Found in the second millennium and even in the whole period to some extent. Earlier, Stuiver and Quay (1981) had (Schove, 1983*d*, p. 194) reported a 133-yr cycle in the first millennium (pp. 97, 193, 378).

200 yr: Found only in the second millennium, but Neftel, Oeschger, and Suess (1981) have found it with phase changes in the B.C. period (Schove, 1983*d*, pp. 378, 379).

Longer cycles may exist but a longer period will need to be analyzed before their significance can be assessed.

SOLAR CYCLES IN TERRESTRIAL WEATHER SINCE 3400 B.C.

My tree-ring evidence for a 200-yr weather cycle (Schove, 1983*d*, pp. 318, 327) was not conclusive in the A.D. period, but Sonett and Suess (1984, p. 142) find an approximate 210-yr cycle in the period 3405 B.C./A.D. 1885. They find also cyclicity in the 80-to-115-yr range.

Fisher (1982) finds cyclicity in ice cores linked with ^{14}C (cross-spectral tests). The following cycles seem to be genuine Sun-weather links:

0-2000 B.P.: 195 yr, 375 yr (i.e., for the A.D. period)
2000-4000 B.P.: 167 yr
0-5300 B.P.: ca. 180 yr, ca. 425yr

The 22-yr cycle in solar amplitude has been important only since 1840 (Schove, 1983*d*, p. 192), but meteorological effects go back at least to the seventeenth century (p. 22). No evidence for the Hale cycle in tree rings before 1600 has been found; the 26-yr quasiperiodicity noted in summer drought in the Hudson Valley by E. Cook (pers. comm.) and Thaler's (Chapter 3) 23-yr winter temperature record may be reflections of the Hale cycle. The curves of summer temperatures in northern Sweden given in Karlén (1984) and based on dated tree rings confirmed by the X-radiography of sediments show no clear cycles and no agreement with the solar curve of Figure 21-2.

11-YR CYCLE AND WEATHER

The 11-yr sunspot cycle and weather are related but not in the simple way once supposed. We need to take a wide span in time and place

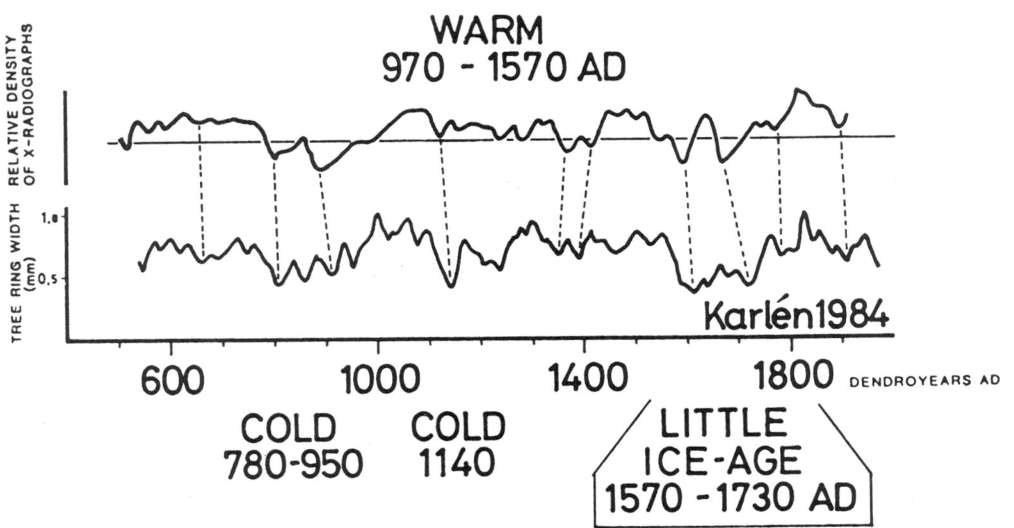

Figure 21-3. Summer temperature proxies for northern Sweden. (Upper curve) Approximately dated sediment densities in a northern Sweden proglacial lake. (Lower curve) Tree-ring widths (31-yr moving means). Cool summers occurred around A.D. 550, 800, 860, 910, 1140, 1240, and during 1570 through 1750. (After Karlén, 1984, p. 268)

before we can demonstrate the effects. These effects can be opposite for strong and weak cycles and even for adjacent years of the sunspot cycle. Nevertheless, there are patterns (Parker, 1976; see also Schove, 1983*d*, pp. 18-19, 229, 291, 248, 326). If we can predict the turning points and amplitude of a solar cycle, we can make successful predictions for the weather of the tropics. My two published predictions were based partly on such factors; my prediction, at the Fairbridge symposium of 1961 (Schove, 1983*d*, p. 257), of a "warm dry wave extending from the old-world tropics to the United States [which] may yet be associated with drought famines" was followed by the Sahel drought of 1964-1972. Professor Fairbridge told me that in Bangladesh in early 1983 I was being blamed for the Indian Ocean droughts because I had predicted in 1980 the usual droughts expected with a negative Southern Oscillation (India, Ethiopia, south-southeast Africa, and Australia) (Schove, 1983*d*, p. 292). The prognosis is now for wetness with floods in India, Australia, and Ethiopia and—despite CO_2—a cooler globe (southeast United States excepted) than in 1982/1983. There will be a risk of further droughts around 1987 (Schove, 1980) as the sunspot minimum approaches, and if the next cycle is as weak as expected, droughts will be more prolonged in the Old World Tropics in the 1990s. A worksheet for use with proxy data since A.D. 1500 is available in *Annales Geophysicae* (1983, **1**:390-396; cf. Schove, 1983*d*, p. 280, Table 5).

Gallardo, in the same 1980 symposium on *Sun and Climate* (p. 81), noted that El Niño years since 1763 occurred especially 1 yr before or after a sunspot minimum and 1 yr or 4 yr after a sunspot maximum. Tree-ring or varve data extending back to the Middle Ages would enable the significance of this suggestion to be tested. El Niño years, and the negative Southern Oscillation often associated with them (ENSO), arise in my view because of pressure discontinuities caused by equatorial inflation (Schove, 1983*d*, p. 275).

My theory of the cause of the El Niño and Southern Oscillation events has been partly published (Schove, 1963), but the theory did not include air-sea interaction. I still believe that the immediate cause lies in persistent cold lows

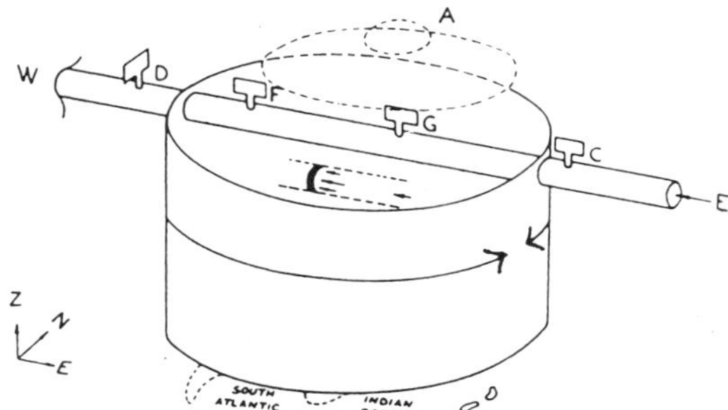

Figure 21-4. The valve model of the Southern Oscillation and El Niño: A three-dimensional view of the upper high over the Indian Ocean countries.

A = The contours of the upper high. E = Equatorial easterlies at 300-100 mb. C = The confluence point over South America as a valve (reflected in a pressure discontinuity at the surface). D = The difluence point over Indonesia as another valve. F, G = Other points off west and east Africa (that may sometimes act as valves). The piston diagram below F and G refers to seasonal changes explained in the text. (From Schove, 1963).

and warm highs at specific longitudes in the subtropics, which interfere with the smooth flow of the equatorial easterlies at the 200-mb level. (This idea was based on pressure anomalies for the year preceding a Southern Oscillation when its value was near zero.) This situation leads to a pressure discontinuity at the surface near the equator. In my model (1963, p. 252) such a discontinuity is regarded as a valve (Fig. 21-4). My main valves over South America and Indonesia seem to be confirmed by Selkirk's (1984) maps, which show that, when a Southern Oscillation has developed, they do separate longitudes with westerly and easterly wind anomalies. Selkirk's analysis is based empirically on actual 200-mb observations; mine was perforce hypothetical as far as the upper air was concerned and needs further elucidation. The relevance of the 1963 model to climatic fluctuations lies in a similar explanation of the superoscillations revealed by the 30-yr means (Schove, 1983*d*, Figs. 17 and 18, cf. pp. 234, 257).

PRESSURE PARAMETER TRENDS

Pressure trends between adjacent 5-yr periods, unlike pressure anomalies for 5-yr means, have a characteristic geographical pattern, or eigenvector (Schove, 1983*d*, pp. 252-254, 287, 288, 320). Temperature trends, for example, in New Zealand and Newfoundland are positively correlated and those in California negatively, but my hope that tree rings would enable us to estimate pressure parameter (PP) values back to A.D. 1-5 has not materialized. Strong solar cycles lead to an increase and weak cycles to a decrease, so New Zealand temperatures are expected to fall as the PP falls. Old World tropical and subtropical droughts and famines increase at such times. A map of the temperature changes associated with a rising pressure parameter is given in my article on sunspot cycles in the *Encyclopedia of Climatology* (Oliver and Fairbridge, 1987), in Professor Fairbridge's series. The different effects on the west and east coasts of the United States have been noted previously (p. 252).

The effects of sunspots on the weather of the temperate zone are complex (cf. Schove, 1983*d*, pp. 226-238), but in former times there was an increase in food prices (pp. 17, 226, 241) and, possibly for this reason, an increase in the number of social and political revolutions (pp. 226-228, 281).

2-YR AND 3-YR (QUASI-BIENNIAL AND QUASI-TRIENNIAL) CYCLES

Corresponding to the Southern Oscillation there is a cycle in stratospheric equatorial winds that averages 2.2 yr. It is often a 2-yr cycle when sunspots are numerous but becomes a 3-yr cycle when they are infrequent (Schove, 1983*d*, pp. 268-273). The cycles are picked up by the troposphere but with a variable delay in each of the four seasons. The fluctuation between the two modes (2 yr or 3 yr) in tree-ring and varve records is a first clue to the cross-dating of those time series. Transatlantic teleconnections have been established in this way in both the Holocene and the Late Glacial, as I explain in the next section.

TELECONNECTIONS AND BIBLICAL DROUGHTS

The bristlecone pine chronology of California is reliably dated, and the pattern of 2-yr and 3-yr cycles matches very well the pattern in a floating dendrochronology from Turkey, based on the timbers of the Midas tombs of the semi-legendary Phrygians (Schove, 1983*d*, p. 364, fig. 3). It has proved possible to date both this tree-ring series and the varves of the USSR, but the latter were found to be nearly 150 yr younger than expected. Droughts north and south of the Black Sea may well have also affected Palestine; the droughts reported in the Books of Kings in the time of Amos (and just after) could well be about 868 B.C. and 847-839 B.C. when tree rings were persistently narrow at Gordion in northwestern Asia Minor.

TELECONNECTIONS AND THE ERUPTION OF THERA (SANTORINI)

There have been several estimates about the date of the eruption of Thera (Santorini) that caused so much damage in Minoan Crete. Weir noticed a year in the Venus Tablets of Ammizaduga in which there were no Venus observations. The date originally given for this year (Schove, 1983*d*, p. 361) was 1637 B.C., but with the revised Weir-Huber dating that year became 1693 B.C., both near the date (1626 B.C.). The X-ray curve of Alpine tree rings (Fig. 21-5) reflects summer temperatures and confirms that around the period suggested by LaMarche, the summers were cold also in Europe, but that particular cold period is what I call Little Ice Age III and it lasted longer than 140 yr from about 1670 B.C. to 1510 B.C. (Schove, 1983*b*, using Renner's floating chronology, as cross-dated with the bristlecone). One particular summer in Renner's Alpine curve is much colder than its neighbors, and I suggest that year may be dated 1479 ± 1 yr B.C. (Fig. 21-5). Could this be the date of the catastrophic destruction of Thera?

THE SUN AND HOLOCENE CLIMATE

The reality of the Suess wiggles in the B.C. period was often doubted at first, but in 1954 I suggested (cf. Schove, 1983*d*, p. 23) that their reality could be tested by a histogram because certain radiocarbon dates b.p. should be especially frequent in a random sample as they would

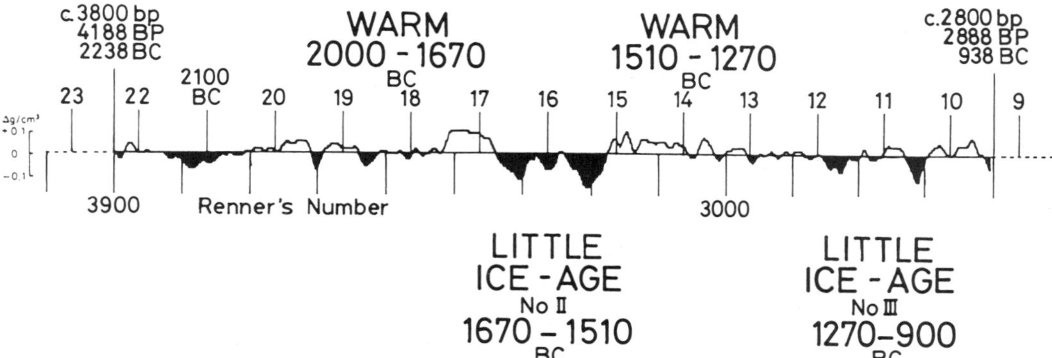

Figure 21-5. Renner's curve compared with Schove dates. Late summer temperature.

Proxy data based on X-ray densities of trees in the Central Alps kindly supplied by F. Renner. The radiocarbon dating of this floating chronology was originally given as about 3895-2610 b.p., and 1662 yr has been deducted to yield sidereal dates B.C. that Schove believes will match the bristlecone pine chronology.

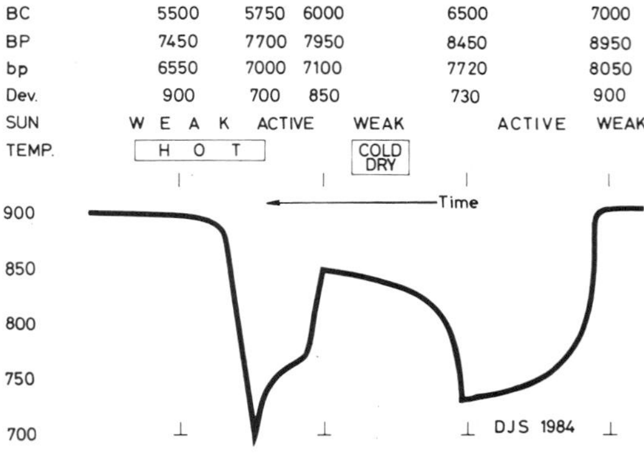

Figure 21-6. Radiocarbon deviation in the so-called Missing Millennium (cf. Schove, 1983*d*, p. 337) based on the ^{14}C dates of European tree rings given by Bruns et al. (1984). SUN relates to solar activity inferred from the ^{14}C anomalies shown on the curve. TEMP refers to the temperature as given in Schove (1983*d*, p. 356).

correspond to perhaps three different true B.C. dates, whereas other b.p. dates would be rare, although they would convert very accurately to a specific sidereal date B.C. (Table 21-4). As standard deviations became smaller and radiocarbon dates more numerous, it was possible to attempt this conversion (Fig. 21-6) (cf. Ottoway and Ottoway, 1974), and tables of each type are published in Schove (1983*a*). These tables (Tables 21-3, 21-4) have been updated slightly with footnotes using information kindly supplied by Pearson and co-workers (1983). Histograms of radiocarbon dates obtained since 1974 where the errors are small might be used to improve on the 1974 table.

LATE GLACIAL TELECONNECTIONS

The methods I used for teleconnections were published in Sweden in 1971, but I understand that in Sweden nobody believes in the possibility of even trans-Baltic (let alone transatlantic) teleconnections such as given by Schove (1983*d*, pp. 335-342). Certainly, at the Stockholm conference of 1983, I did not find anybody who had studied my results or who accepted my synthesis as given in the abstracts (Schove, 1983*b*) published in Denmark. It was probably my fault for not having published the details of some of the tests used, but this I have now done (in 1984*b*). The 1971 chronological table, corrected for obvious errors, is thus reprinted here (Table 21-5). Good radiocarbon dates were not obtainable from the varve areas of either Fennoscandia or New England in 1971. In Sweden it is usually supposed that the Pleistocene-Holocene boundary is about 10,000 yr ago, not only in radiocarbon time b.p. but also in sidereal time B.P.

Table 21-3. Auroral Minima, Radiocarbon Production, and Climate

Auroral Lull (*a*)	*Radiocarbon Production* (*b*)	*Ambiguous Dates* (*c*)	*Radiocarbon Equivalent* (*d*)	*Cold Centuries* A.D. *or* B.P. (*e*)
1646/99		1695/1875	110	(A.D. 1600/50)
1410/1520	1410/1520	1525/1625	320	(ca. A.D. 1450)
1230/1340	1270/1340	1305/90	620 b.p.	(ca. A.D. 1350)
1020/70	990/1070	1040/1150	920	(A.D. 900/930)
880/916		905/975	1100	
630/740	630/730	695/770	1260	
400/70	420/70	435/535	1560	
230/80(?)	250/80	260/330	1720	
105/130	ca. 125	140/210	1830	
25 B.C./A.D. 42	−20/+50	75/120	1895	
ca. 160/135 B.C.				
ca. 440/400 B.C.	−410/−375	−350/220	2230	ca. 2300 IV
(9th-8th cents.)	−840/775	−770/630	2500	ca. 2800 III
(12th cent.)	−1150/−1100	−1250/1125	3000	ca. 3225 II
(ca. 1400 B.C.)				
(ca. 2300 B.C.)				
(33rd cent.)		−3300/3180	4500	4575 I
(34th cent.)	−3400/3350	−3475/3375	4700	
(37th cent.)	−3650/3620	−3625/3500	4775	
(47th/44th cent.)				
(ca. 54th cent.)		5325	6550	ca. 6315
(ca. 59th cent.)		−5850/5750	7000	ca. 7350
(ca. 70th cent.)		−7000/6800	8050	ca. 8250
				ca. 9350
				ca. 10,500
				ca. 10,700

Note: See Pearson et al. (1983) for additional details; for example, 29th century B.C., −2870/2800, −2850/2600, 4100 B.P.
(*a*) Lull in observations of the aurora (Schove, 1983*d*, pp. 149-168). Dates in parentheses are inferences from radiocarbon peaks.
(*b*) Peak in radiocarbon production, as measured by Stuiver or implied by Suess (see Bruns et al., 1983).
(*c*) Ambiguous tree-ring dates, corresponding to the same radiocarbon (b.p.) date in *d*. (Subject to revision.)
(*d*) The corresponding radiocarbon date (suggesting a change from low to high solar activity).
(*e*) Cold (A.D. or b.p.) centuries. These sometimes occur earlier than the auroral minima.
I-IV are Little Ice Ages.

Table 21-4. Radiocarbon Ages that Calibrate to Precise True Dates (Presumably Associated with about a Half-Century of Sunspot Decline)

bp (*a*)	*ad or bc* (*b*)	*AD or BC* (*c*)	(*d*)	(*e*)	(*f*)
250	1700	1650		c	
500	1450	1425	w		c
700	1250	1280	w		c
1060	890	990		w	
1400	550	650		c	
2000	−50	0	w		c
2300	−350	−400		cc	
2600	−650	−800	w		c
4250	−2300	−2900	c		w
4600	−2650	−3350	w		c
5100	−3150	−3900	w		c
5520	−3570	−4400		ww	
5750	−3800	−4560	w		c
6750	−4800	−5650		ww	
7550	−5600	−6375	w		c

Note: See Pearson et al. (1983), whose charts indicate 4325 rather than 4250 b.p. as likely to be one of those rare dates with precise chronological meaning.
(*a*) bp = A: Radiocarbon age that is rare in randomly distributed high-precision samples (cf. Bruns et al., 1983 for oldest dates).
(*b*) ad/bc: Converted but uncalibrated date.
(*c*) AD/BC: Calibrated or tree-ring date.
(*d*) to (*f*): The climatic trend in European summers is given in col. *d* and *f*.
ww = very warm, w = warm, c = cold, cc = very cold. Letters are put in col. *e* when trend is uncertain.

My original reasons for adopting 11,000 sidereal yr (9000 B.C.) as the III/IV boundary (Fig. 21-7) were at that time based only on some uncertain German varves (Fig. 21-8) (p. 343). The question remains open, but careful drilling and paleontological work show that the abrupt change from cold to warm climate indicators falls very close to 10,000 b.p. (Olausson, 1982), which confirms the resolution put forward by Professor Fairbridge at the INQUA congress in Paris in 1969 (see also Fairbridge, 1982).

In North America the careful measurements of varves made by Antevs (1922) (Fig. 21-9) did not lead to further progress because of a general suspicion at that time of the accuracy of varve counting. Antevs's only temporary follower in the United States, Clement Reid, died leaving a large collection of varve measurements for the winter and summer layers; no institution was willing to accept them so they were thrown away (information given to me by Dr. R. Pardi of Queens College, New York). Antevs's measurements, he told me before he died, were in the archives of the American Geographic Society, but I was officially told that they could not be found. As his only follower in Europe, I published my interpretation for Canadian and New England chronology in the reports of the INQUA meeting of 1969. Now, radiocarbon dates are being obtained in North America that are consistent with the dates suggested. (Incidentally, my six large boxes of varve analyses will shortly be available.) Ashley (1972) and Dr. Pardi have satisfied themselves about the accuracy of Antevs's work, but measurements are more difficult now that the clay pits (where the varves were so clearly exposed) have become overgrown with vegetation; future work will require coring.

In Canada the accuracy of the Antevs counting has been confirmed by several workers. Some of the Agterberg-Banerjee measurements were eventually traced and have been kindly supplied by Agterberg for publication (Schove and Fairbridge, 1986). As far as I am aware, they will be the first published results from North America that include both summer and winter measurements. Their importance lies not only in their established teleconnections with the Swedish time scale (Fig. 21-10), but also in the possibility of cross-dating with summer-sensitive tree rings. Such tree rings may be found either in the bristlecone area of California (even under snow) or near the river Danube in Europe, enabling us to test my hypothesis that the beginning of the Holocene, which took place about 10,000 radiocarbon yr b.p., may be dated about 11,000 sidereal yr B.P.—that is, about 9000 yr B.C. They would then bridge the gap between the Holocene tree rings and the Late Glacial varve chronologies and thus make it possible to extend the absolute chronology back to at least 12,500 b.p. (believed to be ca. 11,500 B.C. or ca. 13,500 B.C.).

I am unable to link or supply absolute dates for the several sets of southern New England varves, but help would be appreciated; for example, formulas linking the varve numbers with

Table 21-5. Chronologies from 20,000 to 8,000 B.P.

C14 Stade	Warm or Cold Zone	Varve Dates: Sweden N (All Dates Negative)	Varve Dates: Finland F	Varve Dates: United States A, B	Varve Dates: Canada C or BH	Phase	Absolute Dates B.P.	Absolute Dates B.C.	C14 Error Years (5,568 half-life) Tentative
19,650 b.p. or									
ca. 196	CCCCC								Very large
ca.180	CCCCC								Very large
ca.167	W								
ca.147	W						16,000?		1250?
ca.142±	CCCCCWW			4800?		Vashon/Cary	15,000 ±	13,000±	750?
ca.134	WW			5800?		Raunis/Susáca	14,000±	12,000±	550?
ca.128±	CCCC Ia	10,750?		6300	(−ca.2650)	Early Dryas	13,500±	11,500±	
126	CCCWW	10,400		6630	BH:−				
123	WWWC Ib	10,120			0	Mid-Bölling	ca.13,100	ca.11,150	ca. 750±
122	CWWW Ib	10,025	(−1740)	7010	80	Mid-Bölling	ca.13,000	ca.11,050	ca. 750±
121/0	CCW Ic	9800	(−1575)	7130	200	Middle Dryas	12,850	10,900	ca. 770±
119/8	WCCWWW I/II	9700	−1475	7250	330	Interface	12,740	10,790	ca. 840±
117/6	WW IIa	9600	−1375	7350	430	Early Alleröd	12,640	10,690	ca. 940±
115	CWCWCC IIb	9400	−1175	7550	630	Mid-Alleröd	12,440	10,490	ca. 890±

(continued)

C14 stade: Stades are in conventional C14 years before the present with the last two figures omitted. *Warm or cold:* Temperature sequence indicated by letters: e.g., CCW = very cold then warm. *Zone:* Pollen zones. *Varve dates:* Scales N, F, A, B, C, D as in Fig. 1, p. 363, Schove, 1983*d*, BH = Bracebridge. *Phase:* European terminology. *Absolute dates:* Based on the assumption that Canadian varve 2000 = tree-ring date 9,200 B.P. or 7250 B.C. *C14 error:* This assumption implies an error in the conventional C14 dates of the order of 950 yr, but all these estimates are uncertain.

Table 21-5. *(continued)*

C14 Stade	*Warm or Cold*	*Zone*	*Varve Dates* Sweden N (All Dates Negative)	Finland F	United States A, B	Canada C or BH	*Phase*	*Absolute Dates* B.P.	B.C.	*C14 Error Years (5,568 half-life) Tentative*
ca.113/2	WWCCC	IIc	9200	−1075	7750	(830)	Mid-Alleröd	ca.12,325	10,375	ca.1025±
ca.112/1	WW	IIc	9000	−775	(7950)	(BF?)	Late Alleröd	ca.12,025	10,075	ca. 925±
109/108	WCC	II/III	8900	−659	(8050)	—	Interface	11,900	9950	1000±
107	CCC	IIIa	8800	−550	(8150)	—	Salpausselka I	ca.11,800	ca. 9850	ca.1050±
106/5	CWCWW	IIIb	8500	−300	(8450)	(BF?)	Later Dryas	ca.11,550	ca. 9600	ca. 950±
105/4	CC	IIIc	8300	−100		−C	Salpausselka II	ca.11,350	ca. 9400	ca. 850±
103/2	CCWWW	III/IV	ca.8213N =8140V	0		−52	Late Glacial ends	ca.11,252	9302	ca. 950±
102	WWW		ca.8088	+52		0	Canadian zero	11,200		
95	WW	IV	7600	+540		+500	Preboreal	10,700	8750	ca.1150±
93	C		7200	(+900)		+900		10,300	8350	ca.1000±
90	WW	V	6923 =OD	(+1200)		1166	Döviken Zero	10,033	8083	ca. 983
	WW		(360D)			1528 =0	N. Canada Zero	9,672	7722	
82	C		(6100)			2000	Cockburn	9,200	7250	ca. 950
81	W		(6000)			(2100)				

C14 stade: Stades are in conventional C14 years before the present with the last two figures omitted. *Warm or cold:* Temperature sequence indicated by letters: e.g., CCW = very cold then warm. *Zone:* Pollen zones. *Varve dates:* Scales N, F, A, B, C, D as in Fig. 1, p. 363, Schove, 1983*d*, BH = Bracebridge. *Phase:* European terminology. *Absolute dates:* Based on the assumption that Canadian varve 2000 = tree-ring date 9,200 B.P. or 7250 B.C. *C14 error:* This assumption implies an error in the conventional C14 dates of the order of 950 yr, but all these estimates are uncertain.

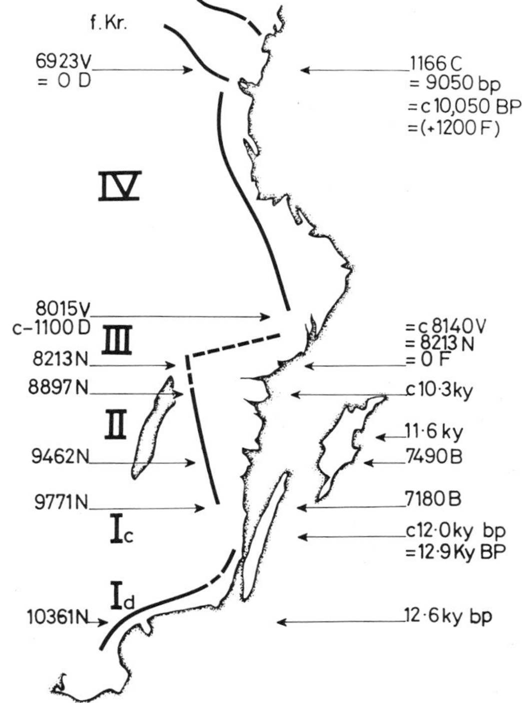

Figure 21-7. Eastern Swedish varve sets as mapped by Kristiansson (1982).

The Swedish dates given by Kristiansson have been labeled V in northern Sweden and N in southern Sweden. 6,923 V, or 8,893 varve yr B.P. (as the Swedish geologists now put it), is the zero datum (OD). Pollen zones Ib to IV are shown approximately. Teleconnected dates (from Schove, 1971) have been added on the right-hand side of the map.

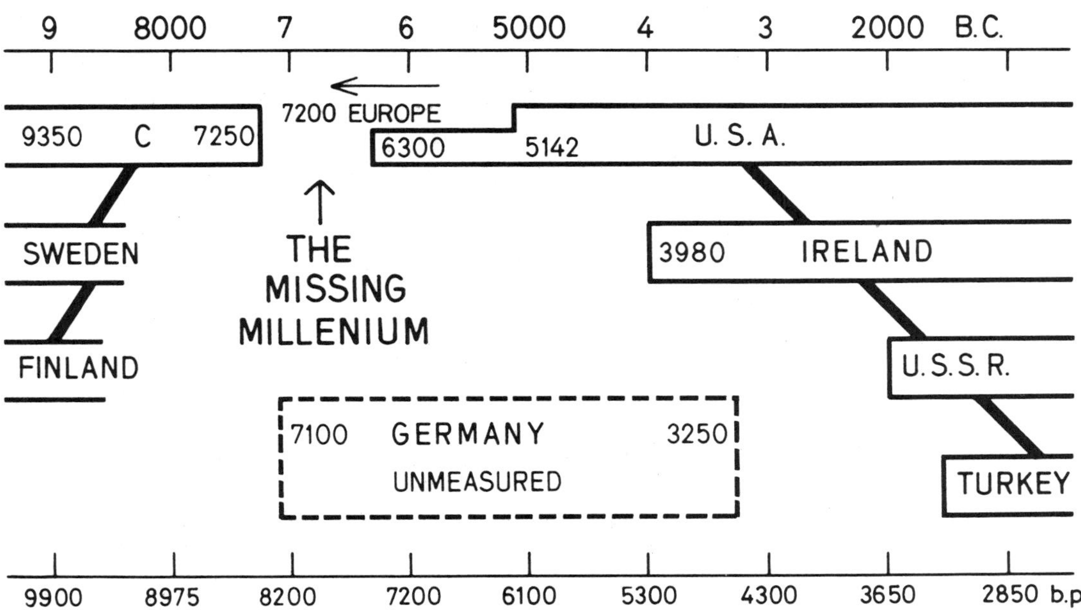

Figure 21-8. The Missing Millennium. This gap has now been effectively filled (cf. Schove, 1983*d*, p. 337, Fig. 2, and p. 364, Fig. 4 for the original version) as indicated in Figure 21-6.

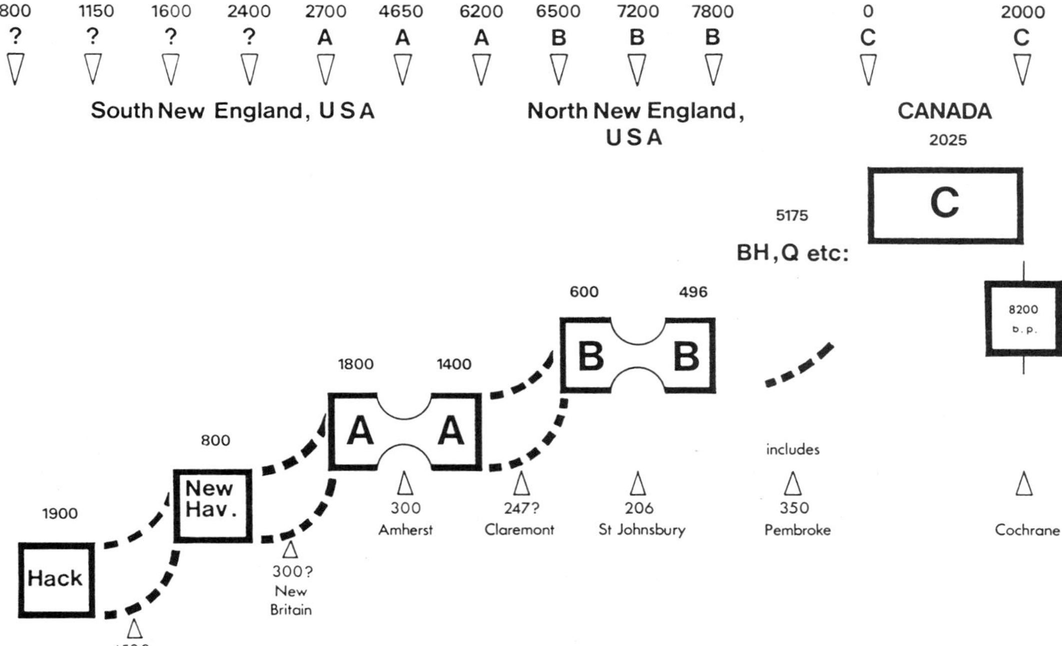

Figure 21-9. North American varve sets in series as Antevs (1922) envisaged them. Hack = Hackensack, New Hav. = New Haven.

approximate ^{14}C dates would be very useful. My present notes are as follows.

Hackensack, N.J.

The earliest long series in the Antevs (1922) time scale are those of Hackensack, N.J., northwest of New York City (40°50′N). These are now believed to lie in the range 16,800-15,300 b.p. There were 1,490 varves in all, but Antevs (1922, pp. 173 and 225) pointed out that, from varve 671 on, there were uncertainties in the counting. I attempted to link the early part of the series with the A series in Centre 1 Connecticut, considering the possibility that varve 1 was about 2,865 A, varve 366 was 3,231 A, and varve 478 was 3,343 A. Varve 1,490, if this were correct, would be about 4,350 A. The drainage date of Lake Hackensack is not known, and the series may be much earlier.

New Haven, Conn. (Haverstraw, N.Y., 41°15′N, and New Haven)

Three overlapping series found by Antevs (1922, pp. 226-227) yielded a series of about 732 varves. They are now believed to lie in the period 16,500-15,500 b.p., but no overlap with the main A series was identified.

Southern New England (Lake Hitchcock)

This is the main A series and appears to cover over 3,000 yr (Fig. 21-11), although Antevs had some reservations in the period of readvance that suggested that the glacier readvanced at 4,800 A and returned to the position it had occupied at 4,450 A. The radiocarbon dates for Lake Hitchcock varves are now believed to lie

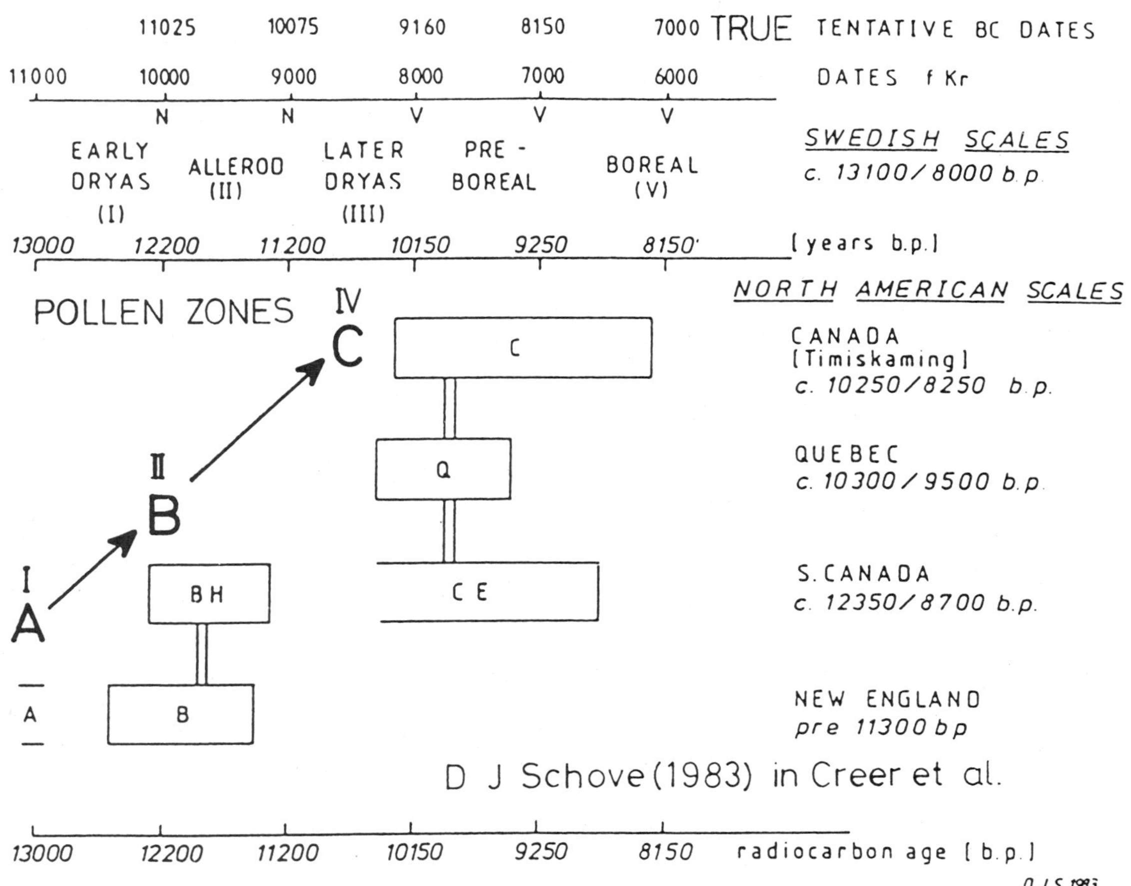

Figure 21-10. North American varve sets in a time scale based on European teleconnections. (Using Schove, 1969/1971)

near 15,700 b.p. (varve ca. 2,700 A), 15,500 b.p. (varve 2,850 A), and the later varves around 13,500 b.p. (5,700 A), so 3,000 varve yr (if correctly counted) correspond to some 2,000 radiocarbon yr. The last varve counted was 6,352 (Antevs, 1922, p. 57) and was placed at about 12,820 b.p. in my 1971 table (Schove, 1971) but without radiocarbon evidence.

Northern New England (Lake Upham)

This is the main B series and covers varves numbered 6,601-7,400 A. The counting was not always certain, and the radiocarbon dates were not available. However, the teleconnections with Sweden enabled me to place it in my 1971 table (Schove, 1971) as beginning at about 12,620 b.p. and continuing to 7,350 A at 11,700 b.p., or 11.7 kyr.

Antevs (1922) may have been correct in assuming the New Haven A and B sets could be placed in series as shown in the diagram. His main error (Fig. 21-12) was to allow over 5,000 yr between the New England and the main Canadian (Timiskaming) sets; the BH (Bracegirdle-Huntsville) and Q (Quebec) sets can be shown (Schove, 1969/1971) to overlap the Timiskam-

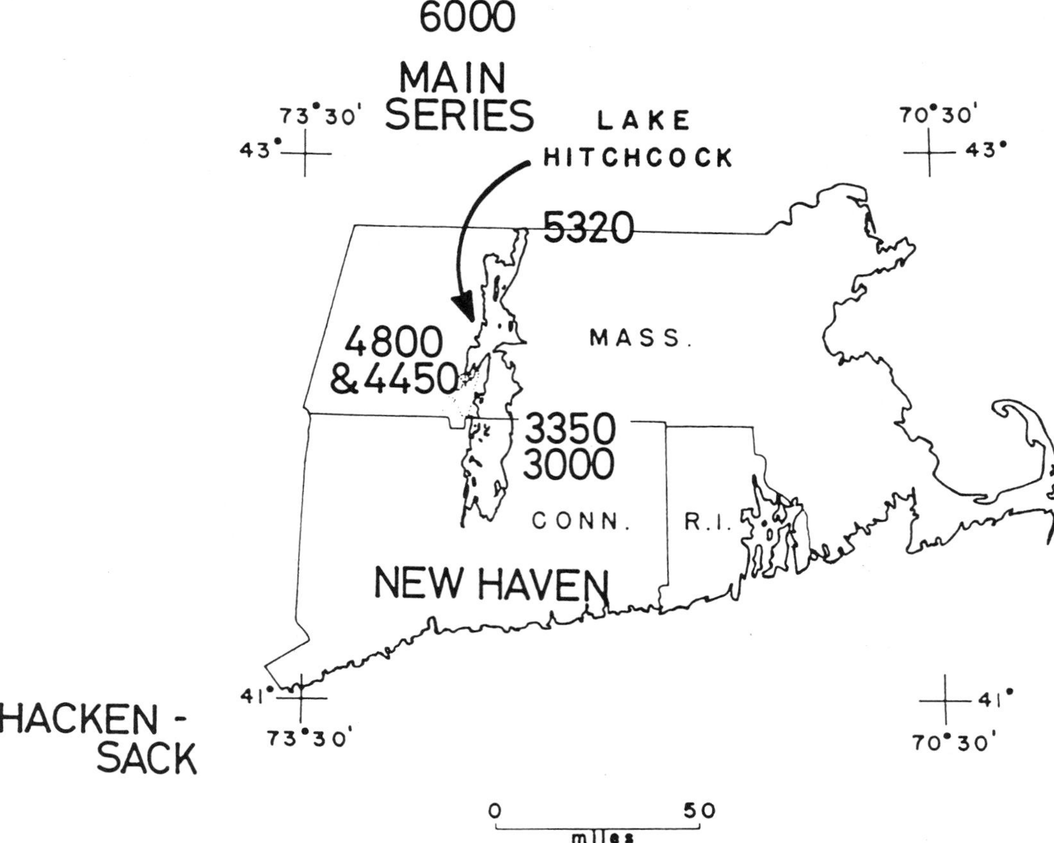

Figure 21-11. The floating varve chronologies of southern New England as Antevs (1922) numbered them. The main series (A) is limited to the radiocarbon time scale by a formula: A + b.p. = 16,000 ± 400. (Base map from Ashley, 1972)

ing set and indeed help to confirm the accuracy of the original counts.

A symposium, *Late Wisconsin Glaciation of New England* (Larson and Stone, 1982), including maps of the ice retreat at 2,000-yr intervals to 12,000 b.p. for moraines are as follows:

- Maximum: 21,750 b.p.
- Beginning of slow retreat: 21,200 b.p.
- Lower Hudson Valley ice-free (Pellets Island Moraine, West of Hudson River, Shenandoah Moraine to East): 17,400 b.p.
- Hudson-Champlain lobe retreats to Glen Falls, New York: After 17,000 b.p.
- Lucerne readvance: ca. 13,200 b.p.
- Champlain Valley may have been ice free: 12,600 b.p.

The notes on transatlantic varve teleconnections given in the INQUA 1969 report (Schove, 1969/1971) are summarized here. In North America, as in Sweden, there was at that time little help from radiocarbon, but no date changes have been made, except for the alternative date for the floating chronology (A) of southern New England and for conversion of dates B.C. to dates b.p. The nomenclature has, however, been updated.

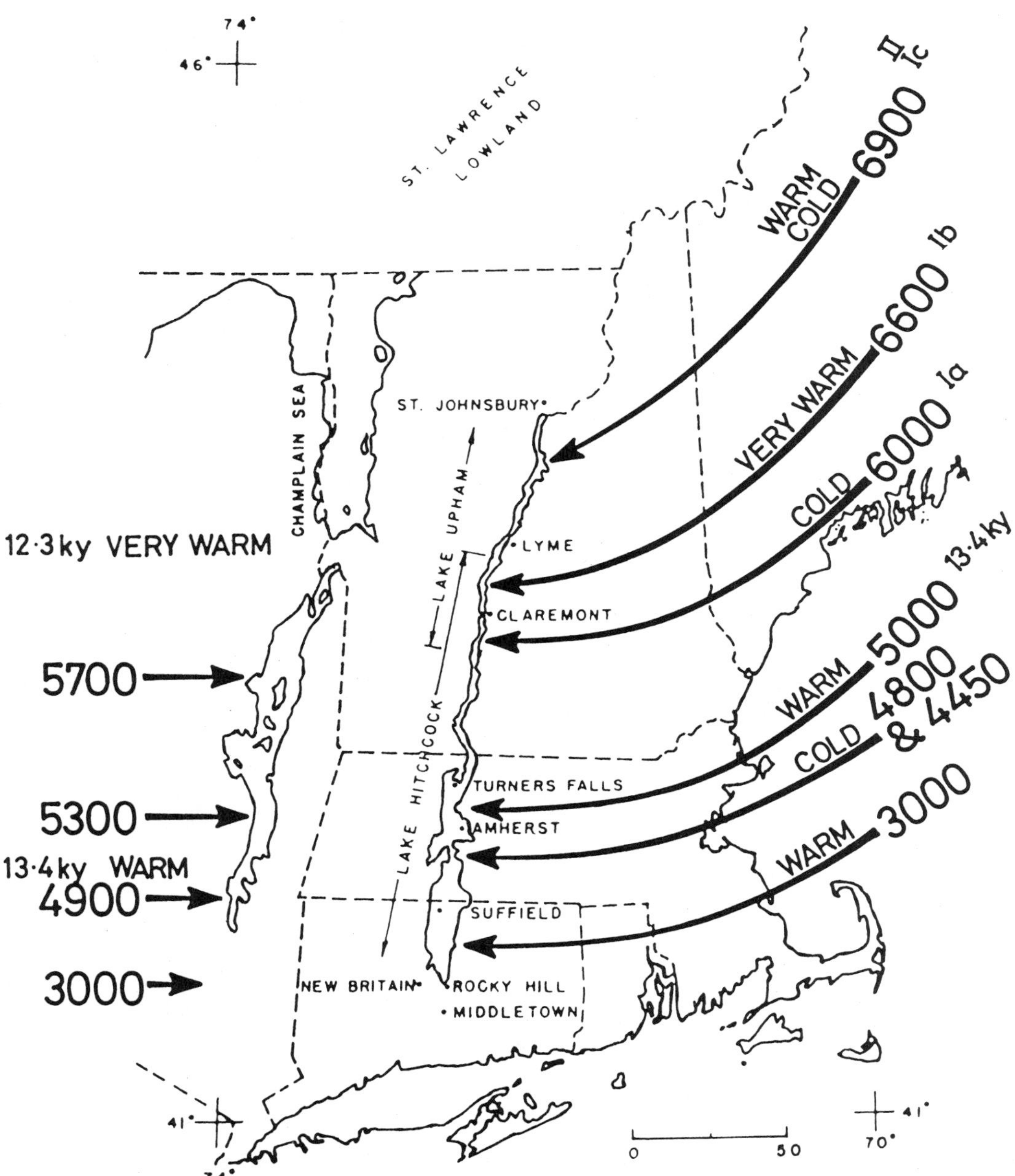

Figure 21-12. The main floating chronology (A to 6,000) and the fixed (B 6,600±) chronology of New England as Antevs numbered them. The dates in kyr b.p. and the European pollen zones I, II are as determined in 1971 by teleconnections. (Base map showing paleolakes from Ashley, 1972)

14 kyr b.p., Rosendale Readvance

The main interruption to the glacial retreat in south-central New England, now known as the Rosendale readvance, began at varve date about 4,500 A, accelerated about 4,580 A and about 4,635 A and culminated about 4,800 A. The alternative formula is

$$A + (b.p.) = 18{,}700 \text{ or } 19{,}400;$$

the later date (Schove, 1983*d*, p. 338) assumes the Rosendale readvance in the Hudson valley was 14,800 b.p. as has since been suggested. The rapid retreat is probably dated around 13,450 b.p. so that a date near 19,300 b.p. may be best, and the 13.4 kyr line in Figure 21-12 could be put farther north.

13.45+ kyr, Susaca Interstade

The term *Susaca* for a short, mild interstade ascribed to about 13,450 b.p. was proposed by Van der Hammen. This interstade was assumed to coincide with the rapid glacial recession in the Merrimac and Connecticut valleys, around 5,700-6,100 A in southern Vermont and southern New Hampshire. However, the formula resulting,

$$A + (b.p.) = 19{,}150,$$

seemed too young, and it was suggested that when bottom varves are reached the most rapid recession may have occurred at earlier A dates than Antevs's maps indicate (Fig. 21-12).

12.65+ kyr, Pre-Bölling Readvance (Claremont)—Zone 1a

The onset of this readvance about 6,220 A, about 6,290 A, and after around 6,360 A presumably has a parallel in events in southern Sweden before 10,700 f.Kr. (B.C.) (in the N scale of southern Sweden, or as the Swedes suppose, before 12,650 varves N, b.p.). The formula given for this floating chronology was

$$A + (b.p.) = 18{,}750.$$

These three formulas should be similar if, as seems likely, there were no errors in the varve counts for this floating chronology and if there were no major wiggles on the ^{14}C curve. A detailed analysis of the nonrandom features in this and the earlier A series is available for comparison with varves in the southern Baltic and USSR.

12.55 kyr, Bölling: Sudden Warmth—Boundary 1a/1b

There is a gap between Antevs's A and B scales estimated by him as between 200 yr and 300 yr. The B scale could, however, be teleconnected with the Swedish varves, and preliminary tests used gave the formula

$$B - N = 17{,}027 + 3.$$

The rapid recession in Sweden at the varve around 10,550 N (12,500 V, B.P.) was thus synchronous with that in New England from 6,580 B. The formula

$$B + b.p. = 19{,}130$$

yields the approximate radiocarbon date. These results are consistent with the latest Swedish results.

12.3 kyr, The Bölling Warmth and Mid-Bölling Moraine—Zone 1b

The recession in Sweden in the one hundred first century f.Kr., about 12,100 N, B.P., was greater than at any other time in the Late Glacial. A Danish botanist, Tauber, suggested that this was the climax of the Bölling and not the Alleröd as some Swedish scientists thought. I suggested that the Swedish varve date 10,050 N or 12,000

N appears to be 300 yr less than the radiocarbon date, and the varve date seems to be correct, although the Swedes admit there is still some confusion in Sweden. (Discussed by Berglund and Mörner, 1984, p. 19.) The Bölling in the strict sense, as charted in Schove (1983*d*, p. 339), consists of just over a century of remarkable warmth 12 kyr, around 12,300 b.p. (my interstade 123/122), with a mid-Bölling retardation that produced a moraine in south Sweden. My date was based especially on summer temperatures inferred from beetle finds by Coope. Recession reached a climax in Vermont about 6,904-6,954 B, and the rate around 6,945 was the fastest observed in New England according to Antevs (1922), but retardation followed around 6,975 B with a renewal of rapid retreat around 7,000 B. The same $B - N = 17{,}027$ formula was applicable up to this date (see Schove, 1983*d*, p. 339).

12.1 kyr, Cold of the Final Early Dryas (Mid-Dryas in Schove, 1983*d*, p. 339)

Details were given by Schove (1969, 1971), but it was pointed out that there was a discrepancy of 70 varves in the teleconnections after this cold phase. I suggested (Schove, 1971, p. 932) that this was "due presumably to extra slump varves in Sweden," but no confirmation of a specific error in the varve counts of this period has come from Sweden, for example, for the paleomagnetically examined varve series of southeastern Sweden (Schove, 1983*d*, p. 340).

12 to 11.3 kyr b.p., Early Alleröd—Zone II

The Alleröd patterns have been illustrated (Schove, 1983*d*, p. 342, fig. 4), and although the dates in that figure refer to the Nilsson time scale f.Kr., the counts are good on both sides of the Atlantic, and the sudden increase of warmth that in northwestern Europe marks the zone boundary I/II corresponds in New England with New England varves around 7,280 B. The transatlantic correlation formula is

$$B - N = 16{,}956 + 5.$$

Final tests have not been made to determine the exact point, inasmuch as small errors may occur in both series. Recession was very rapid in the first 65 yr of the Alleröd.

About 11.25 kyr b.p., Mid-Alleröd—Zone IIb

Some cool phases were dated in both Sweden and Finland (Schove, 1983*d*, pp. 333-357).

About 11.05 kyr b.p., Late Alleröd—Zone IIc

Two centuries of warmth were dated in Sweden, and it was estimated that Nilsson's varve dates were only about 100 yr too young for the ^{14}C scale.

About 10.7 kyr b.p., Later Dryas: Early Phase—Zone IIIa

Two very cold periods with an intervening milder phase are identified in Pollen Zone III in Europe, but varves of this period have not, as far as I know, been identified in North America. The cold of the Early Phase was, in Europe, both sudden and severe. In North America absence of information suggests that recession stopped but readvances were not important.

About 10.4 kyr b.p., Later Dryas: Mild Phase—Zone IIIb

Changes in Fennoscandia are discussed by Schove (1969/1971, p. 933).

About 10.3 kyr b.p., Later Dryas: Second Cold Phase—Zone IIIc

The changes in Europe have again been discussed previously, but I understand that the concept of a single, sudden drainage at Billingen at Nilsson's varve 8,213 is complicated (this varve in my view fits the Finland zero varve). In Sweden some projects to find a bridge between the north and south are still in progress, although my 1971 interpretation had suggested that nearly a century of superfluous varves were included in Nilsson's numeration; 86 such extra varves would bring the Swedish and Finnish series into line. However, the Finnish series have not been rechecked from the work of Niemala (1971).

About 10.1 kyr b.p. (8,100 V), Holocene Commencement

The Canadian series were teleconnected with one another and with the Fennoscandian varves (see formulae, in Schove, 1983*d*, p. 335), and the sequence is described more fully by Schove (1969/1971, p. 933; 1984*b*). The 1971 table is presumably acceptable now that the ^{14}C deviation at about 8,200 b.p. has been confirmed (Fig. 21-8). Parameters derived from specimens of bristlecone from beneath the snow on the slopes of Methuselah Grove are expected to match the patterns on the Canadian varves. The concept of deglaciation in steps was not accepted in the 1970s, but Figure 21-13 shows the implications in North America.

THE 30-M.Y. CYCLE

Professor Fairbridge suggested in 1964 to Professor Schwarzbach that I should be invited to the Geological Conference on Palaeoclimates that he was organizing at Köln, so I felt it might be helpful to prepare a chart of paleomagnetically equatorial temperatures through the Phanerozoic. Earlier, with the aid of Nairn and Opdyke (Schove, Nairn, and Opdyke, 1958), we had used the evidence of corals, paleomagnetism, and paleowinds to view the Permo-Carboniferous globally. Now with the new oxygen-18 work of Bowen (see Schove 1983*d*, pp. 316-332) it was possible to produce a curve with dates of turning points. This curve and the associated table has been reproduced in Schove (1983*d*, p. 328). In the text I stated, "ultimately a spectral analysis of geological time is envisaged. At the time we merely noticed (see Schove, 1983*d*, p. 328, fig. 10) a wave train of about 20 m.y. in the Mesozoic." I excluded this sentence because, several years later (I think Rhodes Fairbridge first noticed it), the Kulp time scale of 1964 had been stretched; the Mid-Triassic has been redated from Kulp's date of around 205-238 m.y. in Harland (1982). In any case even in my chart the cycle looked more than 20 m.y. and 30 m.y. was suggested in 1967 (Schove, 1983*d*, p. 94). Indeed the revised climatic turning points by Frakes (1979) should now be dated roughly as follows:

Warm peaks: 570(?), 483, 448, 438, 420, 400, 238, 213, 188, 156, 105, 83, 51, and 19 m.y.
Cold peaks: 700-600 (?), 475, 443, 420, 408, 300, 275, 265, 220, 200, 175-165, 125, 92, 64, 30, and 1 m.y.

Since 125 m.y. there is a very clear 30-m.y. period. The extinction and cometary impact dates noted by Rampino and Stothers (1984) often fit the turning points but do not keep exactly in step.

REFERENCES

Anderson, R. Y., 1961, Solar-terrestrial climatic patterns in varved sediments, *New York Acad. Sci. Annals* **95**(1):424-439.
Antevs, E., 1922, *The Recession of the Last Ice Sheet in New England (i.e., Lake Hitchcock Varves)*, New York: American Geographical Society.
Ashley, G. M., 1972, *Rhythmic Sedimentation in Glacial Lake Hitchcock, Massachusetts-Connecticut*, Contribution 10, Amherst: Geology Department, University of Massachusetts.
Bain, W. C., 1976, The power spectrum of temperatures in central England, *Royal Meteorol. Soc. Quart. Jour.* **102**:464-466.
Berglund, B., and Mörner, N.-A., 1984, Late Weich-

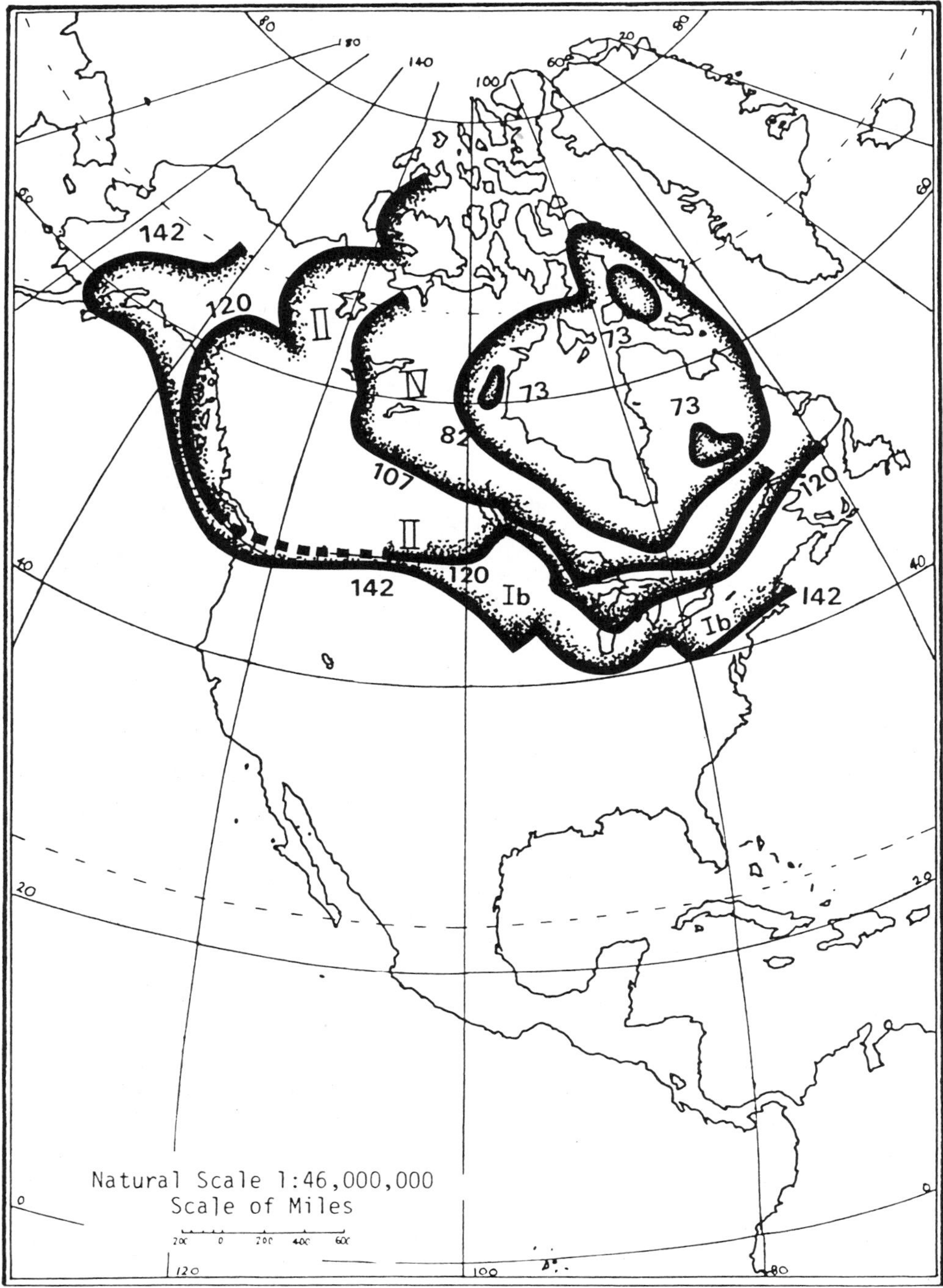

Figure 21-13. Deglaciation of North America idealized from maps by Prest, Bryson, Denton, and Hughes to illustrate Schove's belief in deglaciation in steps (see Schove, 1984*b*).

selian deglaciation and chronostratigraphy of Southern Scandinavia: Problems and present "state of the art," in N.-A. Mörner and W. Karlén (eds.), *Climatic Changes on a Yearly to Millennial Basis*, Dordrecht: Reidel, pp. 17-24.

Bruns, M.; Rhein, M.; Linick, T. W.; and Waterbolk, H. T., 1984, The atmospheric level in the 7th millennium BC, *PACT, Strasbourg* **8:**511-516.

Bryson, R. A., and Dutton, J. A., 1961, Some aspects of the variance spectra of tree rings and varves, *New York Acad. Sci. Annals* **95**(1):580-604.

Dai, N., and Chen, M., 1980, Chronology of historical aurora data in China, Korea and Japan, *Kejishiwenji* **6:**87-146 (in Chinese, translation in progress).

Eddy, J. A., 1976. The Maunder Minimum, *Science* **192:**1189-1202.

Fairbridge, R. W., 1982, The Holocene boundary stratotype: Local and global problems, *Sveriges Geol. Undersökning Arsb.* **C-794:**281-286.

Fairbridge, R. W., and Hameed, S., 1983, Phase coherence of solar cycle minima over two 178-year periods, *Astronomical Jour.* **88:**867-869.

Fisher, D. A., 1982, Carbon-14 production compared to oxygen isotope records from Camp Century, Greenland and Devon Island, Canada, *Climatic Change* **4:**419-426.

Frakes, L. A., 1979, *Climate Throughout Geological Time*, Amsterdam: Elsevier.

Gnevyshev, M. N., 1977, Essential features of the eleven-year solar cycle, *Solar Physics* **51:**175-183.

Harland, W. B., 1982, *A Geologic Time-Scale*, Cambridge: Cambridge University Press.

Jong, A. F. M. de, 1981, Natural ^{14}C Variations, Ph.D dissertation, University of Groningen.

Karlén, W., 1984, Dendrochronology, mass balance and glacier front fluctuations in northern Sweden, in N.-A. Mörner and W. Karlén (eds.), *Climatic Changes on a Yearly to Millennial Basis*, Dordrecht: Reidel, pp. 263-271.

Kristiansson, J., 1982, Varved sediments and the Swedish time scale, *Geol. Fören. Stockholm Förh.* **104:**273-275.

Larson, B. J., and Stone, B. D., 1982, *Late Wisconsin Glaciation of New England*, Dubuque, Iowa: Kendall/Hunt.

Neftel, A.; Oeschger, H.; and Suess, H. E., 1981, Secular non-random variations of cosmogenic carbon-14 in the terrestrial atmosphere, *Earth and Planetary Sci. Letters* **56:**127-147.

Niemalä, J., 1971, Quaternary stratigraphy of the clay layers between Helsinki and Hameenlinna in south Finland, *Finland Geol. Survey Bull. 259.*

Olausson, E. (ed.), 1982, The Pleistocene-Holocene boundary in south-western Sweden, *Sveriges Geol. Undersökning Arsb.* **C-794:**1-288.

Oliver, J. E., and Fairbridge, R. W. (eds.), 1987, *Encyclopedia of Climatology*, New York: Van Nostrand Reinhold.

Ottoway, B., and Ottoway, J. H., 1974, Irregularities in dendrochronological calibration, *Nature* **250:**407-408.

Parker, B. N., 1976, Global pressure variation and the 11-year solar cycle, *Meteorol. Mag.* **105:**33-44.

Pearson, G. W.; Pilcher, J. R.; and Baillie, M. G. L., 1983, High precision ^{14}C measurement of Irish Oaks to show the natural ^{14}C variations from 200 BC to 4000 BC. *Radiocarbon* **25**(2):179-186.

Rampino, M. R., and Stothers, R. B., 1984, Terrestrial mass extinctions, cometary impacts and the Sun's motion perpendicular to the galactic plane, *Nature* **308:**709-712.

Rizzo, P. V., and Schove, D. J., 1962, Early new world aurorae, 1644, 1700, 1719, *British Astron. Assoc. Jour.* **72**(8):396-397.

Schove, D. J., 1961*a*, Solar cycles and the spectrum of time, 250 BC-AD 2000, *New York Acad. Sci. Annals* **95**(1):107-123.

Schove, D. J., 1961*b*, Tree-rings and climatic chronology, *New York Acad. Sci. Annals* **95**(1):605-622.

Schove, D. J., 1963, Models of the Southern Oscillation in the 300-100 mb layer and the basis of seasonal forecasting, *Geofisica Pura e Appl.* **55:**249-261.

Schove, D. J., 1969, Reviews of Swedish palaeoclimatology: Mörner (1969), Berglund (1966), Königsson (1968), *Geog. Jour.* **135:**594-596.

Schove, D. J., 1969/1971, A varve teleconnection project, in M. Ters (ed.), *Etudes sur le Quaternaire dans le Monde*, Paris: INQUA, pp. 927-935.

Schove, D. J., 1971, Varve-teleconnection across the Baltic, *Geog. Annaler*, ser. A, **53:**214-234.

Schove, D. J., 1980, The 200-, 22- and 11-year cycles and long series of climatic data, mainly since AD 200, in *Sun and Climate*, Dordrecht: Reidel, pp. 87-100.

Schove, D. J., 1983*a*, Global oscillations and the absolute dating of varves, ice-cores and tree rings, in *Abstracts of the Second Nordic Symposium on Climatic Changes and Related Problems*, Copenhagen: Det. Danske. Met. Inst.

Schove, D. J., 1983*b*, Recent progress in dendrochronology: A review article, *Inst. Archaeology London Bull.* **20:**191-194.

Schove, D. J., 1983*c*, Sunspot, auroral, radiocarbon and climatic fluctuations since 7000 BC. *Annales Géophysique* **1**(4-5):391-396.

Schove, D. J., 1983*d*, *Sunspot Cycles*, Benchmark Papers in Geology series, vol. 68, New York: Van Nostrand Reinhold, 393p.

Schove, D. J., 1984*a*, Sunspot cycles and global oscillations, in N.-A. Mörner and W. Karlén (eds.), *Climatic Changes on a Yearly to Millennial Basis*, Dordrecht: Reidel, pp. 257-259.

Schove, D. J., 1984*b*, Varves in Canada and the USA: Their relative and absolute chronology, in W. Mahaney (ed.), *Correlation of Quaternary Chronologies*, Norwich: GeoBooks, pp. 395-408.

Schove, D. J., and Frewer, A., 1961, Tree-rings in the

Cairngorms, *Royal Soc. Forestry Scotland Jour.* **15:**63-71.

Schove, D. J.; Nairn, A. E. M.; and Opdyke, N. D., 1958, The climatic geography of the Permian, *Geog. Annaler* **40:**216-231.

Selkirk, R., 1984, Seasonally stratified correlation of the 200 mb tropical wind field to the Southern Oscillation, *Jour. Climatology* **4:**365-382.

Sonett, C. P., and Suess, H. E., 1984, Very long solar periods and the radiocarbon record, *Rev. Geophysics and Space Physics* **22**(2):239-258.

Stuiver, M., 1983, The AD record of climatic and carbon isotope change, *Radiocarbon* **25:**221.

Stuiver, M., and Quay, P. D., 1981, Atmospheric ^{14}C changes resulting from fossil fuel CO_2 release and cosmic flux variability, *Earth and Planetary Sci. Letters* **53:**349-362.

Wittmann, A., 1978, The sunspot cycle before the Maunder Minimum, *Astronomy and Astrophysics* **66:**93-97.

22: Examples and Implications of 18.6- and 11-yr Terms in World Weather Records

Robert Guinn Currie
State University of New York, Stony Brook

Abstract: Analysis of 525 air temperature and pressure records yields worldwide evidence for an 18.6-yr lunar nodal cycle, and a weaker 11-yr solar cycle, tidal forcing of the atmosphere. Waveforms of both signals exhibit bistable flip-flop phasing with respect to both geography and time. The shift in phasing with respect to time occurred principally near nodal epoch 1917.5. Similar behavior has been found for nodal- and solar-induced drought/flood in North America, South America, India, China, and Africa. Depending on phase, the solar drought/flood cycle can enhance or diminish the severity of lunar nodal drought/flood. Both terms appear to be modulated by longer-period forcing. At epochs of maxima in the lunar nodal tide (e.g., 1973.3, 1991.9) and at mid-epochs (e.g., 1982.6), the variability of weather worldwide with respect to drought/flood and consequent problems with food production should be increased but, as a result of nonstationary behavior in time and space for each region, the degree of seriousness cannot be forecast. Between epochs and mid-epochs, weather should be more normal, with optimal prospects for crop yield.

INTRODUCTION

Nearly a century ago, Sir George Darwin investigated the 18.6-yr lunar nodal oceanic tide. Later, in 1907-1909, H. E. Rawson (see Currie, 1984*c*) reported nodal variation in latitudes of subtropical anticyclone belts, as well as nodal-induced drought in South Africa and the Argentine. In the 1970s, Currie (1974, 1976) reported nodal and weaker solar cycle terms in atmospheric parameters, and in 1980 concluded that the well-known tendency for clusters of severe drought years to recur on a time scale of 20 yr in the U.S. great plains and prairies is a periodic phenomenon and therefore predictable.

The procedures used in analysis are prediction-error filtering and maximum entropy spectrum analysis, which is a small facet of probability theory:

Bayes's Theorem (Jaynes, 1979, 1983*a;* Smith and Grandy, 1985):
Laplace, Theory of Probability ca. 1810
Jeffreys, Theory of Probability ca. 1939
Shannon, Information Theory ca. 1949
Jaynes, Principle of Maximum Entropy ca. 1956

In 1939, following earlier work by Lord Keynes (1921), Sir Harold Jeffreys (1939) reinstated the views of Laplace concerning probability theory. Shannon's information theory followed (Shannon and Weaver, 1949), culminating in Jaynes's Principle of Maximum Entropy (1957), which is now widely known in economics, physics, chemistry, statistics, and philosophy (Jaynes, 1983*a*). Appendix 1 contains appropriate references.

Analysis in terms of a spectrum is necessary for establishing the likelihood of signals in data, but the extracted waveforms of such signals have proven to be much more informative. For this reason, almost all examples given in this chapter are presented in the time domain. Currie and Fairbridge (1985) and Currie (1987*a*) provide a nontechnical discussion of the procedures employed in the frequency and time domains.

While a few of the illustrations in this paper are drawn from published work, as noted, the majority are based on new analyses of the *World*

Weather Records of air temperature and air pressure as archived at the National Oceanographic and Atmospheric Administration (NOAA) National Climate Data Center in Asheville, North Carolina. Prior to signal processing all the records were reduced to unit variance which allows direct comparison of the amplitudes of both data sets.

EXAMPLES

Time series over a range of A.D. 632-1984, and spanning a cumulative 150,000 yr of data, have yielded evidence for an 18.6-yr nodal cycle and a generally weaker 11-yr solar cycle in the drought and flood records for North and South America, India, China, and Africa. Figure 22-1 illustrates the solar cycle drought and flood waveform in North America since A.D. 1600 obtained from 42 tree-ring time series (see Currie, 1984*c*, Fig. 8). The centers of solid circles denote epochs of maximum drought correlated to the solar cycle, whereas vertical bars are maxima in sunspot numbers. An important feature is that maxima in drought and sunspot numbers are not normally in phase, and the phase drifts in time. Results for the solar drought cycle in China show it has been out of phase with the solar drought cycle in North America over the past two centuries. This finding implies that a standing-wave pattern exists in the atmosphere with a node somewhere across the Pacific between China and North America (Currie and Fairbridge, 1985).

In Figure 22-1, downward-pointing arrows are epochs of maxima in the 18.6-yr lunar tide, and as Currie (1984*c*) showed, these are highly correlated with maxima in nodal drought in the western United States. These dates in maxima of the lunar nodal cycle are not those of the maximum declination but reflect also the solar

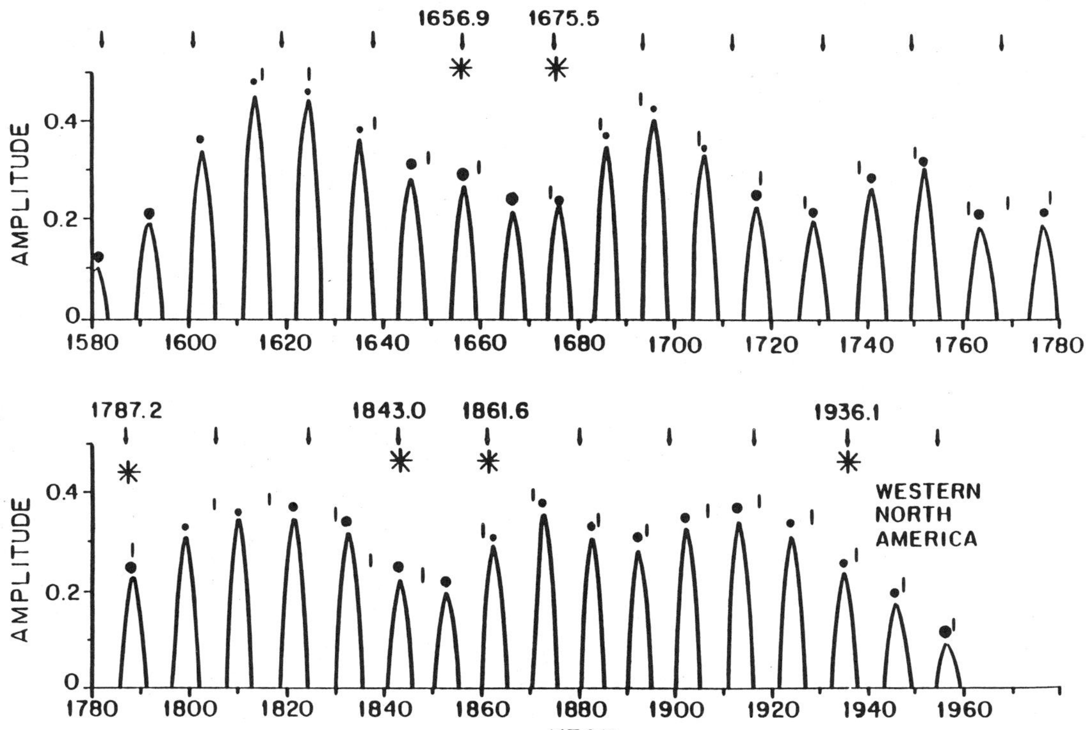

Figure 22-1. Positive polarity waveform of solar-cycle-induced drought maxima in western North America. Symbols and dates are explained in the text (from Currie, 1984*c*, Fig. 8).

tide and are the maxima in the tide-raising force. At epoch 1936.1, nodal and solar drought maxima closely aligned, as indicated by an asterisk, and produced the worst drought of our century, the Dust Bowl. Currie, Hameed, and Handler (1985) found the same lunar nodal and solar cycle effects in U.S. corn production. In addition, Currie (1985, 1987*a*) reported it in wheat yield, acres of sowed crops abandoned, and U.S. production of horses, hogs, chickens (including eggs), and other livestock. For hogs there were three independent time series, all yielding similar results. These results established that, depending on their phase relationships, the solar wave modulates the severity of the 18.6-yr wave.

Figure 22-2 illustrates the 18.6-yr nodal waveform for drought and flood in northern China the past five centuries (Currie and Fairbridge, 1985). From 1500 to 1680, drought maxima were in phase with tidal maxima. The signal was then below noise level (I suspect this is a data-related problem) until 1800 but had experienced a bistable flip-flop—that is, a 180° phase reversal—because, as shown in the lower panel, at epoch 1805.8, the nodal flood maximum had come into phase with the tidal maximum. Another flip-flop occurred at epoch 1843.0, and the latest reversal occurred at 1936.1. Other work (Currie, 1983, 1984*a*; Hameed and Currie, 1985; Currie, 1987*b*) shows that the same reversal, and in the same sense, occurred in tree-ring data from South America, flood-area data from India, and Nile flood data in Africa, but each of them occurred one cycle earlier at epoch 1917.5. Therefore, the nodal drought maxima in these regions have been out of phase with the western United States for most of the twentieth century. Nodal drought maxima in the northeastern United States have also been out of phase with the western U.S. for most of the twentieth century (Currie, 1987*d*).

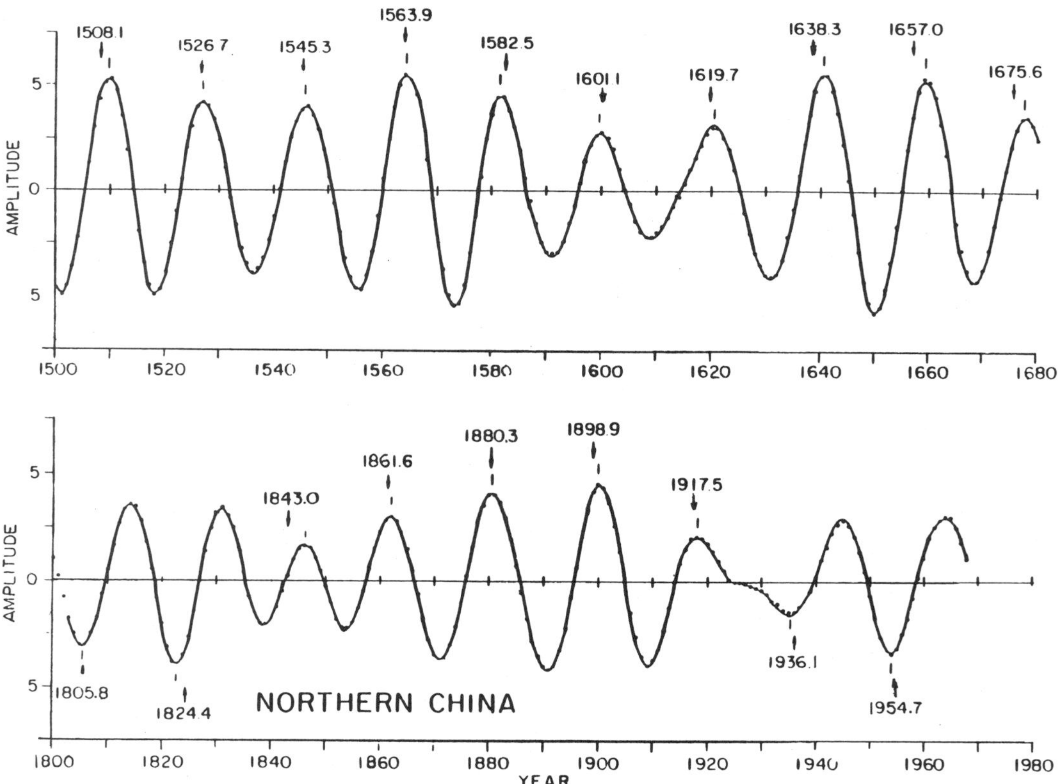

Figure 22-2. Epochs of 18.6-yr nodal drought maxima in northeastern China (from Currie and Fairbridge, 1985).

As is discussed later, these discoveries in each region have serious implications for food production worldwide. Depending on phasing, critical times are at epochs or mid-epochs of the nodal cycle. The most recent nodal epoch occurred at 1973.3, and the next will occur at 1991.9. The most recent mid-epoch was 1982.6, and the next will be 2001.6.

We now turn to examples of 18.6- and 11-yr terms in world records that establish these two phenomena as worldwide. Evidence for the two terms in Earth rotation and in air temperature, air pressure, sea surface temperature (SST), and height of sea level for certain regions has been published.

Out of 330 worldwide air temperature records (based on meteorological data) whose length was 60 yr or more, evidence for the nodal and solar terms was found in 74% and 49% of the records respectively. Results for 195 pressure records was about 10% less successful. Figure 22-3 displays average spectra for pressure in South America and for temperature in Africa, Europe, and the USSR. For illustrative purposes, only those stations that yielded evidence for the 11-yr term were included in the average, so

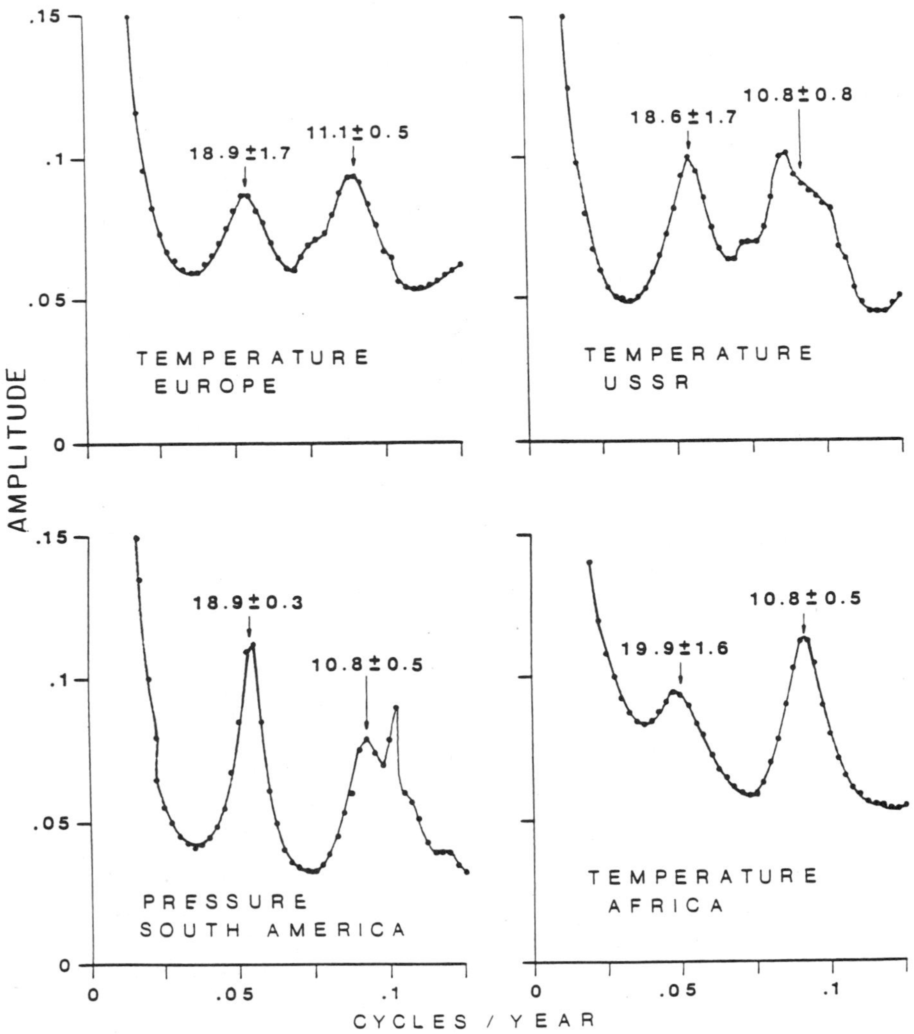

Figure 22-3. Ensemble average spectra of air temperature or pressure for three continents.

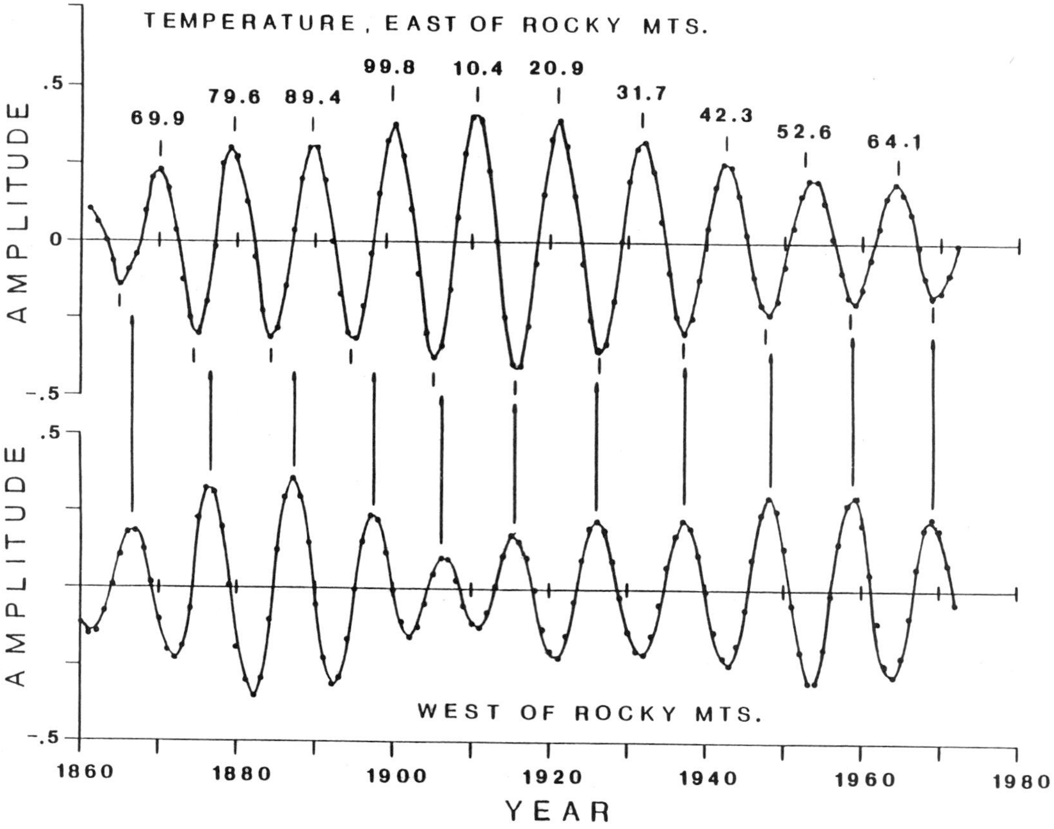

Figure 22-4. Mixed and out-of-phase relationships for solar cycle term in air temperature west and east of the Rocky Mountains.

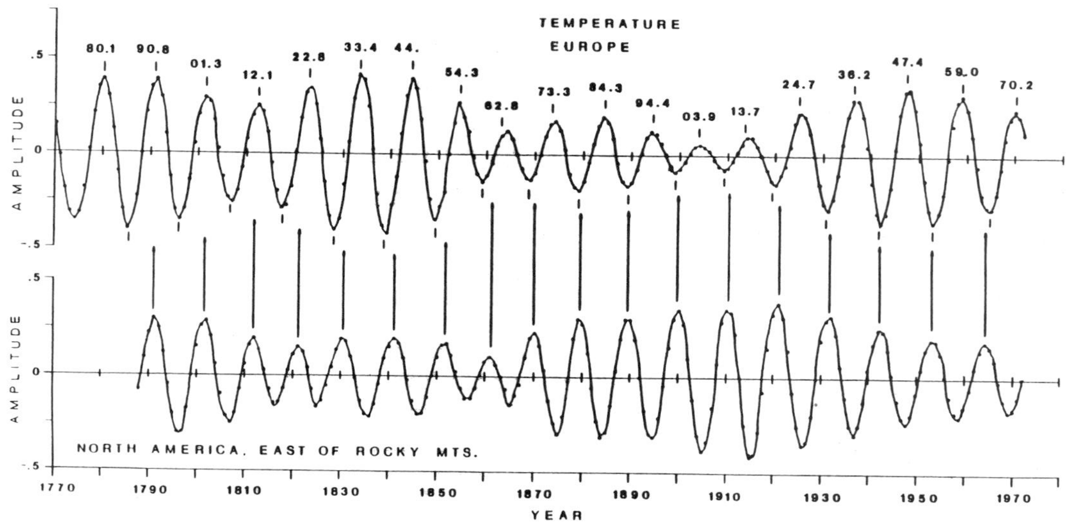

Figure 22-5. Mixed and out-of-phase relationships for solar cycle term in air temperature for the eastern United States and Europe.

this term appears larger, whereas in fact, it is smaller. Papers have been published on the two terms in North American and Japanese data (Currie, 1979, 1981*b*, 1981*d*, 1982). We shall now discuss solar cycle waveforms and later give examples for the 18.6-yr wave.

Solar Component

Figure 22-4 displays the average solar cycle air temperature waves east and west of the Rocky Mountains. From the upward-pointing vertical arrows, it is apparent that since 1900, the signals are 180° out of phase and that these mountains correspond approximately to a node in the global wave pattern. Prior to 1900, a systematic mixed phase difference of about 3 yr exists. This is the first indication, among many in the world records, that at about the year 1900, atmospheric dynamics experienced changes that profoundly influenced the subsequent global pattern of 18.6-yr-induced drought and flood.

Figure 22-5 shows the solar wave for Europe vis-à-vis North America east of the Rockies. Since about 1870, the solar cycle component in the temperature data from Europe is out of phase with eastern North America, so a node must have existed somewhere over the Atlantic Ocean. Prior to 1870, a systematic mixed phase difference exists. Now, where over the North Atlantic does this node lie?

Figure 22-6 provides a clue. The lowest panel is for two stations in the Azores Islands; these are out of phase with Europe in the middle panel. The node thus lies in the eastern Atlantic quite close to the continent of Europe. The upper two panels for Europe and the western USSR are closely in phase with a mean discrepancy of only 0.2 ± 0.5 years for seven epochs.

We now have evidence for two nodes, one along the Rockies and another in the Atlantic near Europe. Figure 22-7 shows results for central versus eastern USSR. Without going into details, these results suggest that a third node exists over western Asia, although it may have a progressive component and be broadly diffuse in longitude. In summary, for the twentieth century, the mean wave number is three for the solar cycle thermal wave in the middle latitudes of the Northern Hemisphere. The evidence also clearly shows that the wave in the atmosphere is quasistanding. This wave number is the simplest triple form of the Rossby wave, commonly expressed by the Northern Hemisphere jet stream.

There are not enough land stations to establish the distribution of nodes in any detail, but on available evidence, the pattern of these nodes is not simple. Figure 22-8 displays results for two locations in Greenland. These show a bistable switch near 1920. Ignoring polarity, however, the mean discrepancy between the stations is only -0.2 ± 0.8 years for seven solar epochs.

Figure 22-9 presents examples of the solar term for Bombay and Madras, India, using pressure data. Note the high amplitude at the mid-nineteenth century followed by rapid attenuation to low amplitude and a phase switch at 1900. India displays great complexity because other stations, near the two shown, show waves that are quite different. Such differences are characteristic of oceanic tides and apparently this holds true in some regions for atmospheric tides.

In summary, by inference from ground-based data, an 11-yr thermal tide seems to be present in the atmosphere, although it has not been measured in situ. Currie (1979) estimated that a modulation in solar luminosity of 0.1% would suffice to induce the observed phenomenon.

Lunar Component

By Newton's law and the celestial mechanics of Laplace, the 18.6-yr atmospheric tidal constituent is the twelfth largest in equilibrium theory; moreover, it is enhanced above equilibrium by orders of magnitude in some regions (Currie, 1982; Campbell, 1983). This tide also appreciably modulates the amplitude and phase of three principal short-period progressive tides as well as the principal 13.7-day standing wave (Currie, 1984*b*). Few in atmospheric science seem to have realized the potential importance of these phenomena as regards weather.

Currie (1979) accidentally detected the nodal

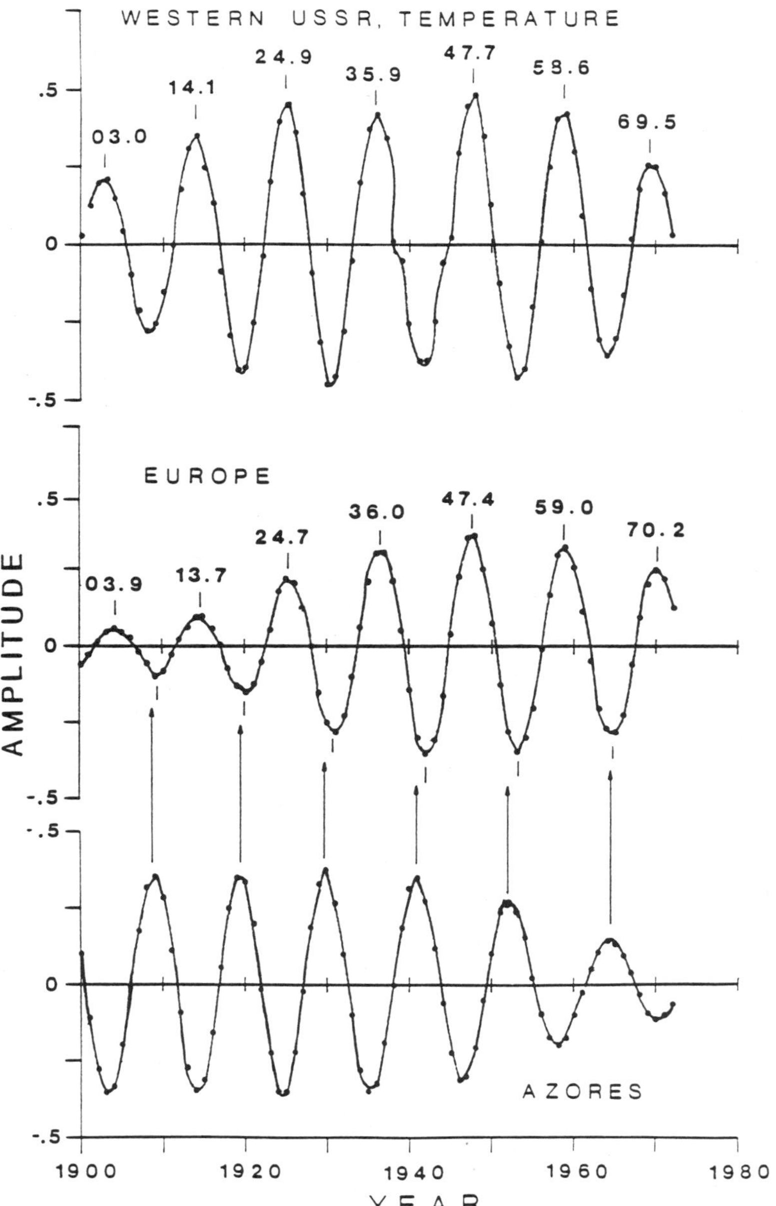

Figure 22-6. Comparison of phase relationships for solar cycle term in air temperature for the Azores and Europe and for Europe vis-à-vis the western USSR.

term in North American temperature records and later reported it in detail (Currie, 1981*d*). Phase measurements on the temperature series showed the response of the atmosphere to tidal forcing is quite different in western Canada compared to the Midwest and northeastern portions of the continent. This difference is reflected at ground level by another remarkable phenomenon, a nodal-induced drought and flood pattern that in western Canada is out of phase with that in the western United States (Currie, 1984*c;* Hameed and Currie, 1986).

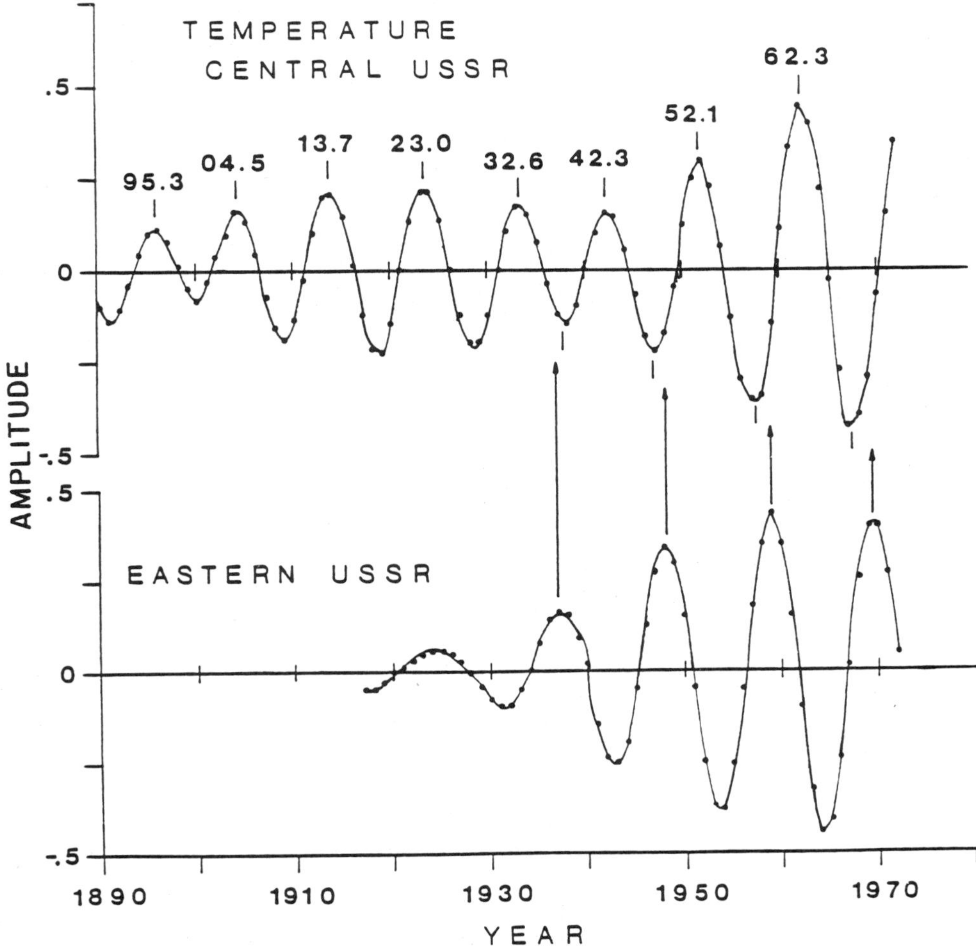

Figure 22-7. Comparison of phase relationships for solar cycle term in air temperature for central and eastern USSR.

Figure 22-10 shows new results for the nodal air temperature (from meteorological data) in South America at Santiago, Chile, near 40° latitude and at Punta Arenas near the Strait of Magellan. One may note a systematic attenuation in amplitude from 1880 to 1918 followed by a flip-flop. The upper panel is for a Peruvian station and two interior locations, but they did not experience the bistable switch. These three records are unfortunately very short; indeed, probably one-third of the 525 records are effectively less than 70 yr in length.

Figure 22-11 displays the nodal pressure term at Santiago, Chile, west of the Andes Mountains versus Buenos Aires, Argentina, to the east. For Santiago, note the large amplitude prior to about 1910 followed by a flip-flop after 1918 and much smaller amplitude. Buenos Aires does not show the bistable switch.

Figure 22-12 shows that nodal pressure at Rio de Janeiro, Brazil, is out of phase with stations in the interior. Because Rio is in phase with Buenos Aires, a node must exist somewhere to the west. Such phase reversals would be expected for a standing wave in the atmosphere. Why Santiago, Chile, to the west of the Andes displays a bistable flip-flop in time is more perplexing.

Moving to another continent, Figure 22-13 displays nodal pressure for Kimberley in the

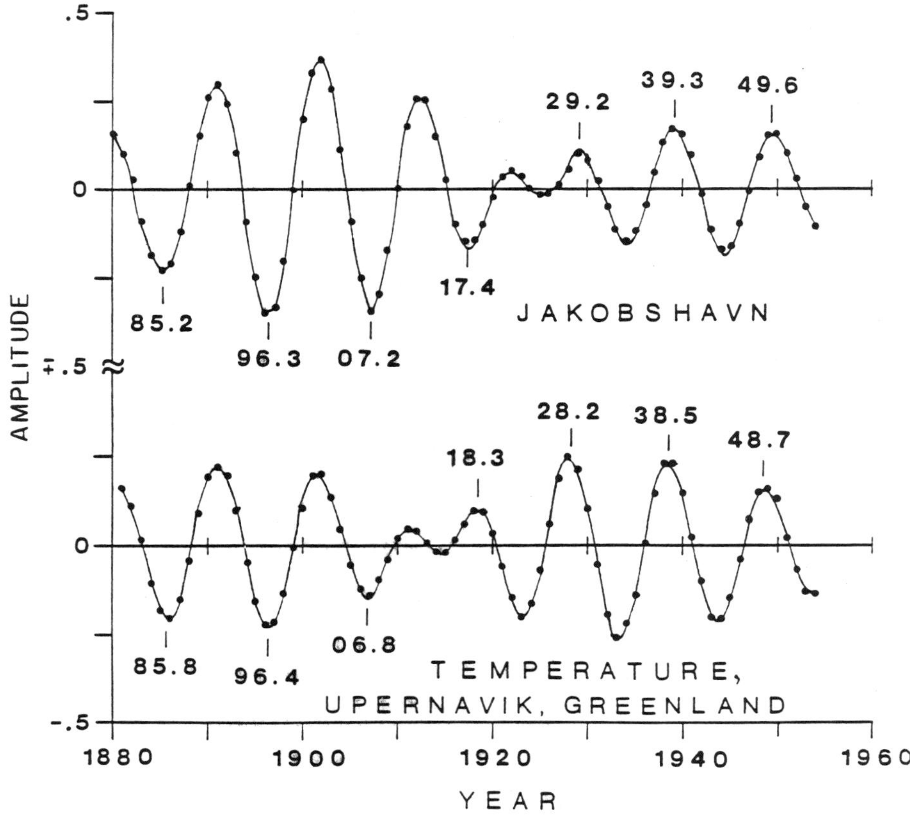

Figure 22-8. A display of bistable polarity switches in solar cycle air temperature for two locations in Greenland.

interior high plains of southern Africa. We note it is out of phase with other locations, mainly coastal, in the same region. Figure 22-14 shows the phasing for nodal temperature from seven stations in Africa where minimum temperature is in phase with maximum tidal force.

The circumstances are reversed for some regions of India, as Figure 22-15 illustrates, because maximum temperature is in phase with maximum lunar tide. As noted earlier, the solar term in India is complex and perplexing in terms of geography. This is also apparent for the nodal term in Figure 22-16 for Columbo and Trincomalee in Sri Lanka off the southern tip of the subcontinent. Although these two cities are not far separated, they display remarkably different behavior. We are reminded again of oceanic tides in this respect.

Most of the proxy data on drought and flood related to the nodal and solar cycles are four to five centuries in length; in one instance, for the River Nile, the span of records is nearly continuous for about ten centuries (Currie, 1987*b*). The flood-area index for India is, in contrast, very short, as Figure 22-17 illustrates in the bottom panel. Large values are large-scale annual monsoonal floods, whereas small values correspond to widespread drought. Bhalme and Mooley (1981), and later Campbell (1983), fit 22-yr and 18.6-yr waves, respectively, to the India flood index, and both were found to be significant.

Currie (1984*a*) reinterpreted the data as a flip-flop at epoch 1918 as shown in the top panel of Figure 22-17. Since then, maximum flood is approximately in phase with tide, the latest epoch being 1973.3. This interpretation implies that India should experience widespread

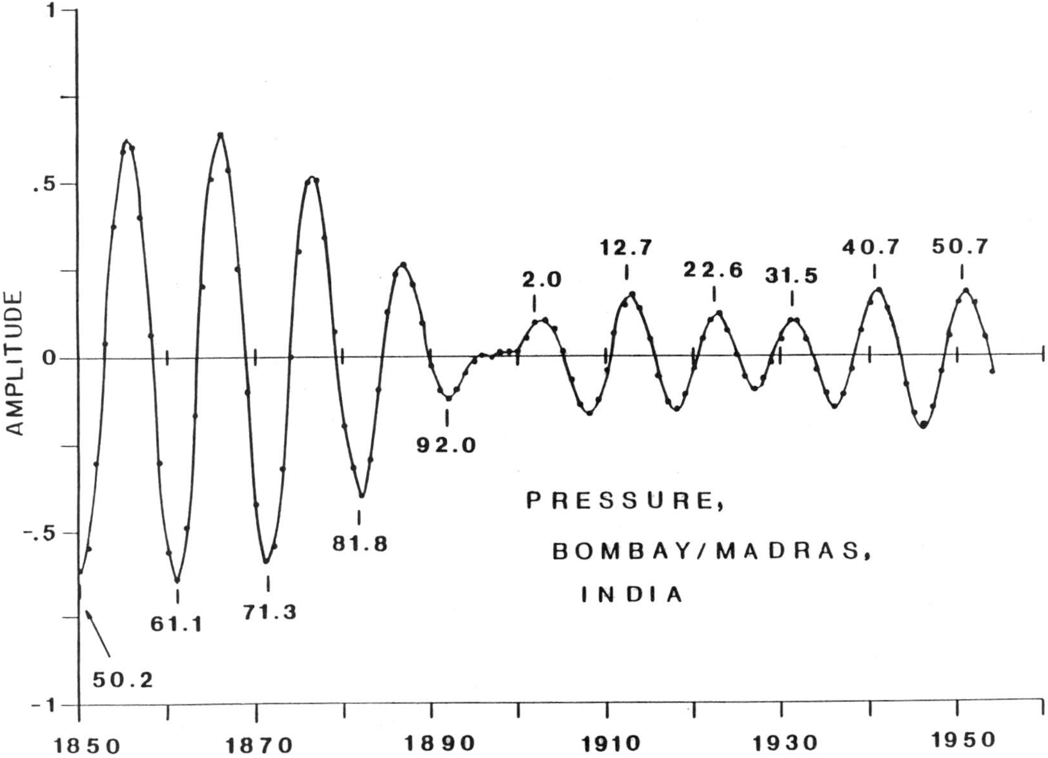

Figure 22-9. A presentation of bistable polarity switches in solar cycle air pressure for two locations in India.

dryness in an interval ±2-3 yr centered at mid-epoch 1982.6. We cannot forecast individual years in the interval nor how serious the drought will be. The previous mid-epoch occurred at 1964; in 1965 and 1966 the monsoon failed, there was a crisis in world grain markets, and the United States gave, as a gift, 20% of its entire wheat crop to India for two years running. The 1983 mid-epoch drought in India was mild, whereas drought was severe over the African continent.

Moving to Australia, Figure 22-18 gives the temperature wave for the coastal cities of Adelaide and Sydney, which are in phase with tide. The record from Cloncurry is from the continental interior and is out of phase. We earlier saw an example of this situation in southern Africa.

Figure 22-19 shows results for several stations in New Zealand and New Guinea. The epoch 1918 is again the pivot because polarity flip-flops in a bistable fashion. We have numerous other examples but insufficient space to mention them. Suffice it to say the evidence in toto indicates that the 18.6- and 11-yr phenomena are worldwide. Moreover, in the spectral range 0.1-0.05 cycle/yr, the atmosphere exhibits two oscillations that are characteristic of electronic parametric amplifiers. One of my colleagues in theoretical physics considered the first-cut dynamics so trivial that he worked briefly on the equations of motion. Within a week he told me the solution does involve a class of mathematical functions (Mathieu functions) that do exhibit bistable modes of oscillation when coefficients are subjected to periodic forcing (Stoker, 1950).

The upper panel of Figure 22-20 shows the ensemble-averaged nodal wave for 50 out of 60 tree-ring chronologies in what is termed the *Arizona network*. This network includes large portions of the western United States and the northern border region of Mexico. It shows that

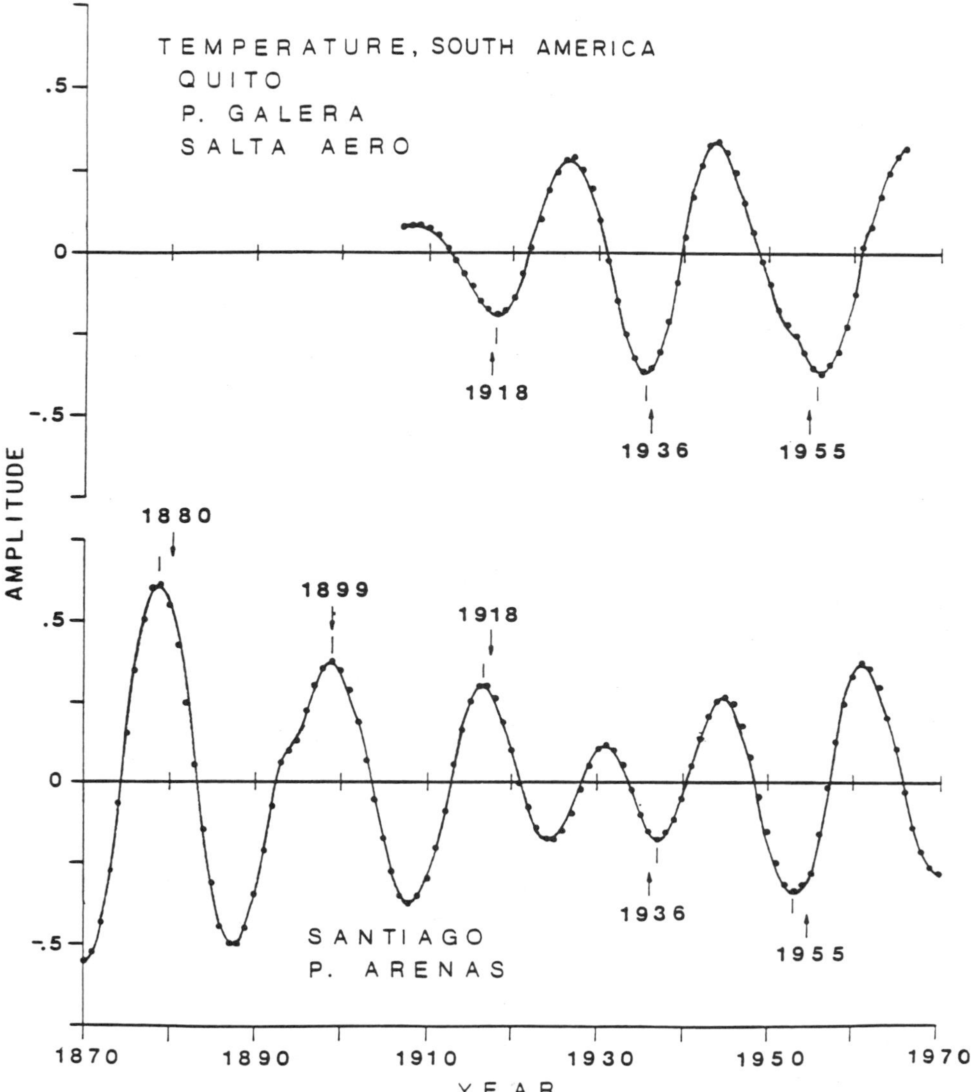

Figure 22-10. An example of the 18.6-yr wave in air temperature for two regions in South America, one of which displays a bistable phase switch.

since 1800, nodal drought epochs have been closely in phase with maxima in the nodal tide with a mean discrepancy of 0 ± 1.3 yr. A mysterious phase discontinuity near 1800 is indicated by an asterisk. Nodal drought in western Canada is out of phase with the tide shown in Figure 22-20, and there is no 1800 discontinuity (Currie, 1984*c*; Hameed and Currie, 1986).

IMPLICATIONS

Borchert (1971) extensively documented the serious impact on the U.S. economy of the prolonged widespread droughts of the 1910s, 1930s, and 1950s and warned that another serious clustering of droughts might be imminent, a warning echoed by other agronomists

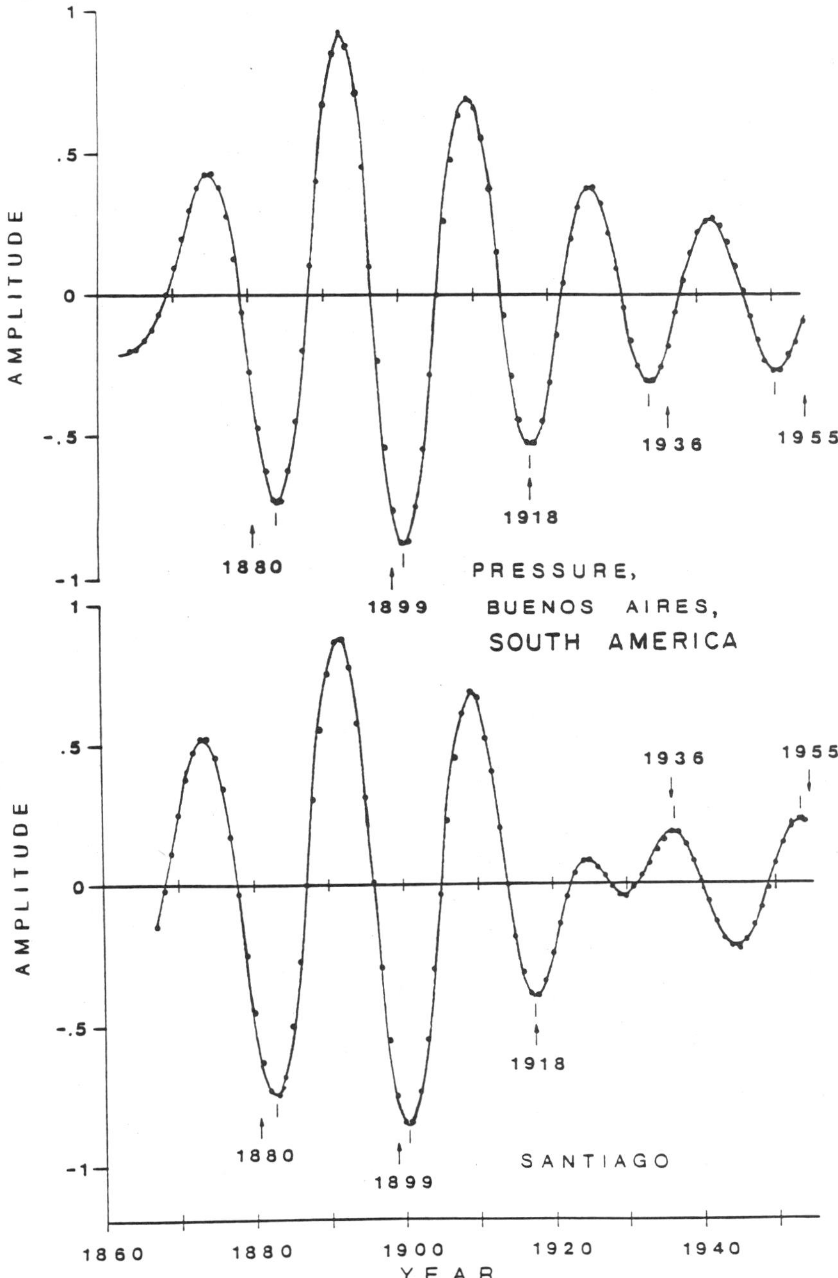

Figure 22-11. An example of the 18.6-yr wave in air pressure for two regions in South America, one of which displays a bistable phase switch.

(Text continues on page 394.)

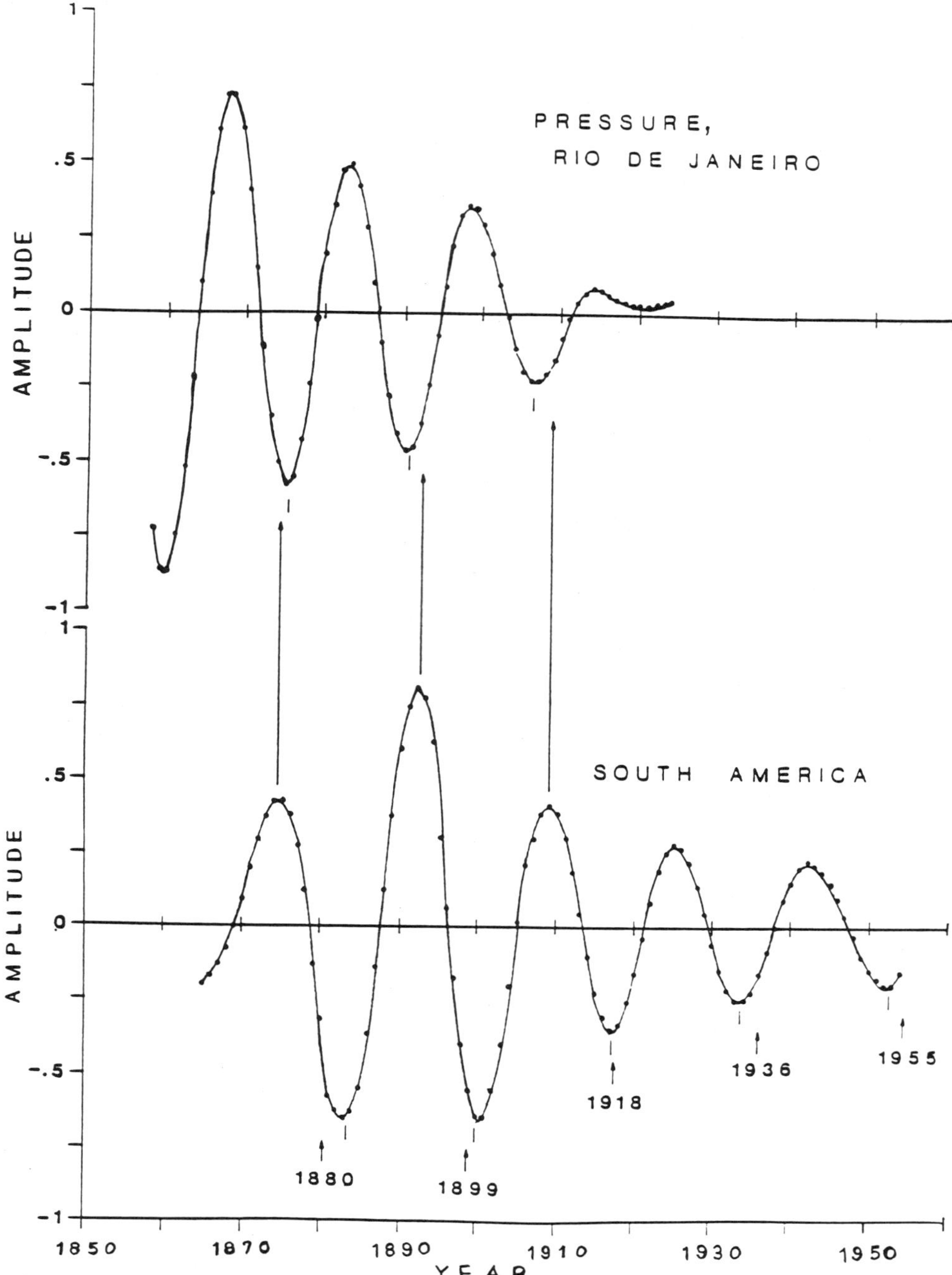

Figure 22-12. A display of the out-of-phase relationship between air pressure at Rio de Janeiro and other South American stations.

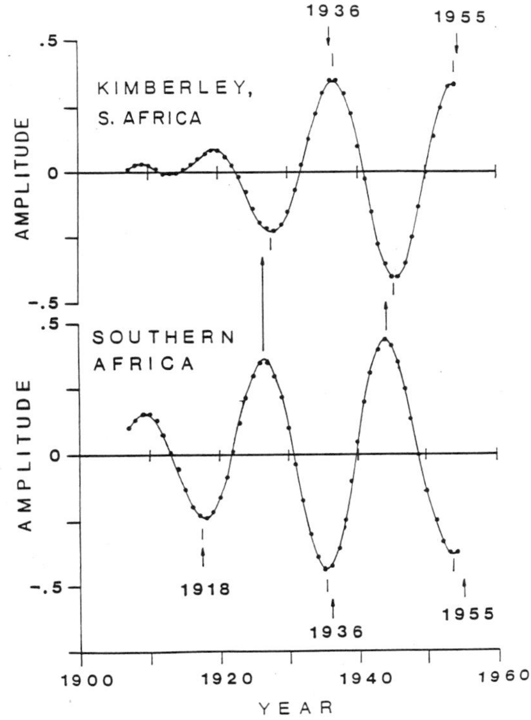

Figure 22-13. A display of the out-of-phase relationship between air pressure at Kimberley in the interior and coastal sites in southern Africa.

Figure 22-14. Lunar nodal 18.6-yr air temperature wave for some sites in Africa.

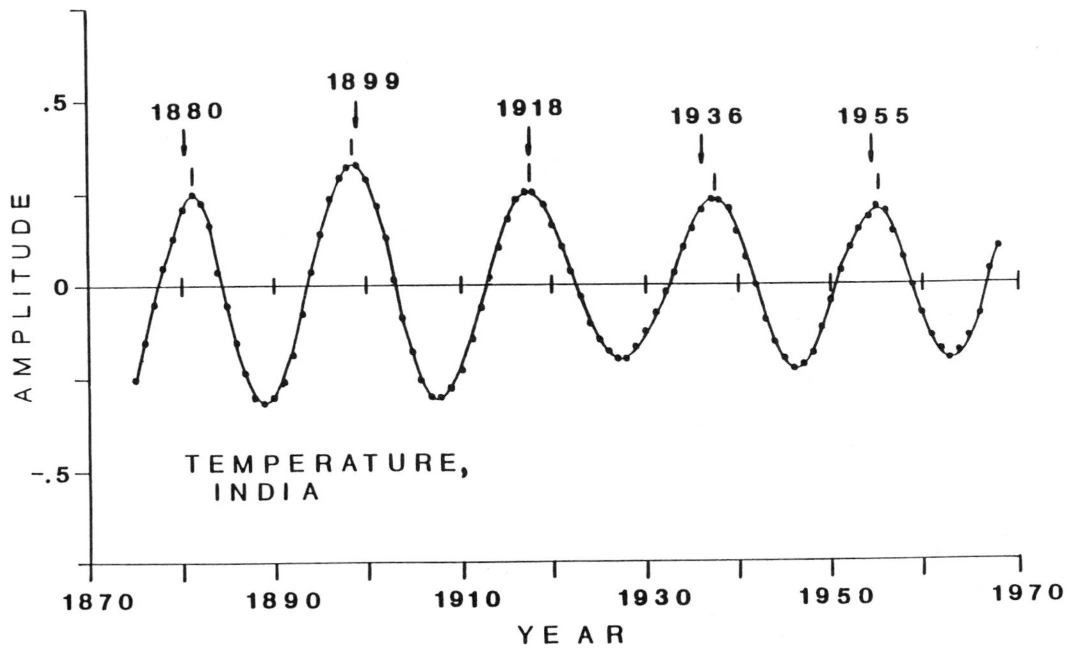

Figure 22-15. Lunar nodal 18.6-yr air temperature wave for some sites in India.

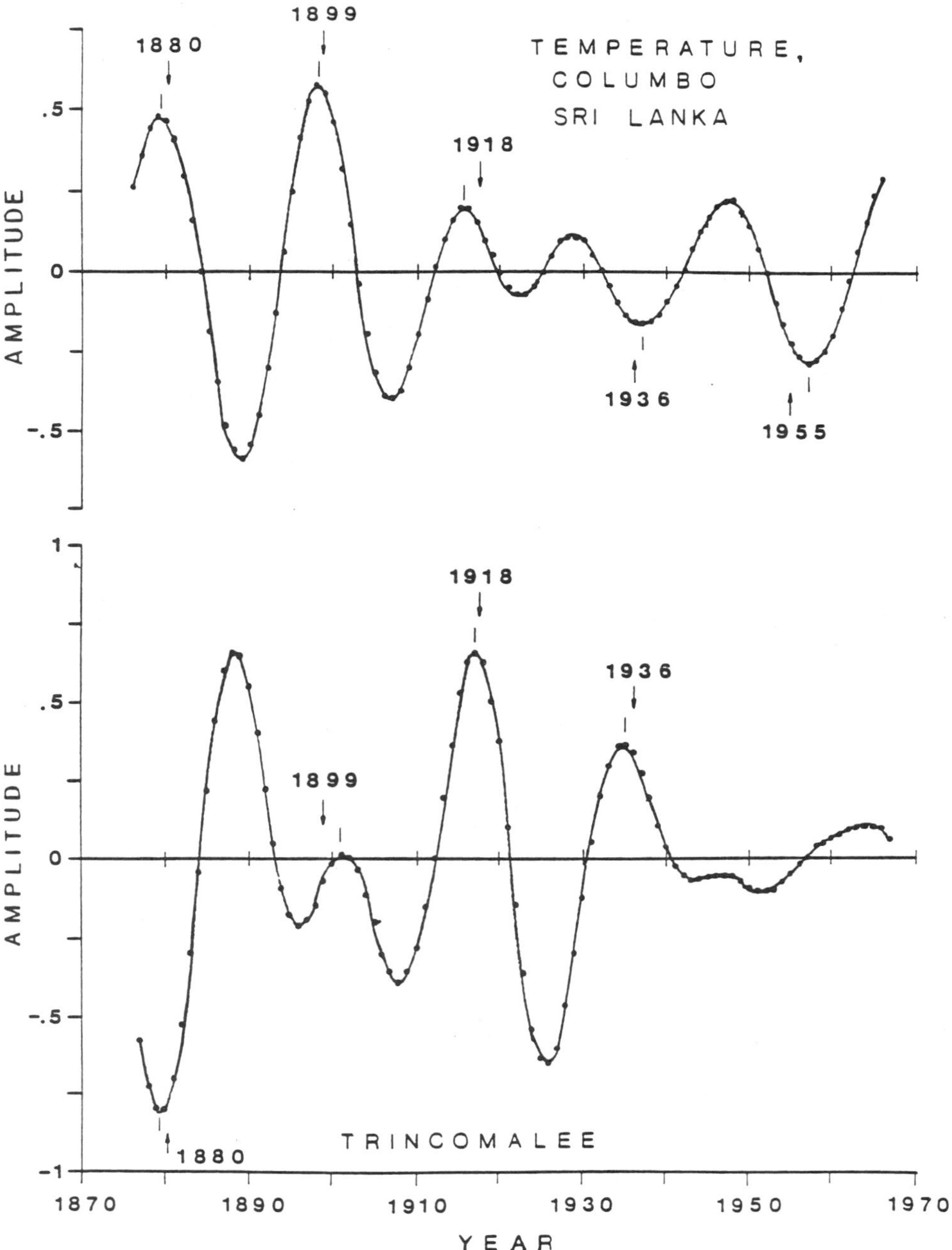

Figure 22-16. A comparison of the lunar nodal air temperature wave for two cities in Sri Lanka.

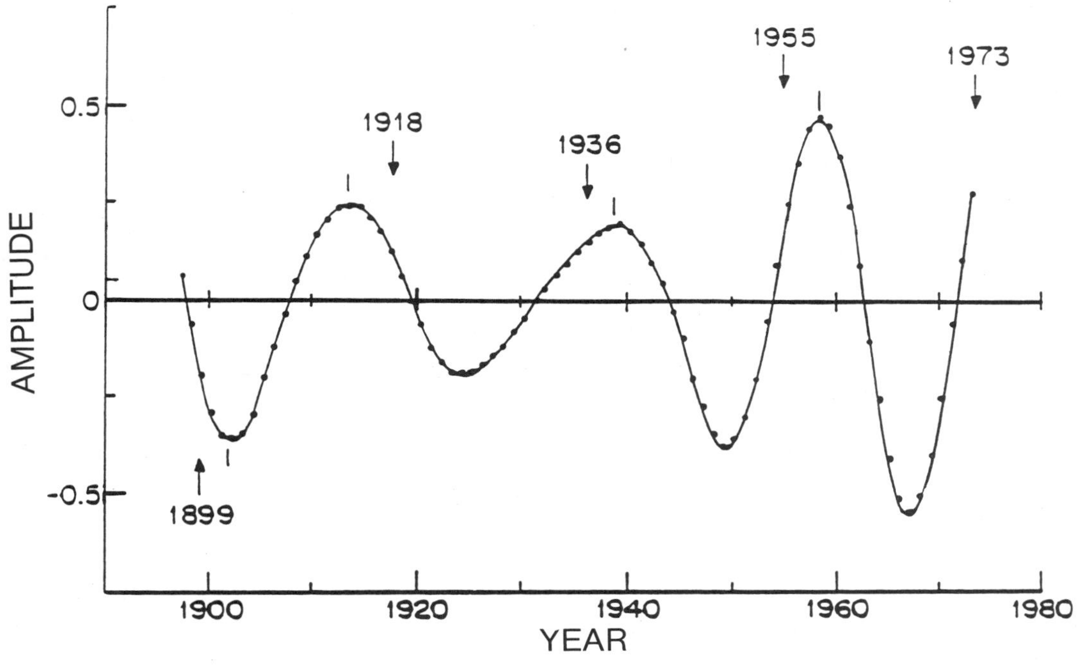

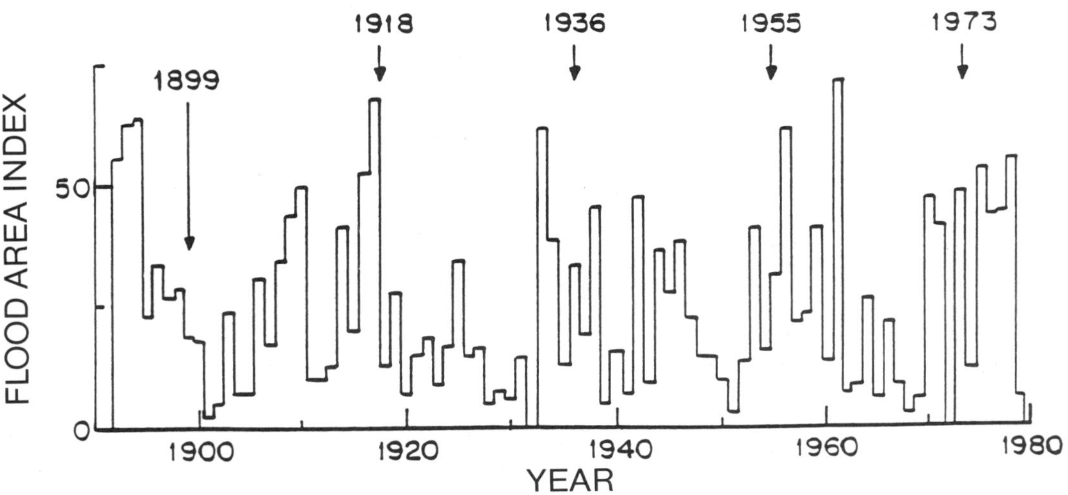

Figure 22-17. Results for a flood-area index in India. (From Currie, 1984*a*, Fig. 2)

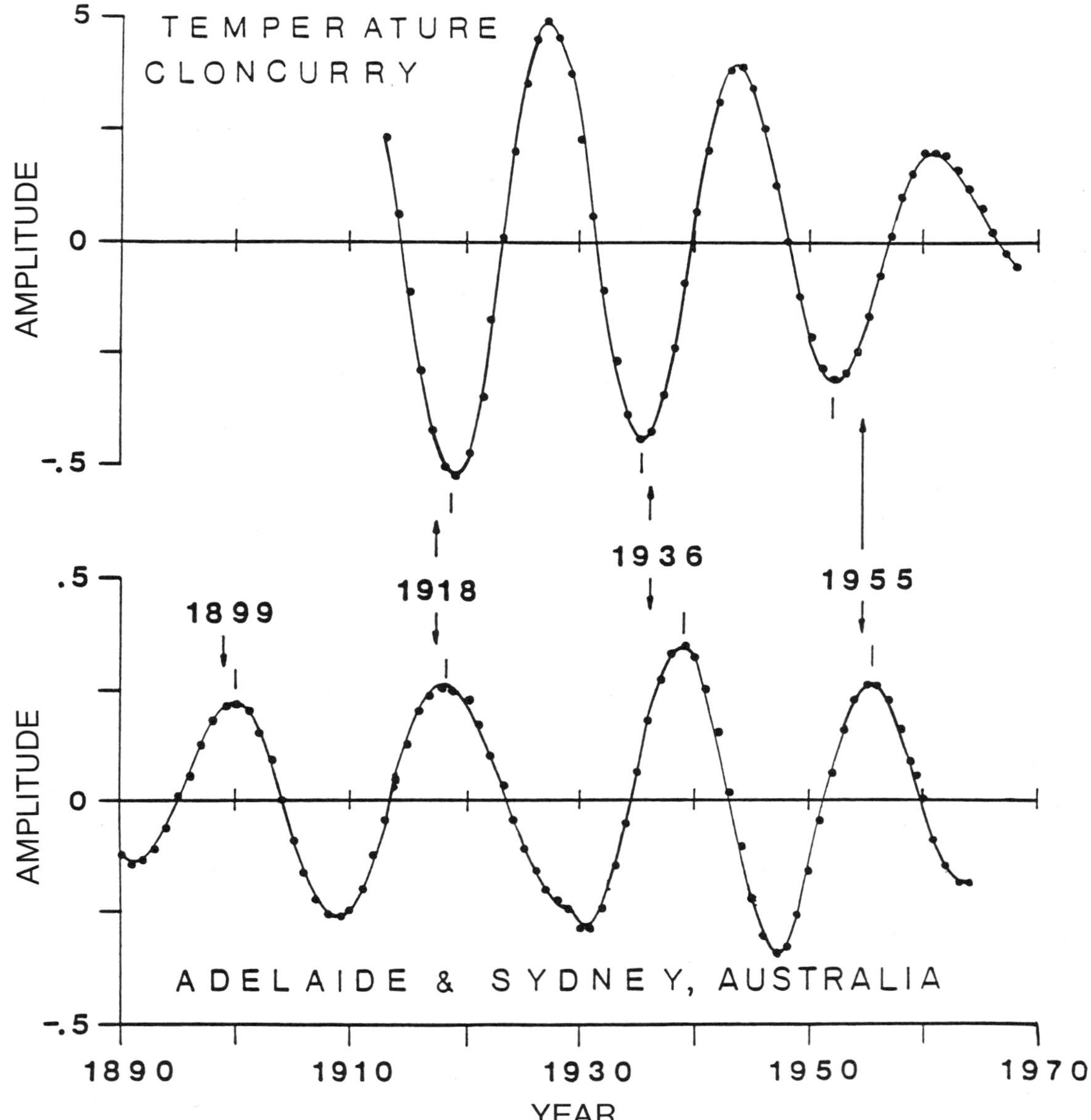

Figure 22-18. An example of out-of-phase relationship between air temperature at Cloncurry in the interior and two coastal cities in Australia.

in the Midwest. This drought actually began in 1970 in northern Mexico and southwest Texas (Thompson, 1973), advanced northward, expanding in space with growing severity, and abruptly ended in 1977 (Rosenberg, 1978). These droughts of record have represented a catastrophic change in the resource base that has directly affected agriculture and indirectly stimulated changes in patterns of migration, expenditures on water development, and a variety of other actions particularly important to people living in the Great Plains and prairies (see Borchert, 1971; Currie, 1984*b*, 1984*c*).

Let us digress a moment because, indeed,

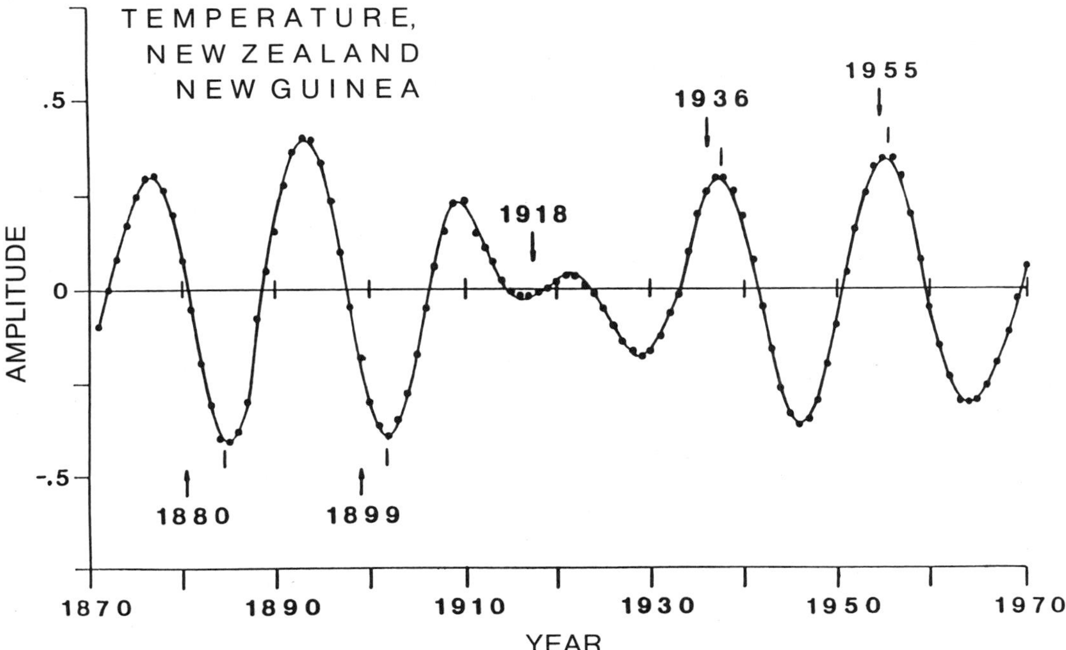

Figure 22-19. Examples of bistable polarity switches in air temperature for locations in New Zealand and New Guinea.

there is a cycle in the economy of the United States, and in other nations, a period of roughly 20 yr, named the Kuznets cycle in honor of Simon Kuznets, Nobel Laureate in economics; selected references are given in Appendix 2. Currie, Hameed, and Handler (1985), Currie and Hameed (1986), and Currie (1987*a*) concluded that the modulation of agricultural output by the lunar nodal tide is likely to be a major determinant of Kuznets cyclic behavior in economics.

We also think modulation of agricultural output by the 10-to-11-yr wave is likely to be a major determinant of Jevons' cyclic behavior in economics (the so-called harvest cycle associated with the nineteenth-century economist W. S. Jevons). Lord Keynes (1936) defended this hypothesis in his celebrated 1936 work *General Theory of Employment, Interest, and Money* and was always opposed to dogmatic approaches to science that overemphasize axioms and postulates and ignore ordinary thought.

Keynes (1939, p. 562) dismissed the methods of G. Yule, E. Slutsky, and J. Tinbergen (low-order auto- and multivariate correlation and regression modeling in economics to explain business cycles) as "puzzles for children where you write down your age, multiply, add this and that, subtract something else, and eventually end up with the number of the Beast in Revelation" (see Currie, 1987*a*).

Mitchell (1927, p. 265), a distinguished American economist, sounded the alarm much earlier by writing:

> Once started upon this career of transforming time series into new shapes for comparison, statisticians have before them a limitless field for the exercise of ingenuity. They are beginning to think of the original data, coming to them in a shape determined largely by administrative convenience, as concealing uniformities which it is theirs to uncover. With more emphasis upon statistical technique than upon rational hypothesis, they are experimenting with all sorts of data, recast in all sorts of ways. Starting with two series having little resemblance in their original shape, they can often transmute one series into "something new and strange," which agrees closely with the

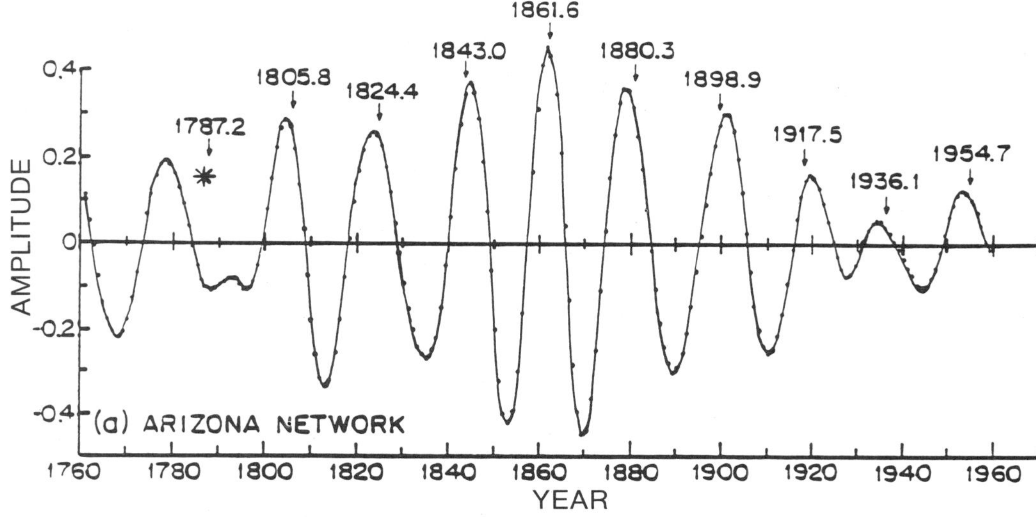

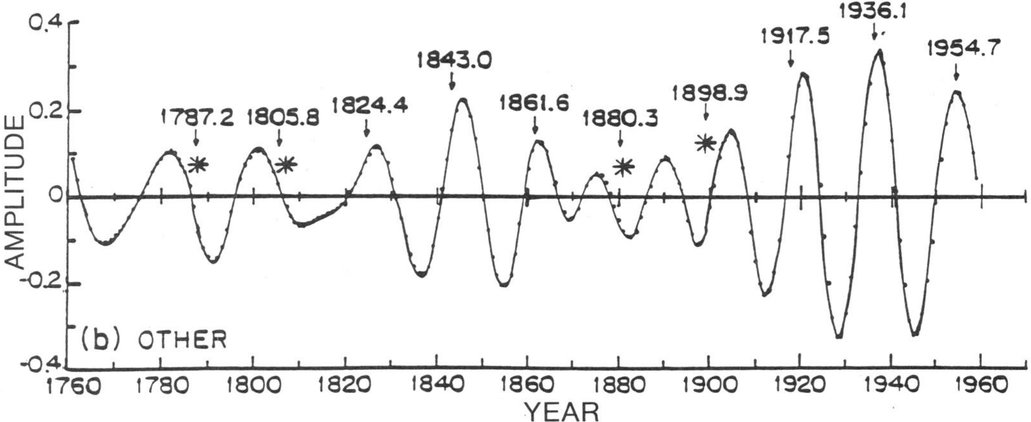

Figure 22-20. Lunar nodal 18.6-yr wave for tree-ring chronologies in western North America. (From Currie, 1984*c*, Fig. 4)

other series. In work of this type, they rely upon the coefficient of correlation to test the degree of relationship between the successive transformations.

Mitchell (1927, pp. 266-270) then gives examples of searches made for the "number of the Beast in Revelation." The methods employed by scientists noted in Appendix 2 were thus aggressively disputed by orthodox mathematical statisticians for decades until the economic shocks of the 1970s caused P. Volcker (1978), chairman of the Federal Reserve Board, to announce the rediscovery of the business cycle (Currie, 1987*a*).

Returning now to the main theme, the droughts of the 1930s struck when all geographical regions, except Europe, possessed net grain surpluses, and despite the population explosion that began after World War II and the gradual emergence of North America as grain supplier to the world, the 1950s drought was less severe, particularly in the Corn Belt region, so food scares did not occur. Another factor is that development of the major remaining world croplands was underway. For example, a single area in western Siberia the size of Argentina was put into food production in a short time span of 4 yr.

The 1960s were a close race between food production and population increase. Only one world food crisis occurred and that happened near the nodal mid-epoch 1964.0 when world food reserves fell to only 59 days (Brown, 1975). The principal factor in this world crisis was two consecutive failures of the Indian monsoon in 1965 and 1966. Crop shortfalls also happened in the USSR and China and a dramatic famine in the Indonesian Islands in 1964 (Borgstrom, 1966; Paddock and Paddock, 1967; Cochrane, 1969; Marx, 1975; Akroyd, 1975).

By 1970, almost all geographical regions were grain deficit and imported a portion of their food from North America, Argentina, and Australia. That year (1970) foreshadowed the nodal epoch 1973.3 and passed with food reserves (as days of world consumption) standing at 90 days. By 1974, these reserves had fallen to less than 30 days (Brown, 1975). In November 1974, some 130 sober nations gathered at an emergency meeting in Rome to discuss the world food crisis (Staff of the New York Times, 1975; Vicker, 1975). Fifty million acres of idled U.S. cropland, the only major reserves left, were put into production but could not stem the tide. In three out of four years, as a result of domestic political pressure, the United States limited exports (Brown, 1975, 1976).

With assigned probability greater than .9, Currie (1984*b*) forecast a return of prolonged widespread drought to the continental interior of North America, considered as a whole, at the next epoch 1991.9. The converse is that mid-epoch 1982.6 witnessed widespread flooding, which is not as serious an adverse factor in food production as drought. For the other regions—the environs of northern China, India, Africa, and South America in middle latitudes—enhanced prospects existed for widespread dryness in mid-epoch 1982.6. Large areas of northeast Brazil experienced severe drought in the early 1980s, which led to famine conditions in 1983-1984 (McDowell, 1984). A prolonged drought over about 80% of the African continent created, in large measure, a calamity that broke into public view in late 1984. According to *Time* magazine (November 26, 1984), 31 nations and more than 150 million people were threatened by starvation, but in 1985 and 1986 the rains returned. The prolonged dryness in India and China was apparently mild.

What about forecasts for a small region such as the U.S. Corn Belt? In an unpublished study (L. M. Thompson, pers. comm. 1981, 1984) on historical corn yields, probability of .75 for a fair or poor corn crop in 4-yr intervals was assigned, which encompassed all four nodal epochs for the twentieth century (1918, 1936, 1955, and 1973). In the intervening 16-yr intervals, the probability was .75 for good or bumper crops. The 1980 corn crop was adversely affected by a short drought, but farm groups in the Midwest were being advised in March 1981 (L. M. Thompson, pers. comm.) that, despite 1980, probability still favored good or bumper crops for 1981 with assigned probability .75. This was good advice because both 1981 and 1982 were bumper. The 1983 corn crop was adversely affected by a short drought in the critical months of July and August, but the probability for a good or bumper crop in 1984 was .75, as it was for the years 1985 to 1988. The crops in 1984, 1985, and 1986 were good or bumper.

Since early 1985, attention has shifted from climatic data discussed in this chapter to records directly related to economics. Shown in Figure 22-21 is evidence for the two terms in yield of corn for Iowa; Figure 22-22 is the nodal waveform for corn in Missouri. Currie (1985, 1987*a*) notes that only corn and wheat data have been analyzed on a state-by-state basis but that evidence exists for both terms in national aggregate time series of oats, barley, rye, buckwheat, flaxseed, sugar beets, Irish potatoes, sweet potatoes, and sugar cane. He also noted that the number of acres of planted wheat abandoned is cyclicly modulated. Figure 22-23 shows two examples for winter wheat in Colorado and Texas.

Currie (1985, 1987*a*) also showed that the phenomena are evident in U.S. livestock production. Figure 22-24 display spectra for three independent series on hog production where

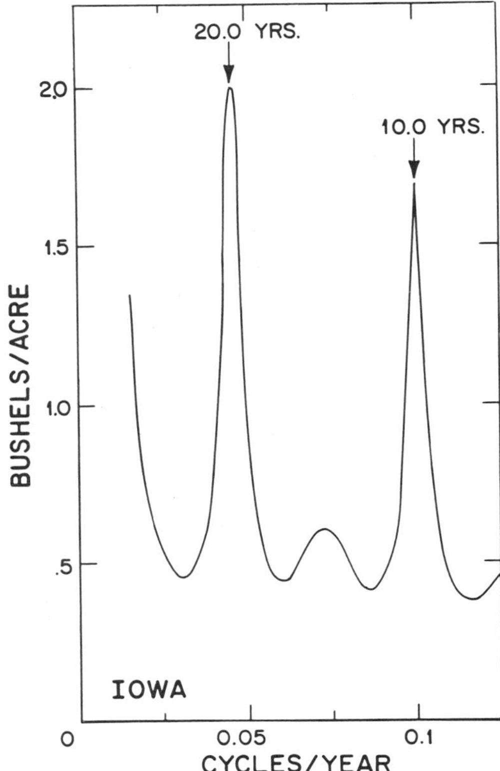

Figure 22-21. Spectrum for Iowa corn production. (From Currie, Hameed, and Handler, 1985)

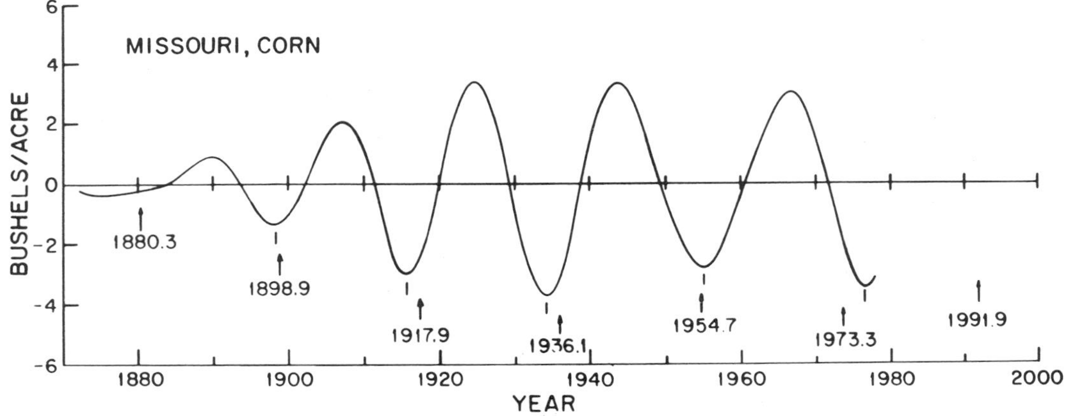

Figure 22-22. Lunar nodal wave for Missouri corn production. (From Currie, Hameed, and Handler, 1985)

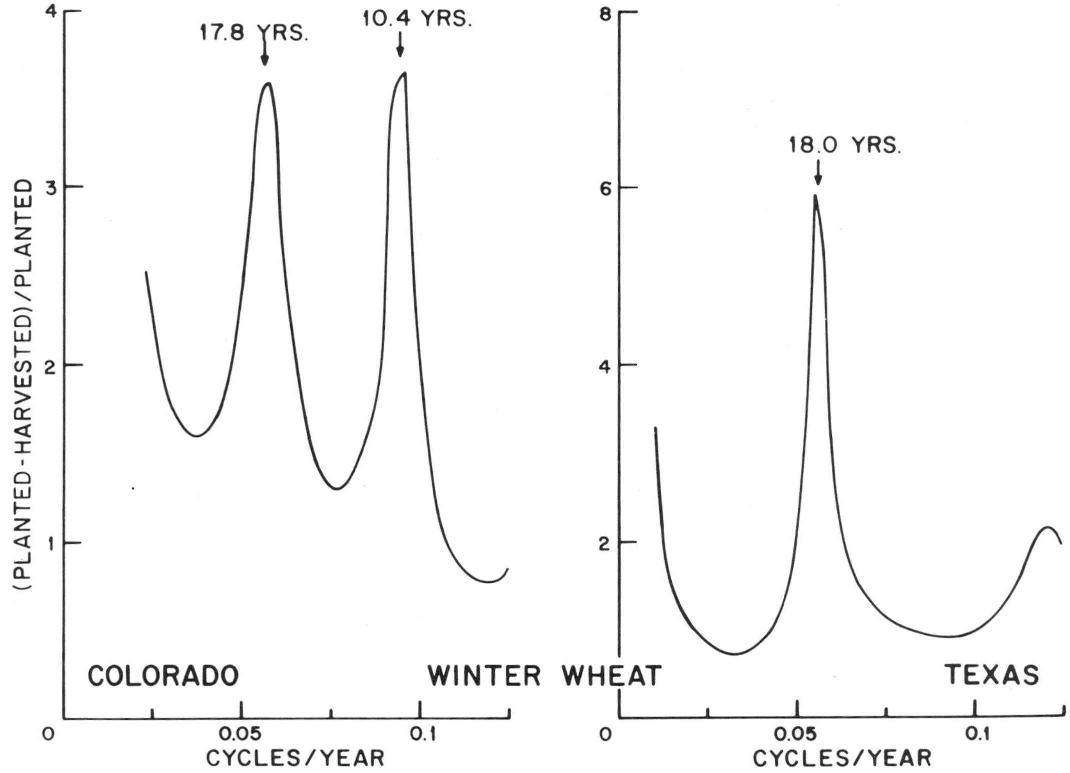

17.8 YRS.
10.4 YRS.
18.0 YRS.
(PLANTED-HARVESTED)/PLANTED
COLORADO
WINTER WHEAT
TEXAS
CYCLES/YEAR
CYCLES/YEAR

Figure 22-23. Spectra for percentage of abandoned planted winter wheat in Colorado and Texas. Note that the scale for Texas is double that for Colorado. (From Currie, 1987*a*)

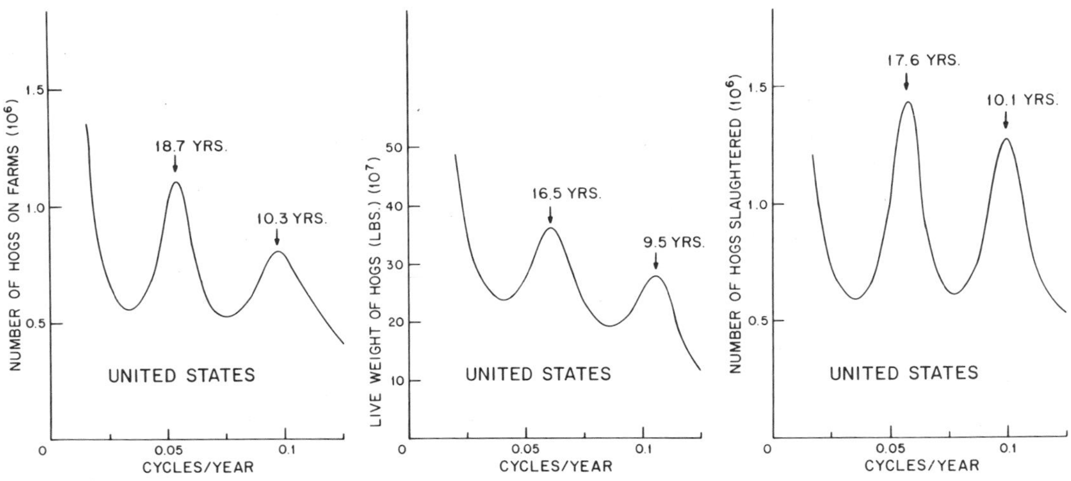

Figure 22-24. Spectra for three independent records of U.S. hog production. (From Currie, 1987*a*)

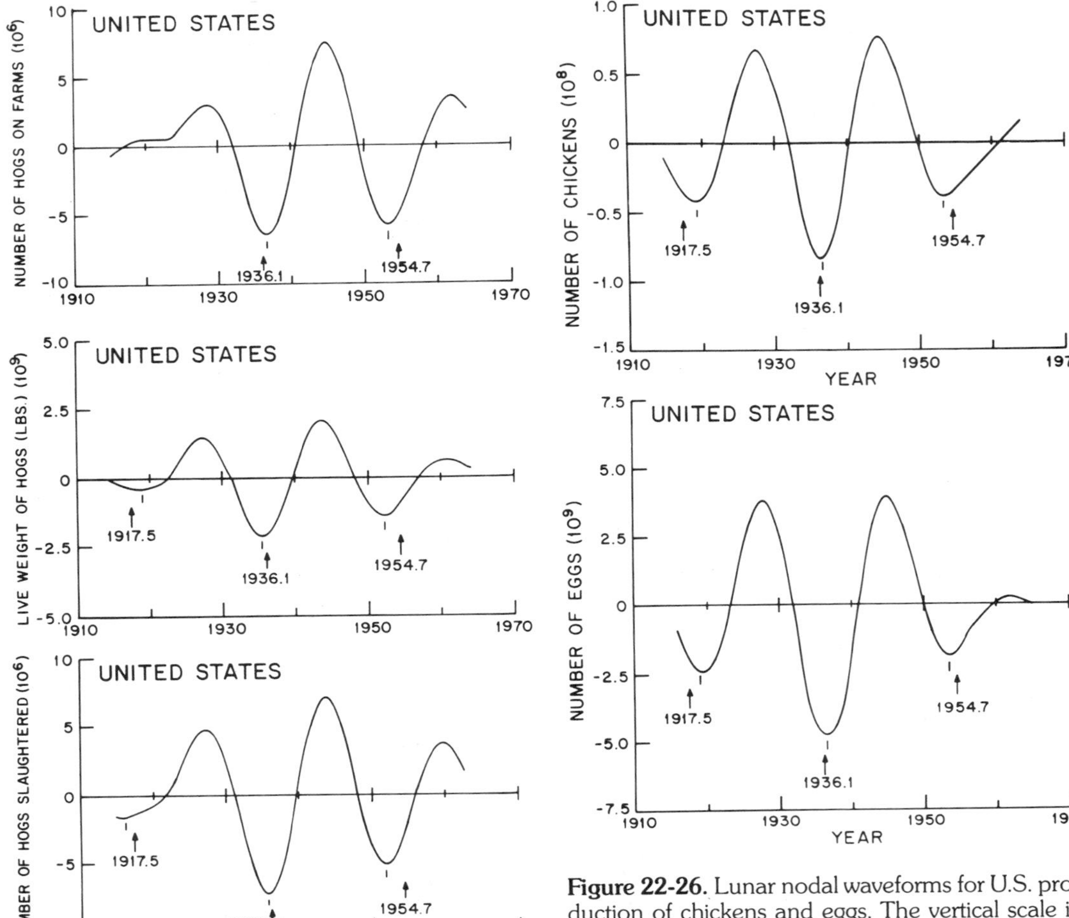

Figure 22-26. Lunar nodal waveforms for U.S. production of chickens and eggs. The vertical scale is amplitude. (From Currie, 1987*a*)

Figure 22-25. Lunar nodal waveforms for U.S. hog production. The vertical scale is amplitude. (From Currie, 1987*a*)

the phenomena appear in each instance. The 18.6-yr nodal waveforms for these three spectra are shown in Figure 22-25. We see that minima in hog production, like minima in corn production, are correlated with maxima in epochs of the nodal tide. Figure 22-26 displays the 18.6-yr wave for the U.S. production of chickens and eggs; both are in phase with production of hogs and corn.

We have presented evidence that the 18.6-yr lunar-nodal and 11-yr solar cycle phenomena are of a global character. Two caveats are offered. First, with regard to drought, individual years cannot be forecast, but the occurrence of extreme 1-yr anomalies does not obviate a forecast over a nominal interval ±2-3 yr centered at epochs or mid-epochs. Second, the degree of severity of nodal drought varies in space and time from one epoch to the next, and this has been established experimentally for North America (Borchert, 1971). But whereas the degree of severity averaged over a nominal interval cannot be properly forecast, the solar drought term is certainly one mechanism causing variable severity from one epoch to the next. Currie (1987*a*) shows that for the twentieth century, the economies of Corn Belt states were most seriously affected in the 1930s and 1950s when both waves in corn production were closely in phase.

REFERENCES

Aykroyd, W. R., 1975, *The Conquest of Famine,* New York: Reader's Digest Press.

Bell, E. P., 1981, Predominant periods in the time series of drought area index for the western high plains AD 1700 to 1962, in S. Sofia (ed.), *Variations of the Solar Constant,* Washington, D.C.: NASA.

Bhalme, H. N., and Mooley, D. A., 1981, Cyclic fluctuations in the flood area and relationship with the double (Hale) sunspot cycle, *Jour. Appl. Meteorology* **20:**1041-1059.

Borchert, J. R., 1971, The dust bowl of the 1970s, *Assoc. Amer. Geogr. Annals* **61:**1-22.

Borgstrom, G. A., 1966, *The Hungry Planet: The Modern World at the Edge of Famine,* New York: Macmillan.

Brown, L. R., 1975, The world food prospect, *Science* **190:**1053-1059.

Brown, L. R., 1976, World population trends: Signs of hope, signs of stress, *Worldwatch Paper 8,* Washington, D.C.: Worldwatch Institute.

Campbell, W. H., 1983, Possible Tidal Modulation of the Indian Monsoon Onset, Ph.D. dissertation, University of Wisconsin, Madison.

Campbell, W. H.; Blechman, J. B.; and Bryson, R. A., 1983, Long-period tidal forcing of Indian monsoon rainfall: An hypothesis, *Jour. Appl. Meteorology* **22:**289-296.

Cochrane, W. W., 1969, *The World Food Problem: A Guardedly Optimistic View,* New York: Crowell.

Currie, R. G., 1974, Solar cycle signal in surface air temperature, *Jour. Geophys. Research* **79:**5657-5660.

Currie, R. G., 1976, The spectrum of sea level from 4 to 40 years, *Royal Astron. Soc. Geophys. Jour.* **46:**513-520.

Currie, R. G., 1979, Distribution of solar cycle signal in surface air temperature over North America, *Jour. Geophys. Research* **84:**753-761.

Currie, R. G., 1980, Detection of the 11-yr sunspot cycle signal in earth rotation, *Royal Astron. Soc. Geophys. Jour.* **61:**131-139.

Currie, R. G., 1981*a,* Solar cycle signal in earth rotation: Nonstationary behavior, *Science* **211:**386-389.

Currie, R. G., 1981*b,* Solar cycle signal in air temperature in North America: Amplitude, gradient, phase and distribution, *Jour. Atmos. Sci.* **38:**808-818.

Currie, R. G., 1981*c,* Amplitude and phase of the 11-yr term in sea level: Europe, *Royal Astron. Soc. Geophys. Jour.* **67:**547-556.

Currie, R. G., 1981*d,* Evidence for 18.6 year signal in temperature and drought conditions in North America since A.D. 1800, *Jour. Geophys. Research* **86:**11055-11064.

Currie, R. G., 1982, Evidence for 18.6 year term in air pressure in Japan and geophysical implications, *Royal Astron. Soc. Geophys. Jour.* **69:**321-327.

Currie, R. G., 1983, Detection of 18.6-year nodal induced drought in the Patagonian Andes, *Geophys. Research Letters* **10:**1089-1092.

Currie, R. G., 1984*a,* On bistable phasing of 18.6 year induced flood in India, *Geophys. Research Letters* **11:**50-53.

Currie, R. G., 1984*b,* Evidence for 18.6 year lunar nodal drought in western North America during the past millennium, *Jour. Geophys. Research* **89:**1295-1308.

Currie, R. G., 1984*c,* Periodic (18.6-year) and cyclic (11-year) induced drought and flood in Western North America, *Jour. Geophys. Research* **89:**7215-7230.

Currie, R. G., 1985, The Trinity wave in climate and economics, Paper presented at the Fifth Workshop on Maximum Entropy and Bayesian Methods in Applied Statistics, University of Wyoming, Laramie, August. (Reprints available on request.)

Currie, R. G., 1987*a,* Climatically induced cyclic variations in United States crop production: Implications in economic and social science, in G. Erickson and C. R. Smith (eds.), *Maximum Entropy and Bayesian Methods in Science and Engineering,* Cambridge: Cambridge University Press, in press.

Currie, R. G., 1987*b,* On bistable phasing of 18.6-year induced drought and flood in the Nile records since A.D. 650, *Jour. Climatology,* in press.

Currie, R. G., 1987*c,* Is Fourier's theorem physically valid? paper presented at Seventh Workshop on Maximum Entropy and Bayesian Methods, Seattle University, Seattle, August.

Currie, R. G., 1987*d,* Periodic 18.6-year signal in precipitation data in the northeastern United States, *Jour. Climatology,* in press.

Currie, R. G., and Fairbridge, R. W., 1985, Periodic 18.6-year and cyclic 11-year induced drought and flood in northeastern China, *Quat. Sci. Rev.* **4:**109-134.

Currie, R. G., and Hameed, S., 1986, Climatically induced cyclic variations in United States corn yield and possible economic implications, *Proceedings of Canadian Hydrology Symposium No. 16-1986,* Regina, June, pp. 661-674.

Currie, R. G.; Hameed, S.; and Handler, P., 1985, Cyclic variations in United States corn production, paper presented at the Federal Reserve Bank, Minneapolis, May. (Reprints available on request.)

Hameed, S., 1984, Fourier analysis of Nile flood levels, *Geophys. Research Letters* **11:**843-845.

Hameed, S., and Currie, R. G., 1985, An analysis of long term variations in the flood levels of the Nile river, *Proceedings, 3rd Conference on Climate Variations and Symposium on Contemporary Climate,* Los Angeles, January, pp. 68-69.

Hameed, S.; Yeh, W. M.; Li, M. T.; Cess, R. D.; and Wang, W. C., 1983, An analysis of periodicities in the 1470 to 1974 Beijing precipitation record, *Geophys. Research Letters* **10**:436-439.

Hameed, S., and Currie, R. G., 1986, Cyclic variations in Canadian and United States drought, *Proceedings of Canadian Hydrology Symposium No. 16-1986*, Regina, June, pp. 113-122.

Jeffreys, H., 1939, *Theory of Probability*, Oxford: Oxford University Press.

Keynes, J. M., 1939, Professor Tinbergen's method, *Economic Jour.* **49**:558-568.

Landsberg, H. E., and Kaylor, R. E., 1977, Statistical analysis of Tokyo winter temperature approximations, 1443-1970, *Geophys. Research Letters* **4**:105-107.

Libby, L. M., 1983, *Past Climates: Tree Thermometers, Commodities, and People*, Austin: University of Texas Press.

Lisitzin, E., 1974, *Sea-Level Changes*, New York: Elsevier.

Loder, J. W., and Garrett, C., 1978, The 18.6-year cycle of sea surface temperature in shallow seas due to variations in tidal mixing, *Jour. Geophys.* Research **83**:1967-1970.

Marx, H. L., Jr. (ed.), 1975, *The World Food Crisis*, New York: Wilson.

McDowell, E., 1984, Brazil: Famine in the backlands, *The Atlantic Monthly* **253**(March):22-28.

Mori, Y., 1981, Evidence of an 11-year periodicity in tree-ring series from Formosa related to the sunspot cycle, *Jour. Climatology* **1**:345-353.

Paddock, W., and Paddock, P., 1967, *Time of Famines: America and the World Food Crisis*, New York: Little Brown.

Rosenberg, N. J., ed., 1978, *North American Droughts* (AAAS Selected Symp. 15), Boulder, Colo.: Westview.

Staff of the New York Times, 1975, *Give Us This Day . . . A Report on the World Food Crisis*, New York: Arno Press.

Stockton, C. W.; Mitchell, J. M., Jr.; and Meko, D. M., 1983, A reappraisal of the 22-year drought cycle, in B. M. McCormac (ed.), *Weather and Climate Responses to Solar Variation*, Boulder: Colorado Associated Universities Press, pp. 507-515.

Stoker, J. J., 1950, *Nonlinear Vibrations in Mechanical and Electrical Systems*, New York: Interscience.

Thompson, L. M., 1973, Cyclical weather patterns in the middle latitudes, *Jour. Soil and Water Conserv.* **28**:87-89.

Tyson, P. D., 1980, Temporal and spatial variation of rainfall anomalies in Africa south of latitude 22° during the period of meteorological record, *Climatic Change* **2**:363-371.

Tyson, P. D., 1981, Atmospheric circulation variation and the occurrence of extended wet and dry spells over Southern Africa, *Jour. Climatology* **1**:115-130.

Vicker, R., 1975, *This Hungry World*, New York: Scribner's.

Vines, R. G., 1980, Analysis of South African rainfall, *South African Jour. Sci.* **76**:404-409.

Vines, R. G., 1982, Rainfall patterns in the western United States, *Jour. Geophys. Research* **87**:7303-7311.

Woodworth, P. L., 1985, A worldwide search for the 11-year solar cycle in mean sea level records, *Royal Astron. Soc. Geophys. Jour.* **80**:743-755.

APPENDIX 1: SIGNAL PROCESSING

Bayes, T., 1783, An essay toward solving a problem in the doctrine of chances, *Royal Soc. London Philos. Trans.*, pp. 330-418.

Burg, J. P., 1967, Maximum entropy spectral analysis, paper presented at the Thirty-Seventh Annual International Meeting of the Society of Exploration Geophysicists, Oklahoma City.

Burg, J. P., 1968, New analysis technique for time series data, paper presented at the NATO Advanced Study Institute on Signal Processing.

Burg, J. P., 1975, Maximum Entropy Spectral Analysis, Ph.D. dissertation, Stanford University, California.

Childers, D. G., 1979, *Modern Spectrum Analysis*, vol. 1, New York: IEEE Press.

Haykin, S. (ed.), 1979, *Nonlinear Methods of Spectral Analysis, Topics in Applied Physics*, vol. 34, Berlin: Springer-Verlag.

Jaynes, E. T., 1957, Information theory and statistical mechanics, *Phys. Rev.* **106**:620-630.

Jaynes, E. T., 1979, Where do we stand on maximum entropy? in R. D. Levine and M. Tribus (eds.), *The Maximum Entropy Formalism*, Cambridge, Mass.: MIT Press, pp. 15-118.

Jaynes, E. T., 1982, On the rationale of maximum entropy methods, *IEEE Proc.* **70**:939-952.

Jaynes, E. T., 1983*a*, *Papers on Probability, Statistics, and Statistical Physics*, R. D. Rosenkrantz (ed.), Hingham, Mass.: D. Reidel.

Jaynes, E. T., 1983*b*, Bayesian spectrum and chirp analysis, paper presented at the Third Workshop on Maximum Entropy and Bayesian Methods in Applied Statistics, Department of Physics, University of Wyoming, Laramie, August.

Jaynes, E. T., 1985, Where do we go from here? in C. R. Smith and W. T. Grandy, Jr. (eds.), *Maximum Entropy and Bayesian Methods in Inverse Problems*, Dordrecht, The Netherlands: Reidel, pp. 21-58.

Justice, J. H. (ed.), 1986, *Maximum Entropy and*

Bayesian Methods in Applied Statistics, London: Cambridge University Press.

Kay, S. M., 1987, *Modern Spectral Estimation,* Englewood Cliffs: Prentice-Hall.

Kessler, S. B. (ed.), 1986, *Modern Spectrum Analysis,* vol. 2, New York: IEEE Press.

Keynes, J. M., 1921, *A Treatise on Probability,* London: Macmillan.

Laplace, P. S. de, 1820, *Theorie Analytique des Probabilities,* 3rd ed., Paris.

Marple, S. L., Jr., 1987, *Digital Spectral Analysis with Applications,* Englewood Cliffs: Prentice-Hall.

Morf, M.; Viera, A.; Lee, D. T. L.; and Kailath, T., 1978, Recursive multichannel maximum entropy spectral estimation, *IEEE Trans. Geosci. Electronics* **GE-16:**85-95.

Shannon, C., and Weaver, W., 1949, *The Mathematical Theory of Communication,* Urbana: University of Illinois Press.

Smith, C. R., and Grandy, W. T., eds., 1985, *Maximum-Entropy and Bayesian Methods and Inverse Problems,* Hingham, Mass.: D. Reidel.

Ulrych, T., and Bishop, T. N., 1975, Maximum entropy spectral analysis and autoregressive decomposition, *Rev. Geophysics and Space Physics* **13:**183-200.

Ulrych, T.; Smylie, D. E.; Jensen, O. G.; and Clark, G. K. C., 1973, Predictive filtering and smoothing of short records by using maximum entropy, *Jour. Geophys. Research* **78:**4959-4964.

APPENDIX 2: ECONOMICS

Abramovitz, M., 1958, *Long Swings in United States Economic Growth,* New York: National Bureau of Economic Research.

Abramovitz, M., 1960, *Long Swings in Economic Growth in the United States,* New York: National Bureau of Economic Research.

Abramovitz, M., 1964, *Evidences of Long Swings in Aggregate Construction Since the Civil War,* New York: National Bureau of Economic Research.

Burns, A. F., 1934, *Production Trends in the United States Since 1870,* New York: National Bureau of Economic Research.

Dewey, E. R., 1970, *Cycles: Selected Writings,* Pittsburgh, Penn.: Foundation for the Study of Cycles.

Dewey, E. R., and Dakin, E. F., 1947, *Cycles—The Science of Prediction,* New York: Holt and Co.

Easterlin, R. A., 1960, *Long Swings in the Growth of Population and Labor Force,* New York: National Bureau of Economic Research.

Easterlin, R. A., 1968, *Population, Labor Force, and Long Swings in Economic Growth: The American Experience,* New York: National Bureau of Economic Research.

Gottlieb, M., 1964, *Estimates of Residential Building, United States, 1840-1939,* New York: National Bureau of Economic Research.

Hoffmann, W. G., 1955, *British Industry, 1700-1950,* New York: Kelly & Millman.

Kuznets, S. S., 1961, *Capital in the American Economy: Its Formation and Financing,* Princeton, N.J.: Princeton University Press.

Kuznets, S. S., 1967, *Secular Movements in Production and Prices: Their Nature and Their Bearing Upon Cyclical Fluctuations,* New York: Kelley.

Kuznets, S. S., and Rubin, E., 1954, *Immigration and the Foreign Born,* New York: National Bureau of Economic Research.

Lewis, J. P., 1965, *Building Cycles and Britain's Growth,* London: Macmillan.

Long, C. D., 1940, *Building Cycles and the Theory of Investment,* Princeton, N.J.: Princeton University Press.

Matthews, R. C. O., 1959, *The Business Cycle,* Chicago: University of Chicago Press.

Mitchell, W. C., 1927, *Business Cycles: The Problem and Its Setting,* New York: National Bureau of Economic Research.

Soper, J. C., 1978, *The Long Swing in Historical Perspective,* New York: Arno Press.

Thomas, B., 1954, *Migration and Economic Growth,* Cambridge: Cambridge University Press.

Volcker, P. A., 1978, *The Rediscovery of the Business Cycle,* New York: Free Press.

Warren, G. F., and Pearson, F. A., 1937, *World Prices and the Building Trades: Index Numbers of Prices of 40 Basic Commodities for 14 Countries in Currency and in Gold, and Material on the Building Industry,* New York: Wiley.

23: Climatic Responses to Variable Solar Activity—Past, Present, and Predicted

Hurd C. Willett
Massachusetts Institute of Technology

Abstract: A brief look at the temperature climate of the middle latitudes of the Northern Hemisphere during the past few centuries relative to the long secular (Gleissberg) sunspot cycle establishes a clear pattern of long-term solar climatic relationships: prolonged periods of minimum sunspot activity are periods of minimum terrestrial temperature, whereas periods of maximum temperature tend to precede periods of maximum sunspot activity. The realization by the author as early as 1945 of this basic solar climatic relationship and its expression in the patterns of the general circulation led to the prediction of a long-term climatic trend first published in the *Journal of Meteorology* (Willett, 1951). The remarkable accuracy of the first quarter century of that prediction was attested to by Lamb (1977). The accuracy of the prediction for that period stands in sharp contrast to the ensuing failure of the prediction in strict correspondence to a sudden unexpected reversal of the solar Gleissberg cycle.

The solar climatic record of the twentieth century, most notably of the recent unexpected break in the cycle, is examined in some detail in light of a solar-wind index and a U.S. temperature index, additional data not available for the previous centuries. Some unique features of the recent record, since 1975, are stressed. Whether the climatic trend continues toward colder weather during the next two decades, as predicted in 1951, or whether the unexpected warm trend since 1979 persists, will depend on whether the recent break in the solar Gleissberg cycle is merely a temporary aberration or a permanent break in the cyclical pattern.

INTRODUCTION

The statistics of solar climatic relationships is so extensive and so involved that a brief chapter like this one is perforce restricted primarily to long-term outstanding features of the patterns. By contrast, the vocal antisolar front insists that the evidence is inconclusive to the point that solar climatic relationships must be placed on a firmer statistical footing before major research funding of the problem can be justified. Yet a 10-yr effort to find funding for just that purpose has uncovered only very meagre support, which nonetheless has produced highly significant statistical results (Prohaska and Willett, 1983). However, the purpose of this chapter is to report only some outstanding features of the long-term solar climatic relationships that the author finds to be conclusive evidence of a primary causal role of variable solar activity in climatic fluctuations.

SOLAR CLIMATIC PREDICTIONS OF 1951

Figure 23-1 contains the observational solar climatic record, on which the 1951 predictions were based, as far back as it extends in reliable detail. However, the preceding three centuries constituted a period of highly significant solar climatic relationships, which demands preliminary mention.

The three centuries from approximately A.D. 1450 to 1750 were the coldest and stormiest in the middle latitudes of the Northern Hemisphere (40°-65°N) since the climatic optimum

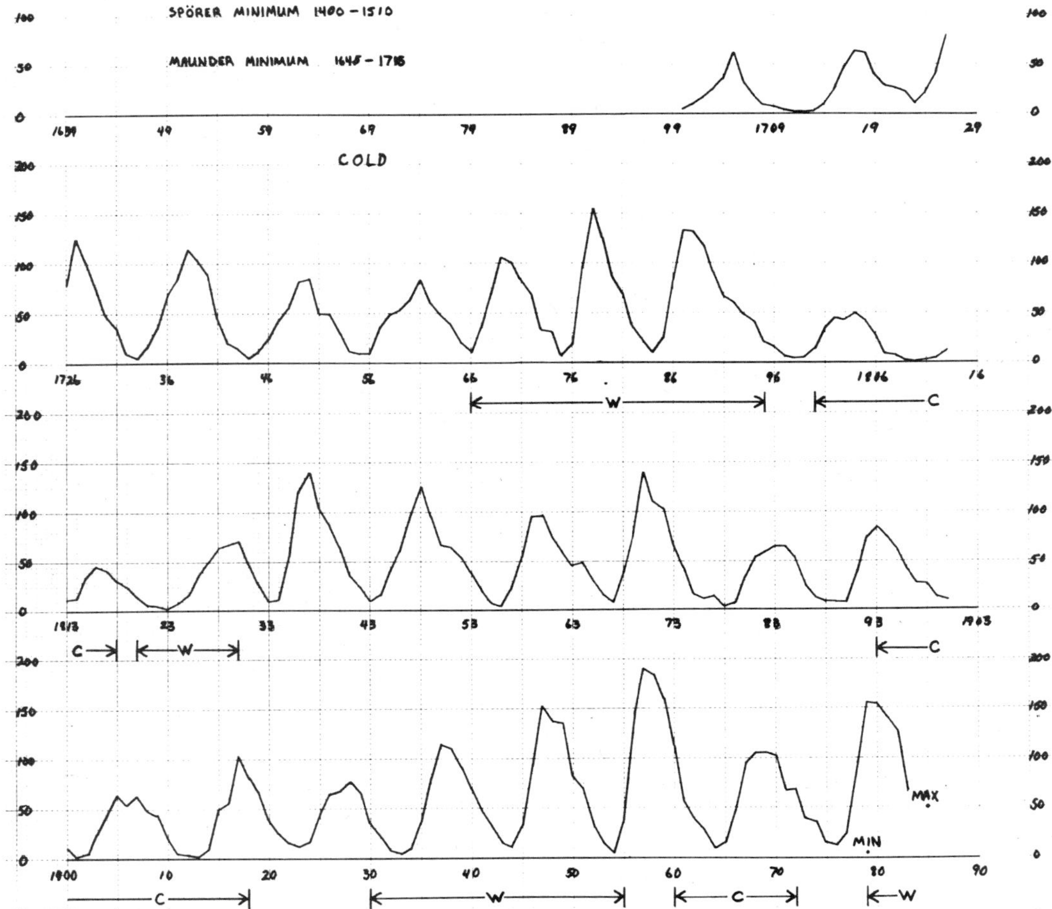

Figure 23-1. Curve of annual mean sunspot number, Relative Sunspot Number (RSS), and current International Sunspot Number (ISS), with indication of cold (C) and warm (W) periods for the past four Gliessberg cycles, 1639-1990.

of some 5,000 yr ago, obviously a period of predominantly low-latitude zonal circulation, responsible for the glacial expansion that earned the name Little Ice Age for the period. Particularly to be noted is the further fact that all available sunspot, auroral, and ^{14}C data indicate the three centuries A.D. 1425-1725 to have been predominantly a period of the outstandingly lowest sunspot activity of the last millennium, phenomenally so during the Spörer Minimum (1400-1510) and the Maunder Minimum (1645-1715) (Eddy, 1977). These two minimum periods coincided with the two coldest periods of the Little Ice Age. Certainly the Little Ice Age implicates a quiet Sun as a primary cause of terrestrial coldness.

The period of record covered by Figure 23-1 is divided into the last four Gleissberg (G) cycles, as dated by Feynman (1982*a*) and others. In 1950, the Gleissberg cycle was not recognized as such, but rather an unnamed secular cycle of approximately 80 yr was used in secular trend prediction. The sunspot numbers graphed in Figure 23-1 are annual mean Zurich Relative Sunspot Numbers (RSS), replaced recently by its successor, the International Sunspot Numbers (ISS). The periods in Figure 23-1 designated as warm (W) and cold (C) previous to

1875 are derived from Brueckner's (1890) masterful treatise, based on his curves of 5-yr mean temperature departures for the continents of Europe, Asia, and North America separately.

Of the four Gleissberg (G) cycles represented in Figure 23-1, the record of the first one, dominated by the cold Maunder Minimum, is largely missing in detail; the alternate second and fourth cycles, as happens typically also with the shorter 11-yr cycles, were similar in pattern; the intermediate third cycle differed in that cycles two and four had a high and steadily climbing level of sunspot activity with prolonged warm periods in mid-portions of the cycles. The years 1766-1790 have been exceeded only by the years 1930-1955 as the warmest quarter-centuries of the past 500 yr. Note that the third cycle had no such steady climb to a late very high peak and only a brief period of less extreme warmth preceding the early sunspot peak of 1834. In all cases before 1950, prolonged periods of coldness were restricted to times of minimum sunspot activity. It was the realization of this basic solar climatic relationship, together with extensive knowledge of the synoptic details of how the low-latitude zonal circulation patterns tend to affect specific continental and maritime regions, that formed the basis of the 1951 climatic trend predictions. The primary uncertainty is the prediction in detail of the continuing G cycle. In 1950 two such errors were involved: the first a minor one, in predicting the sunspot downturn one 11-yr cycle too soon (thinking in terms of an 80-yr instead of 90-yr secular cycle), which had little effect on the weather prediction, but the second, a vital one, the sudden unexpected upturn of sunspot activity following 1975, completely reversed the predicted trend of temperature. In broad terms, the 1951 forecast called for a cooling trend to start in the mid-1950s, bottoming out in a first cold period during the 1960s, then recovering to slightly higher levels during the 1970s, followed during the 1980s and 1990s by the coldest weather since the first two decades of the nineteenth century, considered to be the analogous period.

Figure 23-1 shows that the predicted trend was followed very closely to the mid-1970s, at which time there commenced a sudden reversal of trend of sunspot activity to very high levels, introducing the onset of a markedly warm period instead of a continuing trend to record coldness. The very low minimum and maximum points, indicated, respectively, in 1979 and 1985 at the end of the fourth cycle in Figure 23-1, are located where strict correspondence to the analog period would have placed the actual observed minimum of 1976 (highest minimum on record) and maximum of 1979 (second highest maximum on record), corresponding, respectively, to 1798 and 1804, which were the second lowest minimum and the lowest maximum on record.

Before we look more closely at the sudden reversal of the solar climatic cycle since 1975 and its predictive portent, it is equally important to look briefly at the detailed accuracy of the 1951 trend predictions for the first 25 yr as an impressive example of the possible performance of the solar climatic technique when the cyclical pattern continues undisturbed. Fortunately this sizable task has been performed by an impartial observer whose concise statements can be quoted verbatim, H. H. Lamb (1977). Lamb points out that the downturn of sunspot number was 10 yr (one short cycle) later than predicted, but he summarizes the weather forecasts and their verification in the following words:

> The temperature level over much of the world will fall significantly during the next fifteen years, probably reaching a first minimum level during 1960-1965. This temperature fall will be sharpest where the anomalous warmth of the past twenty-five years has been most extreme. The greatest cooling, then, should occur in the Spitzbergen-Greenland-Iceland area, while significant cooling will occur in northern Europe, the eastern United States, and in middle and lower latitudes of both hemispheres. Exceptions to this downward trend of temperature may be noted in the Antarctic, the interior of western Canada, and the northwestern United States, and particularly from the northeastern Mediterranean northeastward into Russia. . . . The rainfall in lower middle latitudes, south of 50°N, will be substantially higher during the next twenty years. . . . The general recent recession of glaciers in all regions is due to be reversed in the very near future. The change of trend should be most pronounced during the next 20 years. [p. 700]

After commenting on the failure of sunspot numbers to turn down as sharply as predicted, Lamb (1977, p. 700) went on to say: "The statements about temperature and rainfall were, however, all completely right. Glaciers generally have been slower to respond to the change of temperature trend than Willett expected."

This is a rather sweeping verification of such an extended and detailed set of weather predictions. The same set of predictions, continuing beyond 1975, called for the brief reversal during the 1970s of the general downward trend of temperature with a further renewal of the cooling trend during the 1980s and 1990s to the lowest levels of the century, followed by a general significant warming trend to set in at about the turn of the century. The two coldest decades at the close of the century were to be analogous to the two coldest decades of record at the beginning of the nineteenth century, also the two decades of lowest sunspot number of record since 1700. To look more closely now at what has happened to solar activity and weather since 1975, and its portent for the future, we turn first to Figure 23-2 and the added information it presents.

BREAKDOWN OF THE SOLAR CLIMATIC CYCLES

Figure 23-2 presents a highly comprehensive solar climatic record since 1875 for which the observational data were not available during earlier years. The bottom curve in Figure 23-2 is the sunspot number graph taken from Figure 23-1, with the designation of the alternate positive (+) and negative (−) polarity of the Sun's magnetic field with the alternate major and minor 11-yr maxima of the double (Hale, 22-yr) sunspot cycle.

The curve above that of sunspot number in Figure 23-2 is of the annual mean value of Mayaud's aa index of geomagnetic disturbance. On the basis of the observational record, this index is accepted by the author as the best indicator of solar-flare (solar-wind) disturbance of the geomagnetic field and of the temperature and wind fields of the upper atmosphere. A quiet Sun favors the establishment of a strong cold circumpolar cyclonic vortex—that is, a predominantly low-latitude zonal circulation pattern. A strong solar wind penetration of the geomagnetic field and upper atmosphere permits strong spot (auroral zone) heating of the atmosphere in high latitudes, hence destruction of the circumpolar symmetry of the temperature and wind fields and a rapid breakdown of the zonal into a cellular blocking pattern of the circulation—that is, advective heating of higher latitudes, advective cooling of lower, with strong meridional thermal contrasts and localized areas of excess storminess and rain and of excess drought in middle to lower latitudes.

The top two vertical lines in Figure 23-2 contain a uniquely informative index series of winter season, Winter Continental Temperatures (TCW), and summer season, Summer Continental Temperatures (TCS), departures from normal of temperature in the continental United States that deserves some extended comment and explanation. This particular index of the fluctuation of terrestrial temperature during the past century was selected for this study of solar climatic relationships and the prediction of trend primarily for three reasons:

1. The continental United States is a large, relatively homogeneous continental area in lower middle latitudes that must reflect to the fullest variable direct solar radiational influence on terrestrial temperature—notably, in any contrast between the summer and winter seasons.
2. This is probably the largest area for which a relatively complete homogeneous record of temperature for the past century is readily available.
3. This is the area with which we have worked most extensively both in the processing of climatic data and in the analysis and prognosis of synoptic weather patterns.

The derivation and content of the TC indexes is explained by Figure 23-3 and Table 23-1. Figure 23-3 outlines the eight relatively homogeneous climatic districts into which the country is divided and the locations of the 65 first-order weather bureau stations by which the monthly mean departure from normal of temperature is

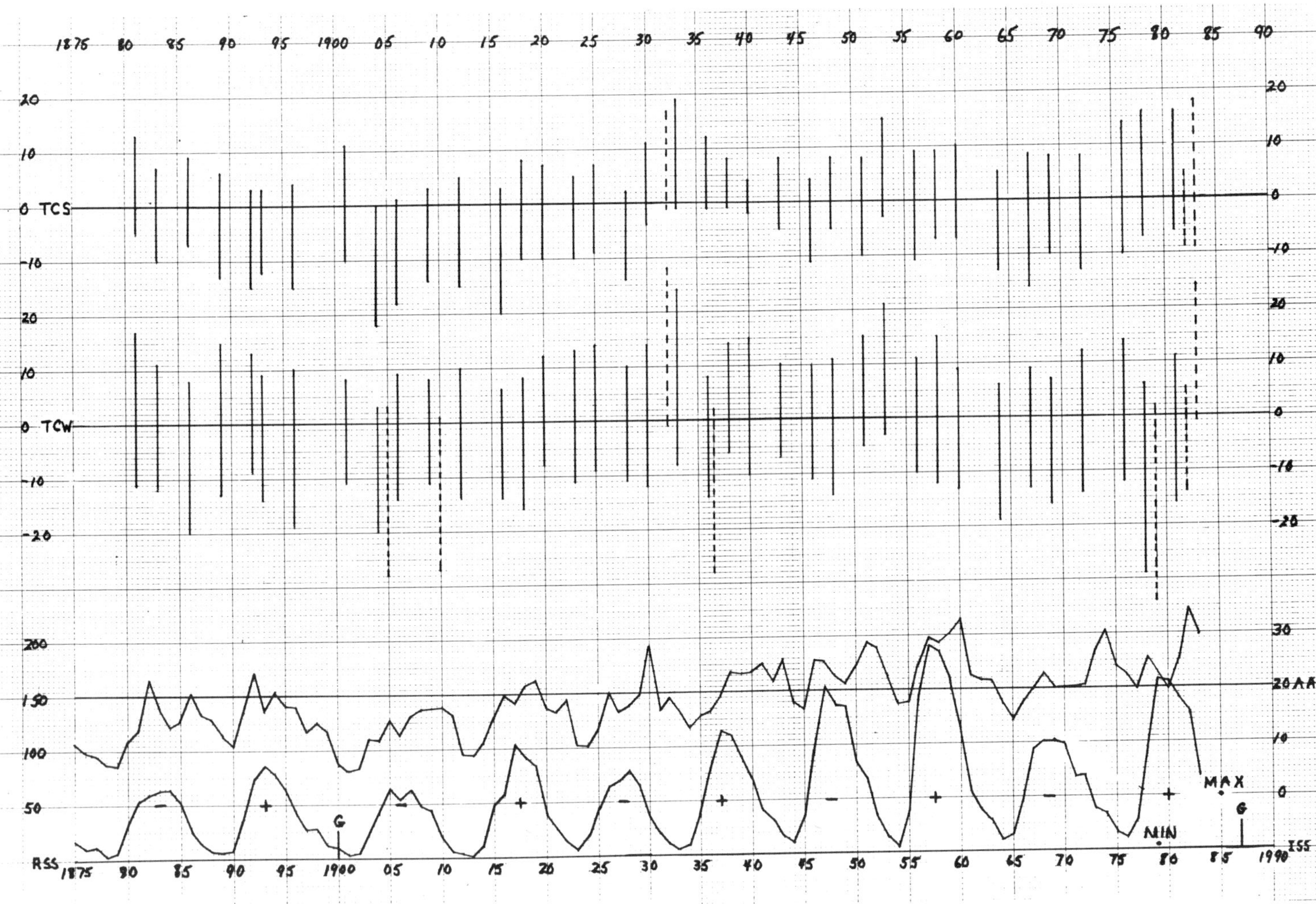

Figure 23-2. Record of annual sunspot number (RSS and ISS), of geomagnetic disturbance (aa index), and of winter (TCW) and summer (TCS) seasonal departures of temperature in the continental United States since 1875.

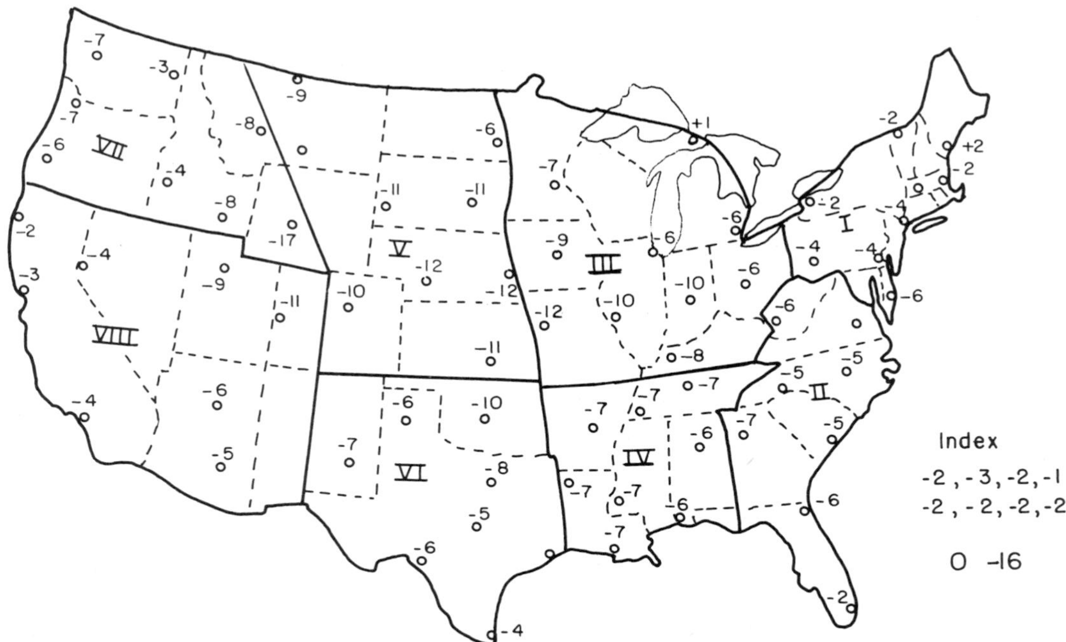

Figure 23-3. Sample monthly map from the 60-yr file used in the computation of the TC indexes, including boundaries of the eight unit climatic districts. The eight negative numbers are the mean temperature departures as given in Table 23-1 for the eight climatic districts. Zero is the sum of all the positive numbers and −16 is the sum of the negative numbers.

Table 23-1. Ratings of Climatic Regions and Their Equivalent Monthly Temperature Departures

	Departure Category[a]						
	+3	+2	+1	0	−1	−2	−3
Winter months[b]	>10	10 to 6	6 to 2	±2	−2 to −6	−6 to −10	<−10
Summer months[b]	>6	6 to 3	3 to 1	±1	−1 to −3	−3 to −6	<−6

Note: For results for the month of December 1909, see Figure 23-3.
[a] From extremely warm (+3) to extremely cold (−3).
[b] Given in °F.

determined for each of the eight climatic districts. Each district is then rated for the month in a departure category ranging from −3 through 0 to +3, by the degree limits listed in Table 23-1, for the winter months (December to February) and for all other months, including the summer months (June to August). The eight district categories for the sample month in Figure 23-3 are listed to the lower right of the figure. The TC index for the month is the pair of numbers of which the first is the sum of all of the positive district category rating numbers and the second is the sum of all of the negative; those listed in Figure 23-3 add up, respectively, to 0 and −16. There has been only one colder winter month in the United States since 1875, January 1977,

0 and −17. The particular merit of the TC index is that in addition to mean temperature it expresses regional contrast, a significant feature of the weather pattern. The winter (TCW) or summer (TCS) seasonal TC index is merely the numerical sum of the three constituent monthly TC indexes.

It should be noted that from 1900 through 1960, the monthly TC indexes were computed exactly as represented in Figure 23-3. Previous to 1900, a few of the 65 regional observational stations dropped out of record, probably with negligible effect on the regional departures. Since 1960, the regional departures have been determined from the monthly mean temperature departure charts in the *Weekly Weather and Crop Bulletin*, probably with no perceptible effect on the regional values. The TC index series is considered strictly homogeneous from 1900 through 1983.

The TC index values plotted in Figure 23-2 are all seasonal, TCW or TCS; the positive component of the index is indicated by the length of the vertical line above the axis, the negative by the length of the line below the axis, the year by the setting of the vertical line on the time scale. All the solid-line index values plotted in Figure 23-2 are 3-yr winter or summer seasonal averages, set on the time scale for the mid-year of the 3-yr index period. A few single-year seasonal indexes of special interest are entered as dashed lines for the indicated year. The 3-yr periods for which average index values are plotted are the successive phase periods of the eight phases of the 22-yr Hale sunspot cycle. Accordingly, the 3-yr average indexes are not always separated by exactly 3 yr. Phase-period index averages were used because the appreciable tendency for some phases of the solar cycle to have distinctive temperature characteristics makes trend tendencies stand out more clearly than they do in the greater confusion of all the single-year seasonal values.

Before examining in Figure 23-2 the breakdown of the current G cycle since 1975, we might comment briefly on a few interesting solar climatic features of the cycle since its beginning in 1900.

1. The first 15 yr of the cycle, 1900-1915, were, solarwise, with respect both to sunspot number and solar wind, the quietest years of the century. They were also, notably so in summer as might be expected, the coldest years of the century. In fact, for the six successive summers of 1903-1908, not one of the eight climatic districts of the country averaged in a positive temperature category for a single summer season, a truly remarkable uniformity of summer continental coolness that certainly must be of direct solar origin.
2. As sunspots and solar wind became most active during the middle years of the G cycle (1930-1960), so also occurred the warmest part of the century. Note that peaks of warmth occurred near or following the sunspot minima that preceded the high positive maxima of 1937, 1957, and 1979. These peak warm periods were the severe drought periods in the middle west—that is, climatic stress periods of blocking circulation patterns.
3. Periods of relative coolness occurred near or following the sunspot minima that preceded the negative maxima of 1947 and 1969. These were periods of wetness in the middle west—that is, of predominantly zonal circulation as opposed to the blocking circulation of the warm drought periods near the alternating sunspot minima.

Consider now in some detail the solar climatic events near the end of the G cycle, following 1975, the years during which the climatic trend predictions of 1951 failed after 25 yr of impressive performance. As noted earlier, that prediction for 1975-2000 was based on the cyclically analogous period 1795-1820, two G cycles previous. That analog (see Fig. 23-1), called for a very low sunspot minimum in 1979 and a very low maximum in 1985. The two decades from 1800 to 1820 were the coldest since 1700 and wet in middle latitudes, typical of strong low-latitude zonal circulation.

To note in some detail the observed solar climatic sequence from 1975 to 1983 (see Fig. 23-2), the following comments are most pertinent. The sunspot minimum count, instead of decreasing to almost zero in 1979, stopped at the highest of record (12) already in 1976. The maximum count, instead of being one of the lowest of reliable record in 1985, reached the second highest of record already in 1979. The length of the 11-yr sunspot cycle tends to vary inversely as the level of activity.

The aa index of geomagnetic disturbance, presumably disturbance by the solar wind, remained surprisingly low during the sunspot maximum of 1979, the only time during the period of record in Figure 23-2 that the two curves crossed. Three years later, in 1982, the aa index rose sharply to the highest level in its 118 yr of record, remaining quite high in 1983, an even more extreme departure than that of sunspot number from the level of activity expected for this period by cyclical analogy.

Only brief mention of significant highlights of the weather of the past 8 yr is possible; any documentation of the remarkable weather events of the period lies far beyond the scope of this chapter. With respect to the U.S. temperature record in Figure 23-2, the following facts should be noted: First, the unpredicted high sunspot maximum of 1979 was accompanied by a 6-yr period of peak warmth on the TCS curve as to be expected, equally unpredicted. Note on the TCW curve that the three winters from 1977 to 1979 were the coldest of the century; 1979 was the coldest single winter of the century, and January 1977, was the coldest month of the century. But only the comparatively sunless season was cold. Of the other nine seasons of this 3-yr period, the TC index was not negative for any of them. It was a warm period as required by insolational heating. Then why the cold winters? The relatively very low level of solar-flare (solar-wind) activity as indicated by the aa index is most favorable to maximum insolational heating (with high sunspots) during the sunny seasons and most favorable to maximum radiational cooling of the higher latitudes in winter in the expanded cold circumpolar vortex in the absence of meridional disturbance of the general circulation.

Second, suddenly, in February 1982, the aa index jumped to double its level of the preceding months, its highest of record, and remained at fluctuating high levels through 1982 and 1983. The whole pattern of weather action changed. The summer of 1982 was the first cool wet one in years (by TCS), thanks to record heavy rains in the central part of the country. After the cold winter of 1982, with its strong zonal circulation and rapidly moving snow storms across the northern half of the country, the winter of 1983 was the second warmest on record, second only to that of 1932, which also followed an analogously sharp high peak of the aa index (see Fig. 23-2).

The outstanding feature of the weather picture of 1983 and 1984, as unpredicted as the record high levels of the aa solar-wind index of which it is a predictable result, is the extreme degree of blocking (climatic stress) breakdown of the general circulation. In one region after another, particularly over the eastern half of the oceans in middle and lower latitudes, persistent deep troughs sweep intensely strong branches of the jet stream at abnormally low latitudes around against blocking ridges to the east. Individual storms develop, peak, and die out in the lower latitudes, blocked from northward movement into the customary high-latitude centers of maximum cyclogenesis, bringing record-breaking storminess and precipitation to the eastern side of the deep troughs. Drought may occur on the eastern side of the blocking ridges, but it is the repeated occurrence of extensive areas of persistent record-breaking precipitation in middle and lower latitudes that characterizes the recent period of record high aa index. The more impressive of these are noted briefly, as follows:

Summer 1982: One of the coldest and wettest on record in eastern Europe, notably Russia.

September 1982: Record-shattering precipitation totals from most of California eastward across the Southern Plateau and central Rockies.

Autumn 1982: Precipitation totals more than twice normal in a broad zone from California northeastward to the Dakotas.

November 1982 through June 1983: An unorthodox and unprecedentedly strong El Niño (undoubtedly physically related to the September 1982 and autumn 1982 items), which brought precipitation totals for the 8-mo period in much of Colombia and northern Peru to amounts more than 50% greater than ever recorded (Rasmussen and Wallace, 1983).

December 1982 through January 1983: Severe to record heavy rains and flooding in parts of the Mississippi Valley and southern plains.

Spring 1983: Record flooding in central China

followed by severe rains and flooding in much of the summer monsoon rain area of India.
Spring 1983: Record heavy rains in the north Atlantic coastal states of the United States, very heavy also in coastal California. Also record storminess and rain in western Europe, with London reporting a record-shattering 31 consecutive days with rain in May.
Late fall and early winter 1983-1984: Record heavy snows and avalanching in the European Alps.
November-December 1983: Record heavy snow over much of the Plateau and central Rockies of the western United States.

From the evidence discussed in brief outline in this section, we can make three summarizing statements without reasonable contradiction:

1. Since 1975, the pattern of variable solar activity has taken off on a tangent not at all predicted from its preceding pattern of cyclical behavior.
2. Equally certainly the trend of climatic change of both temperature and flooding storminess has taken off on a tangent from the course predicted from preceding solar climatic cycles.
3. The fact that this unpredicted solar trend and climatic trend mutually conform to the preceding statistical pattern of solar climatic relationships suggests, but by no means proves, that the former is a primary cause of the latter.

PREDICTION AND RESEARCH

The author's conviction is that the further trend of climatic change will be determined by the further trend of variable solar activity. Accordingly, the basic question of prediction is whether the recent unpredicted reversal of trend of variable solar activity represents a permanent change or merely a temporary aberration of the past cyclical behavior of the Sun.

The cyclical behavior of the Sun's activity is obviously too little understood at the present time to justify any real discussion of this question. A guess must be made, however, if only by reason of practical need.

When in June 1982 I made as usual my first seasonal forecast of the 1982-1983 winter, I was struck by the analogy of the 1957-1960 sunspot and aa maxima (Fig. 23-2) with those of 1979-1982, except that the aa maximum was sharper in 1982, corresponding to the observed sharper blocking storminess of the world weather patterns. The 1979-1982 maxima might be looked at either as a temporary flareback in the descending phase of the current G cycle, or possibly the 1969 maximum might be looked at as a mirror point in the 700-(720?) yr cycle whose glacial (low-latitude zonal) peak was reached during the Maunder Minimum and whose warm climatic stress phase last peaked during the stressful fourteenth century. In the latter case the 1980s and 1990s would tend to be on the warm side. However, initial and final solar inactivity and coldness have been such consistent phases of the last three G cycles that the decision was made to interpret the 1979-1982 peak of activity as a temporary flareback on the downtrend from the peak of 1957-1960. Accordingly, the winter of 1962 was selected as the analog weather pattern for 1984, modified in the expectation that 1984 would be colder throughout because of its later position in the descending phase of the G cycle. The forecast of a cold winter was couched specifically in terms of the expectation of the rapid decrease of solar activity to low levels, as occurred during the analog period of 1961 (see Fig. 23-2). December and January were predicted to be very cold months, coldest in the central part of the country, with some moderation in February, except in the northeast.

What happened? Sunspot number fell off sharply to comparatively low levels in November and December 1961, while the aa solar-wind index settled down to about normal levels. But then in January and February, sunspot number suddenly doubled its recent level, and the aa index returned to an extremely active state. And the weather? Already in the late summer severe cold appeared in the Soviet Arctic and moved down to give northern Europe and much of Asia bitterly cold weather in October and November, which were cold months also across Canada. In December, the bitter cold swept into most of the central United States much more severely even than predicted for December and the first part of January, while it moderated in Europe. The moderation predicted for much of the United States in February set in strongly in January and, by February, swept practically the entire country to produce a very warm month. The TC index of U.S. temperature rose from 1, −14 in December to 12, 0 in February. Not only across North America but also in Eurasia, the bitter coldness of the early season had receded to the polar latitudes and largely disappeared.

No clear trend appears toward the quiet Sun and severe cold of the 1951 prediction for the past two decades of the century. That is, to some extent, called for during the 1984-1987 period in the normal course of the double sunspot cycle, but two cold decades will materialize only if the solar G cycle runs through a normal quiet termination. Our expectation, with no very logical basis, is that it will.

This conclusion brings us back full circle to the opening paragraph of the chapter. It is obvious that the most effective use of solar climatic relationships for the specific prediction of large-scale trends of weather and climate requires an extensive comprehensive program of research to establish a firm statistical and physical basis. This basic research must be planned and executed in three distinct and consecutive stages:

1. The establishment with the greatest possible statistical precision of the pattern of atmospheric weather response to specific solar disturbance.
2. The selection of numerous key-day instances of specific solar disturbance and the extensive day-to-day synoptic analysis of preceding and following atmospheric conditions to establish the physical mechanism from initial solar disturbance to final weather reaction.
3. By the study of solar activity, both cyclical and sudden burst, to establish any possible predictability of pertinent solar disturbance. This problem lies entirely in the province of the solar physicist who has long been engaged in it and not of the meteorologist.

From the very preliminary research that available financing has permitted us to carry out under stage 1 (Prohaska and Willett, 1983), certain results should be mentioned briefly if only because of the prevailing skepticism about the reality of solar weather relationships. Eigenanalysis was applied to fields of 70-yr linear correlation between monthly mean departures of temperature from normal at a network of stations over the United States and southern Canada and each of three selected indexes of solar activity, lagging the index at from 0 yr to 23 yr. The eigenvectors obtained from the temperature-solar index correlation matrices were compared in several ways with eigenvectors obtained in a Monte Carlo test replacing the solar-index series with random numbers. In all tests, consistent significance of the solar-index eigenvector series compared with the random number series was demonstrated statistically. This result confirms the physical reality of solar climatic relationships.

On the other hand, the percentage of the total variance of the temperature fields that is predictable by the eigenvectors is consistently small, less than 10%. This finding is not surprising, because if it were much higher the predictability would have been picked up long ago. Furthermore, there is little doubt but that most solar weather relationships are highly nonlinear, that a rather high threshold of solar disturbance is needed to produce measurable weather reaction—hence, linear correlation is blurred by the one-sidedness of the relationship and the lack of any relationship much of the time. This consideration rather requires that our next effort to get a true picture of solar climatic relationships must be through nonlinear (contingency) correlation. And there is the further well-established fact that atmospheric response to sudden solar wind disturbance is very much a function of the atmosphere's initial state, whether strongly zonal or blocking in pattern.

There is certainly much to learn, and much that can be learned, to improve the application of solar climatic relationships to the prediction of climatic trend. In the meantime, my prediction stands for a markedly quiet termination of the current G cycle. This entails the prediction of the two coldest decades of the century coming up in middle to lower middle latitudes, before the next prolonged period of substantial warming. The Hale sunspot cycle sets the years from 1984 to 1990 to be the coldest period of the next two decades. The strongly low-latitude zonal pattern of the general circulation, which is a necessary accompaniment of broad middle latitude coldness, promises well-distributed above-normal precipitation in lower middle latitudes, in contrast to the alternate meridional zones of flood and drought of 1982 and 1983. Prolonged deficiency of precipitation is favored primarily on continental east coasts, notably on the middle and north Atlantic coast of the United States.

REFERENCES

Brueckner, E., 1890, Klimaschwankungen seit 1700, *Geogr. Abh.* **4**(2):153-484.

Eddy, J. A., 1977, Climate and the changing sun, *Climate Change* **1:**173-190.

Feynman, J., 1982*a*, Geomagnetic and solar wind cycles, *Jour. Geophys. Research* **87**(A8):6153-6162.

Feynman, J., 1982*b*, Solar cycle and long term changes in the solar wind, *Rev. Geophys. Space Physics* **21:**338-348.

Lamb, H. H., 1977, *Climate Present, Past and Future,* vol. 2, *Climatic History and the Future,* London: Methuen & Company, 700p.

Prohaska, J. T., and Willett, H. C., 1983, Dominant Modes of Correlation between Solar Activity and the Month to Month Sequences of the U.S. Temperature Field, Final Report NASA contract NAS5-26218.

Rasmussen, E. M., and Wallace, J. M., 1983, Meteorological aspects of the El Niño/Southern Oscillation, *Science* **222:**1195-1202.

Willett, H. C., 1951, Extrapolation of sunspot-climate relationships, *Jour. Meteorology* **8**(1):1-6.

24: Long-Range Prediction of Solar Activity

Jane B. Blizard
University of Colorado, Boulder

Abstract: Although solar activity has been cyclic for millions of years, the amplitudes and durations of the cycles have varied. Solar activity originates in internal oscillations, but these may be modified by the dynamics of the solar system. The strongest correlation between sunspot number and dynamic factors exists for the Sun's orbit around the barycenter of the solar system. A statistically significant variation of solar radius has been found with an 11-yr period. An increase in radius implies a change in spin momentum, which apparently varies inversely with the Sun's orbital momentum about the barycenter. In addition, other solar system effects correlate with sunspot number. One is the Sun-centered Coriolis effect, having discontinuities when the Sun is close to the barycenter, as it was in 1951 and will be again in 1990. Another is a precessional effect of the planets, which are in orbits inclined to the solar equator, and this effect produces a torque on the oblate Sun. Displacements of more than 500 km may be caused by horizontal tides on the Sun induced by the planets.

Motion of the Sun about the barycenter can account for 49% of the sunspot number. The Sun-centered Coriolis force could contribute 9%; precessional planet torque, 4%; and horizontal tides, another 4% of the sunspot number. Many of these effects should be evaluated in the solar-equatorial coordinate system. Thus, modulations resulting from motions of the Sun and of the solar system could, when added, explain one-half to two-thirds of the sunspot number. Solar activity and its relationship to climate should be studied by examining both the Sun's internal constitution and its surrounding solar system.

INTRODUCTION

When they make geophysical forecasts, geophysicists recognize the importance of solar activity. It is well known that solar activity influences the geomagnetic field and the Earth's ionosphere. If climate is related to solar activity, then long-range solar activity forecasts make possible improved climate forecasts. At present, only statistical methods have been used for making forecasts of solar activity. To raise the level and scope of solar activity forecasts, it is essential to study the nature of solar activity using physical and mathematical models.

Climate signals commonly have been noted as displaying regular cycles of 20 yr (Hibler and Johnson, 1979), 22 yr, or multiples of these (Dicke, 1979; Fairbridge and Hillaire-Marcel, 1977). Other data may reflect the variable solar cycle. For example, the sunspot cycle, whose nominal period is cited as 11 yr, actually varies in length from 7 yr to 17 yr, and its amplitude ranges from 4 to 1, sometimes almost vanishing (Eddy, Gilman, and Trotter, 1976), and at such times, the Earth experiences little ice ages.

In this chapter I discuss several possible solar system effects on sunspots and solar activity and emphasize that the celestial mechanics of the solar system can be computed far into the future. The solar cycle may be driven by nonlinear interactions between interior oscillations (g-modes; Wolff, 1976) and external perturbations related to the solar system.

A survey of variable-flare stars having ages close to that of the Sun (Wilson, 1978) showed a range of regular cycles with periods from 7 yr to 14 yr. Moreover, as with the Sun's cycle, the rise times are shorter than decline times. The flare activity of the Sun is also similar to that of RS Can Van stars. I discuss some reasons why the Sun's cycle may be highly irregular.

SUN'S MOTION AROUND THE BARYCENTER OF THE SOLAR SYSTEM

The Sun is a typical lower main sequence G-2-type star that is presumed to have lost most of its rotational velocity to the solar wind and the planets. Although 99.9% of the mass of the solar system is contained in the Sun, 98% of the remaining angular momentum is located in the planets (Table 24-1). Depending on the solar model used, the Sun's axial rotation accounts for only 1.5-2% of the total angular momentum of the solar system.

Before I discuss the entire solar system, it will help to begin by considering some two-body systems. In the Earth-Moon system, tidal friction is thought to slow the Earth's spin rate; the resulting change in angular momentum is presumed to be transferred to the Moon's orbital motion. Thus, the angular momentum of the total system is constant, but such momentum may be transferred from spin to orbit or from one body to another. In a hypothetical two-body solar system, consisting only of the Sun and Jupiter, both would orbit the barycenter with a period of 11.9 yr.

In a three-body system consisting only of the Sun, Jupiter, and Saturn, the Sun's orbit would vary between loops having a large radius when Jupiter and Saturn are on the same side and loops having a small radius when Jupiter and Saturn are on opposite sides. In such a three-body system, the Sun's orbital angular momentum would vary by an order of magnitude in less than a decade. What becomes of the lost or gained angular momentum?

In celestial mechanics, point masses are used for the Sun and the planets. By convention, the change in solar orbital momentum would be transferred entirely by perturbations in the planetary orbits. By contrast, in the solar system, in addition to its orbital momentum, each body possesses an axial-spin momentum. One might speculate that angular momentum could be transferred from the Sun's orbit to the Sun's spin momentum (either by changes in the Sun's radius or its angular velocity). If the angular momentum were transferred from orbit to spin, then during an 11-yr period it could make a difference of up to 7% in solar equatorial radius or velocity or both.

Variations in the solar radius having periods of 11 yr and 76 yr have been found (Gilliland, 1981). The large solar radius has been shown to correspond to small orbital momentum, implying a momentum transfer (Prokudina, 1981). Equatorial rotation rates of the Sun have been shown to be periodic. Anomalies have been found during the Maunder Minimum (Eddy, Gilman, and Trotter, 1976, 1977), and sudden glitches have been found during alignment of the Sun, Jupiter, and the solar system's barycenter (Howard, 1978; Landscheidt, 1981). Because the Sun is a fluid body, variations in radius or rotation rate could exert significant effects on the formation of sunspots and on solar activity.

An exhaustive statistical study has been made by matching 21 variables of the solar system with monthly sunspot numbers for the interval 1749-1964 (Pimm and Bjorn, 1969). One of the largest correlation coefficients found was between solar orbital momentum around the

Table 24-1. Angular Momentum in the Solar System

Planet	*Spin Angular Momentum* $gm/cm^2/sec$	*Orbit Angular Momentum* $gm/cm^2/sec$
Sun	6 to 1.45×10^{48}	4×10^{47} max
Jupiter	41.9×10^{44}	193.4×10^{48}
Saturn	8.3×10^{44}	78.35×10^{48}
Uranus	0.21×10^{44}	16.95×10^{48}
Neptune	0.21×10^{44}	25.65×10^{48}

Sources: Based on data from Alfvén and Arrhenius, 1976; Dole, 1970.
Note: Values for other planets are less than 2% of Jupiter's values.

barycenter of the solar system and sunspot number. The variance in sunspot number is found by squaring the correlation coefficient $(0.6992)^2 = 0.489$, or 48.9% of the variance could be related to the motion of the Sun around the barycenter. Because solar activity may result from other possible solar system influences, the results of Pimm and Bjorn's multiple correlations are discussed in a following section.

SUN-CENTERED CORIOLIS ACCELERATION

Because the positions of the planets keep changing, the center of the Sun is forced to make an orbit consisting of a series of looping paths within a region of about two solar radii (R_{Sun}) from the barycenter (Jose, 1965; Landscheidt, 1981). If one assumes that the barycenter is fixed in inertial space, then the vector R from the barycenter (also the origin of the coordinate system) to the center of the Sun was almost zero in 1951, but the displacement from one year to the next is never zero. The center of the Sun is always moving; the chief motive force is supplied by Jupiter, with Saturn next, followed by Uranus and Neptune.

The center of the Sun moves relatively to the fixed axes but also to axes rotating about the barycenter, with the distance to the center of the Sun, R. The Sun acts like a rotating body on a rotating platform, located at a variable distance R from the axis of the rotating platform and moving relative to the platform's rotating axes. The acceleration of the center of the Sun changes as the Sun moves around the barycenter, depending on the planet-determined velocities and positions of the Sun. The acceleration viewed by an observer fixed in inertial space (at the barycenter, e.g.), A_f, is related to the acceleration A_r viewed by a rotating observer (at the center of the Sun) by the equation

$$A_f = A_r + \text{centrifugal acceleration} + \text{Coriolis acceleration.}$$

If the Sun were moved by only a single planet—say, Jupiter—then in an inertial frame, the Sun's orbit would be approximately circular with a mean radius of $1.06 R_{Sun}$. At the center of the Sun, the only acceleration would be centrifugal acceleration in the direction of Jupiter. In a three-body system consisting of Jupiter, Saturn, and the Sun, the Sun's path would describe an epitrochoid (three-lobed rosette). The solar acceleration would not be perpendicular to the radius vector R; hence, a large Coriolis disturbance is present as a result of the superimposition of Saturn's revolution on a rotating platform of the Sun and Jupiter. Uranus and Neptune would have less effect here (about 3.7%) than for displacement (23.4%) or for angular momentum (13.6%).

An equation can be written for this Sun-centered Coriolis acceleration A_{SCC} referred to axes rotating about the barycenter with the Sun's radius vector. One finds that the radial and tangential V_θ components of the velocity of the Sun about the barycenter are both important, as well as the distance from the barycenter; thus (Pimm and Bjorn, 1969),

$$A_{SCC} = 2\, V_\theta V_r / R,$$

where V_r is the radial component of the Sun's velocity about the barycenter. The correlation coefficient of this function with sunspot number is 0.3125, with alternate cycles plotted as positive and negative (to show the ~22-yr magnetic cycle; Pimm and Bjorn, 1969). Because the variance depends on the square of the correlation coefficient, the proportion of the sunspot number variance contributed by this factor is $(0.3125)^2 = 0.0977$ (rounds to 9.8%). Large discontinuities appeared in this function in 1951 and can be calculated for 1990 when the center of the Sun makes a close approach to the barycenter (nuclear transit of Landscheidt, 1981).

PRECESSIONAL EFFECTS ON AN OBLATE SUN

Because of its axial rotation, the Sun's equatorial radius must exceed its polar radius; that is, the Sun is an oblate spheroid. The magnitude of the Sun's oblateness on its visible edge has been measured by Hill and Stebbins (1975). Here, I describe the effects of the planets on an oblate Sun and show that these effects can contribute to some part of the cyclic solar activity.

Before I consider the solar system, I discuss the case of the Moon, the Sun, and the oblate Earth. Two effects become apparent. First is the luni-solar precession of the equinoxes. Not so well known is the effect on the Earth's fluid cover, the oceans. Precessional effects on a fluid differ from those on a solid body. Precessional action of the Moon on the Earth's ocean water has actually been observed in a study of the Gulf Stream through the Straits of Florida. Twice during each lunar month, at high and low declination of the Moon, flow characteristics change. Following the changes in the Moon's declination, Gulf Stream flow at a given point is either accelerated or decelerated.

In its equatorial belt, the Earth's surface is not wholly fluid. Nevertheless, the effect of precessional action on the ocean is present. Similarly, the planets could cause precessional action (fluid motion) on the Sun's equatorial belt, which could contribute to sunspots and solar activity. A precessional effect could occur because of the Sun's oblateness and the inclination to the Sun's equator of the orbital plane of a planet. Such an effect attempts to bring into line the planet's orbital plane and the plane of the Sun's equator, but it never succeeds. Because the Sun is not rigid, its axis of rotation would not exhibit conical motion. Instead, the fluid equilibrium near the Sun's equator could be affected.

The precessional torque of a planet on the Sun would tend to shift fluid of the equatorial belt down toward the equator on one side and up toward the equator on the opposite side. The centrifugal force resulting from the Sun's rotation would try to maintain the equatorial belt in the central position. The interaction of these forces could produce fluid disturbances.

Most of the planetary orbits make angles of from 3° to 8° with the plane of the Sun's equator. Pluto is an exception; its plane makes an angle of 12°, but Pluto is so tiny that its effect on the Sun is presumed to be negligible (Table 24-2). The magnitude of the precessional torque is approximately

$$T = M \sin 2\theta / R^3,$$

where M is the mass of the planet, R is the mean distance of the planet from the Sun, and θ is the angle of inclination (planet orbit to plane of Sun's equator). The relationship is derived from McCullagh's theorem for gravitational potential.

During each revolution, the precessional effect becomes zero twice (at the nodes) and reaches maximum values twice (at points midway between the nodes, assuming a circular orbit). If eccentricity is significant, then the greater effect is on the side of the perihelion position. As can be seen from Table 24-2 (relative magnitudes of planetary torques), the most significant torques are exerted by Jupiter and Venus; the Earth ranks third.

The ascending nodes (see Table 24-2) of all planets except Mercury and Pluto are clustered within 12° on either side of heliographic longi-

Table 24-2. Orbital Elements and Precessional Torques of Planets on the Sun, in the Solar-Equatorial Coordinate System

Planet	*Inclination to Solar Equator*	*Longitude Ascending Node*	$M \sin 2\theta / R^3$ *Mean Torque*
Mercury	3°21′	327°9′	0.33
Venus	3°47′	253°2′	1.07
Earth	7°10′	254°4′	1
Jupiter	6°	249°2′	1.87
Saturn	5°25′	237°9′	0.08

Source: Based on data from Blizard, 1969, 1983.
Note: Torque values for other planets are less than 1% of Earth's torque.

tude 250°. Thus, reinforcement could take place when two or more planets are located at right angles to the nodes of this coordinate system. The transformation in coordinates from ecliptic to solar equatorial is shown elsewhere (Blizard, 1969).

A sinusoidal variation of sunspot number, with predicted values at the 1% confidence level, and in phase with the Earth's period, has been detected (Vassilyeva, Schpitalnaya, and Petrova, 1980). In the statistical study of sunspot numbers and solar system parameters, previously mentioned, a correlation coefficient of 0.2048 was found between precessional torque and sunspot number (Pimm and Bjorn, 1969). Because the variance in sunspot number depends on the square of the correlation coefficient, precessional torque may influence sunspot number and solar activity by 4.2%.

TIDAL EFFECT ON SOLAR ACTIVITY

Of all the proposed ideas of planetary influence on solar activity, the hypothesis of planetary tidal action has been the most controversial. Much has been written on the subject (reviewed by Kuklin, 1976). Despite rejection in the past, however, some recent developments tend to lend credence to planetary tidal effects.

Tidal influences are both vertical and horizontal. The vertical effect can be shown to be negligible; most critics have shown that all planets acting together could raise a solar tide of only 1 mm. In contrast, the horizontal tidal effect may be significant. In a 14-day period, half a solar rotation, the horizontal displacement of planetary tide would be 560 km and its velocity, 0.93 m/sec (Opik, 1972; Wood, 1972). The relative importance of planets in creating tidal effects on the Sun is about in the same order as for precessional effects: Jupiter, Venus, Earth, Mercury.

Correlation coefficients computed between sunspot number and tidal effects bear out the relative importance of horizontal as contrasted with vertical (=radial) tides. The correlation between sunspot number and horizontal tide is 0.2043; that between sunspot number and vertical tide is 0.1253 (Pimm and Bjorn, 1969). Still, the direct influence on sunspot number (based on correlation coefficient squared) is only 4.17% for horizontal, 1.57% for vertical, and 4.40% for total. In addition, the four planets mentioned are the same ones that show strong correlation with sunspots and solar flares (Blizard, 1969; Brier, 1979). Other factors include possible transfer of angular momentum from the outer planets (Mörth and Schlamminger, 1979) and resonance between planetary periods and solar oscillations (Brier, 1979).

CONCLUSION

Although four possible external effects on solar activity have been discussed, it remains to be determined whether these effects are statistically independent. Two (motion about the barycenter and Sun-centered Coriolis effect) correlate with the 22-yr cycle and may not be independent. The other two (precessional torque and planetary tides on the Sun) correlate with the 11-yr cycle and thus appear to be independent of the first two (yet they may not be independent of each other). Thus, depending on assumptions made, several multiple correlation coefficients can be obtained. If the first two are computed together, one obtains a multiple correlation coefficient of 0.72, resulting in a variance of 52% of the sunspot number. If all four are combined, the correlation coefficient is 0.74 and the variance is 54% of the sunspot number.

Finally, the influence of factors external to the solar system cannot be ignored. Up to 10% dependence of sunspot number on heliocentric longitude has been found as a result of the Sun's motion toward the solar apex (Trellis, 1967) and with the state of the interplanetary field (Vassilyeva and co-workers, 1980). The possibility of effects by a tenth planet or by a solar companion has not been ruled out.

In conclusion, the motions of the Sun and those related to the solar system may contribute to solar activity and may explain in part the variation in amplitudes and durations of the basic internal cycles of the Sun. If half or more of the variance in sunspot number could be related to parameters inherent in the solar system, then it would be possible to use these parameters computed for years into the future to make predictions of long-range solar activity.

ACKNOWLEDGMENTS

I thank J. P. Cox, professor of astrophysics, University of Colorado, for encouragement, many helpful comments, and review of the manuscript. I also thank K. D. Wood, professor of aerospace at the University of Colorado, and H. H. Sargent, III, of the Environmental Research Laboratories, NOAA, Department of Commerce.

REFERENCES

Alfvén, H., and Arrhenius, G., 1976, *Evolution of the Solar System,* National Aeronautics and Space Administration, NASA SP, 345, Washington, D.C.: U.S. Government Printing Office, 599p.

Blizard, J. B., 1969, Long range flare prediction, Contract NAS8-21436, Final Report to NASA, Marshall Space Flight Center, Huntsville, Ala. (Appendix).

Blizard, J. B., 1983, Precessional effects on an oblate sun, *Am. Astronomical Assoc. Bull.* **15:**959.

Brier, G., 1979, Use of difference equations for predicting sunspot numbers, in B. M. McCormac and T. A. Seliga (eds.), *Solar-Terrestrial Influences on Weather and Climate,* Dordrecht: D. Reidel Publishing Co., pp. 209-213.

Dicke, R. H., 1979, Solar luminosity and the sunspot cycle, *Nature* **280:**24-27.

Dole, S. H., 1970, *Habitable Planets for Man,* 2nd ed., New York: American Elsevier, 158p.

Eddy, J. A.; Gilman, P. A.; and Trotter, D. E., 1976, Solar rotation during the Maunder Minimum, *Solar Physics* **46**(1):3-14.

Eddy, J. A.; Gilman, P. A.; and Trotter, D. E., 1977, Anomalous solar rotation during the seventeenth century, *Science* **198:**824-829.

Fairbridge, R. W., and Hillaire-Marcel, C., 1977, An 8,000-year paleoclimatic record of the "Double-Hale" 45-yr solar cycle, *Nature* **268:**413-416.

Gilliland, R., 1981, Solar radius variations over the past 265 years, *Astrophys. Jour.* **258:**1144-1155.

Hibler, W. D., III, and Johnson, S. J., 1979, The 20-year cycle in Greenland ice core records, *Nature* **280:**481-483.

Hill, H. A., and Stebbins, R. T., 1975, The intrinsic visual oblateness of the sun, *Astrophys. Jour.* **200:**471-483.

Howard, R., 1978, The rotation of the sun, *Rev. Geophysics and Space Physics* **16:**721-732.

Jose, P. D., 1965, Sun's motion and sunspots, *Astron. Jour.* **70:**193-200.

Kuklin, G. V., 1976, Cyclical and secular variations of solar activity, in V. Bumba and J. Kleczek (eds.), *Basic Mechanisms of Solar Activity,* Dordrecht, The Netherlands: D. Reidel Publishing Co., pp. 147-190.

Landscheidt, T., 1981, Swinging sun, 79-year cycle and climate change, *Jour. Interdisciplinary Cycle Research* **12:**3-19.

Mörth, H. T., and Schlamminger, L., 1979, Planetary motion, sunspots, and climate, in B. M. McCormac and T. A. Seliga (eds.), *Solar-Terrestrial Influences on Weather and Climate,* Dordrecht, The Netherlands: D. Reidel Publishing Company, pp. 193-207.

Öpik, E., 1972, Planetary tides and sunspots, *Irish Astron. Jour.* **10:**293-301.

Pimm, R. S., and Bjorn, T., 1969, Prediction of smoothed sunspot number using dynamic relations between the sun and planets, Contract NAS8-21445, Final Report to NASA, Lockheed-Huntsville Research Center, Huntsville, Ala.

Prokudina, V. S., 1981, Variations of the solar radius and the motion of the sun about the barycenter, *Astron. Tsirk., no. 1179,* pp. 6-8 (in Russian).

Trellis, M., 1967, Influence of the sun's motion towards the apex and origins of centers of activity, *Astrophys. Letters* **1:**57-58.

Vassilyeva, G. I.; Schpitalnaya, A. A.; and Petrova, N. S., 1980, Prediction of solar activity taking into account its extraneous conditionality, in R. F. Donnelly (ed.), *Solar and Terrestrial Predictions, Proceedings,* vol. 3, Washington, D.C.: U.S. Government Printing Office, pp. A45-A57.

Wilson, O. C., 1978, Chromospheric variations in main-sequence stars, *Astrophys. Jour.* **226:**379-396.

Wood, K. D., 1972, Sunspots and planets, *Nature* **240:**91-93.

Wood, R. M., 1975, Comparison of sunspot periods with planetary synodic period resonances, *Nature* **255:**312-313.

Wood, R. M., and Wood, K. D., 1965, Solar motion and sunspot comparison, *Nature* **208:**129-130.

Wolff, C., 1976, Timing of solar cycles by rigid internal rotations, *Astrophys. Jour.* **205:**612-621.

25: Long-Range Forecasts of Solar Cycles and Climate Change

Theodor Landscheidt
Schroeter Institute for Research in Cycles of Solar Activity, F.R. Germany

Abstract: The secular cycle of solar activity, which seems to be connected with climatic change, volcanism, and the ozone column, is correlated with a wave pattern formed by secular variations in impulses of the torque driving the sun's oscillatory motion about the center of mass of the solar system. Information about the epoch, phase, and amplitude of secular maxima and minima of sunspot activity and concomitant solar-terrestrial effects can be read from this secular wave.

The low-pass-filtered supersecular variation in the energy of the secular wave runs parallel with the supersecular sunspot cycle, characterized by excursions like the Maunder Minimum and the Medieval Maximum, and with the central England air temperature time series. The future course of both waves, accessible to computation, points to a secular minimum of sunspot activity past 1990 and a subsequent supersecular minimum of the Maunder Minimum type around 2030. In view of the relationship between secular and supersecular sunspot minima and climate, a prolonged phase of cooling and glacier advance is to be expected from 1990 to 2070.

The cross-correlation function of 256 yearly absolute values of the time rate of change of the torque dT/dt, related to impulses of the torque, and respective sunspot numbers R_M, covering the years 1705-1960, yields a significant correlation at 11 yr, 22 yr, and 31 yr. The latter quasiperiod fits the Sahelian drought cycle of 31 yr found by Faure and Gac. Extrema of the function dT/dt form a quasicycle with a mean period of 13.3 yr, which emerges in the spectral analysis of surface temperature, atmospheric pressure, and zonal westerlies. Maximum entropy analysis yields a strong relationship of the 13.3-yr cycle and its harmonics to energetic solar eruptions, which are connected with weather. This cyclicity has opened up the possibility of a dependable flare forecast that was successfully put to the test of experience. The fact that climate is the integral of weather should lead to an improved understanding of variations in climate that are subject to solar forcing.

SECULAR AND SUPERSECULAR CYCLES IN SUNSPOTS AND CLIMATE

Climate is a cyclic phenomenon. The sun that drives weather and climate shows cyclic behavior, too. The 11-yr cycle of sunspots is the best known feature of the sun's activity. Thus, it is natural that research in solar-terrestrial relationships in this field was fixed on the 11-yr cycle for many decades. But 100 yr of controversial results seems to be enough to show that the effects looked for must be of minor practical importance. Since 1976 Eddy (1976) has focused attention on the finding that a curve connecting the 11-yr peaks, the long-term envelope of sunspot activity, reflects changes in climate much better than the 11-yr variations. This modulation that corresponds to decades of higher or lower solar activity is known in astronomy as the secular, or 80-yr, cycle of sunspot activity. According to Gleissberg (1958, 1976) this cycle showed maxima around 1630,

1700, 1780, 1850, and 1950, while minima occurred around 1670, 1740, 1810, and 1900. Similar results can be deduced from Figure 25-1 after Schönwiese (1979), which shows yearly sunspot numbers subjected to a Gaussian low-pass filter. The Maunder Minimum is identical with an extremely deep secular minimum around 1670, as well as the Spörer Minimum about 1500. Both grand minima mark the Little Ice Age in Europe and similar periods of cold in North America, as derived from records of mid-latitude glacier advance (Eddy, 1977). Such advance, though to a lesser extent, was also observed around the secular minima about 1810 and 1900 (Rudloff, 1967; Dronia, 1975), while secular maxima went along with times of warmer climate and glacier retreat or at least a stop in advance. As will be shown, the secular modulation is again modulated by a supersecular cycle, which outlines excursions of the Maunder Minimum type, or corresponding maxima of extreme height like the Medieval Maximum of the twelfth century, which coincided with the Medieval Climatic Optimum.

MAGNETIC STORMS AND CIRCULATION

This long-term connection, if it is real, points to a solar signal or a complex array of output factors that build up a sufficient potential of climatic change by steady accumulation, or lead to an effect by their constant lack during a period of several decades. Some promising candidates exist for such a relationship. Small variations in the solar constant, which were observed during the Solar Maximum Mission in 1980, do not offer an explanation. There are only negative deviations that, according to Hoyt and Eddy (1982), are the result of temporary blocking of solar radiation by the strong magnetic fields in sunspots. This blocking should result in a small drop in temperature around an 11-yr maximum, while secular maxima favor rising temperatures. But Bucha (1983) has demonstrated that, in Europe and North America, the prevalence of zonal- or meridional circulation is dependent on the incidence of magnetic storms. Thirteen to twenty days after sudden commencements, caused by solar activity, a pronounced increase in temperature occurs due to prevailing zonal circulation, while lull periods in geomagnetic activity lead to meridional flow and drops in temperature induced by Arctic air.

If Bucha's (1983) observation is valid, secular maxima with intense 11-yr maxima and frequent magnetic storms should generate an excess of zonal circulation over meridional flow, while secular minima with weak 11-yr maxima and extended lulls in strong magnetic activity should produce a surplus of meridional circulation. This situation matches the positive correlation between secular maxima and rising temperatures and between secular minima and cooling. A smoothed index of geomagnetic activ-

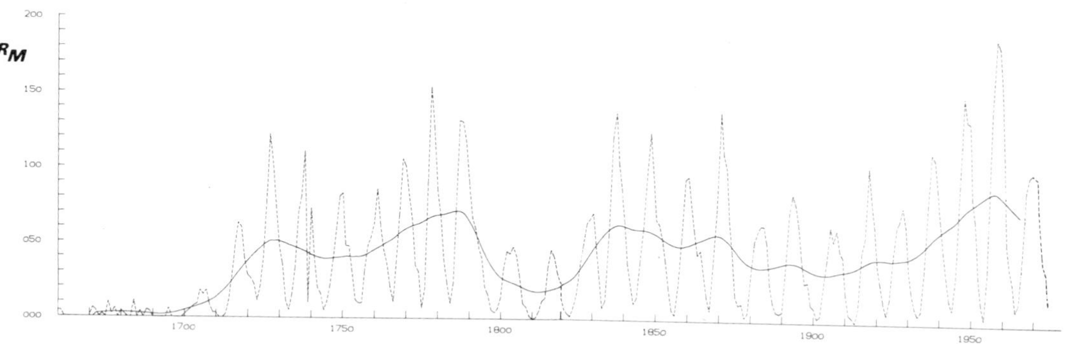

Figure 25-1. Eleven-year sunspot cycle 1670-1975 (hatched line) subjected to a Gaussian low-pass filter (solid line). The smoothed outline sets off the secular minima about 1670, 1740, 1810, and 1900. (From Schönwiese, 1979)

ity shows a maximum around 1950 and a minimum about 1900.

Researchers are trying to explain effects like that observed by Bucha. Neubauer (1983) has discovered that sudden commencements of geomagnetic storms, induced by solar eruptions, are related to displacements in the 70-mb polar vortex in the lower stratosphere, which again influence the polar vortex in the troposphere. The shift in the 70-mb polar vortex is caused by a highly localized sudden stratospheric warming. Of 66 magnetic storms observed during the winters 1978/1979 and 1979/1980, as many as 61 were accompanied by a sudden stratospheric warming.

Here is a brief summary of Neubauer's (1983) argumentation: Solar particles enter the cusp of the magnetosphere. They penetrate to the mesosphere and to the upper stratosphere where they cause significant increases in NO and other ozone-destroying molecules and ions. A localized UV (ultraviolet) window is created by the sudden reduction of ozone above 30 km at the base of the cusp. UV radiation, which is absorbed above 30 km at other locations, passes through the UV window at the base of the cusp and is absorbed in the lower stratosphere. A localized sudden stratospheric warming results. The polar vortex in the lower stratosphere is displaced. Eventually, the tropospheric polar vortex and jet stream are similarly displaced, which enhances vorticity. This effect is strongest in winter. Bucha (1983) has explained this quantitatively.

SOLAR ACTIVITY, VOLCANISM, AND CLIMATE

A rather promising possibility is indicated by an anticorrelation between volcanic activity and secular variations in solar activity. Bray (1974) could show that nearly all significant glacier advances of the past 40,000 yr coincided with enhanced volcanic eruptions. This finding is valid also for glacier expansions since 1500 (Bray, 1974) that can be compared with the Dust Veil Index of Lamb (1970). It should be noted that they all coincide with minima in the secular cycle of solar activity. Wexler (1953), Lamb (1982), and Mitchell (1970) found relationships between the eruptive activity of volcanoes and periods of low temperature in the Northern Hemisphere. These relationships were corroborated by analysis of the core drilled from the Greenland Ice Sheet. Hammer, Clausen, and Dansgaard (1980) found a strong correlation between energetic volcanic explosions, indicated by a higher degree of acidity in the respective ice layers, and periods of cooling. An investigation into relationships of the global temperature of the Northern Hemisphere with the Dust Veil Index of Lamb by Miles and Gildersleeves (1977) yielded the correlation $r = -0.72$. Antarctic ice cores showed similar results. Between 1450 and 1850 the ice layers were marked by a double concentration of sulfates (Allard, 1981). This finding corresponded with the Little Ice Age in the Northern Hemisphere.

Figure 25-2 presents the observed mean values of the optical depth of the aerosol in the Northern Hemisphere for 1880-1980. This quantity measures the degree of the reduction of the transparency of the atmosphere by the aerosol. The epochs of the secular minimum of solar activity about 1900 and the secular maximum around 1950 are marked by thick arrows. It is obvious that the envelope of the peaks of the optical depth shows a negative correlation with the secular cycle. The optical depth is still growing. After the eruptions of Alaid and Pagan in 1981 and especially of El Chichón in 1982, the curve reached an ascent (Reiter and Heck, 1983), the extrapolation of which indicates a new maximum around 1990. As is shown later, this is expected to be the epoch of the next secular minimum.

Bryson (1982) has developed a model yielding the optical depth of an aerosol generated by explosive eruptions of volcanoes that blow sulfuric gases high up into the stratosphere, where they can stay for years. The variation in the optical depth of both aerosols, the observed one and that derived from Bryson's model for the period 1880-1980, show a strong correlation ($r = .95$). This means that only 5% of the decrease in solar radiation by diminishing transmission cannot be attributed to energetic volcanic eruptions. If this could be confirmed by

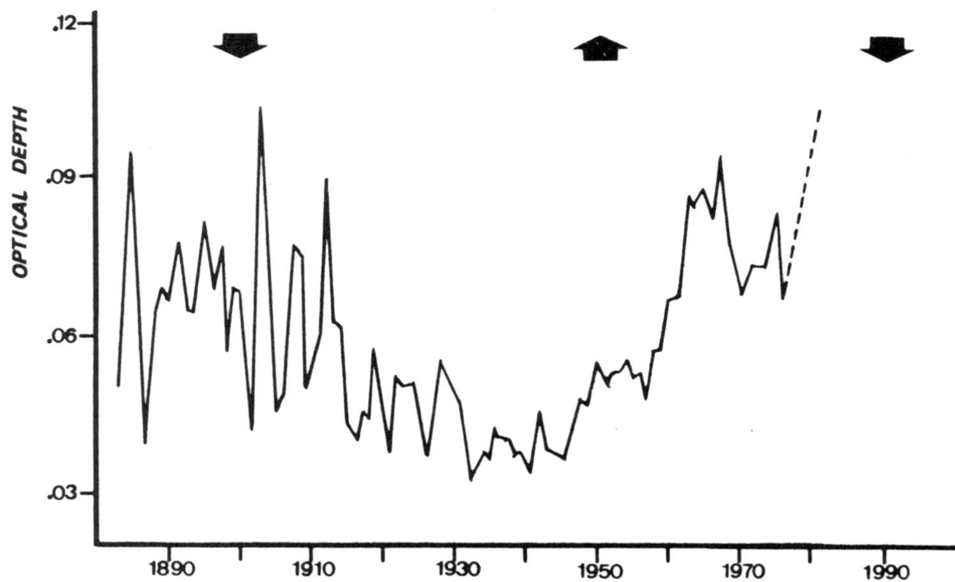

Figure 25-2. Variations in the observed mean values, 1880-1980, of the optical depth of the aerosol in the Northern Hemisphere show an anticorrelation with secular extrema in sunspot activity. The secular minimum about 1900 and the secular maximum around 1950 are indicated by thick arrows. The values of the optical depth show a steep ascent after 1980. An extrapolation indicates a new maximum of smoothed values about 1990. This finding fits a secular minimum in sunspots at this epoch. (After Bryson, 1982)

data before 1880, the inverse relationship between the course of the secular cycle and volcanic activity would get practical importance. Flohn (1984) has explained how a cluster of strong volcanic eruptions, or a prolonged lull of volcanic activity like that in the decades before 1950, can trigger pronounced cooling or warming, if the effect is maintained over a period of decades.

The regulation of volcanic activity by secular variations in the sun's activity is not beyond physical speculation. The solar wind, the velocity of which is dependent on the level of solar activity, is coupled with the Earth's magnetic field, the origin of which is in the interior of the Earth. This coupling can result in small jumps in the Earth's rotation on its axis. Danjon (1959) observed a deceleration of the Earth's rotational velocity after the very energetic solar eruptions on February 23, 1956, and July 15, 1959. The day became about 2 msec longer. This deceleration was observed again after the huge flare on August 7, 1972 (Gribbin and Plagemann, 1972). A systematic investigation by Challinor (1971), covering the period 1956-1969 from one 11-yr maximum to the next one, yielded a relationship between variations in solar activity and a slow change in the length of day. Munk and MacDonald (1960) discussed the possible consequences of an interaction between the solar wind and the geomagnetic field of the Earth and concluded that the interaction can cause variations in the rotation.

On the basis of plate tectonics, it may be assumed that jumps or less conspicuous decelerations or accelerations in the Earth's rotation can influence the relative movements between quasirigid plates, to which most of the Earth's volcanism is related. Eighty percent of active volcanoes lie near subduction zones where oceanic plates plunge downward, 16% on mid-ocean ridges, and 4% near the boundaries of colliding plates or on hot spots in the midst of plates. It seems to be consistent to surmise that variations in the degree of coupling of the solar wind and the Earth's magnetic field influence

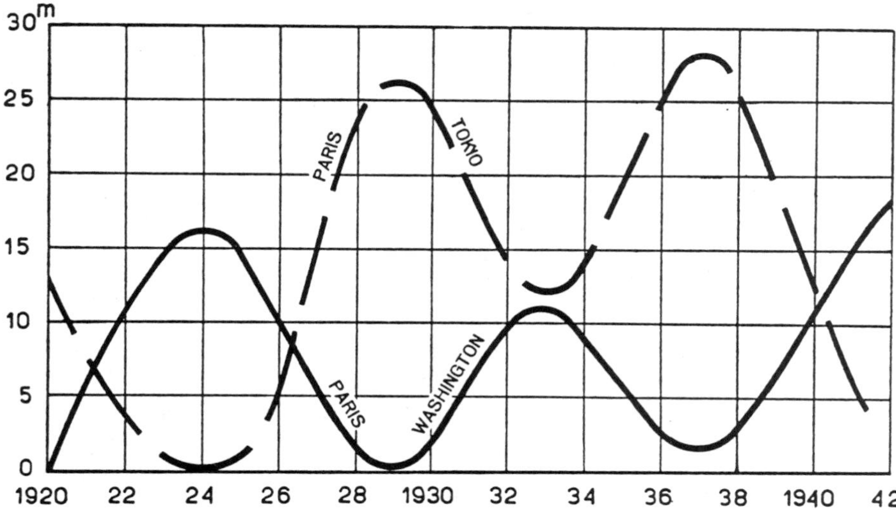

Figure 25-3. Differences in longitude between Paris on the one side, and Tokyo and Washington, D.C., on the other side. The extrema of the emergent oscillations match extrema of the 11-yr sunspot cycle. Sunspot maxima occurred in 1928 and 1937 and minima in 1923 and 1933. (From Stoyko, in Bachmann, 1965)

explosive volcanic activity via changes in the Earth's rotational velocity and the motion of plates with respect to one another. There is some evidence that these motions follow solar activity, which varies the torque of the solar wind on the Earth. Geodesic measurements by 71 stations evaluated by Stoyko, director of the Bureau International de l'Heure in Paris, showed a pulsation in the distance from Paris to Washington, D.C., and to Tokyo that ran parallel with the 11-yr cycle (see Bachmann, 1965). These results can be read from Figure 25-3. Eleven-yr maxima occurred in 1928 and 1937, and minima in 1923 and 1933. Stoyko (see Bachmann, 1965) drew the conclusion that the continents swing back and forth, while the 11-yr sunspot cycle goes up and down. In any case, the parallel course of the change in the volcanic aerosol and the secular cycle of solar activity seems to indicate that conditions for the development of a volcanic aerosol are best when the solar activity is near a secular minimum going along with a special state of magnetic coupling. It is interesting in this respect that, according to Benioff (1951), global strain, caused by tectonic forces, reached a maximum around 1900, came to a minimum after a 50-yr strain release in 1950, and then began to grow again (Richter, 1958).

SOLAR VARIATIONS, OZONE COLUMN, AND HEAT TRANSPORT

Another approach is based on changes in ozone concentration in the stratosphere, induced by variations in the solar UV flux. Since 1953 UV radiation varied on the order of a factor of 2 from sunspot minimum to maximum. This variation is valid, too, for the radiation within 170-210 nm that generates ozone. According to Heath and Thekaekara (1977), the solar flux about 170 nm is a factor of approximately 2.5 greater at solar maximum than at solar minimum, and at 210 nm there is still a factor of 1.9. These variations have been calculated to change the global surface temperature. High-resolution maps of ozone distribution, based on Nimbus 7 observations, show clearly that ozone follows the course of planetary waves and that strong ozone concentration gradients are related to jet streams (Aimedieu, 1981). Avery and Geller (1978) calculated that the wave structure in the troposphere and stratosphere is sensitive to changes in the strength of the polar night jet. This jet is a direct result of the meridional temperature gradient in the stratosphere. Thus, it is possible that changes in the ozone abundance influence intensity and location of the polar jet, which again modify the momentum and heat transport of planetary waves in the stratosphere and troposphere (Bates, 1977). Schwentek (1971) observed a strong correlation between mean temperature in the stratosphere and sunspot number *R*. An increase in *R* from 5 to 160 induces an increase in stratospheric temperature of 5 K at 25 km, of 16 K at 30 km, of 27 K at 35 km, and of 40 K at 40 km. Thus, the temperature gradient should be strong enough to release the assumed effect, at least when integrated over several decades around a secular maximum. As Fairbridge and Hillaire-Marcel (1977) have pointed out, planetary albedo, too, would be affected. During secular minima the decrease in ozone production would be intensified by an additional effect. Chamberlain (1976) holds that long periods of low solar activity and therefore enhanced cosmic ray production of NO_x in the stratosphere would reduce the ozone concentration. Secular minima are excellent examples of such periods.

The Arosa total ozone data (Rowland, 1978), presented in Figure 25-4, are the longest and most complete records available. Their trend follows the secular cycle with a maximum around 1950. The curve plots the deviation of yearly mean values from the 50-yr mean. As Hill and Sheldon (1975) have shown, the Arosa record has only a poor relationship to the 11-yr sunspot cycle. Sunspots are no appropriate crite-

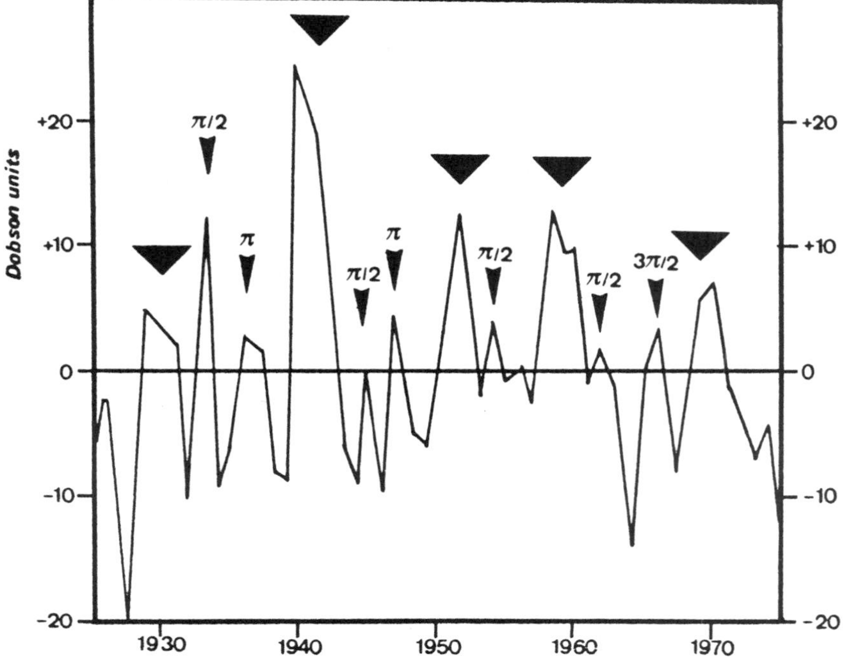

Figure 25-4. Arosa long-term record of ozone concentration for the years 1925-1975. The curve plots the deviation of annual mean values from the 50-yr mean. Triangles point to the epochs of heliocentric conjunctions of Jupiter and the center of mass (CM) of the solar system. Harmonics of cycles formed by consecutive conjunction events are indicated by arrowheads. Peaks in the ozone concentration coincide with periods of energetic solar eruptions that are connected with conjunctions of Jupiter and the CM.

rion in this respect, however. Periods of enhanced eruptional activity with strong UV flux do not stick to the 11-yr maximum. About 1942, for instance, 2 yr before the 11-yr minimum, there was a period of very energetic eruptions, which is reflected in the Arosa ozone record. The triangles in Figure 25-4, pointing to such excursions in the curve, indicate the epochs of heliocentric conjunctions of Jupiter and the center of mass of the solar system (CM), which are a more effective criterion of solar-terrestrial interactions than sunspots. Even harmonics of cycles formed by consecutive conjunction events, especially at $\pi/2$ and π radians, seem to be effective in solar-terrestrial interaction. Such harmonics are indicated in Figure 25-4 by arrowheads. Conjunctions of Jupiter and the CM are such a dependable indicator of energetic eruptional activity that they opened a possibility to predict X-ray bursts >X9 (Landscheidt, 1984). Such bursts are rare events. Only two of this category occurred from 1970 to 1981. According to a long-range forecast issued in January 1982, events >X9 were expected April 29 to May 5, May 22 to June 9, and September 27 to the middle of December 1982. Bursts X12, X12.9, and X10 were observed on June 6, December 15, and December 17, 1982. In addition, several X7, X8, and X9 events were observed. The forecast was checked by the Space Environment Services Center, Boulder, and by the astronomers Gleissberg, Pfleiderer, and Woehl. The discussion section shows that the conjunction of Jupiter and CM could be looked at as a case of spin-orbit coupling involving transfer of angular momentum.

IMPULSES OF THE TORQUE IN THE SUN'S MOTION ABOUT THE CENTER OF MASS

There seems to be some evidence that the secular and supersecular cycles of solar activity are related with climate in the Northern Hemisphere. The practical application of such knowledge, however, depends on the feasibility of a dependable forecast of long-term solar activity. Knowledge of the mean length of solar cycles is no real help in this respect. According to Gleissberg (1975) the 80-yr cycle varies within 40-120 yr. Even the mean period of the supersecular cycle has not been clearly defined. It is necessary to find a set of predictable time series running parallel with long-term solar variation and climatic change. The sun's oscillations about the CM of the solar system seem to meet these requirements (Fairbridge, 1983; Landscheidt, 1976, 1980, 1981*c*, 1983, 1984, 1986).

Figure 25-5 shows this relative motion in the ecliptic plane, seen from the center of the sun (CS). The distance of both centers varies from 0.01 to 2.19 solar radii (Jose, 1965). One revolution around the CM takes 9-14 yr. It is conspicuous that just in 1951, when the sun's orbit showed high rates of curvature, the secular sunspot cycle reached a maximum. This is no fortuitous fit. Closer examination shows that impulses of the torque $\Delta L = \int_{t_0}^{t_1} T(t)dt$, which drive the sun's motion around the CM, are the special quantitative criterion of relationships with the secular and supersecular cycles of solar activity.

GRAND MINIMA AND QUANTITATIVE THRESHOLDS OF IMPULSES OF THE TORQUE

Rare periods of extremely weak activity like the Maunder Minimum as a rule occur only when two consecutive impulses of the torque of opposite direction are so strong that the strength of each of them and the sum of both transgresses a very high threshold defined quantitatively (Landscheidt, 1981*c*). The plot in Figure 25-6 shows measured radiocarbon deviations since 5300 B.C. derived from the analysis of dated tree rings by Damon (1977), with a smoothed curve of sinusoidal variation in the Earth's magnetic moment from Lin et al. (1975). The excursions modulating the sine curve, which reflect fluctuations in cosmic ray flux, are attributed to supersecular variations in sunspots regulating the strength of the solar wind. When the sun is quiet and the solar wind is weak, more radio-

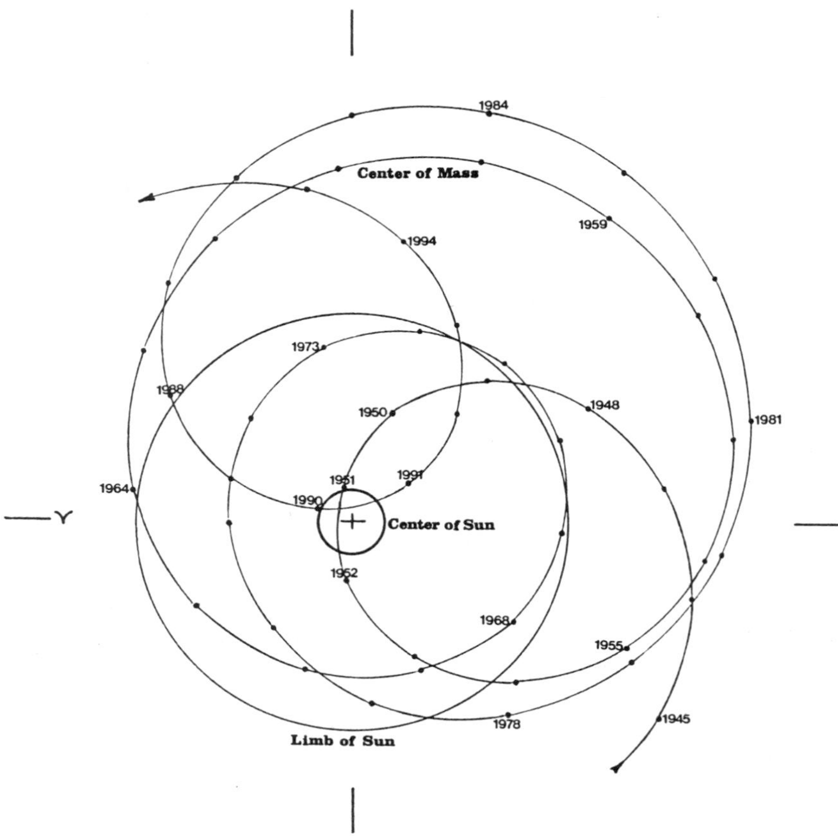

Figure 25-5. Position of the CM of the solar system in the ecliptic plane relative to the sun's center for the years 1945-1995.

carbon is generated by the impact of galactic cosmic rays on the atmosphere, while less radiocarbon is produced when the sun is active and the shielding effect of its extended magnetic field against cosmic rays becomes strong. The graph by Eddy (1976) allows for the inverse relationship of cosmic ray flux with solar activity. Eddy (1978) has focused attention on the finding that grand minima in sunspot activity can be read from the positive radiocarbon excursions. These are the Maunder Minimum M, the Spörer Minimum S, the Greek Minimum G, the Homeric Minimum H, the Egyptian Minimum E, and a Sumerian Minimum S.

The long arrows in Figure 25-6 represent those epochs when extremely strong impulses of the torque met the defined quantitative criterion. Since 5300 B.C. only eight such events occurred, which all marked grand minima. A Pearson test of this result, making use of Yates's correction, yields the value 27 for 1 degree of freedom ($P < .00001$). It should be noted that, according to Eddy (1976) also, all grand minima prior to the Maunder Minimum and the Spörer Minimum coincided with periods of prolonged cooling and glacier advance. The shorter arrows indicate impulses of the torque meeting a medium threshold criterion. The precise function of this category can be judged near the minimum of the sine curve displaying amplified radiocarbon features. The last event of this category occurred about 1810. This was a period of relatively cold climate going along with glacier advance (Rudloff, 1967). Remember that all

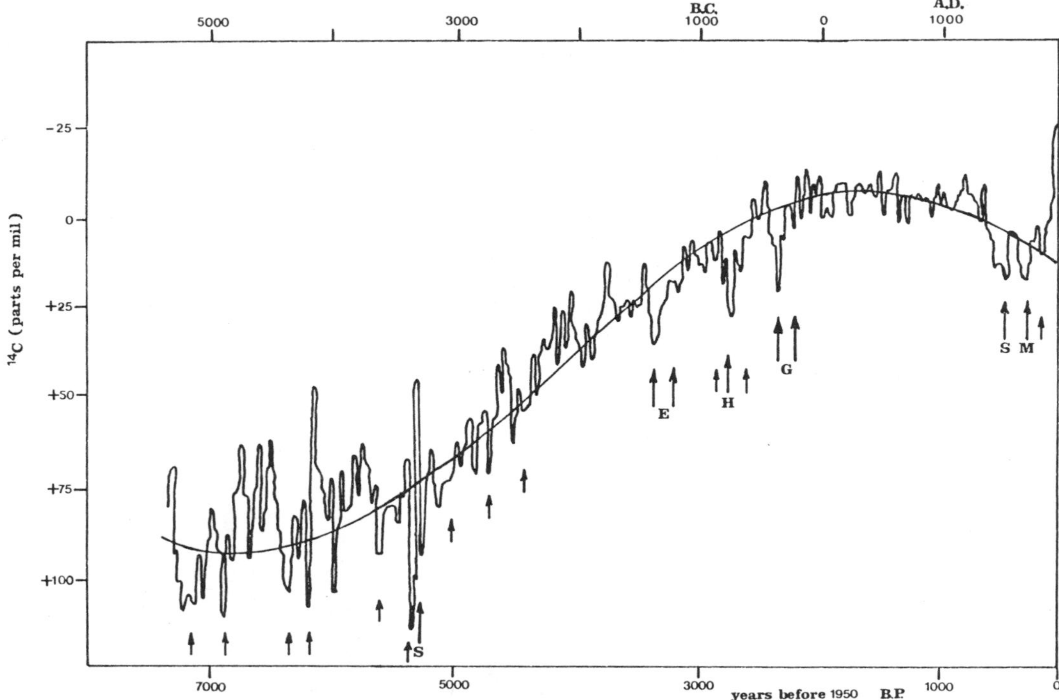

Figure 25-6. Radiocarbon deviations since 5300 B.C. derived from analysis of dated tree rings by Damon (1977), with a smoothed curve of sinusoidal variation in the Earth's magnetic moment from Lin et al. (1975). The graph by Eddy (1976) allows for the inverse relationship of cosmic ray flux with solar activity. The excursions modulating the sine curve are attributed to supersecular variations in sunspots. Six grand minima (Maunder Minimum M, Spörer Minimum S, Greek Minimum G, Homeric Minimum H, Egyptian Minimum E, Sumerian Minimum S) coincide with rare impulses of the torque meeting a severe quantitative threshold criterion. These eight events, which all fit grand minima, are indicated by long arrows. Shorter arrows represent impulses of the torque of a medium category.

these excursions occurred at epochs of secular minima that were obviously subjected to a low-frequency modulation.

EXTREMA OF THE TIME RATE OF CHANGE OF THE TORQUE, ZONAL WESTERLIES, AIR TEMPERATURE, QBO, AND LOCUST PLAGUES

The definition of impulses of the torque [$\int_{t_0}^{t_1} T(t)dt$; t_0 when $T = 0$; $t_1 = 300$ days after t_0] makes evident that single impulses of the torque are cumbersome to compute. It is more convenient to use the continuous function $d^2L/dt^2 = dT/dt$, the maxima and minima of which coincide with positive and negative impulses of the torque and are proportional to their strengths. These extrema are henceforth called $+/-A$. Figure 25-7 presents a plot of the function dT/dt covering A.D. 1900-2000. Extrema $+/-A$, representing impulses of the torque of varying strength, are marked in Figure 25-7 by short pointers. An analysis covering more than 7,600 yr yields a mean interval of 6.65 yr from one extremum $+/-A$ to the next one. Schönwiese (1980) found this quasiperiod and periods of about double, fourfold, and sixteenfold length in the central England air temperature time series of 1659-1978. The mean interval from

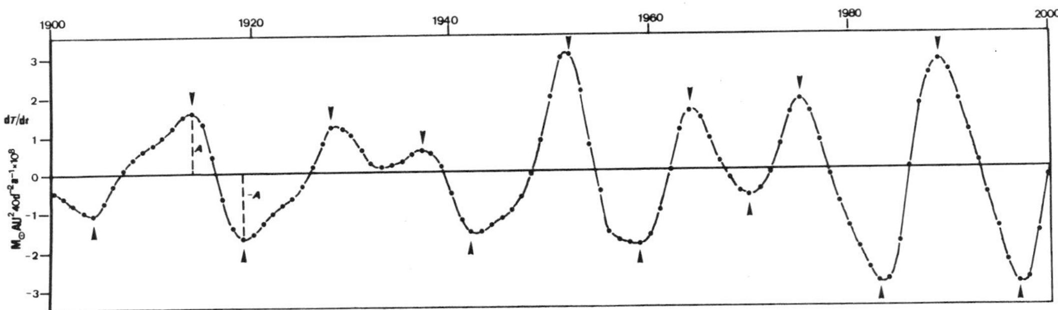

Figure 25-7. Time rate of change of the torque dT/dt, plotted for A.D. 1900-2000. Extrema +/−A are marked by pointers. (Mathematical expressions are explained in the text.)

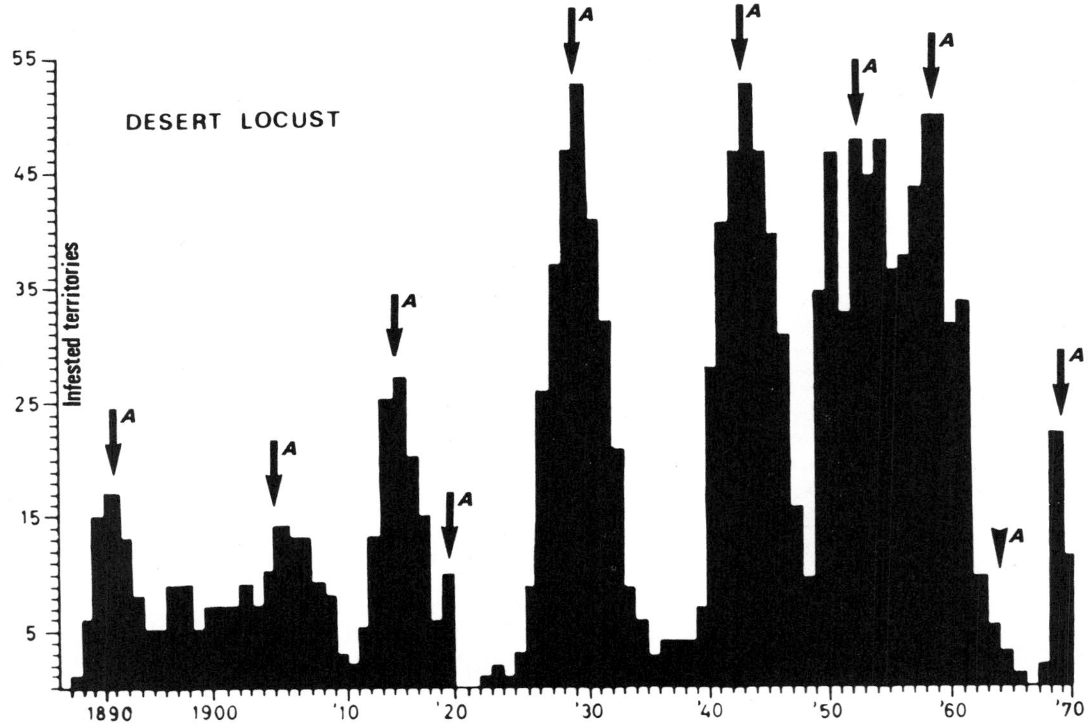

Figure 25-8. Outbreaks of the desert locust correlated with extrema +/−A in the time rate of change of the torque dT/dt. (The mathematical relationships are explained in the text.)

one maximum +A to the next maximum, and from one minimum −A to the next minimum, is 13.3 yr. This quasiperiod matches a significant sharp peak emerging in a periodogram analysis of zonal westerlies in the Northern Hemisphere elaborated by Brier (1968). A maximum entropy spectral analysis of four combined time series of atmospheric pressure from 1881 to 1975, carried out by Junk (1982), yielded a relatively strong period of 13.3 yr when 200 coefficients were chosen. Gilliland (1983) found a significant peak at 12-13 yr in the analysis of residuals from volcanic, CO_2, and solar climate modeling; he holds that this peak is of solar-induced origin and cannot be looked at as an ill-determined 11-yr peak. The fourth harmonic of 13.3 yr, 3.3 yr, often appears in meteorological time series (Visser, 1959; Carrea, 1978; Schove, 1983). The sixth harmonic, 2.2 yr, has the length of the period of the quasibiennial oscillation (QBO).

Figure 25-8 presents indirect evidence of a correlation among the secular cycle, impulses of the torque, and climate. Locust plagues are dependent on rainfall in certain desert areas. The numbers of countries infested by swarms of the desert locust (Haskell, Ashall, and Hemming, 1966) show peaks of more or less massive invasions. The envelope connecting these peaks reflects the secular maximum around 1950 and the secular minimum around 1900. The foregoing secular maximum about 1850 was again going along with huge migratory swarms of locusts. The single peaks in Figure 25-8 correlate with the epochs of extrema $+/-A$. Only in 1963 did an amplitude A not coincide with an outbreak. A Pearson test with Yates's correction gives out the value 15.2 for 1 degree of freedom ($P < .0001$). Historical records of plagues going back to the sixth century confirm this relationship.

TORQUE VARIATIONS AND CYCLES OF FLARES

Sunspots constitute only potentials of solar activity that are actually released by solar eruptions. These are the hallmark of solar-terrestrial interaction. Single energetic flares and periods of enhanced eruptional activity seem to be related to weather. This finding is valid for the quality of weather forecasts (Scherhag, 1952; Reiter, 1979), atmospheric circulation changes (Schuurmans, 1979), rainfall (Markson, 1983), and thunderstorm incidence (Bossolasco, 1972). There are models that explain this effect (Roberts and Olson, 1973; Markson, 1983; Neubauer, 1983; Bucha, 1983). These connections between solar eruptions and weather pose the problem of the prediction of flares. Astronomers hold that flares show a stochastic distribution, but closer examination discloses cycles of solar flares with mean periods of 9 yr, 2.25 yr, and 3.3 mo (Landscheidt, 1984).

Other flare cycles in the range of months are related to variations in dT/dt. The last 13.3-yr cycle $-A$ to $-A$ (1970-1982) had a length of 153.7 mo. Exact harmonics of this cycle represent most of the variance in power spectra of energetic flares covering this period. Figure 25-9 shows the result of a maximum entropy spectral analysis that uses the Burg algorithm (Burg, 1975). The most prominent amplitudes represent the torque cycle itself, the harmonic 4.8 mo, the harmonic 1.2 mo with a neighboring peak at 1.1 mo, and a strong amplitude at 2.8 mo that seems to drop out of this sequence, as

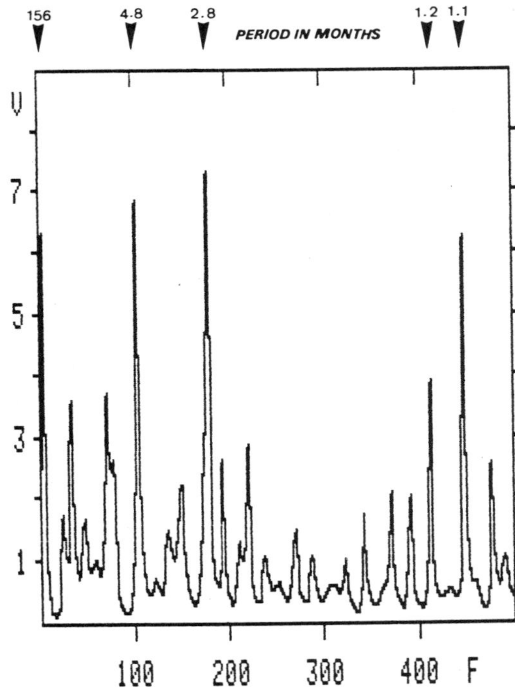

Figure 25-9. Maximum entropy spectrum (Burg algorithm, 312 data, 84 filter coefficients) of energetic solar eruptions.

The analysis consists of X-ray bursts ≥ X1 observed from 1970 to 1982, the number and weight of which were integrated over periods of half-months. P = relative power. Frequency, F, in millicycles per half-month. The prominent amplitudes reflect the quasicycle formed by extrema $-A$ in the time rate of change of the torque in the sun's oscillations about the CM and harmonics of this cycle.

2.4 mo would be the fitting harmonic between 4.8 mo and 2.4 mo. But this shift is the result of the interference with another cycle in the same range. A strong 100-day cycle is formed by the change in the angular acceleration of the vector of the tidal forces of the planets Venus, Earth, and Jupiter that shows a very strong relationship to energetic X-ray bursts ≥ X1 (Landscheidt, 1984). The mean length of this cycle of 3.367 mo and the harmonic of the torque cycle of 2.4 mo, when combined, yield the mean value 2.88 mo, which is near the strong amplitude at 2.8 mo. The maximum entropy spectral analysis is based on 82 filter coefficients. The frequency is measured in millicycles per sampling interval of half a month. Subject of the analysis are X-ray bursts ≥ X1 observed from 1970 to 1982; their numbers and weights were summed up within each sampling interval. A Blackman-Tukey power spectrum yields similar results. A maximum entropy spectrum of the same sample based on 104 filter coefficients shows a finer resolution, but there is no shift in the frequencies. Long-range forecasts by means of such cycles showed a high degree of reliability (Landscheidt, 1981*a*, 1981*b*, 1981*c*, 1983, 1984). The cycles in question are subject to strong variation. The torque cycle 1982-1998 will have a length of 15.72 yr. So the former 4.8-mo cycle will be 5.9 mo long. Without knowledge of these changes, a forecast based on a statistical analysis of previous data will go astray.

SECULAR CYCLE OF 83 YR

A Fourier analysis of yearly values of dT/dt, covering the millennium A.D. 1000-2000, yielded a period of 83 yr related to the frequency component containing the greatest amount of energy of all harmonics. Wolf had found such a period already at the end of the nineteenth century (Brunner-Hagger, 1957). Brier (1979) found exactly this period in the unsmoothed cosine transform of 2,148 autocorrelations of 2,628 monthly sunspot numbers. Just the period of 83 yr comes out again, when all extrema $+/-A$, representing those intriguing impulses of the torque, are taken to constitute a smoothed time series covering more than 7,600 yr. The interval is from 5259 B.C. to A.D. 2347. Smoothing was done over five consecutive values, attributing half filter weight to wing data. Gaussian curve smoothing yielded the same result. Cascade filtering had to be applied to get a really smooth curve. The low-pass-filtered wave in Figure 25-10 represents a section of the time series covering A.D. 1100-2100. The wave has a mean length of 166 yr, but each extremum, whether positive or negative, is related to a maximum in the secular sunspot cycle, while minima occur when the curve is near zero values. The mean period, measured from one extremum $+/-A_S$ to the next one, is 83 yr and varies within 47-118 yr. Gleissberg (1975) found a corresponding range from 40 yr to 120 yr in the secular sunspot cycle. The thick arrows in Figure 25-10 indicate maxima in the secular cycle fixed by Gleissberg (1958), making use of data by Schove (1955). They are in phase with extrema $+/-A_S$. The smoothed curve reveals that there is an excess of positive or negative impulses of the torque for many decades. This points to a cumulative effect.

The wave in Figure 25-10 also contains information about the energy in a secular sunspot maximum. The intensity of the highest 11-yr maximum within the secular period is directly proportional to the amplitude A_S and inversely proportional to the surface integral SI measuring the area above or below the time axis. The following regression equation, based on these quantities, gives out the yearly mean of the relative sunspot number R_M of the highest 11-yr maximum:

$$R_M\,(+/-8) = 902\, A_S/SI - 104.$$

This yields, for example, $R_M = 85$ for the weak secular maximum in A.D. 1377 and $R_M = 240$ for A.D. 1128, when the secular maximum reached the highest value since 5259 B.C.

Figure 25-11 represents corresponding results for the period A.D. 300-1100. The curve elaborated by Gleissberg (1958) is based on data by Schove (1955). These data may be regarded as reliable at least as to the main features of the secular cycle; remaining deviations and uncertainties cannot have great weight because the data were subjected to secular smoothing. D_M stands for intervals between consecutive 11-yr maxima, which follow the secular cycle; minima of D_M correspond to intense sunspot maxima and vice versa. The flat triangles indicate the epochs of secular maxima and minima derived from A_S data. The two data sets are in phase. An evaluation of the total result for the years A.D. 300-1960 confirms the high degree of synchronism of the Gleissberg data and the calculated A_S epochs. A Pearson test yields the value 54.5 (1 degree of freedom; $P << .00001$) for the maxima and 22.3 (1 degree of freedom; $P < .00001$) for the minima. When subsets are formed, the results prove to be homogeneous.

SUPERSECULAR CYCLE OF 391 YR

The supersecular cycle of sunspots, characterized by features like the Maunder Minimum and the Medieval Maximum, seems to be related to the varying energy of the secular wave proportional to squared values of $+/-A_S$. The plot in the middle of Figure 25-12 shows the change in smoothed values of A_S^2 for A.D. 1000-2000. The data were smoothed over three consecutive values, attributing half weight to wing data. There are only two maxima and minima within the millennium investigated. They are synchronous with the supersecular sunspot excursions on top of Figure 25-12, representing a section of Figure 25-6. M stands for Maunder Minimum, S for Spörer Minimum, and O for Medieval Maximum or Medieval Optimum in climate. The coefficient of correlation between the two sets of data amounts to $r = .774$. When the data plotted on top of Figure 25-12 are smoothed, the positive correlation with A_S^2 increases ($r = .8$).

The plot at the bottom of Figure 25-12 shows the time series of the central England air temperature from 1659 to 1978. The dashed line is based on estimated values by Lamb (1969) and the solid line on observed data (Manley, 1974), which are smoothed. The strong positive correlation between these data and A_S^2 is reflected by the coefficient of correlation $r =$ .866. The fact that A_S^2 is also in good agreement with the solar excursions on top of Figure 25-12 is further evidence of the statement by Gates and Mintz (1975) that in the last millennium the climate in the Northern Hemisphere has undergone changes that correspond closely in date, phase, and amplitude with the sun's long-term variations. But there is a third conforming time series now, the wave pattern of the variation of impulses of the torque, that can be computed. Thus, the parallel course of the predictable values of A_S^2 and the supersecular sunspot cycle seems to offer new aspects in the long-range forecast of climate. The continuing steep ascent in the A_S^2 plot, going beyond the year 2000, seems to indicate a new supersecular maximum. If this were right, the secular minimum, which according to the course of the secular wave in Figure 25-10 and additional calculations is to be expected past 1990, should release only a reduced cooling effect. With respect to climatic change, however, a millennium is a rather short period. Therefore, the A_S^2 analysis is extended back to the sixth millennium B.C.

FORECAST OF IMMINENT MINIMA IN THE SECULAR AND SUPERSECULAR CYCLES

Figure 25-13 gives a survey of the energy in the secular wave for more than 7,600 yr. The deepest point in the plot, covering the period 5259 B.C. to A.D. 2347, is at 4847 B.C., and the highest point is reached at A.D. 1128. If this is half an ultralong cycle, the complete cycle should cover about 12,000 yr. Black points represent extrema $+/-A_S$ numbers -64 to $+28$. Black triangles indicate maxima of the supersecular sunspot cycle, and white triangles represent minima. Grand minima and the Medieval Maximum are designated as in Figures 25-6 and 25-12. The extrema properly reflect all marked peaks and troughs of the excursions modulating the sine curve in Figure 25-6. This result can be checked by means of Figure 25-14, presenting a plot from Eddy (1978). The thick arrows, added to the plot, indicate epochs of supersecular extrema derived from the varying energy in the secular

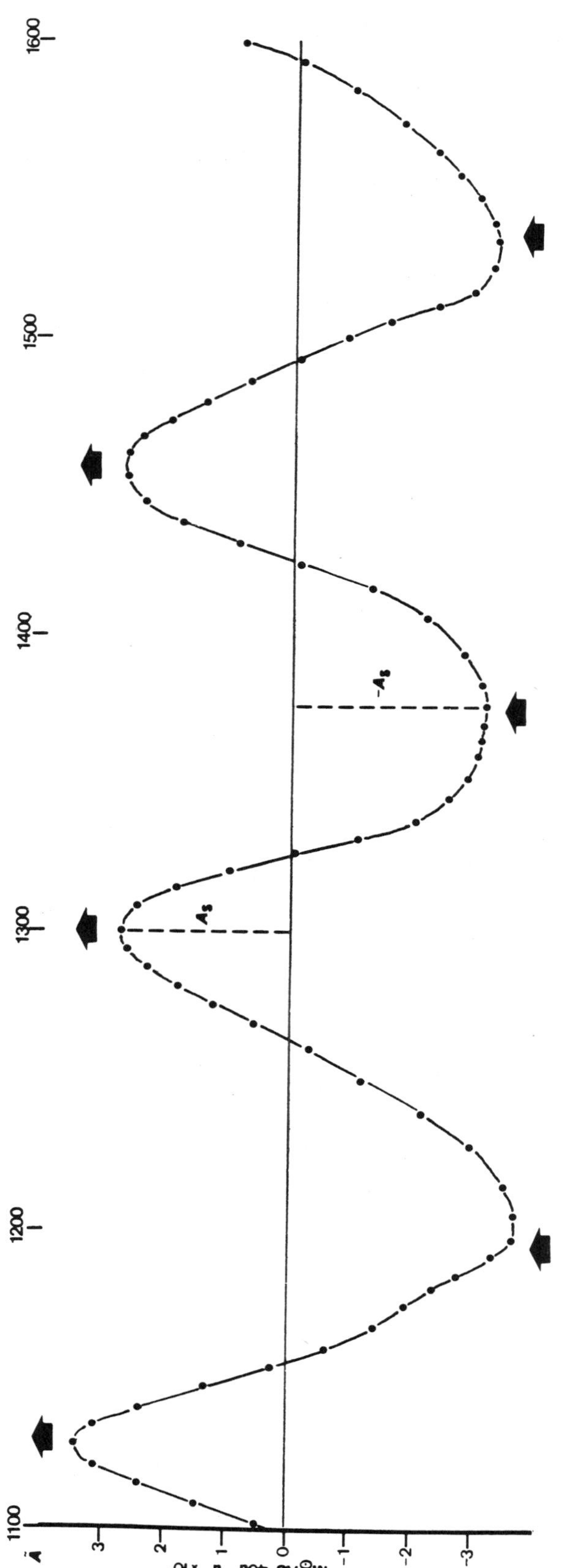
1100
1200
1300
1400
1500
1600
3
2
1
0
-1
-2
-3
A_s
$-A_s$

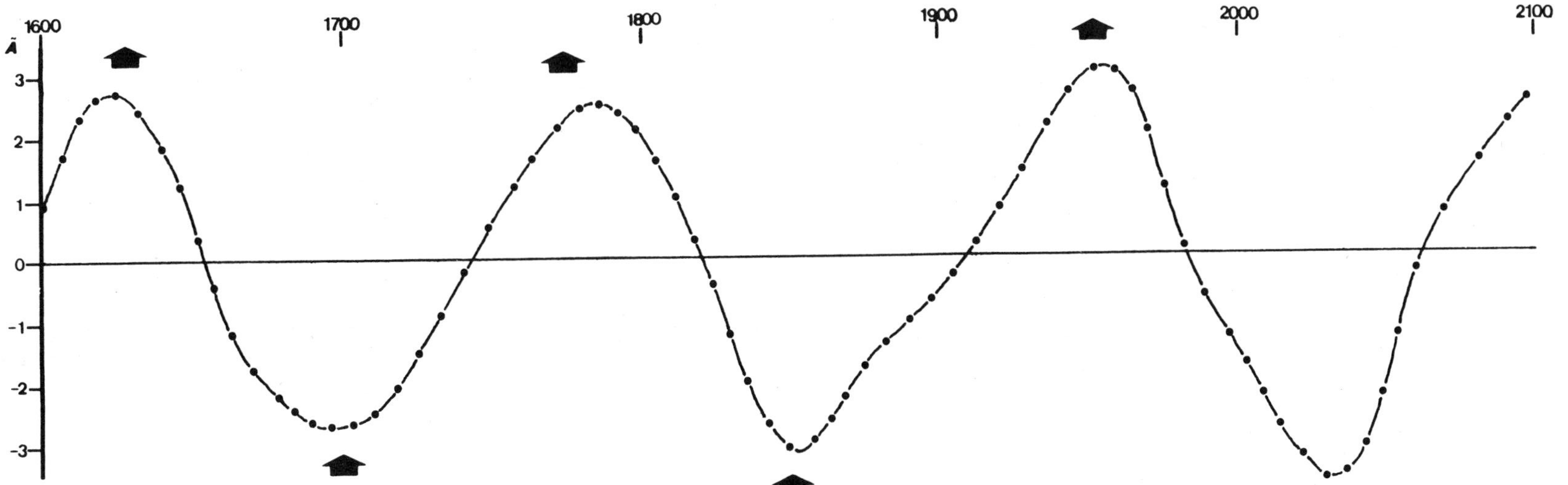

Figure 25-10. Plot of the time series A.D. 1100-2100 of the variation in smoothed extrema $+/-A$, proportional to the strength of impulses of the torque. The secular wave has a mean length of 166 yr. Each extremum $+/-A_S$, appearing at mean intervals of 83 yr, is related to a maximum in the secular sunspot cycle. These maxima, defined by Gleissberg, are marked by thick arrows. Minima occur when the wave is near zero values.

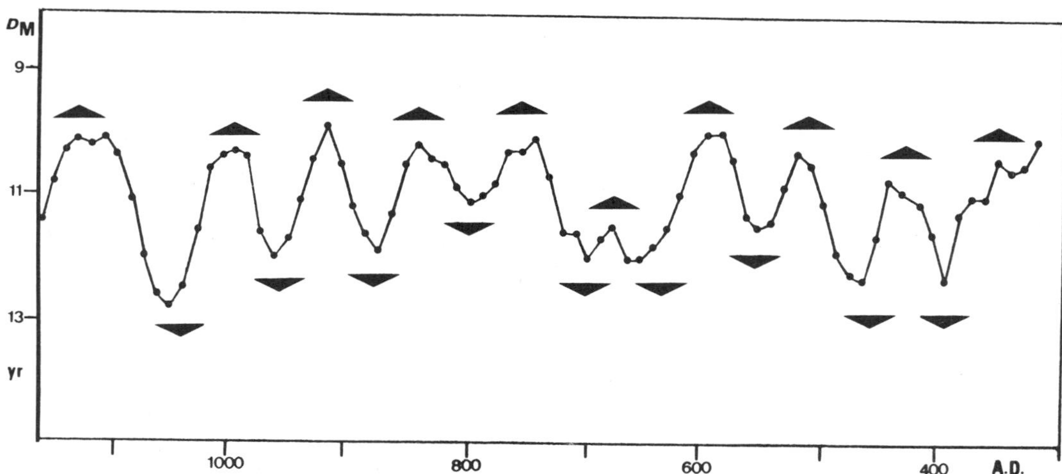

Figure 25-11. Secular cycle of sunspots for A.D. 300-1100. Flat triangles indicate epochs of secular maxima and minima derived from A_S data. (After Gleissberg, 1958; Schove, 1955)

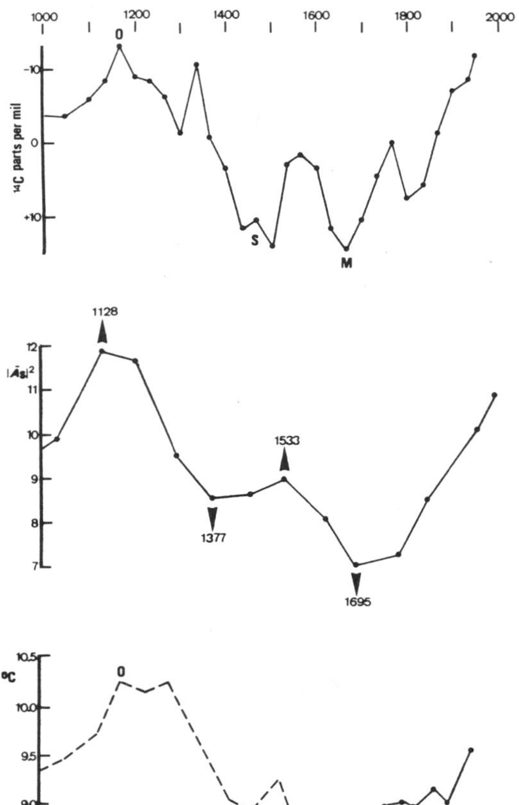

Figure 25-12. Supersecular variation in the energy of the secular wave, proportional to A_S^2 (middle), phase locked with the supersecular sunspot change derived from radiocarbon data (top), and the central England air temperature time series (bottom) (hatched line, estimated values by Lamb, 1969; solid line, observed data). (O, S, and M are explained in the text.)

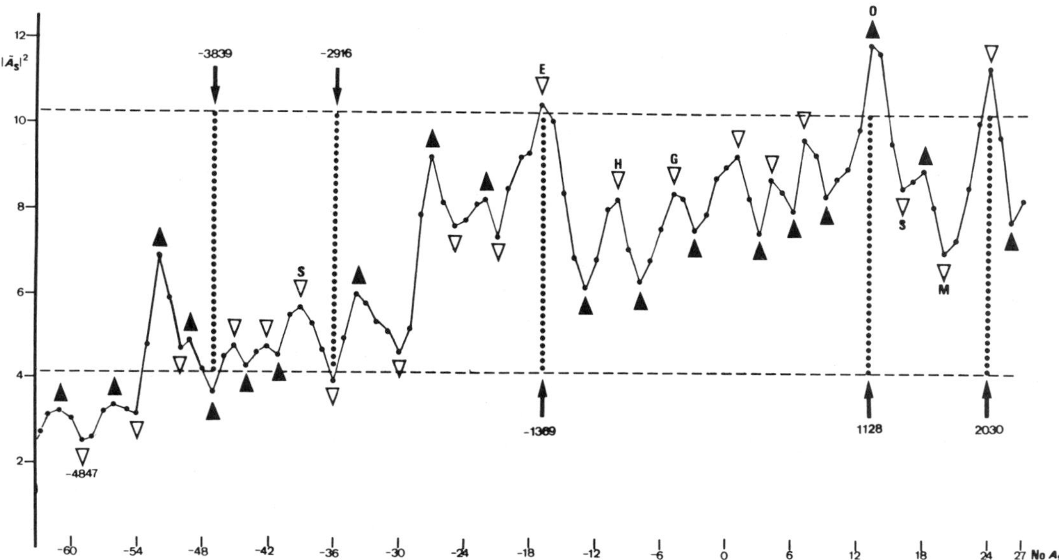

Figure 25-13. Supersecular variation of the energy in the secular wave 5259 B.C. to A.D. 2347. Points in the curve represent epochs of extrema A_S numbers −64 to +28. The mean length of the cycle is 391 yr. Black triangles point at maxima in the corresponding supersecular sunspot cycle; open triangles indicate minima. Grand minima and the Medieval Maximum are designated as in Figures 25-6 and 25-12. When the energy goes beyond quantitative thresholds, indicated by hatched horizontal lines, a phase jump occurs in the correlated supersecular sunspot cycle. These critical phases are marked by vertical dotted lines. A new phase jump is imminent about 2030. It points to a supersecular minimum, comparable with the Egyptian Minimum (E) around 1369 B.C., a prolonged period of distinct cooling and glacier advance.

wave shown in Figure 25-13. Only around A.D. 1 is there some divergence. The A_S^2 values indicate a more differentiated development as indicated by smaller markers. The supersecular cycle has a mean length of 391 yr, which varies within 166-665 yr. This quasiperiod is near 400 yr assumed by Link (1964) and not far from 405 yr derived from the core drilled from the Greenland Ice Sheet by Dansgaard et al. (1969). The 200-yr cycle found by Suess (Schove, 1968) and confirmed by Dewey (1960) and Schove (1961) seems to be a harmonic of the 391-yr cycle.

The plot in Figure 25-13 shows two quantitative thresholds indicated by dashed horizontal lines. When the energy in the wave surmounts the upper line or falls beneath the lower line, a phenomenon like a cutoff potential or a phase jump emerges in the correlated cycle of solar activity. The thresholds seem to work like sharply drawn waveguide walls. At the crucial points, which are set off in Figure 25-13 by dotted vertical lines, a supersecular maximum is followed by another maximum and a minimum by a further minimum. A_S number +23, indicated by a black dot just below the upper horizontal line in Figure 25-13, represents the secular maximum about 1950. After the ensuing secular minimum past 1990, indicated by values near zero in the secular wave in Figure 25-10, the curve will again climb beyond the upper threshold. Therefore, another phase jump is to be expected. The next extremum about 2030 should turn out to be a supersecular minimum. The situation may be compared with that around 1369 B.C., when a comparable phase jump occurred. A_S number −17, about 1369 B.C., marked the Egyptian Minimum, which was characterized by cold climate and a marked glacier expansion (Denton and Karlén, 1973). Thus, a prolonged period of cooling, affecting the Northern Hemisphere, will presumably be initiated by the secular minimum past 1990, reach a climax

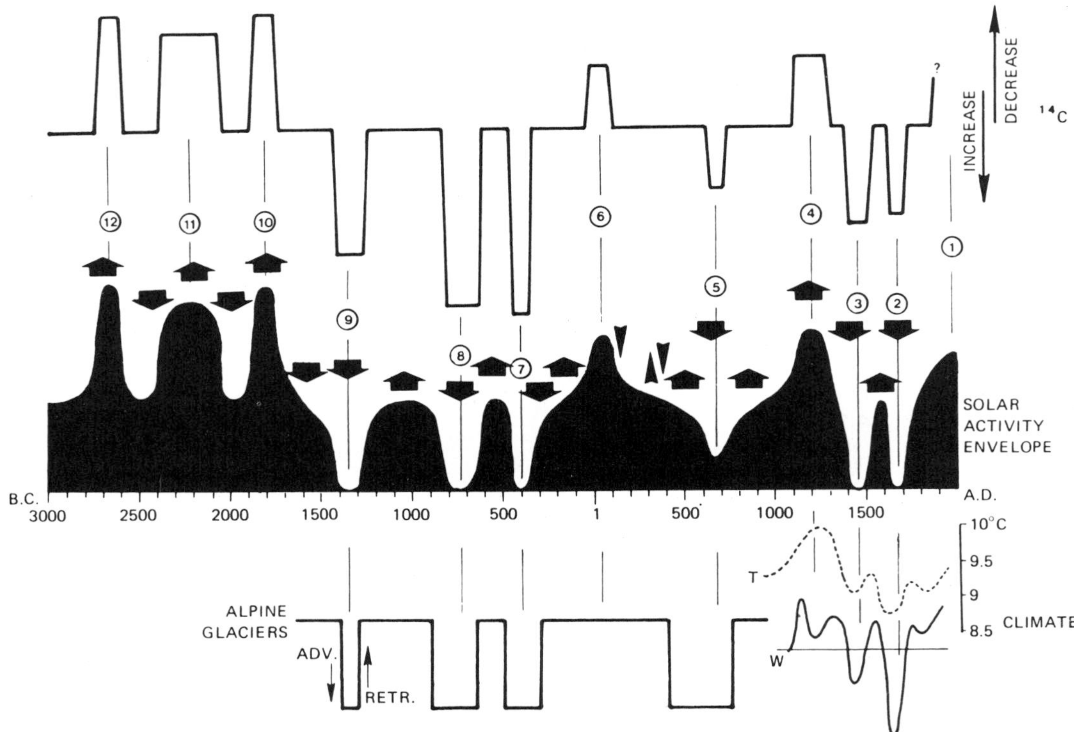

Figure 25-14. Plot from Eddy (1978) showing radiocarbon excursions, solar activity envelope, European climate (T = mean annual temperature; W = severity of northern European winters), and advance and retreat of alpine glaciers.

Thick arrows represent supersecular maxima and minima from Figure 25-13, derived from variations in the energy of the secular wave. Only around A.D. 1 is there some divergence; A_S^2 values delineate a more differentiated course, as indicated by smaller markers. Two consecutive minima or maxima, representing phase jumps marked in Figure 25-13 by vertical dotted lines, change the general level of activity. This is evident from the Egyptian Minimum (No. 9) and the Medieval Maximum (No. 4).

around 2030, and come to an end about 2070. This conforms, as to the next decades, with Lamb's (1982) expectations.

THE SUN'S MOTION AND THE 11-YR CYCLE

The secular minimum past 1990 should go along with cooler climate, reaching at least the level of the secular minimum around 1900. The high level of sunspot activity in the current 11-yr cycle is not opposed to this forecast. The very weak 11-yr maximum in 1805 (R = 49) was the immediate successor of the energetic maximum in 1788 (R = 141). The distance between these maxima, however, was 17 yr, time enough for a gradual decay of activity. Just such a wide distance can be expected after the last 11-yr peak at the end of 1979. The maximum of cycle number 22 should not be reached before 1995/1996 and will be weak ($R < 60$). This forecast is based on the outcome of a cross-correlation analysis of 256 absolute yearly values of the time rate of change of the torque dT/dt with corresponding sunspot numbers R_M. The analysis covers the period 1705-1960. One-year lags go up to +/−34. The estimated cross-correlation function, plotted in Figure 25-15, is not symmetrical about zero. The well-

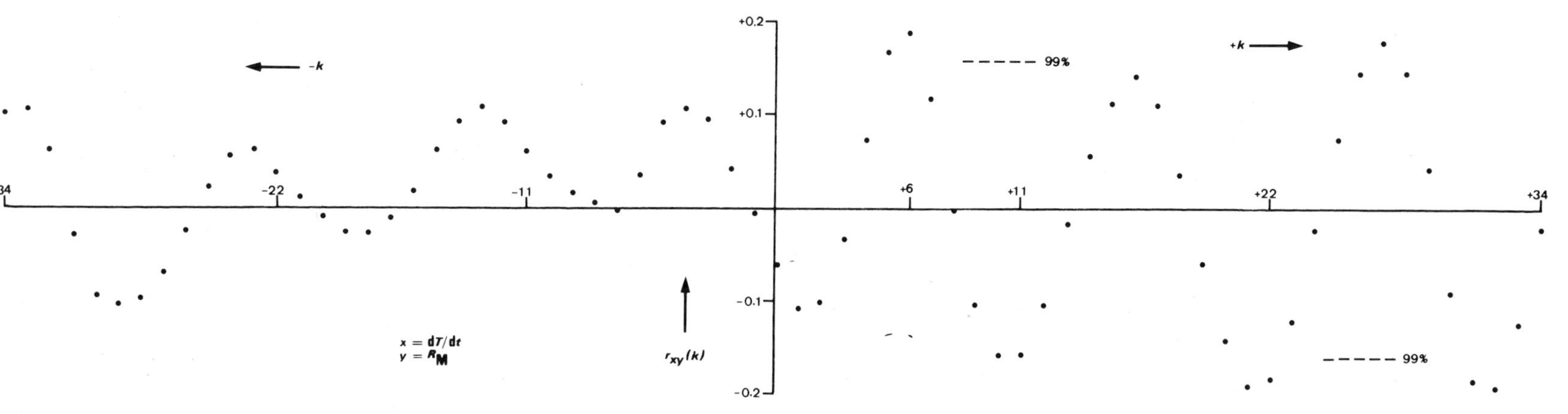

Figure 25-15. Cross-correlation function of 256 absolute yearly values of dT/dt from 1705 to 1960, and corresponding yearly sunspot means R_M, with 1-yr lags k up to $+/-34$. Lack in symmetry indicates that sunspots lag 5.5 yr behind dT/dt. Negative correlation at 11 yr, 22 yr, and 31 yr turns into positive correlation, when a shift is applied that compensates the lag.

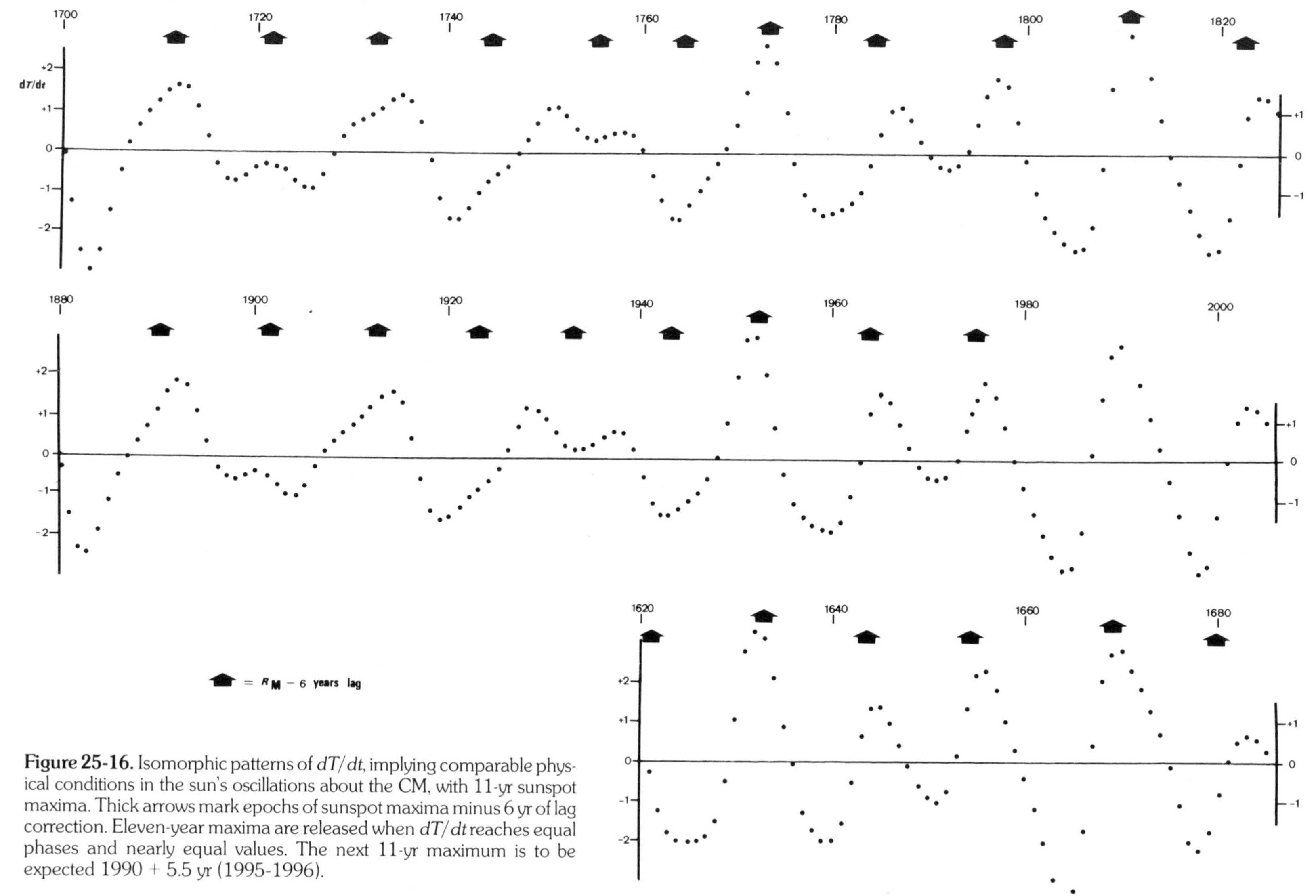

Figure 25-16. Isomorphic patterns of dT/dt, implying comparable physical conditions in the sun's oscillations about the CM, with 11-yr sunspot maxima. Thick arrows mark epochs of sunspot maxima minus 6 yr of lag correction. Eleven-year maxima are released when dT/dt reaches equal phases and nearly equal values. The next 11-yr maximum is to be expected 1990 + 5.5 yr (1995-1996).

defined peak at $k = +6$ shows that sunspots lag 6 yr behind dT/dt. Closer examination yields a lag of 5.5 yr. The negative cross-correlation near 11 yr, 22 yr, and 31 yr, distinctly visible in Figure 25-15, turns into a significant positive correlation at 11 yr, 22 yr, and 31 yr, when a shift is applied that compensates the lag of 5.5 yr.

Figure 25-16 demonstrates the result of a shift of 6 yr. The data—dT/dt, and R minus 6 yr—are arranged such that they allow for the 179-yr recurrence pattern, detected by Jose (1965), which also characterizes variations in the torque. In comparison with the preceding 179-yr period, 11-yr maxima occur when the time rate of change of the torque dT/dt reaches equal phases and nearly equal values. The plot at the bottom of Figure 25-16 demonstrates that this relationship is not dependent on the 179-yr period. What matters is the repetition of nearly the same physical situation indicated by nearly the same phase and value in the change of the torque. It is easy to see that the next extremum fitting the pattern will occur in 1990. Since the lag of 5.5 yr has to be added, the next 11-yr maximum should be released in 1995/1996, 16-17 yr after the last maximum.

THE 31-YR CYCLE AND SAHELIAN DROUGHT

The 31-yr cycle, emergent from the cross-correlation function beside the well-known 11-yr cycle and the magnetic Hale cycle, is a surprise. But it seems to be real, as it is related to climatic data. Sir Francis Bacon held as long ago as 1625 that such a cycle existed in climate. Brueckner (1890) derived a cycle near 34 yr from many types of European weather data including some records centuries long. A power spectrum of Sirén's (1961) Lapland tree-ring index from 1463 to 1960, which is believed to register mean temperatures of the high summer period in northern Finland, shows a prominent peak around 32 yr. As Fairbridge (1984) pointed out at the Second Nordic Symposium on Climatic Changes and Related Problems (Stockholm, 1983), the cycle also has a complicated relationship to the Nile floods. Faure and Gac (1981) derived a 31-yr cycle of Sahelian drought from runoff data of the river Senegal. They predicted an end to the drought in 1985 with full wet conditions reestablished in 1992. It is interesting that Faure and Gac used 7-yr running means of the runoff data because they showed the best fit. Seven years are near the mean interval of consecutive extrema $+/-A$ of 6.65 yr.

Every fifth of these extrema, which followed the rhythm of the 31-yr cycle, exactly fit the drought maxima in 1913, 1941, and 1975. This was also correct for data in the nineteenth century. This time series is not long enough to judge the reliability of this relationship definitely. If it is real, the next Sahelian drought maximum is to be expected about 2010. A humid period should reach a first maximum around 1986. Because the relationship is based on smoothed values, a certain shift may occur.

DISCUSSION

There is some evidence that wave patterns, derived from variations in impulses of the torque, correlate well with solar cycles, and to some extent with climatic change. Thus, it seems possible now to calculate the date, phase, and amplitude of extrema in secular and supersecular cycles of solar activity and climate. Magnetic storms, volcanism, and the ozone column may also be involved. This is a step ahead in practice. Theoretically, it is qualitatively plausible that impulses of the torque should cause disturbances in the symmetry of the sun's plasma flow. But quantitatively the weakness of single impulses of the torque poses a problem. Even the strongest impulses do not go beyond $3\ M_{\odot}\ AU^2/d \times 10^9$. Unfortunately, threshold values for disturbances capable of triggering turbulence and ensuing macroscopic variations in plasma flow cannot be derived from the theory of solar activity because it is still in a rudimentary state of development. $\alpha\omega$ dynamos do not explain the secular cycle, the supersecular cycle, or even variations in length and intensity of the 11-yr cycle. But observation of the sun under natural conditions in the space-time laboratory of the solar system shows definitely that impulses of the torque are strong enough when their effects cumulate over decades or centuries.

The sun oscillating around the CM is, among other impacts, subject to centrifugal force. Plasma exposed to centrifugal force shows convection (Grycz, 1966). Besides, it should be considered that plasma, the elasticity of which allows the excitation and propagation of diverse oscillations and waves, is very sensitive to kinetic and magnetohydrodynamic instabilities (Frank-Kamenetzki, 1972). Hydromagnetic turbulence, generated by microscopic small disturbances, can build up macroscopic magnetic fields and considerable plasma flow (Kadomtsev, 1965). Analogous to weather, turbulence plays an important role in solar activity. Research by Ruelle (1980), Takens (1971), Hofstadter (1981), Gollub and Swinney (1975), and Graham (1982) has shown that contrary to former views the onset of turbulence does not presuppose the involvement of millions or even billions of degrees of freedom; a third oscillation, joining two already existing, can release the transition from laminar to turbulent flow. So-called attractors, mathematical structures, the existence of which in the equations of celestial mechanics was already known to Poincaré, are involved in this transition to turbulence in Navier-Stokes flow. Ruelle (1980) thinks that the Earth's magnetism could be an example of a strange attractor. This would explain magnetic reversals and other instabilities. The sun's motion, too, seems to meet the conditions of such an attractor. The sun is a dissipative system, subjected to quasiperiodic driving force, which makes it oscillate about the CM. Thus, the sun's swing around the CM could play the part of the third varying period in Gollub's turbulence experiments, while the sun's rotation on its axis and the cyclic variations in magnetic fields, generated by the solar dynamo, represent the first two periods.

Another approach considers spin-orbit coupling. The sun revolving around the CM and the sun rotating on its axis could be looked at as coupled oscillators, capable of internal resonance resulting in slight positive or negative accelerations in the sun's spin, which according to John (1932), Perepelkin (1933), Belopolsky (1933), and Howard (1975) are actually observed. A speeding up or slowing down of the sun's rotation rate is liable to influence the sun's activity. Slower rotation is related to enhanced activity and faster rotation to weak activity (Eddy, 1977; Howard, 1975). It is noteworthy in this respect that in 1911/1912, 1929/1930, 1967, and 1970, when the rate of rotation diminished conspicuously, Jupiter was in conjunction with the CM as seen from the sun's center. Jupiter holds 71% of the total mass of the planets and 61% of the total angular momentum. Coupling of both oscillators could result from the sun's motion through its own magnetic fields. Thus, transfer of angular momentum could occur. The four giant planets Jupiter, Saturn, Uranus, and Neptune, which regulate the sun's motion about the CM, could be involved in such transfer as they dispose of more than 99% of the angular momentum of the solar system. The sun's orbital angular momentum is not negligible in comparison with its spin momentum; the relationship is 1 in 10.

With respect to the lack of an elaborated theory of solar activity, it will be difficult to develop magnetohydrodynamic models explaining the observed effects. But this difficulty does not impair the heuristic importance of the results. The stage of gathering data and establishing morphological relationships always precedes the emergence of elaborated theories. As the results cover the fields of astronomy, astrophysics, geophysics, climatology, and meteorology, interdisciplinary cooperation should lead to an improved understanding of the interconnections involved.

REFERENCES

Aimedieu, P., 1981, Les menaces sur l'ozone se confirment, *La Recherche* **12:**493.

Allard, P., 1981, Les Cendres du Mont Saint-Helens peuvent-elles voiler durablement le soleil? *La Recherche* **12:**368.

Avery, S. K., and Geller, M. A., 1979, as quoted in B. M. McCormac and T. A. Seliga (eds.), *Solar-Terrestrial Influences on Weather and Climate,* Dordrecht: Reidel Publishing Company, pp. 20-21.

Bachmann, E., 1965, *Wer hat Himmel und Erde gemessen,* Zurich: Buechergilde Gutenberg, 296p.

Bates, J. R., 1977, Dynamics of stationary ultralong waves in middle latitudes, *Royal Meteorol. Soc. Quart. Jour.* **103:**397.

Belopolsky, A., 1933 (article title not provided), *Zeitschr. Astrophys.* **7:**357.

Benioff, H., 1951, Earthquakes and rockcreeps, *Seismological Soc. America Bull.* **41:**31.

Bossolasco, M.; Dagnino, I.; Elena, A.; and Flocchini, G., 1972 (article title not provided), *Napoli Univ. Ist. Meteorol. e Oceanog.* **1:**213.

Bray, J. R., 1974, Volcanism and glaciation during the past 40 Millennia, *Nature* **252:**679.

Brier, G. W., 1968, Long-range prediction of zonal westerlies and some problems in data analysis, *Rev. Geophysics* **6:**525.

Brier, G. W., 1979, Use of the difference equation methods for predicting sunspot numbers, in B. M. McCormac and T. A. Seliga (eds.), *Solar-Terrestrial Influences on Weather and Climate,* Dordrecht: Reidel Publishing Company, pp. 209-214.

Brueckner, E., 1890, *Klimaschwankungen seit 1700, nebst Bemerkungen ueber die Klimaschwankungen der Diluvialzeit,* Wein: Hoelzel.

Brunner-Hagger, W., 1957, Die Entdeckung der grossen Sonnenfleckenperiode von 83 bzw. 178 Jahren, durch Rudolf Wolf, in den Jahren 1861-1889, *Orion, Schweizer. Astron. Gesell. Mitt.* **55.**

Bryson, R. A., 1982, Volcans et climat, *La Recherche* **13:**845.

Bucha, V., 1983, Direct relations between solar activity and atmospheric circulation, Its Effect on weather and climate, *Studia Geophys. et Geod.* **27:**19.

Burg, J. P., 1975, Maximum Entropy Analysis, Ph.D. dissertation, Stanford University, Palo Alto, California.

Carrea, G., 1978, Sun-weather relationships at Oxford, *Weather* **33:**179.

Challinor, R. A., 1971, Variations in the rate of rotation of the Earth, *Science* **172:**1022.

Chamberlain, J. W., 1976, *A Mechanism for Inducing Climatic Variations through Ozone Destruction: Screening of Galactic Cosmic Rays by Solar and Terrestrial Magnetic Fields,* NASA-CR-148695, 28p.

Damon, P. E., 1977, Solar induced variations at one AU., in O. R. White (ed.), *The Solar Output and Its Variation,* Boulder, Colo.: Colorado Associated University Press, pp. 429-445.

Danjon, A., 1959, Solar flares and changes in the length of day, *Acad. Sci. Comptes Rendus* **B249:**2254; **250:**1399.

Dansgaard, W.; Johnson, S. J.; Miller, J.; and Langway, C. C., Jr., 1969, One thousand centuries of climatic record from Camp Century on the Greenland ice sheet, *Science* **166:**377-381.

Denton, G., and Karlén, W., 1973, Holocene climatic variations: Their pattern and possible cause, *Quat. Research* **3:**155-205.

Dewey, E. R., 1960, The 200-year cycle in the length of the sunspot cycle, *Jour. Cycle Research* **7:**70.

Dronia, H., 1975, Veraendern wir unser Klima? *Bild d. Wissenschaft* **12:**52.

Eddy, J. A., 1976, The Maunder Minimum, *Science* **192:**1189.

Eddy, J. A., 1977, The case of the missing sunspots, *Sci. American* **236**(5):80.

Eddy, J. A., 1978, Evidence for a changing sun, in J. A. Eddy (ed.), *The New Solar Physics,* Boulder, Colo.: Westview Press, pp. 11-33.

Fairbridge, R. W., 1983, Climate Change: An Exogenetic Model, unpublished manuscript.

Fairbridge, R. W., 1984, The Nile Floods as a global climatic solar proxy, in N. A. Mörner and W. Karlén (eds.), *Proceedings of the Second Nordic Symposium on Climatic Change and Related Problems,* Dordrecht: Reidel Publishing Company, pp. 181-190.

Fairbridge, R. W., and Hillaire-Marcel, C., 1977, An 8000-year paleoclimatic record of the "Double-Hale" 45-year solar cycle, *Nature* **268:**413.

Faure, H., and Gac, J. Y., 1981, Will the Sahelian drought end in 1985? *Nature* **291:**475.

Flohn, H., 1984, Possible geophysical mechanism of abrupt climatic change, in N. A. Mörner and W. Karlén (eds.), *Proceedings of the Second Nordic Symposium on Climatic Change and Related Problems,* Dordrecht: Reidel Publishing Company, pp. 521-531.

Frank-Kamenetski, D. A., 1972, *The Fourth State of Matter,* New York: Plenum Press.

Gates, W. L., and Mintz, Y., 1975, *Understanding Climatic Change,* Appendix A, Washington, D.C.: National Academy of Sciences.

Gilliland, R. L., 1983, Climate change as a test of the influence of external perturbations, in B. M. McCormac (ed.), *Weather and Climate Responses to Solar Variations,* Boulder, Colo.: Colorado Associated University Press, pp. 273-281.

Gleissberg, W., 1958, The 80-year sunspot cycle, *British Astron. Assoc. Jour.* **68:**150.

Gleissberg, W., 1975, Gibt es in der Sonnenfleckenhaeufigkeit eine 179-jaehrige Wiederholungstendenz? *Frankfurt Univ. Astron. Inst. Veroeff.* **57:**2.

Gleissberg, W., 1976, Das juengste Maximum des 80-jaehrigen Sonnenfleckenzyklus, *Kleinheubacher Berichte* **19:**661.

Gollub, J., and Swinney, H. L., 1975, Onset of turbulence in a rotating fluid, *Phys. Rev. Letters* **35:**927-930.

Graham, R., 1982, Ein Stueck unberechenbarer Natur: Die Turbulenz, *Bild d. Wissenschaft* **4:**68.

Gribbin, J., and Plagemann, S., 1972, Discontinuous

change in Earth's spin following great solar storm of August 1972, *Nature* **243:**26.

Grycz, B., 1966, *Fourth State of Matter,* London: Iliffe Books.

Hammer, C. U.; Clausen, H. B.; and Dansgaard, W., 1980, Greenland ice sheet evidence of post-glacial volcanism and its climatic impact, *Nature* **288:**230-235.

Haskell, P. T.; Ashall, C.; and Hemming, C. F. (eds.), 1966, *Locust Handbook,* London: Antilocust Research Center.

Hays, D. H., 1977, Climatic change and the possible influence of variations of solar input, in O. R. White (ed.), *The Solar Output and Its Variation,* Boulder, Colo.: Colorado Associated University Press, pp. 73-90.

Heath, D. F., and Thekaekara, M. P., 1977, The solar spectrum between 1200 and 3000 Å., in O. R. White (ed.), *The Solar Output and Its Variation,* Boulder, Colo.: Colorado Associated University Press, pp. 193-212.

Hill, W. J., and Sheldon, P. N., 1975, Statistical modeling of total ozone measurements with an example using data from Arosa, Switzerland, *Geophys. Research Letters* **2:**541-544.

Hofstadter, D. R., 1981, Strange attractors; Mathematical patterns delicately poised between order and chaos, *Sci. American* **245:**16.

Howard, R., 1975, The rotation of the Sun, *Sci. American* **232**(4):106-114.

Hoyt, D. V., and Eddy, J. A., 1982, *An Atlas of Variations in the Solar Constant Caused by Sunspot Blocking and Facular Emissions from 1874 to 1981,* Boulder, Colo.: National Center for Atmospheric Research.

John, S., 1932 (article title not provided), *Internat. Astron. Union Trans.* **4:**43.

Jose, P. D., 1965, Sun's motion and sunspots, *Astron. Jour.* **70:**193-200.

Junk, H. P., 1982, Die Maximum-Entropie-Spektral-Analyse und ihre Anwendung auf meteorologische Zeitreihen, Diplomarbeit, Meteorologisches Institut der Universität, Bonn, 96p.

Kadomtsev, B. B., 1965, *Plasma Turbulence,* London: Academic Press.

Lamb, H. H., 1969, Climatic fluctuations, in H. E. Landsberg and H. Flohn (ed.), *World Survey of Climatology,* vol. 2, Amsterdam: Elsevier.

Lamb, H. H., 1970, Volcanic dust in the atmosphere; with a chronology and assessment of its meteorological significance, *Royal Astron. Soc. Philos. Trans.* **A266:**425.

Lamb, H. H., 1982, *Climate, History, and the Modern World,* London and New York: Methuen, 387p.

Landscheidt, T., 1976, Beziehungen zwischen der Sonnenaktivitaet und dem Massenzentrum des Sonnensystems, *Nachrichten der OlbersGesellschaft Bremen, no. 100,* p. 2.

Landscheidt, T., 1980, Saekularer Tiefpunkt der Sonnenaktivitaet-Ursache einer Kaelteperiode um das Jahr 2000? *Jahrbuch der Wittheit zu Bremen* **24:**189.

Landscheidt, T., 1981*a*, Cycles of solar flares, in D. Overdieck, J. Mueller, and H. Lieth (eds.), *Ninth International Congress of Biometeorologie, Abstract Volume,* Universität Osnabrueck.

Landscheidt, T., 1981*b*, Solar oscillations, flare pattern, rise of leucopenia, and cycle of locust plagues, in D. Overdieck, J. Mueller, and H. Lieth (eds.), *Ninth International Congress of Biometeorologie, Abstract Volume,* Universität Osnabrueck.

Landscheidt, T., 1981*c*, Swinging sun, 79-year cycle, and climatic change, *Jour. Interdisciplinary Cycle Research* **12:**3.

Landscheidt, T., 1983, Solar oscillations and climatic change, in B. M. McCormac (ed.), *Weather and Climate Responses to Solar Variation,* Boulder, Colo.: Colorado Associated University Press, pp. 293-308.

Landscheidt, T., 1984, Cycles of solar flares and weather, in N. A. Mörner and W. Karlén (eds.), *Proceedings of the Second Nordic Symposium on Climatic Change and Related Problems,* Dordrecht: Reidel Publishing Company.

Landscheidt, T., 1986, Long-range forecast of sunspot cycles, in G. Heckman and M. A. Shea (eds.), *Solar Terrestrial Predictions,* Boulder, Colo.: National Oceanographic and Atmospheric Administration, pp. 48-57.

Lin, Y. C.; Fan, C. Y.; Damon, P. E.; and Wallick, E. J., 1975 (article title not provided), *14th Internat. Cosmic Ray Conf. (Muenchen) Proc.* **3:**995.

Link, F., 1964, Manifestations de l'activité solaire dans le passé historique, *Planetary and Space Sci.* **12:**333.

Manley, G., 1974, Central England temperatures: Monthly means 1659-1973, *Royal Meteorol. Soc. Quart. Jour.* **100:**389.

Markson, R., 1983, Solar modulation of fair-weather and thunderstorm electrification and a proposed program to test an atmospheric electrical sun-weather mechanism, in B. M. McCormac (ed.), *Weather and Climate Responses to Solar Variation,* Boulder, Colo.: Colorado Associated University Press, pp. 323-343.

McCormac, B. M. (ed.), 1983, *Weather and Climate Response to Solar Variations,* Boulder, Colo.: Colorado Associated University Press.

McCormac, B. M., and Seliga, T. A. (eds.), 1979, *Solar-Terrestrial Influences on Weather and Climate,* Dordrecht: Reidel Publishing Company.

Miles, M. K., and Gildersleeves, P. B., 1977, A statistical study of the likely causative factors in the climatic fluctuations of the last 100 years, *Meteorol. Mag.* **106:**314.

Mitchell, J. M., 1970, *Global Effect of Environmental Pollution,* Holland: Singer.

Mörner, N. A., and Karlén, W. (eds.), 1984, *Pro-*

ceedings of the Second Nordic Symposium on Climatic Change and Related Problems (Stockholm, May, 1983), Dordrecht: Reidel Publishing Company.

Munk, W. H., and MacDonald, G. J. F., 1960, *The Rotation of the Earth*, Cambridge: Cambridge University Press.

Neubauer, L., 1983, The sun-weather connection—Sudden stratospheric warmings correlated with sudden commencements and solar proton events, in B. M. McCormac (ed.), *Weather and Climate Responses to Solar Variations*, Boulder, Colo.: Colorado Associated University Press, pp. 395-397.

Perepelkin, E. J., 1933 (article title not provided), *Zeitschr. Astrophys.* **6:**121.

Reiter, R., 1979, Influences of solar activity on the electrical potential between the ionosphere and the Earth, in B. M. McCormac and T. A. Seliga (eds.), *Solar-Terrestrial Influences on Weather and Climate*, Dordrecht: Reidel Publishing Company, pp. 243-251.

Reiter, R., and Heck, H. D., 1983, Die geheimnisvolle Wolke, *Bild d. Wissenschaft* **20:**32.

Richter, C. F., 1958, *Elementary Seismology*, San Francisco and London: W. H. Freeman.

Roberts, W. O., and Olson, R. H., 1973, New evidence for effects of variable solar corpuscular emission on the weather, *Rev. Geophysics and Space Physics* **11:**371.

Rowland, F. S., 1978, Stratospheric ozone: Earth's fragile shield, in *Encyclopaedia Britannica 1979 Yearbook of Science and the Future*, The University of Chicago, pp. 170-191.

Rudloff, H. V., 1967, *Die Schwankungen und Pendelungen des Klimas in Europa seit dem Beginn der regelmaessigen Instrumentenbeobachtungen*, Vieweg: Braunschweig.

Ruelle, D., 1980, Les attracteurs étranges, *La Recherche* **11:**132.

Ruelle, D., and Takens, F., 1971, On the nature of turbulence, *Commun. in Math. Phys.* **20:**167-192.

Ruelle, D., and Takens, F., 1971, Note concerning our paper "On the nature of turbulence," *Commun. in Math. Phys.* **23:**343-344.

Scherhag, R., 1952, Die explosionsartigen Stratosphaerenerwaermungen des Spaetwinters 1951/52, *Berichte des Deutschen Wetterdienstes US-Zone, no. 38*, p. 51.

Schoenwiese, C. D., 1979, *Klimaschwankungen*, Berlin, Heidelberg, New York: Springer-Verlag.

Schoenwiese, C. D., 1980, Climatic change predictability and realization by means of statistical filter models, *Meteorol. Rundschau* **33:**75.

Schove, D. J., 1955, The sunspot cycle 649 BC to AD 2000, *Jour. Geophys. Research* **60:**127.

Schove, D. J., 1968, Solar, Auroral, and Climatic Cycles since 4000 BC, paper presented at the Solar-Terrestrial Conference 1968, Brussels.

Schove, D. J., 1961, Auroral numbers since 500 BC, *British Astron. Assoc. Jour.* **72:**31.

Schove, D. J., 1983, Sunspot cycles and global oscillations in N. A. Mörner and W. Karlén (eds.), *Proceedings of the Second Nordic Symposium on Climatic Changes and Related Problems*, Dordrecht: Reidel Publishing Company, pp. 257-259.

Schuurmans, C. J. E., 1979, Effects of solar flares on the atmospheric circulation, in B. M. McCormac and T. A. Seliga (eds.), *Solar-Terrestrial Influences on Weather and Climate*, Dordrecht: Reidel Publishing Company, pp. 105-118.

Schwentek, H., 1971, The sunspot cycle 1958/1970 in ionospheric absorption and stratospheric temperatures, *Jour. Atmos. and Terrest. Physics* **33:**1839.

Sirén, G., 1961, Skogsgraenstallen som indikator foer klimatfluktuationerna i norra Fennoskandien under historisk tid, *Commun. Inst. Forest. Fenniae* **54**(2):1-66.

Visser, S. W., 1959, On the connection between the 11-year sunspot period and the periods of about 3 and 7 years in world weather, *Geofisica Pura e Appl.* **43:**302-318.

Wexler, H., 1953, Radiation balance of the Earth as a factor in climate change, in H. Shapley (ed.), *Climate Change*, Cambridge, Mass.: Harvard University Press.

White, O. R. (ed.), 1977, *The Solar Output and Its Variation*, Boulder, Colo.: Colorado Associated University Press.

26: The Sun's Orbit, A.D. 750–2050: Basis for New Perspectives on Planetary Dynamics and Earth–Moon Linkage

Rhodes W. Fairbridge and John E. Sanders
Columbia University

Abstract: The Sun is forced into an orbit around the center of mass (barycenter) of the solar system because of the changing mass distributions of the planets. Irregularities in the solar orbit generate cycles that have periods ranging from a few years to several hundred years. Much of the solar orbit is a response to the movements of Jupiter and Saturn, which change angular relationships by 90° in just under 5 yr. A fundamental rhythm is the Saturn-Jupiter Lap (SJL) cycle of 19.859 Earth yr. Solar-planetary-lunar dynamic relationships form a new basis for understanding cyclic solar forcing functions on the Earth's climate.

INTRODUCTION

This paper brings to the attention of scientists studying the Earth's climate and solar-terrestrial relationships some aspects of celestial mechanics that we think are significant and that seem to have been either ignored by astronomers or cast in the completely wrong terms of a Sun-centered system. The central topics are usually considered to be astronomy; indeed, when we raise these topics with geologists, they wonder why we are bothering about such astronomical details.

We think that our results are vital to understanding solar-terrestrial relationships. We show the physical basis for many cycles, some of which appear to possess highly specific periods and thus are amenable to mathematical analysis. We caution, however, that nearly every cycle within our solar system and galaxy is subject to built-in variations that complicate projection of these cycles through long time periods. In contrast, certain other periodicities lack any such apparently neat simplicity even on short time scales, but they nonetheless are real. For these, the term *recurrence tendency* seems more appropriate than *cycle*.

We present some of the first results of our comprehensive analysis of the Sun's orbit during the period A.D. 750-2050. We explore thereafter some new perspectives made possible by our results. We conclude with a discussion of solar-lunar-terrestrial relationships.

We have used the NASA-JPL Long Ephemeris DE-102 to iterate the Cartesian coordinates of the Sun at 200-day intervals for the period A.D. 750-2050. We have plotted the successive positions of the Sun on a series of two-dimensional diagrams. Our diagrams display 64 solar orbits of the barycenter with periods that range from 15 yr to 26 yr and average 19.9 yr.

We have organized the orbit diagrams into

The reader is directed to the Selected Bibliography on Sun-Earth Relationships and Cycles Having Periods of Less Than 10,000 Years, also by J. E. Sanders and R. W. Fairbridge, for the references cited in this paper.

eight classes, designated by letters of the alphabet from A through G. Using the patterns of orbital classes and the distance of the center of the Sun from the barycenter, we have diagnosed seven repeat cycles having periods of approximately 178-179 yr. For each orbit diagram we recognize a geometrically determined axis of symmetry (AXSYM). From one major orbital loop to another, the AXSYM undergoes a clockwise (hence, retrograde) motion of an average of 121.07°. After nine solar orbits, the AXSYM net motion is 9.63° of arc, corresponding to an Orbital Symmetry Progression (OSP) cycle of 178.73 yr.

The Sun's orbit is locked onto the 19.857 SJL cycle. The 9-orbit repeat cycle corresponds approximately with a Saturn-Jupiter Repeat (SJR) cycle of 177.92 yr (formed by 15 orbits of Jupiter and 6 orbits of Saturn). An Outer-Planets Restart (OPR) cycle of 4,448 yr is defined by Stacey (1963). The approximate commensurabilities among all the cycles re-emphasize the dynamic unity of the solar system. Other important geophysical variables that are possibly linked to the Sun's orbit include changes in the Sun's oblateness, diameter, and spin rate, involving both the spin rate of the Sun's surface and the differential spin rate between the Sun's core and its surface.

The Earth's spin rate also fluctuates in changes that seem to be linked to variables connected with both the Sun and the Moon. Accelerations are accompanied by amplified zonal circulation patterns, mild winters in the Northern Hemisphere, and stronger monsoons in the tropics. By contrast, spin deceleration is associated with blocking circulation patterns, Northern Hemisphere winters that are colder than normal, and weaker or delayed monsoons. In addition, spin rate changes are accompanied by fluctuations of the intensity and of the secular drift of the geomagnetic field.

Both the Earth and the Sun respond to the variable forces introduced by the Moon's orbital fluctuations. Lunar cycles range in length from the apsides cycle of 8.849 yr and the nodal cycle of 18.6134 yr to the Perigee-Syzygy (P-S) cycle of 31.008 yr, the apsides-perihelion cycle of 62.013 yr, the nodal-perihelion cycle of 93.020 yr, to the progression of the lunar perigee (PLP) cycle of 556.027 yr and the parallactic tidal cycle of 1,843.346 yr (F. J. Wood, 1985).

Individual solar orbits do not correspond to the sunspot cycle (SSC), but approximately 16 SSCs correspond to one 178.71 yr APS. The repeat cycles as well as the SJL cycle of about 20 yr coincide with like-period terrestrial climate cycles including those inferred from isotopic studies of the Greenland ice cores and from other data.

RESULTS

Solar-Orbit Diagrams

Our joint investigation grew out of an attempt by Sanders to determine if the Maunder Sunspot Minimum could be explained in terms of the changing ellipticity of the Sun's orbit. Accordingly, the first iteration of the computer runs was at 200-Julian-day intervals for a 500-yr period beginning in A.D. 1516. We later extended this series back to A.D. 750. Our diagrams are two-dimensional maps of the positions of the Sun in the X-Y plane of the Cartesian coordinates of the NASA-JPL Ephemeris, whose origin is the center of mass of the solar system. This X-Y plane is the Earth's equatorial plane extended in all directions, scarcely the most appropriate one for our purposes. We have not followed the method of Jose (1965) whereby the X-Y plane was shifted so it became the invariable plane of the solar system (mean plane of all planetary orbits). Such a shifting of the X-Y plane would doubtless introduce some small modifications to our results. However, we do not think serious changes would be required. Future studies of the effects of the Z direction should be made.

In plotting the diagrams such as shown in Figures 26-1 and 26-2, we have adopted the method of showing only one orbit at a time. We have defined each orbit as extending from a given closest approach of the center of the Sun to the barycenter (for convenience, abbreviated as B or BAC) to the next following closest approach after an excursion to the remotest

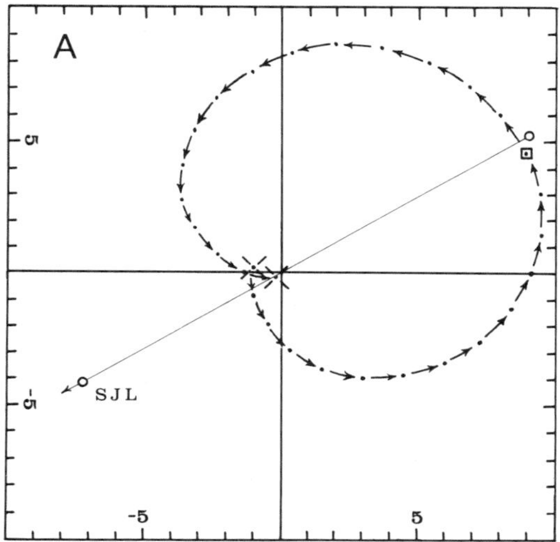

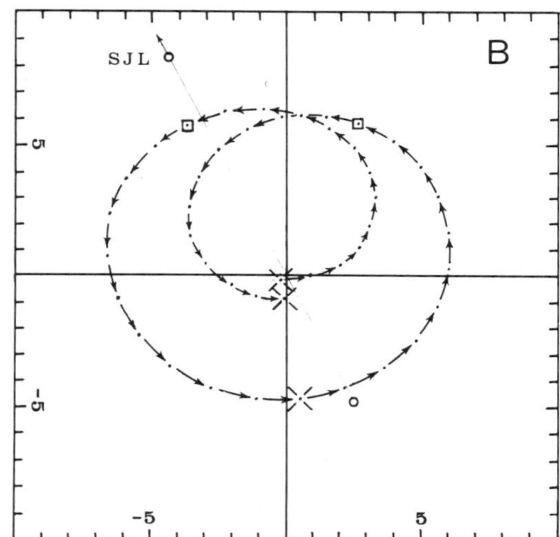

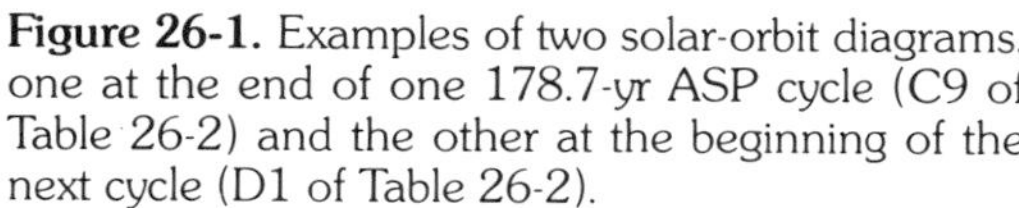

Figure 26-1. Examples of two solar-orbit diagrams, one at the end of one 178.7-yr ASP cycle (C9 of Table 26-2) and the other at the beginning of the next cycle (D1 of Table 26-2).

The lengths of the sides of the square frames of these diagrams are 20×10^{-3} A. U.; the scale marks are at 10^{-3} A. U.; the horizontal and vertical axes intersect at the barycenter (BAC); and the positive X axis is the vernal equinox of 1950.0. Dots mark the positions of the center of the Sun at 200-day intervals; arrows show the direction of the Sun's motion. Apobacs, dots within squares; peribacs, diagonal crosses. Line from orbit path through barycenter to Saturn-Jupiter lap positions (SJL) marked by small circles; arrow points to the direction of the SJL.

(A) In this class C orbit, the distance from the center of the Sun to the barycenter (=heliocenter-barycenter separation) at the apobac of A.D. 1306 was 1.459×10^6 km. On Ocotber 31, 1304, an All-Planets Synod took place in the "southwest" quadrant, opposite the Sun's apobac position; for several years thereafter, all the major planets lay in this same quadrant. The Saturn-Jupiter lap (SJL) happened in about 1306; it nearly coincided with the time of the Sun's apobac of this orbit. At the Sun's peribac position of this orbit, Jupiter and Saturn were in opposition with respect to the barycenter; the heliocenter-barycenter separation was then only 49,800 km, an approach close enough to qualify as a nuclear transit of Landscheidt (1976). The sunspot minimum of A.D. 1299, the maximum of A.D. 1307, and another minimum in A.D. 1311 do not coincide with any of the key points of this orbit.

(B) In this class E orbit, the next-succeeding one after that of Figure 26-1*(a)*, the Sun traces an epitrochoid with both a major loop and a minor loop. The heliocenter-barycenter separation at the A.D. 1320 apobac was only 1.017×10^6 km; that at the A.D. 1331 minor maximum was 0.942×10^6 km. The SJL of 1326 followed 6 yr after the Sun's 1320 apobac.

These two solar orbits spanned most of the Wolf Sunspot Minimum of A.D. 1280-1350. The two sunspot maxima at around 1318 and 1327 were very weak. As with other such sunspot minima, radiocarbon flux on the Earth was very high. In northern Europe, starting in 1310, the climate deteriorated drastically (Lamb, 1972, 1977). In the winter of 1322-1323, the Baltic Sea was totally frozen over. Catastrophic droughts or dry spells occurred in north Africa and in the American southwest.

position on that orbit. We then start a new diagram. Thus, the time of the end point of one orbit map becomes the starting time of the next orbit map. With small orbits, we modified this procedure by incorporating them into the following larger orbit to make a closed figure having a clearly defined AXSYM (see Figs. 26-3 and 26-4). Our method of showing only single orbits contrasts with that used by Jose (1965, Fig. 1, p. 194), in which many orbits were superimposed.

We have found that we needed some new terms for referring to the Sun's position with respect to the barycenter. By analogy with standard astronomic terms (such as *perigee* and *apogee* for the Moon's orbit around the Earth,

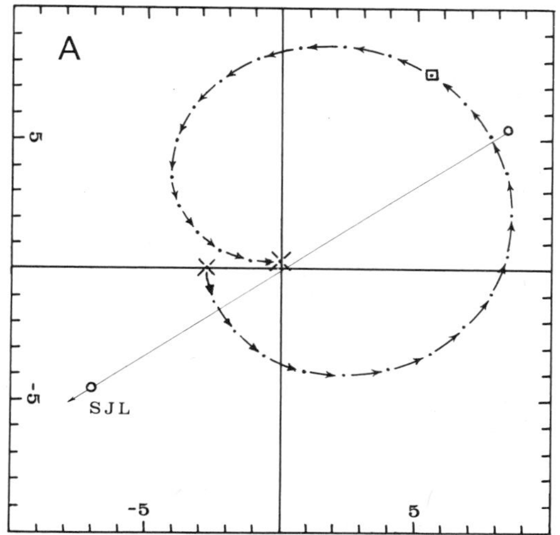

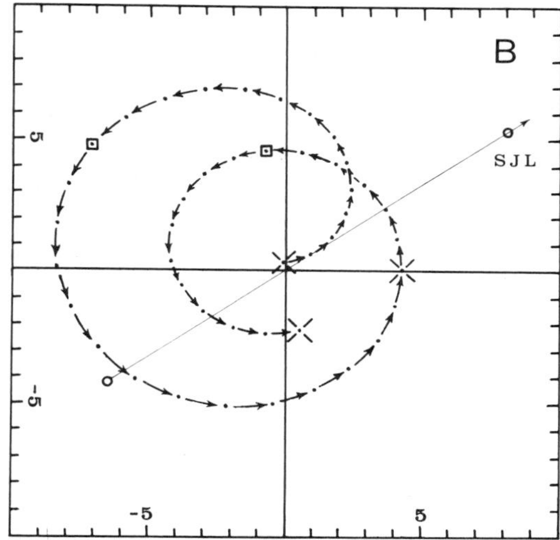

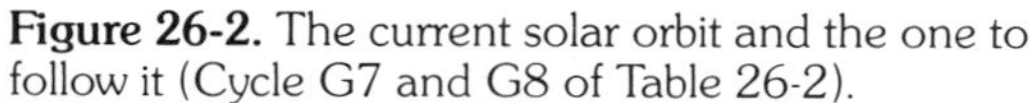

Figure 26-2. The current solar orbit and the one to follow it (Cycle G7 and G8 of Table 26-2).

(A) In the current solar orbit, of class G, the solar apobac was reached in A.D. 1983, about a year following the SJL of A.D. 1982. The Sun is moving toward its projected peribac of 1990, at which time it will be close enough to the barycenter to qualify as a nuclear transit. This event will mark the close of the first of two 39-yr segments in a Landscheidt cycle of 178 yr that is subdivided into one segment of 100-yr duration and two successive segments of 39 yr each.

(B) The next following orbit, another of Class E (compare with Figure 26-1*b*), begins with a nuclear transit, includes a major apobac on the large loop (projected for A.D. 1997) and a minor one on the smaller loop, and ends with a peribac projected for A.D. 2014.

and *perihelion* and *aphelion* for the planets' orbits around the Sun), we propose to define the minimum Sun-barycenter distance as *peribac* and the maximum Sun-barycenter distance as *apobac.* We have delineated each orbit of the Sun around the barycenter (SOB) on the basis of peribac to peribac.

The path of the Sun describes complex figures consisting of major loops and minor loops (see Figs. 26-1 and 26-2). This configuration has been interpreted as the result of the Sun's trajectory in moving around an orbital ellipse of large eccentricity at the same time as the semimajor axis of this ellipse (its so-called longitude of peribac) is progressing very rapidly—notably, at the orbital period of Saturn—with one focus serving as the pivot point (Sanders, 1981). When the orbits of Jupiter and Saturn place them on the same side of the Sun, the Sun generates major loops (swinging around the orbital ellipse on the side opposite the pivot focus). When the orbits of Jupiter and Saturn place them on opposite sides of the Sun, the Sun's path makes minor loops (Sun curving around its orbital ellipse on the same side as the pivot focus). The single major loops approximate *cardioids,* or heart-shaped figures; the double loops, *epitrochoids* (Blizard, 1981).

In reviewing the 64-orbit diagrams, we noticed a curious repetition of distinctive patterns. The orbital diagrams conform to eight classes, defined geometrically and designated arbitrarily by us as A to H (Fig. 26-3). They constitute empirical modes. The eight categories merge with one another, and the last ones (G and H) approach the form of the first one (class A).

The major loop in each class partly resembles a cardioid, the two bulges of which are connected by a crossover leading to the minor loop. We found that a line drawn at a tangent to the two bulges of the cardioid could be plotted, together with a line intersecting at 90° and passing through the crossover to extend to the midpoint on the major loop. This projection

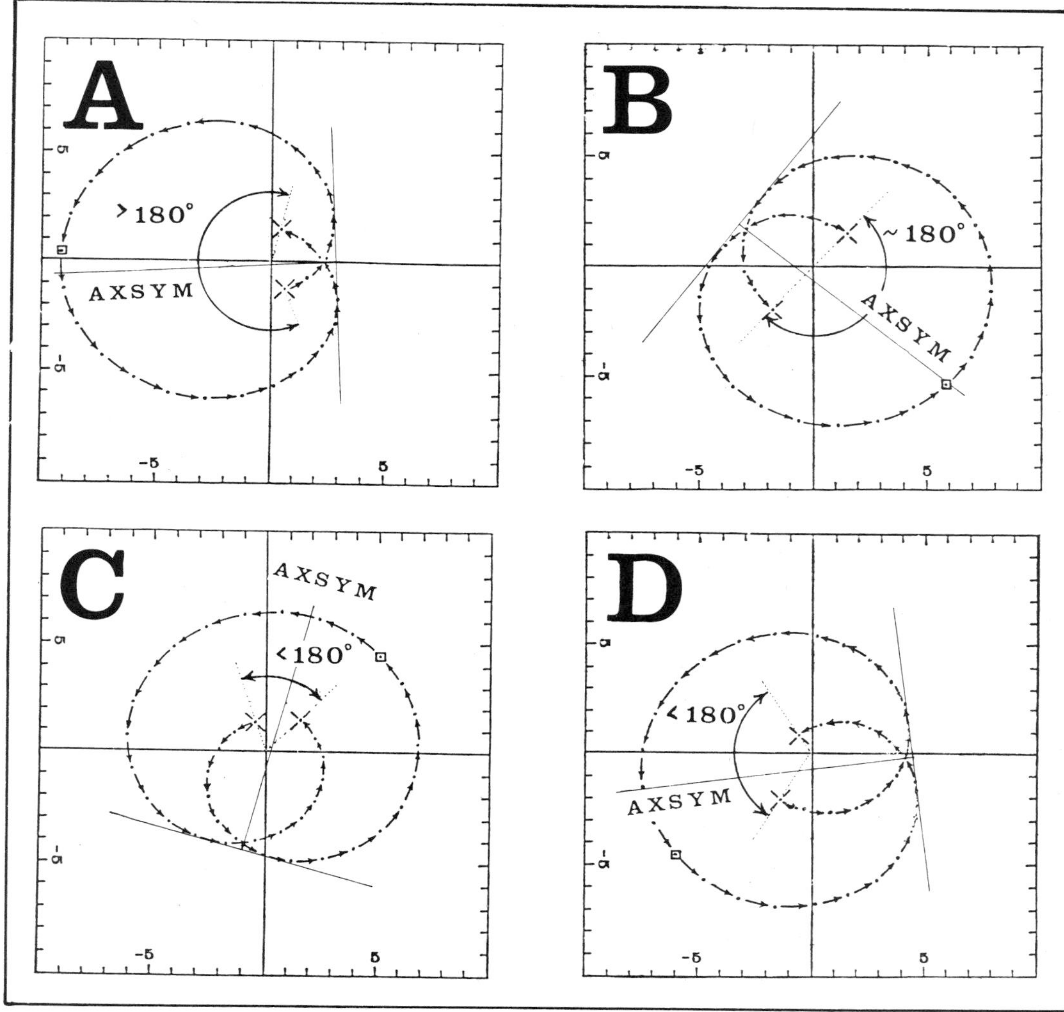

Figure 26-3. Examples of the eight solar-orbital classes.

Tangent-to-curvature lines are basis for drawing visual "axes of symmetry" (AXSYMS; lines perpendicular to tangent-to-curvature lines). Thin dotted lines connect peribac positions and the barycenter. Further explanation can be found in the text.

If the solar orbit is an ellipse, then the thin dotted lines define the instantaneous positions of the semi-major axis of the ellipse and the peribac-barycenter separation, the distance form the pivot focus to the trace of the ellipse.

A, 1516-1533; *B*, 1733-1751; *C*, 1712-1733; *D*, 1573-1593; *E*, 1671-1694; *F*, 1929-1951; *G*, 1616-1632; *H*, 800-816.

corresponds to an AXSYM. In the most regular examples, the AXSYM comes close to intersecting also the barycenter (B) and near a point on the outer loop that corresponds in time to the SJL (a term we shall use as a synonym to the barycentric synod of Saturn and Jupiter) as well as being close to the apobac.

Solar-orbital classes A through D are characterized by considerable degrees of symmetry. Class A is the most regular. In it, the barycenter is found between the two peribacs on lines that intersect at an angle of more than 180°. Class B is similar but with the barycenter close to being on a straight line (around 180°) between

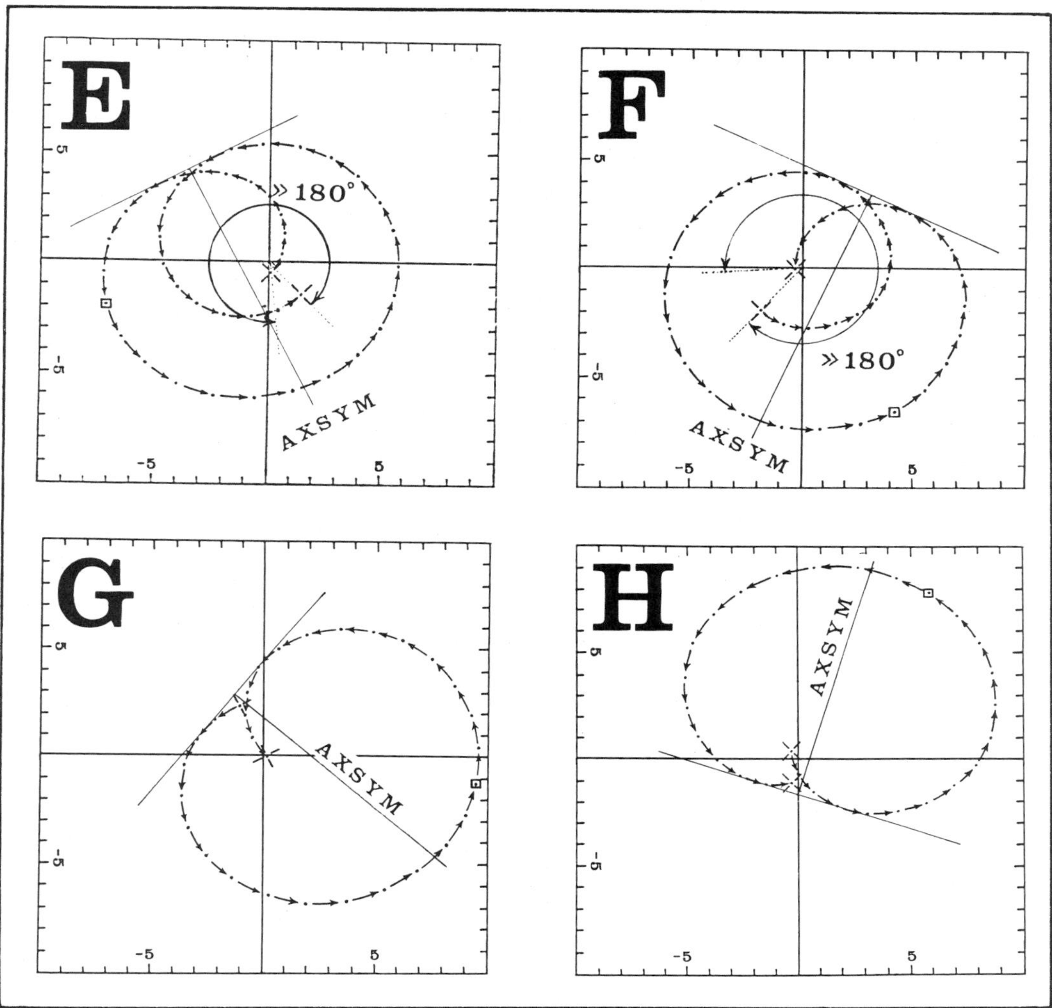

the two peribacs. C and D are less than 180° and somewhat less symmetrical, C having a shorter initial phase and D a longer one. Solar-orbital classes E through H are collectively less regular, E and F with peribacs almost enclosed (approaching 360°), E shorter in the initial phase, F longer. Finally, G and H are truncated forms without crossovers but with otherwise similar asymmetry.

Through the 13 centuries covered by this study, categories A and B are generally followed by C, D, E, or F; the end members, G and H, are unusual but similar in habit to A and B. No category is ever followed by another in the same class, except for the case of Class C, which appears twice in a pair (A.D. 869-888 and A.D. 1186-1207).

The durations of the orbits range from 15 yr to 26 yr; the average length is about 19.9 yr. It corresponds to the mean time required for Jupiter to gain one lap on Saturn (the SJL cycle; Table 26-1). According to the newest ephemerides, this value is 19.859 (±0.88) tropical yr of the Earth, averaged over 7,000 yr.

Sun-Barycenter Separation Distances

Although in plotting the orbit diagrams we have used only the positions of the Sun in the

Table 26-1. Orbital Periods, Yearly Angular Motions, Lap Periods, and Angular Motion per Lap of the Four Major Outer Planets of the Solar System

OUTER PLANETS:	*JUPITER*	*SATURN*	*URANUS*	*NEPTUNE*
Orbital Period (yr.)	11.86223	29.4577	84.0139	164.793
Angular Motion (degrees/yr)	30.348	12.221	4.285	2.185

	SATURN/ JUPITER LAP	URANUS/ SATURN LAP	NEPTUNE/ URANUS LAP	URANUS/ JUPITER LAP
Beat Frequency:	19.859	45.387	171.3921	13.8115
Barycentric Phase:	1940.9	1942.21	1993.3	1941.4
Angular Motion (per lap):	(360)+242.68°	↓ 194.48° (194.5°)	(360)+14.49°	

Jupiter: 15 × 360 = 5400 + 24°; Saturn 2160 + 24°

[24 × 15 = 360°; Resonance Drift Cycle: 15 × 178.71 = *2680.65* yr ≅ 135 SJL ≅ 59 USL]

SUN'S ORBITAL SYMMETRY PROGRESSION CYCLE (OSP)
178.73 yr (±0.27)

Sunspot Cycle (SSC): 11.12 yr (± 6 yr); 16.07 per OSP; 8400 per APR
SJL: 19.859 × 9 = 178.73 SJL × 16 = USL; × 7 = *317.74 yr* [USJL]
OSP Triad Cycle: 59.57 yr
SJL/USL/OSP: 5719.4 yr (32 × OSP)
Triple Synod (TS-USJR): 2859.7 yr (16 × OSP)
USJR (2859.7) = 63 × USL (45.387)
USJL (317.74) × 3 = 953.22; × 9 = 2859.7 yr
NUL/OSP: ≅ 4111 yr ≅ 24 OSP; 13 × NUL ≅ 2224 yr
Outer Planet Restart: Quadruple Synod (NUSJ): *4448 yr* (~ 2224 yr)
NUSJ/ASP RESTART: 40,032 yr = 9 × NUSJ = 45 × 889.6
(= 224 × OSP = 2016 × SJL = 882 × USL = 233 × NUL)
Note: Last OPR (NUSJ in phase approx.) AD 1306

Other Commensurabilities:

ALL-PLANET RESTART [APR] 93,408 yr (STACEY, 1967)	ORBITAL SYMMETRY PROGRESSION CYCLE – OSP: 178.73 yr = 3 TRIADS
21 × OPR (4448 yr)	3 × SJL = 1 Triad = 59.571 yr
545 × NUL (171.39 yr)	TRIAD-LENGTH CYCLE (TLC):
2058 × USL (45.387 yr)	24 SJL (476.57 yr) *OR* 21 SJL (417 yr)
1568 × OSP-TRIADS (59.57 yr)	TLC-5 PHASE REPEAT: 13,940 yr
522⅔ × OSP (178.73 yr)	= 30 TLC = 234 Triads
4704 × SJL (19.859 yr)	= 702 SJL = 78 OSP

Note: Also shown are the commensurability relationships of planetary periods and lap periods to the OPR cycle of 4,627 years.

X-Y plane of the Cartesian coordinates of the DE-102 Ephemeris, we have not altogether omitted the Z direction. We have used it indirectly by plotting the computed resultant distance *R*, which is the true distance in three dimensions from the center of the Sun to the barycenter (Fig. 26-4).

One might suppose that because of the great mass of the Sun compared to that of the planets, the barycenter would always lie within the body of the Sun and so close to the center of the Sun that it need not be separately recognized. A brief consideration of the significance of the positions of the planets with respect to the barycenter of the solar system will show why the center of the Sun is not the barycenter of the solar system and why the position of the center of the Sun with respect to the barycenter keeps shifting. At times these two points are so far apart that the barycenter lies outside the body of the Sun (Figs. 26-5 and 26-6).

A scan of Figure 26-4 shows that the Sun-barycenter separation distances at the apobac positions (points inside triangles for major orbital loops but not including intermediate fluctuations, which are shown by dots only in both apobac and peribac positions) range from slightly more than 6 units to a little more than 10 units on the vertical scale (10^{-3} A.U., or about 149,600 km). By contrast, the distances at the peribac positions fall within the range of less than 1 to a little more than 5 units.

If the Sun's orbit were an ellipse departing only slightly from a circle, like the Earth's orbit, for example, then the Sun-barycenter separation distance at apobac and peribac would be nearly the same (which they approach during the first intermediate fluctuation shown, A.D. 782-790). If the Sun's orbit were an extremely flat ellipse, nearly approaching a straight line, then the Sun-barycenter separation distances would increase as the value of the eccentricity of the ellipse increases. For example, the greatest ranges of distance are displayed by the orbits having peaks at about A.D. 950 and A.D. 1130 (far right, top two rows). In these two orbits, the heliocenter-barycenter separation at apobac is more than 10 units yet the separation at the next following peribac position is less than 1 unit.

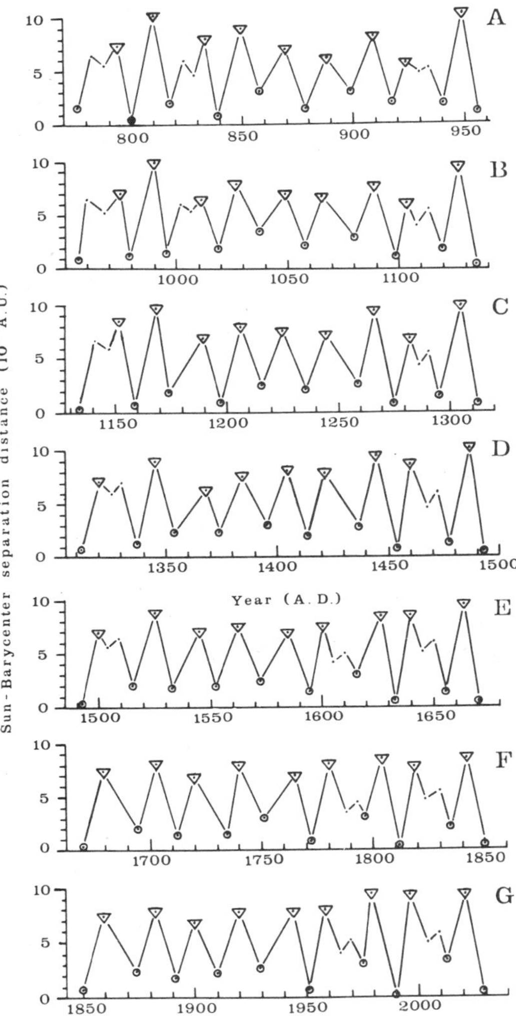

Figure 26-4. Sun-barycenter distance plots, schematized as straight lines connecting peribac positions (dots inside circles) with apobac positions (dots inside triangles) shown with time axis (horizontal) segmented by Saturn-Jupiter repeat cycles (SJR) of approximately 178.7 Earth yr.

Minor orbital loops ("intermediate fluctuations"), shown by dots only, are related to Landscheidt cycles but shift their positions in the various ASP-APS cycles (labeled by letters, beginning with A at A.D. 777). Times of sunspot minima and cold climates in the Northern Hemisphere (such as Little Ice Ages) span the times from the ends of ASP-APS cycles C (Wolf Minimum) and E (Maunder Minimum) and into the beginnings of the following 178.7-yr cycles. If this pattern holds, then a comparable Little Ice Age can be expected to begin near the end of cycle G, early in the twenty-first century.

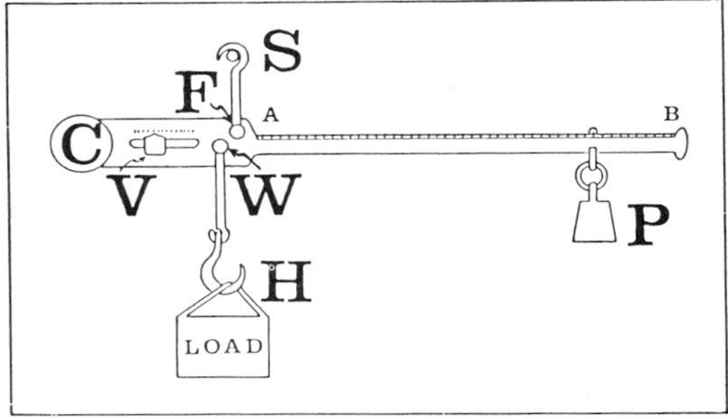

Figure 26-5. Sketch of an old-time Roman steelyard, such as used by butchers and agricultural merchants.

S, suspension hook, attached to the balance arm at F, the fulcrum; W, point of attachment of the load, which is placed on the weighing hook, H; C, counter-balance weight; V, the vernier (a very small movable weight next to its graduated scale); P, the poise, is a larger movable weight that slides along the calibrated arm AB. When P is at point A and no load is present, the steelyard is brought into equilibrium by adjusting the vernier. The object to be weighed would then be placed on hook H and P moved along the graduated lever arm toward B until a new equilibrium balance became established.

The steelyard provides a simple analogy from statics that may help illustrate the role of the barycenter within the solar system. The Sun would be at H (attached to the arm at W); the collective planets would be the poise, P; and the barycenter, the fulcrum, F. Because the solar system is dynamic, however, the vector mass distribution of the planets is always shifting. Thus, following the steelyard analogy, P would be moving back and forth. To keep the arm in balance when P is moved toward B, for example, the distance from W to F would have to increase but can be compensated for by moving V to the left. When P moves toward A, then W must also move toward F. (A more realistic model might be constructed by suspending the load from W attached to a movable V.) However, the solar system involves more than the balancing around the barycenter of all the forces exerted by the orbiting bodies. In addition, the spin rates of the orbiting bodies (Sun and planets) also change during the orbits. (Further explanation can be found in the text.)

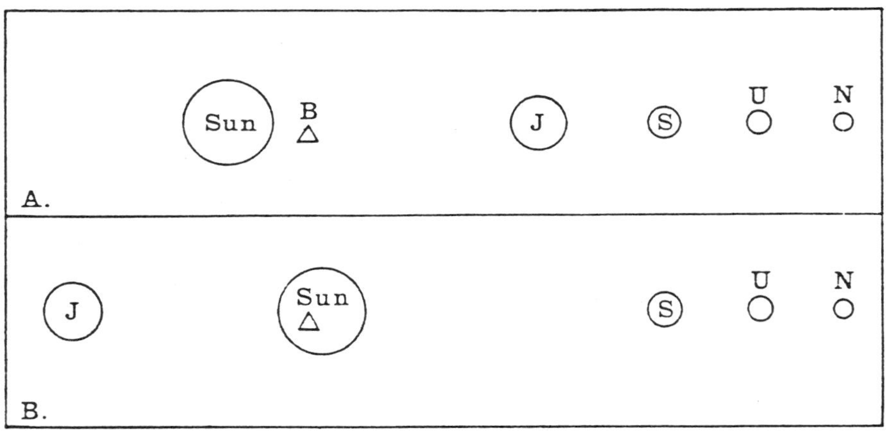

Figure 26-6. Diagrammatic sketches to illustrate the Sun's relationship to the barycenter (apex of triangle at B = center of mass) of the solar system.

A: When Jupiter *(J)*, Saturn *(S)*, Uranus *(U)*, and Neptune *(N)* assume positions on one side of the barycenter, the Sun swings out about 1.5×10^6 km (a little more than one solar diameter) away from B on the opposite side. In so doing, its orbital angular velocity increases and its spin rate is believed to slow down by 4-5%. Because of their relatively larger masses, Jupiter and Saturn together on the same side of the Sun bring about nearly the same effect. Because Jupiter gains 90° on Saturn in just under 5 yr, Jupiter "laps" Saturn in the SJL cycle every 19.859 yr. *B:* The minimum distance from B to the center of the Sun is achieved when Jupiter and Saturn come into opposition and the other planets lie on the side of Saturn. (Sizes of the Sun and planets and distances from the Sun to the planets are not to scale. The shift of the Sun with respect to the barycenter is in proportion to the size of the Sun, that is, about one solar diameter.)

The arrangement of the rows of Figure 26-4 coincides with that of Table 26-2. Each row represents a cycle of about 178 yr.

Attention is directed to the "intermediate fluctuations." These result when Jupiter and Saturn lie on opposite sides of the barycenter. (By contrast, the larger separations between apobac and peribac coincide with times when Jupiter and Saturn lie together on one side of the barycenter, and the center of the Sun, therefore, must lie well away from the barycenter on the opposite side.) After being present in several successive cycles, the intermediate fluctuations are seen to die out, while new ones pick up. In contrast to these comings and goings, notice the presence of the intermediate fluctuation in all rows of cycle 8 (counting from the left).

We do not know what, if any, physical significance may be attached to the turning points of these Sun-barycenter distance-separation cycles. Two important points about them are as follows: (1) They are caused by the changing mass distributions of the planets and thus represent a kind of single resultant function reflecting the instantaneous situation within the solar system, and (2) they are predictable.

Orientations of the AXSYM of the Orbital Traces

In plotting the orbit diagrams, we noticed that not only the patterns were being repeated but that orientations in the X-Y plane were being rotated. To try to determine exactly what this rotation amounted to, we measured and recorded the azimuth bearing of each AXSYM in geometric terms, and then calculated the differences. As might be expected from the shifting locations of Jupiter and Saturn in their lap positions, the AXSYM is found to progress about 120° for each orbit of the Sun. Because of the changeable orbits, however, the angular variation is quite appreciable. From A.D. 790 to A.D. 1445, the range usually lies between 100° and 125° and commonly alternated high and low, with four phase jumps. Quite unexpectedly, an altogether distinctive regime appears from A.D. 1445 to 2021 with a range from 65° to 165°, four times larger than at any time in the preceding seven centuries. Three phase changes occur in cycles beginning A.D. 1564, 1742, and 1921.

We conclude, therefore, that the solar orbit around the barycenter is demonstrably responding to planetary gravitational dynamics, as predicted by Newton. Furthermore, we conclude that the two largest planets, Jupiter and Saturn, control the system. The other planets exert minor modulations but never to the extent of shifting the date of the AXSYM by more than 10% of the mean periodicity (19.9 yr).

These conclusions may hardly be news in the world of celestial mechanics, but the demonstrated pattern of motion of the AXSYM, which we have studied for some 13 centuries, is an exercise that does not appear to have been attempted previously.

Effects of Jupiter and Saturn

Orientation plots of each AXSYM disclose a relatively close coincidence between it and the times of the SJL. These laps were established from the JPL Ephemeris data for the period since A.D. 1500. The maximum difference between the date calculated from the mean periodicity and that based on ephemeris coordinates is ±2 yr, but 70% of the dates fall within the same year (data iterated at 200-day intervals). Prior to A.D. 1500, for the years A.D. 750-1500, the mean SJL periodicity was used.

The SJR cycle of 177.92 yr, formed by 15 orbits of Jupiter and 6 orbits of Saturn, is displayed by the near-sine curves generated by plotting the distances between the centers of these planets and the barycenter of the solar system (iterated as R in the computer printouts). In recent centuries, every third SJL closely matches in time the peribac of Jupiter in its third, eighth, and thirteenth orbit in the SJR, whereas one SJR cycle later, every third SJL roughly matches Saturn's second, fourth, and sixth orbits (lagging about 1 yr). Thus, a cycle of about 356 yr is created. Based on this general pattern, it is convenient to count the starts of the SJR cycles as the points in time when the

Table 26-2. Systematic Table of Solar Cycles, A.D. 777–2030

	SOB								
	CYCLE 1	*CYCLE 2*	*CYCLE 3*	*CYCLE 4*	*CYCLE 5*	*CYCLE 6*	*CYCLE 7*	*CYCLE 8*	*CYCLE 9*
Cycle A	777 (F) 777 782 787 792 796 794 (789)	800 (H) 802 806 813 808 (809)	816 (F) 819 824 827 835 833 (829)	839 (B) 839 846 851 856 847 (849)	857 (C) (856) 861 868 872 870 (869)	878 (C) .880 886 891 896 887 (889)	899 (B) 900 904 910 916 908 (909)	917 (F) 922 925 933 937 925 (928)	940 (A) 946? 953 948 (948)
Cycle B	956 (F) 958 962 970 974 973 (968)	979 (H) 981 986 990 987 (988)	995 (F) 996 1000 1003 1011 1015 1011 (1008)	1018 (B) 1024 1028 1026 (1028)	1037 (D) 1036 1041 1048 1054 1048 (1048)	1057 (C) 1061 1067 1072 1077 1065 (1067)	1078 (B) 1083 1087 1093 1087 (1087)	1096 (F) 1097 1105 1111 1115 1103 (1107)	1119 (A) 1119 1124 1128 1127 (1127)
Cycle C	1134 (E) 1134 1137 1144 1149 1141 (1147)	1158 (A) 1161 1168 1172 1166 (1167)	1174 (C) 1180 1184 1189 1193 1189 (1187)	1197 (C) 1198 1202 1209? 1205 (1206)	1216 (D) 1215 1222? 1227 1235 1227 (1226)	1236 (C) 1239 1246 1251 1256 1244 (1246)	1258 (A) 1261 1269 1267 (1266)	1275 (E) 1275 1281 1286 1289 1282 (1286)	1298 (A) 1297 1298 1308 1312? 1305 (1306)
Cycle D	1313 (E) 1318 1322? 1327 1334	1337 (A) 1340 1347 1352	1353 (C) 1358 1361 1368	1376 (B) 1378 1382 1387 1391	1394 (D) 1398 1404 1411	1415 (E) 1416 1422 1428 1434	1437 (G) 1439 1445 1450	1454 (E) 1455 1460 1468	1477 (A) 1474 1480 1485 1489?

	1321 (1326)	1345 (1345)	1367 (1365)	1385 (1385)	1405 (1405)	1422 (1425)	1446 (1445)	1460 (1465)	1485 (1484)
Cycle E	1493 (E) 1493 1501 1506 1513 1499 (1504)	1516 (A) 1517 1524 1528 1524 (1526)	1533 (C) 1534 1537 1543 1547 1543 (1544)	1555 (B) 1554 1558 1567 1571 1563 (1564)	1573 (D) 1578 1581 1587 1585 (1584)	1593 (E) 1593 1598 1604 1609 1614 1601 (1604)	1616 (F) 1620 1625 1625 (1623)	1632 (E) 1633 1639 1645 1650 1655 1639 (1643)	1656 (A) 1661 1666 1664 (1663)
Cycle F	1671 (E) 1673 1679 1685 1689 1678 (1684)	1694 (A) 1694 1699 1705 1703 (1703)	1712 (C) 1712 1718 1723 1727 1721 (1722)	1733 (B) 1734 1738 1745 1750 1742 (1743)	1751 (D) 1755 1761 1766 1769 1764 (1762)	1772 (E) 1775 1778 1784 1788 1780 (1782)	1796 (G) 1798 1805 1810 1804 (1802)	1811 (E) 1816 1823 1829 1833 1818 (1822)	1835 (G) 1837 1843 1848 1842 (1842)
Cycle G	1850 (E) 1856 1860 1867 1870 1858 (1862)	1873 (A) 1878 1883 1889 1882 (1882)	1891 (C) 1894 1901 1907 1900 (1901)	1912 (A) 1913 1917 1923 1928 1920 (1921)	1929 (F) 1933 1937 1944 1947 1943 (1941)	1951 (E) 1954 1957 1964 1968 1958 (1961)	1975 (G) 1976 1979 1983 (1981)	1990 (E) [1990] 1997 (2001)	2013 (A) 2021 (2021)

Note: Table is based on the Sun's OPS Cycles shown by the horizontal rows and individual solar-orbit cycles (SOB, defined by successive peribac positions, neglecting intermediate fluctuations, and shown by the vertical columns). Within boxes, circled letters designate orbital class; large numbers at upper left are calendar dates (rounded) of times of peribac positions; smaller numbers at lower left are calendar dates (rounded) of apobac positions; smallest numbers indicate dates of sunspot minima and sunspot maxima (underlined); numbers in parentheses at bottom show calendar dates (rounded) of SJLs (note frequent coincidences with times of solar apobacs).

peribac of Jupiter most closely matches the peribac of Saturn—for example, A.D. 1501, 1678, 1856, and 2034.

Only a study of the long-term values discloses that the SJR (177.92-yr) and the APS (or 178.73-yr) cycles are not the same. At the start of the SJR cycle in 1678, we find this Jupiter peribac coinciding with the solar apobac, but for the SJR cycle starting in 1856, the solar apobac does not appear until 1858, and for the SJR starting in 2034, the solar apobac is calculated to be in 2037. Going back in time, the SJR cycle starting in 1501 fails to match the solar apobac of 1499. In the same way with the SJL cycle (to which the AXSYM cycle is locked), one may note that the SJLs of 1663, 1702, 1762, and 1821 are all within 0.5 yr of matching Jupiter's peribacs, but in the twentieth century they are noticeably drifting apart.

Although we observe that the apobac positions of the Sun (from which a line drawn to the barycenter locates the instantaneous position of the semimajor axis of the Sun's presumed orbital ellipse) correspond to the 19.9-yr SJL approximately, during the 13 centuries of data studied, the correspondence may lag up to 4 yr. Matches or near matches (±1 yr) of the SJL and Sun's apobac positions are noted at the following SJLs: A.D. 769.6, 808.4, 868.9, 908.7, 948.4, 988.1, 1027.8, 1047.7, 1127.1, 1166.8, 1206.5, 1266.1, 1305.8, 1563.9, 1583.5, 1663.6, 1702.6, 1742.8, 1842.2, and 1901.7. (Because of our 200-day spacing of data points, the correctness of these dates is only to the nearest two-tenths.)

The Solar-Orbital and Saturn-Jupiter Cycles

As we have shown, the period of the Sun's orbit around the barycenter (SOB), measured from peribac to peribac, ranges between about 14 yr and 26 yr. The mean period is 19.9 ± 6.0 yr. Short SOB cycles of about 15-17 yr are usually followed by longer ones of 22-24 yr.

The Sun's apobac position usually corresponds within ±1 yr of an SJL, but the times of these two may differ by as much as about 6 yr. The SJL was first physically observed by Kepler in 1603. Stacey (1963, 1967*a*) pointed out that the SJL period corresponds to the mean time required for Jupiter to gain one lap on Saturn. Thus, it is simply a synodic interval as seen from the barycenter. We regard this SJL cycle of 19.859 yr as "the pulse of the solar system." Its long-term predictability is very precise and seems to be commensurable with the OPR period of about 4,448 yr (19.859 × 224 = 4,448.4 yr). Because of orbital variability, however, individual epochs display a variation of about ±0.8 yr.

Determination of the mean length of the SJL deserves the closest attention. The standard mathematical procedure for calculating a beat frequency is straightforward, as has been demonstrated for all planetary ratios by Mörth and Schlamminger (1979). It requires only the best available long-term means of the orbital periods concerned. The commonly cited value is 19.857 yr (Stacey, 1963), but we have employed newer orbital periods to conclude that 19.859 is better, and, for general purposes, a value of 19.86 is adequate.

A second approach is to begin with the period of the beat frequencies of all the outer planets that are commensurable at a period of 4,627.25 (±0.05) yr. Using this approach and excluding the SJL component, we find that 233 × SJL = 4,627.25, where SJL = 19.85944 yr.

A third procedure is to take the astronomically observed and mathematically calculated Saturn/Jupiter barycentric synods. Over the course of the present century, the mean value is 19.8496 (courtesy of our colleague, James Shirley), with a range from 19.18 yr to 20.44 yr. For other centuries and longer time spans this range is somewhat larger. For the last 7,000 yr the range is 18.98 yr to 20.62 yr. From the foregoing we conclude that the best value for the SJL at this time is 19.86 ± 0.88 yr.

The last two SJL epochs (barycentric conjunctions) were A.D. 1961, April 11, or 1961.280 (J.D. 2437405.8) and A.D. 1981, April 15, or 1981.284 (J.D. 2444711.2). The next Saturn/Jupiter opposition phase will be in A.D. 1990, December 27, or 1990.99 (J.D. 2447887.5). The conjunction of 1961.28 serves as a datum year for quick calculations of approximate SJL dates to within ±0.65 yr during the historical record. Stacey (1963) recognized 1961 as a mid-phase "stellium" of several long-term cycles.

The SJL half-cycle (Saturn and Jupiter in opposition) of 9.9295 yr is strongly represented in spectral analyses of sunspot numbers (Cohen and Lintz, 1974; Wittmann, 1978; Currie, 1979; Lomb and Andersen, 1980). The effects of this cycle become especially noticeable at times when Jupiter is alone on one side of the barycenter and the other three major planets lie within a narrow sector on the opposite side. This almost balanced alignment brings the Sun's core close to the barycenter, a condition that Landscheidt (1976) names a *nuclear transit.*

NEW PERSPECTIVES

The Orbital Progression Cycle and All-Planet Synod

With each epitrochoidal or cardioidal orbit, the axis of symmetry of the major loop (solar AXSYM) undergoes a clockwise (hence, retrograde) motion of a little more than 120° (the average is 121.07°). Every third orbit, it returns to within 3.21° of a complete circle, creating a three-lobed rosette pattern. Every third solar orbit also more or less coincides with the return of the SJL position to about the same celestial longitude (a result of the relationship whereby Jupiter gains about 90° on Saturn every 5 yr). The periodicity of this SJL return position ("triad") is 59.577 yr. It is very close to being commensurable with a Venus cycle (96 × 0.615 = 59.04 yr), which has been incorporated into the Mayan calendar (Schove, 1977). Also, it has been noticed in sunspot numbers and named the Wolf Cycle.

Blizard (1983) pointed out that because of the Sun's oblateness and the tilt of its axis of rotation to the invariable plane, the whole system undergoes a slow retrograde motion that leads to a longer cycle. A near-symmetry in the SOB is reached at the end of every 9 orbital periods. This is the ASP cycle. Thus, after 9 solar orbits, the AXSYM undergoes a net retrograde motion of 9.63°, which is also 9 SJLs. Because of their 5 : 2 commensurability (details in the following section), Jupiter and Saturn develop a repeat cycle (SJR) of nearly the same period (177.93 yr), which drifts progressively with respect to the OSP-ASP cycles.

The OSP and APS cycles correspond closely, but they are not identical. The APS was defined by Ren and Li (1980) as a quasiperiodic interval averaging 178.71 yr marked by "the assemblage of all planets and the Sun on the same side of the Earth, with the subtending angle of the two outermost planets at a minimum." Frequently they come to lie within a 90° quadrant, except for the Earth.

By contrast, the periodicity of the OSP cycle is very regular. The orbital patterns of the SOB paths conform to it in a systematic, but not completely uniform, way. The peribac dates defining the starts of each of the OSP cycles are optional but here chosen as follows: A.D. 777, 956, 1134, 1314, 1493, 1671, 1950, and 2030. In contrast, the APS is defined on a geocentric basis. These synodic dates are commonly close to, but not identical with, the start dates of the OSP cycles. The AP synodic dates are A.D. 768, 949, 1126, 1304, 1483, 1665, 1844, 1982, and 2163. An alternative system for defining the 178.73 yr cycle was to choose sunspot data, minimum to minimum (Fairbridge and Hameed, 1983).

Each OSP, as defined here, begins and ends with a solar peribac in which the center of the Sun comes close to the barycenter, at about 1-2 × 10^{-3} AU (a Landscheidt nuclear transit). Similar transits also occur at other times, but infrequently. One such event is expected for A.D. 1990.3; it will be the first since 1454. The SOB patterns of no two OSP cycles are ever quite identical, but the rule locking them into the SJL seems never to be broken (9 × 19.859 = 178.73 yr).

Long-Term Commensurabilities

When Kepler visualized the planetary periods and separations as being arranged in a harmonic series, he used as unity the mean Earth-Sun separation (defined as the astronomic unit, AU, commonly approximated as 149.5 × 10^{-6}

km, but according to Newhall, Standish, and Williams (1983), a more precise value is 149,597,870.68 ± 0.03 km). Although the AU definition is specific enough, in reality the instantaneous distance between the Earth and the Sun varies during each orbit of the Earth-Moon pair and during each orbit of the Sun (a type of orbit not known to Kepler). The change during each solar orbit may exceed 1.5×10^6 km, a variation of about 1%.

Astronomers have long realized that the classic calculations of the Titius-Bode Law, made on the basis of 1 A.U., were weak approximations (Nieto, 1972). A unit of distance subject to fewer variations than the astronomic unit is the mean Saturn-Jupiter separation. Using the mean Saturn-Jupiter separation distance as unity, Stacey (1967*b*, p. 745) recalculated the Titus-Bode sequence. His result shows the following mean distance ratios among the planets from Venus to Neptune: 1 : 16, 1 : 8, 1 : 4, 1 : 2, 1, and 2. The end members are repetitive.

Stacey's (1967*b*) ratios provide a clue to the key roles of Saturn and Jupiter in controlling solar system dynamics. The Saturn : Jupiter commensurability (2 : 5) and the SJR cycles (6 orbits of Saturn to 15 orbits of Jupiter) of 177.9 yr very nearly approximate the ASP and APS cycles of 178.73 yr. In Stacey's words (pers. comm.), the fundamental "tone" is "set by Jupiter each time it gains a lap on Saturn and, figuratively speaking, plucks the cord that unites Saturn to the Sun."

The dynamic and commensurability systems exemplified by our solar-orbital plots are thus established by the various interplanetary lap rates (Mörth and Schlamminger, 1979). The commensurabilities *(C)* are easily calculated by the formula

$$C = \frac{P_o \times P_i}{P_o - P_i}$$

where P = planetary period, i = inner planet, and o = outer planet. Among the outer planets, the key lap rates are SJL = 19.859 yr, USL = 45.363 yr, NUL = 171.392 yr (see Table 26-1).

Every 233 SJLs, 102 USLs, and 27 NULs, the outer planets return to a starting configuration. Using the gear-theory language of Stacey (1963), this configuration brings about an OPR cycle of about 4,627.25 yr.

The significant inner-planet lap periods are Jupiter-Mars = 2.235 yr, Mars-Earth = 2.135 yr, Jupiter-Earth = 1.092 yr, and Jupiter-Venus = 0.6488 yr. The importance of these short lap cycles is shown by their appearance in spectral analyses of sunspot numbers and in various terrestrial time series.

Spin Rates of the Sun and the Earth

The current mean spin rate of the visible surface of the Sun is about 27 Earth day (Bruzek and Durrant, 1977), but it varies from solar equator to solar polar regions. It is not uniform in time, and its value depends on whether one measures it as seen from the Earth or from a fixed point in space. The spin rate is measured from differential displacement of sunspots. These appear first at high solar latitudes and migrate equatorward to the maximum of the solar cycle, a phenomenon known as Spörer's Law, after its chief early investigator. The spin rate seems to range from about 25 day at the solar Equator to 31 day near the poles. The mean rate as seen from the Earth (synodic rate) is 27.275 day, but as seen from a fixed point in space (sidereal rate), it is 25.38 day.

Solar rotation was first precisely measured by R. C. Carrington (1863). He began a series of rotation numbers (Carrington Numbers), beginning with number 1 in November 1853 at the central meridian. The effect of the Carrington rotation, combined with the equatorward expansion of sunspot clusters, a phenomenon that is paired in the two solar hemispheres, yields a pattern resembling two wings. When these have been plotted, they form what has been called a "butterfly diagram" (of E. W. Maunder, 1921).

The observed oblateness of the Sun is appreciably greater than it should be if the core rotated at the same rate as the surface. Accordingly, it is inferred that the core spins at a considerably higher rate than the solar surface (Steenbeck and Krause, 1969).

According to the standard precepts of celestial mechanics, during the 4.6 b.y. of the solar system's existence, the Sun has been progressively losing angular momentum from its outer spheres. This presumed loss should be reflected in a progressively diminishing spin rate. Discovery of evidence for the existence of the 11-yr sunspot period in varved sedimentary rocks of Late Precambrian age (G. E. Williams, 1981), however, indicates that no significant change in the 11-yr cycle has taken place within the past billion years or so. Therefore, the 11-yr cycle would seem not to be connected to solar spin rate (or if it is so connected, the spin rate has not changed as thought).

In planetary trajectories, orbital velocity and spin rate vary systematically around the orbit. At perihelion, for example, the velocity of a planet and its spin rate reach maximum values. In contrast, the Sun's velocity reaches maximum values at apobac. The minimum solar velocity is reached at peribac; at such times its spin rate and oblateness reach their maxima.

Landscheidt (1981, 1984, paper 25) has suggested that solar activity is related to the Sun's spin rate. Slowing of the spin rate is thought to be marked by increasing emissions; acceleration, by decreased activity. He sees a coupling between the Sun spinning on its own axis and the Sun revolving around the barycenter as twin oscillations that develop a resonance that results in spin-rate variations. Two kinds of resonant spin rates for celestial bodies have been found (Goldreich and Peale, 1966). One is exemplified by the Moon and the other by Venus (but these authors did not consider the rotation of the sun).

Other aspects of solar resonance have been discussed by Nohonoj (1971) and Blizard (1983), who kindly provided us with a copy of her translation of Nohonoj's paper (published in Russian). According to them, Jupiter-Sun-barycenter conjunctions create the resonance effect. Such conjunctions appear with a periodicity having a long-term mean of about 9 yr and a range from 2 yr to 15 yr. During each ASP and APS cycle, 20 such conjunctions take place; thus the mean is 8.9355 yr. Dates of recent conjunctions, each coinciding with a rise in solar flares and X-ray activity, are 1942, 1951, 1959, 1967, 1970, 1974, and 1982 (Landscheidt, 1981, 1984). These dates do not coincide with the recent sunspot maxima of 1948, 1958, 1969, and 1979.

Statistical analyses between the 11-yr solar cycle and the Earth's spin rate (length of day, LOD) for the period 1838-1970 using the maximum entropy method (MESA) have disclosed a coherence of 0.82 (Currie, 1980). The changes in the Earth's spin rate lag the peak sunspot number by 3-3.5 yr. Terrestrial climatic reactions can be expected to result from changes in the Earth's spin rate. Increased zonality of circulation accompanies accelerations, and the contrasting style, that of blocking, prevails during decelerations. For example, during the fastest spin accelerations within the past 200 yr, Great Britain experienced the mildest winter on record (1868-1869), some 8° above the coldest (mean for December, January, and February). Winters nearly as mild match the high-spin epochs in the mid-1830s and 1930s. By contrast, the last spin decelerations coincided with one of Great Britain's coldest winters (1962-1963) and with several seasons of general freeze-up in the Baltic. During the winter of 1837-1838, the North Sea was frozen from Denmark to Norway (Lamb, 1977). After 1840, the year of the low-spin minimum, China also recorded extremely cold episodes.

Another factor that may be closely connected with changes of the Earth's spin rate is geomagnetic secular variation. Three abrupt decelerations in spin rate during the past 200 yr, around 1830-1840, 1895-1910, and 1960-1970, were paralleled by regionally variable accelerations of the geomagnetic secular drift.

The 11- and 22-yr Sunspot and Heliomagnetic Cycles

Several early workers have suggested that the sunspot cycle was linked in some way to the motions of the planets (de la Rue, Stewart, and Loewy, 1872; Schuster, 1911; Sanford, 1936; Johnson, 1946, e.g.), but this viewpoint has often been challenged (e.g., Okal and Anderson, 1975). Recent support for the importance of the planets comes from the work of Mörth and Schlamminger (1979).

In the first attempt to use a computer-based ephemeris to analyze the Sun's orbit, Jose (1965) showed that no simple relationship exists between the Sun's orbit and numbers of sunspots. He did show the existence of an orbital cycle of 178.71 yr, but he did not extend his analysis beyond a single such cycle. In fact, Jose was not able to convince others that his 178-yr cycle was a reality even though it matched a comparable-length cycle that some investigators had claimed to have found within sunspot numbers (King-Hele, 1966). The existence of such a cycle of 178 yr has been challenged (Cohen and Lintz, 1974; Okal and Anderson, 1975). In connection with their attempt to discredit the planetary hypothesis of sunspots, Okal and Anderson (1975, p. 512) wrote, with respect to the spectral analysis of sunspot numbers performed by Cohen and Lintz, "In addition they demonstrated that the longer period, ~180-yr cycle proposed for the solar sunspot spectrum arises from the beat of the 11- and 9.8-yr cycles and is not an intrinsic periodicity." What we think that these authors have "discovered" is just another link in the long chain of commensurability within the solar system that goes far back—to the Titius-Bode Law (Stacey, 1967*b;* Mörth and Schlamminger, 1979). Our orbital plots clearly demonstrate the validity of a 178-yr cycle.

In a comprehensive statistical analysis between sunspot numbers and various solar-orbital parameters based on the computerized ephemeris, Pimm and Bjorn (1969) calculated that the proportion of the statistical variance ranged from 10% to 50%. They found that the correlation coefficient between sunspot number and the factor we have emphasized, distance from the center of the Sun to the barycenter of the solar system, was −0.5547, which accounted for 31% of the variance. They found the greatest correlation (coefficient of −0.7057) between sunspot numbers and the radius of curvature of the Sun's path; this accounted for 50% of the variance of the sunspot numbers. Other factors computed by Pimm and Bjorn included angular momentum about center of curvature, velocity about center of mass, rate of change of radius of curvature, angular momentum, acceleration, and Sun-centered Coriolis force.

Around the middle of the nineteenth century, when Schwabe (1843) announced his discovery of the sunspot cycle, its periods were rather short; he reckoned a period of about 10 yr. The duration of this cycle is variable, however, and its period changes with time. Later workers established a period of about 11 yr. Although we do not think much is to be gained by seeking a single number for the sunspot cycle, we review the various attempts to refine its duration. Schove (1967) determined 11.1 yr; Stacey (1967*a*) computed 11.12 yr (and he thought it coincided with his "anomalistic year of the Sun," a concept we have rejected as explained previously); Wittmann (1978) decided on 11.135 yr; and Dicke (1978), 11.14 yr. Kanda (1933) pointed out that systematic changes through time cause its period to vary between 10.38 yr and 11.28 yr, in addition to the range of variation of individual cycles (maximum to maximum) lying between 7 yr and 17 yr. If one uses the spacing between minima, the variation is much less.

Using a climatically based proxy to confirm the periodicity of solar activity within the Maunder Minimum, Hameed and Wyant (1982) extended the baseline of the astronomically counted cycle of sunspot numbers to cover two complete 178.7-yr-long cycles. Using the material, Fairbridge and Hameed (1983) demonstrated that within each of these long cycles, the periods of the SSC were short at the beginnings and ends but longer in the middle parts, approximately as Kanda had shown. The existence of these systematic changes in SSC periods provides the basis for a new approach to the problem of predicting future SSCs. It is now possible to weight the statistical formulae toward the appropriate corresponding epoch of the 178-yr cycles.

Using our chronology of the SOB cycles (see Table 26-2), we have made a preliminary study of the SSCs as related to the solar orbits. Each SOB (which averages 19.9 yr) contains one to three SSCs. The distribution of the SSCs within each SOB is not tightly constrained by the solar orbit.

Initially, we identified the SSC chronology over our entire 13-century data base. We used SSC minima and maxima dates from Schove (1983), with cross-checks from Wittmann (1978) and the auroral series (Siscoe, 1980). As Table 26-2 shows, each 178-yr OSP cycle contains about 16 SSCs of variable lengths (averaging 11.12 ± 0.12 yr and with an average range between 11.00 yr and 11.25 yr) but the sunspot numbers do not so much reflect the major planets as the inner ones-plus-Jupiter (Landscheidt, 1979, 1984).

Our study of data from the outer planets at 200-day intervals sheds no light on the subjects of either the individual variation of amplitudes or the periods of the SSC. In this connection, several researchers have proposed the concept that the so-called tidal planets (Jupiter, Venus, Earth, and Mercury) develop torques on the Sun having periods that coincide with peaks on spectral diagrams of sunspot numbers (Bigg, 1967; Wood, 1975). Furthermore, the torque crescendos must develop a tidal resonance (horizontal, or tractive, tides) leading to rupture that would go a long way toward explaining the solar-flare and X-ray eruptions (Landscheidt, 1981, 1983, 1984, paper 25). Landscheidt visualizes that an exchange of torques takes place among the Sun and its tidal planets and infers that this exchange is responsible for fluctuations in the spin rates of all planets. The variation of the tidal potential exerted by the planets on the Sun is appreciable, although the actual force is small. If the Earth-Moon potential is taken as 1, the maximum planetary tide is more than 7 and the minimum less than 3 (Bollinger, 1968).

Although the amplitudes of the planetary tide-raising forces on the Sun are slight, their values change rapidly. Several authors have denigrated the importance of tides, while studiously avoiding various important papers, for example, of Schuster (1911) who indicated that particles accelerated by explosions within the solar photosphere will come under external tidal influences as they escape from the Sun's gravitational field. Bollinger has calculated these tidal values in a detailed set of tables for the years 1900-1980. The principal cycle is the Jupiter-Venus hemisynodic period of about 4 mo (118.15617 day) with peaks about every 12 ± 0.05 yr (average 11.9693 yr). Based on an Earth-tide force (m/r^3) at a mean distance = 1.0, Bollinger's minimum tide peak during the present century was June 30, 1904 (±4 day), with a value of 2.9. His maximum Sun-tide peak (determined for seven planets out to Uranus) was November 5, 1941, with a value of 7.53 (six planets were almost in conjunction; only Mercury was in opposition). Other peaks included December 1905, December 1917, April 1929, May 1953, December 1965, and December 1977. The exceptional years, 1929 and 1953, fell on Jupiter-Earth conjunctions in December and with Venus in opposition. Each year having a peak tidal period, lasting 1 or 2 mo, is accompanied by a very similar preceding and following year.

Bollinger has pointed out that extreme fluctuations of the Sun tide are to be expected in view of the eccentric orbits—in particular, of Jupiter, Earth, and Mercury but to some extent of Saturn and Neptune. The perihelion positions of all lie in the same heliocentric quadrant with Jupiter at 13°35′ and Earth at 102°09′. The tide-raising effect of Venus changes little; its orbit is nearly circular. Through torque (Landscheidt, 1984), the planetary tides exerted on the Sun affect both the Sun's radiation and also modulate the lunar influences on Earth. Nile flood peaks (Fairbridge, 1984) show a remarkably high tendency to follow the Sun-tide maxima by 1-2 yr. The lunar nodal cycle is often seen, but typically the sunspot cycle is not represented.

An important point that emerges from the foregoing is that although long-term sunspot periodicity is commensurable with the all-planets cycles of the solar system, the individual timing and amplitudes of the SSCs relate primarily to motions of the tidal planets.

Linkage with the Earth-Moon Pair

That variations in the orbit of the Moon around the Earth (more exactly, the orbits of both about their common barycenter) seem to exert some effect on the Sun was first proposed by Otto Pettersson (1912, 1914*b*, 1915*b*). This concept remains controversial. But no doubt exists that the Moon's effects on the Earth vary with the nutation of its orbital plane in a mean nodal cycle of 18.6134 yr and by the progression within this plane of the longitude of lunar perigee, yielding the mean apsides cycle of 8.849 yr. Combinations of these cycles with others control mutual alignments at perigee-syzygy and/or perihelion.

The effects of the various lunar cycles have been found in many terrestrial phenomena. These include nutation of about 9 sec of arc in the motion of the Earth's pole of rotation *(Astronomical Almanac);* variations in atmospheric pressure (Chapman, 1919*b*, 1951; Bryson, 1948; Currie, 1982); precipitation (Landsberg, 1976; Currie, 1981*b*, 1984*a;* Campbell, Blechman, and Bryson, 1983); sea level (Pettersson, 1912, 1915*a*, 1930, 1934; Currie, 1976); tidal flooding (Wood, 1978, 1985); sea-ice conditions (Otto Pettersson, 1912; Maksimov and Sleptsov-Shevlevich, 1970); tidal currents (Herlineaux, 1957); currents in submarine canyons (Shepard, Sullivan, and Wood, 1981); sea-surface temperatures (Loder and Garrett, 1978); geyser eruptions (Rinehart, 1972*a*, 1972*b*); volcanic eruptions (Bell, 1981; Shirokov, 1983); earthquakes (Kilston and Knopoff, 1983); tree-ring widths (Currie, 1984*b*); and zoological growth series (Hurt et al., 1979).

Using harmonic analysis of Stockholm temperature records, Stromberg (1913) showed a correlation between lunar-solar tidal maxima and maximum range of monthly averages, especially in winter. Stromberg also calculated longer-term lunar cycles, and Pettersson (1915*a*) elaborated their gravitational effects. Otto Pettersson (1912, 1914*a*, 1914*b*) related these effects to various climatic time series and climatic proxies. Pettersson showed that the maximum tide-raising force coincided with times when the perihelion-node-apsides line points directly at the Sun. This arrangement occurs only infrequently. Pettersson used an older ephemeris to compute the last date at A.D. 1433. This date has been widely cited in the literature. Based on newer ephemeris data, this maximum has been recalculated as January 8, 1340 (F. J. Wood, 1985), which is 93 yr (5 × 18.6 yr) earlier. An approximation of this peak tidal alignment takes place at various seasons about every 62.013 yr. One such alignment took place in the present century on January 16, 1961 (621 yr, or 10 × 62.1 yr after January 8, 1340).

Within the present millennium, almost all the closest perihelion dates fall in January (midwinter in the Northern Hemisphere). This seasonal position is the result of the approximately 25,800-yr gyroscopic precession of the Earth's pole of rotation, which affects the position of the equinox (and hence both the calendar year and the tropical year), being in turn governed by the torques developed (chiefly by the Moon and the Sun) on the Earth's equatorial bulge. This axial precession cycle also involves a tilt cycle of 8,600 anomalistic yr, during which the angle between the Earth's pole of rotation and the pole normal to the plane of the ecliptic makes its maximum departure and then returns (3 × 8,600 = 25,800).

The apsides cycle of 8.849 yr comes into phase with the nodal regression cycle about once every 186 yr (10 × 18.6; 21 × 8.849). Approximately 0.21% of the average tidal variation is connected to the apsides cycle. About 0.2% is related to a node-perihelion cycle of 93 yr and to an apse-perihelion cycle of 62 yr.

Pettersson (1915*a*) compared the variability of the tide-raising force to a long-period oceanic swell; he supposed that the heights of all the secondary wave trains would be amplified toward the swell crests. The total envelope of variance is therefore considerable, which from the terrestrial climate viewpoint, is very important. It means that the year-to-year variance is much greater than usual during lengthy epochs near the swell crests. Such a relationship might help explain the remarkable climatic fluctuations that marked the fourteenth and fifteenth centuries (A.D. 1340 ± 100 yr). Although in general the weather during these two centuries was remarkably warm, the winters of A.D. 1432-1439 were extremely cold (Lamb, 1977). Examples of comparably and remarkably unstable climatic epochs are found in the long-term proxy series from the Nile, Japan, the Crimea, and elsewhere (Schove, 1983; Fairbridge, 1984*b*).

Two principal causes of lunar-terrestrial climatic variables are (1) changes in the gravitational field and (2) nutational changes in insolation. Both are modulated by the Moon's changing orbital eccentricity, proximity to Earth, and the angle and direction between the Moon's orbital plane and the plane of the ecliptic. These are the kinds of orbital variables that Milankovitch used as the basis for his explanation of Pleistocene climatic swings from glacial to interglacial. As with the Sun, however, these lunar variables operate in short time spans.

The changing Earth-Moon separation distance (356,355-406,691 km) constitutes a large ellipticity variable. It is important because the tide-producing force is inversely proportional to the cube of this separation distance. The fact that during the past 400 yr, because all the minimum lunar perigee positions occurred during the perihelion passage of the Earth (thus during the winter season in the Northern Hemisphere), the lunar cycles tend to be reinforced by the solar (F. Wood, 1985, pp. 150, 218).

During the past four centuries, the closest approach was on January 4, 1912. The times of exact perigee and syzygy were separated by only 6 min. F. Wood (1985) has designated these close alignments as "proxigee-syzygy" phases. Solar and lunar declination cycles may fall in unison. When the Moon at equinox crosses the plane of the ecliptic, in which the Sun lies, the Sun's semidiurnal component adds 27% to the lunar to create the so-called "equinoctial tides." Furthermore, at times of close proxigee-syzygy, the centers of the Sun, Earth, and Moon lie along a line passing through the Earth-Moon barycenter, with the Sun's declination being typically (but not always) negative and the Moon's positive. Further, the apsides cycle is important in that it causes variations in the Sun-Earth separation. The retrograde apsides line rotation is commensurable with the direct (counterclockwise) lunar nodal cycle. Thus, 164.7/8.849 = 18.612 tropical yr. The declination maxima correspond only on average to apsides cycles; they are approximately in phase every 186 yr (18.6×10; 8.849×21). The solar-lunar tidal maxima precede declination maxima by an average of 4 yr—thus, A.D. 1340, 1433, 1526, 1619, 1712, 1805, 1891, 1992, a Quintuple Tidal-Nodal cycle (93 yr).

Another lunar periodicity discovered by Wood (1978) is the P-S cycle of 31.008 yr. It develops a Triple P-S (gravitational) cycle of 93.020 yr, which is closely in phase with the Quintuple Tidal-Nodal (nongravitational) cycle of 93.067 yr; thus, A.D. 1340, 1433, 1526, 1619, 1712, 1805, 1898, 1992. This tidal periodicity, not the simple declination, is most important in terrestrial climate cycles (Currie, 1984*b*). The date in the Northern Hemisphere winter season in which this cycle reaches its maxima creeps up gradually. In 1340, it happened on January 8. In 1992, the date will be January 19. The 93.020-yr node-perihelion cycle is only approximately in phase with the Apsides cycle in solar alignment on rare occasions (e.g., 1526, 1712). At such times, tidal effects are especially enhanced.

How do these lunar periodicities relate to terrestrial climates? It is easier to demonstrate apparent correlations than to explain the linkage mechanism(s). Further study is needed for all the dynamic systems involved. In the late nineteenth century, Ekholm and Arrhenius (1898) showed that a relationship exists between lunar motions and the frequency of thunderstorms and the behavior of the aurora borealis. The subject has been developed further by Chapman (1964*a*, 1964*b*) and by others. Clues to possible linkage mechanisms may also be gleaned from the kinds of weather parameters involved.

Most obvious of the longer-term variables is the 18.6-yr cycle in precipitation or in its inverse, the drought incidence in semiarid to marginally humid regions. This cycle has been demonstrated in North America in tree rings and in other series (Currie, 1981*c*, 1982, 1984*b*). The long-term Chinese historical records of rainfall at Beijing have given the same result (Hameed et al., 1983; Currie and Fairbridge, 1985). In connection with solar cycles, the 78-to-80-yr

Gleissberg cycle is represented in the Chinese data, but the signals from 11- and 22-yr periods are weak.

In North Africa, scattered documentary records of droughts (or famines) cover seven centuries. The maxima of 18.6-yr tidal cycles are matched by droughts, usually in groups of 3 yr or so, at the following times: 1303, 1340, 1396, 1508, 1601, 1619, 1694, 1713, 1731, 1749, 1787, 1861, and 1973. About half of these tidal peaks match times of sunspot maxima, but a few coincide with sunspot minima (Fairbridge, 1984*b*). As with the Beijing rainfall records, spectral analysis of the Nile floods shows strong peaks for the 18.6-yr tidal signal and for the Gleissberg cycle but weak representation of the 11- and 22-yr cycles (Hameed, 1984).

Studies of the Indian monsoon show both solar cycle and lunar cycle effects. The 22-yr solar cycle is felt especially at high latitudes, possible because of the greater flux of solar particles to the north magnetic pole. Its peaks (every second sunspot minimum) are marked by strong zonal circulation and a warm Northern Hemisphere (Bhalme and Mooley, 1981). At such times, the signal from the lunar-tidal effect is weak. In contrast, during epochs when blocking (meridional) circulation prevails, the effect of lunar forcing is two to three times stronger (Campbell, Blechman, and Bryson, 1983). During the twentieth century in the Northern Hemisphere, blocking circulation was predominant at first and peaked about 1910. Around 1920, the circulation crossed over to zonal, which peaked around 1930 and then diminished to another crossover around 1940 (Dzerdzeevski, 1969). This same pattern appears in Nile floods. Low levels, which correspond to droughts in north Africa, begin in 1911, 1941, and 1973. The epochs of meridional circulation are punctuated by strong lunar signals (associated with droughts). The levels of Lake Victoria (one of the sources of the White Nile) rose and fell with the SSC from 1890 to 1930, but then this pattern shifted (Lawrence, 1965). The lunar factor is supported by the extreme perigee-syzygy alignment of March 4, 1931, a season that was marked in coastal areas by 4 mo of perigean spring-tidal flooding (F. J. Wood, 1985).

In mid-latitude continental areas, especially in mid-winter, when the subtropical high-pressure area is weak, high precipitation correlates with the lunar-tidal maxima (Brier and Bradley, 1964). The lunar correlation is more prominent when numbers of sunspots are low. The lunar linkage mechanism has not been established and evidently needs research. As mentioned previously, a connection has been inferred via the mechanism of the atmospheric tide.

In very high latitudes (>70°), climatic series strongly reflect solar cycles and the lunar effects are weakly represented. Between latitudes 40° and 60°, these two forcing potentials show irregular, alternating phases (Guiot, 1984). In these regions, under the influence of maritime westerly air masses, the contradictory signals of solar and lunar origins have tended to become obscured in the white noise.

It is apparent that the extreme proxigee-syzygy events fall in line with maximum-declination phases over an approximate 168-yr periodicity (tables in F. J. Wood, 1985, p. 201.20). Examples are known from the winters of 1725 and 1893. These extreme proxigee-syzygy events alone are not completely regular, but during the past four centuries, on average (66 cases) they have occurred once every 5.90 yr (F. J. Wood, 1985).

A long-term cycle, the Progression of Lunar Perigee (PLP), displays a period of 556.027 tropical yr. This cycle has been claimed to match the times of major climatic change during the Holocene (Stacey, 1963, 1967*c*), but no causative relationship has been demonstrated. An overall variation in the Earth-Moon separation distance, as defined by the observed parallax, has been called the parallactic tide of "nearly 1850 yr" (Otto Pettersson, 1930). Actually, its period seems to be 1843.346 yr (F. J. Wood's "Cycle K": exactly 209 apsides cycles and 99.033 nodal cycles; the latter is almost exactly commensurable with the apside—the difference is only 1.8 day). Other long-term lunar periods identified by Wood (1985) include one of 222.103 yr that controls episodes of extreme perigee-syzygy.

The dynamics of the lunar-tidal/atmosphere linkage mechanisms have not been as thoroughly investigated as they deserve to be. The principal worker after Pettersson has been the Russian oceanographer, Maksimov. The variations in the speed of the Gulf Stream are particularly illuminating. Although its semiannual variation has attracted considerable attention (Stommel, 1958; and others too numerous to mention), its long-term changes have been largely neglected. Maksimov and Smirnov (1965) calculated that toward the peak of the 18.6-yr nodal tide cycle, the static height of the ocean surface undergoes a south-to-north tilting. This tilting is thought to increase the speed of the Gulf Stream and to raise the temperature of the water at the sea surface in the North Atlantic between Iceland and Great Britain by 0.5-1°C. A small statistical climatic signal has been found in British winter temperatures (the odd decades of Lamb, 1977, p. 484). The Icelandic low center of action oscillates east and west from the low of a nodal cycle to a high. During that cycle, the locus of vertical lunar passage shifts by more than 1,000 km north and south (in each hemisphere). Consideration of west coast currents, which set from north to south in the Northern Hemisphere, indicates phase reversal. Tide-gauge records suggest that whereas nodal-cycle rise in tide range and temperature of seawater would be amplified by the speedup of the Gulf Stream (by both Coriolis and steric effects), such effects would be negative for the California Current.

DISCUSSION

The following discussion of solar-terrestrial relationships is rather long, but individual topics are treated briefly. Despite the study in great detail of certain individual aspects of it, the overall subject is not well understood. Our discussion is intended to remind ourselves of the subject headings and to cite references to the literature where further information and ideas are available.

Many important points we emphasize have been published previously in well-known professional publications, for example, the *Journal of Geophysical Research* (Arriaga, 1955) and *the Astronomical Journal* (Jose, 1965). Nevertheless, our subject matter is evidently not, however, considered by astronomers to be very important and seems to have been totally ignored by solar physicists and others concerned with solar-terrestrial relationships. It was not mentioned in the recent report by the Geophysical Study Committee of the U.S. National Research Council (Eddy, 1982). In astronomy textbooks, little or no space is devoted to the Sun's orbit around the center of mass (barycenter) of the solar system. Moreover, the term *barycenter* rarely appears in the indexes of astronomy textbooks.

Some exceptions to these points are contained in books on celestial mechanics (e.g., Kurth, 1959; Krogdahl, 1962) and in a recent volume about the solar system (Jones, 1984). Jones's Figure 1-5 illustrates two bodies moving about their common center of mass. The following text describes the relationship: "If the one body has negligible mass compared to the other then the centre of mass of the system lies at the centre of mass of the more massive body, and this is also where the focus lies. This is nearly the case of the Sun and planets" (Jones, 1984, p. 9). We consider that this statement is not correct (see Figs. 26-5 and 26-6).

Two other aspects of our subject merit special emphasis. First, although most solar physicists and climatologists acknowledge the existence of a solar cycle, they display little respect for the many papers that have been published in which correlations have been sought between solar variation (e.g., as expressed by mean annual numbers of visible sunspots) and aspects of the Earth's weather. A number of physical scientists have accused those seeking correlations between temperature, rainfall, or other climatic parameters and sunspot numbers as indulging in autosuggestion (Pittock, 1979). At an early stage, we set out to test the conclusion published in 1963 by the late Clyde M. Stacey that the Sun's orbit could be employed to define an "anomalistic year of the Sun" whose period he thought was equal to that of the mean sunspot period. Although we support many of Stacey's conclusions, we find that his concept of the relationship between the sunspot cycle and an anomalistic year of the Sun is not correct.

Second, the Sun is forced to orbit the solar

system's barycenter because of the changing mass distributions of the planets. This means that planetary configurations are fundamental physical factors—not simply the nostrums of astrologers. This key point was stated clearly long ago by Sir Isaac Newton but seems to have been ignored or misunderstood by many subsequent workers. Newton wrote, "Since the centre of gravity [of the solar system] is constantly at rest, the sun, according to the various positions of the planets, must continually be moved every way, but will never recede far from that centre" (1687).

Besides the pioneering papers by Arriaga (1955) and Jose (1965), several authors subsequently have taken into account the basic fact of the Sun's orbit and that the center of the Sun is not the same point as the barycenter of the solar system (Wood and Wood, 1965; Pimm and Bjorn, 1969; Landscheidt, 1976, 1981, 1983, 1984, paper 25; Mörth and Schlamminger, 1979; Blizard, 1981, 1983, paper 24; Ren and Li, 1980). In some of these papers, the notion of a Sun-centered system is perpetuated by illustrations in which the barycenter is shown as orbiting the Sun. In reality, of course, the converse is true.

Our efforts to understand the importance of this topic has been uniquely aided by data from the computer-based barycentric ephemeris from NASA's Jet Propulsion Laboratory, Pasadena, California. An even longer ephemeris than the one we used is now available (Newhall, Standish, and Williams, 1983). Our copy of the ephemeris tapes and the work enabling them to be used locally resulted from the courtesy of Dr. Robert Jastrow, former director, Goddard Institute of Space Studies in New York City, and through the efforts of Dr. Stephen Unger.

We fully appreciate that Earth-bound astronomers cannot calibrate their telescopes to an invisible point, the barycenter of the solar system. In the past, celestial observations, essential for classical navigators, required a geocentric base. Ephemerides, such as those contained in the *Astronomical Almanac* and solar system catalogs (e.g., Meeus and Victor, 1983), are constructed on a geocentric (or in some cases, a heliocentric) framework. Kepler's famous laws of planetary orbits, memorized by every astronomy student, are based on a heliocentric system. And, because Newton had not been born until 1630, 12 yr after Kepler had died, the importance of the Sun's motion around the barycenter of the solar system has never been connected with Kepler's expressions. Our "solar orbit" can be described as "non-Keplerian."

For their convenience, astronomers use the Ptolemaic Earth-centered coordinate system for pointing their telescopes into space. However, as mentioned by Blizard (1981), in using such a system of geocentric or even one having heliocentric coordinates for describing planetary orbits, astronomers simply find it convenient to ignore the Sun's orbit, in so doing, they give a totally wrong basis for understanding what may be important geophysical cyclic variations.

Any star orbited by one or more natural satellites (such as planets) will behave in the same manner as our Sun. That is, the orbit of the star around the barycenter of its system will make that star appear to oscillate or wobble when viewed from a distance. Astrophysicists interpret the oscillations of distant stars (e.g., Barnard's star) as a basis for inferring that such stars are being orbited by planetary bodies that are not visible to telescopic observation (D. C. Black, 1982; Black and Scargle, 1982).

We have shown that during 64 orbits, the Sun's variable distance from the barycenter of the solar system changes in systematic ways that define at least two cycles. These are (1) the orbital period of the Sun, based on the time between successive peribac positions and ranging between 15 yr and 26 yr but with an average of about 19.9 yr, and (2) a cycle of approximately 178-180 yr, which is an orbital-repeat cycle. These two cycles are controlled by the positions of the major planets. The first is tied closely to the Saturn-Jupiter lap (SJL) cycle of 19.859 yr and the second, to the Orbital Symmetry Progression (OSP) cycle of 178.73 yr and the All-Planets Synod (APS) of 178.71 yr, which are closely related to the Saturn-Jupiter Repeat cycle (SJR) of 177.93 yr.

Because these solar cycles are repetitive and predictable, they open up new areas for research into solar-terrestrial relationships. Clearly, these (and possibly other) orbital cycles constitute important but previously ignored physical mechanisms that must be taken into account in any attempts to understand the Sun's behavior. Other periodicities that may be related to them include changes in the rate of the Sun's rotation and changes in the surface-to-surface separation distance between the Earth and the Sun. Indeed, solar cyclicity must now be considered as being a fact of life.

What needs to be done is to determine if these solar cycles are in any way(s) connected with other kinds of time-series information about the Sun (e.g., numbers of sunspots) or about the Earth's geophysical variables (e.g., changing spin rate as indicated by variable lengths of the day and variations in the geomagnetic field) or climate (e.g., temperature or rainfall). In the past, references to the solar cycle meant the sunspot cycle. However, recent studies have shown that not even the associated phenomena all coincide. Spectral analyses of sunspot numbers hint at the existence of many cycles, some of which we discuss in light of our work.

The possibility that long-term predictability may now be at hand leads to some interesting philosophic implications. For example, opinion seems to be solidifying behind the concept that external celestial-mechanical factors of the kind recognized by Milankovitch can affect the Earth's climate on time scales ranging from tens of thousands of years to hundreds of thousands of years. Such Milankovitch-type cycles are based on variables tied to planetary orbits. In particular, Milankovitch emphasized the predictably varying distances from the Earth to the Sun as the ellipticity of the Earth's orbit varies and as the angles of tilt and of direction of the Earth's axis that affect seasonality undergo regular oscillations. These cyclically varying parameters bring about important changes in the Earth's climate even if the Sun and other factors remain constant. This presumed constancy of the Sun and other factors may not be valid. Be that as it may, what we think we have demonstrated is that the Sun's orbit undergoes the same kinds of celestial-mechanical variables emphasized by Milankovitch but with two major differences: (1) the time scales of many of these solar variations are in the ranges of a few years to a few hundred years, and (2) the Sun's motions vary in cycles having identical periods with some of these planetary-orbital periods and to their commensurabilities.

The time has come to stop searching for a single solar cycle. Solar cycles are numerous; they are real, demonstrably cyclic, and reproducible. The period of one of the most important SOB cycles is about 20 yr. When such a peak appears in the spectral analysis of a terrestrial time series (as in Hibler and Johnsen, 1979, e.g.), it should be considered as being valid and not written off as some kind of average between other cycles such as those having periods of 18.6 yr and 22.4 yr (as, e.g., Roberts, 1979).

Another point that our work highlights is that many practices of long-term statistical averaging can suppress what may be a valid signal. Because the forcing mechanisms of solar cycles are related to planetary motions, they are both periodic and slightly irregular. At some times, they may reinforce one another and build up to crescendos. At other times, they may cancel out and thus drop out of sight for a time. Searches for signals from such cycles in smoothed terrestrial time series should not obscure these characteristics. The cycles must be treated chronologically because the so-called 11-yr cycle is almost a figment of the statistician's art.

When it comes to climate prediction on the planet Earth, one is faced with monumental difficulties involving variable inputs, feedbacks, and dynamic complexities. Not the least of prediction problems is psychological. It stems from a curious paradox among certain climatologists who, in their alleged searches for solar-terrestrial climatic links, homogenize their basic data. For example, the WMO-approved, world-wide or nationwide system of year-round averaging in 30-yr blocks combines data that clearly should be kept separate. Three examples include the following: (1) West coast data should not be mixed with east coast data, nor arctic with tropical; maritime air masses are subject to long-term water mass memories; high latitudes are much more responsive to climatic factors, on the order of 20 : 1 (i.e., a factor causing a 1°C change in the tropics over a decade or more is likely to be matched by a 20°C shift in the high arctic). (2)

Year-round averages commonly suppress summer and winter departures whose signs may be opposite. Daily and monthly means are often less illuminating than the envelope of extremes. The biologic effects of extremes typically are more significant than those of averages (e.g., widths of tree rings). Many trees do not grow during the winter. Thus, growth season data may be much more meaningful than yearly averages. (3) The standard twentieth-century 30-yr basis for averaging weather statistics fails to recognize the long-term alternation between zonal and meridional circulation modes (Dzerdzeevski, 1961; Lamb, 1977).

A different approach for climatological research should include an aggressive search for causality, using attempts to dissect (as in biology) the complex mix of signals contained in day-to-day weather. Useful parallels can be drawn from the work of coastal specialists dealing with sea level. The readings from tide gauges must be diligently and imaginatively analyzed to recognize quantitatively the components resulting from many variables. These include the tidal factors (lunar, solar, and planetary), the steric factor, atmospheric pressure, wind and fetch, density, effects of rivers, current dynamics, and the Coriolis effect. Not to be omitted are the vertical motions of the land surface resulting from neotectonics, compaction, and removal of interstitial fluids. The net result of these factors is the actual measured water level. Comparable complexities characterize weather.

By *aggressive approach* we mean that searches should concentrate on critical areas. For example, if one wants to test for the effects of the sunspot cycle on temperature, precipitation, or wind, one should concentrate on records from continental, high-pressure, low-cloud stations. Examples include Chihuahua, Khartoum, or Alice Springs. More than a century ago, Köppen (1873) reported an 11-yr correlation from low-latitude stations. As we would expect, however, he did not find such a relationship elsewhere. In very high-latitude localities, we anticipate seeing the strong signals that relate to particulate radiation. These particles tend to be funneled in at the magnetic poles. On the short term, the major solar-flare events are dramatically expressed in the Earth's weather data (Bucha, 1977, 1980). Other solar effects are related to the solar-sector boundaries (Wilcox, 1976). On the long term, we expect signs of the great sunspot dearths, the Maunder, Spörer, Wolf, and other minima, to appear in climatic proxy series that lack the time control provided by radiocarbon or by beryllium-10, for example. In the ice cores from glaciers on Greenland and Baffin Island, the oxygen-18 and deuterium analyses display a clear 180-yr signal (Johnsen, Dansgaard, and Clausen, 1970). The harmonics of 90 yr and 45 yr are also displayed.

In the Hudson Bay region, a summer storminess cycle of about 45 yr has been dramatically recorded by 183 uniform beach ridges. These form an uninterrupted time series that reaches back more than 8,000 yr (Fairbridge and Hillaire-Marcel, 1977).

Rhodes Fairbridge is chairman of a working group (WG 8-D) of the International Quaternary Association (INQUA) Holocene Commission that has been instructed to study cyclical trends during the past 10,000 yr that are identifiable from sedimentary and sea-level data. The present tasks of this INQUA group consist mainly of identifying field evidence, cataloging and analyzing the data, and refining the chronological base. Particularly useful in the last category are the methods of dendrochronology, varve chronology, and ice-core analysis, each of which aims at 1-yr precision. Radiocarbon flux rates, in combination with tree-ring counting, tell us something about the Sun (Eddy, 1980).

That great and abrupt environmental changes have occurred repeatedly during the past 10,000 yr is now widely accepted by many paleoclimatologists. Not only have the Suess wiggles been amply confirmed, but also in closely analyzed tree rings, de Jong, Mook, and Becker (1979) have demonstrated a series of approximately 180-yr cycles in which the amount of radiocarbon rises steeply and fades gradually. This relationship implies that solar activity decays abruptly and builds up gradually. This is the same relationship that has been noticed in the pattern of sunspot maxima within successive cycles, usually in 4-8-16 groups (45, 90, 180 yr), although within single cycles, asymmetry varies with intensity (Schove, 1983).

We address a fervent plea to historical climatologists (and to historians interested in climate) to pay special attention to climatic singularities. By these, we mean unusual climatic events of even a few days' duration or of seasonal extent. Some of these singularities have been connected to veils of volcanic dust (Lamb, 1970), but others, particularly precipitation events, may be connected with meteor showers (Bowen, 1953, 1956) or with lunar-solar tidal peaks (Brier and Bradley, 1964; Currie, 1984*b*). During the past 1,000 yr or so, historical documents help to identify such special events (e.g., Ludlum, 1968; LeRoy Ladurie, 1971). These events are important even if their causes are not always easily understood.

Our analysis of the Sun's orbit and its remarkably uneven behavior forms a firm chronometric basis against which solar-related terrestrial events can be matched. The Sun's position and other attributes of its orbit can now be firmly related to planetary motions. Inasmuch as solar emission activity conforms to solar motions as well as to many aspects of terrestrial climate, the door is now open for fundamental new efforts on terrestrial predictions.

We contend that the new planetary dynamics carries a profound philosophical message. The course of history, as well as the future, is highly deterministic. The Sun and planets are not "walking randomly." We think it is highly probable that terrestrial climate is strongly dependent on extraterrestrial cycles having periods of from 1 yr to 1,000 yr. In this sense, nature's dice really are somewhat loaded.

ACKNOWLEDGMENTS

This study would not have been possible without the generous assistance and cooperation of Dr. Robert Jastrow, former director, Goddard Institute of Space Studies (GISS) in New York City, who arranged for us to obtain a copy of the tape of the JPL Ephemeris and made available time on the institute computer for us to make the calculations that formed the basis of our work. Dr. Stephen Unger wrote the programs that enabled the GISS computer to read the JPL Ephemeris tape.

Parts or all of this manuscript have been reviewed and improved by Jane Blizard, Robert G. Currie, Sultan Hameed, the late Derek Justin Schove, James H. Shirley, and Fergus J. Wood. Many others have helped with useful discussions and reprints of their papers.

VII

Appendix

Selected Bibliography on Sun–Earth Relationships and Cycles Having Periods of Less than 10,000 Years

John E. Sanders and Rhodes W. Fairbridge
Columbia University

Aaby, B., 1976, Cyclic variations in climate over the past 5,500 yr. reflected in raised bogs, *Nature* **263:**281-284.

Abbe, C., 1903, Australian droughts and the Moon, *Monthly Weather Rev.* **30:**525-526.

Abbe, C., 1904, Periods in solar radiation and terrestrial temperatures, *Monthly Weather Rev.* **31:**595.

Abbe, C., 1906*a*, Drought and atmospheric electricity, *Monthly Weather Rev.* **34:**121-122.

Abbe, C., 1906*b*, Benjamin Franklin as meteorologist, *Am. Philos. Soc. Proc.* **45:**117-128.

Abbot, C. G., 1904, Recent studies of the solar constant of radiation, *Monthly Weather Rev.* **31:**587-592.

Abbot, C. G., 1911*a*, The Sun's energy-spectrum and temperature, *Astrophys. Jour.* **34:**197-208.

Abbot, C. G., 1911*b*, *The Sun*, New York and London: Appleton, 448p.

Abbot, C. G., 1918, On periodicity in solar variation, *Smithsonian Misc. Colln.* **69**(6):8p.

Abbot, C. G., 1923, Values of the solar constant, *Monthly Weather Rev.* **51:**71-81.

Abbot, C. G., 1925*a*, Solar variation and forecasting, *Smithsonian Misc. Colln.* **77**(5):27p.

Abbot, C. G., 1925*b*, Solar variation and the weather, *Am. Meteorol. Soc. Bull.* **6**(7):100-105.

Abbot, C. G., 1927, A group of solar changes, *Smithsonian Misc. Colln.* **80**(2):16p.

Abbot, C. G., 1931, Weather dominated by solar changes, *Smithsonian Misc. Colln.* **85**(1):18p.

Abbot, C. G., 1933*a*, Sun spots and weather, *Smithsonian Misc. Colln.* **87**(18):10p.

Abbot, C. G., 1933*b*, Forecasts of solar variation, *Smithsonian Misc. Colln.* **89**(5):5p.

Abbot, C. G., 1935*a*, Solar radiation and weather studies, *Smithsonian Misc. Colln.* **94**(10):89p.

Abbot, C. G., 1935*b*, Mount St. Katherine, an excellent solar-radiation station, *Smithsonian Misc. Colln.* **94**(12):11p.

Abbot, C. G., 1935*c*, Rainfall variations, *Royal Meteorol. Soc. Quart. Jour.* **61**(258):90-92.

Abbot, C. G., 1936*a*, The dependence of terrestrial temperatures on the variations of the Sun's radiation, *Smithsonian Misc. Colln.* **95**(12):15p.

Abbot, C. G., 1936*b*, Further evidence of the dependence of terrestrial temperatures on the variations of solar radiation, *Smithsonian Misc. Colln.* **95**(15):4p.

Abbot, C. G., 1936*c*, Cycles in tree-ring widths, *Smithsonian Misc. Colln.* **95**(19):5p.

Abbot, C. G., 1940, The variation of the Sun and weather, *Am. Meteorol. Soc. Bull.* **21**(12):407-416.

Abbot, C. G., 1941, An important weather element hitherto generally disregarded, *Smithsonian Misc. Colln.* **101**(1):34p.

Abbot, C. G., 1944*a*, *The Sun and the Welfare of Man*, New York: Smithsonian Institution Series, Inc., 384p.

Abbot, C. G., 1944*b*, A 27-day period in Washington precipitation, *Smithsonian Misc. Colln.* **104**(3).

Abbot, C. G., 1944*c*, Weather predetermined by solar variation, *Smithsonian Misc. Colln.* **104**(5):44p.

Abbot, C. G., 1945*a*, Correlations of solar variation with Washington weather, *Smithsonian Misc. Colln.* **104**(13).

Abbot, C. G., 1945*b*, Solar variation and weather, in *Annual Report for 1944*, Washington, D.C.: Smithsonian Institution, pp. 119-154.

Abbot, C. G., 1946, The Sun makes the weather, *Sci. Monthly* **62:**201-210, 341-348.

Abbot, C. G., 1947*a*, The Sun's short regular variation and its large effect on terrestrial temperatures, *Smithsonian Misc. Colln.* **107**(4):33p.

Abbot, C. G., 1947*b*, Precipitation affected by solar variation, *Smithsonian Misc. Colln.* **107**(9):4p.

Abbot, C. G., 1947*c*, A revised analysis of solar-constant values, *Smithsonian Misc. Colln.* **107**(10):9p.

Abbot, C. G., 1948*a*, Solar variation attending West Indian hurricanes, *Smithsonian Misc. Colln.* **110**(1):7p.

Abbot, C. G., 1948*b*, 1947-1948 report on the 27.0074-day cycle in Washington precipitation, *Smithsonian Misc. Colln.* **110**(4):2p.

Abbot, C. G., 1948*c*, Magnetic storms, solar radiation, and Washington temperature departures, *Smithsonian Misc. Colln.* **110**(6):12p.

Abbot, C. G., 1949, Short periodic solar variations and the temperatures of Washington and New York, *Smithsonian Misc. Colln.* **111**(13):8p.

Abbot, C. G., 1952*a*, Periodicities in the solar-constant measures, *Smithsonian Misc. Colln.* **117**(10):31p.

Abbot, C. G., 1952*b*, Important interferences with normals in weather records associated with sunspot frequency, *Smithsonian Misc. Colln.* **117**(11):3p.

Abbot, C. G., 1952*c*, Solar variations and precipitation at Peoria, Illinois, *Smithsonian Misc. Colln.* **117**(16):18p.

Abbot, C. G., 1953*a*, Solar variations and precipitation at Albany, New York, *Smithsonian Misc. Colln.* **121**(5):16p.

Abbot, C. G., 1953*b*, Long-range effects of the Sun's variation on the temperature of Washington, D.C., *Smithsonian Misc. Colln.* **122**(1):14p.

Abbot, C. G., 1953*c*, Solar variation, a leading weather element, *Smithsonian Misc. Colln.* **122**(4):35p.

Abbot, C. G., 1955*a*, Sixty-year weather forecasts, *Smithsonian Misc. Colln.* **128**(3):22p.

Abbot, C. G., 1955*b*, Periodic solar variation, *Smithsonian Misc. Colln.* **128**(4):20p.

Abbot, C. G., 1956, Periods related to 273 months or 22-3/4 years, *Smithsonian Misc. Colln.* **134**(1):17p.

Abbot, C. G., and Fowle, F. E., 1913*a*, Volcanoes and climate, *Smithsonian Misc. Colln.* **60**(29):24p.

Abbot, C. G., and Fowle, F. E., 1913*b*, The variation of the Sun, *Astrophys. Jour.* **38:**181-186.

Abbot, C. G.; Fowle, F. E.; and Aldrich, L. B., 1913, Die Solarkonstante und ihre Schwankungen, *Zeitsch. Meteorologische* **30:**257-261.

Abbot, C. G., and McCandlish, N. M., 1939, The weekly period in Washington precipitation, *Smithsonian Misc. Colln.* **98**(21):4p.

Adderley, E. E., 1963, The influence of the moon on atmospheric ozone, *Jour. Geophys. Research* **68:**1405-1408.

Adderley, E. E., and Bowen, E. G., 1962, Lunar component in precipitation data, *Science* **137:**749-750.

Adem, J., 1964, On the physical basis for numerical prediction of monthly and seasonal temperature in the troposphere-ocean-continent system, *Monthly Weather Rev.* **92:**91-103.

Adem, J., 1970, On the prediction of monthly mean temperatures, *Tellus* **22**(4):410-430.

Afanasieva, V. I., and Laptukhov, A. I., 1982, Periodical oscillations of geomagnetic and solar activity, *Geomagnetism and Aeronomy (Moscow)* **22**(4).

Agee, E. M., 1980, Present climatic cooling and a proposed causative mechanism, *Am. Meteorol. Soc. Bull.* **61**(11):1356-1367.

Ahlmann, H. W., 1948, The present climatic fluctuation, *Geog. Jour.* **112:**165-195.

Ahlmann, H. W., 1953, *Glacier Variations and Climatic Fluctuations*, American Geographical Society, Bowman Memorial Lecture Series, 51p.

Aitken, John, 1908, Ions as nuclei of condensation, *Symons' Meteorol. Mag. (London)* **43:**118.

Aitken, M. J., and Weaver, G. H., 1965, Recent archaeomagnetic results in England, *Jour. Geomagnetism and Geoelectricity* **17:**391.

Akasofu, S.-I., and Meng, C.-I., 1969, A study of polar magnetic substorms, *Jour. Geophys. Research* **74**(1):293-313.

Alfven, H., 1958, On the theory of magnetic storms and aurorae, *Tellus* **10:**104-116.

Alfven, H., and Arrhenius, G., 1970, Structure and evolutionary history of the solar system, I, *Astrophys. Space Sci.* **8:**338.

Allan, D. W., 1958, Reversals of the Earth's magnetic field, *Nature* **182:**469-470.

Alldredge, L. R., 1975, A hypothesis for the source of impulses in geomagnetic secular variation, *Jour. Geophys. Research* **80**(11):1571-1578.

Alldredge, L. R., 1982, Geomagnetic models and the solar cycle effect, *Rev. Geophysics and Space Physics* **20**(4):965-970.

Alldredge, L. R., 1983, Varying geomagnetic anomalies and secular variation, *Jour. Geophys. Research* **88B**(11):9443-9451.

Alldredge, L. R., and Stearns, C. O., 1969, Dipole model of the sources of the Earth's magnetic field and secular change, *Jour. Geophys. Research* **74**(27):6583-6593.

Alldredge, L. R.; Van Voorhis, G. D.; and Davis, T. M., 1963, A magnetic profile around the world, *Jour. Geophys. Research* **68**(12):3679-3692.

Allen, C. W., 1944, Relation between magnetic storms and solar activity, *Royal Astron. Soc. Monthly Notices* **104:**13-21.

Allen, C. W., 1958, Solar radiation, *Royal Meteorol. Soc. Quart. Jour.* **84**(362):307-318.

Allen, C. W., 1960, A sunspot cycle model, *Observatory* **80:**94-98.

Alt, E., 1909, Die Doppeloszillation des Barometers, inbesondere im arktischen Gebiete, *Zeitsch. Meteorologische (Braunschweig)* **26:**145-164.

Alter, D., 1922*a*, A rainfall period equal to one-ninth the sunspot period, *Kansas Univ. Sci. Bull.* **13**(11).

Alter, D., 1922*b*, Possible period in mean Sun spottedness, *Popular Astronomy*, February, p. 104.

Alter, D., 1924, Application of Schuster's periodogram to long rainfall records, beginning 1748, *Monthly Weather Rev.* **52:**479-487.

Alter, D., 1927, A group or correlation periodogram with application to the rainfall of the British Isles, *Monthly Weather Rev.* **55**(6):263-266.

Alter, D., 1928, A new analysis of the sunspot numbers, *Monthly Weather Rev.* **56**(10):399-401.

Alter, D., 1929, A critical test of the planetary hypothesis of sun spots, *Monthly Weather Rev.* **57**(4):143-146.

Alter, D., 1933*a*, Some results of cycle analysis of rainfall, *Am. Meteorol. Soc. Bull.* **14**(11):256-257.

Alter, D., 1933*b*, Correlation periodogram investigation of English rainfall, *Monthly Weather Rev.* **61:**345-350.

Ammons, R.; Ammons, A.; and Ammons, R. B., 1983, Solar activity-related quasi-cycles in Tertiary tree-ring records: Evidence and methodological studies, in B. M. McCormac (ed.), *Weather and Climate Response to Solar Variation,* Boulder: Colorado Associated University Press, pp. 535-543.

Anders, H. S., 1902*a*, The relation of sunshine to the prevalence of influenza, Meteorological Study No. 2, *Balneol. and Climatol. Soc. (London) Jour.* **6:**264-270.

Anders, H. S., 1902*b*, Atmospheric pressure and epidemic influenza in Philadelphia, Meteorological Study No. 3, *Balneol. and Climatol. Soc. (London) Jour.* **6:**270-275.

Andersen, A. P., and Lomb, N. R., 1980, Yet another look at two classical time series, in O. D. Anderson (ed.), *Time Series: Proceedings of an International Conference* (Nottingham University, March 1979), Amsterdam and New York: North Holland Publishing Company, pp. 5-20.

Andersen, B. N., and Maltby, P., 1983, Has rapid solar-core rotation been observed?, *Nature* **302:**808-810.

Anderson, C. N., 1939, A representation of the sunspot cycle, *Jour. Geophys. Research* **44:**175-179.

Anderson, C. N., 1954, Notes on the sunspot cycle, *Jour. Geophys. Research* **59**(4):455-461.

Anderson, R. Y., 1961, Solar-terrestrial climatic patterns in varved sediments, *New York Acad. Sci. Annals* **95:** Article 1, 424-439.

Andersson, G., 1909, The climate of Sweden in the late Quaternary period. Facts and theories, *Sveriges Geol. Undersökning Årsb.* **3**(1):88p.

Andrews, J. T., and Ives, J. D., 1972, Late- and postglacial events (>10,000 B.P.) in the eastern Canadian Arctic, with particular reference to the Cockburn moraines and break-up of the Laurentide Ice Sheet, in Y. Vasari, H. Hyvarinen, and S. Hicks (eds.), *Climatic Changes in Arctic Areas during the Last Ten-Thousand Years,* Oulu, Finland: Universitatis Oulensis, pp. 149-176.

Angell, J. K., and Korshover, J., 1963, Harmonic analysis of the biennial zonal-wind and temperature regimes, *Monthly Weather Rev.* **91:**537-548.

Angell, J. K., and Korshover, J., 1974, Quasi-biennial and long-term fluctuations in the centers of action, *Monthly Weather Rev.* **102:**669-678.

Angell, J. K., and Korshover, J., 1976, Global analysis of recent total ozone fluctuations, *Monthly Weather Rev.* **104:**63-75.

Angell, J. K., and Korshover, J., 1978*a*, Global ozone variations: An update into 1976, *Monthly Weather Rev.* **106:**725-737.

Angell, J. K., and Korshover, J., 1978*b*, Recent rocket sonde derived temperature variations in the western hemisphere, *Jour. Atmos. Sci.* **35:**1758-1764.

Angot, A., 1903, Sur les variations simultanees des taches solaires et des temperatures terrestres, *Acad. Sci. Comptes Rendus* **136:**1186-1189.

Angot, A., 1906, Etudes sur le climat de la France. Pression atmospherique, *Paris, Bureau Central Meteorologique, Ann.* **1:**83-249.

Angstrom, A., 1935, Teleconnections of climatic changes in present time, *Geog. Annaler* **17:**242-258.

Angstrom, A., 1939, The change of temperature climate in present time, *Geog. Annaler* **21:**119-131.

Angstrom, A., 1949, Atmospheric circulation, climatic variations and continentality of climate, *Geog. Annaler* **31:**316-320.

Angstrom, A., 1970, Apparent solar constant variations and their relation to the variability of atmospheric transmission, *Tellus* **22**(2):205-218.

Anonymous, 1961, *Explanatory Supplement to the Astronomical Ephemeris,* London: Her Majesty's Stationery Office; and Washington, D.C.: U.S. Government Printing Office, 533p.

Anonymous, 1983, *The Astronomical Almanac (for 1983),* Washington, D.C.: U.S. Naval Observatory; and London: Royal Greenwich Observatory.

Antalova, A., and Gnevyshev, M. N., 1965, Principal characteristics of the 11-year solar activity cycle, *Soviet Astronomy* **9:**198-201.

Arago, F., 1865, Sur l'état thermometrique du globe terrestre, in *Oevres Complètes,* vol. 8, Paris: Gide, pp. 184-487.

Arakawa, H., 1937, Fujiwara on five centuries of freezing dates of Lake Suwa in central Japan, *Archiv Meteorologie, Geophysik, u. Bioklimatologie,* Series B, **6:**152-166.

Arakawa, H., 1957, Climatic change as revealed by the data from the Far East, *Weather* **12**(2):46-51.

Arakawa, H., and Tsutsumai, K., 1956, A decrease in

the normal incidence radiation values for 1953 and 1954 and its possible cause, *Japan Meteorol. Agency Geophys. Mag.* **27:**205-208.

Archibald, E. D., 1879*a*, Barometric pressure and sun-spots, *Nature* **20:**28-29.

Archibald, E. D., 1879*b*, Barometric pressure and temperature in India, *Nature* **20:**626-627.

Archibald, E. D., 1879*c*, The weather and the Sun, *Nature* **20:**626-627.

Archibald, E. D., 1880, Sunshine cycles, *Nature* **21:**393-394.

Archibald, E. D., 1881*a*, Abnormal barometric gradient between London and St. Petersburg in the sun-spot cycle, *Nature* **23:**618-619.

Archibald, E. D., 1881*b*, On the connection between solar phenomena and climatic cycles, part I, *The Scientific Roll*, pp. 17-37. (Reprinted in A. Ramsay, 1898, *A Bibliography, Guide, and Index to Climate*, London: Swan Sonnenschein Company, 449p.)

Archibald, E. D., 1882*a*, Variations in the Sun's heat, *Nature* **25:**316.

Archibald, E. D., 1882*b*, Conservation of solar energy, *Nature* **25:**504.

Archibald, E. D., 1885, The eleven-year meridional oscillation of the auroral zone, *Nature* **33:**52-53.

Archibald, E. D., 1890, Cyclical periodicity in meteorological phenomena, *Am. Meteorol. Jour.* **7:**289-295.

Archibald, E. D., 1896, The long period weather forecasts of India, *Nature* **55:**85-88.

Archibald, E. D., 1900, Droughts, famines, and forecasts in India, *Monthly Weather Rev.* **28:**246-248.

Arcimis, A., 1904, Telegraphic disturbances in Spain on October 31, 1903, *Nature* **69:**29.

Arctowski, H., 1907, De l'influence de la lune sur la vitesse du vent aux sommets du Santis, du Sonnblick et du Pike's Peak, *Bruxelles, Soc. Astron. Bull.*, pp. 388-398.

Arctowski, H., 1908*a*, Variations de la repartition de la pression atmospherique à la surface du globe, *Bruxelles Soc. Astron. Bull.*, pp. 161-163.

Arctowski, H., 1908*b*, Les anomalies de la repartition de la pression atmospherique aux Etats-Unis, *Bruxelles Soc. Astron. Bull.*, pp. 200-201.

Arctowski, H., 1908*c*, Recherches sur la periodicité des phénomènes meteorologiques à Bruxelles, *Bruxelles Soc. Astron. Bull.*, pp. 226-231.

Arctowski, H., 1908*d*, Les variations seculaires du climat de Varsovie, *Bruxelles Soc. Astron. Bull.*, pp. 301-326.

Arctowski, H., 1910*a*, Correlations of climatic changes, *Science* (new series) **31:**25-26.

Arctowski, H., 1910*b*, On changes of atmospheric presssure in North America, *Science* (new series) **31:**427-428.

Arctowski, H., 1910*c*, Sur les anomalies de la repartition de la pression atmospherique aux Etats-Unis, *Acad. Sci. Comptes Rendus* **150:**753-754.

Arctowski, H., 1910*d*, Studies on climate and crops. 1. Variations in the distribution of atmospheric pressure in North America, *Am. Geog. Soc. Bull.* **42:**270-282.

Arctowski, H., 1910*e*, Studies on climate and crops. 2. The yield of wheat in the United States and in Russia during the years 1891 to 1900, *Am. Geog. Soc. Bull.* **42:**481-495.

Arctowski, H., 1912*a*, Studies on climate and crops. 3. The "solar constant" and the variations of atmospheric temperature at Arequipa and some other stations, *Am. Geog. Soc. Bull.* **44:**598-606.

Arctowski, H., 1912*b*, Studies on climate and crops. 4. Corn crops in the United States, *Am. Geog. Soc. Bull.* **44:**745-760.

Arctowski, H., 1914*a*, Zur Dynamik der Klimaanderungen, *Zeitschr. Meteorologische (Braunschweig)* **31:**417-426.

Arctowski, H., 1914*b*, A study of the changes of the distribution of temperature in Europe and North America during the years 1900 to 1909, *New York Acad. Sci. Annals* **24:**39-113.

Arctowski, H., 1917, Sunspots, magnetic storms and rainfall, *Monthly Weather Rev.* **45**(11):538.

Arctowski, H., 1940, Researches on temperature changes from day to day and solar constant variations, *Am. Meteorol. Soc. Bull.* **21**(6):257-261.

Arora, B. R., and Sastri, N. S., 1977, Some features of the variability associated with solar and lunar daily variations, *Royal Astron. Soc. Geophys. Jour.* **50**(1):235-241.

Arrhenius, S., 1904, On the electric equilibrium of the Sun, *Royal Soc. [London] Proc.* **73:**496-499.

Arrhenius, S., 1909, *Die physikalischen Grundlagen der Kohlensauretheorie der Klimaveranderungen*, Stuttgart: Centralblatt für Mineralogie, pp. 481-491.

Arriaga, N., 1955, Relations between solar activity and the center of gravity of the planetary system, *Jour. Geophys. Research* **60:**535-536.

Ashley, A. McC., 1901, Long range seasonal forecasts for the Pacific Coast states, *Monthly Weather Rev.* **29:**16-19.

Ashworth, J. R., 1904, A source of the ionisation of the atmosphere, *Nature* **70:**454.

Atwater, M. A., 1970, Planetary albedo changes due to aerosols, *Science* **170:**64-66.

Austin, L. W., 1927*a*, Long-wave radio measurements at the Bureau of Standards in 1926, with some comparisons of solar activity and radio phenomena, *Inst. Radio Engineers Proc.* **15**(10):825-836.

Austin, L. W., 1927*b*, Radio atmospheric disturbances and solar activity, *Inst. Radio Engineers Proc.* **15**(10):837-842.

Axmann, 1905, Physiologie des Wetters, *Himmel und Erde* **17:**219-234.

Babcock, H. W., 1961, The topology of the Sun's magnetic field and 22-year cycle, *Astrophys. Jour.* **133:**572-587.

Babcock, H. W., and Babcock, H. D., 1955, The Sun's magnetic field, *Astrophys. Jour.* **121:**349.

Bagby, J. P., 1973, Further evidence of tidal influence of earthquake incidence, *The Moon* **6**(3/4):398-404.

Bagby, J. P., 1975, Sunspot cycle periodicities, *Nature* **253:**482.

Bailey, S. I., 1901, Recent total eclipses of the Sun, *Popular Science Monthly* **60**(2):240-252.

Baker, D. G., and Skaggs, R. H., 1985, The Minnesota long-term temperature record, *Clim. Change* **7:**225-236.

Balasubrahmanyan, V. K.; Boldt, E.; and Palmeira, R. A. R., 1967, Solar modulation of galactic cosmic rays, *Jour. Geophys. Research* **72**(1):27-36.

Balling, R. C., Jr., and Lawson, M. P., 1982, Twentieth century changes in winter climatic regions, *Clim. Change* **4:**57-69.

Bandeen, W. R., and Maran, S. P., eds., 1974, *Symposium on Possible Relationships between Solar Activity and Meteorological Phenomena Proceedings,* Greenbelt, Md.: NASA Goddard Space Flight Center, NASA SP-366, 263p.

Bangs, N. H., 1929, Is the semidiurnal barometric variation an electric phenomenon?, *Monthly Weather Rev.* **57:**383.

Banks, R. J., 1969, Geomagnetic variations and the electrical conductivity of the upper mantle, *Royal Astron. Soc. Geophys. Jour.* **17**(5):457-487.

Barber, K. E., 1982, Peat-bog stratigraphy as a proxy climate record, in A. F. Harding (ed.), *Climatic Change in Later Prehistory,* Edinburgh: Edinburgh University Press, pp. 103-113.

Barnes, A. A., 1919, Rainfall in England; The true long-average as deduced from symmetry, *Royal Meteorol. Soc. Quart. Jour.* **45:**209-232.

Barnes, J. A.; Sargent, H. H., III; and Tryon, P. V., 1980, Sunspot cycle simulation using random noise, in R. O. Pepin, J. A. Eddy, and R. B. Merrill (eds.), *Ancient Sun: Proceedings of a Conference on the Ancient Sun: Fossil Record in the Earth, Moon and Meteorites,* New York: Pergamon Press, pp. 159-163.

Barnes, R. T. H.; Hide, R.; White, A. A.; and Wilson, C. A., 1983, Atmospheric angular momentum fluctuations, length-of-day changes and polar motion, *Royal Soc. [London] Proc.* **387A:**31-73.

Barnett, J. J., 1975, Large sudden warming in the Southern Hemisphere, *Nature* **255:**387-389.

Barnett, T. P., 1978, The role of the oceans in the global climate system, in J. R. Gribbin (ed.), *Climatic Change,* Cambridge: Cambridge University Press, pp. 157-177.

Barraclough, D. R.; Harwood, J. M.; Leaton, B. R.; and Malin, S. R. C., 1975, A model of the geomagnetic field at epoch 1975, *Royal Astron. Soc. Geophys. Jour.* **43**(3):645-659.

Barrell, J., 1914, The status of hypotheses of polar wanderings, *Science* (new series) **40:**333-340.

Barry, R. G., and Chorley, P. J., 1968, *Atmosphere, Weather, and Climate,* London: Methuen and Company, 379p.

Barta, G., 1956, A 40-50 year period in the secular variation of the geomagnetic field, *Acta Geol. Budapest* **4:**15-42.

Bartels, J., 1932, Terrestrial magnetic activity and its relation to solar phenomena, *Terr. Magnetism and Atmos. Electricity* **37:**1-52.

Bartels, J., 1934, Twenty-seven day recurrences in terrestrial-magnetic and solar activity, 1923-1933, *Terr. Magnetism and Atmos. Electricity* **39:**201-202.

Bartels, J., 1935, Random fluctuations, persistence, and quasi-persistence in geophysical and cosmical periodicities, *Terr. Magnetism and Atmos. Electricity* **40:**1-60.

Bartels, J., 1963, Behauptete Einflusse des Mondes auf der Erdmagnetische Unruhe und ihr statistischer Hintergrunde, *Naturwissenschaften* **50:**592.

Bartels, J., 1964, Statistische Hintergrunde für geophysikalische Synchronisierungs-versuche und Kritic behaupteten Mond-einflussen auf die erdmagnetische Aktivität, *Akad. Wiss. Göttingen Nachr. Math.-Phys. Kl.* **2**(23):333-356.

Barton, C. E., and Merrill, R. T., 1983, Archaeo- and palaeosecular variation and long-term asymmetries of the geomagnetic field, *Rev. Geophysics and Space Physics* **21:**603-614.

Barton, C. E.; Merrill, R. T.; and Barbetti, M., 1979, Intensity of the Earth's magnetic field during the last 10,000 years, *Physics Earth and Planetary Interiors* **20:**96-110.

Bauer, L. A., 1910*a*, Beginning and propagation of the magnetic disturbance of May 8, 1902, and of some other magnetic storms, *Terr. Magnetism and Atmos. Electricity* **15:**9-20.

Bauer, L. A., 1910*b*, Analysis of the magnetic disturbance of January 26, 1903, and general considerations regarding magnetic changes, *Terr. Magnetism and Atmos. Electricity* **15:**21-30.

Bauer, L. A., 1910*c*, Solar activity and terrestrial magnetic disturbances, *Am. Philos. Soc. Proc.* **49:**130-144.

Bauer, L. A., 1923, Solar and terrestrial correlations. Discussion, report of a conference on cycles, *Geog. Rev.* **13**(4) (Spec. Supp.):671-672.

Bauer, L. A., 1924*a*, Correlations between solar activity and atmospheric electricity, *Terr. Magnetism and Atmos. Electricity* **29:**23-32; 161-186.

Bauer, L. A., 1924*b*, Correlations between solar activity and atmospheric electricity, *Phys. Rev.* **23:** 23-32.

Bauer, L. A., 1925, Regarding atmospheric electricity and its relation with solar activity, *Terr. Magnetism and Atmos. Electricity* **30:**17-23.

Bauer, L. A., 1926*a*, Activity of the Sun and of atmospheric electricity on land and sea, 1916-1920, *Terr. Magnetism and Atmos. Electricity* **31:**31.

Bauer, L. A., 1926*b*, Sunspot and annual variations of atmospheric electricity with special reference to the Carnegie observations, 1915-1921, *Carnegie Inst. Washington Pub. 175* **5:**361-386.

Bauer, L. A., and Duvall, C. R., 1925, Studies concerning the relation between the activity of the Sun and of the Earth's magnetism, I, *Terr. Magnetism and Atmos. Electricity* **30:**191-213.

Bauer, L. A., and Duvall, C. R., 1926, Studies concerning the relation between the activity of the Sun and of the Earth's magnetism, II, *Terr. Magnetism and Atmos. Electricity* **31:**37-47.

Baur, F., 1922, Periodic oscillations of annual temperatures in Germany, *Monthly Weather Rev.* **50:**199-208.

Baur, F., 1925*a*, The 11-year period of temperature in the northern hemisphere in relation to the 11-year sunspot cycle, *Monthly Weather Rev.* **53:**204-208.

Baur, F., 1925*b*, The 3 to 3½ year periodic pressure oscillation in the free atmosphere, *Monthly Weather Rev.* **53:**392-394.

Baur, F., 1932*a*, Zur Frage der Realitat der Schwankungen der Solar Konstante, *Zeitschr. Meteorologische.*

Baur, F., 1932*b*, Changes in the solar constant of radiation, *Monthly Weather Rev.* **60:**242-246.

Baur, F., 1937, *Einfuhrung in die Grosswetterforschung*, Leipzig and Berlin: B. G. Teubner, 51p.

Baur, F., 1948, *Einfuhrung in die Grosswetterkunde*, Wiesbaden.

Baur, F., 1949*a*, Beziehungen des Grosswetters zu kosmischen Vorgange, in *Haannsuring Lehrbuch der Meteorologie*, (5 Auflage), Teil 8, Leipzig: Hirzel.

Baur, F., 1949*b*, Die doppelte Schwankung der atmospharischen Zirkulation in der gemassigten Zone innerhalb des Sonnenfleckenzyklus, *Meteorol. Rundschau* **2.**

Baur, F., 1951, Extended-range weather forecasting, in T. F. Malone (ed.), *Compendium of Meteorology*, Boston, Mass.: American Meteorological Society, pp. 814-833.

Baur, F., 1953, Die Abhangigkeit des Groswetters von solaren Erscheinungen, *Archiv Meteorologie, Geophysik, u. Bioklimatologie*, Series A **6**(2):129-152.

Baur, F., 1956, *Physikalische-Statistische Regeln als Grundlagen für Wetter- und Witterungsvorhersagen, I*, Frankfurt-am-Main: Akademische Verlagsgesellschaft.

Baur, F., 1958, *Physikalische-Statistische Regeln als Grundlagen für Wetter- und Witterungsvorhersagen. II*, Frankfurt-am-Main: Akademische Verlagsgesellschaft.

Baur, F., 1959, *Die sommerniederschlage Mitteleuropas in den letzten 1½-Jahrhunderten und ihre Beziehungen zum Sonnenfleckenzyklus*, Leipzig: Akademische Verlag, Geest & Portig K.-G.

Baur, F., 1963, Beziehungen irdischer Erscheinungen zu Vorgangen auf der Sonne, *Sterne u. Weltraum (Mannheim)* **2**(7/8):155-158.

Baur, F., 1964, Ist die sogenannte Solarkonstante wirklich konstant? (Bericht uber die III. Meteorologische Fortbildungstagung für Grosswetterkunde und langfristige Witterungsvorhersage), *Meteorol. Rundschau* **17:**19-25.

Baxendell, J., 1925, Meteorological periodicities of the order of a few years, *Royal Meteorol. Soc. Quart. Jour.* **51:**371-392.

Baxter, M. S., and Walton, A., 1971, Fluctuations of atmospheric carbon-14 concentrations during the past century, *Royal Soc. [London] Proc.* **321A:**105-127.

Beals, E. A., 1911, Variations in rainfall, *Monthly Weather Rev.* **39:**1448-1452.

Beals, E. A., 1927, The northeast trade winds of the North Pacific, *Monthly Weather Rev.* **55**(5):211-226.

Beard, D. B., 1964, The effect of an interplanetary magnetic field on the solar wind, *Jour. Geophys. Research* **69**(7):1159-1168.

Bell, B., 1970, The oldest records of the Nile floods, *Geog. Jour.* **136**(4):569-573.

Bell, B., and Defouw, R. T., 1964, Concerning a lunar modulation of geomagnetic activity, *Jour. Geophys. Research* **69:**3169-3174.

Bell, B., and Menzel, D. H., 1972, *Toward the Observation and Interpretation of Solar Phenomena*, Cambridge, Mass.: Air Force Cambridge Research Laboratories, AFCRL F19628-69-C0077 and ADCRL-TR-74-0357, pp. 8-12.

Bell, G. J., 1977, Changes in sign of the relationship between sunspots and pressure, rainfall and the monsoon, *Weather* **32:**26-32.

Bell, P. R., 1981, The combined solar and tidal influence in climate, in S. Sofia (ed.), *Variations in the Solar Constant*, Washington, D.C.: National Aeronautics and Space Administration, NASA Conference Publication 2191, pp. 241-255.

Benedicks, C., 1940, Influence of lunar periodicity on climate according to O. Pettersson, *Arkiv Matematik, Astronomi och Fysik (Stockholm)* **27A**(7):1-15.

Bennett, J. B., 1914, The practical value of long-period rainfall observations, *Scottish Meteorol. Soc. Jour.* **16:**320-328.

Berendsen, H. J. A., 1984, Quantitative analysis of radiocarbon dates of the perimarine area in The Netherlands, *Geologie en Mijnbouw* **63:**343-350.

Berger, A., ed., 1981, *Climatic Variations and Variability: Facts and Theories*, Dordrecht, The Netherlands: D. Reidel, 795p.

Berger, A., 1984, Accuracy and frequency stability of the Earth's orbital elements during the Quaternary, in A. Berger, J. Imbrie, J. D. Hays, G. Kukla, and B. Saltzman (eds.), *Milankovitch and Climate. Understanding the Response to Astronomical Forcing*, vol. 1, Dordrecht, The Netherlands: D. Reidel, pp. 3-39.

Berger, A.; Imbrie, J.; Hays, J. D.; Kukla, G.; and Saltzman, B., eds., 1984, *Milankovitch and Climate. Understanding the Response to Astronomical Forcing*, 2 vols. Dordrecht, The Netherlands: D. Reidel, vol. 1, 510p; vol. 2, 384p.

Bergsten, F., 1928, On periods in the water-height of Lake Vänern, *Geog. Annaler* **10:**140-144.

Bergthorsson, P., 1962, Preliminary notes on past climates of Iceland in *Proceedings, Conference on the Climate of the 11th and 16th Centuries*, Aspen, Colorado, 16-24 June 1962, National Center for Atmospheric Research and Air Force Cambridge Research Laboratory, (unpublished mimeographed notes).

Berlage, H. P., 1927, Ueber den Erhaltungstrieb gewisser langperiodischer Schwankungen des Luftdruckes und der Temperatur, *Zeitschr. Meteorologische* **44:**91-94.

Berlage, H. P., 1929, Ueber die Ursache der dreijahrigen Luftdruckschwankung, *Zeitschr. Meteorologische (Braunschweig)* **46:**249-259. (Review by C. F. Brooks, Ocean currents the probable cause of the 3-year pressure cycle of the tropical South Pacific, *Monthly Weather Rev.* **57**(9):384-385.)

Berlage, H. P., 1957*a*, Fluctuations of the general atmospheric circulation of more than one year, their nature and prognostic value, *Koninkl. Nederlands Meteorol. Inst. De Bilt, Med. Verh. No. 69*, 152p.

Berlage, H. P., 1957*b*, Is the Southern Oscillation in meteorology becoming the classic example of a terrestrial cycle steered by the solar cycle?, *Koninkl. Nederlands Geol. Mijnbou Genoot. Verh. Geol. Ser.* **18:**13-21.

Berlage, H. P., 1959, On the extension of the Southern Oscillation throughout the world during the period July 1, 1949 up to July 1, 1957, *Geofisica Pura e Applicata* **44:**287.

Berlage, H. P., 1961, Variations in the general atmospheric and hydrospheric circulation of periods of a few years, affected by variations of solar activity, *New York Acad. Sci. Annals* **95**(1):354-367.

Berlage, H. P., 1966, The Southern Oscillation and world weather, *Koninkl. Nederlands Meteorol. Inst. De Bilt, Med. Verh. no. 88*, 152p.

Berlage, H. P., and deBoer, H. J., 1960, On the Southern Oscillation, its way of operation and how it affects pressure patterns in the higher latitudes, *Geofisica Pura e Applicata* **46:**329-351.

Bernard, E.-A., 1962*a*, Theorie astronomique des pluviaux et interpluviaux du Quaternaire africain, *Acad. Royale Sci. Outre-Mer, Cl. Sci. Nat. et Med.* (nouvelle serie) **12**(1).

Bernard, E.-A., 1962*b*, Le caractère tropical des paleoclimates à cycles conjoints de 11 et 21000 ans et ses causes: Migration des poles ou derive des continents, *Acad. Royale Sci. Outre-Mer Cl. Sci. Nat. et Med.* (nouvelle serie) **13**(6).

Bernard, E.-A., 1964, The laws of physical palaeoclimatology and the logical significance of palaeoclimatic data, in A. E. M. Nairn (ed.), *Problems in Palaeoclimatology*, New York: John Wiley & Sons, pp. 309-321.

Bernheimer, W. J., 1929, Radiation and temperature of the Sun, *Monthly Weather Rev.* **57**(10):412-417.

Berson, F. A., and Kulkarni, R. H., 1968, Sunspot cycle and quasi-biennial stratospheric oscillation, *Nature* **217:**1133-1134.

Beug, H.-J., 1982, Vegetation history and climatic changes in central and southern Europe, in A. F. Harding (ed.), *Climate Changes in Later Prehistory*, Edinburgh: Edinburgh University Press, pp. 85-102.

Beveridge, W. H., 1927, Weather and harvest cycles, *Econ. Jour.*, December, pp. 429-452.

Bezrukova, A. Y., 1962, On the epoch of the maxima of the eleven year cycles 20, 21, and 22, *Soln. Dannye*, pp. 69-74.

Bhadani, M. M., and Ludlow, N. G. T., 1961, Precipitation of sub-micron dust in still air by cloud-size water droplets, *Nature* **190:**974-976.

Bhalme, H. N., and Mooley, D. A., 1981, Cyclic fluctuations in the flood area and relationship with the double (Hale) sunspot cycle, *Jour. Appl. Meteorology* **20:**1041-1059.

Bhatnagar, V. P., and Jakobsson, T., 1978, Lack of effects of solar magnetic sector crossings on the troposphere, *Geophys. Research Letters* **5:**180-182.

Bigelow, F. H., 1891, The causes of the variation of the magnetic needle, *Am. Jour. Sci.* (3rd series) **42**(249):253-256.

Bigelow, F. H., 1894, Inversion of temperatures in the 26.68 day solar magnetic period, *Am. Jour. Sci.* (3rd series) **48**(288):435-451.

Bigelow, F. H., 1898, *Solar and Terrestrial Magnetism*, U.S. Department of Agriculture, Weather Bureau, Bulletin No. 21.

Bigelow, F. H., 1902*a*, A contribution to cosmical meteorology, *Monthly Weather Rev.* **30**(7):347-354.

Bigelow, F. H., 1902*b*, Studies on the meteorological effects of the solar and terrestrial physical processes, *Monthly Weather Rev.* **30:**559-567.

Bigelow, F. H., 1903*a*, Synchronous changes in the solar and terrestrial atmospheres, *Monthly Weather Rev.* **31:**9-18.

Bigelow, F. H., 1903*b*, Sun spots and the weather conditions on the Earth, *Monthly Weather Rev.* **31:**474.

Bigelow, F. H., 1903*c*, Studies on the circulation of the atmospheres of the Sun and of the Earth. I. The circulation of the Sun's atmosphere, *Monthly Weather Rev.* **31**(10):459-466.

Bigelow, F. H., 1903*d*, Studies on the circulation of the atmospheres of the Sun and of the Earth. II. Synchronism of the variations of the solar prominences with the terrestrial barometric pressures and the temperatures, *Monthly Weather Rev.* **31**(11):509-516.

Bigelow, F. H., 1903*e*, Synchronism of the variation of the solar prominences with terrestrial barometric pressures and the temperatures, *Monthly Weather Rev.* **31:**509-516.

Bigelow, F. H., 1904, Studies on the circulation of the atmospheres of the Sun and of the Earth, *Monthly Weather Rev.* **32:**15-20, 71-78, 166-169, 212-216, 260-263.

Bigelow, F. H., 1908, The relations between the meteorological elements of the United States and the solar radiation, *Am. Jour. Sci.* (4th series) **25**(149):413-430.

Bigelow, F. H., 1909, Important problems in climatology, *Monthly Weather Rev.* **37:**979-982.

Bigelow, F. H., 1910, Studies on the general circulation of the Earth's atmosphere. A discussion of the departures and the residuals of temperature and precipitation in climatology, *Am. Jour. Sci.* (4th series) **29:**277-292.

Bigelow, F. H., 1915, *A Meteorological Treatise on the Circulation and Radiation in the Atmospheres of the Earth and of the Sun,* New York: John Wiley & Sons, 431p.

Bigg, E. K., 1963*a*, A lunar influence on ice nucleus concentrations, *Nature* **197:**172-173.

Bigg, E. K., 1963*b*, The influence of the Moon on geomagnetic disturbances, *Jour. Geophys. Research* **68**(5):1409-1413.

Bigg, E. K., 1963*c*, Lunar and planetary influences on geomagnetic disturbances, *Jour. Geophys. Research* **68**(13):4099-4104.

Bigg, E. K., 1964, Lunar influences on the frequency of magnetic storms, *Jour. Geophys. Research* **69**(23):4971-4974.

Bigg, E. K., 1967, Influence of the planet Mercury on sunspots, *Astron. Jour.* **72**(4):463-466.

Bingham, D. K., and Stone, D. B., 1972, Secular variation in the Pacific Ocean region, *Geophys. Jour.* **28:**337.

Binnie, Sir A. R., 1913, *Rainfall Reservoirs and Water Supply,* London: Constable and Co., 157p.

Birkeland, K., 1896, Sur les rayons cathodiques sous l'action de forces magnetiques intenses, *Archives Sci.* **1:**497-512.

Birkeland, K., 1899, Recherches sur les taches du soleil et leur origine, *Norske Vidensk. Christiana, Skr. Mat.-Naturv. Kl. No. 1,* 173p.

Birkeland, K., 1911, The simultaneity of certain abruptly-beginning magnetic storms, *Nature* **87:**483-484.

Birkeland, K., 1914, A possible connection between magnetic and meteorologic phenomena, *Monthly Weather Rev.* **42**(4):211.

Bjerknes, J., 1937, Theorie der aussertropischen Zyklonenbildung, *Zeitschr. Meteorologische (Braunschweig)* **54:**462-466.

Bjerknes, J., 1961, El Niño: Study based on analysis of ocean surface temperatures, 1935-1957, *Internat. Tropical Tuna Comm. Bull.* **5:**217-303.

Bjerknes, J., 1969, Atmospheric teleconnections from the equatorial Pacific, *Monthly Weather Rev.* **97**(3):163-172.

Black, D. C., 1982, The detection of other planetary systems, *Mercury* **9**(5):105-111.

Black, D. C., and Scargle, J. D., 1982, On the detection of other planetary systems by astrometric techniques, *Astrophys. Jour.* **263**(2) (part 1):854-869.

Black, D. I., 1967, Cosmic ray effects and faunal extinctions at geomagnetic field reversals, *Earth and Planetary Sci. Letters* **3:**225.

Black, J. N., 1956, The distribution of solar radiation over the Earth's surface, *Archiv Meteorologie, Geophysik u. Bioklimatologie* **7B:**165-189.

Blackman, R. B., and Tukey, J. W., 1958, The measurement of power spectra from the point of view of the communications engineer, *Bell System Tech. Jour.* **37:**185-282.

Blackman, R. B., and Tukey, J. W., 1959, *The Measurement of Power Spectra,* New York: Dover Books.

Blackmon, M. L.; Wallace, J. M.; Lau, N. C.; and Mullen, S. L., 1977, An observational study of the northern hemisphere wintertime circulation, *Jour. Atmos. Sci.* **34:**1040-1053.

Blackshear, W. T., and Tolson, R. H., 1978, High correlation between monthly averages of solar activity and total atmospheric ozone, *Geophys. Research Letters* **5:**921-924.

Blake, D., and Lindzen, R. S., 1973, Effect of photochemical models on calculated equilibria and cooling rates in the stratosphere, *Monthly Weather Rev.* **101:**783-802.

Blanford, H. F., 1875*a*, Solar heat and sun-spots, *Nature* **12:**147-148.

Blanford, H. F., 1875*b*, Solar radiation and sun-spots, *Nature* **12:**188-189 (also *Zeitschr. Meteorologische* **10:**261-264).

Blanford, H. F., 1880, On the barometric see-saw between Russia and India in the sun-spot cycle, *Nature* **21:**477-482.

Blanford, H. F., 1884*a*, Variations of rainfall in northern India during the sun-spot period, *Bengal Asiatic Soc. Proc. 1884,* pp. 165-167.

Blanford, H. F., 1884*b,* Periodicity of the sun-spot cycles, and their effect on meteorological and other phenomena, *Bengal Asiatic Soc. Proc. 1884,* pp. 168-169.

Blanford, H. F., 1884*c,* On the connexion of the Himalayan snowfall with dry winds and seasons of drought in India, *Royal Soc.* [*London*] *Proc.* **37:**3-22.

Blanford, H. F., 1884*d,* The theory of the winter rains of northern India, *Nature* **30:**304-305.

Blanford, H. F., 1887*a,* The eleven-year periodical fluctuation of the Carnatic rainfall, *Nature* **36:**227-229.

Blanford, H. F., 1887*b,* Remarks on some recent evidence on the subject of the variation of rainfall of the Carnatic and Northwest Himalayas with the sun-spot period, *Bengal Asiatic Soc. Proc. 1887,* pp. 116-121.

Blanford, H. F., 1889, *The Climates and Weather of India, Ceylon, Burma and the Storms of the Indian Seas,* London.

Blanford, H. F., 1891*a,* The paradox in the sunspot cycle in meteorology, *Nature* **43:**583-587.

Blanford, H. F., 1891*b,* The forecast of the Indian monsoon rains, *Nature* **44:**225-227.

Blanford, H. F., 1891*c,* Rain-making in Texas, *Nature* **44:**473-475.

Blasing, T. J., and Duvick, D., 1984, Reconstruction of the precipitation history in North American corn belt using tree rings, *Nature* **307:**143-145.

Bliss, G. S., 1913, *Forecasting the Weather,* Washington, D.C.: U.S. Department of Agriculture, Weather Bureau, Bulletin No. 42, 34p.

Blizard, J. B., 1969, *Long Range Solar Flare Prediction,* Washington, D.C.: U.S. National Aeronautics and Space Administration, Contractor's Report, NASA CR-61316.

Blizard, J. B., 1981, Solar activity and the irregular motions of the Sun (abstract), *Am. Astron. Soc. Bull.* **13**(4):876.

Blizard, J. B., 1983, Precessional effects of an oblate Sun (abstract), *Am. Astron. Soc. Bull.* **15**(4):959.

Block, L. H., 1934, The record-breaking drought, heat and dust storms of 1934, *Am. Meteorol. Soc. Bull.* **15**(12):300-307.

Bohme, W., 1967, Eine 26-monatige Schwankung der Haufigkeit meridionaler Zirkulationsformen uber Europa, *Zeitschr. Meteorologie* **19:**113-115.

Bollinger, C. J., 1935, The relation of solar radiation to sun-spot cycles, *Oklahoma Acad. Sci. Proc.* **15.**

Bollinger, C. J., 1945, The 22-year pattern of rainfall in Oklahoma and Kansas, *Am. Meteorol. Soc. Bull.* **26**(9):376-383.

Bollinger, C. J., 1960, *Atlas of Planetary Solar Climate, with Sun-Tide Indices of Solar Radiation and Global Insolation,* vol. 1 (1900-1959), Norman, Okla.: Battenburg Press, 121p.

Bollinger, C. J., 1962, *Atlas of Planetary Solar Climate, with Sun-Tide Indices of Solar Radiation and Global Insolation,* vol. 2 (1960-1980), Norman, Okla.: Battenburg Press, 51p.

Bollinger, C. J., 1964, *Atlas of Planetary Solar Climate.* Vol. 4: *Planetary Periodicities in Sun-Tide Cycles and Climate Variation,* Norman, Okla.: Battenburg Press, 44p.

Bollinger, C. J., 1968, Sun tides: An unexplored astronomical approach to climatic trends and cycles, *Tellus* **20**(3):412-416.

Bonov, A. D., 1968, Prognosis of the 11-year cycles No. 20, 21, and 22 of the solar activity, *Soln. Danknye,* pp. 68-73.

Borchert, J. R., 1971, The dust bowl in the 1970s, *Assoc. Am. Geographers Annals* **61**(1):1-22.

Borisenkov, Y. P.; Tsvetkov, A. V.; and Agaponov, S. V., 1983, On some characteristics of insolation changes in the past and in the future, *Clim. Change* **5:**237-244.

Borucki, W. J., and others, 1980, The influence of solar UV variations on climate, in R. O. Pepin, J. A. Eddy, and R. B. Merrill (eds.), *Ancient Sun: Proceedings of a Conference on the Ancient Sun: Fossil Record in the Earth, Moon and Meteorites,* New York: Pergamon Press, pp. 513-522.

Bos, G. J., 1972, Possible relationship between sun spot cycles and fluctuations in frequency of Mongolism, *Jour. Interdisc. Cycle Research* **3**(3-4):267-268.

Bossolasco, M.; Dagnino, I.; Elena, A.; and Flocchini, G., 1973*a,* The thunderstorm activity over the Mediterranean area, *Riv. Italiana Geofis.* **22**(1-2):21-26.

Bossolasco, M.; Dagnino, I.; Elena, A.; and Flocchini, G., 1973*b,* Thunderstorm activity and interplanetary magnetic field, *Riv. Italiana Geofis.* **22**(5-6):293-295.

Bossolasco, M., and Elena, A., 1960, On the lunar semidiurnal variation of the D and F2 layers, *Geofisica Pura e Applicata* **46:**167-172.

Botley, C. M., 1957, Some great tropical aurorae, *British Astron. Assoc. Jour.* **67:**188-191.

Botley, C. M., 1967, Unusual auroral periods, *British Astron. Assoc. Jour.* **77:**328-330.

Bouma, J. J., and Tromp, S. W., 1972, Daily, monthly and yearly fluctuations in total number of suicides and suicide attempts in the western part of The Netherlands, *Jour. Interdisc. Cycle Research* **3**(3-4):269-270.

Bowen, E. G., 1953, Influence of meteoric dust on rainfall, *Australian Jour. Physics* **6**(4):490-497.

Bowen, E. G., 1956, A relation between meteor showers and the rainfall of November and December, *Tellus* **8**(3):394-402.

Bowen, E. G., 1963, A lunar effect on the incoming meteor rate, *Jour. Geophys. Research* **68**(5):1401-1403.

Bowen, E. G., 1964, Lunar and planetary tails in the solar wind, *Jour.. Geophys. Research* **69**(23):4969-4970.

Bowen, E. G., 1974, Kidson's relation between sunspot number and the movement of high pressure systems in Australia, in W. R. Bandeen and S. P. Maran (eds.), *Symposium on Possible Relationships between Solar Activity and Meteorological Phenomena, Proceedings,* Greenbelt, Md.: NASA Goddard Space Flight Center, NASA SP-366, pp. 56-59.

Bowie, E. H., 1909, Rivers and floods, *Monthly Weather Rev.* **37:**399-400.

Bowie, E. H., 1929, The long dry season of 1929 in the far West, *Monthly Weather Rev.* **57**(11):449-451.

Bowman, I., 1935, Our expanding and contracting "desert," *Geog. Rev.* **25**(1):43-61.

Boyden, C. J., 1963, Development of the jet stream and cut-off circulations, *Meteorol. Mag.* **92:**287-299.

Braak, C., 1913, Ueber die Ursache langperiodischer Barometer- und Temperaturschwankungen, *Zeitschr. Meteorologische (Braunschweig)* **30**(2):49-58.

Braak, C., 1919, Atmospheric variations of short and long duration in the Malay Archipelago and the possibility to forecast them, *Koninkl. Magnet. Meteorol. Observatorium, Batavia, Verh. No. 5,* 57p.

Bradley, D. A.; Woodbury, M. A.; and Brier, G. W., 1962, Lunar synodical period and widespread precipitation, *Science* **137:**748-749.

Bradley, R. S., and Miller, G. H., 1972, Recent climatic change and increased glacierization in the Eastern Canadian Arctic, *Nature* **237:**385-387.

Bradley, R. S., and Miller, G. H., 1972, Recent climatic change and increased glacerization in the Eastern Canadian Arctic, *Nature* **237:**385-387.

Brasseur, G., and Nicolet, M., 1973, Chemospheric processes of nitric oxide in the mesosphere and stratosphere, *Planetary and Space Sci.* **21:**939-961.

Brasseur, G., and Simon, P. C., 1981, Stratospheric chemical and thermal response to long-term variability in solar UV irradiance, *Jour. Geophys. Research* **86:**7343-7362.

Bray, J. R., 1966, Atmospheric carbon-14 content during the past three millennia in relation to temperature and solar activity, *Nature* **209:**1065-1067.

Bray, J. R., 1967, Variation in atmospheric carbon-14 activity relative to sunspot-auroral solar index, *Science* **156:**640-642.

Bray, J. R., 1968, Glaciation and solar activity since the fifth century B.C., and the solar cycle, *Nature* **220:**672-674.

Bray, J. R., 1971, Vegetational distribution, tree growth and crop success in relation to recent climatic change, *Advances in Ecol. Research* **7:**177-233.

Bray, J. R., 1972, Cyclic temperature oscillations from 0-20,300 yr B.P., *Nature* **237:**277-279.

Bray, J. R., 1974, Glacial advance relative to volcanic activity since 1500 A.D., *Nature* **248:**42-43.

Bray, J. R., 1978, Volcanic eruptions and climate during the past 500 years, in A. B. Pittock, L. A. Frakes, D. Jenssen, J. A. Peterson, and J. W. Zillman (eds.), *Climatic Change and Variability: A Southern Perspective,* London and New York: Cambridge University Press, pp. 256-262.

Bray, J. R., 1982, Alpine glacial advance in relation to a proxy summer temperature index based mainly upon wine harvest dates A.D. 1453-1973, *Boreas* **11**(1):1-10.

Bray, J. R., and Loughhead, R. E., 1964, *Sunspots,* London: Chapman and Hall, 364p.

Bray, J. R., and Struik, G. J., 1963, Forest growth and glacial chronology in eastern British Columbia and their relation to recent climatic trends, *Canadian Jour. Botany* **41:**1245-1271.

Brewer, A. W., and Wilson, A. W., 1965, Measurements of the solar ultra violet radiation in the stratosphere, *Royal Meteorol. Soc. Quart. Jour.* **91:**451-461.

Briden, J. C., 1966, Variation of the intensity of the geomagnetic field through geological time, *Nature* **212:**246-247.

Brier, G. W., 1947, 40-year trends in Northern Hemisphere surface pressure, *Am. Meteorol. Soc. Bull.* **28:**237-247.

Brier, G. W., 1961, Some aspects of long-term fluctuations in solar and atmospheric phenomena, *New York Acad. Sci. Annals* **95**(Article 1):173-187.

Brier, G. W., 1968, Long-range prediction of the zonal westerlies and some problems in data analysis, *Rev. Geophysics* **6:**525-551.

Brier, G. W., 1978, The quasi-biennial oscillation and feedback processes in the atmosphere-ocean-Earth system, *Monthly Weather Rev.* **106:**938-946.

Brier, G. W., and Bradley, D. A., 1964, The lunar synodical period and precipitation in the United States, *Jour. Atmos. Sci.* **21:**386-395.

Britton, C. E., 1937, A meteorological chronology to A.D. 1450, *Royal Meteorol. Office (Great Britain), Geophys. Mem.* **8**(70).

Broecker, W. S., 1966, Absolute dating and the astronomical theory of glaciation, *Science* **151:**299-304.

Broecker, W. S., 1968, In defense of the astronomical theory of glaciation, *Am. Meteorol. Soc. Meteorol. Mon.* **8**(30):139-141.

Brooks, C. E. P., 1914*a,* The meteorological conditions of an ice-sheet and their bearing on the desiccation of the globe, *Royal Meteorol. Soc. London Quart. Jour.* **40:**53-68.

Brooks, C. E. P., 1914*b*, Climatic change, *Nature* **93:**532.

Brooks, C. E. P., 1918, Continentality and temperature—second paper: The effect of latitude on the influence of continentality on temperature, *Royal Meteorol. Soc. Quart. Jour.* **44:**253-269.

Brooks, C. E. P., 1919, A mechanism of climatic cycles, *Meteorol. Mag.* **55:**205-206 (also in *Monthly Weather Rev.* **48**(10):596-597.)

Brooks, C. E. P., 1921, Secular variation in climate, *Geog. Rev.* **11**(1):120-135.

Brooks, C. E. P., 1923*a*, Variations in the levels of the central African lakes Victoria and Albert, *Royal Meteorol. Office (Great Britain), Geophys. Mem.* **2**(20):337-344.

Brooks, C. E. P., 1923*b*, Weather influences in the British Isles, *Nature* **112:**386.

Brooks, C. E. P., 1925, The problem of warm polar climates, *Royal Meteorol. Soc. Quart. Jour.* **51:**83-91.

Brooks, C. E. P., 1926, *Climate through the Ages; A Study of the Climatic Factors and Their Variations*, London and New Haven, Conn.: Yale University Press, 439p.

Brooks, C. E. P., 1927*a*, Planets and periodicities, *Nature* **119:**298.

Brooks, C. E. P., 1927*b*, Non-linear relations with sunspots, *Royal Meteorol. Soc. Quart. Jour.* **53:**68-71.

Brooks, C. E. P., 1927*c*, Performance in long-range weather forecasting, *Monthly Weather Rev.* **55**(9):390-395.

Brooks, C. E. P., 1928*a*, The problem of the varves, *Royal Meteorol. Soc. Quart. Jour.* **54:**64-70.

Brooks, C. E. P., 1928*b*, Periodicities in the Nile floods, *Royal Meteorol. Soc. Mem.* **2**(12):9-26.

Brooks, C. E. P., 1928*c*, Sunspots and the distribution of pressure over western Europe, *United Kingdom Air Ministry Meteorological Office Professional Notes No. 49*, 6p.

Brooks, C. E. P., 1928*d*, Arctic ice and British weather, *Monthly Weather Rev.* **56:**416-417.

Brooks, C. E. P., 1930, The role of the oceans in the weather of Western Europe, *Royal Meteorol. Soc. Quart. Jour.* **56**(234):131-140.

Brooks, C. E. P., 1934, The variation of the annual frequency of thunderstorms in relation to sunspots, *Royal Meteorol. Soc. Quart. Jour.* **60**(254):153-166.

Brooks, C. E. P., 1949*a*, *Climate through the Ages*, 2nd ed., London: Benn, 395p.

Brooks, C. E. P., 1949*b*, Post-glacial climatic changes in the light of recent glaciological research, *Geog. Annaler* **31**(1-4):21-24.

Brooks, C. E. P., 1951, Geological and historical aspects of climatic change, in T. F. Malone (ed.), *Compendium of Meteorology*, Boston, Mass.: American Meteorological Society, pp. 1004-1018.

Brooks, C. E. P., 1957, Annotated bibliography on solar relations with weather, *Meteorol. Abs. and Bibliography* **8**(1):95-119.

Brooks, C. E. P., and Glasspoole, J., 1922, The drought of 1921, *Royal Meteorol. Soc. Quart. Jour.* **48**(202):139-168.

Brooks, C. E. P., and Glasspoole, J., 1928, *British Floods and Droughts*, London: Benn.

Brooks, C. E. P., and Quennel, W., 1928, On the influence of Arctic ice on the subsequent distribution of pressure over the eastern North Atlantic and western Europe, *United Kingdom Air Ministry Meteorological Office Geophys. Mem.* **4**(41). (Review by A. J. Henry, 1929, *Monthly Weather Rev.* **57**(3):99-102.)

Brooks, C. E. P., and Thorman, G. L., 1928, The distribution of heat and maxima and minima of temperature over the globe, *United Kingdom Air Ministry Meteorological Office Geophys. Mem.* **4**(44).

Brooks, C. F., 1918, Ocean temperatures in long-range forecasting, *Monthly Weather Rev.* **46**(11):510-512.

Brooks, C. F., 1935, *Why the Weather?* New York: Harcourt, Brace and Company, 295p.

Brooks, C. F., and Bangs, N. H., 1929, Severe winter in Europe 1928-29, *Monthly Weather Rev.* **57**(2):58-60.

Brooks, C. F., and Schell, I. I., 1939, Certain synoptic antecedents of severe cold waves in southern New England, *Am. Meteorol. Soc. Bull.* **20**(12):439-444.

Brouwer, D., and van Woerkom, A. J. J., 1950, The secular variations of the orbital elements of the principal planets, Astronomical papers prepared for the use of *The American Ephemeris and Nautical Almanac*, vol. 13, part 2, Washington, D.C.: U.S. Government Printing Office.

Brown, E. W., 1896, *Introductory Treatise on the Lunar Theory*, Cambridge: Cambridge University Press (also New York, Dover reprint).

Brown, E. W., 1898, On the mean motions of the lunar perigee and node, *Royal Astron. Soc. Monthly Notices* **57:**332-341, 566-567.

Brown, E. W., 1899, Theory of the motion of the Moon; containing a new calculation of the expressions for the coordinates of the Moon in terms of time, *Royal Astron. Soc. Mem. 53*, pp. 39-116, 163-202.

Brown, E. W., 1900, A possible explanation of the sun-spot period, *Royal Astron. Soc. Monthly Notices* **60:**599-606.

Brown, G. M., 1974, A new solar-terrestrial relationship, *Nature* **251:**592-594.

Brown, G. M., and John, J. I., 1979, Solar cycle

influences in tropospheric circulation, *Jour. Atmos. and Terr. Physics* **42:**43.

Brown, R. M., 1912, The Mississippi River flood of 1912, *Am. Geog. Soc. Bull.* **44:**645-657.

Brückner, E., 1890, Klimaschwankungen seit 1700 nebst Bemerkungen uber die Klimaschwankungen der Diluvialzeit, *Geog. Abh. Herausgeben von A. Penck* **4**(2):325p.

Brückner, E., 1912, Klimaschwankungen und Volkerwanderungen, *Akad. Wiss. Alman. Wien* **62:**421-445.

Brunner, W., 1939, Monthly sunspot numbers, *Terr. Magnetism and Atmos. Electricity* **44:**84.

Brunner-Hagger, W., 1955, Beziehungen zwischen Sonnentatigkeit, komischer Strahlung, Meteorstromen und Zurich Niederschlagen, *Schweizer. Naturf. Gesell. Verh.* **135:**112-113.

Brunt, Sir D., 1919, A periodogram analysis of the Greenwich temperature records, *Royal Meteorol. Soc. Quart. Jour.* **45:**323-338.

Brunt, Sir D., 1925, Periodicities in European weather, *Royal Soc. London Philos. Trans.* **225A:**247-302.

Brunt, Sir D., 1927, An investigation of periodicities in rainfall, pressure, and temperature at certain European stations, *Royal Meteorol. Soc. Quart. Jour.* **53**(221):1-30.

Brunt, Sir D., 1937, Climatic cycles, *Geog. Jour.* **89**(3):214-239.

Brunt, Sir D., 1954, Some aspects of human ecology in hot tropical regions, in *Biology of Deserts*, London: Institution of Biology, pp. 213-218.

Bruzek, S., and Durrant, C. J., eds., 1977, *Illustrated Glossary for Solar and Solar-Terrestrial Physics*, Dordrecht, The Netherlands: D. Reidel, 204p.

Bryson, R. A., 1948, On a lunar bi-fortnightly tide in the atmosphere, *Am. Geophys. Union Trans.* **29:**473-475.

Bryson, R. A., 1972, Climatic modification by air pollution, in N. Polunin (ed.), *The Environmental Future*, New York, Macmillan, pp. 133-154.

Bryson, R. A., 1973, Drought in Sahelia, Who or what is to blame? *Ecologist* **3:**366-371.

Bryson, R. A., 1974, A perspective on climatic change, *Science* **184:**753-760.

Bryson, R. A., 1975, The lessons of climatic history, *Environmental Conservation* **2**(3):163-170.

Bryson, R. A., and Dutton, J. A., 1961, Some aspects of the variance spectra of tree rings and varves, *New York Acad. Sci. Annals* **95**(article 1):580-604.

Bryson, R. A., and Murray, T. J., 1977, *Climates of Hunger. Mankind and the World's Changing Weather*, Madison and London: University of Wisconsin Press, 171p.

Bryson, R. A., and Padoch, C., 1981, On the climates of history, in R. I. Rotberg and T. K. Rabb (eds.), *Climates and History. Studies in Interdisciplinary History*, Princeton, N.J.: Princeton University Press, pp. 3-17.

Bryson, R. A., and Swain, A. M., 1981, Holocene variation of monsoon rainfall in Rajasthan, *Quat. Research* **16**(2):135-145.

Bucha, V., 1965, Results of archaeomagnetic research in Czechoslovakia for the epoch from 4400 B.C. to the present, *Jour. Geomagnetism and Geoelectricity* **17:**407-412.

Bucha, V., 1970, Influence of the Earth's magnetic field on radiocarbon dating (with discussion), in I. U. Olsson (ed.), *Radiocarbon Variations and Absolute Chronology*, Stockholm: Almqvist and Wiksell, pp. 501-511.

Bucha, V., 1976*a*, Variations of the geomagnetic field, the climate and weather, *Studia Geophysica et Geodetica* **20:**149.

Bucha, V., 1976*b*, Changes in the geomagnetic field and solar wind—Causes of changes of climate and atmospheric circulation, *Studia Geophysica et Geodetica* **20:**346.

Bucha, V., 1976*c*, Effect of changes in the Earth's magnetic field on weather, climate, and glaciations, pp. 161-187 in International Geological Correlation Programme, Project 73/1/24, *Quaternary Glaciations in the Northern Hemisphere*, Report No. 3 on the session in Bellingham, Washington, 6-19 September 1975, Bellingham-Prague 1976 [edited in collaboration with the International Union for Quaternary Research (INQUA)].

Bucha, V., 1977, Mechanism of solar-terrestrial relations and changes of atmospheric circulation, *Studia Geophysica et Geodetica* **21**(Suppl. 416):350-360.

Bucha, V., 1980, Mechanism of the relations between the changes of the geomagnetic field, solar corpuscular radiation, atmospheric circulation, and climate, *Jour. Geomagnetism and Geoelectricity* **32:**217-264.

Bucha, V., 1984, Mechanism for linking solar activity to weather-scale effects, climatic changes and glaciations in the northern hemisphere, in N.-A. Mörner and W. Karlén (eds.), *Climatic Changes on a Yearly to Millennial Basis*, Dordrecht, The Netherlands: D. Reidel, pp. 415-448.

Buchan, A., 1901, The rainfall in Scotland in relation to sunspots, *Scottish Meteorol. Soc. Jour.* **12:**117.

Buchanan, J. Y., 1911, Fish and drought, *Nature* **88:**107-110.

Budyko, M. I., 1969, The effect of solar radiation variations on the climate of the Earth, *Tellus* **21:**611-619.

Budyko, M. I., 1972, Future climate, *EOS* **54:**868-874.

Budyko, M. I., 1978, The heat balance of the Earth, in J. R. Gribbin (ed.), *Climatic Change*, Cambridge: Cambridge University Press, pp. 85-113.

Buettner, K., 1952, Human bioclimatology and energy in arid and humid tropical zones, in *Proceedings, Mid-Southwest Conference on Tropical Housing and Building*, Austin, Texas, April 1952, pp. 37-48.

Bullard, E. C., 1967, The removal of trend from

magnetic surveys, *Earth and Planetary Sci. Letters* **2:**293-300.

Bullard, E. C.; Freedman, C.; Gellman, H.; and Nixon, J., 1950, The westward drift of the Earth's magnetic field, *Royal Soc. London Philos. Trans.* **243A:**67-92.

Bullard, Sir E. C., and Gellman, H., 1954, Homogeneous dynamos and terrestrial magnetism, *Royal Soc. London Philos. Trans.* **247A:**213-278.

Bulmer, M. G., 1974, A statistical analysis of the 10-year cycle in Canada, *Jour. Animal Ecology* **43:**701-718.

Bumba, V., and Kleczek, J., eds., 1976, *Basic Mechanisms of Solar Activity: International Astronomical Union, Symposium 71,* Dordrecht, The Netherlands: D. Reidel, 481p.

Bunting, A. H.; Dennett, M. D.; Elston, J.; and Milford, J. R., 1975, Seasonal rainfall forecasting in West Africa, *Nature* **253:**622-623.

Burchuladze, A. A.; Pagava, S. V.; Povenic, P.; Togonidza, G. I.; and Ulsacev, S., 1980, Radiocarbon variations with the 11-year solar cycle during the last century, *Nature* **287:**320-322.

Bureau, R. A., and Craine, L. B., 1970, Sunspots and planetary orbits, *Nature* **228:**984.

Burns, G. B.; Bond, F. R.; and Cole, K. D., 1980, An investigation of the southern hemisphere vorticity response to solar sector boundary crossings, *Jour. Atmos. and Terr. Physics* **42:**765-769.

Bushby, F. H., and Timpson, M. S., 1967, A 10-level atmospheric model and frontal rain, *Royal Meteorol. Soc. Quart. Jour.* **93:**1-17.

Buss, A. A., 1902, Synchronism of solar storms and terrestrial magnet disturbances, *Knowledge (London)* **25:**277-278.

Buss, A. A., 1903, Solar activity and terrestrial magnetism, *Knowledge (London)* **26:**155-156.

Cahill, L. J., Jr., 1964, The geomagnetic field, in D. P. LeGalley and A. Rosen (eds.), *Space Physics,* New York: John Wiley & Sons, pp. 301-349.

Cairnes, J. L., and LaFond, E. C., 1966, Periodic motions of the seasonal thermocline along the Southern California coast, *Jour. Geophys. Research* **71**(16):3903-3915.

Callendar, G. S., 1960, Discussion of a paper by E. B. Kraus on "Synoptic and dynamic aspects of climate change," *Royal Meteorol. Soc. Quart. Jour.* **86:**572-573.

Callendar, G. S., 1961, Temperature fluctuations and trends over the Earth, *Royal Meteorol. Soc. Quart. Jour.* **87:**1-12.

Callis, L. B.; Natarajan, M.; and Nealy, J. E., 1979, Ozone and temperature trends associated with the 11-year solar cycle, *Science* **204:**1303-1306.

Callis, L. B., and Nealy, J. E., 1978, Solar UV variability and its effect on stratospheric thermal structure and trace constituents, *Geophys. Research Letters* **5:**249-252.

Camp, J. McK., 1979, A drought in the late 8th century BC, *Hesperia* **48:**397-411.

Campbell, W. H., 1983, Possible tidal modulation of the Indian Monsoon onset, Ph.D. dissertation, University of Wisconsin, Department of Meteorology.

Campbell, W. H.; Blechman, J. B.; and Bryson, R. A., 1983, Long-period tidal forcing of Indian monsoon rainfall: An hypothesis, *Jour. Climate and Appl. Meteorology* **22:**289-296.

Cane, M. A., 1983, Oceanographic events during El Niño, *Science* **222:**1189-1195.

Capon, R. S., 1914, The influence of the Sun on terrestrial magnetism, *Observatory (London)* **37:**341-344.

Carapiperis, L. N., 1945, Sur la fréquence et la périodicité des étésiens à Athènes, *Akad. Athénón Praktika* **20:**126-134.

Carapiperis, L. N., 1960, On the variation of the Etesians within the sunspot cycle, *Geofisica Pura e Applicata* **46:**190-192.

Carapiperis, L. N., 1962, The Etesian winds. III. Secular changes and periodicity of the Etesian winds, *Upomnemata Tou Ethnikon Asteroskopeion Athenon,* Series II (Meteorology), no. 11.

Carlisle, N., and Carlisle, M., 1980, *Where to Live for Your Health,* New York: Harcourt, Brace Jovanovich.

Carrington, R. C., 1854, On a method of observing the positions of spots on the Sun, with an example of its application, *Royal Astron. Soc. Monthly Notices* **14:**153-158.

Carrington, R. C., 1858, On the evidence which the observed motions of the solar spots offer for the existence of an atmosphere surrounding the Sun, *Philos. Mag.* **15:**444-452.

Carrington, R. C., 1859, On the distribution of the Solar spots in latitude since the beginning of the year 1854, *Royal Astron. Soc. Monthly Notices* **19:**1-4.

Carrington, R. C., 1863, On certain phenomena in the motions of Solar Spots: Elements of position of the Solar Equator, *Royal Astron. Soc. Monthly Notices* **22:**300-301.

Cartwright, D. E., and Tayler, R. J., 1971, New computations of the tide-generating potential, *Royal Astron. Soc. Geophys. Jour.* **24:**45-74.

Castagnoli, G., and Lal, D., 1980, Solar modulation effects in terrestrial production of carbon-14, *Radiocarbon* **22:**133-158.

Cato, I., 1985, The definitive connection of the Swedish geochronological time scale with the present, and the new date of the zero year in Doviken, northern Sweden, *Boreas* **14**(2):117-122.

Caviedes, C. N., 1973, Secas and El Niño: Two simultaneous climatical hazards in South America, *Assoc. Am. Geog. Proc.* **5:**44-49.

Chalmers, J. A., 1967, *Atmospheric Electricity,* 2nd ed., New York: Pergamon Press, 515p.

Chamberlain, J. W., 1961, *Physics of the Aurora and Airglow*, New York: Academic Press, 704p.

Chandler, S. C., 1891*a*, On the variation of latitude, I, *Astron. Jour.* **11:**59-61.

Chandler, S. C., 1891*b*, On the variation of latitude, II, *Astron. Jour.* **11:**65-70.

Chandler, S. C., 1891*c*, On the variation of latitude, III, *Astron. Jour.* **11:**75-79.

Chandra, S., 1983, A study of solar activity-ozone relationship from Nimbus-4 BUV data, in B. M. McCormac (ed.), *Weather and Climatic Responses to Solar Variations*, Boulder, Colo.: Colorado Associated University Press, pp. 137-144.

Chandra, S., and Maeda, K., 1980, A search for correlation between geomagnetic activity and stratospheric ozone, *Geophys. Research Letters* **7:**757-760.

Chapman, S., 1919*a*, The solar and lunar diurnal magnetic variations, *Royal Soc. London Philos. Trans.* **218A:**1-118.

Chapman, S., 1919*b*, The lunar tide in the Earth's atmosphere, *Royal Meteorol. Soc. Quart. Jour.* **45:**113-139.

Chapman, S., 1925, The lunar diurnal magnetic variation at Greenwich and other observatories, *Royal Soc. London Philos. Trans.* **225A:**49-91.

Chapman, S., 1951, Atmospheric tides and oscillations, in T. F. Malone (ed.), *Compendium of Meteorology*, Boston, Mass.: American Meteorological Society, pp. 510-530.

Chapman, S., 1956, The electrical conductivity of the ionosphere: A review, *Nuovo Cimento* (Supp.) **9**(10):1385.

Chapman, S., 1961, Sun storms and the Earth: The Aurora Polaris and the space around the Earth, *Am. Scientist* **49:**249-284.

Chapman, S., 1964*a*, *Solar Plasma, Geomagnetism and Aurora*, New York: Gordon and Breach, Science Publishers, 141p.

Chapman, S., 1964*b*, Aurora and geomagnetic storms, in D. P. LeGalley and A. Rosen (eds.), *Space Physics*, New York: John Wiley & Sons, pp. 226-269.

Chapman, S., and Bartels, J., 1940, *Geomagnetism*, 2 vols., Oxford: Clarendon Press, 1049p.

Chapman, S., and Falshaw, E., 1922, The lunar atmospheric tide at Aberdeen, 1869-1919, *Royal Meteorol. Soc. Quart. Jour.* **48:**246-250.

Chapman, S., and Lindzen, R., 1970, *Atmospheric Tides; Thermal and Gravitational*, Dordrecht, The Netherlands: D. Reidel, 200p.

Charlier, C. V. L., 1901*a*, Ueber die astronomische Erklarung einer Eiszeit, *Deutsch. Naturf. Gesell. Verh.* **73:**10-15.

Charlier, C. V. L., 1901*b*, Die astronomische Erklarung der Eiszeit, *Umschau* **5:**802-805.

Charlson, R. J.; Harrison, H.; and Witt, G., 1972, Aerosol concentrations: Effect on planetary temperatures, *Science* **175:**95-96.

Chernovsky, E. J., 1966, Double sunspot-cycle variation in terrestrial magnetic activity 1884-1963, *Jour. Geophys. Research* **71**(3):965-974.

Chernovsky, E. J., and Hagan, M. P., 1958, The Zurich sunspot number and its variations, 1700-1957, *Jour. Geophys. Research* **63:**775-788.

Chervin, R. M., 1978, The limitations of modelling: The question of statistical significance, in J. R. Gribbin (ed.), *Climatic Change*, Cambridge: Cambridge University Press, pp. 191-201.

Chistoni, C., 1901, Ueber den Regenfall in Modena, 1830-1896, *Zeitschr. Meteorologische* **18:**93.

Chittenden, F. H., 1901, Insects and the weather during the season of 1900, *U.S. Dept. Agriculture Entomology Div. Bull.* (new series), no. 30, pp. 63-75.

Chree, C., 1901, Terrestrial magnetism and atmospheric electricity, *Nature* **64:**151-153.

Chree, C., 1903*a*, Solar and magnetic disturbances, *Nature* **69:**6.

Chree, C., 1903*b*, Preliminary note on the relationships between sun-spots and terrestrial magnetism, *Royal Soc. [London] Proc.* **71:**221-224.

Chree, C., 1904, Nature of the relationship between sun-spot frequency and terrestrial magnetism, *Royal Astron. Soc. Mem.* **55**(Appendix II):151-187.

Chree, C., 1905*a*, Magnetic storms and aurora, *Nature* **73:**101, 173.

Chree, C., 1905*b*, Review of Maunder's recent investigations on the cause of magnetic disturbances, *Terr. Magnetism and Atmos. Electricity* **10:**9-14.

Chree, C., 1907*a*, Magnetic storms and aurora on February 9-10, 1907, *Nature* **75:**367.

Chree, C., 1907*b*, Atmospheric electricity and fog, *Nature* **77:**343.

Chree, C., 1909, Magnetic storms and solar eruptions, *Nature* **81:**456.

Chree, C., 1912*a*, Wireless telegraphy and terrestrial magnetism, *Nature* **90:**37.

Chree, C., 1912*b*, Studies of aurora, *Nature* **90:**38-40.

Chree, C., 1912*c*, *Studies in Terrestrial Magnetism*, London: Macmillan, 206p.

Chree, C., 1912*d*, Some phenomena of sun-spots and of terrestrial magnetism at Kew Observatory, *Royal Soc. London Philos. Trans.* **212A:**75-116.

Chree, C., 1912*e*, Magnetic storms, *Royal Inst. [London] Proc.* **19:**772-787.

Chree, C., 1913*a*, Some phenomena of sunspots and of terrestrial magnetism. Part 2, *Royal Soc. London Philos. Trans.* **213A:**245-277.

Chree, C., 1913*b*, Magnetic storms and solar phenomena, *Nature* **92:**19.

Chree, C., 1914*a*, Time measurements of magnetic disturbances and their interpretation, *Phys. Soc. [London] Proc.* **26:**137-153.

Chree, C., 1914*b*, The 27-day period in magnetic phenomena, *Royal Soc. [London] Proc.* **90A:**583-599.

Chree, C., 1924, Periodicities, solar and meteorological, *Royal Meteorol. Soc. Quart. Jour.* **50**(210):87-97.

Christie, A. D., 1973, Secular or cyclic changes in ozone, *Pure and Applied Geophysics* **105–108:** 1000-1009.

Christie, W. H. M., 1882, Sun-spots and magnetic storms, *Nature* **25:**55-56.

Chu, C.-C., 1926, Climatic pulsations during historic time in China, *Geog. Rev.* **16**(2):274-282.

Cirera, D., 1909, Perturbation magnetique du 25 Septembre 1909 (et relation avec l'activité solaire), *Acad. Sci. Comptes Rendus* **149:**1035-1036.

Cirera, D., and Balcelli, M., 1907, Etude des rapports entre l'activité solaire et les variations magnetiques et electriques enregistrées à Tortose (Espagne), *Acad. Sci. Comptes Rendus* **144:**959-961.

Claiborne, R., 1970, *Climate, Man, and History,* New York: W. W. Norton, 444p.

Clark, D. H., and Stephenson, F. R., 1978, An interpretation of the pretelescopic sunspot records from the Orient, *Royal Astron. Soc. Quart. Jour.* **19:**387-410.

Clark, J. E., 1903, The study of sunspot cycles, *Symons' Meteorol. Mag. (London)* **38:**136.

Clark, J. E., 1904, A danger in "smoothing" rainfall values, *Symons' Meteorol. Mag. (London)* **39:**83-84.

Clark, J. E., 1907*a,* York rainfall records and their possible indication of relation to solar cycles, *British Assoc. Advance. Sci., Rept. for 1906,* pp. 500-501.

Clark, J. E., 1907*b,* A relation between rainfall at York and solar cycles, *Symons' Meteorol. Mag. (London)* **42:**32-33.

Clark, J. E., 1911, Relations of phenological and climatic variation, *Nature* **86:**192.

Clark, J. E., 1912, The abnormal summers of 1910 and 1911, *Croydon Nat. History and Sci. Soc.,* pp. 54-56.

Clark, J. E., 1913, Effect upon atmospheric transparency of the Alaskan eruptions, *Royal Meteorol. Soc. Quart. Jour.* **39:**219-220.

Clayton, H. H., 1901*a,* The eclipse cyclone and the diurnal cyclones, results of meteorological observations in the solar eclipse of May 28, 1900, *Am. Acad. Arts and Sci. Proc.* **36:**307-318.

Clayton, H. H., 1901*b,* The influence of rainfall on commerce and politics, *Popular Sci. Monthly* **60**(2):158-165.

Clayton, H. H., 1903, The 27-day period in auroras and its connection with sunspots, *Science* (new series) **18:**632.

Clayton, H. H., 1904, The study of sunspot cycles, *Symons' Meteorol. Mag. (London)* **39:**8.

Clayton, H. H., 1907, A proposed new method of weather forecasting by analysis of atmospheric conditions into waves of different lengths, *Monthly Weather Rev.* **35:**161-167.

Clayton, H. H., 1908, The meteorology of total solar eclipses, including the eclipse of 1905, *Harvard College Observatory Annals* **58:**192-216.

Clayton, H. H., 1909, Diurnal and semi-diurnal atmospheric variations, *Nature* **79:**397-398.

Clayton, H. H., 1917, On the effect of short period variations of solar radiation on the Earth's atmosphere, *Smithsonian Misc. Colln.* **68**(3):18p.

Clayton, H. H., 1923, *World Weather, Including a Discussion of the Influence of Variations in Solar Radiation on the Weather and of the Meteorology of the Sun,* New York: Macmillan, 393p.

Clayton, H. H., 1925, Solar radiation and weather, or forecasting weather from observations of the Sun, *Smithsonian Misc. Colln.* **77**(6):64p.

Clayton, H. H., 1926, Solar activity and long-period weather changes, *Smithsonian Misc. Colln.* **78**(4):62p.

Clayton, H. H., 1927, World weather records, *Smithsonian Misc. Colln.* **79:**1199p.

Clayton, H. H., 1930, The atmosphere and the Sun, *Smithsonian Misc. Colln.* **82**(7):49p.

Clayton, H. H., 1934*a,* World weather and solar activity, *Smithsonian Misc. Colln.* **89**(15):52p.

Clayton, H. H., 1934*b,* World weather records, *Smithsonian Misc. Colln.* **90:**616p.

Clayton, H. H., 1936, Long-range weather changes and methods of forecasting, *Monthly Weather Rev.* **64:**359-376.

Clayton, H. H., 1939, The sunspot period, *Smithsonian Misc. Colln.* **98**(2):18p.

Clayton, H. H., 1940, The 11-year and 27-day solar periods in meteorology, *Smithsonian Misc. Colln.* **99**(5):20p.

Clayton, H. H., 1946, Sunspot changes and weather changes, *Smithsonian Misc. Colln.* **104**(19):29p.

Clemence, G. M., 1953, Coordinates of the center of mass of the Sun and the five outer planets, *Am. Ephemeris and Nautical Almanac Astron. Pap.* **13**(pt. 4).

Clemence, G. M., 1954, Perturbations of the five outer planets by the four inner ones, *Am. Ephemeris and Nautical Almanac Astron. Pap.* **13**(pt. 5).

Clements, F. E., 1921, Drought periods and climatic cycles, *Ecology* **2**(3):181-188.

Clements, H., 1903, *Weather Discoveries and Forecasts for the Coming Winter, 1904,* London: Wilkes and Company, 20p.

Clements, H., 1904*a, Clements's Nine Astronomical Weather Disturbances. Weather Prediction for the Year 1904 by an Entirely New and Scientific Method,* East Dulwich.

Clements, H., 1904*b, How to Predict the Weather, Wind and Magnetic Storms and Sunspots,* Dulwich, 26p.

Clements, H., and Disby, W., 1902, *Natural Law in Terrestrial Phenomena. A Study in the Causation of Earthquakes, Volcanic Eruptions, Temperature—*

With a Record of Evidence, London: W. Hutchinson and Company, 370p.

Clough, H. W., 1905, Synchronous variations in solar and terrestrial phenomena, *Astrophys. Jour.* **22:** 42-75.

Clough, H. W., 1920, An approximate seven-year cycle in terrestrial weather, with solar correlation, *Monthly Weather Rev.* **48**(10):593-596.

Clough, H. W., 1924, A systematically varying period with an average length of 28 months in weather and solar phenomena, *Monthly Weather Rev.* **52:**421-441.

Clough, H. W., 1933, The 11-year sun-spot period, secular periods of solar activity, and synchronous variations in terrestrial phenomena, *Monthly Weather Rev.* **60**(4):99-108.

Clough, H. W., 1934, The weather of the present century—Past and prospective, *Am. Meteorol. Soc. Bull.* **15.**

Clough, H. W., 1939, The effect of climatic changes upon the price of wheat, *Am. Meteorol. Soc. Bull.* **20**(11):399-400.

Clough, H. W., 1943, The long-period variations in the length of the 11-year solar period, and concurrent variations in terrestrial phenomena, *Am. Meteorol. Soc. Bull.* **24**(4):154-163.

Coakley, J. A., 1979, A study of climate sensitivity using a simple energy balance climate model, *Jour. Atmos. Sci.* **36:**260-269.

Cobb, W. E., 1967, Evidence of solar influence on the atmospheric electrical elements at Mauna Loa Observatory, *Monthly Weather Rev.* **95:**905-911.

Cobb, W. E., 1977, Atmospheric electric measurements at the South Pole, in H. Dolezalek and R. Reiter (eds.), *Electrical Processes in Atmospheres,* Darmstadt: Steinkopff Verlag, pp. 161-167.

Coeurdevache, P., 1903*a,* Frequence et variation diurne du vent a Perpignan, *Soc. Meteorol. France Annuaire* **51:**50-52.

Coeurdevache, P., 1903*b,* Pluie a Paris suivant le cycle lunaire, *Soc. Meteorol. France Annuaire* **51:**106-107.

Coffey, H. E., and Gilman, P. A., 1969, Sunspot motion statistics for 1965-67, *Solar Physics* **9:**423-426.

Cohen, C. J.; Hubbard, E. C.; and Oesterwinter, C., 1972, Elements of the outer planets for one million years, *Am. Ephemeris and Nautical Almanac Astron. Pap.* **22**(pt. 1).

Cohen, C. J.; Hubbard, E. C.; and Oesterwinter, C., 1973, Planetary elements for 10,000,000 years, *Celestial Mechanics* **7**(4):438-448.

Cohen, T. J., and Lintz, P. R., 1974, Long term periodicities in the sunspot cycle, *Nature* **250:**398-400.

Cohen, T. J., and Sweetser, E. I., 1975, The "spectra" of the solar cycle and of data for Atlantic tropical cyclones, *Nature* **256:**295-296.

Cohn, B., 1901, Die Sonnenflecke und das Wetter, *Natur (Halle)* **50:**436-438.

Cole, H. P., 1975, An investigation of a possible relationship between the height of the low-latitude tropopause and the sunspot number, *Jour. Atmos. Sci.* **32:**998.

Cole, K. D., 1976, Physical argument and hypothesis for Sun-weather relationships, *Nature* **260:**229.

Cole, T. W., 1973, Periodicities in solar activity, *Solar Physics* **30:**103-110.

Coleman, J. M., 1982, Recent seasonal rainfall and temperature relationships in peninsular Florida, *Quat. Research* **18**(2):144-151.

Colombo, G., 1965, Rotational period of the planet Mercury, *Nature* **208:**575.

Colombo, G., and Shapiro, I. I., 1966, The rotation of the planet Mercury, *Astrophys. Jour.* **145:**296.

Conrad, V., 1940, Investigations into periodicity of the annual range of air temperature at State College, Pennsylvania, *Pennsylvania State College Studies* **8.**

Conway, V. M., 1948, Von Post's work on climatic rhythms, *New Phytology* **47:**220-237.

Cook, J. W.; Brueckner, G. E.; and van Hoosier, M. E., 1980, Variability of the solar flux in the far ultraviolet 1175-2000 A, *Jour. Geophys. Research* **85:**2257-2268.

Cornish, E. A., 1936, On the secular variation of the rainfall at Adelaide, South Australia, *Royal Meteorol. Soc. Quart. Jour.* **62:**481-492.

Cornish, E. A., 1954, On the secular variation of rainfall at Adelaide, Australia, *Australian Jour. Physics* **7:**334-346.

Cortie, A. L., 1902*a,* Minimum sunspots and terrestrial magnetism, *British Assoc. Advance. Sci. Rept. for 1902,* pp. 522-523.

Cortie, A. L., 1902*b,* Minimum sunspots and terrestrial magnetism, *Astrophys. Jour.* **16:**203-210.

Cortie, A. L., 1903, Solar prominences and terrestrial magnetism, *British Assoc. Advqnce. Sci. Rept. for 1903,* pp. 574-575.

Cortie, A. L., 1904, Magnetic storms and associated sun-spots, *Royal Astron. Soc. Monthly Notices* **65:**197-205.

Cortie, A. L., 1910*a,* The sunspots and associated magnetic storms of September-October, 1909, *Royal Astron. Soc. Monthly Notices* **70:**19-23.

Cortie, A. L., 1910*b,* The recent magnetic disturbances and the Sun's activity, *Observatory (London)* **33:**100-101.

Cortie, A. L., 1912, Sunspots and terrestrial magnetic phenomena, 1898-1911: The cause of the annual variation in magnetic disturbances, *Royal Astron. Soc. Monthly Notices* **73:**52-60.

Cortie, A. L., 1913*a*, Magnetic disturbances, sunspots, and the Sun's corona, *British Assoc. Advance. Sci. Rept. 1912*, p. 411.

Cortie, A. L., 1913*b*, Sunspots and terrestrial magnetic phenomena, 1898-1911: The greater magnetic storms. Second paper, *Royal Astron. Soc. Monthly Notices* **73:**148-155.

Cortie, A. L., 1913*c*, Sunspots and terrestrial magnetic phenomena, 1898-1911: Sunspot areas, magnetic storms, and the Sun's corona. Third paper, *Royal Astron. Soc. Monthly Notices* **73:**431-436.

Cortie, A. L., 1913*d*, The mode of propagation of the Sun's influence in magnetic storms, *Royal Astron. Soc. Monthly Notices* **73:**539-543.

Cortie, A. L., 1914*a*, An area of long-continued solar disturbance, and the associated magnetic storms, *Royal Astron. Soc. Monthly Notices* **74:**670-678.

Cortie, A. L., 1914*b*, Solar and terrestrial magnetic disturbances, *British Assoc. Advance. Sci. Rept. 1913*, pp. 394-395.

Coughlan, M. J., 1983, A comparative climatology of blocking action in the two hemispheres, *Australian Meteorol. Mag.* **31:**3-13.

Courtillot, V., and leMouel, J. L., 1984, Geomagnetic secular variation impulses, *Nature* **311:**709-716.

Courtillot, V.; leMouel, J. E.; Dueruix, J.; and Cazenave, A., 1982, Magnetic secular variation as a precursor of climatic change, *Nature* **297:**386-387.

Cowling, T. G., 1975, Sunspots and the solar cycle, *Nature* **255:**189-190.

Cox, A., 1969, Geomagnetic reversals, *Science* **163:**237.

Cox, A.; Dalrymple, G. B.; and Doell, R. R., 1967, Reversals of the Earth's magnetic field, *Sci. American* **216**(2):44.

Cox, A.; and Doell, R. R., 1960, Review of paleomagnetism, *Geol. Soc. America Bull.* **71:**645.

Cox, J. F., 1951, On some possibilities offered by the study at planetary scale of terrestrial phenomena, *Am. Geophys. Union Trans.* **32:**536-537.

Cox, R. A.; Eggleton, A. E. J.; Derwent, R. G.; Lovelock, J. E.; and Pack, D. H., 1975, Long-range transport of photochemical ozone in northwestern Europe, *Nature* **255:**118-121.

Coy, L., 1979, A usually large westerly amplitude of the quasi-biennial oscillation, *Jour. Atmos. Sci.* **36:**174-176.

Craddock, J. M., 1965, The analysis of meteorological time series for use in forecasting, *The Statistician* **15**(2):167-190.

Craig, R. A., 1951*a*, Solar variability and meteorological anomalies, *Am. Acad. Arts and Sci. Proc.* **79**(4):280-290.

Craig, R. A., 1951*b*, Atmospheric pressure changes and solar activity, *New York Acad. Sci. Trans.* (Series 2) **13**(7):280-282.

Craig, R. A., and Willett, H. C., 1951, Solar energy variations as a possible cause of anomalous weather changes, in T. F. Malone (ed.), *Compendium of Meteorology*, Boston, Mass.: American Meteorological Society, pp. 379-390.

Cram, L. E., and Thomas, J. H., 1981, The physics of sunspots, *Nature* **293:**101-102.

Creer, K. M., 1981, Long-period geomagnetic secular variations since 12,000 yr B.P., *Nature* **292:**208-212.

Creer, K. M.; Hogg, T. E.; Readman, P. W.; and Reynaud, C., 1980, Palaeomagnetic secular variation curves extending back to 13,400 years B.P. recorded by sediments deposited in Lac de Joux, Switzerland, *Jour. Geophysics* **48:**139-147.

Croll, J., 1875, Climate and time in their geological relations, London: Daldy and Isbister (New York: Appleton), 577p.

Cronin, J. F., 1971, Recent volcanism and the stratosphere, *Science* **172:**847-849.

Crowe, C., 1958, Carbon-14 activity during the past 5000 years, *Nature* **182:**470-471.

Crutzen, P. J., 1977, The stratosphere-mesosphere, in O. R. White (ed.), *The Solar Output and Its Variation*, Boulder: Colorado Associated University Press, pp. 13-16.

Crutzen, P. J.; Isaksen, I.; and Reid, G., 1975, Solar proton events: Stratospheric sources of nitric oxide, *Science* **189:**457-459.

Cullen, C., 1980, Was there a Maunder Minimum?, *Nature* **283:**427-428.

Curott, D. R., 1966, Earth deceleration from ancient solar eclipses, *Astron. Jour.* **71:**264-269.

Currie, R. G., 1966, The geomagnetic spectrum—40 days to 5.5 years, *Jour. Geophys. Research* **71:**4579-4598.

Currie, R. G., 1968, Geomagnetic spectrum of internal origin and lower mantle conductivity, *Jour. Geophys. Research* **73:**2779-2786.

Currie, R. G., 1974*a*, Period and *Qw* of the Chandler wobble, *Royal Astron. Soc. Geophys. Jour.* **38:**178-185.

Currie, R. G., 1974*b*, Solar cycle signal in surface air temperature, *Jour. Geophys. Research* **79:**5657-5660.

Currie, R. G., 1976, The spectrum of sea level from 4 to 40 years, *Royal Astron. Soc. Geophys. Jour.* **46:**513-520.

Currie, R. G., 1979, Distribution of solar cycle signal in surface air temperature over North America, *Jour. Geophys. Research* **84:**753-761.

Currie, R. G., 1980, Detection of the 11-year sunspot cycle signal in Earth rotation, *Royal Astron. Soc. Geophys. Jour.* **61:**131-139.

Currie, R. G., 1981*a*, Solar-cycle signal in Earth rotation: Nonstationary behavior, *Science* **211:**386-389.

Currie, R. G., 1981*b*, Solar cycle signal in air temperature in North America: Amplitude gradient, phase and distribution, *Jour. Atmos. Sci.* **38:**808-818.

Currie, R. G., 1981c, Evidence for 18.6 year signal in temperature and drought conditions in North America since A.D. 1800, *Jour. Geophys. Research* **86:**11055-11064.

Currie, R. G., 1982, Evidence for 18.6 year term in air pressure in Japan and geophysical implications, *Royal Astron. Soc. Geophys. Jour.* **69:**321-327.

Currie, R. G., 1984*a*, On bistable phasing of 18.6 year induced flood in India, *Geophys. Research Letters* **11:**50-53.

Currie, R. G., 1984*b*, Evidence for 18.6 year lunar nodal drought in western North America during the past millennium, *Jour. Geophys. Research* **89:**1295-1308.

Currie, R. G., 1984*c*, Periodic (18.6-year) and cyclic (11-year) induced drought and flood in western North America, *Jour. Geophys. Research* **89**(D5):7215-7230.

Currie, R. G., and Fairbridge, R. W., 1985, Periodic 18.6-year and cyclic 11-year induced drought and flood in Northeastern China and some global implications, *Quater. Sci. Rev.* **4:**109-134.

Dallas, W. L., 1902, Meteorological history of the seven monsoon seasons, 1893-1899, in relation to the Indian rainfall, *Indian Meteorol. Mem.* **12**(Part 4):409-484.

Damboldt, T., 1972, The 22-year cycle in the recurrence tendency of geomagnetic activity, *Jour. Interdisc. Cycle Research* **3**(3/4):365-371.

Damon, P. E., 1977, Solar induced variations of energetic particles at one A.U., in O. R. White (ed.), *The Solar Output and Its Variations,* Boulder, Colo.: Colorado Associated University Press, pp. 429-448.

Damon, P. E., 1970, Climatic versus magnetic perturbation of the atmospheric C14 reservoir (with discussion), in I. U. Olsson (ed.), *Radiocarbon Variations and Absolute Chronology,* Stockholm: Almqvist and Wiksell, pp. 571-593.

Damon, P. E., 1977, Solar induced variations of energetic particles at one A. U., in O. R. White (ed.), *The Solar Output and Its Variations,* Boulder, Colo.: Colorado Associated University Press, pp. 429-448.

Damon, P. E.; Lerman, J. C.; and Long, A., 1978, Temporal fluctuations of atmospheric ^{14}C: Causal factors and implications, in F. A. Donath, F. G. Stehli, and G. W. Wetherill (eds.), *Annual Review of Earth and Planetary Sciences 1978,* vol. 6, Palo Alto, Calif.: Annual Reviews, pp. 457-494.

Damon, P. E.; Long, A.; and Grey, D. C., 1966, Fluctuations of atmospheric C14 during the last six millennia, *Jour. Geophys. Research* **71**(4):1055-1071.

Dansgaard, W.; Johnsen, S. J.; Clausen, H. B.; and Langway, C. C., 1971 Climatic record revealed by the Camp Century ice core, in K. K. Turekian (ed.), *Late Cenozoic Glacial Ages,* New Haven, Conn.: Yale University Press, pp. 37-56.

Dansgaard, W.; Johnsen, S. J.; Moller, J.; and Langway, C. C., 1969, One thousand centuries of climate record from Camp Century on the Greenland Ice Sheet, *Science* **166:**377-381.

Dansgaard, W., and Tauber, H., 1969, Glacier oxygen-18 content and the Pleistocene ocean temperatures, *Science* **166:**499-502.

Dauvillier, A., 1963, *Les hypotheses cosmogeniques: Theories des cycles cosmiques et des planètes jumelles,* Paris: Masson et Cie, 258p.

Dauvillier, A., 1970, Sur la rotation differentielle des corps celestes, *Soc. Royale des Sci. (Liège) Bull.* **39**(9/10):486-507.

Davis, G. E., and McCarthy, T. L., 1932, Twenty-nine months of solar radiation at Tucson, Arizona, *Monthly Weather Rev.* **60**(12):237-242.

Davis, H. T., 1941, *The Analysis of Economic Time Series,* Bloomington, Ind.: Principia Press.

Dawson, E., and Newitt, L. R., 1982, The magnetic poles of the Earth, *Jour. Geomagnetism and Geoelectricity* **34**(4):225-240.

de Boer, H. J., 1951, Tree ring measurements and weather fluctuations in Java from A.D. 1514, *Koninkl. Nederlandse Akad. Wetensch. Proc.* **54B:**194.

de Boer, H. J., 1967, Meteorological cycles, in R. W. Fairbridge (ed.), *The Encyclopedia of Atmospheric Sciences and Astrogeology,* Encyclopedia of Earth Sciences, vol. 2, New York: Reinhold, pp. 564-572.

de Boer, H. J., and Berlage, H. P., 1962, The relation between the mean barometric pressure anomaly at Jakarta and the length of the period of the Southern Oscillation, *Geofisica Pura e Applicata* **53:**198-207.

de Boer, H. J., and Euwe, W., 1949, On long-periodical temperature variations, *Koninkl. Magnet. Meteorol. Observatorium Verh. No. 35,* pp. 1-16.

Dechevrens, M., 1901, The Moon and rainfall, *Symons' Meteorol. Mag. (London)* **36:**183-184.

Deeley, R. M., 1930, Sunspots and pressure distribution, *Nature* **126:**401.

Defant, A., 1912, Die Veranderungen in der allgemeinen Zirkulation der Atmosphare in den gemassigten Breiten der Erde, *Akad. Wiss. Wien Sitzungsberichte* **121**(Series 2a):379.

Defant, A., 1921, Die Zirkulation der Atmosphare in den gemassigten Breiten der Erde, *Geog. Annaler* **3:**209-265.

DeGeer, G., 1908, On late Quaternary time and climate, *Geol. Fören. Stockholm Förh.* **30:**459-464.

DeGeer, G., 1910*a*, A geochronology of the last 12,000 years, *Internat. Geol. Congr. 11th, Comptes Rendus* **1:**245-246.

DeGeer, G., 1910*b*, A thermographical record of the late quaternary climate, *Internat. Geol. Congr. 11th, Symposium: Die Veranderungen des Klimas*, pp. 303-310.

DeGeer, G., 1921, Correlation of late glacial annual clay varves in North America with the Swedish time scale, *Geol. Fören. Stockholm Förh.* **43:**420.

DeGeer, G., 1927, Geochronology as based on solar radiation, *Science* **66:**458-460.

DeGeer, G., 1934, Equatorial Palaeolithic varves in East Africa, *Geog. Annaler* **16:**75-96.

Degens, E. T.; Michaelis, W.; Mopper, K.; and Kempe, S., 1978, Warven-chronologie holozaner Sedimente des Schwarzen Meeres, *Neues Jahrb. Geologie u. Paläontologie Monatsch.*

Degens, E. T.; Stoffers, P.; Golubic, S.; and Dickman, M. D., 1978, Varve chronology: Estimated rates of sedimentation in the Black Sea deep basin, pp. 499-508 in *Initial Reports of the Deep Sea Drilling Project*, vol. 42, part 2.

Dehsara, M., and Cehak, K., 1970, A global survey on periodicities in annual mean temperatures and precipitation totals, *Archiv Meteorologie, Geophysik, u. Bioklimatologie* **18B:**269-278.

deJong, A. F. M., and Mook, W. G., 1981, Natural C-14 variations and consequences for sea-level fluctuations and frequency analysis of periods of peat growth, *Geologie en Mijnbouw* **60:**331-336.

deJong, A. F. M.; Mook, W. G.; and Becker, B., 1979, Confirmation of the Suess wiggles: 3200-3700 B.C., *Nature* **280:**48-49.

Delany, M. J., and Waterhouse, F. L., 1956, International Society of Bioclimatology and Biometeorology, *Nature* **178:**966.

de la Rue, W.; Stewart, B.; and Loewy, B., 1869, Researches on solar physics, *Royal Soc. London Philos. Trans.* **159:**1-110.

de la Rue, W.; Stewart, B.; and Loewy, B., 1860, Researches on solar physics, *Royal Soc. London Philos. Trans.* **160:**389-496.

de la Rue, W.; Stewart, B.; and Loewy, B., 1872, Further investigation on planetary influence upon solar activity, *Royal Soc. [London] Proc.* **20:**210-218.

DeLury, R. E., 1922, Meteorological and astronomical pulses, *Am. Meteorol. Soc. Bull.* **3**(3):38-39.

Dennett, F. C., 1905, Magnetic storms and aurorae, *Nature* **73:**152.

Dennett, F. C., 1909, Great magnetic storms, *Knowledge and Science News (London)* **6:**455-457.

Denning, W. F., 1907, Summer snow and frost: Climatic vagaries, *Symons' Meteorol. Mag. (London)* **42:**111.

Dermott, S. F., 1968, On the origin of commensurabilities in the solar system—II. The orbital period relation, *Royal Astron. Soc. Monthly Notices* **141:**363-376.

de Rop, W., 1971, A tidal period of 1800 years, *Tellus* **23**(3):261-262.

Devoy, R. J. N., 1979, Flandrian sea level changes and vegetational history of the lower Thames estuary, *Royal Soc. London Philos. Trans.* **285B:** 355-410.

Dewey, E. R., 1970, The Sun's peregrinations, *Cycles* **21**(3):60.

Dicke, R. H., 1966, The secular acceleration of the Earth's rotation, in B. G. Marsden and A. G. W. Cameron (eds.), *The Earth-Moon System*, New York: Plenum Press, pp. 98-164.

Dicke, R. H., 1969, Average acceleration of the Earth's rotation and the viscosity of the deep mantle, *Jour. Geophys. Research* **74**(25):5895-5902.

Dicke, R. H., 1976, Evidence for a solar distortation rotating with a period of 12.2 days, *Solar Physics* **47:**475-515.

Dicke, R. H., 1978, Is there a chronometer hidden deep in the Sun?, *Nature* **276:**676-680.

Dicke, R. H., 1979, Solar luminosity and the sunspot cycle, *Nature* **280:**24-27.

Dicke, R. H., 1981, Seismology and geodesy of the sun: Solar geodesy, *U.S. Natl. Acad. Sci. Proc.* **78:**1309-1312.

Dicke, R. H., 1982, A magnetic core in the Sun? The solar rotator, *Solar Physics* **78:**3-16.

Dicke, R. H., and Goldberg, H. M., 1967, Solar oblateness and general relativity, *Phys. Rev. Letters* **18:**313.

Dicke, R. H., and Goldberg, H. M., 1974, The oblateness of the Sun, *Astron. Jour. Supp.* **27:**131-182.

Dickson, H. N., 1901, The mean temperature of the atmosphere and the causes of glacial periods, *Symons' Meteorol. Mag. (London)* **36:**145-146.

Dines, J. S., 1912, Some long-period fluctuations in the trade winds of the Atlantic, *Royal Meteorol. Soc. Quart. Jour.* **38:**265-270.

Dines, W. H., 1901, Weekly death-rate and temperature curves, 1890-1899, *Royal Meteorol. Soc. Quart. Jour.* **27:**69-75.

Dines, W. H., 1910, Atmospheric tides, *Symons' Meteorol. Mag. (London)* **45:**51.

Dines, W. H., 1911, Planetary rainfall, *Symons' Meteorol. Mag. (London)* **46:**179.

Dines, W. H., 1912, Winter in the seventeenth century, *Symons' Meteorol. Mag. (London)* **47:**56-57.

Ding, Y.; Lou, B.; and Feng, Y., 1982, The ancient solar activity maxima epochs of various periods, *Acta Astron. Sinica* **23**(3):287-298.

Ditchani, G., 1911, The Moon and the weather, *Knowledge (London)* **8:**47-48.

Doake, C. S. M., 1977, A possible effect of ice ages on the Earth's magnetic field, *Nature* **267:**415-416.

Doberitz, R., 1967, Statistical investigations of the

climatic anomalies of the equatorial Pacific, *Bonner Meteorol. Abh.* **7.**

Doberitz, R., 1968, Cross spectrum analysis of rainfall and sea temperature at the equatorial Pacific Ocean, *Bonner Meteorol. Abh.* **8.**

Doberk, W., 1882, Sun-spots and Markree rainfall, *Nature* **25:**366-367.

Dobson, G. M. B., 1929, Summary of the present state of our knowledge of the distribution of ozone in the upper atmosphere, *Monthly Weather Rev.* **57**(2):56-57.

Dobson, G. M. B.; Harrison, D. N.; and Lawrence, J., 1929, Ozone and geomagnetic activity, *Royal Soc. (London) Proc.* **122A:**456.

Doell, R. R., and Cox, A., 1971, Pacific geomagnetic secular variation, *Science* **171:**248-254.

Doell, R. R., and Cox, A., 1972, The Pacific geomagnetic secular variation anomaly and the question of lateral uniformity in the lower mantle, in E. C. Robertson (ed.), *The Nature of the Solid Earth,* New York, McGraw-Hill, pp. 245-284.

Dolezalek, H., 1972, Discussion of the fundamental problem of atmospheric electricity, *Geofisica Pura e Applicata* **100:**8-43.

Donnelly, R. F., 1983, Solar UV spectral irradiance variations, in B. M. McCormac (ed.), *Weather and Climate Responses to Solar Variations,* Boulder: Colorado Associated University Press, pp. 43-55.

Donnelly, R. F., ed., 1980, *Solar-Terrestrial Predictions Proceedings,* Boulder, Colorado, U.S. Department of Commerce, National Oceanic and Atmospheric Administration.

Dorman, L. I., and others, 1982, The energy spectrum of 11-year modulation of galactic cosmic rays, *Geomagnetism and Aeronomy* **22**(4).

Douglass, A. E., 1909, Weather cycles in the growth of big trees, *Monthly Weather Rev.* **37:**225-237.

Douglass, A. E., 1918, A method of estimating rainfall by the growth of trees, *Carnegie Inst. Washington Pub. 192.*

Douglass, A. E., 1919*a*, Climatic cycles and tree-growth; A study of the annual rings of trees in relation to climate and solar activity, *Carnegie Inst. Washington Pub. 289* **1:**127p.

Douglass, A. E., 1919*b*, The relations of weather and business, *Monthly Weather Rev.* **47**(12):867.

Douglass, A. E., 1928, Climatic cycles and tree-growth; A study of the annual rings of trees in relation to climate and solar activity, *Carnegie Inst. Washington Pub. 289* **2:**166p.

Douglass, A. E., 1936, A study of cycles, *Carnegie Inst. Washington Pub. 289* **3:**171p.

Dressler, A. J., 1974, Some problems in coupling solar activity to meteorological phenomena, in W. R. Bandeen and S. P. Maran (eds.), *Symposium on Possible Relationships between Solar Activity and Meteorological Phenomena, Proceedings,* Greenbelt, Md.: NASA Goddard Space Flight Center, NASA SP-366, pp. 187-197.

Droste, B., 1924, Die 11-jahrige Sonnenfleckenperiode und die Temperaturschwankungen auf die nordlichen Halbkugel im jahrzeitlicher und regionaler Differenzierungen, *Zeitschr. Meteorologische,* pp. 261-268.

Duell, B., and Duell, G., 1948, The behavior of barometric pressure during and after solar particle invasions and solar ultraviolet invasions, *Smithsonian Misc. Colln.* **110**(8):34p.

Duggal, S. P.; Forbush, S. E.; and Pomerantz, M. A., 1970, The variation with a period of two solar cycles in the cosmic ray diurnal anisotropy for the nucleonic component, *Jour. Geophys. Research* **75**(7):1150-1156.

Dull, B., 1933, Uber den Einfluss der solaren Tatigkeit auf die Erdatmosphare, *Beitr. Geophysik* **39:**1-9.

Duncombe, R. L., 1973, Planetary ephemerides, in B. D. Tapley and V. Szebehely (eds.), *Recent Advances in Dynamical Astronomy,* Dordrecht, The Netherlands: D. Reidel, pp. 289-308.

Dunham, D. W.; Sofia, S.; Fiala, A. D.; Herald, D.; and Muller, P. M., 1980, Observation of a possible change in the solar radius between 1715 and 1979, *Science* **210:**1243-1245.

Dunn, E. B., 1902, *The Weather and Practical Methods of Forecasting It,* New York: Dodd, Mead, 356p.

Dunn, J. R.; Fuller, M.; Ito, H.; and Schmidt, V. A., 1971, Paleomagnetic study of a reversal of the Earth's magnetic field, *Science* **172:**840-845.

Dütsch, H. U., 1971, Photochemistry of atmospheric ozone, in H. E. Landsberg and J. Van Mieghen (eds.), *Advances in Geophysics,* vol. 15, New York: Academic Press, pp. 219-322.

Dütsch, H. U., 1979, The search for solar-cycle-ozone relationships, *Jour. Atmos. and Terr. Physics* **41:**771-785.

Dyson, F. W., 1905, Magnetic storms and the solar rotation, *Observatory (London)* **28:**176-179.

Dzerdzeevski, B. L., 1961, The general circulation of the atmosphere as a necessary link in the Sun-climatic variations chain, *New York Acad. Sci. Annals* **95**(article 1):188-199.

Dzerdzeevski, B. L., 1966, Some aspects of dynamic climatology, *Tellus* **18:**751-760.

Dzerdzeevski, B. L., 1969, Climatic epochs in the twentieth century and some comments on analysis of past climates, in H. E. Wright (ed.), *Quaternary Geology and Climate,* Washington, D.C.: U.S. National Academy of Science, Publication 1701, pp. 49-60.

Easton, C., 1918, Periodicity of winter temperatures in western Europe since A.D. 760, *Koninkl. Nederlandse Akad. Wetensch. Proc.* **20A:**1092.

Easton, C., 1928, *Les hivers dans l'Europe occidentale,* Leyden: E. J. Brill, 210p.

Ebdon, R. A., 1975, The quasi-biennial oscillation and its association with tropospheric circulation patterns, *Meteorol. Mag.* **104:**282-297.

Eckert, W. J.; Brouwer, D.; and Clemence, G. M., 1951, Coordinates of the five outer planets 1653-2060, *Am. Ephemeris and Naut. Almanac Astron. Paper* **12:**327p.

Eckhardt, W. R., 1909, *Das Klimaproblem der geologischen Vergangenheit und historischen Gegenwart*, Braunschweig: F. Vieweg und Sonne, 183p.

Eckhardt, W. R., 1911, Die Veranderungen des Klimas seit dem Maximum der letzten Eiszeit, *Zeitschr. Geog.* **17:**378-386.

Eddy, J. A., 1975, A new look at solar-terrestrial relations, *Am. Astron. Soc. Bull.* **7:**410.

Eddy, J. A., 1976, The Maunder Minimum, *Science* **192:**1189-1202.

Eddy, J. A., 1977*a*, Climate and the changing Sun, *Clim. Change* **1:**173-190.

Eddy, J. A., 1977*b*, Historical evidence for the existence of the solar cycle, in O. R. White (ed.), *The Solar Output and Its Variation*, Boulder: Colorado Associated University Press, pp. 51-71.

Eddy, J. A., 1977*c*, The case of the missing sunspots, *Sci. American* **236**(5):80-92.

Eddy, J. A., 1979, *A New Sun: The Solar Results from Skylab*, Washington, D.C.: National Aeronautics and Space Administration, NASA SP-402.

Eddy, J. A., 1980, Climate and the role of the Sun, *Jour. Interdisc. History* **10:**725-747.

Eddy, J. A., 1981, Climate and the role of the Sun, in R. I. Rotberg and T. K. Rabb (eds.), *Climate and History. Studies in Interdisciplinary History*, Princeton, N.J.: Princeton University Press, pp. 145-167.

Eddy, J. A., 1983, The Maunder Minimum: A reappraisal, *Solar Physics* **89:**195-207.

Eddy, J. A., and Boornazian, A. A., 1979, Secular decrease in the solar diameter, 1863-1953, *Am. Astron. Soc. Bull.* **11:**437.

Eddy, J. A.; Gilman, P. A.; and Trotter, D. E., 1976, Solar rotation during the Maunder Minimum, *Solar Physics* **46**(1):3-14.

Eddy, J. A.; Gilman, P. A.; and Trotter, D. E., 1977, Anomalous solar rotation in the early 17th century, *Science* **198:**824-829.

Edholm, O. G., 1954, Physiological effects of cold environments on man, in J. L. Cloudsley-Thompson (ed.), *Biology of Deserts*, London: Institute of Biology, pp. 207-212.

Edwards, D. P., 1982, Solar heating by ozone in the tropical stratosphere, *Royal Meteorol. Soc. Quart. Jour.* **108:**253-262.

Egyed, L., 1961, Temperature and magnetic field, *New York Acad. Sci. Annals* **95**(Article 1):72-77.

Ekholm, N., 1901, On the variations of the climate of the geological and historical past and their causes, *Royal Meteorol. Soc. Quart. Jour.* **27:**1-27.

Ekholm, N., 1902, On the meteorological conditions of the Pleistocene epoch, *Geol. Soc. London Quart. Jour.* **58:**37-45.

Ekholm, N., Periodicitat der Sonnenthatigkeit, *Kongl. Svenska Vetenskaps-Akademiens Forhandl.* **26**(5).

Ekholm, N., and Arrhenius, S., 1898, Ueber den Einfluss des Mondes auf die Polarichter und Gewitter, *Kongl. Svenska Vetenskaps-Akademiens Handlingar* **31**(2).

Ellis, H. T., and Pueschel, R. F., 1971, Solar radiation: Absence of air pollution trends at Mauna Loa, *Science* **172:**845-846.

Ellis, W., 1901, Sun-spots and magnetic disturbance, *Royal Astron. Soc. Monthly Notices* **61:**537-541.

Ellis, W., 1902, The Moon and rainfall, *Symons' Meteorol. Mag. (London)* **37:**142-143.

Ellison, M. A., 1955, *The Sun and Its Influences*, London: Roulledge and Kegan Paul, 235p.

Ellison, M. A., 1958, Magnetic activity following a solar flare, *Jour. Atmos. and Terr. Physics* **12:**214-215.

Ellison, M. A., 1968, *The Sun and Its Influence; An Introduction to the Study of Solar-Terrestrial Relations*, 3rd ed. (revised by Patrick Moore), New York: Elsevier, 240p.

Ellison, T. H., 1970, Discussion meeting on long-range weather forecasting: Introductory remarks, *Royal Meteorol. Soc. Quart. Jour.* **96:**326-328.

Elsasser, W. M., 1941, A statistical analysis of the Earth's magnetic field, *Phys. Rev.* **60:**876.

Elsasser, W. M., 1950, The Earth's interior and geomagnetism, *Rev. Modern Physics* **22:**1-35.

Elsasser, W. M., 1951, Hydromagnetic dynamo theory, *Rev. Modern Physics* **28:**135-163.

Emiliani, C., 1972*a*, Quaternary paleotemperatures and the duration of the high-temperature intervals, *Science* **178:**398-401.

Emiliani, C., 1972*b*, Interglacial high sea levels and the control of Greenland ice by the precession of the Equinoxes, *Science* **178:**398-401.

Emiliani, C., and Shackleton, N. J., 1974, The Brunhes epoch: Isotopic paleotemperatures and geochronology, *Science* **183:**511-514.

Enger, I., 1959, Optimum length of record for climatological estimates of temperatures, *Jour. Geophys. Research* **64**(7):779-787.

Epstein, S., and Yapp, C. J., 1976, Climatic implications of the D/H ratio of hydrogen in C-H groups in tree cellulose, *Earth and Planetary Sci. Letters* **30:**252-261.

Evans, J. W., 1955, Solar influence of the Earth, in *Smithsonian Institution Annual Report for 1954*, Washington, D.C., pp. 189-200.

Everest, R., 1834, On the influence of the Moon

upon atmospherical phenomena, *Asiatic Soc. Bengal Jour.* **3:**345-359, 631-634.

Evershed, J. R. T., and Sitaramo Ayyer, S., 1913, The apparent effect of planets on the distribution of prominences, *Kodiakanal Bull.* **35:**37-43.

Eythorsson, J., 1949, Temperature variations in Iceland, *Geog. Annaler* **31**(1-4):36-55.

Eythorsson, J., 1952, Polar ice at the coasts of Iceland, *Jökull* **2:**32.

Fairbridge, R. W., ed., 1967*a*, *The Encyclopedia of Atmospheric Sciences and Astrogeology*, Encyclopedia of Earth Sciences, vol. 2, New York: Reinhold, 1200p.

Fairbridge, R. W., 1967*b*, El Niño effect, in R. W. Fairbridge (ed.), *The Encyclopedia of Atmospheric Sciences and Astrogeology*, Encyclopedia of Earth Sciences, vol. 2, New York: Reinhold, pp. 352-354.

Fairbridge, R. W., 1977, Global climate change during the 13,500-B.P. Gothenburg geomagnetic excursion, *Nature* **265:**430-431.

Fairbridge, R. W., 1980, Prediction of long-term geologic and climatic changes that might affect the isolation of radioactive waste, in *Underground Disposal of Radioactive Wastes*, vol. 2, Vienna: International Atomic Energy Agency, IAEA-SM-243/43, pp. 385-405.

Fairbridge, R. W., 1983, The Pleistocene-Holocene boundary, *Quat. Sci. Rev.* **1**(3):215-244.

Fairbridge, R. W., 1984*a*, Planetary periodicities and terrestrial climate stress, in N.-A. Mörner and W. Karlén (eds.), *Climatic Changes on a Yearly to Millennial Basis*, Dordrecht, The Netherlands: D. Reidel, pp. 509-520.

Fairbridge, R. W., 1984*b*, The Nile floods as a global climatic/solar proxy, in N.-A. Mörner and W. Karlén (eds.), *Climatic Changes on a Yearly to Millennial Basis*, Dordrecht, The Netherlands: D. Reidel, pp. 181-190.

Fairbridge, R. W., and Hameed, S., 1983, Phase coherence of solar cycle minima over two 178-year periods, *Astron. Jour.* **88**(6):867-869.

Fairbridge, R. W., and Hillaire-Marcel, C., 1977, An 8,000-yr palaeoclimatic record of the "double-Hale" 45-yr solar cycle, *Nature* **268:**413-416.

Fairbridge, R. W., and Krebs, O. A., Jr., 1962, Sea level and the Southern Oscillation, *Royal Astron. Soc. Geophys. Jour.* **6**(4):532-545.

Faris, R. L., 1910*a*, Times of abruptly beginning magnetic disturbances, as recorded at the Coast and Geodetic Survey magnetic observatories, *Terr. Magnetism and Atmos. Electricity* **15:**33-35.

Faris, R. L., 1910*b*, On precursors of magnetic storms, *Terr. Magnetism and Atmos. Electricity* **15:**209-210.

Faris, R. L., 1911*a*, On the non-simultaneity of suddenly beginning magnetic storms, *Nature* **87:**78.

Faris, R. L., 1911*b*, The peculiar magnetic disturbances of December 28-31, 1908, *Terr. Magnetism and Atmos. Electricity* **16:**13-24.

Farthing, E. D., 1955*a*, A possible relationship between the solar corona and weather conditions in the central midwest, *Am. Meteorol. Soc. Bull.* **36**(9):427-435.

Farthing, E. D., 1955*b*, Progress report on solar-weather research utilizing a daily computed index of solar coronal activity as the solar parameter, *Am. Meteorol. Soc. Tech. Bull. 55-7*, 27p.

Farthing, E. D., 1956, Utilizing the sun to forecast the weather, *Shell Aviation News* **214:**19-24.

Fast, J., 1979, *Weather Language*, New York: Simon and Schuster.

Faure, H., and Gac, J. Y., 1981, Sahelian drought to end in 1985?, *Nature* **291:**475-478.

Feldman, W. C.; Asbridge, J. R.; Bame, S. J.; and Gosling, J. T., 1977, Plasma and magnetic fields from the sun, in O. R. White (ed.), *The Solar Output and Its Variations*, Boulder: Colorado Associated University Press, pp. 351-382.

Fenner, J. A., 1975, The winter of 1974-5: The problem of extrapolation, *Weather* **30**(8):272-273.

Fenyi, J., 1905, Ueber Temperatur-Erniedrigungsinfolge erhohter Insolation, *Zeitschr. Meteorologische* **22:**311-313.

Ferguson, C. W., 1969, A 7104-year annual tree-ring chronology, *Tree-Ring Bull.* **29**(3-4):3-29.

Fermar, L. L., 1939, Varved sediments in Malaya, *Geol. Mag.* **76:**473-478.

Ferris, G. A., 1969, Planetary influences on sunspots, *British Astron. Jour.* **79:**385-388.

Feshenfeld, F. C.; Ferguson, E. E.; Streit, G. E.; and Albritton, D. L., 1976, Stratospheric ion chemistry and the 11-year variation in polar ozone, *Science* **194:**544-545.

Fessenden, R. A., 1907, Atmospheric absorption of wireless signals, *Nature* **76:**444.

Feynman, J., 1983, Solar cycle and long term changes in the solar wind, *Rev. Geophysics and Space Physics* **21**(2):338-348.

Figee, S., 1905, The influence of the Moon on the magnetic needle at Batavia from observations made from April 1883 up to March 1899, *Royal Magnetical and Meteorol. Observatory at Batavia Folio* **26:**183-285.

Finch, H. F., and Leaton, B. M., 1957, The Earth's main magnetic field—epoch 1955.0, *Royal Astron. Soc. Monthly Notices* (Geophys. Supp.) **7:**314.

Fincham, G. H. H., 1905, The effect of the sun spot period on the daily variation of the magnetic elements at the Cape of Good Hope, *British Assoc. Advance. Sci., Rept. 1905*, pp. 338-339.

Fincham, G. H. H., 1906, On the nature of the effect of the sun-spot frequency on the variation of the magnetic elements at the Cape of Good Hope, *South African Philos. Soc. Trans.* **16:**301-312.

Finn, W., 1903, Influences of the sunspots upon the electrical and magnetic forces of the Earth, *Sci. American* (Supp.) **56:**23351-23352.

Fischer, H.-J., and Muhleisen, R., 1972, Variationen des Ionosphären potentials und der Weltgewittertätigkeit im 11 jährigen solaren Zyklus, *Meteorol. Rundschau* **25:**6-10.

Fisher, D. A., 1982, Carbon-14 production compared to oxygen isotope records from Camp Century, Greenland and Devon Island, Canada, *Clim. Change* **4:**419-426.

Fisher, R. A., 1929, Tests of significance in harmonic analysis, *Royal Soc. [London] Proc.* **125A:**54-59.

Flammarion, C., 1907, Le magnetism solaire, *Soc. Astron. France Bull.*, pp. 159-165.

Flint, R. F., 1951, Climatic implications of glacier research, in T. F. Malone (ed.), *Compendium of Meteorology*, Boston, Mass.: American Meteorological Society, pp. 1019-1023.

Flohn, H., 1958, Bemerkungen zum Problem der globalen Klimaschwankungen, *Archiv Meteorologie, Geophysik, u. Bioklimatologie* **9B:**1-13.

Flohn, H., 1963, Klimaschwankungen und grossraumige Klimabeeinflussung, Köln und Oplade: Westdeutscher Verlag.

Flohn, H., 1974, Background of a geophysical model of the initiation of the next glaciation, *Quat. Research* **4:**385-404.

Fodor, G., 1910, Die Sonnenflecken und die magnetischen Sturme des Jahres 1909, *Natur (Leipzig)*, pp. 141-143.

Folland, C. K., 1977, Recent work on some quasi-cyclic fluctuations of meteorological parameters affecting British climate, *Weather* **32:**336-342.

Folland, C. K.; Parker, D. E.; and Kates, F. E., 1984, Worldwide marine temperature fluctuations 1856-1981, *Nature* **310:**670-673.

Follin, J. W., Jr.; Grey, E. P.; and Yu, K., 1977, The connection between cosmic rays and thunderstorms, *EOS* **58:**110.

Forbes, J. D., 1861, Inquiries about terrestrial temperature, *Royal Soc. Edinburgh Trans.* **22**(pt. 1):75-100.

Forbush, S. E., 1954, World-wide cosmic-ray variations, 1937-52, *Jour. Geophys. Research* **59:**525-542.

Forbush, S. E., 1966, Time variation of cosmic rays, in J. Bartels (ed.), *Handbuch der Physik, Geophysik*, vol. 3, New York: Springer-Verlag, pp. 159-274.

Fowle, F. E., 1929, Atmospheric ozone: Its relation to some solar and terrestrial phenomena, *Smithsonian Misc. Colln.* **81**(11):27p. (Review by W. W. Kimball, 1929, *Monthly Weather Rev.* 57(2):58.)

Frank, F. C., and Alldredge, L. R., 1968, A peculiar property of the geomagnetic field, *Jour. Geophys. Research* **73:**677-682.

Frech, F., 1903, Ueber Eiszeiten und das Klima der geologischen Vergangenheit, *Weltall (Berlin)* **3:**193-198.

Frech, F., 1906, Studien uber das Klima der geologischen Vergangenheit, *Zeitschr. Gesell. Erdkunde (Berlin)*, pp. 533-553.

Frederick, J. E., 1977, Chemical response of the middle atmosphere to changes in the ultraviolet flux, *Planetary and Space Sci.* **25:**1-5.

Freeman, J. C., and Portig, W. H., 1965, Planetary effects on the sunspot cycle, in *Investigation of the Relation Between Solar Variations and Weather on Earth*, NASA CR-77306.

Freeman, O. W., 1929, Evidence of prolonged droughts on the Columbia Plateau prior to white settlement, *Monthly Weather Rev.* **57**(6):250-251.

Freier, G. D., 1961, Auroral effects on the Earth's electric field, *Jour. Geophys. Research* **66:**2695-2702.

Frenzel, B., 1966, Climatic change in the Atlantic-Sub-boreal transition on the northern hemisphere, in J. S. Sawyer (ed.), *World Climate from 8000-0 B.C.*, London: Royal Meteorological Society, pp. 99-123.

Frenzel, B., 1973, Climatic fluctuations of the Ice Age, Cleveland: The Press of Case-Western Reserve University, 306p. (trans. from German by A. E. M. Nairn).

Frenzel, F., ed., 1977, *Dendrochronologie und Postglaziale Klimaschwankungen in Europa*, Wiesbaden: F. Steiner Verlag GmbH.

Fricke, W., and Teleki, G., eds., 1982, *Sun and Planetary System*, Dordrecht, The Netherlands: D. Reidel, 538p.

Friese, F. W., 1938, The drought region of northeastern Brazil, *Geog. Rev.* **28:**363-378.

Fritsche, H., 1902, *Die tagliche Periode der erdmagnetischen Elemente*, St. Petersburg.

Fritsche, H., 1906, Die jahrliche und tagliche Periode der erdmagnetischen Elemente, *Zeitschr. Physikalische (Leipzig)* **7:**130-133.

Fritts, H. C., 1966, Growth-rings of trees, their correlation with climate, *Science* **114:**973-979.

Fritts, H. C., 1976, *Tree Rings and Climate*, New York: Academic Press, 567p.

Fritts, H. C.; Blasing, T. J.; Hayden, B. P.; and Kutzbach, J. E., 1971, Multi-variate techniques for specifying tree-growth and climate relationships and for reconstructing anomalies in palaeoclimate, *Jour. Appl. Meteorology* **10**(5):845-864.

Fritts, H. C.; Lofgreau, G. R.; and Gordon, G. A., 1979, Variations in climate since 1602 as reconstructed from tree-rings, *Quat. Research* **12:**18-46.

Fritz, H., 1878, *Ueber die Beziehungen der Sonnenflecken zu den meteorologischen und magnetischen Erscheinungen der Erde*, Haarlem: De Erven Loosjes.

Fritz, H., 1879, *Die wichtigsten periodischen Erscheinungen der Meteorologie und Kosmologie*, Zurich.

Fritz, H., 1883, Die Sonnenfleckenperioden und die Planetenstellen, *Zurich Naturf. Gesell. Vierteljahrschrift* **27.**

Fritz, H., 1893, Die Perioden solarer und terrestricher Erscheinungen [trans. by W. W. Reed, in *Monthly Weather Rev.* **56**(10):401-407].

Fritz, H., 1889, *Die Wichtigsten Periodischen Erscheinungen,* Leipzig.

Fritz, S., 1949, The albedo of the planet Earth and of clouds, *Jour. Meteorology* **6:**277-282.

Fromm, E., 1985, Chronological calculation of the varve zero in Sweden, *Boreas* **14**(2):123-126.

Gage, K. S., and Reid, G. C., 1981, Solar variability and the secular variation in the tropical tropopause, *Geophys. Research Letters* **8:**187-190.

Gager, C. S., 1912, Wheat rusts and sunspots, *Science* (new series) **35:**74-75.

Gallant, R., 1963, Changes in the Earth's axis during large meteorite collisions, *Nature* **200:**414-415 (reply by F. Dachille, pp. 415-416).

Gani, J., 1975, The use of statistics in climatological research, *Search* **6:**504-508.

Gannett, H., 1906, Certain relations of rainfall and temperature in tree growth, *Am. Geog. Soc. Bull.* **38:**424-434.

Garretson, L. T., 1904, Deflection of thunderstorms with the tides, *Monthly Weather Rev.* **32:**520.

Garrido, R., 1909, Les variations de l'activité solaire pendant l'année 1908, *Cosmos* **60:**228-229.

Garriott, E. B., 1904*a, Long-range Weather Forecasts,* Washington, D.C.: U.S. Department of Agriculture, Weather Bureau, pp. 38-42.

Garriott, E. B., 1904*b, Long-range Weather Forecasts,* Washington, D.C.: U.S. Department of Agriculture, Weather Bureau Bulletin No. 35, 68p.

Garriott, E. B., 1905, Rivers and floods, *Monthly Weather Rev.* **33**(4):129.

Garriott, E. B., 1906, A possible extension of the period of weather forecasts, *Monthly Weather Rev.* **34:**22-23.

Garriott, E. B., 1907, Equinoctial storms, *Monthly Weather Rev.* **35:**276-277.

Gasjukov, P. S., and Smirnov, N. P., 1967, Pressure-field oscillations over the northern hemisphere within the 11-year cycle of solar activity, *Akad. Nauk. SSSR Doklady* **173:**567-569 (in Russian).

Gast, P. R., 1929, A correlation between solar radiation intensities and relative humidities, *Monthly Weather Rev.* **57**(11):464-465.

Gast, P. R., and Stickel, P. W., 1929, Solar radiation and relative humidity in relation to Duff moisture and forest fire hazard, *Monthly Weather Rev.* **57**(11):466-468.

Gedeonov, A. D., 1972, Some manifestations of cyclic fluctuations of the climate of the Northern Hemisphere, *Jour. Interdisc. Cycle Research* **3**(3/4):345-348.

Geinitz, E., 1913, Die grossen Schwankungen der norddeutschen Seen, *Naturwissenchaften* **1:**665-670.

Geller, M. A., and Alpert, J. C., 1980, Planetary-wave coupling between the troposphere and the middle atmosphere as a possible Sun-weather mechanism, *Jour. Atmos. Sci.* **37:**1197-1215.

Gerard, V. B., 1959, The propagation of world-wide sudden commencements of magnetic storms, *Jour. Geophys. Research* **64**(6):593-596.

Gerety, E. J., 1978, Reply to Xanthakis, Possible Sun-weather correlation, *Nature* **275:**775.

Gerety, E. J.; Olson, R. H.; and Roberts, W. O., 1978, Analysis of a possible sun-weather correlation, *Nature* **272:**231-232.

Gerety, E. J.; Wallace, J. M.; and Zerefos, C. S., 1977, Sunspots, geomagnetic indices and the weather: A cross-spectral analysis between sunspots, geomagnetic activity and global weather data, *Jour. Atmos. Sci.* **34:**673-678.

Gerson, N. C., 1960, The ionosphere and solar activity, *Geofisica Pura e Applicata* **45:**117-122.

Geyh, M. A., 1971, Middle and young Holocene sea level changes as global contemporary events, *Geol. Fören. Stockholm Förh.* **93:**679-692.

Geyh, M. A., 1980, Holocene sea-level history: Case study of the statistical evaluation of 14-C dates, *Radiocarbon* **22:**695-704.

Geyh, M. A.; Merkt, J.; and Muller, H., 1971, Sedimentological pollen-analytical and isotopic studies of annually laminated sediments in the central part of the Schleinsee, Germany, *Archiv Hydrobiologie* **69:**355-399.

Ghazi, A., 1977, Global variation of atmospheric ozone, *Jour. Interdisc. Cycle Research* **8:**183-188.

Ghazi, A.; Ramanathan, V.; and Dickinson, R. E., 1979, Acceleration of upper atmosphere radiative damping: Observational evidence, *Geophys. Research Letters* **6:**437-440.

Gheury, M. E. J., 1907, Sur l'influence meteorologique de la lune, *Soc. Belge Astronomie Bull.* **12:**163-165.

Giannitrapani, L., 1910, Il presente periodo climatico e le variazioni periodiche di Bruckner, *Soc. Geog. Italiana Boll.* **11**(2):1345-1350.

Gilette, H. P., 1938, Coincidence of some climate and sea-level cycles, *Pan-American Geologist* **70:**279.

Gilliland, R. L., 1981, Solar radius variations over the past 265 years, *Astrophys. Jour.* **248:**1144-1155.

Gilliland, R. L., 1982, Solar, volcanic, and CO_2 forcing of recent climatic changes, *Clim. Change* **4:**111-131.

Gilman, D. L., 1982, The nature of climatic variability, in J. A. Eddy (ed.), *Solar Variability, Weather, and Climate,* Washington, D.C.: National Academy Press, pp. 53-63.

Gilman, P. A., 1968, Thermally driven Rossby-mode dynamo for solar magnetic-field reversals, *Science* **160:**760-763.

Girardin, P., 1910, Les oscillations des glaciers de Savoie, particulierement de 1902 à 1909, *Assoc. Français pour l'Avancement des Sci. Comptes Rendus* **39:**285-287.

Glass, B., and Heezen, B. C., 1967, Tektites and geomagnetic reversals, *Nature* **214:**372.

Glass, B., and Heezen, B. C., 1967, Tektites and geomagnetic reversals, *Sci. American* **217**:32-38.

Gleissberg, W., 1944, A table of secular variations of the solar cycle, *Terr. Magnetism and Atmos. Electricity* **49**:243.

Gleissberg, W., 1955, The 80-year sunspot cycle, *British Astron. Assoc. Jour.* **68**:148-152.

Gleissberg, W., 1956, Zur Konstanz der Skala der Sonnenfleckenrelativzahlen, *Naturwissenschaften* **43**:196.

Gleissberg, W., 1965, The eighty-year cycle in auroral frequency numbers, *British Astron. Assoc. Jour.* **75**:227-231.

Gleissberg, W., 1971, The probable behavior of Sunspot Cycle 21, *Solar Physics* **21**:240-245.

Gleissberg, W., 1972, The 80-year solar cycle and its use for solar-activity forecasting, *Jour. Interdisc. Cycle Research* **3**:391-394.

Gleissberg, W., 1973, The 11-year and 80-year cycles in the frequency of sunspots easily visible to the naked eye, *Jour. Interdisc. Cycle Research* **4**:313-318.

Gleissberg, W., and Damboldt, T., 1979, Reflections on the Maunder Minimum of sunspots, *British Astron. Assoc. Jour.* **89**(5):440-449.

Glock, W. S., 1955, Growth rings and climate, *Bot. Rev.* **21**:73-188.

Gnevyshev, M. N., 1977, Essential features of the eleven-year solar cycle, *Solar Physics* **51**:175-183.

Godden, W., 1912, Is our winter becoming less severe?, *Symons' Meteorol. Mag. (London)* **47**:101.

Godwin, H., 1946, The relationship of bog stratigraphy to climatic change and archeology, *Prehistorical Soc. Proc.* **12**:1-11.

Goekel, A., 1908, Klimaschwankungen, *Natur u. Kultur (Munchen)* **5**:353-357.

Gold, E., 1910, Periodic variation in the velocity of the centers of high and low pressure, *U.S. Dept. Agriculture, Mt. Weather Observatory, Bull. No. 2,* pp. 193-195.

Gold, E., 1913, Secular desiccation of the Earth, *Nature* **92**:435.

Gold, T., 1959, Motions in the magnetosphere of the Earth, *Jour. Geophys. Research* **64**(9):1219-1224.

Gold, T., 1962, Magnetic storms, *Space Sci. Rev.* **1**:100-114.

Goldberg, R. A., 1982, A review of reported relationships linking solar variability to weather and climate, in J. A. Eddy (ed.), *Solar Variability, Weather and Climate,* Washington, D.C.: National Academy Press, pp. 19-32.

Goldreich, P., 1965, An explanation of the frequent occurrence of commensurable mean motions in the solar system, *Royal Astron. Soc. Monthly Notices* **130**:159-181.

Goldreich, P., and Peale, S., 1966, Spin-orbit coupling in the solar system, *Astron. Jour.* **71**:425-438.

Goldthwait, R. P., 1966, Evidence from Alaskan glaciers of major climatic changes, in J. S. Sawyer (ed.), *World Climate from 8000-0* B.C., London: Royal Meteorological Society, pp. 40-53.

Golovanov, Y., 1960, Can climate be changed?, *Priroda,* no. 2, pp. 124-128.

Goodall, W. M., 1939, The solar cycle and the F2 region of the ionosphere, *Inst. Radio Engineers Proc.,* pp. 701-703.

Gorczynski, W., 1904, Sur la diminution de l'intensité du rayonnement solaire en 1902 et 1903, *Acad. Sci. Comptes Rendus* **138**:255-258.

Gordon, A. H.; Byron-Scott, R. A. D.; and Bye, J. A. T., 1982, A note on QBO-SO interaction, the Quasi-Triennial Oscillation and the sunspot cycle, *Jour. Atmos. Sci.* **39**(9):2083-2087.

Gordon, G. A., 1982, Verification of dendroclimatic reconstructions, in M. K. Hughes, P. M. Kelly, J. R. Pilcher, and V. C. LaMarche, Jr. (eds.), *Climate from Tree-Rings,* Cambridge: Cambridge University Press, pp. 58-61.

Gordon, H. B., and Davies, D. R., 1975, Mathematical prediction of climatic change, *Nature* **253**:419-420.

Gornitz, V.; Lebedeff, S.; and Hansen, J., 1982, Global sea level trend in the last century, *Science* **215**:1611-1614.

Gough, D. I., 1973, The geophysical significance of geomagnetic variation anomalies, *Physics Earth and Planetary Interiors* **7**:379.

Goutereau, C., 1904, Sur l'affaiblissement du rayonnement solaire en 1902-1903, *Soc. Meteorol. France Annuaire* **52**:189-195.

Graber, M. A., 1976, Polar motion spectra based upon Doppler, I.P.M.S. and B.I.H. data, *Royal Astron. Soc. Geophys. Jour.* **46**:75-85.

Graham, W. V., 1912*a*, Winters in the seventeenth century, *Symons' Meteorol. Mag. (London)* **47**:30.

Graham, W. V., 1912*b*, The weather in the seventeenth century, *Symons' Meteorol. Mag. (London)* **47**:101-102.

Gravelius, H., 1901, Die jahrliche Periode der Regenmenge zu Marburg a Lahn, *Zeitschr. Gewasserkunde* **4**:99-103.

Gravelius, H., 1902, Methodische-Bemerkung zür Discussion von Periodicitaten in der Klimatologie, *Isis Naturw. Gesell. (Dresden) Sitzungsber.,* January-July, pp. 24-28.

Gravelius, H., 1905, Zur Abhangigkeit des Regenfalls von der Meereshohe. Die Hyetographische Kurve, *Zeitschr. Gewasserkunde* **7**:129-145.

Greaves, W. M. H., and Newton, H. W., 1928, Large magnetic storms and large sunspots, 1874-1927, *Royal Astron. Soc. Monthly Notices* **88**:556-567.

Green, F. H. W., 1975, The February-June weather relationship in north-west Europe, *Nature* **253**:522-523.

Green, J. S. A., 1974, No consensus yet on climate, *Nature* **252**:343.

Gregg, D. P., 1984, A nonlinear solar cycle model with potential for forecasting on a decadal time scale, *Solar Physics* **90:**185-194.

Gregg, W. R., 1908, *Recent Auroral Displays and Magnetic Disturbances,* Washington, D.C.: U.S. Department of Agriculture, Mt. Weather Observatory, Bulletin No. 1, pp. 232-236.

Gregg, W. R., 1909, *Auroral Displays and Magnetic Disturbances at Mount Weather during September, 1908,* Washington, D.C.: U.S. Department of Agriculture, Mt. Weather Observatory, Bulletin No. 2, pp. 12-23.

Gregory, J. W., 1905, The Southern Ocean and its climatic control over Australasia, *Australian Assoc. Advance. Sci. Repts.* **10:**328-347.

Gregory, J. W., 1914, Is the Earth drying up?, *Geog. Jour.* **43:**148-172, 293-313.

Gregory, R. A., 1912, Cycles of the Sun and weather, *Nature* **89:**147-149.

Gregory, Sir R., 1930, Weather recurrences and weather cycles, *Royal Meteorol. Soc. Quart. Jour.* **56**(234):103-120.

Grey, D. C., 1969, Geophysical mechanisms for C14 variations, *Jour. Geophys. Research* **74**(26):6333-6340.

Gribbin, J. R., 1973, Planetary alignments, solar activity and climatic change, *Nature* **246:**453-454.

Gribbin, J. R., 1978*a, What's Wrong with Our Weather? The Climatic Threat of the 21st Century,* New York: Charles Scribner's Sons, 174p.

Gribbin, J. R., ed., 1978*b, Climatic Change,* Cambridge: Cambridge University Press, 280p.

Gribbin, J. R., 1978*c,* Long-term effects, in J. R. Gribbin (ed.), *Climatic Change,* Cambridge: Cambridge University Press, pp. 133-138.

Gribbin, J. R., 1978*d,* The search for cycles, in J. R. Gribbin (ed.), *Climatic Change,* Cambridge: Cambridge University Press, pp. 139-149.

Gribbin, J. R., 1978*e,* Short-term effects, in J. R. Gribbin (ed.), *Climatic Change,* Cambridge: Cambridge University Press, pp. 150-154.

Gribbin, J. R., 1981*a,* Geomagnetism and climate, *New Scientist* **91:**350-353.

Gribbin, J. R., 1981*b,* The Sun, the Moon and the weather, *New Scientist* **91:**754-757.

Gribbin, J. R., 1982, Stand by for bad winters, *New Scientist* **93:**220-223.

Gribbin, J. R., and Lamb, H. H., 1978, Climatic change in historical times, in J. R. Gribbin (ed.), *Climatic Change,* Cambridge: Cambridge University Press, pp. 68-82.

Gribbin, J. R., and Plagemann, S. H., 1973, Discontinuous change in the Earth's spin rate following great solar storm of August 1972, *Nature* **243:**26-27.

Gribbin, J. R., and Plagemann, S. H., 1976, *The Jupiter Effect. The Planets as Triggers of Devastating Earthquakes,* rev. ed., New York: Vintage Books-Random House, 178p.

Gribbin, J. R., and Plagemann, S. H., 1976, *The Jupiter Effect Reconsidered,* New York, Vintage Books, 182p.

Griggs, R., 1937, Timber lines as indicators of climatic trends, *Science* **85:**251-255.

Grobler, J., 1912, Wetter und Krankheit, *Deutsche Rev. Stuttgart* **37:**304-322.

Groissmayr, F., 1927, Die Nilflut und der Folgewinter in Zentraleuropa, *Zeitschr. Meteorologische* **44:**292-296.

Grosse, 1912, Astrometeorologie, *Weltall* **12:**172-177.

Groveman, B. S., and Landsberg, H. E., 1979*a, Reconstruction of Northern Hemisphere Temperature: 1579-1880,* Bethesda: University of Maryland, Meteorological Program, Publication 79-181.

Groveman, B. S., and Landsberg, H. E., 1979*b,* Simulated Northern Hemisphere temperature departures 1579-1880, *Geophys. Research Letters* **6:**767-769.

Groves, G. W., 1971, Dynamics of the Earth-Moon system, in Z. Kopal (ed.), *Physics and Astronomy of the Moon,* 2nd ed. New York: Academic Press, 302p.

Grunsky, C. E., 1927, The improbability of rainfall cycles, *Monthly Weather Rev.* **55**(2):66-68.

Gubbins, D., 1974, Theories of the geomagnetic and solar dynamos, *Rev. Geophysics and Space Physics* **12:**137.

Guinot, B., 1970, Work of the Bureau international de l'heure on the rotation of the Earth, in L. Mansinha, D. E. Smylie, and A. E. Beck (eds.), *Earthquake Displacement Fields and the Rotation of the Earth,* Dordrecht, The Netherlands: D. Reidel, New York: Springer-Verlag, pp. 54-62.

Gunther, J. J., 1834, Witterungslauf vor und wahrend der letzten Erscheinung der Halleyschen Cometen im Jahre 1759, wo er am 13 März seine Sonnennahe erreichte, *Kastner Archiv Naturf.* **26:**31-35.

Gutch, J. W. G., 1842, Some of the remarkable meteorological phenomena that have occurred in the last thousand years, *Quart. Jour. Meteorol.* **1:**134-148.

Hachey, H. B., and McLellan, H. J., 1948, Trends and cycles in surface temperatures of the Canadian Atlantic, *Canadian Fisheries Research Board Jour.* **7:**351-362.

Hale, G. E., 1908, On the probable existence of magnetic fields in sunspots, *Astrophys. Jour.* **28:**315-343.

Hale, G. E., 1914, The Earth and Sun as magnets, *Smithsonian Inst. Rept. for 1913,* pp. 145-158.

Hale, G. E., 1924, Sunspots as magnets and the periodic reversal of their polarity, *Nature* **113:**105.

Hale, G. E., and Nicholson, S. B., 1925, The law of sunspot polarity, *Astrophys. Jour.* **62:**270.

Hale, L. C., and Croskey, C. L., 1979, An auroral effect on the fair weather electrical field, *Nature* **278:**239-241.

Hale, L. C.; Croskey, C. L.; and Mitchell, J. D., 1981, Measurements of middle-atmosphere electrical fields and associated electrical conductivities, *Geophys. Research Letters* **8:**927.

Hall, G. W., 1837, On the connexion of the weather with the tide, *British Assoc. Advance. Sci. Rept. 1836, Notices,* p. 41.

Hall, M., 1906, A method of predicting the movement of tropical cyclones, *Monthly Weather Rev.* **34:**165-167.

Hameed, S., 1984, Fourier analysis of Nile flood levels, *Geophys. Research Letters* **11:**843-845.

Hameed, S., and Cess, C. D., 1983, Impact of a global warming on biospheric sources of methane and its climatic consequences, *Tellus* **35**(B):1-7.

Hameed, S., and Currie, R. G., 1985, An analysis of long term variations in flood levels of the Nile River, in *Proceedings of the Third Conference on Climatic Variations and Symposium on Contemporary Climate,* January 1985, Los Angeles, pp. 68-69.

Hameed, S., and Wyant, P., 1982, Twenty-two year cycle in surface temperatures during the Maunder Minimum, *Geophys. Research Letters* **9**(1):83-86.

Hameed, S.; Yeh, W. M.; Li, M. T.; Cess, R. D.; and Wang, W. C., 1983, An analysis of periodicities in the 1470 to 1974 Beijing precipitation record, *Geophys. Research Letters* **10:**436-439.

Hamilton, W. L., 1973, Tidal cycles of volcanic eruptions: Fortnightly to 19 yearly periods, *Jour. Geophys. Research* **78**(17):3363-3375.

Hammer, C. U.; Clausen, H. B.; and Dansgaard, W., 1980, Greenland ice sheet evidence of post-glacial volcanism and its climatic impact, *Nature* **288:**230-235.

Hammond, A. L., 1973, Research progress on a broad front, earth and planetary science, *Science* **182:**1329-1331.

Handler, P., and Handler, F., 1983, Climatic anomalies in the tropical Pacific Ocean and corn yields in the United States, *Science* **220:**1155-1156.

Hann, J., 1902, Die Schwankungen der Niederschlagsmengen in grosseren Zeiträumen, *Akad. Wiss. Wien Sitzungsber.* **111**(IIa):67-186.

Hann, J., 1903*a*, *Handbook of Climatology.* Part I: *General Climatology,* New York: Macmillan, 437p. (translated with additional references and notes by R. deC. Ward).

Hann, J., 1903*b*, Ergebnisse 43 jahriger Regenmessungen auf der Insel Malta, *Zeitschr. Meteorologische* **20:**180-181.

Hann, J., 1903*b*, Tagliche und jahrliche Periode der Sturme auf dem Ben Nevis, *Zeitschr. Meteorologische* **20:**223-224.

Hann, J., 1904*a*, Die jahrliche Periode der magnetischen Deklination, *Zeitschr. Meteorologische* **21:**129-131.

Hann, J., 1904*b*, Die tagliche und jahrliche Gang der magnetischen Inklination, *Zeitschr. Meteorologische* **21:**131-135.

Hann, J., 1905*a*, Regenfall zu Greenwich 1815-1903, *Zeitschr. Meteorologische* **22:**30-32.

Hann, J., 1905*b*, The anomalies of the weather in Iceland, 1851-1900, and their relation to the simultaneous weather anomalies in North-Western Europe, *Royal Meteorol. Soc. Quart. Jour.* **31:**152-163.

Hansen, R. T., 1959, Recurrent geomagnetic storms and solar prominences, *Jour. Geophys. Research* **64**(1):23-25.

Hanzlik, S., 1930, Die Luftdruckeffekt der Sonnenfleckenperiode. I Mitteilung: Jahresmittel, *Gerlands Beitr. Geophysik* **28:**114-125.

Hanzlik, S., 1936, Der Niederschlagseffekt der Sonnenfleckenperiode, *Gerlands Beitr. Geophysik* **47**(1/2):15-30.

Hanzlik, S., 1937, The Hale double solar-cycle rainfall in western Canada, *Am. Meteorol. Soc. Bull.* **18**(2):60.

Harding, A. F., ed., 1982, *Climatic Change in Later Prehistory,* Edinburgh: Edinburgh University Press, 210p.

Hartman, D. L., 1974, Time spectral analysis of mid-latitude disturbances, *Monthly Weather Rev.* **102:**348-362.

Hartman, W. K., and Larson, S. M., 1967, Angular momenta of planetary bodies, *Icarus* **7:**257-260.

Hartmann, R., 1971, A new representation of the 80-year cycle in sunspot frequency, *Solar Physics* **21:**246-248.

Hassan, F. A., 1981, Historical Nile floods and their implications for climate change, *Science* **212:**1142-1145.

Hassard, A. R. J. F., 1909, Sun-spots, *Popular Astronomy* **17:**120-121.

Hasselman, K., 1976, Stochastic climate models, Part I, Theory, *Tellus* **28**(6):473-485.

Hastenrath, S., and Heller, L., 1977, Dynamics of climatic hazards in north-east Brazil, *Royal Meteorol. Soc. Quart. Jour.* **103:**77-92.

Hastenrath, S., and Kaczmarczyk, E. B., 1981, On spectra and coherence of tropical climate anomalies, *Tellus* **33:**453-462.

Hastenrath, S., and Kutzbach, J. E., 1983, Paleoclimatic estimates from water and energy budgets of East African lakes, *Quat. Research* **19**(2):141-153.

Hastenrath, S., and Rosen, A., 1983, Patterns of India monsoon rainfall anomalies, *Tellus* **35A:**324-331.

Hastenrath, S.; Wu, M.; and Chus, P., 1984, Towards the monitoring and prediction of north-east Brazil droughts, *Royal Meteorol. Soc. Quart. Jour.* **110:**411-425.

Hatcher, D. A., 1984, Simple formulae for Julian day numbers and calendar dates, *Royal Astron. Soc. Quart. Jour.* **25:**53-55.

Haurwitz, B., 1946, Relations between solar activity

and the lower atmosphere, *Am. Geophys. Union Trans.* **27**(2):161-163.

Haurwitz, B., 1964, Atmospheric tides, *Science* **144:**1415-1422.

Haurwitz, M. W., and Brier, G. W., 1981, A critique of the superposed epoch analysis method, *Am. Meteorol. Soc. Bull.* **60:**1344-1345.

Hays, P. B., and Roble, R. G., 1979, A quasi-static model of global atmospheric electricity. I. The lower atmosphere, *Jour. Geophys. Research* **84:**3291-3305.

Heath, D. F., 1973, Space observations of the variability of solar irradiance in the near and far ultraviolet, *Jour. Geophys. Research* **78:**2779-2792.

Heath, D. F.; Krueger, A. J.; and Crutzen, P. J., 1977, Solar proton event: Influence on stratospheric ozone, *Science* **197:**886-889.

Heath, D. F., and Thekaekara, M. P., 1977, The solar spectrum between 1200 and 3000 Å, in O. R. White (ed.), *The Solar Output and Its Variation*, Boulder: Colorado Associated University Press, pp. 193-212.

Hecht, A. D., 1985, *Paleoclimatic Analysis and Modeling*, New York: Wiley-Interscience, 445p.

Hegyfoky, J., 1909, Sonnenflecke und Regen, *Zeitschr. Meteorologische* **26:**228-229.

Heidel, K., 1972, Turbidity trends at Tucson, Arizona, *Science* **177:**882-883.

Heimbrod, G., 1904, Results of harmonic analysis of the diurnal variation at the Cape of Good Hope and at Hobart, *Terr. Magnetism and Atmos. Electricity* **9:**9-14.

Heimbrod, G., 1905, Results of harmonic analysis of the diurnal variation at the Cape of Good Hope and at Hobart, *Terr. Magnetism and Atmos. Electricity* **10:**131-141.

Helland-Hansen, B., 1949, Remarks on some variations in atmosphere and sea, *Geog. Annaler* **31**(1-4):75-82.

Helland-Hansen, B., and Nansen, F., 1920, Temperature variations in the North Atlantic Ocean and in the atmosphere, *Smithsonian Misc. Colln.* **70**(4):408p. (English translation of original published in 1917 in German, in Oslo, Norway.)

Hellmann, G., 1906, *Die Niederschläge in den norddeutschen Strombegieten* (Precipitation in the North German River Basins), 3 vols., Berlin.

Hellmann, G., 1908, Ueber die extremen Schwankungen des Regenfalls, *Zeitschr. Gesell. Erdkunde (Berlin)*, pp. 605-613.

Hellmann, G., 1909, Untersuchungen uber die Schwankungen der Niederschläge, *Konigl. Preussischen Meteorologischen Inst. Berlin, Abh.* **3**(1) (1909-1910), 81p. (Veroff. No. 207).

Henkel, F. W., 1911, Cyclones and the Sun's rotation period, *Symons' Meteorol. Mag. (London)* **45**(1910): 232-235.

Henkel, R., 1972, Evidence for an ultra-long cycle of solar activity, *Solar Physics* **25:**498-499.

Hennig, R., 1904, Katalog bemerkenswerter Witterungsereignisse von den altesten Zeiten bis zum Jahre 1800, *Kongl. Preussischen Meteorologischen Inst. Berlin, Abh.* **2:**93p.

Hennig, R., 1912, Sonnenflecken und Witterung, *Wetter* **29:**20-24.

Henry, A. J., 1902, Average annual precipitation in the United States [and Canada] for the period 1871-1901, *Monthly Weather Rev.* **30:**207-213.

Henry, A. J., 1906, Weather forecasting from synoptic charts, *Franklin Inst. Philadelphia Jour.* **162:** 297-316.

Henry, A. J., 1914, Secular variation of precipitation in the United States, *Am. Geog. Soc. Bull.* **46:**192-201.

Henry, A. J., 1919, The distribution of maximum floods, *Monthly Weather Rev.* **47**(12):861-867.

Henry, A. J., 1921*a*, Seasonal forecasting of precipitation—Pacific coast, *Monthly Weather Rev.* **49:**213-219.

Henry, A. J., 1921*b*, A review of some of the literature on sunspot-pressure relations, *Monthly Weather Rev.* **49:**281-284.

Henry, A. J., 1926, The Bruckner cycle in the United States, *Monthly Weather Rev.* **54:**507.

Henry, A. J., 1927*a*, The Bruckner cycle of climatic oscillations in the United States, *Assoc. Am. Geographers Annals* **17:**60-71.

Henry, A. J., 1927*b*, Abnormal summers in the United States, *Monthly Weather Rev.* **55**(8):349-353.

Hepner, H., 1903, Luftdruck und Wetterprognose, *Weltall* **3:**182-186.

Heppner, J. P., 1977, Empirical models of high-latitude electric fields, *Jour. Geophys. Research* **82:**1115-1125.

Hepworth, M. W. C., 1906, Cold weather and Atlantic ice, *Symons' Meteorol. Mag. (London)* **41:**108.

Hepworth, M. W. C., 1911, The recurrence of warm and cold periods, *Symons' Meteorol. Mag. (London)* **46:**12-13.

Hepworth, M. W. C. 1912, The effect of the Labrador current upon the surface temperature of the North Atlantic; and of the latter upon air temperature and pressure over the British Isles, *British Meteorol. Office, Geophys. Mem. No. 1*, 10p.

Herget, P., 1953, Solar coordinates 1800-2000, *Am. Ephemeris and Naut. Almanac Astron. Papers* **14.**

Herget, P., 1955, Coordinates of Venus 1800-2000, *Am. Ephemeris and Naut. Almanac Astron. Papers* **15**(pt. 3).

Herlineaux, R. H., 1957, On tidal currents and properties of the sea water along the British Columbia coast, *Canada Fisheries Research Board, Pacific Coast Station, Progress Report*, no. 108, pp. 7-9.

Her Majesty's Nautical Office, 1958, *Planetary Coordinates for the Years 1960-1980 Referred to the Equinox of 1950.0*, London: H. M. Stationery Office, 160p (reprinted, corrected ed. 1962; heliocentric coordinates).

Herman, J. R., and Goldberg, R. A., 1978*a*, Initiation of non-tropical thunderstorms by solar activity, *Jour. Atmos. and Terr. Physics* **40:**121-134.

Herman, J. R., and Goldberg, R. A., 1978*b*, *Sun, Weather, and Climate*, Washington, D.C.: U.S. National Aeronautics and Space Administration, NASA SP-426, 350p.

Hermann, E., 1914, System der Einwirkung von Sonne und Mond auf de atmospharischen Vorgange und seine Auswertung, *Annalen Hydrographie u. Maritimen Meteorologie* **42:**121-141.

Herr, R. B., 1978, Solar rotation determined from Thomas Harriot's sunspot observations of 1611 to 1613, *Science* **202:**1079-1081.

Herring, R. S., 1978, Hydrology and chronology: The Rodah Nilometer as an aid in dating interlacustrine history, in J. B. Webster (ed.), *Chronology, Migration and Drought in Interlacustrine Africa*, New York: Africana Publishing Company, Dalhousie University African Studies Series, pp. 39-86.

Herron, M. M., and Herron, S. L., 1983, Past atmospheric environments revealed by polar ice core studies, *Hydrol. Sci. Jour.* **28**(1):139-153.

Herron, T. J.; Tolstoy, I.; and Kraft, D. W., 1969, Atmospheric pressure background fluctuations in the mesoscale range, *Jour. Geophys. Research* **74**(6):1321-1329.

Hesselberg, Th., and Birkeland, B. J., 1940, Sakulare Schwankungen des Klimas von Norwegen. Die Lufttemperatur, *Geofys. Publikasj.* **14**(4).

Hesselberg, Th., and Birkeland, B. J., 1941, Sakulare Schwankungen der Klimat von Norwegen. Der Niederschlag, *Geofys. Publikasj.* **14**(5).

Hesselberg, Th., and Birkeland, B. J., 1943, Sakulare Schwankungen des Klimas von Norwegen. Luftdruck und Wind, *Geofys. Publikasj.* **14**(6).

Heusser, C. J., 1966, Polar hemispheric correlation: Palynological evidence from Chile and the Pacific north-west of America, in J. S. Sawyer (ed.), *World Climate from 8000-0 B.C.*, London: Royal Meteorological Society, pp. 124-141.

Heyworth, A., 1978, Submerged forests around the British Isles: Their dating and relevance as indicators of Post-glacial land and sea level changes, in J. Fletcher (ed.), *Dendrochronology in Europe*, British Archeological Reports, International Series 51, pp. 279-288.

Hibler, W. D., III, and Johnsen, S. J., 1979, The 20-yr cycle in Greenland ice core records, *Nature* **280:**481-483.

Hide, R., 1966, Free hydromagnetic oscillations of the Earth's core and the theory of the geomagnetic secular variation, *Royal Soc. London Philos. Trans.* **259A:**615.

Hide, R., and Roberts, P. H., 1961, The origin of the main geomagnetic field, in L. H. Ahrens, F. Press, K. Rankama, and S. K. Runcorn (eds.), *Physics and Chemistry of the Earth*, vol. 4, New York: Pergamon Press, pp. 27-98.

Hide, R.; Birch, N. T.; Morrison, L. V.; Shea, D. J., and White, A. A., 1980, Atmospheric angular momentum fluctuations and changes in the length of day, *Nature* **286:**114-117.

Hildebrandsson, H. H., 1903, Bericht uber die Errichtung von Observatorien in der Nahe der Aktionscentsren der Atmosphare, in *Bericht des Internationale Meteorologische Komitees*, Versammelung zu St. Petersburg 1899, Berlin, pp. 64-66.

Hildebrandsson, H. H., 1904, Sur la circulation general de l'atmosphere, *British Assoc. Advance. Sci. Repts.* for 1903, pp. 562-565.

Hill, H. A., 1978, Seismic sounding of the Sun, in J. A. Eddy (ed.), *The New Solar Physics*, Boulder, Colo.: Westview Press, pp. 135-214.

Hill, L., 1919, Atmospheric conditions which affect health, *Royal Meteorol. Soc. Quart. Jour.* **45**(191): 189-207.

Hill, T. W., and Wolf, R. A., 1977, Solar-wind interaction, in *The Upper Atmosphere and Magnetosphere*, Washington, D.C.: U.S. National Research Council Geophysical Study Committee, National Academy of Sciences, pp. 25-41.

Hines, C. O., 1965, Wind-induced magnetic fluctuations, *Jour. Geophys. Research* **70**(7):1758-1760.

Hines, C. O., 1974, A possible mechanism for the production of Sun-weather correlations, *Jour. Atmos. Sci.* **31:**589-591.

Hines, C. O., and Halevy, I., 1975, Reality and nature of a Sun-weather correlation, *Nature* **258:**313-314.

Hines, C. O., and Halevy, I., 1977, On the reality and nature of a certain Sun-weather correlation, *Jour. Atmos. Sci.* **34:**382-404.

Hinsdale, G., 1903, Climates and health resorts in the Dominion of Canada, *Balneol. and Climatol. Soc. London Jour.* **7:**245-261.

Hinsdale, G., 1936*a*, Climate and disease with special reference to heat, humidity, sunlight, heliotherapy and seasonal influence. I., *Am. Meteorol. Soc. Bull.* **17**(10):275-284.

Hinsdale, G., 1936*b*, Climate and disease. II. Humidity, *Am. Meteorol. Soc. Bull.* **17**(12):371-374.

Hinsdale, G., 1937*a*, Climate and disease. III. Climate indoors and out (air conditioning); effects of heat and cold; kidney disease; heat and sunstroke, *Am. Meteorol. Soc. Bull.* **18**(2):53-60.

Hinsdale, G., 1937*b*, Climate and disease. IV. Pneumonia; influenza; diarrhoea in children; Asiatic cholera, *Am. Meteorol. Soc. Bull.* **18**(6):226-232.

Hinsdale, G., 1937*c*, Climate and disease. V. Tropical diseases in temperate zones; tropical acclimatization, *Am. Meteorol. Soc. Bull.* **18**(12):383-388.

Hinsdale, G., 1938, Climate and disease. VI. Pain; rheumatism; the aged, *Am. Meteorol. Soc. Bull.* **19**(12):424-430.

Hinsdale, G., 1939, Climate and disease. VII. Climatic stations for tuberculosis, *Am. Meteorol. Soc. Bull.* **20**(12):428-433.

Hinsdale, G., 1940, Climate and disease. VIII. Heliotherapy for tuberculosis, *Am. Meteorol. Soc. Bull.* **21**(12):417-424.

Hirshberg, J., 1973, The solar wind cycle, the sunspot cycle, and the corona, *Astrophysics and Space Sci.* **20:**473-481.

Hirschboek, K. K., 1980, A new worldwide chronology of volcanic eruptions, *Palaeogeography, Palaeoecology, and Palaeoclimatology* **29:**223-241.

Hissink, C. W., 1909, Zonnevlekken en temperatuur (Sunspots and temperature), *Hemel en Dampkring's Gravenhage* **7:**7-8.

Hoinkes, H. C., 1968, Glacier variations and weather, *Jour. Glaciology* **7:**3-19.

Holmes, R. L., 1907, Phenomenal rainfall in Suva Fiji, August 8, 1906, *Royal Meteorol. Soc. Quart. Jour.* **33:**201-205.

Holton, J. R., 1968, A note on the propagation of the biennial oscillation, *Jour. Atmos. Sci.* **25:**519-521.

Holton, J. R., 1979, *An Introduction to Dynamic Meteorology*, 2nd ed., New York: Academic Press, 391p.

Holton, J. R., 1982, The nature and origin of weather variability, in J. A. Eddy (ed.), *Solar Variability, Weather, and Climate*, Washington, D.C.: National Academy Press, pp. 48-52.

Holton, J. R., and Lindzen, R. S., 1972, An updated theory for the quasibiennial cycle in the tropical stratosphere, *Jour. Atmos. Sci.* **29:**1076-1080.

Holton, J. R., and Tan, H. C., 1979, The influence of the equatorial quasibiennial oscillation on the global circulation at 50 mb, *Jour. Atmos. Sci.* **37:**2200-2208.

Holtzworth, R. H., 1980, High latitude stratospheric electrical measurements in fair and foul weather under various solar conditions, *Jour. Geophys. Research* **85:**2200-2208.

Holtzworth, R. H., and Mozer, F. S., 1979, Direct evidence of solar flare modification of stratospheric electrical fields, *Jour. Geophys. Research* **84:**363-367.

Hones, E. H., Jr., 1979, Solar wind-magnetosphere-ionosphere coupling, in B. M. McCormac and T. A. Seliga (eds.), *Solar-terrestrial Influences on Weather and Climate*, Dordrecht, The Netherlands: D. Reidel, pp. 83-100.

Hooker, R. H., 1922, The weather and the crops in eastern England, 1885-1921, *Royal Meteorol. Soc. Quart. Jour.* **48**(202):115-138.

Hope-Simpson, R. E., 1978, Sunspots and flu: A correlation, *Nature* **275:**86.

Horton, R. E., 1899, Report on the run-off and water power of Kalamazoo River, in A. C. Lane (ed.), *Water Resources of the Lower Peninsula of Michigan*, Washington, D.C.: U.S. Geological Survey Water-Supply and Irrigation Paper No. 30, pp. 22-38.

Horton, W. H., 1903, The Sun's apparent electrical influence on the Earth, *Western Electrician (Chicago)* **33:**275.

Hosmark, G., 1925, Solar radiation and the weekly weather forecast of the Argentine Meteorological Service, *Smithsonian Misc. Colln.* **77**(7).

Houben, H.; Gierasch, P. J.; and Turcotte, D. L., 1975, Can the geomagnetic dynamo be driven by the semidiurnal tides? (abstract), *EOS* **56:**356.

Houghton, D. M., 1958, Heat sources and sinks at the Earth's surface, *Meteorol. Mag.* **67:**132-143.

Houghton, H. G., 1954, On the annual heat balance of the northern hemisphere, *Jour. Meteorology* **11:**1-9.

Hovmoller, E., 1947, Climate and weather over the coast-land of northeast Greenland and the adjacent sea, *Medd. Grönland* **144**(1):Appendix 1.

Howard, Robert, ed., 1970, *Solar Magnetic Fields*, International Astronomic Union Symposium No. 43, Dordrecht, The Netherlands: D. Reidel, 782p.

Howard, R., 1981, Global velocity fields of the Sun and the activity cycle, *Am. Scientist* **69**(1):28-36.

Hoyanagi, M., 1965, Sand-buried ruins and shrinkage of rivers along the Old Silk Road region in the Tarim Basin, *Tokyo Geog. Soc. Jour. Geography* **74:**1-12, 55-75.

Hoyt, D. V., 1979*a*, Variations in sunspot structure and climate, *Clim. Change* **2:**79-92.

Hoyt, D. V., 1979*b*, Smithsonian Astrophysical Observatory solar constant program, *Rev. Geophysics and Space Physics* **17:**427-458.

Hoyt, W. G., and others, 1936, *Studies of Relations of Rainfall and Run-off in the United States*, U.S. Geological Survey Water-Supply Paper No. 772, 301p.

Hudson, H. S., 1983, Variations of the solar radiation input, in B. M. McCormac (ed.), *Weather and Climatic Responses to Solar Variations*, Boulder: Colorado Associated University Press, pp. 31-41.

Hughes, D. W., 1980, Solar size variation, *Nature* **286:**439-440.

Hughes, M. K.; Kelly, P. M.; Pilcher, J. R.; and LaMarche, V. C., Jr., eds., 1982, *Climate from Tree Rings*, Cambridge: Cambridge University Press, 223p.

Hughes, O. L., 1965, Surficial geology of part of the Cochrane District, Ontario, *Geol. Soc. America Spec. Paper 84*, pp. 535-565.

Humphreys, W. J., 1910, Solar disturbances and terrestrial temperatures, *Astrophys. Jour.* **32**(2):97-111.

Humphreys, W. J., 1913, Volcanic dust and other factors in the production of climatic changes, and their possible relation to ice ages, *Mount Weather Observatory Bull. No. 6*, pp. 1-34.

Humphreys, W. J., 1926, *Rain Making and Other Weather Vagaries*, Baltimore, Md.: Williams and Wilkins, 157p.

Humphreys, W. J., 1931, How droughts occur, *Am. Meteorol. Soc. Bull.*

Humphreys, W. J., 1937, The greenhouse effect of volcanic dust, *Monthly Weather Rev.* **65**(7):261-262.

Hundhausen, A. J., 1972, *Coronal Expansion and the Solar Wind,* Berlin: Springer-Verlag, 238p.

Hundhausen, A. J., 1979, Solar activity and the solar wind, *Rev. Geophysics and Space Physics* **17:**2034-2048.

Hunt, B. G., 1978, Atmospheric vacillations in a general circulation model I: The large-scale energy cycle, *Jour. Atmos. Sci.* **35:**1133.

Hunt, B. G., 1981, An evaluation of a Sun-weather mechanism using a general circulation model of the atmosphere, *Jour. Geophys. Research* **86**(C2):1233-1245.

Hunt, H. A., 1919, A basis for seasonal forecasting in Australia, *Royal Meteorol. Soc. Quart. Jour.* **55**(232):323-334.

Huntington, E., 1908, Coincident activities of the Earth and the Sun, *Popular Sci. Monthly* **72:**492-502.

Huntington, E., 1914*a, The Climatic Factor as Illustrated in Arid America* (with contributions by Charles Schuchert, A. E. Douglass, and C. J. Kullmer), Washington, D.C.: Carnegie Institution of Washington Publication No. 192, 341p.

Huntington, E., 1914*b,* The solar hypothesis of climatic changes, *Geol. Soc. America Bull.* **25:**477-590.

Huntington, E., 1917, Solar activity, cyclonic storms, and climatic changes, *2nd Pan-Am. Sci. Congr., 2nd, Proc.* (series II) **2:**411-432.

Huntington, E., 1918*a,* Solar disturbances and terrestrial weather: I. Extreme barometric gradients compared with sun spots, *Monthly Weather Rev.* **46:**123-141.

Huntington, E., 1918*b,* Solar disturbances and terrestrial weather: II. Sun spots compared with changes in the weather, *Monthly Weather Rev.* **46:**168-177.

Huntington, E., 1918*c,* Solar disturbances and terrestrial weather: III. Faculae and the solar constant compared with barometric gradient, *Monthly Weather Rev.* **46:**269-277.

Huntington, E., 1920, The control of pneumonia and influenza by the weather, *Ecology* **1:**6-23 (review by J. B. Kincer in *Monthly Weather Rev.* **48**(9):501-505).

Huntington, E., 1923*a, Earth and Sun. A Hypothesis of Weather and Sunspots,* New Haven, Conn.: Yale University Press, 296p.

Huntington, E., 1923*b,* Cycles of health. Report of a conference on cycles, *Geog. Rev.* **13**(4, Spec. Supp.):662-664.

Huntington, E., 1924, *Civilization and Climate,* 3rd ed., New Haven, Conn.: Yale University Press, 453p (reprinted 1977, Hamden, Conn.: Archon Books).

Huntington, E., and Visher, S. S., 1922, *Climatic Changes,* New Haven, Conn.: Yale University Press, 329p.

Huntington, E., and others, 1930, *Weather and Health: A Study of Daily Mortality in New York City,* Washington, D.C.: U.S. National Research Council Bulletin 75.

Hustich, I., 1947, On variations in climate, in crop of cereals and in growth of pine in northern Finland, 1890-1931, *Fennia* **70:**2.

Hustich, I., 1948, The Scotch Pine in northernmost Finland and its dependence on the climate in the last decades, *Acta Bot. Fennica* **42.**

Hustich, I., 1949, On the correlation between growth and the recent climatic fluctuation, *Geog. Annaler* **31**(1-4):90-105.

Hustich, I., 1956, Correlation of tree-ring chronologies of Alaska, Labrador and northern Europe, *Acta Geog.* **15:**1-26.

Imbrie, J., and Imbrie, K. P., 1979, *Ice Ages: Solving the Mystery,* New York: Macmillan, and Short Hills, N.J.: Enslow Publications, 224p.

Ingram, M. J.; Underhill, D. J.; and Wigley, T. M. L., 1978, Historical climatology, *Nature* **276:**329-334.

Intriligator, D. C., 1974, Evidence of solar-cycle variations in the solar wind, *Astrophys. Jour.* **188:**L23-L26.

Israel, H., 1973, *Atmospheric Electricity,* vols. 1 and 2 (trans. from German), Jerusalem: Israel Program for Scientific Translation.

Ives, R. L., 1948, Recent climatic fluctuations in the Great Basin region of the United States, *Weather* **3:**374.

Jacobs, W. C., 1951, The energy exchange between the sea and the atmosphere and some of its consequences, *Scripps Inst. Oceanography Bull.* **6**(2):27-122.

Jacques-Felix, H., 1947, La vie et la mort du Lac Tschad, *[France] Outre-Mer Bull. Agriculture* **3** (review in *Geog. Rev.* January 1949).

Jagannathan, P., and Bhalme, H. N., 1973, Changes in pattern of distribution of southwest monsoon rainfall over India associated with sunspots, *Monthly Weather Rev.* **101**(9):691-700.

Jagannathan, P., and Parthasarathy, B., 1973, Trends and periodicities of rainfall over India, *Indian Jour. Meteorology, Hydrology, and Geophysics* **4:**371-375.

Jansen, E., and Bjorklund, K. R., 1985, Surface ocean circulation in the Norwegian Sea 15,000 B.P. to present, *Boreas* **14**(3):243-258.

Jardine, W. G., 1975, Chronology of Holocene marine transgression and regression in south-western Scotland, *Boreas* **4:**173-196.

Jeffreys, Sir H., 1954, Dynamics of the Earth-Moon system, in G. P. Kuiper (ed.), *The Earth as a Planet,* Chicago, Ill.: University of Chicago Press, pp. 42-56.

Jenkins, G. M., and Watts, D. G., 1968, *Spectral*

Analysis and Its Applications, San Francisco, Calif.: Holden-Day, 525p.

Johansson, O. V., 1931, Die Hauptcharakeristika des jahrlichen Temperaturganges, *Gerlands Beitr. Geophysik* **33:**406-428.

Johnsen, S. J.; Dansgaard, W.; and Clausen, H. B., 1970, Climatic oscillations 1200-2000 A.D., *Nature* **277:**482-483.

Johnson, M. O., 1946, *Correlation of Cycles in Weather, Solar Activity, Geomagnetic Values, and Planetary Configurations,* San Francisco, Calif.: Phillips and Van Orden, 149p.

Johnston, M. J., and Mauk, F. J., 1972, Earth tides and the triggering of eruptions from Mt. Stromboli, Italy, *Nature* **239:**266-267.

Jones, H. S., compiler, 1955, *Sunspots and Geomagnetic-Storm Data Derived from Greenwich Observations, 1874-1954,* London: H.M. Stationery Office.

Jones, W. B., 1984, *The Solar System,* Oxford: Pergamon Press, 336p.

Joos, M., 1982, Swiss midland-lakes and climatic changes, in A. F. Harding (ed.), *Climatic Change in Later Prehistory,* Edinburgh: Edinburgh University Press, pp. 44-51.

Jose, P. D., 1965, Sun's motion and sunspots, *Astron. Jour.* **70**(1):193-200.

Julian, P. R., and Chervin, R. M., 1978, A study of the Southern Oscillation and Walker circulation phenomenon, *Monthly Weather Rev.* **106:**1433-1451.

Julian, P. R.; Washington, W. M.; Hembree, L.; and Ridley, C., 1970, On the spectral distribution of large-scale atmospheric kinetic energy, *Jour. Atmos. Sci.* **27:**376-387.

Jurdy, D. M., and Van der Voo, R., 1975, True polar wander since the Early Cretaceous, *Science* **187:**1193-1196.

Kahle, A. B.; Ball, R. H.; and Cain, J. C., 1969, Prediction of geomagnetic secular change confirmed, *Nature* **223:**165.

Kals, W. S., 1982, *Your Health, Your Moods, and the Weather,* New York: Doubleday.

Kanda, S., 1933, Ancient records of sunspots and auroras in the Far East and the variation of the period of solar activity, *Imperial Acad. Japan Proc.* **9:**293-296.

Karlén, W., 1973, Holocene glacier and climatic variations, Kebnekajse Mountains, Swedish Lapland, *Geog. Annaler* **55A:**29-63.

Karlén, W., 1976, *Holocene Climatic Fluctuations Indicated by Glacier and Tree-Limit Variations in Northern Sweden,* Forskningsrapport 23, Stockholm University, Naturgeografiska Institutionen, 9p.

Karlén, W., ed., 1982, Holocene glacial fluctuations in Scandinavia, *Striae* **18:**26-34.

Karlstrom, T. N. V., 1961, The glacial history of Alaska: Its bearing on paleoclimatic theory, *New York Acad. Sci. Annals* **95**(article 1):290-340.

Karlstrom, T. N. V., 1976, Quaternary and Upper Tertiary time-stratigraphy of the Colorado Plateau, continental correlation, and some paleoclimatic implications, in W. C. Mahaney (ed.), *Quaternary Stratigraphy of North America,* Stroudsburg, Penn.: Dowden, Hutchinson and Ross, pp. 275-282.

Kasemir, H. W., 1972, Atmospheric electric measurements in the Arctic and Antarctic, *Pure and Appl. Geophysics* **100:**70-80.

Kasemir, H. W., 1977, Theoretical problems of the global electric circuit, in H. Dolezalek and R. Reiter (eds.), *Electrical Processes in Atmospheres,* Darmstadt, West Germany: Steinkopff, pp. 423-439.

Kassatrine, T. J., 1928, The microsynoptic structure of rainfall, *Royal Meteorol. Soc. Quart. Jour.* **54:**227.

Kassner, C., 1906, Mond und Gewitter, *Wetter* **23:**237-238.

Katz, E. J., and Garzoli, S., 1982, Response of the western Equatorial Atlantic Ocean to an annual wind cycle, *Jour. Marine Research* **40**(Supp.):307-327.

Kaula, W. M., 1968, *An Introduction to Planetary Physics: The Terrestrial Planets,* New York: John Wiley & Sons, 490p.

Kawai, N., 1972, The magnetic control on the climate in the geologic time, *Japan Acad. Sci. Proc.* **48**(9):687-689.

Kayer, E., 1913, *Ueber die Arrheniussche Theorie der Eiszeit,* Stuttgart: Centralblatt fur Mineralogie, Geologie, und Palaontologie, pp. 769-771.

Keating, G. M., 1978, Relation between monthly variations in global ozone and solar activity, *Nature* **274:**873-874.

Keating, G. M., 1981, The response of ozone to solar activity variations: A review, *Solar Physics* **74:**321-347.

Keen, F. P., 1937, Climatic cycles in eastern Oregon as indicated by tree rings, *Monthly Weather Rev.* **65**(5):175-188.

Keith, L. B., 1963, *Wildlife's Ten-Year Cycle,* Madison: University of Wisconsin Press, 201p.

Kelley, J. B., 1956, Heat, cold and clothing, *Sci. American* **194**(2):109-116.

Kellogg, W. W., and Schneider, S. H., 1974, Climate stabilization: For better or for worse?, *Science* **186:**1163-1172.

Kelly, P. M., 1979, Solar influence on North Atlantic mean sea level pressure, in B. M. McCormac and T. A. Seliga (eds.), *Solar-Terrestrial Influences on Weather and Climate,* Dordrecht, The Netherlands: D. Reidel, pp. 297-298.

Kelly, P. M., and Jones, P. D., 1983, The evidence for cycles on solar time scales in climate data, in B. M. McCormac (ed.), *Weather and Climate Response to Solar Variation,* Boulder: Colorado Associated University Press, pp. 581-590.

Kelly, P. M.; Jones, P. D.; Sear, C. B.; Cherry, B. S. G.; and Tavakol, R. K., 1982, Variations in surface air temperatures: Part 2. Arctic regions, 1881-1980, *Monthly Weather Rev.* **110:**71-83.

Kelvin, Lord, 1899, The age of the Earth as an abode fitted for life, *Philos. Mag.* **47:**66.

Kempe, S., 1977, Hydrographie, Warven-Chronologie und organische Geochimie des Van Sees, Ost-Turkei, *Geologische u. Palaontologische Inst. (Hamburg University) Mitt.* **47:**125-228.

Kempe, S., and Degens, E. T., 1978, Lake Van varve record: The past 10,420 years, in E. T. Degens and F. Kurtman (eds.), *The Geology of Lake Van*, Ankara, Turkey: Maden Tetkik ve Arama, pp. 56-63.

Kempe, S., and Degens, E. T., 1979, Varves in the Black Sea and in Lake Van (Turkey), in C. Schluchter (ed.), *Moraines and Varves*, Rotterdam, The Netherlands: A. A. Balkema, pp. 309-318.

Kennett, J. P., and Watkins, N. D., 1970, Geomagnetic polarity change, volcanic maxima and faunal extinction in the South Pacific, *Nature* **227:**930-934.

Kepler, J., 1619, *Harmonices mundi, libri V*, Austriae: Lincii.

Kerner, F., 1905, Thermoisodromen; Versuch einer kartographischen Darstellung des jahrlichen Ganges der Lufttemperatur, *Koninkl. Geog. Gesell. Wien* **6**(3).

Kerr, R. A., 1978, Climate control: How large a role for orbital variations? *Science* **201:**144-146.

Kesslitz, W., 1902, Magnetische Storung in Pola wahrend der Eruption des Mont Pelee am 8. Mai 1902, *Zeitschr. Meteorologische* **19:**316-317.

Khromov, S. P., 1973, Solar cycles and climate, *Meteorology and Hydrology* **9:**90-124 (in Russian in *Meteorol. Gidrol,* **14:**90-112).

Kidson, E., 1925, Some periods in Australian weather, *Australia Bur. Meteorology Bull. No. 17.*

Kilston, S., and Knopoff, L., 1983, Lunar-solar periodicities of large earthquakes in southern California, *Nature* **304:**21-25.

Kimball, H. H., 1901, Sun-spots and the weather, *Monthly Weather Rev.* **29:**248-249.

Kimball, H. H., 1928, Amount of solar radiation that reaches the surface of the Earth on the land and on the sea, and methods by which it is measured, *Monthly Weather Rev.* **56:**393-398.

Kincer, J. B., 1931, The drought of 1930, *Am. Meteorol. Soc. Bull.* **12**(1):14-17.

Kincer, J. B., 1937, *Is the Climate Changing?*, Springfield, Ill.

Kincer, J. B., 1940, Relation of recent glacier recessions to prevailing temperatures, *Monthly Weather Rev.* **68**(6):158-160.

Kincer, J. B., 1941*a*, Climate and weather data in the United States, in *Climate and Man*, Washington, D.C.: U.S. Department of Agriculture, Yearbook.

Kincer, J. B., 1941*b*, Some pressure-precipitation trend relations, *Monthly Weather Rev.* **69:**232-235.

Kincer, J. B., 1946, Our changing climate, *Am. Geophys. Union Trans.* **27:**342-347.

King, J. W., 1973, Solar radiation changes and the weather, *Nature* **245:**443-446.

King, J. W., 1974, Weather and the Earth's magnetic field, *Nature* **247:**131-134.

King, J. W., 1975, Sun-weather relationships, *Aeronautics and Astronautics* **13**(4):10-19.

King, J. W.; Hurst, E.; Slater, A. J.; Smith, P. A.; and Tamkin, B., 1974, Agriculture and sunspots, *Nature* **252:**2-3.

King, J. W.; Slater, A. J.; Stevens, A. D.; Smith, P. A.; and Willis, D. M., 1977, Large-amplitude stationary planetary waves induced in the troposphere by the Sun, *Jour. Atmos. and Terr. Physics* **39:**1357-1367.

King, J. W., and Willis, D. M., 1974, Magnetometeorology: Relationships between the weather and the Earth's magnetic field, in W. R. Bandeen and S. P. Maran (eds.), *Symposium on Possible Relationships between Solar Activity and Meteorological Phenomena Proceedings*, Greenbelt, Md.: NASA, Goddard Space Flight Center, pp. 52-55.

King, J. W.; Slater, A. J.; Stevens, A. D.; Smith, P. A.; and Willis, D. M., 1977, Large-amplitude standing planetary waves induced in the troposphere by the Sun, *Jour. Atmos. and Terr. Physics* **39:**1357-1367.

King-Hele, D. G., 1966, Predictions of the dates and intensities of the next two sunspot maxima, *Nature* **209:**285-286.

Kingsmill, T. W., 1906, A 300-year climatic and solar cycle, *Nature* **73:**413-414.

Kirk, B. L., and Rust, B. W., 1983, The solar cycle effect on atmospheric carbon dioxide levels, in B. M. McCormac (ed.), *Weather and Climate Response to Solar Variation*, Boulder: Colorado Associated University Press, pp. 129-136.

Klages, K. H., 1930, Geographical distribution of variability in the yields of field crops in the states of the Mississippi Valley, *Ecology* **11**(2).

Klein, W. H., 1957, *Principal Tracks and Mean Frequencies of Cyclones and Anticyclones in the Northern Hemisphere*, Washington, D.C.: U.S. Weather Bureau, Research Paper No. 40, 60p.

Klejmenova, E. P., 1967, On the variation of thunderstorm activity in the solar cycle, *Leningrad Glav. Uprav. Gidromet. Sluzb., Meteorol. Gidrol, no. 8,* pp. 64-68 (in Russian).

Knight, J. W., and Sturrock, P. A., 1976, Solar activity geomagnetic field and terrestrial weather, *Nature* **264:**239-240.

Knopoff, L., 1964, Earth tides as a triggering mechanism for earthquakes, *Seismol. Soc. America Bull.* **54:**1865-1870.

Knopoff, L., 1970, Correlation of earthquakes with lunar orbital motions, *The Moon* **2:**140-143.

Knott, C. G., 1901, Solar radiation and Earth temperatures, *Royal Soc. Edinburgh Proc.* **23:**296-311.

Koch, L., 1945, The East Greenland ice, *Medd. Grönland* **130**(3):374p.

Komintz, M. A., and Pisias, N. G., 1979, Pleistocene climate: Deterministic or stochastic?, *Science* **204:**171-173.

Kondratyev, K. Y., 1969, *Radiation in the Atmosphere,* New York: Academic Press, 912p.

Kondratyev, K. Y., 1985, *Changes in Global Climate: A Study of the Effect of Radiation and Other Factors during the Present Century,* Rotterdam, The Netherlands: Balkema, 288p.

Kondratyev, K. Y., and Nikolsky, G. A., 1970, Solar radiation and solar activity, *Royal Meteorol. Soc. Quart. Jour.* **96:**509-522.

Kondratyev, K. Y., and Nikolsky, G. A., 1979, The stratospheric mechanism of solar and anthropogenic influences on climate, in B. M. McCormac and T. A. Seliga (eds.), *Solar-Terrestrial Influences on Weather and Climate,* Dordrecht, The Netherlands: D. Reidel, pp. 317-322.

Kono, M., 1971, Intensity of the Earth's magnetic field during the Pliocene and Pleistocene in relation to the amplitude of Mid-Ocean Ridge magnetic anomalies, *Earth and Planetary Sci. Letters* **11:**10-17.

Koppe, H., 1952, Solareaktivitat und Atmosphare, *Zeitschr. Meteorologie* **6**(12):369-378.

Köppen, W., 1873, Uber mehrjahrige Perioden der Witterung, inbesondere uber die 11-jahrige Periode der Temperature, *Zeitschr. Meteorologische* **8:**241-248, 257-267.

Köppen, W., 1901, Ueber Periodicitat in meteorologischen Zahlenreihen, *Annalen Hydrographie Maritimen Meteorologie,* **29:**135-136.

Köppen, W., 1914, Lufttemperaturen, Sonnenflecken, und Vulkanausbrucke, *Zeitschr. Meteorologische* **31:**305-328.

Köppen, W., 1918, Eine 89 jahrige Periode in der Witterung, *Zeitschr. Meteorologische* **35:**98.

Köppen, W., and Wegener, A., 1924, *Die Klimate der geologischen Vorzeit,* Berlin: Gebrudern Borntrager, 256p.

Kornblum, J. J., 1969, Concentration and collection of meteoric dust in the atmosphere, *Jour. Geophys. Research* **74**(8):1908-1919.

Kosters, J. J., and Murcray, D. G., 1979, Change in the solar constant between 1868 and 1978, *Geophys. Research Letters* **6:**382-384.

Kozlowski, T. T., ed., 1962, *Tree Growth,* New York: Ronald Press, 442p.

Kozlowski, T. T., 1979, *Tree Growth and Environmental Stresses,* Seattle: University of Washington Press, 192p.

Kraus, E. B., 1954, Secular changes in the rainfall regime of S. E. Australia, *Royal Meteorol. Soc. Quart. Jour.* **80:**591-601.

Kraus, E. B., 1955*a,* Secular changes of tropical rainfall regimes, *Royal Meteorol. Soc. Quart. Jour.* **81:**198-210.

Kraus, E. B., 1955*b,* Secular changes of east coast rain-fall regimes, *Royal Meteorol. Soc. Quart. Jour.* **81:**430-439.

Kraus, E. B., and Hanson, H. P., eds., 1982, International symposium on climatic variability in the oceans, *Prog. Oceanography* **11**(2):61-218.

Krause, E. H. L., 1910, Die veranderungen des Klimas seit der letzten Eiszeit, *Deutsch. Geol. Gesell. Zeitschr.* **62:**123-128.

Krause, E. H. L., 1913, Das europaische Klima im lezten vorchristlichen Jahrtausend, *Naturw. Wochenschr.* **28:**689-693.

Krebs, W., 1904*a,* Sonnenflecken und erdmagnetische Ungewitter im Jahre 1903, *Weltall* **4:**362-367.

Krebs, W., 1904*b,* Sonnenflecken und erdmagnetische Ungewitter im Jahre 1903, *Weltall* **5:**99-101.

Krebs, W., 1907*a,* Atmospheric see-saw phenomenon and the occurrence of typhoon storms, *Nature* **75:**560.

Krebs, W., 1907*b,* Qualitative analysis of curve diagrams, *Symons' Meteorol. Mag. (London)* **42:**27-28.

Krebs, W., 1908, Der Mond als Sonnenuhr zur Zeitbestimmung fur Erdkatastrophen, *Weltall* **8:**105-112.

Krebs, W., 1909, Sonnentatigkeit und Witterung: *Natur u. Kultur* **6:**606-608, 637-639, 667-669, 700-701, 735-736; **7:**31-32, 92-93, 125-126, 189-191.

Krebs, W., 1910, Sonnentatigkeit und Witterung, *Natur u. Kultur* **7:**349-350, 382-383, 412-414, 445-447, 476-477, 509-511, 540-542, 573-574, 604-607, 635-638, 669-670, 701-703, 732-734, 766-767; **8:**25-37, 88-91.

Krebs, W., 1911, Sonnentatigkeit und Witterung, *Natur u. Kultur* **8:**154-156, 211-221, 347-348, 411, 474-477, 539-541, 603-605, 664-666; **9:**55-58, 91-93, 153-154.

Kritzinger, H. H., 1924, Grundlagen der periodischen Schwankungen der Sonnenflecken und des Klimas, *Zeitschr. Meteorologische* **41**(1):21-23.

Krogdahl, W. S., 1962, *The Astronomical Universe,* 2nd ed., New York: Macmillan, 585p.

Kruger, G., 1910, Sonnenflecken und Witterung, *Naturw. Rundschau* **25:**301-303.

Kuhn, F. R., 1927, Mehrjahrige periodische Schwankungen des Luftdruckes in Ost- und Nordeuropa, ein Beitrag zur Erforschung der Periodizitat klimatischer Elemente, *Zeitschr. Meteorologische* **44:**307-308.

Kuliyeva, R. N., 1975*a,* Relation between the sector structure of the interplanetary magnetic field and zonal circulation indexes, *Geomagnetism and Aeronomy* **15:**278-280.

Kuliyeva, R. N., 1975*b,* Effect of the sector structure of the interplanetary magnetic field on atmospheric circulation, *Geomagnetism and Aeronomy* **15:**438.

Kullmer, C. J., 1914, The shift of the storm track, in E. Huntington (ed.), *The Climatic Factor as Illustrated in Arid America,* Washington, D.C.: Carnegie Institution of Washington, Publication No. 192, pp. 193-205.

Kullmer, C. J., 1933, The latitude shift of the storm track in the 11-year solar period, *Smithsonian Misc. Colln.* **89**(2):34p.

Kullmer, C. J., 1943, A remarkable reversal in the distribution of storm frequency in the United States in the double Hale solar cycles, of interest in long-range forecasting, *Smithsonian Misc. Colln.* **103**(60):20p.

Kupetskiy, V. N., 1970, Trend in recent marine and continental glacierization in relation to solar activity, *Sov. Hydrol. Sel. Pap. No. 3* (1969), pp. 304-308.

Kurth, R., 1959, *Introduction to the Mechanics of the Solar System,* New York: Pergamon Press, 177p.

Kutzbach, J. E., 1976, The nature of climate and climatic variations, *Quat. Research* **6:**471-480.

Kutzbach, J. E., 1981, Monsoon climate of the early Holocene: Climate experiment with the Earth's orbital parameters for 5000 years, *Science* **214:**59-61.

Kutzbach, J. E., 1983, Monsoon rains of the late Pleistocene and early Holocene: Patterns, intensity, and possible causes of changes, in A. Street-Perrott, M. Beran, and R. Ratcliffe (eds.), *Variations in the Global Water Budget,* Dordrecht, The Netherlands: D. Reidel, pp. 371-389.

Kutzbach, J. E., and Bryson, R. A., 1974, Variance spectrum of the Holocene climatic fluctuations in the North Atlantic sector, *Jour. Atmos. Sci.* **31:**1958-1963.

Kutzbach, J. E.; Bryson, R. A.; and Shen, W. C., 1968, An evaluation of the thermal Rossby number in the Pleistocene, *Am. Meteorol. Soc. Meteorol. Mons.* **8**(30):134-138.

LaBonte, B. J., and Howard, R., 1981, Measurement of solar radi change, *Science* **214:**907-909.

Labrijn, A., 1945, Het Klimaat van Nederland gedurende de laast twee an en half Eeuw, *Koninkl. Nederlands Met. Inst., s'Gravenhage Meded en Verh., No. 49,* 114p.

LaMarche, V. C., Jr., 1974, Paleoclimatic inferences from long tree-ring records, *Science* **183:**1043-1048.

LaMarche, V. C., Jr., 1978, Tree-ring evidence of past climatic variability, *Nature* **276:**334-338.

LaMarche, V. C., Jr., and Fritts, H. C., 1971, Tree rings, glacier advance and climate in the Alps, *Zeitschr. Gletscherkunde u. Glazialgeologie* **7:**125-131.

LaMarche, V. C., Jr., and Hirschboeck, K. K., 1984, Frost rings in trees as records of major volcanic eruptions, *Nature* **307:**121-126.

Lamb, H., 1911, On atmospheric oscillations, *Royal Soc. [London] Proc.* **84A:**551-572.

Lamb, H. H., 1965, The early medieval warm epoch and its sequel, *Palaeogeography, Palaeoclimatology, and Palaeoecology* **1:**13-37.

Lamb, H. H., 1967, Britain's changing climate, *Geog. Jour.* **133:**445-468.

Lamb, H. H., 1970, Volcanic dust in the atmosphere; with a chronology and assessment of its meteorological significance, *Royal Soc. London Philos. Trans.* **266A:**425-533.

Lamb, H. H., 1972, *Climate Present, Past and Future.* Vol. 1: *Fundamentals and Climate Now,* London: Methuen and Company, Ltd.; New York: Barnes and Noble Books, 613p.

Lamb, H. H., 1974*a*, Fluctuations in climate, *Nature* **251:**568.

Lamb, H. H., 1974*b*, Climate, vegetation and forest limits, *Royal Soc. London Philos. Trans.* **276A:**195-230.

Lamb, H. H., 1977, *Climate Present, Past and Future.* Vol. 2: *Climate History and the Future,* London: Methuen and Company, Ltd.; New York: Barnes and Noble Books, 835p.

Lamb, H. H., 1979, Climatic variation and changes in the wind and ocean circulation: The Little Ice Age in the northeast Atlantic, *Quat. Research* **11**(1):1-20.

Lamb, H. H., 1982*a*, *Climate, History and the Modern World,* London and New York: Methuen and Company, 387p.

Lamb, H. H., 1982*b*, Reconstruction of the course of postglacial climate over the world, in A. F. Harding (ed.), *Climate Change in Later Prehistory,* Edinburgh: Edinburgh University Press, pp. 11-32.

Lamb, H. H., and Johnson, A. I., 1959, Climatic variation and observed changes in the general circulation, Parts I and II, *Geog. Annaler* **41:**94-134.

Lamb, H. H., and Johnson, A. I., 1961, Climatic variation and observed changes in the general circulation, Part III. Investigation of long series of observations and circulation changes in July, *Geog. Annaler* **43**(3-4):363-400.

Lamb, H. H., and Johnson, A. I., 1966, Secular variations of the atmospheric circulation since 1750, *H. M. Meteorol. Off. Geophys. Mem. No. 110.*

Lamb, H. H.; Lewis, R. P. W.; and Woodroffe, A., 1966, Atmospheric circulation and the main climatic variables between 8000 and 0 B.C.: Meteorological evidence, in J. S. Sawyer (ed.), *World climate from 8000-0* B.C., London: Royal Meteorological Society, pp. 174-217.

Lamb, H. H., and Morth, H. T., 1978, Arctic ice, atmospheric circulation and world climate, *Geog. Jour.* **144:**1-22.

Lamb, H. H., and Woodroffe, A., 1970, Atmospheric circulation during the Last Ice Age, *Quat. Research* **1**(1):29-58.

Lamb, P. J., 1981, Do we know what we should be trying to forecast—climatically?, *Am. Meteorol. Soc. Bull.* **62:**1000-1001.

Lambeck, K., 1975, Effects of tidal dissipation in the oceans on the Moon's orbit and the Earth's rotation, *Jour. Geophys. Research* **80:**2917-2925.

Lambeck, K., 1980*a*, Changes in the length of day and atmospheric circulation, *Nature* **286:**104-105.

Lambeck, K., 1980*b*, The Earth's variable rotation, *New Scientist* **88:**426-429.

Lambeck, K., 1980*c*, *The Earth's Variable Rotation*, New York: Cambridge University Press, 450p.

Lambeck, K., and Cazenave, A., 1973, The Earth's rotation and atmospheric circulation—I. Seasonal variations, *Geophys. Jour.* **32:**79.

Lambeck, K., and Cazenave, A., 1974, The Earth's rotation and atmospheric circulation—II. The continuum, *Geophys. Jour.* **38:**49.

Lambeck, K., and Cazenave, A., 1976, Long term variations in the length of day and climatic change, *Royal Astron. Soc. Geophys. Jour.* **46:**555-573.

Landsberg, H. E., 1949, Climatic trends in the series of temperature observations at New Haven, Connecticut, *Geog. Annaler* **31**(1-4):125-132.

Landsberg, H. E., 1960, Note on the recent climatic fluctuations in the United States, *Jour. Geophys. Research* **65:**1519-1525.

Landsberg, H. E., 1962, Biennial pulses in the atmosphere, *Beitr. Physik Atmosphare* **35:**184-194.

Landsberg, H. E., 1969, *Weather and Health: An Introduction to Biometeorology*, New York: Doubleday, 148p.

Landsberg, H. E., 1976, Spectral analysis of long meteorological series, *Jour. Interdisc. Cycle Research* **7**(3):237-243.

Landsberg, H. E., and Albert, J. M., 1974, The summer of 1816 and volcanism, *Weatherwise* **27:**63-66.

Landsberg, H. E., and Jacobs, W. C., 1951, Applied climatology, in T. F. Malone (ed.), *Compendium of Meteorology*, Boston, Mass.: American Meteorological Society, pp. 976-992.

Landsberg, H. E., and Kaylor, R. E. 1977, Statistical analysis of Tokyo winter temperature approximations, 1443-1970, *Geophys. Research Letters* **4:**105-107.

Landsberg, H. E.; Mitchell, J. M., Jr.; Crutcher, H. L.; and Quinlan, F. T., 1963, Surface signs of the biennial atmospheric pulse, *Monthly Weather Rev.* **91:**549-556.

Landscheidt, T., 1976, Beziehungen zwischen dem Sonnenaktivitat und dem Massenzentrum des Sonnensystems, *Olbers-Gesell. Bremen Nachricht.*, no. 100, pp. 2-19.

Landscheidt, T., 1981, Swinging Sun, 79-year cycle, and climatic change, *Jour. Interdisc. Cycle Research* **12**(1):3-19.

Landscheidt, T., 1983, Solar oscillations and climatic change, in B. M. McCormac (ed.), *Weather and Climatic Responses to Solar Variations*, Boulder: Colorado Associated University Press, pp. 293-308.

Landscheidt, T., 1984, Cycles of solar flares and weather, in N.-A. Mörner and W. Karlén (eds.), *Climatic Changes on a Yearly to Millennial Basis*, Dordrecht, The Netherlands: D. Reidel, pp. 473-481.

Langley, S. P., 1901, Note [on solar changes of temperature and variations in rainfall in the region surrounding the Indian Ocean, by Sir Norman Lockyer], *Smithsonian Inst. Rept. for 1900*, pp. 183-184.

Langley, S. P., 1903, The "solar constant" and related problems [with bibliography], *Astrophys. Jour.* **17:**89-99.

Langley, S. P., 1904, On a possible variation of the solar radiation and its probable effect on terrestrial temperature, *Astrophys. Jour.* **19:**305-321.

Lanzerotti, L. J., 1977, Measures of energetic particles from the Sun, in O. R. White (ed.), *The Solar Output and Its Variation*, Boulder: Colorado Associated University Press, p. 383.

Lanzerotti, L. J., and Raghavan, R. S., 1981, Solar activity and solar neutrino flux, *Nature* **293:**122-124.

Laplace, P. S., 1798-1823, Traite de Mecanique celeste, tome 5, livre 13, Paris: J. B. M. Duprat, pp. 149-167, 237-243.

Larsen, M. F., and Kelley, M. C., 1977, A study of an observed and forecasted meteorological index and its relation to the interplanetary magnetic field, *Geophys. Research Letters* **4:**337-340.

Lawrence, D. B., 1950*a*, Glacier fluctuation for six centuries in southeastern Alaska and its relation to solar activity, *Geog. Rev.* **40:**191-223.

Lawrence, D. B., 1950*b*, Estimating dates of recent glacier advances and recession rates by studying tree growth layers, *Am. Geophys. Union Trans.* **31:**243-248.

Lawrence, D. B., and Lawrence E. G., 1959, Recent glacier variations in southern South America, *Am. Geog. Soc. Tech. Rept.*

Lawrence, E. N., 1965, Terrestrial climate and the solar cycle, *Weather* **20:**334-343.

Layzer, D.; Rosner, R.; and Doyle, H. T., 1979, On the origin of solar magnetic field, *Astrophys. Jour.* **229:**1126-1137.

Leighly, J. B., 1934, Graphic studies in climatology. III. A graphic interpolation device for dating the extremes of the annual temperature cycle, *Univ. California Pub. Geography* **6**(5):173-190.

Leighton, R. B., 1969, A magneto-kinematic model of the solar cycle, *Astrophys. Jour.* **156:**1-26.

Leith, C. E., 1975, Numerical weather prediction, *Rev. Geophysics and Space Physics*, **13:**681.

Leith, C. E., 1978, Objective methods of weather prediction, *Annual Rev. Fluid Mechanics* **10.**

Lejenas, H., 1984, Characteristics of southern hemisphere blocking as determined from a time series of observational data, *Royal Meteorol. Soc. Quart. Jour.* **110:**967-979.

Le Mouel, J. L.; Madden, T. R.; Ducruix, J.; and Courtillot, V., 1981, Decade fluctuations in geomagnetic westward drift and Earth rotation, *Nature* **290:**763-765.

Lempfert, R. G. K., 1913, British weather forecasts: Past and present, *Royal Meteorol. Soc. Quart. Jour.* **39:**173-184.

LeRoy Ladurie, E., 1967, *Histoire du climat depuis l'an mil*, Paris: Flammarion, 377p.

LeRoy Ladurie, E., 1971, *Times of Feast, Times of Famine*, London: George Allen and Unwin Ltd.; Garden City, N.Y.: Doubleday (trans. Barbara Bray), 428p.

Lethbridge, M. D., 1979, Thunderstorm frequency and solar sector boundaries, in B. M. McCormac and T. A. Seliga (eds.), *Solar-Terrestrial Influences on Weather and Climate*, Dordrecht, The Netherlands: D. Reidel, pp. 253-257.

Leverrier, V. J., 1843, Recherches sur l'orbite de Mercure et sur les perturbations, *Jour. Mathematics Pure Appl.* **13:**87p.

Lewis, R. G., 1935, Correspondence: Pleistocene chronology of central Europe, *Geol. Mag.* **72:**431-432.

Leyst, E., 1912, Luftdruck und Sonnenflecken, *Soc. Imperiale Naturalistes Moscou Bull.* **25:**93-158.

Li, Zhi-sen, and others, 1978, Correlation between the short anomalies of residuals of astronomical time and latitude and the major earthquakes around the observatories, *Acta Geophys. Sinica* **16**(2):226-256.

Libby, L. M., 1983, *Past Climates, Tree Thermometers, Commodities, and People*, Austin: University of Texas Press, 143p.

Libby, L. M., and Pandolfi, L. J., 1974, Calibration of isotope thermometers in an oak tree using official weather records, in J. Labeyrie (ed.), *Les methodes quantitatives d'etude des variations du climat au cours du Pleistocene*, Paris: Editions du Centre national de la Recherche Scientifique, pp. 299-310.

Liljequist, G. H., 1943*a*, The severity of the winters at Stockholm 1757-1942, *Statens Meteorol. Anstalt*, serien Uppsatser, no. 46.

Liljequist, G. H., 1943*b*, The severity of the winters at Stockholm 1757-1942, *Geog. Annaler* **25**(1-2):81-104.

Liljequist, G. H., 1949, On fluctuations of the summer mean temperature in Sweden, *Geog. Annaler* **31**(1-4):159-178.

Limb, C., 1912, Electrisation par la pluie d'une antenne de telegraphie sans fil, *Acad. Sci. Comptes Rendus* **154:**625-626.

Lincoln, J. V., 1970, Smoothed observed and predicted sunspot numbers, cycle 20, *Solar-Geophys. Data*, September, no. 313, part I, p. 9.

Lindgren, S., and Neumann, J., 1981, The cold and wet year 1695—A contemporary German account, *Clim. Change* **3:**173-187.

Lindqvist, R., 1932, *A Treatise on Reliable Predictions of Water Conditions*, Stockholm.

Lindzen, R. S., 1970, Internal gravity waves in atmospheres with realistic dissipation and temperature, Part I, *Geophys. Fluid Dynamics* **1:**303-355.

Lindzen, R. S., and Farrell, B., 1977, Some realistic modifications of simple climate models, *Jour. Atmos. Sci.* **34:**1387-1501.

Lindzen, R. S., and Holton, J. R., 1968, A theory of quasi-biennial oscillation, *Jour. Atmos. Sci.* **25:**1095-1107.

Lingenfelter, R. E., and Ramagy, R., 1970, Astrophysical and geophysical variations in C14 production (with discussion), in I. U. Olsson (ed.), *Radiocarbon Variations and Absolute Chronology*, Stockholm: Almqvist and Wiksell, pp. 513-537.

Link, F., 1958, Kometen, Sonnentatigkeit und Klimaschwankungen, *Die Sterne* **34:**129-140.

Link, F., 1964, Manifestations de l'activité solaire dans le passe historique, *Planetary and Space Sci.* **12:**333-348.

Link, F., 1968, Auroral and climatic cycles in the past, *British Astron. Assoc. Jour.* **78:**195-205.

Link, F., 1977, Solar and climatic cycles in the past, *Jour. Interdisc. Cycle Research* **8**(3/4):199-204.

Linsley, R. K., 1951, The hydrologic cycle and its relation to meteorology—River forecasting, in T. F. Malone (ed.), *Compendium of Meteorology*, Boston, Mass.: American Meteorological Society, pp. 1048-1056.

Lisitzin, E., 1974, *Sea-level changes*, Amsterdam: Elsevier Scientific Publishing Company, 286p.

Lockwood, G. W., and Thompson, D. J., 1979, A relationship between solar activity and planetary albedos, *Nature* **280:**43-45.

Lockwood, J. G., 1979, *Causes of Climate*, New York: John Wiley & Sons, 260p.

Lockyer, Sir N., 1902*a*, La relation entre les protuberances solaires et le magnetism terrestre, *Acad. Sci. Comptes Rendus* **135:**364-365.

Lockyer, Sir N., 1902*b*, The West Indian eruption and solar energy, *Science* **15:**915-916.

Lockyer, Sir N., and Lockyer, W. J. S., 1902, On some phenomena which suggest a short period of solar and meteorological changes, *Royal Soc. [London] Proc.* **70:**500-504.

Lockyer, Sir N., and Lockyer, W. J. S., 1902-1903*a*, On the similarity of the short period pressure variation over large areas, *Royal Soc. [London] Proc.* **71:**134-145.

Lockyer, Sir N., and Lockyer, W. J. S., 1902-1903*b*,

The relation between solar prominences and terrestrial magnetism, *Royal Soc. [London] Proc.* **71:**244-250.

Lockyer, W. J. S., 1901a, A long period sunspot variation, *Nature* **64:**196-197.

Lockyer, W. J. S., 1901*b*, The solar activity, 1833-1900, *Royal Soc. [London] Proc.* **68:**285-300.

Lockyer, W. J. S., 1902*a*, Die Sonnentatigkeit 1833-1900, *Zeitschr. Meteorologische* **19:**59-71.

Lockyer, W. J. S., 1902*b*, Mont Pelee eruption and dust falls, *Nature* **66:**53.

Lockyer, W. J. S., 1902*c*, The similarity of the short-period barometric pressure variations over large areas, *Nature* **67:**224-226.

Lockyer, W. J. S., 1902*d*, Solar prominences and terrestrial magnetism, *Nature* **67:**377-379.

Lockyer, W. J. S., 1903, The solar and meteorological cycle of thirty-five years, *Nature* **68:**8-10.

Lockyer, W. J. S., 1905, The flow of the River Thames in relation to British pressure and rainfall changes, *Royal Soc. [London] Proc.* **76:**494-506.

Lockyer, W. J. S., 1906, Barometric pressures of long duration over large areas, *Royal Soc. [London] Proc.* **78:**43-60.

Loder, J. W., and Garrett, C., 1978, The 18.6-year cycle of sea surface temperature in shallow seas due to variations in tidal mixing, *Jour. Geophys. Research* **83**(C4):1967-1970.

Lodge, Sir O., 1914, The electrification of the atmosphere, natural and artificial (Fifth Kelvin Lecture), *Inst. Elec. Engineers Jour.* **52:**333-352.

Loewe, F., 1937, A period of warm winters in western Greenland and the temperature see-saw between western Greenland and Central Europe, *Royal Meteorol. Soc. Quart. Jour.* **63:**365-372.

Lomb, N. R., and Andersen, A. P., 1980, The analysis and forecasting of the Wolf sunspot numbers, *Royal Astron. Soc. Monthly Notices* **190:**723-732.

London, J., 1956, Solar eruptions and the weather, *New York Acad. Sci. Trans.* **19:**138-146.

London, J.; Bjarnason, G. G.; and Rottman, G. J., 1984, 18 months of UV irradiance observations from the Solar Mesosphere Explorer, *Geophys. Research Letters* **11:**54-56.

London, J., and Haurwitz, M. W., 1963, Ozone and sunspots, *Jour. Geophys. Research* **68**(3):795-801.

London, J., and Kelly, J., 1974, Global trends in total atmospheric ozone, *Science* **184:**987-989.

London, J., and Oltmans, S. J., 1973, Further studies of ozone and sunspots, *Geofisica Pura e Applicata* **105-108:**1302-1307.

London, J., and Reber, C. A., 1979, Solar activity and total atmospheric ozone, *Geophys. Research Letters* **6:**869-872.

London, J.; Ruff, I.; and Tick, L. J., 1959, The relationship between geomagnetic variations and the circulation at 100 mb, *Jour. Geophys. Research* **64**(11):1827-1833.

Longman, I. M., 1959, Formulas for computing the tidal acceleration due to the Moon and the Sun, *Jour. Geophys. Research* **64**(12):2351-2355.

Loomis, E., 1868, *A Treatise on Meteorology*, New York: Harpers, 305p.

Loper, D. E., 1975, Torque balance and energy budget for the precessionally driven dynamo, *Physics Earth and Planetary Interiors* **11:**43.

Lorenz, E. N., 1968, Climatic determinism, *Meteorol. Mon.* **8**(30):1-3.

Lorenz, E. N., 1969, The future of weather forecasting, *New Scientist*, **42:**290-291.

Lorius, C.; Merlival, L.; Jouzel, J.; and Pourchet, M., 1979, A 30,000-yr isotope climatic record from Antarctic ice, *Nature* **280:**644-648.

Lovelius, N. V., 1972, Reconstruction of the course of meteorological processes on the basis of the annual tree rings along the northern and altitudinal forest boundaries, in F. E. Wielgolaski and Th. Rosswall (eds.), *Tundra Biome: Proceedings of the Fourth International Meeting on the Biological Production of Tundra*, Stockholm: Swedish International Biological Program Committee, pp. 248-260.

Lowe, E. J., 1880, *The Coming Drought, or the Cycle of the Seasons*, London.

Lowes, F. J., and Runcorn, S. K., 1951, The analysis of the geomagnetic secular variation, *Royal Soc. London Philos. Trans.* **243A:**525-546.

Ludlum, D. M., 1968, *Early American Winters*, vol. 2, *1821-1870*, History of American Weather, Boston, Mass.: American Meteorological Society, 257p.

Ludlum, D. M., 1973-1976, The weather of American independence, *Weatherwise* **26:**152-159; **27:**162-168; **28:**118-121, 147, 172-176; **29:**236-240, 288-290.

Lundqvist, J., 1985, The 1984 symposium on clay-varve chronology in Sweden, *Boreas* **14**(2):97-100.

Lyons, H. G., 1905, The Nile flood and its variation, *Geog. Jour.* **26:**249-272, 395-411.

Lysegaard, L., 1937, Anderungen des Klimas von Danemark seit 1800, *Zeitschr. Meteorol.* **54:**109-112.

Lysegaard, L., 1949, Recent climatic fluctuations, *Folia Geog. Danica* **5:**215p.

Maack, R., 1956, Ueber Waldverwustung und Bodenerosion im Staate Parana, *Die Erde, no. 3/4*, pp. 191-228.

McCormac, B. M., ed., 1983, *Weather and Climate Response to Solar Variation*, Boulder: Colorado Associated University Press, 626p.

McCormac, B. M., and Seliga, T. A., eds., 1979, *Solar-Terrestrial Influences on Weather and Climate*, Dordrecht, The Netherlands: D. Reidel, 346p.

Macdonald, A. G., 1922, Meteorology in medicine: With special reference to the occurrence of malaria in Scotland, *Royal Meteorol. Soc. Quart. Jour.* **48:**11-28.

Macdonald, N. J., and Roberts, W. O., 1960, Further evidence of a solar corpuscular influence on large-scale circulation at 300 Mb, *Jour. Geophys. Research* **65:**529-534.

Macdonald, N. J., and Woodbridge, D. D., 1959, Relation of geomagnetic disturbances to circulation changes at the 30,000-foot level, *Science* **129:**638-639.

MacDowall, A. B., 1901*a*, Gradual change in our climate, *Knowledge (London)* **24:**39-40.

MacDowall, A. B., 1901*b*, Sunspots and winters, *Knowledge (London)* **24:**156-157.

MacDowall, A. B., 1901*c*, The Moon and wet days, *Nature* **64:**424-425.

MacDowall, A. B., 1901*d*, The Moon and rainfall, *Symons' Meteorol. Mag. (London)* **36:**165-166.

MacDowall, A. B., 1901*e*, Does the Moon affect rainfall?, *Knowledge (London)* **24:**276.

MacDowall, A. B., 1901*f*, Die Sonnenfleckenperiode und der Charakter der Winter von Wien, *Zeitschr. Meteorol.* **18:**588-589.

MacDowall, A. B., 1901-1902, The Moon and thunderstorms, *Nature* **65:**367.

MacDowall, A. B., 1902*a*, Bruckner's cycle and the variation of temperature in Europe, *Nature* **66:**77-78.

MacDowall, A. B., 1902*b*, Sunspots and wind, *Nature* **66:**320.

MacDowall, A. B., 1902*c*, The Moon and rainfall, *Symons' Meteorol. Mag. (London)* **37:**107-108.

MacDowall, A. B., 1902-1903, Sunspots and winter heat, *Nature* **67:**247.

MacDowall, A. B., 1903*a*, Our rainfall in relation to Bruckner's cycle, *Nature* **68:**56.

MacDowall, A. B., 1903*b*, Sun-spots and phenology, *Nature* **68:**389-390.

MacDowall, A. B., 1903*c*, Our winters in relation to Bruckner's cycle, *Nature* **68:**600.

MacDowall, A. B., 1903*d*, Retardation von Perioden, *Zeitschr. Meteorol.* **20:**88.

MacDowall, A. B., 1903-1904, Sunspots and temperature, *Nature* **69:**607-618.

MacDowall, A. B., 1904, Some weather prophets, *Symons' Meteorol. Mag. (London)* **2:**80.

MacDowall, A. B., 1904-1905, The Moon and the barometer, *Nature* **71:**320.

MacDowall, A. B., 1905*a*, Solar changes and weather, *Nature* **72:**175.

MacDowall, A. B., 1905*b*, Sonnenflecken, Luftdruckabweichungen zu Stykkisholm und Frosttage zu Greenwich, *Zeitschr. Meteorologische* **22:**462-463.

MacDowall, A. B., 1905*c*, Der Mond und die Kalten Tage, *Zeitschr. Meteorologische* **22:**463.

MacDowall, A. B., 1905*d*, Sonnenflecken und Luftdruck, *Zeitschr. Meteorologische* **22:**565-566.

MacDowall, A. B., 1907*a*, Rothesay rainfall and the sun-spot cycle, *Nature* **77:**488.

MacDowall, A. B., 1907*b*, Mondphasen und niedrigen Barometerstand, *Zeitschr. Meteorologische* **24:**87-88.

MacDowall, A. B., 1907*c*, Sonnenflecken und Regenfall zu Rothesay (Schottland) 1804 bis 1904, *Zeitschr. Meteorologische* **24:**514.

MacDowall, A. B., 1908*a*, Forecasting seasons, *Knowledge and Sci. News (London)* **5:**82-83.

MacDowall, A. B., 1908*b*, Weather prediction, *Symons' Meteorol. Mag. (London)* **43:**193-194.

MacDowall, A. B., 1909*a*, The question of sunspot influence, *Symons' Meteorol. Mag. (London)* **44:**10-11.

MacDowall, A. B., 1909*b*, Warm months in relation to sunspot numbers, *Nature* **79:**367-368.

MacDowall, A. B., 1909*c*, Sunspots and sunshine, *Symons' Meteorol. Mag. (London)* **44:**164.

MacDowall, A. B., 1912*a*, The question of sun-spot influence, *Nature* **88:**449.

MacDowall, A. B., 1912*b*, Variations in our climate, *Symons' Meteorol. Mag. (London)* **47:**622.

McElhinny, M. W., and Merrill, R. T., 1975, Geomagnetic secular variation over the past 5 million years, *Rev. Geophysics and Space Sci.* **13:**687.

McEwen, G. F., 1918, Oceanic circulation and its bearing upon attempts to make seasonal weather forecasts, *Univ. California, Scripps Inst. Biol. Research Bull. No. 7.*

McEwen, G. F., 1922, Forecasting seasonal rainfall from oceanic temperature, *Am. Meteorol. Soc. Bull.* **3**(10):135.

McEwen, G. F., 1923, How the Pacific Ocean affects Southern California's climate. Seasonal rainfall for 1923-34 indicated by ocean temperatures, *Am. Meteorol. Soc. Bull.* **4**(10):142-148.

McEwen, G. F., 1924, Forecasting seasonal rainfall from ocean temperatures. Indications for the 1924-25 season in Southern California, *Am. Meteorol. Soc. Bull.* **5**(10):137-139.

McEwen, G. F., 1925, Ocean temperatures and seasonal rainfall in Southern California, *Monthly Weather Rev.* **53**(11):483-489.

McEwen, G. F., 1934, Methods of seasonal weather forecasting at the Scripps Institution of Oceanography, *Am. Meteorol. Soc. Bull.* **15**(11):249-256.

McFadgen, B. G., 1975, Long term cycles in the variation of atmospheric radiocarbon, related to changes in Holocene climate, *Search* **6:**509-511.

McIntyre, M. E., 1982, How well do we understand the dynamics of stratospheric warmings?, *Meteorol. Soc. Japan Jour.* **60.**

Mackie, S. F., 1901, The relation between the level of Great Salt Lake and the rainfall, *Monthly Weather Rev.* **29:**57-61.

McMurray, H., 1941, Periodicity of deep-focus earthquakes, *Seismol. Soc. America Bull.* **31:**33-56.

McNish, A. G., and Lincoln, J. R., 1949, Prediction of sunspot numbers, *Am. Geophys. Union Trans.* **30**(5):673-685.

McPherson, J. G., 1907, The weather and influenza, *Knowledge and Science News (London)* **4:**85.

Madden, R. A., 1976, Estimates of the natural variability of time-averaged sea level pressures, *Monthly Weather Rev.* **104:**942-952.

Madden, R. A., 1977, Estimates of the autocorrelations and spectra of seasonal mean temperatures over North America, *Monthly Weather Rev.* **105:**9-18.

Madden, R. A., and Julian, P. R., 1972, Description of global scale circulation cells in the tropics with a 40-50 day period, *Jour. Atmos. Sci.* **29:**1109-1123.

Madden, R. A., and Williams, J., 1978, Correlation between temperature and precipitation in the United States and Europe, *Monthly Weather Rev.* **106:**142-147.

Maddren, A. G., 1905, Smithsonian exploration in Alaska in 1904, in search of mammoth and other fossil remains, *Smithsonian Misc. Colln.* **49:**117p.

Maeda, K.-I., 1983, Lunar modulations of the equatorial electrojet, *Jour. Atmos. and Terr. Physics* **45**(4):245-254.

Maeda, K.-I., and Heath, D. F., 1980/1981, Stratospheric ozone response to a solar proton event: Hemispheric asymmetries, *Geofisica Pura e Applicata* **119:**1-8.

Magny, M., 1982, Atlantic and Sub-boreal: Dampness and dryness?, in A. F. Harding (ed.), *Climatic Change in Later Prehistory,* Edinburgh: Edinburgh University Press, pp. 33-43 (article trans. by L. E. Forbes and A. F. Harding from Magny, 1979, *Rev. Archeol. l'Est et du Centre-Est* **30**(1-2):57-65).

Mahany, W. C., ed., 1981, *Quaternary Palaeoclimate,* Norwich, U.K.: GeoBooks, 464p.

Mahlmann, W., 1840, Ueber die 9- und 19-jahrige Witterungsperiode, *Gesell. Erdkunde (Berlin) Monats.* **1:**102-104.

Maksimov, I. V., 1952, On the 80-year cycle of climatic variations, *Akad. Nauk. S.S.S.R. Doklady* **86**(5):917-920 (in Russian; English trans. by E. R. Hope, Canada Defense Research Board, 1953).

Maksimov, I. V., and others, 1967, Nutational migration of the Iceland low pressure, *Akad. Nauk. S.S.S.R. Doklady* **177:**88-91.

Maksimov, I. V.; Saruhanjan, E. I.; and Smirnov, N. P., 1970, On the relation between deformation force and the movements of the atmosphere's centres of action, *Akad. Nauk. S.S.S.R. Doklady* **190:**1095-1097 (in Russian).

Maksimov, I. V., and Sleptsov, B. A., 1963, *Study of Eleven-Year Variations in Atmospheric Pressure in the Antarctic,* Moscow: Soviet Antarctic Expedition No. 43 (in English).

Maksimov, I. V., and Sleptsov-Shevlevich, B. A., 1970, Long-term changes of tidal forces of Moon and ice conditions of Arctic seas, *Morsk. Ryb. Khoz. Ok. (PINRO), Polyar Nauchno-Issled. Inst. Trudy* **27:**21-39.

Maksimov, I. V., and Smirnov, N. P., 1965, A contribution to the study of the causes of long-period variations in the activity of the Gulf Stream, *Moscow Oceanology* **5**(2) (trans. from Russian by American Geophysical Union).

Maley, J., 1981, Etudes palynologiques dans le bassin du Tchad et paleoclimatologie de l'Afrique nord-tropicale de 30,000 ans a l'epoque actuelle, *Travaux et Document de l'Orstom, no. 129,* pp. 1-586.

Maley, J., 1982, Dust, clouds, rain types, and climatic variations in tropical North Africa, *Quat. Research* **18**(1):1-16.

Malin, S. R. C., 1969, Geomagnetic secular variation and its changes, 1942.5 to 1962.5, *Royal Astron. Soc. Geophys. Jour.* **17**(4):415-441.

Malin, S. R. C., and Clark, A. D., 1974, Geomagnetic secular variation, 1962.5 to 1967.5, *Royal Astron. Soc. Geophys. Jour.* **36:**11-20.

Malkus, W. V. R., 1963, Precessional torques as the cause of geomagnetism, *Jour. Geophys. Research* **68:**2871-2886.

Malkus, W. V. R., 1968, Precession of the Earth as the cause of geomagnetism, *Science* **160:**259-264.

Manley, G., 1953, The mean temperature of central England, 1698-1952, *Royal Meteorol. Soc. Quart. Jour.* **79:**242-261.

Manley, G., 1959, Temperature trends in England, 1698-1959, *Archiv Meteorologie, Geophysik, u. Bioklimatologie* **9:**413-433.

Manley, G., 1961, Meteorological factors in the great glacial advance (1690-1720), *Internat. Assoc. Sci. Hydrology Pub. 54,* pp. 388-391.

Manley, G., 1966, Problems of the climatic optimum: The contribution of glaciology, in J. S. Sawyer (ed.), *World Climate 8000-O* B.C., London: Royal Meteorological Society, pp. 34-39.

Manley, G., 1974, Central England temperatures: Monthly means 1659-1973, *Royal Meteorol. Soc. Quart. Jour.* **100:**389-405.

Mansinha, L., and Smylie, E. W., 1967, Effect of earthquakes on the Chandler wobble and the secular polar shift, *Jour. Geophys. Research* **72**(18):4731-4743.

Mansinha, L., and Smylie, E. W., 1968, Earthquakes and the Earth's wobble, *Science* **161:**1127-1129.

Mansinha, L., Smylie, E. W.; and Chapman, C. H., 1979, Seismic excitation of the Chandler wobble revisited, *Royal Astron. Soc. Geophys. Jour.* **59:**1.

Manson, M., 1903*a, The Evolution of Climates,* Minneapolis, Minn.: Franklin Printing Company, 86p.

Manson, M., 1903*b,* Rainfall on the Pacific coast of North and South America and the factors of water supply in California, *Assoc. Eng. Soc. Jour.* **30:**104-117.

Manson, M., 1914, The bearing of the facts revealed by Antarctic research upon the problems of the ice age, *Geog. Jour.* **43:**706-708.

Manson, M., 1924, The physical and geological traces of the cyclone belt across North America, *Monthly Weather Rev.* **76.**

Maple, E., 1959*a*, Geomagnetic oscillations at middle latitudes. I. The observational data, *Jour. Geophys. Research* **64**(10):1395-1404.

Maple, E., 1959*b*, Geomagnetic oscillations at middle latitudes. II. Sources of the oscillations, *Jour. Geophys. Research* **64**(10):1405-1409.

Marchand, E., 1901, Sur les relations des phenomenes solaires avec ceux de la physique du globe terrestre, *Cong. Internat. Meteorol. (Paris)*, pp. 148-175.

Marchand, E., 1909*a*, Quelques remarques sur la grande perturbation magnetique du 25 september 1909 et les phenomenes solaires concomitants, *Acad. Sci. Comptes Rendus* **149:**616-619.

Marchand, E., 1909*b*, La perturbation magnetique du 25 september 1909 et les phenomenes terrestres et solaires concomitants, *Soc. Astron. France Bull.* **23:**551-553.

Marchand, E., 1909*c*, Relations des phenomenes solaires avec les phenomenes de l'atmosphere terrestre. Essai d'application à la prevision du temps, *Assoc. Française Avancement Sci. Comptes Rendus* **38:**387-393.

Marchesi, D. D., 1906, *La luna e le sue influenze sull'agricoltura e sul tempo*, Bologne: Treves, 232p.

Marcz, F., 1976, Links between atmospheric electricity and ionospheric absorption due to extraterrestrial influences, *Jour. Geophys. Research* **81**(25):4566-4570.

Markham, C. G., 1974, Apparent periodicities in rainfall at Fortaleza, Ceara, Brazil, *Jour. Appl. Meteorology* **13:**176-179.

Markson, R., 1971, Considerations regarding solar and lunar modulation of geophysical parameters, atmospheric electricity and thunderstorms, *Geofisica Pura e Applicata* **84:**161-202.

Markson, R., 1975, Solar modulation of atmospheric electrification through variation of the conductivity over thunderstorms, in W. R. Bandeen and S. P. Maran (eds.), *Possible Relationships between Solar Activity and Meteorological Phenomena*, Washington, D.C.: National Aeronautics and Space Administration, NASA SP-366, pp. 171-178.

Markson, R., 1976, Ionospheric potential variations obtained from aircraft measurements of potential gradient, *Jour. Geophys. Research* **81**(12):1980-1990.

Markson, R., 1978, Solar modulation of atmospheric electrification and possible implications for the Sun-weather relationship, *Nature* **273:**103-109.

Markson, R., 1979, Atmospheric electricity and the Sun-weather problem, in B. M. McCormac and T. A. Seliga (eds.), *Solar-Terrestrial Influences on Weather and Climate*, Dordrecht, The Netherlands: D. Reidel, pp. 215-232.

Markson, R., 1983, Solar modulation of fair-weather and thunderstorm electrification and a proposed program to test an atmospheric Sun-weather mechanism, in B. M. McCormac (ed.), *Weather and Climate Response to Solar Variation*, Boulder: Colorado Associated University Press, pp. 323-343.

Markson, R., and Muir, M., 1980, Solar wind control of the Earth's electrical field, *Science* **208:**979-990.

Marriott, R. A., 1914, *The Change in the Climate and Its Cause*, London: E. Marlborough and Company, 94p.

Marsden, B. G., and Cameron, A. G. W., eds., 1966, *The Earth-Moon System*, New York: Plenum Press, 288p.

Marsden, R. S., 1906, Scarlatina and certain other diseases in relation to temperature and rainfall, *Royal Sanitary Inst. London Jour.* **27:**397-399.

Marshall, J. R., 1972, Precipitation patterns of the United States and sunspots, Ph.D. dissertation, University of Kansas.

Martin, D., and Hawkins, H. F., Jr., 1950, Forecasting the weather: The relationship of temperature and precipitation over the United States to the circulation aloft, *Weatherwise* **3:**16-19, 40-43, 65-67, 89-92, 113-116, 138-141.

Marvin, C. F., 1923, Concerning normals, secular trends, and climatic changes, *Monthly Weather Rev.* **51:**383-390.

Marvin, C. F., 1925, Concerning normals, secular trends, and climatic changes, *Monthly Weather Rev.* **53:**301.

Marvin, C. F., 1927, The Wolfer sunspot numbers analyzed as frequency distributions, *Am. Meteorol. Soc. Bull.* **8**(5):79-84.

Mason, B. J., 1962, *Clouds, Rain and Rainmaking*, Cambridge and New York: Cambridge University Press, 189p.

Mason, B. J., 1976, Towards the understanding and prediction of climatic variations, *Royal Meteorol. Soc. Quart. Jour.* **102:**473-499.

Mass, C., and Schneider, S. H., 1977, Statistical evidence on the influence of sunspots and volcanic dust on long-term temperature records, *Jour. Atmos. Sci.* **34:**1995-2004.

Mathieu, G., 1902, La lune et le temps, *Ami Cultivat. Liege*, p. 180.

Matthews, J. A., 1977, Glacier and climatic fluctuations inferred from tree-growth variations over the last 250 years, central southern Norway, *Boreas* **6**(1):1-24.

Mattice, W. A., 1935, Weather and corn yields, *Monthly Weather Rev.*

Maunder, E. W., 1903, The sunspots of 1903, October, *Knowledge (London)* **26:**275-278.

Maunder, E. W., 1903-1904*a*, The "great" magnetic storms, 1875 to 1903, and their association with sun-spots, as recorded at the Royal Observatory, Greenwich, *Royal Astron. Soc. Monthly Notices* **64:**205-222.

Maunder, E. W., 1903-1904*b*, Further note on the "great" magnetic storms, 1875-1903, and their association with sun-spots, *Royal Astron. Soc. Monthly Notices* **64:**222-224.

Maunder, E. W., 1904-1905*a*, Magnetic disturbances, 1882 to 1903, as recorded at the Royal Observatory, Greenwich, and their association with sun-spots, *Royal Astron. Soc. Monthly Notices* **65:**2-34.

Maunder, E. W., 1904-1905*b*, Magnetic disturbances as recorded at the Royal Observatory, Greenwich, and their association with sun-spots (second paper), *Royal Astron. Soc. Monthly Notices* **65:**538-559.

Maunder, E. W., 1905*a*, Early suggestions of the indication by magnetic disturbances of the solar rotation-period, *Observatory (London)* **28:**100-104.

Maunder, E. W., 1905*b*, Magnetic disturbances as recorded at the Royal Observatory, Greenwich, and their association with sun spots (third paper), *Royal Astron. Soc. Monthly Notices* **65:**666-681.

Maunder, E. W., 1905*c*, The solar origin of the disturbances of terrestrial magnetism, *Astron. Nachrichte* **167:**177-182.

Maunder, E. W., 1905*d*, The solar origin of terrestrial magnetic disturbances, *Astrophys. Jour.* **21:**101-115.

Maunder, E. W., 1905*e*, The solar origin of terrestrial magnetic disturbances, *Popular Astronomy* **13:**52-64.

Maunder, E. W., 1909-1910, Note on the cyclones of the Indian Ocean, 1856-1867, and their association with the solar rotation, *Royal Astron. Soc. Monthly Notices* **70:**49-62.

Maunder, E. W., 1913, The Sun-spots, *Scientia (Bologna)* **13:**1-9 (supp. pp. 3-11).

Maunder, E. W., 1921-1922, The prolonged sun-spot minimum, 1645-1715, *British Astron. Assoc. Jour.* **32**(4):140-145.

Maunder, S. S. D., 1903-1904, Suggested connection between sun-spots activity and the secular change in magnetic declination, *Royal Astron. Soc. Monthly Notices* **64:**224-228.

Maunder, S. S. D., 1907, An apparent influence of the Earth on the numbers and areas of sun-spots in the cycle 1889-1901, *Royal Astron. Soc. Monthly Notices.* **67:**451-476.

Maurer, J., 1909, Gletscherschwankungen und Variationen der Sommerwarme, *Zeitschr. Meteorologische* **26:**181-183.

Maurer, J., 1924, *Zeitschr. Meteorologische* **61**(3) (trans. by W. W. Reed, Severe winters in southern Germany and Switzerland since the year 1400 determined from severe lake freezes, *Monthly Weather Rev.* **52**(4):222).

Maurer, J., 1927, Les catastrophes meteorologiques et l'activité solaire, Génève, Materiaux pour l'étude des Calamités, *Année* **4:**178-180.

Mayaud, P. N., 1977, On the reliability of the Wolf number series for estimating long-term periodicities, *Jour. Geophys. Research* **82**(7):1271-1272.

Mead, G. D., 1964, Deformation of the geomagnetic field by the solar wind, *Jour. Geophys. Research* **69**(7):1181-1195.

Mead, G. D., and Beard, D. B., 1964, Shape of the geomagnetic field solar wind boundary, *Jour. Geophys. Research* **69**(7):1169-1179.

Meadows, A. J., 1974, *Early Solar Physics*, London: Pergamon Press, 312p.

Meadows, A. J., 1975, A hundred years of controversy over sunspots and weather, *Nature* **256:**95-97.

Mears, E. G., 1942/1943, The ocean current called "The Child," *Smithsonian Inst. Rept.*, pp. 245-251.

Meeus, J., 1975, Comments on The Jupiter Effect, *Icarus* **26:**257-267.

Meeus, J., 1978, Planetes et activité solaire, *Ciel et Terre* **94:**19-27.

Meeus, J., 1982, Planetary quadrants and minimum sectors, 0 to 3000 (letter), *Sky and Telescope*, January, pp. 5-6.

Meeus, J., and Victor, R. C., 1983, *Astronomical Tables of the Sun, Moon and Planets*, Richmond, Va.: Willmann-Bell, Inc.

Meier, M., 1984, Contribution of small glaciers to global sea level, *Science* **226:**1418-1421.

Meinardus, W., 1904, Ueber Schwankungen der nordatlantischen Zirkulation und ihre Folgen, *Annalen Hydrographie u. Maritimen Meteorologie* **32:**353-362.

Meinardus, W., 1905, Ueber Schwankungen der nordatlantischen Circulation und damit zusammenhangende Erscheinungen, *Zeitschr. Meteorologische* **22:**398-412.

Meinardus, W., 1906*a*, Variations in the circulation of the North Atlantic and the phenomena connected therewith (trans. from the German by R. H. Scott), *Royal Meteorol. Soc. Quart. Jour.* **32:**53-65.

Meinardus, W., 1906*b*, Periodische Schwankungen der Eisdrift bei Island, *Annalen Hydrographie u. Maritimen Meteorologie* **34:**148-162, 227-239, 278-285.

Meinardus, W., 1908, Zu den Beziehungen zwischen den Eisverhaltnissen bei Island und der nordatlantischen Zirkulation, *Annalen Hydrographie u. Maritimen Meteorologie* **36:**318-321.

Meinel, A. B., and Meinel, M. P., 1964, Height of the glow stratum from the eruption of Agung on Bali, *Nature* **201:**657-658.

Meinel, A. B., and Meinel, M. P., 1967, Volcanic sunset-glow stratum: Origin, *Science* **155:**189.

Meissner, O., 1906, Einfluss des Mondes auf die Erdbebenhaufigkeit, *Himmel u. Erde* **18:**278-279.

Meissner, O., 1908, Der Mond und die Wolken, *Wetter* **25:**91-92.

Meissner, O., 1911, Der Einfluss der Sonnenfleckenhaufigkeit auf das Klima von Berlin, *Astronomische Nachrichten* **189:**371-374.

Meissner, O., 1912, Niederschlag und Sonnenfleckenperiode, *Wetter* **29:**225-226.

Meissner, O., 1943, Vergleichende Betrachtungen

der Mitteltemperatur von Berlin, Leipzig, Prag und Wien, *Annalen Hydrographie u. Maritimen Meteorologie* **71.**

Melchior, P., and Yumi, S., eds., 1972, *Rotation of the Earth*, Dordrecht, The Netherlands: D. Reidel, 244p.

Michelson, A. A., and Gale, H. G., 1919, The rigidity of the Earth, *Astrophys. Jour.* **50:**330-345.

Mielke, J., 1913, Die temperaturschwankungen 1870-1910 in ihrem verhäitnis 3[u] der 11 jährigen Sonnenfleckenperiode, *Duet. Sewarte Archiv* **36**(3).

Mikola, P., 1952, The effect of recent climatic variations on forest growth in Finland, *Fennia* **75:**69-76.

Mikola, P., 1962, Temperature and tree growth near the northern timber line, in T. T. Kozlowski (ed.), *Tree Growth*, New York: Ronald Press, pp. 265-287.

Milham, W. I., 1924, The year 1816—the causes of abnormalities, *Monthly Weather Rev.* **52:**563-570.

Miller, P. R., and Clark, F. E., 1955, Water and the micro-organisms, in *Water, The Yearbook of Agriculture*, Washington, D.C.: U.S. Department of Agriculture, pp. 25-35.

Mills, C. A., 1938, Weather and health, *Am. Meteorol. Soc. Bull.* **19**(4):141-152.

Mills, C. A., 1941, Some possible relationships of planetary configurations and sunspots to world weather, *Am. Meteorol. Soc. Bull.* **22:**167-173.

Mills, C. A., 1955, Why the weather gets you down, *Sunday Star, This Week Magazine*, 11 September 1955, pp. 116-118.

Minina, L. S.; Petrosyants, M. A.; and Portnyagin, Y. I., 1977, Circulation systems in the Northern Hemisphere at heights of 80-100 km, *Meteorology and Hydrology, No. 3*, p. 15.

Mitchell, C. L., and Wexler, H., 1941, How the daily forecast is made, in *Climate and Man*, Yearbook of Agriculture, Washington, D.C.: U.S. Department of Agriculture, pp. 579-599.

Mitchell, J. M., Jr., 1976, An overview of climatic variability and its causal mechanisms, *Quat. Research* **6:**481-493.

Mitchell, J. M., Jr., 1983, Empirical modeling of effects of solar variability, volcanic events, and carbon dioxide on global-scale average temperature since A.D. 1880, in B. M. McCormac (ed.), *Weather and Climate Response to Solar Variation*, Boulder: Colorado Associated University Press, pp. 265-272.

Mitchell, J. M., Jr.; Stockton, C. W.; and Meko, D. M., 1979, Evidence of a 22-year rhythm of drought in the western United States related to the Hale solar cycle since the 17th century, in B. M. McCormac and T. A. Seliga (eds.), *Solar-Terrestrial Influences on Weather and Climate*, Dordrecht, The Netherlands: D. Reidel, pp. 125-144.

Mobbs, S. D., 1982, External principles for global climate models, *Royal Meteorol. Soc. Quart. Jour.* **108:**535-550.

Mock, S. J., and Hibler, W. D., III, 1976, The 20-year oscillation in eastern North America temperature records, *Nature* **261:**484-486.

Moffett, R. J.; Murphy, J. A.; and Bailey, G. J., 1975, Storm-time increases in the ionospheric total electron content, *Nature* **253:**330-331.

Molchanov, A. M., 1968, The resonant structure of the solar system, *Icarus* **8:**203-215.

Molchanov, A. M., 1969*a*, Resonances in complex systems, a reply to critiques, *Icarus* **11:**95-103.

Molchanov, A. M., 1969*b*, The reality of resonances in the solar system, *Icarus* **11:**104-110.

Monin, A. S., 1972, *Weather Forecasting as a Problem in Physics*, Cambridge, Mass.: MIT Press, 199p.

Monin, A. S., 1974, *Earth's Rotation and Climate*, Delhi, India: Radok-Radhakrishna (trans. from Russian), 140p.

Monin, A. S., and Vulis, I. J., 1971, On the spectra of long-period oscillations of geophysical parameters, *Tellus* **23:**337-345.

Moore, G. W., and Giddings, J. L., 1962, Record of 5000 years of Arctic wind direction recorded by Alaskan beach ridges, *Geol. Soc. America Spec. Paper 68*, 232p.

Moran, P. A. P., 1949, The statistical analysis of the sunspot and lynx cycles, *Jour. Animal Ecology* **18:**115-116.

Moran, P. A. P., 1953, The statistical analysis of the Canadian lynx cycle, *Australian Jour. Zoology* **1:**163-173, 291-298.

Moran, P. A. P., 1954, The logic of the mathematical theory of animal populations, *Jour. Wildlife Management* **l18:**60-66.

Morgan, W. J.; Stoner, J. O.; and Dicke, R. H., 1961, Periodicity of earthquakes and the invariance of the gravitational constant, *Jour. Geophys. Research* **66**(11):3830-3843.

Morikofer, W., 1953, Die Klimakurorte der Schweizer Alpen, *Schweizer. Jour.*

Mörner, N.-A., 1969, The Late Quaternary history of the Kattegatt Sea and the Swedish westcoast, *Sveriges Geol. Undersökning Arsb.* **640C:**1-487.

Mörner, N.-A., 1973, Climate cycles during the last 35,000 years, *Jour. Interdisc. Cycle Research* **4:**189-192.

Mörner, N.-A., 1976*a*, Eustasy and geoid changes, *Jour. Geol.* **84:**123-151.

Mörner, N.-A., 1976*b*, Eustatic changes during the last 8000 years in view of radiocarbon calibration and new information from the Kattegat region and other northwestern European areas, *Palaeogeography, Palaeoclimatology, and Palaeoecology* **19:**63-85.

Mörner, N.-A., 1977*a*, Paleoclimate and short period changes of the core/mantle coupling and interface, *Jour. Interdisc. Cycle Research* **8:**207-210.

Mörner, N.-A., 1977*b*, Palaeoclimatic records from South Scandinavia, global correlations, origin and cyclicity, in S. Horie (ed.), *Paleolimnology of Lake Biwa and the Japanese Pleistocene*, vol. 4, Otsu,

Japan: Otsu Hydrobiological Station, pp. 499-528.

Mörner, N.-A., 1980, A 10,700 year's paleotemperature record from Gotland and Pleistocene/Holocene boundary events in Sweden, *Boreas* **9:**283-287.

Mörner, N.-A., and Karlén, W., eds., 1984, *Climatic Changes on a Yearly to Millennial Basis*, Dordrecht, The Netherlands: D. Reidel, 667p.

Mörner, N.-A., and Wallin, B., 1977, A 10,000 year temperature record from Gotland, Sweden, *Palaeogeography, Palaeoclimatology, and Palaeoecology* **21:**113-138.

Mörth, H. T., and Schlamminger, L., 1979, Planetary motion, sunspots and climate, in B. M. McCormac and T. A. Seliga (eds.), *Solar-Terrestrial Influences on Weather and Climate*, Dordrecht, The Netherlands: D. Reidel, pp. 193-207.

Mosetti, F., 1963, Sull' esistenza di un ritmo con periodo di 45 anni in talune fluttuazioni geofisiche, *Boll. Geofisica Teor. ed Appl.* **5**(18):127-138.

Mosley-Thompson, E., and Thompson, L. G., 1982, Nine centuries of microparticle deposition at the South Pole, *Quat. Research* **17**(1):1-13.

Moulton, F. R., 1914, *An Introduction to Celestial Mechanics*, 2nd ed., New York: Macmillan, 437p.

Mozer, F. S., 1971, Balloon measurement of vertical and horizontal atmospheric electrical fields, *Geofisica Pura e Applicata* **84:**32-45.

Mozer, F. S., and Serlin, R., 1969, Magnetospheric electrical field measurements with balloons, *Jour. Geophys. Research* **74:**4739-4754.

Mueller, I. I., 1969, *Spherical and Practical Astronomy, as Applied to Geodesy*, New York: F. Ungar Publishing Company, 615p.

Mukherjee, A. K., and Singh, B. P., 1978, Trends and periodicities in annual rainfall in monsoon areas over the northern hemisphere, *Indian Jour. Meteorology, Hydrology, and Geophysics* **29:**441-447.

Mulheisen, R., 1971, Neue Ergebnisse und Probleme in der Luftelektrizitat, *Zeitschr. Geophysik* **37:**759-793.

Mulheisen, R., 1977, The global circuit and its parameters, in H. Dolezalek and R. Reiter (eds.), *Electrical Processes in Atmospheres: Proceedings of the Fifth International Conference on Atmospheric Electricity*, Darmstadt, West Germany: Steinkopff Verlag, pp. 467-473.

Munk, W. H., and Hassan, E. S. M., 1961, Atmospheric excitation of the Earth's wobble, *Geophys. Jour.* **4:**339-358.

Munk, W. H., and MacDonald, G. J. F., 1960, *The Rotation of the Earth: A Geophysical Discussion*, Cambridge: Cambridge University Press, 323p.

Munk, W. H., and Miller, R., 1950, Variations in the Earth's angular velocity resulting from fluctuations in atmospheric and oceanic circulation, *Tellus* **2:**93-101.

Murphy, C. H., 1969, Seasonal variation in ionospheric winds over Barbados, West Indies, *Jour. Geophys. Research* **74**(1):339-347.

Murphy, R. C., 1926, Oceanic and climatic phenomena along the west coast of South America during 1925, *Geog. Rev.* **6**(1):26-54.

Myerson, R. J., 1970, Long-term evidence for the association of earthquakes with the excitation of the Chandler wobble, *Jour. Geophys. Research* **75:**6612-6617.

Myrbach, O., 1926, Der Temperaturverlauf in Wien im Winter 1925/26 im Zusammenhang mit Sonnenfleckenkulminationen und Mondphase (abstract), *Zeitschr. Meteorologische* **43:**485-486.

Myrbach, O., 1927*a*, Der Temperaturverlauf zu Wien im Winter 1925/26 im Zusammenhang mit Sonnenfleckenkulminationen und Mondphase, *Wetter* **44:**31-37, 55-56, 74-78.

Myrbach, O., 1927*b*, Die Schwankungen der Grosswetterlage in ihrer Abhangigkeit von der Sonnentatigheit nebst einen Anhang uber die Alteration dieser Beziehungen durch die Mondphase, *Jour. Geophysics and Meteorology* **4**(2):217-231.

Myrbach, O., 1935, Sonnenfleckenzyklus und Gewitterhaufigkeit in Wien, Kremsmunster und Bayern, *Zeitschr. Meteorologische* **52:**225-227.

Nagata, T., 1965, Main characteristics of recent geomagnetic secular variation, *Jour. Geomagnetism and Geoelectricity* **17:**263-276.

Namias, J., 1953, Thirty-day forecasting: A review of a ten-year experiment (with a foreword by H. C. Willett), *Am. Meteorol. Soc. Meteorol. Mon.* **2**(6):83p.

Namias, J., 1959, Recent seasonal interactions between North Pacific waters and the overlying atmospheric circulation, *Jour. Geophys. Research* **64**(6):631-646.

Namias, J., 1963, Large-scale air-sea interactions over the North Pacific from summer 1962 through the subsequent winter, *Jour. Geophys. Research* **68**(22):6171-6186.

Namias, J., 1965, Short-period climatic fluctuations, *Science* **147:**696-706.

Namias, J., 1966, A weekly periodicity in eastern United State precipitation and its relation to hemispheric circulation, *Tellus* **18**(4):731-744.

Namias, J., 1968, Long-range weather forecasting—History, current status, and outlook, *Am. Meteorol. Soc. Bull.* **49**(5):438-470.

Namias, J., 1969, Seasonal interactions between the North Pacific Ocean and the atmosphere during the 1960's, *Monthly Weather Rev.* **97**(3):173-192.

Namias, J., 1970, Macroscale variations in sea surface temperatures in the North Pacific, *Jour. Geophys. Research* **75**(3):565-582.

Namias, J., 1972*a*, Experiments in objectively predicting some atmospheric and oceanic variables for the winter of 1971-1972, *Jour. Appl. Meteorology* **11:**1164-1174.

Namias, J., 1972*b*, Large-scale and long-term fluctuations in some atmospheric and oceanic variables, in D. Dyrssen and D. Jagner (eds.), *The Changing Chemistry of the Oceans: Proceedings of the Twentieth Nobel Symposium*, New York: John Wiley & Sons; Stockholm: Almquist and Wiksell, pp. 27-48.

Namias, J., 1978, Persistence of U.S. seasonal temperatures up to one year, *Monthly Weather Rev.* **106:**1557-1567.

Namias, J., 1979, The enigma of drought—A challenge for terrestrial and extraterrestrial research, in B. M. McCormac and T. A. Seliga (eds.), *Solar-Terrestrial Influences on Weather and Climate*, Dordrecht, The Netherlands: D. Reidel, pp. 41-43.

Namias, J.; and Born, R., 1970, Temporal coherence in North Pacific sea-surface temperature patterns, *Jour. Geophys. Research* **75**(30):5952-5955.

Nansen F., 1918, Changes in oceanic and atmospheric temperatures and their relation to changes in the Sun's activity, *Washington Acad. Sci. Jour.* **8:**135-138 (abstract in *Monthly Weather Rev.* **46**(4):177-178).

Nash, W. C., 1902, Greenwich rainfall, 1841-1902, *Observatory (London)* **26:**414-416.

Nash, W. C., 1903, Greenwich rainfall, 1841-1902, *Symons' Meteorol. Mag. (London)* **38:**177-179.

Nash, W. C., 1904*a*, Monthly rainfall at the Royal Observatory, Greenwich, 1815-1903, *Royal Meteorol. Soc. Quart. Jour.* **30:**291-302.

Nash, W. C., 1904*b*, Sun-spots and hot summers, *Observatory (London)* **27:**319-320.

Nash, W. C., 1910, Daily rainfall at the Royal Observatory, Greenwich, 1841-1903, *Royal Meteorol. Soc. Quart. Jour.* **36:**309-322.

Nastrom, G. D., and Belmont, A. G., 1980, Evidence for a solar cycle signal in tropospheric winds, *Jour. Geophys. Research* **85**(C1):443-452.

Nastrom, G. D., and Belmont, A. G., 1982, Evidence for a solar cycle signal in tropospheric winds, *Jour. Geophys. Research* **88:**11025-11030.

National Defense University (jointly with U.S. Department of Agriculture), 1980, *Crop Yields and Climate Change to the Year 2000: A Survey of Expert Opinion*, Alexandria, Va.: Defense Documentation Center, Superintendent of Documents, No. D5402:C 61/2/vl.

National Research Council (NRC), 1979*a*, *Stratospheric Ozone Depletion by Halocarbons: Chemistry and Transport*, Washington, D.C.: U.S. National Academy of Sciences, 238p.

National Research Council (NRC), 1979*b*, *Carbon Dioxide and Climate: A Scientific Assessment*, Washington, D.C.: U.S. National Academy of Sciences, 22p.

Neher, H. V., 1971, Cosmic rays at high latitudes and altitudes covering four solar maxima, *Jour. Geophys. Research* **76**(7):1637-1651.

Nelson, J. H.; Hurwitz, L.; and Knapp, D. G., 1962, *Magnetism of the Earth*, Washington, D.C.: U.S. Government Printing Office, U.S. Department of Commerce, Coast and Geodetic Survey, Publication 40-1, 79p.

Nelson, J. H., 1951, Shortwave radio propagation correlation with planetary positions, *RCA Rev.* **12**(1):26-34.

Nelson, J. H., 1952, Planetary position effect on short wave signal quality, *Elec. Eng.* **71**(5):421-424.

Nelson, J. H., 1962, Do the planets cause sunstorms?, *Saturday Review*, October 6, 1962, pp. 64-66.

Nelson, J. H., 1963, Circuit reliability, frequency utilization, and forecasting in the high frequency communications band, in G. J. Gassman (ed.), *The Effect of Disturbances of Solar Origin on Communications*, New York: Macmillan, pp. 293-301.

Nelson, J. H., 1974, *Cosmic Patterns, Their Influence on Man and His Communication*, Washington, D.C.: American Federation of Astrology, 76p.

Neuberger, L., 1983, The Sun-weather connection—Sudden stratospheric warmings correlated with sudden commencements and solar proton events, in B. M. McCormac (ed.), *Weather and Climate Response to Solar Variation*, Boulder: Colorado Associated University Press, pp. 395-397.

Neuwirth, R., 1956, Wetterempfindlichkeit und aussere Einflusse auf physikalisch-chemische Systeme, *Angew. Meteorologie* **2**(10):311-317.

Newcomb, S., 1895, The elements of the four inner planets and the fundamental constants of astronomy, *Am. Ephemeris and Nautical Almanac Astron. Papers* **5**(pt. 4).

Newcomb, S., 1897, A new determination of the precessional constant with the resulting precessional motions, *Am. Ephemeris and Nautical Almanac Astron. Papers* **8**(pt. 1):73.

Newcomb, S., 1903, On sunspots and magnetic storms, Address to Astronomical and Astrophysical Society of America, *Science*.

Newcomb, S., 1908, A search for fluctuations in the Sun's thermal radiation through their influence on terrestrial temperature, *Am. Philos. Soc. Trans.* (new series) **21:**309-387.

Newell, R. E., 1982, The Southern Oscillation, *Jour. Climatology* **2**(4):357-373.

Newell, R. E.; Kidson, J. W.; Vincent, D. G.; and Boer, G. J., 1972, *The General Circulation of the Tropical Atmosphere and Interactions with Extratropical Latitudes*, vol. 1, Cambridge, Mass.: MIT Press, 258p.

Newhall, X. X.; Standish, E. M., Jr.; and Williams, J. G., 1983, DE 102: A numerically integrated ephemeris of the Moon and planets spanning forty-four centuries, *Astronomy and Astrophysics* **125:**150-167.

Newkirk, G. A., Jr., 1982, The nature of solar variability, in J. A. Eddy (ed.), *Solar Variations, Weather, and Climate*, Washington, D.C.: National Academy Press, pp. 33-47.

Newkirk, G. A., Jr., and Frazier, K., 1982, The solar cycle, *Physics Today*, April 1982, pp. 25-34.

Newman, E., 1965, Statistical investigation of anomalies in the winter temperature record of Boston, Massachusetts, *Jour. Appl. Meteorology* **4:**706-713.

Newman, J. E., and Pickett, R. C., 1974, World climates and food supply variations, *Science* **186:**877-881.

Newton, H. W., 1928, The Sun's cycle of activity, *Royal Meteorol. Soc. Quart. Jour.* **54**(227):161-173.

Newton, I., 1687, *Mathematical Principles of Natural Philosophy and System of the World* (trans. by A. Motte (1729) and rev. by F. Cajori), Berkeley: University of California Press, 680p.

Newton, R. R., 1969, Secular accelerations of the Earth and Moon, *Science* **166:**825-831.

Newton, R. R., 1972, *Medieval Chronicles and the Rotation of the Earth,* Baltimore, Md.: The Johns Hopkins University Press, 825p.

Ney, E. P., 1959, Cosmic radiation and the weather, *Nature* **183:**451-452.

Nicholls, N., 1978, Air-sea interaction and the quasi-biennial oscillation, *Monthly Weather Rev.* **106:** 1505-1508.

Nichols, E., 1920, Climate and its relation to acute respiratory conditions, *Monthly Weather Rev.* **48**(9):499-501.

Nicholson, S. E., and Flohn, H., 1980, African environment and climatic changes and the general atmospheric circulation in late Pleistocene and Holocene, *Clim. Change* **2:**313-348.

Nicolet, M., 1983, Changes in atmospheric chemistry related to solar flux variations, in B. M. McCormac (ed.), *Weather and Climate Response to Solar Variation,* Boulder: Colorado Associated University Press, pp. 117-128.

Nieto, M. M., 1972, *The Titius-Bode Law of Planetary Distances: Its History and Theory,* Oxford: Pergamon Press, 161p.

Nilsson, E., 1935, Traces of ancient changes of climate in East Africa, *Geog. Annaler* **17:**1-21.

Nilsson, E., 1938, Pluvial lakes in East Africa, *Geol. Fören. Stockholm Förh.* **60.**

Nilsson, E., 1940, Ancient changes of climate in British East Africa and Abyssinia, *Geog. Annaler* **22:**1-79.

Nilsson, E., 1949, The pluvials of East Africa. An attempt to correlate Pleistocene changes of climate, *Geog. Annaler* **31**(1-4):204-211.

Nippoldt, A., 1904, On the investigation of simultaneous occurrences in the solar activity and terrestrial magnetism, *Astrophys. Jour.* **20:**202-206.

Nishida, A., and Jacobs, J. A., 1962, World-wide changes in the geomagnetic field, *Jour. Geophys. Research* **67**(2):525-540.

Nodon, A., 1907, Observations sur l'action electrique du soleil et de la lune, *Acad. Sci. Comptes Rendus* **145:**521-523.

Nodon, A., 1909*a,* L'activité solaire et les phenomenes terrestres, *Soc. Meteorol. France Annuaire* **57:**223-225.

Nodon, A., 1909*b,* L'origine solaire des cyclones et des tempetes, *Soc. Belge Astronomie Bull.,* pp. 121-123.

Nodon, A., 1909*c,* La prevision du temps, *Soc. Belge Astronomie Bull.,* pp. 472-481.

Nodon, A., 1909*d,* Origine electrique des cyclones et des tempetes, *Cosmos* **60:**8.

Nodon, A., 1910*a,* Le role de l'electricite en meteorologie, *Ciel et Terre* **31:**353-358.

Nodon, A., 1910*b,* Recherches sur le magnetism terrestre, *Acad. Sci. Comptes Rendus* **150:**1711-1713.

Nodon, A., 1911, Les cyclones et les perturbations solaires, *Ciel et Terre* **32:**138-155.

Nohonoj, D., 1971, Irregularities in solar rotation and variations of the radius, *Sol. Dannye* **6:**93-97 (trans. courtesy of J. B. Blizard).

Nordmann, C., 1903*a,* La periode des taches solaires et les variations des temperatures moyennes annuelles de la Terre, *Acad. Sci. Comptes Rendus* **136:**1047-1050; also *Cosmos (Paris)* **52:**675-676. (Translated in Monthly Weather Rev. 31(8):371.)

Nordmann, C., 1903*b,* Theorie electro-magnetique des aurores boreales et des variations et perturbations du magnetism terrestre, *Electricien* **26**(Series 2):280-281.

Nordmann, C., 1904, The sun-spot period and the variations of the mean annual temperature of the Earth, *Smithsonian Inst. Rept. for 1903,* pp. 139-140 (translation of an article from *Rev. Général Sci.* 1903, pp. 803-808).

North, G. R.; Mengel, J. G.; and Short, D. A., 1983, Climatic response to a time varying solar constant, in B. M. McCormac (ed.), *Weather and Climate Response to Solar Variation,* Boulder: Colorado Associated University Press, pp. 243-255.

Nowack, J. F., 1908, A simple method of forecasting storms, *Symans' Meteorol. Mag. (London)* **43:**95-96.

Nowroozi, A. A.; Kuo, J.; and Ewing, Maurice, 1969, Solid Earth and oceanic tides recorded in the ocean floor off the coast of North Carolina, *Jour. Geophys. Research* **74**(2):605-614.

Nupen, W., and Kageorge, M., 1958, Bibliography on solar-weather relationships, *Am. Meteorol. Soc. Meteorol. Abs. and Bibliography,* 248p.

Nyberg, A., 1975, An experiment in forecasting monthly mean temperatures in Stockholm, *Tellus* **27**(1):34-38.

O'Brien, K., 1979, Secular variations in the production of cosmogenic isotopes in the Earth's atmosphere, *Jour. Geophys. Research* **84:**423-431.

O'Connell, M. D., 1909, Climate (meteorological environment) as a possible cause of pyrexia, *Jour. Tropical Medicine (London),* 15 March.

Oeschger, H.; Messerli, B.; and Svilar, M., eds., 1980, *Das Klima. Analysen und Modelle, Geschichte und Zukunft*, Berlin: Springer Verlag, 296p.

Oesterwinter, C., and Cohen, C. J., 1972, New orbital elements for Moon and planets, *Celestial Mechanics* **5:**317.

O'Gallagher, J. J., 1969, Analysis of changes in the modulated cosmic-ray spectrum near solar minimum, *Jour. Geophys. Research* **74**(1):43-52.

Ogilvie, A. E. J., 1984, The past climate and sea-ice record from Iceland, Part 1: Data to A.D. 1780, *Clim. Change* **6:**131-151.

Okal, E., and Anderson, D. L., 1975, On the planetary theory of sunspots, *Nature* **253:**511-513.

Olive, P., 1972, La region du Lac Leman depuis 10,000 ans: Donnees paleoclimatiques et prehistoriques, *Rev. Géographie Phys. et Géologie Dynam.* **16**(3):253-264.

Olson, D. E., 1971, The evidence for auroral effects on atmospheric electricity, *Geofisica Pura e Applicata* **84:**118-138.

Olson, R. H., 1979, Comment on "Stratospheric temperatures during solar cycle 20" by R. S. Quiroz, *Jour. Geophys. Research* **84:**7898.

Olson, R. H.; Roberts, W. O.; and Zerefos, C. S., 1975, Short-term relationships between solar flares, geomagnetic storms and tropospheric vorticity patterns, *Nature* **257:**113-115.

O'Mahony, G., 1965, Rainfall and moon phase, *Royal Meteorol. Soc. Quart. Jour.* **91:**196-208.

Omori, F., 1904, Note on the relation between earthquake frequency and atmospheric pressure, *Tokyo Math. and Phys. Soc. Brief Repts.* **2:**113-117.

Öpik, E. J., 1967, Climatic changes, in S. K. Runcorn (ed.), *The Application of Modern Physics to the Earth and Planetary Interiors*, New York: John Wiley & Sons, pp. 139-145.

Öpik, E. J., 1972, Planetary tides and sunspots, *Irish Astron. Jour.* **10:**298-301.

Oppenheim, S., 1910, Uber die Bestimmung der Periode einer periodischen Erscheinung mit Anwendung auf die Theorie des Erdmagnetismus, *Zeitschr. Meteorologische* **27:**266-270.

Orville, R. E., and Spencer, D. W., 1979, Global lightning flash frequency, *Monthly Weather Rev.* **107:**934-943.

Otaola, J. A., and Zenteno, G., 1983, On the existence of long-term periodicities in solar activity, *Solar Physics* **89:**209-213.

Paetzold, H. K., 1969, Variation of the vertical ozone profile over middle Europe from 1951 to 1968, *Annales Geophysique* **25**(1):347-349.

Paetzold, H. K., 1973, The influence of solar activity on the stratospheric ozone layer, *Geofisica Pura e Applicata* **105–108:**1309-1311.

Paetzold, H. K.; Piscalar, F.; and Zschorner, H., 1972, Secular variation of the stratospheric ozone layer over middle Europe during the solar cycles from 1951-1972, *Nature* **240:**106-107.

Page, D. E., 1983, Heliomagnetism, geomagnetism and the Earth's atmosphere, in B. M. McCormac (ed.), *Weather and Climate Response to Solar Variation*, Boulder: Colorado Associated University Press, pp. 345-363.

Page, J., 1906, Has the Gulf Stream any influence on the weather of New York City?, *Monthly Weather Rev.* **34:**465.

Page, L. F., 1937, Temperature and rainfall changes in the United States during the past 40 years, *Monthly Weather Rev.* **65**(2):46-54.

Palmer, A. H., 1920, Economic results of deficient precipitation in California, *Monthly Weather Rev.* **48**(10):586-589.

Panella, G., 1972, Paleontological evidence on the Earth's rotational history since the early pre-Cambrian, *Astrophysics and Space Sci.* **16:**212-237.

Panofsky, H. A., 1967, Meteorological applications of cross-spectrum analysis, in B. Harris (ed.), *Advanced Seminar on Spectral Analysis of Time Series*, New York: John Wiley & Sons.

Park, C. G., 1976*a*, Solar magnetic sector effects on the vertical atmospheric electric field at Vostok, Antarctica, *Geophys. Research Letters* **3:**475-478.

Park, C. G., 1976*b*, Downward mapping of high-latitude electric fields to the ground, *Jour. Geophys. Research* **81:**168-174.

Parker, B. C.; Zeller, E. J.; and Gow, A. J., 1982, Nitrate fluctuations in Antarctic snow and firn: Potential sources and mechanisms of formation, *Annals Glaciology* **3:**243-248.

Parker, B. N., 1976, Global pressure variation and the 11-year solar cycle, *Meteorol. Mag.* **105:**33-44.

Parker, D. E., 1985, Climatic impact of explosive volcanic eruptions, *Meteorol. Mag.* **114:**149-161.

Parker, E. N., 1979, *Cosmical Magnetic Fields: Their Origin and Their Activity*, Oxford: Clarendon Press; New York: Oxford University Press, 841p.

Parkinson, J. H.; Morrison, L. V.; and Stephenson, F. R., 1980, The constancy of the solar diameter over the past 250 years, *Nature* **288:**548-551.

Parkinson, W. D., 1983, *Introduction to Geomagnetism*, Scottish Academic Press/Elsevier Scientific, 433p.

Parthasarathy, B., and Dhar, O. N., 1976, A study of the trends and periodicities in the seasonal and annual rainfall of India, *India Jour. Meteorology, Hydrology, and Geophysics* **27:**23-28.

Parthasarathy, B., and Mooley, D. A., 1978, Some features of long homogeneous series of Indian summer monsoon rainfall, *Monthly Weather Rev.* **106**(6):771-781.

Paul, A. K., 1985, Analysis and prediction of sunspot numbers, *Geophys. Research Letters* **12**(12): 833-834.

Peck, J.; and Snow, E. C., 1913, The correlation of

rainfall, *Royal Meteorol. Soc. Quart. Jour.* **39:**307-313.

Pedersen, G. P. H., and Rochester, M. G., 1972, Spectral analyses of the Chandler wobble, in P. J. Melchior and S. Yumi (eds.), *Rotation of the Earth,* Dordrecht, The Netherlands: D. Reidel, pp. 33-38.

Pekeris, C. L., 1937, Atmospheric oscillations, *Royal Soc. [London] Proc.* **158A:**650-671.

Penck, A., 1914, The shifting of the climatic belts, *Scottish Geog. Mag. (Edinburgh)* **30**(6):281-293.

Penner, J. E., and Chang, J. S., 1978, Possible variations in atmospheric ozone related to the eleven-year solar cycle, *Geophys. Research Letters* **5:**817-820.

Peppler, A., 1931, Energieschwankungen der nordatlantischen Zirkulation und Sonnenflecken 1881-1923, *Gerlands Beitr. Geophysik* **29**(2):187-200.

Perkins, J. A., and Sims, J. D., 1983, Correlation of Alaskan varve thickness with climatic parameters, and use in paleoclimatic reconstruction, *Quat. Research* **20:**308-321.

Peroche, J., 1904, Le mouvement de nos temperatures et la precession des equinoxes, *Rev. Sci. France et de l'etranger (Paris)* **1**(Series 5):579-583.

Petterssen, S., 1956, *Weather Analysis and Forecasting,* 2nd ed., vol. 1, New York: McGraw-Hill, 428p.

Pettersson, O., 1904, On the influence of ice-melting upon oceanic circulation, *Geog. Jour.* **24:**285-333.

Pettersson, O., 1907, On the influence of ice-melting upon oceanic circulation, *Geog. Jour.* **30:**273-295, 671-675.

Pettersson, O., 1912, The connection between hydrographical and meteorological phenomena, *Royal Meteorol. Soc. Quart. Jour.* **38:**173-191.

Pettersson, O., 1914*a,* Climatic variations in historic and prehistoric time, *Svenska Hydrogr. Biol. Komm., Skriften, No. 5,* 26p.

Pettersson, O., 1914*b,* On the occurrence of lunar periods in solar activity and the climate of the Earth, A study in geophysics and cosmic physics, *Svenska Hydrogr. Biol. Komm., Skriften.*

Pettersson, O., 1915*a,* Long periodical variations of the tide-generating force, *Conseil Permanente Internationale pour Exploration de la Mer (Copenhagen), Pub. Circ. No. 65,* pp. 2-23.

Pettersson, O., 1915*b,* Om Solflacksfenomenets Periodicitet och des Sambad med Klimatets Forandringar. En Studie i kosmisk Fysik, *Kungl. Svenska Vetenskapsakademiens, Handligar* **53**(1):64p.

Pettersson, O., 1930, The tidal force, *Geog. Annaler* **12:**261-322.

Pettersson, O., 1934, Tidvattnets problem, IV. Der interna parallaktiska tidvattnet, en studie i geofysik, *Arkiv Matematik, Astronomi och Fysik* **25A**(1):11p.

Pettersson, W. J., 1929, The past cold winter and the possibility of long-range weather forecasting, *Monthly Weather Rev.* **57**(6):256-257.

Pettit, E., 1927, Ultra-violet solar radiation, *Nat. Acad. Sci. Proc.* **13:**380-387.

Pettit, E., 1932, Measurements of ultra-violet solar radiation, *Astrophys. Jour.* **75:**185-221.

Philander, S. G. H., 1983, El Niño Southern Oscillation phenomena, *Nature* **302:**295-301.

Phillips, W. F. R., 1907, Relation of temperature, humidity and winds to chronic nephritis, *British Balneol. and Climatol. Soc. Jour.* **11:**280-285.

Pickard, G. W., 1927*a,* The correlation of radio reception, solar activity, and terrestrial magnetism, I, *Inst. Radio Engineers Proc.* **15**(2):83-97.

Pickard, G. W., 1927*b,* The correlation of radio reception, solar activity, and terrestrial magnetism, II, *Inst. Radio Engineers Proc.* **15**(9):749-766.

Pickering, W. H., 1903, Relation of the Moon to the weather, *Popular Astronomy* **11:**327-328.

Pickering, W. H., 1920, The relation of prolonged tropical droughts to sunspots, *Monthly Weather Rev.* **48**(100):589-592.

Piddington, J. H., 1960, Geomagnetic storm theory, *Jour. Geophys. Research* **65:**93-106.

Piddington, J. H., 1978, The flux-rope-fibre theory of solar magnetic fields, *Astrophys. and Space Sci.* **55:**401-425.

Pilcher, J. R., and Gray, B., 1982, The relationship between oak tree growth and climate in Britain, *Jour. of Ecology* **70:**297-304.

Pilcher, J. R., and Hughes, M., 1982, The potential of dendrochronology for the study of climatic change, in A. F. Harding (ed.), *Climatic Change in Later Prehistory,* Edinburgh University Press, pp. 75-84.

Pilgrim, L., 1903, Der Einfluss der Schwankungen der Schiefe der Ekliptik und der Exzentrizitat der Erdbahn auf das Klima mit besonderer Berucksichtigung des Eiszietproblems, *Wurttemberg Mathematisch-Naturw. Ver. Mathematisch-Naturw. Mitt. (Stuttgart)* **5**(Series 2):33-62.

Pimm, R. S., and Bjorn, T., 1969, *Prediction of Smoothed Sunspot Number Using Dynamic Relations between the Sun and Planets,* Huntsville, Alabama, Lockheed Missiles and Space Company, Huntsville Research and Engineering Center, Final Report, Contract NAS 8-21445 (study of mathematical analysis and prediction of solar activity), variously paginated.

Pines, D., and Shaham, J., 1973, Seismic activity, polar tides and the Chandler wobble, *Nature* **245:**77-81.

Pisias, N. G., 1978, Paleoceanography of the Santa Barbara Basin during the last 8,000 years, *Quat. Research* **10:**366-384.

Pittock, A. B., 1975, Climatic change and the pattern of Australian rainfall, *Search* **6:**498-504.

Pittock, A. B., 1978, A critical look at long-term sun-weather relationships, *Rev. Geophysics and Space Physics* **16:**400-420.

Pittock, A. B.; Frakes, L. A.; Jenssen, D.; Peterson, J. A.; and Zillman, J. W. eds., 1978, *Climatic Change and Variability. A Southern Perspective*, Cambridge: Cambridge University Press, 455p.

Pittock, A. B., 1979, Solar cycles and the weather: Successful experiments in auto-suggestion?, in B. M. McCormac and T. A. Seliga (eds.), *Solar-Terrestrial Influences on Weather and Climate*, Dordrecht, The Netherlands: D. Reidel, pp. 181-191.

Pittock, A. B., 1980*a*, Enigmatic variations, *Nature* **283:**605-606.

Pittock, A. B., 1980*b*, Patterns of climatic variations in Argentina and Chile—I. Precipitation, 1931-60, *Monthly Weather Rev.* **108:**1347-1361.

Pittock, A. B., 1980*c*, Monitoring causality and uncertainty in a stratospheric context, *Geofisica Pura e Applicata* **118:**643-661.

Pittock, A. B., 1983, Solar variability, weather and climate: An update, *Royal Meteorol. Soc. Quart. Jour.* **109**(1):23-55.

Pittock, A. B., and Shapiro, R., 1982, Assessment of evidence of the effect of solar variation on weather and climate, in J. A. Eddy (ed.), *Solar Variability, Weather, and Climate*, Washington, D.C.: National Academy Press, pp. 64-75.

Plassmann, J., 1913, Wetter und Mond, *Ver. von Freunden der Astronomie u. Kosmischen Physik Mitt.* **23:**151-155.

Plumb, R. A., and Bell, R. C., 1982, A model of the quasi-biennial oscillation on an equatorial beta-plane, *Royal Meteorol. Soc. Quart. Jour.* **108**(2):335-352.

Pollack, J. B.; Borucki, W. J.; and Toon, O. B., 1979, Are solar spectral variations a driver for climatic change?, *Nature* **282:**600-603.

Pomerantz, W. A., and Duggal, S. P., 1972, Record-breaking cosmic ray storm stemming from solar activity in August 1972, *Nature* **241:**331-333.

Pomerantz, W. A., and Duggal, S. P., 1974, The Sun and cosmic rays, *Rev. Geophysics and Space Physics* **12:**343-361.

Portig, W. H., and Freeman, J. C., 1965, *Investigation of the Relation between Solar Variations and Weather on Earth*, Washington, D.C.: National Aeronautics and Space Administration, Final Report, NASA Contract No. NASW-724.

Posner, G. S., 1957, The Peru Current, *Bingham Oceanog. Lab. Bull.* **16**(2):106-154.

Prager, M., 1907, Eine Vorhersage der Regenfalle in Indien fur das Jahr 1906, *Wetter* **24:**11-16.

Pramanik, S. K., and Jagaanathan, P., 1953, Climatic changes in India—(I). Rainfall, *Indian Jour. Meteorology and Geophysics* **4:**291-309.

Press, F., and Brace, W. F., 1965, Earthquake prediction, *Science* **152:**1575-1584.

Press, F., and Briggs, P., 1975, Chandler wobble, earthquakes, rotation and geomagnetic changes, *Nature* **256:**270-273.

Priesendorfer, R. W., and Mobley, C. D., 1982, *Climate Forecast Verifications, U.S. Mainland, 1974-1982*, Seattle, Wash.: NOAA Technical Memorandum ERL PMEL-36.

Priestley, C. H. B., 1966, Droughts and wet periods and their association with sea-surface temperature, *Australian Jour. Sci.* **29:**56-57.

Prohaska, J. T., and Willett, H. C., 1983, Dominant modes of relationship between U. S. temperatures and geomagnetic activity, in B. M. McCormac (ed.), *Weather and Climate Response to Solar Variation*, Boulder: Colorado Associated University Press, pp. 489-494.

Prokudina, V. A., 1981, Variations of the solar radius and the motion of the Sun about the barycenter, *Astron. Tsirk., no. 1119*, pp. 6-8 (in Russian, trans. by J. B. Blizard).

Proost, A., 1908, Les taches solaires et la prevision du temps, *Rev. Général Agron. (Louvain)*, pp. 343-346.

Pruppacher, H. R., and Klett, J. D., 1978, *Microphysics of Clouds and Precipitation*, Dordrecht, The Netherlands: D. Reidel, 714p.

Psuty, N. P., 1965, Beach-ridge development in Tabasco, Mexico, *Assoc. Am. Geographers Annals* **55:**112-124.

Quayle, E. T., 1910, On the possibility of forecasting the approximate winter rainfall for Northern Victoria, Australia, *Commonwealth Bur. Meteorology (Melbourne) Bull. No. 5.*

Quayle, E. T., 1925, Australian rainfall in sunspot cycles, *Australia Meteorol. Branch Bull. No. 22.*

Quenisset, F., 1903, Remarques sur le dernier groupe de taches solaires et les perturbations magnetiques, *Acad. Sci. Comptes Rendus* **137:**747-748.

Quensel, P., 1901, Sunspots and terrestrial temperature, *Knowledge (London)* **24:**108-109.

Quiroz, R. S., 1979, Stratospheric temperature during solar cycle 20, *Jour. Geophys. Research* **84**(C5):2415-2420.

Rabot, C., 1915, Recents travaux glaciers dans les Alpes françaises, *La Geographie* **30:**257-268.

Raisbeck, G. M., and Yiou, F., 1980, Temporal variations in cosmogenic ^{10}Be production: Implications for radiocarbon dating, *Radiocarbon* **22:**245-249.

Ralph, E. K., and Michael, H. N., 1967, Problems of the radiocarbon calendar, *Archaeometry* **10:**3-11.

Ramanathan, V., 1980, Climatic effects of ozone change: A review, in A. P. Mitra (ed.), *Low Latitude Aeronomical Processes*, Oxford: Pergamon Press, pp. 223-236.

Ramanathan, V., 1982, Coupling through radiative and chemical processes, in J. A. Eddy (ed.), *Solar Variability, Weather, and Climate*, Washington, D.C.: National Academy Press, pp. 83-91.

Ramanathan, V.; Callis, L. B.; and Boughner, R. E., 1976, Sensitivity of surface temperature and atmospheric temperature to perturbations in the stratospheric concentration of ozone and nitrogen dioxide, *Jour. Atmos. Sci.* **33:**1092-1112.

Ramanathan, V., and Dickinson, R. E., 1979, The role of stratospheric ozone in the zonal and seasonal radiative energy balance of the Earth-troposphere system, *Jour. Atmos. Sci.* **36:**1084-1104.

Rampino, M. R., and Sanders, J. E., 1981, Episodic growth of Holocene tidal marshes in the northeastern United States: A possible indicator of eustatic sea-level fluctuations, *Geology* **9:**63-67.

Rampino, M. R.; Self, S.; and Fairbridge, R. W., 1979, Can rapid climate change cause volcanic eruptions? *Science* **206:**826-829.

Ramsay, A., 1884, *A Bibliography, Guide and Index to Climate*, London: W. Swan Sonnenschein Company, 449p.

Rangarajan, S., 1965, Effect of solar activity on atmospheric ozone, *Nature* **206:**497-498.

Rao, K. S. R., and Harindranathan, N. M. V., 1981, Possible linkage between atmospheric total ozone and solar magnetic sector boundary passage, *Jour. Atmos. and Terr. Physics* **43:**367-372.

Rao, M. B. S., 1902, The rainfall in the city of Madras and the frequency of sun spots, *Monthly Weather Rev.* **30:**438-440.

Rao, U. R., 1972, Solar modulation of galactic cosmic radiation, *Space Sci. Rev.* **12:**719-809.

Rasmusson, E. R., and Wallace, J. M., 1983, Meteorological aspects of the El Niño/Southern Oscillation, *Science* **222:**1195-1202.

Rasool, S. I., 1961, Effets de l'activité solaire sur l'ozone atmospherique et la hauteur de la tropopause, *Geofisica Pura e Applicata* **48:**93-101.

Rasool, S. I., 1964, Global distribution of the net energy balance of the atmosphere from TIROS radiation data, *Science* **143:**567-569.

Rasool, S. I., and Schneider, S. H., 1971, Atmospheric carbon dioxide and aerosols: Effects of large increases on global climate, *Science* **173:**138-141.

Ratcliffe, J. A., 1970, *Sun, Earth and Radio: An Introduction to the Ionosphere and Magnetosphere*, London: World University Library, Weidenfeld and Nicolson, 256p.

Ratcliffe, R. A. S.; Weller, J.; and Collison, P., 1978, Variability in the frequency of unusual weather over approximately the last century, *Royal Meteorol. Soc. Quart. Jour.* **104:**243-255.

Raulin, V., 1902*a*, Variation seculaire du magnetism terrestre, *Annales Chimie et Physique* **25**(Series 7):289-307.

Raulin, V., 1902*b*, Les regimes pluviometriques saisonnaux des isles Britanniques, d'apres les observations des 25 années 1866-1890, *Soc. Meteorol. France Annuaire* **50:**129-139.

Raulin, V., 1903, Sur les observations pluviometriques en Australie, de 1881 à 1900, *Soc. Meteorol. France Annuaire* **51:**121-134.

Raulin, V., 1904, A three years' period in rainfall, *Symons' Meteorol. Mag. (London)* **39:**111-112.

Raulin, V., 1912, Distribution of rain in Mauritius during the decade 1891-1900, *Symons' Meteorol. Mag. (London)* **47:**5-7.

Ravenstien, E. G., 1901, The lake level of Victoria Nyanza, *Geog. Jour.* **18:**403-406.

Rawson, H. E., 1907, Anticyclones as aids to long distance forecasts, *Royal Meteorol. Soc. Quart. Jour.* **33:**309-310.

Rawson, H. E., 1908, The anticyclonic belt of the southern hemisphere, *Royal Meteorol. Soc. Quart. Jour.* **34:**165-188.

Rawson, H. E., 1909, The anticyclonic belt of the northern hemisphere, *Royal Meteorol. Soc. Quart. Jour.* **35:**233-248.

Rawson, H. E., 1910*a*, Periodic changes in the seasonal positions and tracks of anticyclones, *Royal Meteorol. Soc. Quart. Jour.* **36:**65-67.

Rawson, H. E., 1910*b*, The North Atlantic anticyclone: Tracks of centres of high areas, 1882-1883, *Royal Meteorol. Soc. Quart. Jour.* **36:**197-209.

Redway, J. W., 1913, The fluctuating climate of North America, *Geog. Jour.* **41:**75-76.

Reed, C. D., 1914, Droughts at New York City, *Monthly Weather Rev.* **42**(11):629-631.

Reed, C. D., 1932, June temperature indicates corn maturity in Iowa, *Am. Meteorol. Soc. Bull.* **14:**199-202.

Reed, R. J., 1964, A tentative model of the 26-month oscillation in tropical latitudes, *Royal Meteorol. Soc. Quart. Jour.* **90:**441-466.

Reed, R. J., 1965, The present status of the 26-month oscillation, *Am. Meteorol. Soc. Bull.* **46:**374-387.

Reed, R. J., and Rogers, D. G., 1962, The circulation of the tropical stratosphere in the years 1954-1960, *Jour. Atmos. Sci.* **19:**127-135.

Reid, G. C., 1974, Polar-cap absorption—Observations and theory, *Fundamentals Cos. Physics.* **1:**167-200.

Reid, G. C., and Gage, K. S., 1983, Solar variability and the height of the tropical tropopause, in B. M. McCormac (ed.), *Weather and Climate Response to Solar Variation*, Boulder: Colorado Associated University Press, pp. 569-579.

Reid, G. C.; McAfee, J. R.; and Crutzen, P. J., 1978, Effects of intense stratospheric ionization events, *Nature* **275:**489-492.

Reineck, H.-E., 1961, The Orkanflut vom 16. Februar 1962, *Natur u. Mus.* **92**(5):151-172.

Reitan, C. H., 1974, Frequencies of cyclones and cyclogenesis for North America, 1951-1970, *Monthly Weather Rev.* **102:**861-868.

Reiter, R., 1953, Beziehungen zwischen Sonneneruptionen, Wetterablauf und Reaktionen des

Menschen, dargestellt am Verlauf der Infra-Langwellenstorunger, der Geburtenziffer und der Kerkehrsunfallziffer, *Angew. Meteorologie* **1**(10):289-303.

Reiter, R., 1969, Solar flares and their impact on potential gradient and air-earth current characteristics at high mountain stations, *Geofisica Pura e Applicata* **72:**259-267.

Reiter, R., 1971, Further evidence for impact of solar flares on potential gradient and air-earth current characteristics at high mountain stations, *Geofisica Pura e Applicata* **86:**142-158.

Reiter, R., 1972, Case study concerning the impact of solar activity upon potential gradient and air-earth current in the lower troposphere, *Geofisica Pura e Applicata* **94:**218-225.

Reiter, R., 1973, Increased influx of stratospheric air into the lower troposphere after solar H alpha and X ray flares, *Jour. Geophys. Research* **78**(27):6167-6172.

Reiter, R., 1976, The electrical potential of the ionosphere as controlled by the solar magnetic sector structure, *Naturwissenschaften* **63**(Part 4):192.

Reiter, R., 1977, The electrical potential of the ionosphere as controlled by the solar magnetic sector structure, result of a study over a period of a solar cycle, *Jour. Atmos. and Terr. Physics* **39:**95-99.

Reiter, R., 1983, Modification of the stratospheric ozone profile after acute solar events, in B. M. McCormac (ed.), *Weather and Climate Response to Solar Variation,* Boulder: Colorado Associated University Press, pp. 95-116.

Ren, Z., and Li, Z., 1980, Effect of motions of planets on climatic changes in China, *Kexue Tongbao* **25**(5):417-422.

Rex, D. F., 1950*a,* Blocking action in the middle troposphere and its effect on regional climate, I: An aerological study of blocking action, *Tellus* **2:**196-211.

Rex, D. F., 1950*b,* Blocking action in the middle troposphere and its effect on regional climate, II: The climatology of blocking action, *Tellus* **2:**275-297.

Rex, D. F., 1951, The effect of Atlantic blocking upon European climate, *Tellus* **3:**100-112.

Richardson, C., 1919, Australian droughts, *Monthly Weather Rev.* **47**(12):860.

Richardson, R., 1910, On the occurrence of great cold throughout Scotland during November and December, 1909, and January, 1910, *Scottish Meteorol. Soc. Jour.* **15:**158-162.

Richmond, A. D., 1976, Electric field in the ionosphere and plasmasphere on quiet days, *Jour. Geophys. Research* **81**(7):1447-1450.

Richter, C. M., 1902, Sonnenflecken, Erdmagnetismus und Luftdruck, *Zeitschr. Meteorologische* **19:**386-389.

Richter-Bernburg, G., 1964, Solar cycle and other climatic periods in varvitic evaporites, in A. E. M. Nairn (ed.), *Problems in Palaeoclimatology,* New York: John Wiley & Sons, pp. 510-519.

Riehl, H.; El-Bakry, M.; and Meitin, J., 1979, Nile River discharge, *Monthly Weather Rev.* **107:**1546-1553.

Riehl, H., and Meitin, J., 1979, Nile River discharge: A barometer of short-period climate variations, *Science* **206:**1178-1179.

Rikitake, T., 1966*a,* Westward drift of the equatorial component of the Earth's magnetic dipole, *Jour. Geomagnetism and Geoelectricity* **18:**383-392.

Rikitake, T., 1966*b, Electromagnetism and the Earth's Interior,* Amsterdam: Elsevier Publishing Company, 308p.

Rinehart, J. S., 1972*a,* 18.6-year Earth tide regulates geyser activity, *Science* **177:**346-347.

Rinehart, J. S., 1972*b,* Fluctuations in geyser activity caused by variations in Earth tidal forces, barometric pressure and tectonic stresses, *Jour. Geophys. Research* **77:**342-350.

Ringberg, B., and Rudmark, L., 1985, Varve chronology based upon glacial sediments in the area between Karlskrona and Kalmar, southeastern Sweden, *Boreas* **14**(2):107-110.

Roberts, W. O., 1979, Introductory review of solar-terrestrial weather and climate relationships, in B. M. McCormac and T. A. Seliga (eds.), *Solar-Terrestrial Influences on Weather and Climate,* Dordrecht, The Netherlands: D. Reidel, pp. 29-40.

Roberts, W. O., and Olson, R. H., 1973*a,* Geomagnetic storms and wintertime 300-mb trough development in the North Pacific-North Atlantic area, *Jour. Atmos. Sci.* **30:**135.

Roberts, W. O., and Olson, R. H., 1973*b,* New evidence for effects of variable solar corpuscular emission on the weather, *Rev. Geophysics and Space Sci.* **11:**731-740.

Robertson, E. C., ed., 1972, *The Nature of the Solid Earth,* Proceedings of a Symposium Held at Harvard University, April 16-18, 1970, New York: McGraw-Hill, 677p.

Robinson, E., and Robbins, R. C., 1969, Atmospheric CO concentrations on the Greenland Ice Cap, *Jour. Geophys. Research* **74**(8):1968-1973.

Roble, R. G., 1977, The thermosphere, in F. J. Johnson (ed.), *The Upper Atmosphere, and Magnetosphere,* Washington, D.C.: National Academy of Science, pp. 57-71.

Roble, R. G.; Dickinson, R. E.; and Ridley, E. C., 1977, Seasonal and solar cycle variations of the zonal mean circulation in the thermosphere, *Jour. Geophys. Research* **82**(35):5493-5504.

Roble, R. G., and Hays, P. B., 1979, Electrical coupling between the upper and lower atmosphere, in B. M. McCormac and T. A. Seliga (eds.), *Solar-*

Terrestrial Influences on Weather and Climate, Dordrecht, The Netherlands: D. Reidel, pp. 233-241.

Roble, R. G., and Hays, P. B., 1982, Solar-terrestrial effects on the global electrical circuit, in J. A. Eddy (ed.), *Solar Variability, Weather, and Climate,* Washington, D.C.: National Academy Press, pp. 92-106.

Robock, A., 1978, Internally and externally caused climate change, *Jour. Atmos. Sci.* **35:**1111-1122.

Rochester, M. G., 1960, Geomagnetic westward drift and irregularities in the Earth's rotation, *Royal Soc. London Philos. Trans.* **252A:**531-555.

Rochester, M. G., 1976, The secular decrease of obliquity due to dissipative core-mantle coupling, *Geophys. Jour.* **46:**109-126.

Rochester, M. G.; Jacobs, J. A.; Smylie, D. E.; and Chong, K. F., 1975, Can precession power the geomagnetic dynamo?, *Geophys. Jour.* **43:**661-678.

Rochester, M. G., and Smylie, D. E., 1965, Geomagnetic core-mantle coupling and the Chandler wobble, *Geophys. Jour.* **10:**289-315.

Roden, R. B., 1963, Electromagnetic core-mantle coupling, *Geophys. Jour.* **10:**289-315.

Rodenwald, M., 1958, Beitrage zur Klimaschwankung im Meere, 10. Beitrag. Die Anomalie der Wassertemperatur und der Zirkulation in Nordpazifischen Ozean und an der Kuste Perus im Jahre 1955, *Deutsche Hydrographische Zeitschr.* **11.**

Roedel, W., 1980, On the climate-radiocarbon relationship: Nitric oxide and ozone as connecting links between radiation and the Earth's surface temperature, *Radiocarbon* **22:**250-259.

Romanchuk, P. R., 1981, The nature of solar cyclicity, *Soviet Astronomy* **25:**87-92.

Ronai, A., 1984, The development of the Quaternary geology in Hungary, *Acta Geol. Hungarica* **27**(1/2):75-90.

Roosen, R. G.; Harrington, R. S.; Giles, J.; and Browning, I., 1976, Earth tides, volcanoes and climate change, *Nature* **261:**680-682.

Root, C. J., 1937, Deaths during the heat wave of July, 1936, at Detroit, *Am. Meteorol. Soc. Bull.* **18**(6):232-233.

Rose, J. K., 1932, Climate and corn yield in Indiana, 1887-1930, *Indiana Acad. Sci. Proc.* **42:**317-321.

Rosen, R. D., and Salstein, D. A., 1983, Variations in atmospheric angular momentum on global and regional scales and the length of day, *Jour. Geophys. Research* **88**(C-9):5451-5470.

Rosen, R. D.; Salstein, D. A.; Eubanks, T. M.; Dickey, J. O.; and Steppe, J. A., 1984, An El Niño signal in atmospheric angular momentum and Earth rotation, *Science* **225:**411-414.

Rosenberg, R. L., 1970, Unified theory of interplanetary magnetic field, *Solar Physics* **15:**72-78.

Rosenberg, R. L., and Coleman, P. J., Jr., 1974, 27-day cycle in the rainfall at Los Angeles, *Nature* **250:**481-484.

Ross, A. D., 1911, The origin of magnetic storms, *Glasgow Philos. Soc. Proc.* **42:**131-136.

Rossby, C. G.; Allen, R.; Holmboe, J.; Namias, J.; Page, L.; and Willett, H. C., 1939, Relation between variations in the intensity of the zonal circulation of the atmosphere and the displacement of the semi-permanent centers of action, *Jour. Marine Research* **2**(1):38-55.

Rossby, C. G., and Willett, H. C., 1948, The circulation of the upper troposphere and lower stratosphere, *Science* **108:**643-652.

Rowswell, B. T., 1909, The magnetic storm of September 25, [1909], *Symons' Meteorol. Mag. (London)* **44:**163.

Roy, A. E., 1982, *Orbital Motion,* 2nd ed. Bristol: A. Hilger, 495p.

Roy, A. E., and Ovenden, M. W., 1954, On the occurrence of commensurable mean motions in the solar system, *Royal Astron. Soc. Monthly Notices* **114:**234-241.

Rubens, H., and Schwarzschild, K., 1914, Sind im Sonnenspektrum Warmestrahlen von grosser Wellenlange vorhanden?, *Kongl. Preussischen Akad. Wiss. Sistzungsber.,* 1914, pp. 702-708.

Rudel, K., 1904, Sonnenscheindauer und Influenza, *Wetter* **21:**131-132.

Ruderman, M. A., and Chamberlain, J. W., 1975, Origin of the sunspot modulation of ozone: Its implications for stratospheric NO injection, *Planetary and Space Sci.* **23:**247-268.

Ruderman, M. A.; Foley, H. M.; and Chamberlain, J. W., 1976, Eleven year variation in polar ozone and stratospheric-ion chemistry, *Science* **192:**555-557.

Rudolph, W. E., 1953, Weather cycles on the South American west coast, *Geog. Rev.* **43**(4):565-566.

Runcorn, S. K., ed., 1967, *The Application of Modern Physics to the Earth and Planetary Interiors,* New York: John Wiley & Sons, 692p.

Runcorn, S. K., 1970, A possible cause of the correlation between earthquakes and polar motion, in L. Mansinha, D. E. Smylie, and A. E. Beck (eds.), *Earthquake Displacement Fields and the Rotation of the Earth,* Dordrecht, The Netherlands: D. Reidel, pp. 181-187.

Russell, C. T., and McPherron, R. L., 1973, Semiannual variation of geomagnetic activity, *Jour. Geophys. Research* **78:**92-108.

Russell, H. C., 1896, On periodicity of good and bad seasons, *Royal Soc. New South Wales Jour.* **30:**70-115.

Russell, H. C., 1902, The fallacy of assuming that a wet year in England will be followed by a wet year in Australia, *Royal Soc. New South Wales Jour.,* p. 314.

Russell, R. J., 1932, Dry climates of the United States: II. Frequency of dry and desert years 1901-20, *California Univ. Pubs. Geography* **5.**

Rycroft, M. J., and Theobald, A. G., 1978, Estimates of the stratospheric temperature variation in response to changes of the flux of solar UV radiation, *Space Research* **18:**99-102.

Saarnisto, M., 1985, Long varve series in Finland, *Boreas* **14**(2):133-138.

Sagalyn, R. C., and Faucher, G. A., 1957, Time variations of charged atmospheric nuclei, in H. Weickmann and W. Smith (eds.), *The Artificial Stimulation of Rain, Proceedings of the First Conference on the Physics of Cloud and Precipitation Particles,* London: Pergamon Press.

Sakaguchi, Y., 1982, Climatic variability during the Holocene Epoch in Japan and its causes, *Tokyo Univ. Dept. Geog. Bull.* **14:**1-27.

Sakurai, K., 1977, Equatorial solar rotation and its relation to climatic changes, *Nature* **269:**401-402.

Salinger, M. J., 1979, New Zealand climate: The temperature record, historical data and some agricultural implications, *Clim. Change* **2:**109-126.

Salisbury, G. N., 1903, A curious coincidence. Is it accidental or governed by law? [Apparent 3-year cycles in precipitation correlated with annual mean barometer], *Monthly Weather Rev.* **31:**229.

Salter, M. deC. S., and Glasspoole, J., 1923, The fluctuations of annual rainfall in the British Isles considered cartographically, *Royal Meteorol. Soc. Quart. Jour.* **49**(208):207-229.

Sanders, J. E., 1981, *Principles of Physical Geology,* New York: John Wiley & Sons, 624p.

Sandstrom, J. W., 1926, Ueber den Einfluss des Golfstromes auf die Wintertemperatur in Europa, *Zeitschr. Meteorologische* **43:**401-411 (summary by A. Walters, 1928, *Meteorol. Mag.* **63:**61-63).

Sanford, F., 1929, Is the twelve-hour variation in atmospheric pressure an electric phenomenon? *Science* **69:**434-436.

Sanford, F., 1936, Influence of planetary configuration on the frequency of visible sunspots, *Smithsonian Misc. Colln.* **95**(11):5p.

Sao, K., 1967, Correlations between solar activity and the atmospheric potential gradient at the Earth's surface in the polar regions, *Jour. Atmos. and Terr. Physics.* **29:**213-216.

Sargent, H. H., III, 1979, A geomagnetic activity recurrence index, in B. M. McCormac and T. A. Seliga (eds.), *Solar-Terrestrial Influences on Weather and Climate,* Dordrecht, The Netherlands: D. Reidel, pp. 101-104.

Sarukhanyan, E. I., and Smirnov, N. P., 1970, Solar activity, Earth's pressure field, and atmospheric circulation, *Geomagnetism and Aeronomy* **10:**390-392.

Sasajima, S., 1965, Geomagnetic secular variation revealed in the baked earths in West Japan (Part 2). Change of the field intensity, *Jour. Geomagnetism and Geoelectricity* **17:**413.

Sasajima, S., and Maenaka, K., 1969, Variation of the geomagnetic field intensity since the Late Miocene, *Jour. Geophys. Research* **74**(4):1037-1044.

Sassensfeld, M., 1903, Regenfall 1851-1900 zu Trier, *Zeitschr. Meteorologische* **20:**235-237.

Saville, C. M., 1934, Some rainfall variations England and New England, *Royal Meteorol. Soc. Quart. Jour.* **60**(256):313-331.

Sawyer, J. S., 1963, Notes on the response of the general circulation to changes in the solar constant, in *UNESCO Symposium on Changes in Climate,* Rome, 1961, Paris: UNESCO, p. 33.

Sawyer, J. S., 1966*a,* Possible variations of the general circulation of the atmosphere, in J. S. Sawyer (ed.), *World Climate 8000-0* B.C., London: Royal Meteorological Society, pp. 218-229.

Sawyer, J. S., ed., 1966*b, World Climate 8000-0* B.C., London: Royal Meteorological Society, 229p.

Sawyer, J. S., 1971, Possible effects of human activity on world climate, *Weather* **26:**251-262.

Schatten, K.; Goldberg, R.; Mitchell, J. M.; Olson, R.; Schafer, J.; Silverman, S.; Wilcox, J.; and Williams, G., 1979, Solar weather/climate predictions, in R. F. Donnelly (ed.), *Solar-Terrestrial Predictions Proceedings,* vol. 2, Washington, D.C.: U.S. Government Printing Office, pp. 655-668.

Schatzman, E., 1966, Interplanetary torques, in B. G. Marsden and A. G. W. Cameron (eds.), *The Earth-Moon System,* New York: Plenum Press, pp. 12-25.

Schell, I. I., 1943, The Sun's spottedness as a possible factor in terrestrial pressure, *Am. Meteorol. Soc. Bull.* **24**(3):85-93.

Schell, I. I., 1952, On the role of the ice off Iceland in the decadal temperatures of Iceland and some other areas, *Conseil Permanent Internat. pour l'Exploration de la Mer (Copenhagen) Jour.* **18:**1-36.

Schell, I. I., 1956, Interrelations of Arctic ice with the atmosphere and the ocean in the North Atlantic-Arctic and adjacent areas, *Jour. Meteorology* **13:**46-58.

Schell, I. I., 1961*a,* The ice off Iceland and the climates during the last 1200 years approximately, *Geog. Annaler* **43:**354-362.

Schell, I. I., 1961*b,* Recent evidence about the nature of climate changes and its implications, *New York Acad. Sci. Annals* **95**(article 1):251-270.

Schell, I. I., 1965, The origin and possible prediction of the fluctuations in the Peru Current and upwelling, *Jour. Geophys. Research* **70**(22):5529-5540.

Schell, I. I., 1967, Sea ice, climatic changes, in R. W. Fairbridge (ed.), *Encyclopedia of Atmospheric Sciences and Astrogeology,* vol. 2, Encyclopedia of Earth Sciences, New York: Van Nostrand Reinhold, pp. 858-861.

Scheller, A., 1908, Magnetisches Gewitter, *Zeitschr. Meteorologische* **25:**234.

Scherhag, R., 1936, Die Zunahme der atmospharischen Zirkulation in den letzten 25 Jahren, *Annals Hydrography* **65**:397.

Scherhag, R., 1948, Die Wintertemperaturen in Berlin und die Sonnenfleckenrelativzahlen, *Archiv Meteorologie, Geophysik, u. Bioklimatologie* **1A**:233-246.

Scherrer, P. H., 1979, Solar variability and terrestrial weather, *Rev. Geophysics and Space Physics* **17**:724-731.

Schiegl, W. E., 1972, Deuterium content of peat as a palaeoclimate recorder, *Science* **175**:512-513.

Schiegl, W. E., 1974, Climatic significance of deuterium abundance in growth rings of Picea, *Nature* **251**:582-584.

Schmidt, W., 1913, Nachweis von Perioden langer Dauer, *Zeitschr. Meteorologische* **30**:392-394.

Schneider, J., 1907, Ueber den Einfluss des Mondes auf die Windkomponenten zu Hamburg, *Deutsch. Seewarte (Hamburg) Archiv* **30**(2):1-10.

Schneider, S. H., 1974, A new world climate norm? Implications for future world needs, *Am. Acad. Arts and Sci. Bull.* **28**(3):20-35.

Schneider, S. H., 1976, *The Genesis Strategy. Climate and Global Survival*, New York and London: Plenum Press, 414p.

Schneider, S. H., and Dennett, R. D., 1975, Climatic barriers to long-term energy growth, *Ambio* **4**(2):65-75.

Schneider, S. H., and Dickinson, R. E., 1974, Climate modelling, *Rev. Geophysics and Space Physics* **12**:447-493.

Schneider, S. H., and Mass, C., 1975, Volcanic dust, sunspots and temperature trends, *Science* **190**:741-746.

Schoeberl, M. R., and Strobel, D. F., 1978, The response of the zonally averaged circulation to stratospheric ozone reductions, *Jour. Atmos. Sci.* **35**:1751-1757.

Schofield, J. C., 1970, Correlation between sea level and volcanic periodicities of the last millennium, *New Zealand Jour. Geology and Geophysics* **13**:737-741.

Schostakowitsch, W. B., 1927, Die periodischen Schwänkungen der Niederschlagsmenge in Russland und Mittelsibirien und die Sonnenflecken, *Zeitschr. Meteorologische* **44**:347-355.

Schott, C. A., 1876, Atmospheric temperature in the United States, *Smithsonian Inst. Contr. 277.*

Schove, D. J., 1949, Chinese raininess through the centuries, *Meteorol. Mag. (London)* **78**:11-16.

Schove, D. J., 1950, The climatic fluctuation since A.D. 1850 in Europe and the Atlantic, *Royal Meteorol. Soc. Quart. Jour.* **76**:147-167.

Schove, D. J., 1954, Summer temperatures and tree-rings in north-Scandinavia A.D. 1461-1950, *Geog. Annaler* **36**:40-80.

Schove, D. J., 1955, The sunspot cycle, 649 B.C., to A.D. 2000, *Jour. Geophys. Research* **60**:127-146.

Schove, D. J., 1971*a*, Varve teleconnection across the Baltic, *Geog. Annaler* **53A**:214-234.

Schove, D. J., 1971*b*, Biennial oscillations and solar cycles, A.D. 1490-1970, *Weather* **26**:201-209.

Schove, D. J., 1977, African droughts and the spectrum of time, in D. Dalby, R. J. Harrison Church, and F. Bezaz (eds.), *Drought in Africa 2,* London: International African Institute in Association with The Environment Training Program, Afenviron Special Report, pp. 38-53.

Schove, D. J., 1978, Tree-ring and varve scales combined, c. 13,500 B.C., to A.D. 1977, *Palaeogeography, Palaeoclimatology, and Palaeoecology* **25**:209-233.

Schove, D. J., 1979*a*, Sunspot turning points and aurorae since A.D. 1510, *Solar Physics* **63**:423-432.

Schove, D. J., 1979*b*, Varve-chronologies and their teleconnections, 14000-750 B.C., in C. Schluchter (ed.), *Moraines and Varves,* Rotterdam, The Netherlands: Balkema, pp. 319-325.

Schove, D. J., 1980, Aurorae, sunspots and weather, mainly since A.D. 1200, in C. S. Deehr and J. A. Holtet (eds.), *Exploration of the Polar Upper Atmosphere,* NATO Advanced Study Institute, Lillehammer, Norway 5-16 May 1980, Dordrecht, The Netherlands: D. Reidel, pp. 421-430.

Schove, D. J., 1981, The 200-, 22-, and 11-year cycles and long series of climatic data, mostly since A.D. 200, in *Sun and Climate.* Conference on Sun and Climate, Toulouse: Centre National d'Etudes Spatiale, pp. 87-100.

Schove, D. J., ed., 1983, *Sunspot Cycles,* Benchmark Series in Geology, vol. 68, Stroudsburg, Penn.: Hutchinson Ross Publications, 393p. (Note: A serious misprint occurs on p. 369, Sunspot minima A.D. 507-1501. Corrected sheets can be obtained from R. W. Fairbridge).

Schove, D. J., and Fairbridge, R. W., 1983, Swedish chronology revisited, Report on the 2nd Nordic Climate Symposium, *Nature* **304**:583.

Schreiber, 1914, 108 Jahre Beobachtungen der Wasserstande der Elbe an der Augustusbrucke in Dresden, *Wasser* **10**:267-268.

Schreiber, P., 1901, Die Niederschlags- und Abflussverhaltnisse im Gebiet der Weisseritz wahrend der Jahre 1866 bis 1900 und die sich daraus ergebende Einwirkung von Stauanlagen auf die Nutzung des Wassers und die Abflussvorgange, *Kongl. sachsischen meteorologischen Inst. Chemnitz Abh.* **5**:45p.

Schreiber, P., 1903, Die Schwankungen der jahrlichen Niederschlagshohen und deren Beziehungen zu den Relativzahlen für die Sonnenflecken. Untersuchung uber die Periodizitat der Sonnenflecken und des Niederschlages, in *Das Klima des Konigreiches Sachsen*, pp. 22-36.

Schreiber, P., 1904, Kritische Bearbeitung der Luftdruckmessungen im Konigreich Sachsen wahrend der Jahre 1866-1900, *Kongl. sachsischen meteorologischen Inst. Chemnitz Jahrb.* **18:**1-55.

Schroder, W., 1979, Auroral frequency in the 17th and 18th centuries and the 'Maunder Minimum,' *Jour. Atmos. and Terr. Physics.* **41:**445-446.

Schubert, J., 1913, Graphische Darstellung meteorologischer Werte, *Zeitschr. Meteorologische* **30:**545-547.

Schubler, G., 1832, Resultate 60-jahriger Beobachtungen uber den Einfluss des Mondes auf die Veranderungen in unserer Atmosphäre, *Archiv Gesammte Naturlehre* **5**(2):169-212.

Schulman, E., 1942, Centuries-long tree indices of precipitation in the southwest, *Am. Meteorol. Soc. Bull.* **23.**

Schulman, E., 1951, Tree-ring indices of rainfall, temperature, and river flow, in T. F. Malone (ed.), *Compendium of Meteorology,* Boston, Mass.: American Meteorological Society, pp. 1024-1032.

Schultz, L. G., 1912, Weather and the ultraviolet radiations of the Sun, *Nature* **90:**68-70.

Schumm, S. A., and Lichty, R. W., 1965, Time, space and causality in geomorphology, *Am. Jour. Science* **263:**110-119.

Schuster, A., 1897, On lunar and solar periodicities of earthquakes, *Royal Soc. [London] Proc.* **61:**455-465.

Schuster, A., 1898, On the investigation of hidden periodicities with application to a supposed 26-day period of meteorological phenomena, *Terr. Magnetism and Atmos. Electricity* **3**(Mar. 3):13-41.

Schuster, A., 1899, The periodogram of magnetic declination as obtained from the records of the Greenwich Observatory during the years 1874-1895, *Cambridge Philos. Soc. Trans.* **18:**107-135.

Schuster, A., 1901, On magnetic precession, *Phys. Soc. [London] Proc.* **17:**644-655.

Schuster, A., 1904, On sun-spot periodicities—Preliminary note, *Royal Soc. [London] Proc.* **77:**136-140, 141-146.

Schuster, A., 1904-1905, Sun-spots and magnetic storms, *Royal Astron. Soc. Monthly Notices* **65:**186-197.

Schuster, A., 1906, On the periodicities of the sunspots, *Royal Soc. London Philos. Trans.* **206A:**69-100.

Schuster, A., 1911, The influence of planets on the formation of sunspots, *Royal Soc. [London] Proc.* **85A:**309-323.

Schuster, A., 1914, On Newcomb's method of investigating periodicities and its application to Bruckner's weather cycle, *Royal Soc. [London] Proc.* **90A:**349-355.

Schuster, A., and Turner, H. H., 1907, Preliminary note on the rainfall periodogram, *British Assoc. for Advance. Sci. Repts. for 1906*, p. 498.

Schuster, F., 1908, Der Einfluss des Mondes auf unserer Atmosphäre—berechnet und graphische dargestellt, Karlsruhe: F. Gutsch, 31p.

Schuster, F., 1913*a*, Die Gewitterbildung in ihrer Beziehung zu den wichtigsten Mondstellungen, *Zeitschr. Meteorologische* **30:**222-227.

Schuster, F., 1913*b*, Die 18.6 jahrige Mondperiode in meteorologischer Hinsicht, *Zeitschr. Meteorologische* **30:**488-492.

Schuster, F., 1914, Die Anderung des Luftdrucks in mondperiodischen Wellensystemen und deren Interferenz, *Annalen Hydrographie u. Maritimen Meteorologie* **42:**432-438.

Schuster, F., 1916, Die Verschiebung des synodischen Luttdrucksystems unter dem Einfluss der 18.6-jährigen Mondperio de, *Annalen der Hydrographie und Maritimen Meteorologie Hamburg* **44**(8):442-444.

Schuurmans, C. J. E., 1965, Influence of solar flare particles on the general circulation of the atmosphere, *Nature* **205:**167-168.

Schuurmans, C. J. E., 1975, On climate changes related to the 22-year solar cycle, in W. R. Bandeen and S. P. Maran (eds.), *Possible Relationships Between Solar Activity and Meteorological Phenomena*, Greenbelt, Md.: NASA Goddard Space Flight Center, NASA SP-366, pp. 161-162.

Schuurmans, C. J. E., 1978, Influence of solar activity on winter temperatures: New climatological evidence, *Clim. Change* **1:**231-237.

Schuurmans, C. J. E., 1981, Solar activity and climate, in A. Berger (ed.), *Climatic Variations and Variability: Facts and Theories,* Dordrecht, The Netherlands: D. Reidel, pp. 559-575.

Schuurmans, C. J. E., 1983, Differences in tropospheric thermal behavior between alternate solar cycles, in B. M. McCormac (ed.), *Weather and Climate Response to Solar Variation*, Boulder: Colorado Associated University Press, pp. 559-568.

Schuurmans, C. J. E., 1984, Climatic variability and its time changes in European countries, based on instrumental changes, in H. Flohn and R. Fantechi (eds.), *The Climate of Europe: Past, Present, and Future*, Dordrecht, The Netherlands: D. Reidel, pp. 65-100.

Schuurmans, C. J. E., and Oort, A. H., 1969, A statistical study of pressure changes in the troposphere and lower stratosphere after strong solar flares, *Geofisica Pura e Applicata* **75:**233-246.

Schwabe, H., 1843, Die Sonne, *Astron. Nachr.* **20:**213-286.

Sears, P. B., 1947, *Deserts on the March*, 2nd ed., Norman: University of Oklahoma Press, 178p.

Secchi, P., 1872, Observation des variations des diametres solaires; observation des protuberances et de la chromosphere; observation des étoiles filantes; aurore boreale observée à Rome le 10 aout, à 10 heures du matin (letter), *Acad. Sci. Comptes Rendus* **75:**606-613.

Sedgwick, W., 1911, Weather in the seventeenth century, *Symons' Meteorol. Mag. (London)* **46:**61-66, 107-111, 169-172, 213-220.

Sellers, A., and Meadows, A. J., 1975, Long term variations in the albedo and surface temperature of the Earth, *Nature* **254:**44.

Sellers, W. D., 1969, A global climatic model based on the energy balance of the Earth-atmosphere system, *Jour. Appl. Meteorology* **8**(3):392-400.

Septer, E., 1926, Sonnenflecken und Gewitter in Sibirien, *Zeitschr. Meteorologische* **43:**229-231.

Severny, A. B., 1969, Is the Sun a magnetic rotator?, *Nature* **224:**53-54.

Severny, A. B.; Kotov, V. A.; and Tsap, T. T., 1984, Power spectrum of long-period solar oscillations and 160-min pulsations during 1974-82, *Nature* **307:**247-249.

Sewall, H., 1907, The influence of barometric pressure on nephritis, *British Balneol. and Climatol. Soc. [London] Jour.* **11:**108-116.

Shapiro, I. I., 1980, Is the Sun shrinking?, *Science* **208:**51-53.

Shapiro, R., 1953, A planetary-atmospheric response to solar activity, *Jour. Meteorology* **10**(5):350-355.

Shapiro, R., 1956, Further evidence of a solar-weather effect, *Jour. Meteorology* **13**(4):335-340.

Shapiro, R., 1975, The variance spectrum of monthly mean central England temperatures, *Royal Meteorol. Soc. Quart. Jour.* **101:**679-681.

Shapiro, R., 1976, Solar magnetic sector structure and terrestrial atmospheric vorticity, *Jour. Atmos. Sci.* **33:**865-870.

Shapiro, R., 1979, An examination of certain proposed sun-weather connections, *Jour. Atmos. Sci.* **36:**1105-1116.

Shapiro, R., and Stolov, H. L., 1978, A search for solar influence in the skill of weather forecasts, *Jour. Atmos. Sci.* **35:**2334-2345.

Shapiro, R., and Ward, F., 1962, A neglected cycle in sunspot numbers?, *Jour. Atmos. Sci.* **19:**506-508.

Sharpe, A., 1907, Fall in level of Central African Lakes: Lake Nyasa, *Geog. Jour.* **29:**466-467.

Shastakowich, V. B., 1927, Die periodischen Schwankungen der Niederschlagsmenge in Russland und Mittelsibirien und die Sonnenflecken, *Zeitschr. Meteorologische* **44:**347-355.

Shastakowich, V. B., 1931, Die Bedeutung der Untersuchung der Bodenablagerungen der Seen für einige Fragen der Geophysik, *Internat. Ver. Theor. Angew. Limnologie Verh.* **5:**307-317.

Shastakowich, V. B., 1944, An experiment on geochronological analysis of mud deposits of Malinovoie Lake in connection with the uplift of the shore of the White Sea, *Akad. Nauk S.S.S.R. Geog. Obshchestva SSSR Zapiski* **76:**203-206 (in Russian).

Shaw, D., 1965, Sunspots and temperatures, *Jour. Geophys. Research* **70:**4997-4999.

Shaw, W. N., 1901, On the effect of sea temperature upon the seasonal variation of air temperature of the British Isles, *Symons' Meteorol. Mag. (London)* **36:**145.

Shaw, W. N., 1904, *Commission for the Combination and Discussion of Meteorological Observations from the Point of View of Their Relations with Solar Phenomena,* Cambridge, U.K.: International Meteorological Committee, Report of Preliminary Proceedings, 7p.

Shaw, W. N., 1905*a,* Autumn rainfall and yield of wheat, *Symons' Meteorol. Mag. (London)* **40:**10-12.

Shaw, W. N., 1905*b,* On a relation between autumnal rainfall and the yield of wheat of the following year. Preliminary note, *Royal Soc. [London] Proc.* **74:**552-553.

Shaw, W. N., 1905*c,* Seasons in the British Isles from 1878, *Royal Stat. Soc. London Jour.* **68:**247-313.

Shaw, W. N., 1905*d,* A relation between autumnal rainfall and the yield of wheat of the following year. Preliminary note, *Monthly Weather Rev.* **33:**46-47.

Shaw, W. N., 1905-1906, The pulse of the atmospheric circulation, *Nature* **73:**175-177.

Shaw, W. N., 1906*a,* The law of sequence in the yield of wheat for Eastern England, 1885-1904, *Zeitschr. Meteorologische* **23:**208-216.

Shaw, W. N., 1906*b,* Some aspects of modern weather forecasting, *Royal Inst. Great Britain [London] Proc.* **17:**455-457.

Shaw, W. N., 1906*c,* An apparent periodicity in the yield of wheat for Eastern England, 1885-1905, *Royal Soc. [London] Proc.* **78A:**69-79.

Shaw, W. N., 1911, *Forecasting Weather,* London: Constable and Company, 380p.

Shaw, W. N., 1913, On seasons and crops in the east of England, *Scottish Meteorol. Soc. Jour.* **16:**179-183.

Shaw, W. N., 1914, Weather forecasts in England, *Nature* **92:**715-716.

Shaw, W. N., and Cohen, R. W., 1901, Seasonal variation of atmospheric temperatures in the British Isles and its relation to wind-direction and sea temperature, *Royal Soc. [London] Proc.* **69:**61-85.

Shaw, W. N., and Dines, W. H., 1905, The study of the minor fluctuations of atmospheric pressure, *Royal Meteorol. Soc. Quart. Jour.* **31:**39-52.

Sheeley, N. R., Jr., 1964, Solar faculae during the sunspot cycle, *Astrophys. Jour.* **140:**731-735.

Shennan, I., 1982*a,* Problems of correlating Flandrian sea-level changes and climate, in A. F. Harding (ed.), *Climatic Change in Later Prehistory*, Edinburgh: Edinburgh University Press, pp. 52-67.

Shennan, I., 1982*b*, Interpretation of Flandrian sea-level data from the Fenland, England, *Geol. Assoc. Proc.* **83:**53-63.

Shennan, I.; Tooley, M. J.; Davis, M. J.; and Haggart, B. A., 1983, Analysis and interpretation of Holocene sea-level data, *Nature* **302:**404-406.

Shepard, F. P.; Sullivan, G. G.; and Wood, F.J., 1981, Greatly accelerated currents in submarine canyon head during optimum astronomical tide-producing conditions, *Shore and Beach* **49**(1):32-34.

Shirley, J. H., 1979, High latitudes anticyclogenesis and enhanced atmospheric heat transfer processes, in B. M. McCormac and T. A. Seliga (eds.), *Solar-Terrestrial Influences on Weather and Climate*, Dordrecht, The Netherlands: D. Reidel, pp. 323-327.

Shirokov, V. A., 1983, The influence of the 19-year tidal cycle on large-scale eruptions and earthquakes in Kamchatka, and their long-term prediction, in S. A. Fedotov and Y. K. Markhinin (eds.), *The Great Tolbachik Fissure Eruption*, Cambridge: Cambridge University Press.

Shlien, S., 1972, Earthquake-tide correlation, *Royal Astron. Soc. Geophys. Jour.* **28**(1):27-34.

Shuvalov, V. M., 1970, Dependence of the solar cycle on the position of the planets, *Astron. Vestnik* **4:**198-203.

Siebert, M., 1961, Atmospheric tides, *Advances in Geophysics* **7:**105-187.

Simpson, G. C., 1928, Further studies in terrestrial radiation, *Royal Meteorol. Soc. Mem. 3*, pp. 1-26.

Simpson, G. C., 1929, Past climates, *Manchester Literary and Philos. Soc. Mem.* **74:**1-34.

Simpson, G. C., 1934, World climate during the Quaternary Period, *Royal Meteorol. Soc. Quart. Jour.* **60:**425-478.

Simpson, G. C., 1940, Possible causes of changes in climate and their limitations, *Linnaean Soc. London Proc.* **152**(Part 2):190-219.

Simpson, G. C., 1957, Further studies in world climate, *Royal Meteorol. Soc. Quart. Jour.* **83**(358):459-481; invited discussion, pp. 481-485.

Simpson, J. F., 1967, Earth tides as a triggering mechanism for earthquakes, *Earth and Planetary Sci. Letters* **2:**473-478.

Simpson, J. F., 1968, Solar activity as a triggering mechanism for earthquakes, *Earth and Planetary Sci. Letters* **3**(5):417-425.

Siren, G., 1961, Sklogsgranstallen som indikator for klimafluktuationerna i norra Fennoskandien under historisk tid, *Inst. Forest. Fenniae (Helsingfors) Communicationes* **54**(2):66p.

Siren, G., and Hari, P., 1971, *Coinciding Periodicity in Recent Tree Rings and Glacial Clay Sediments*, Kevo, Finland: Kevo Subarctic Research Station, Report, No. 8, pp. 348-352.

Siscoe, G. L., 1976, Minimum effect model of geomagnetic excursions applied to the auroral zone locations, *Jour. Geomagnetism and Geoelectricity* **28**(5):427-436.

Siscoe, G. L., 1978, Solar-terrestrial influences on weather and climate, *Nature* **276:**348-352.

Siscoe, G. L., 1980, Evidence in the auroral record for secular solar variability, *Rev. Geophysics and Space Physics* **18:**647-658.

Siscoe, G. L., and Verosub, K. L., 1983, High medieval auroral incidence over China and Japan: implications for the medieval site of the geomagnetic pole, *Geophys. Research Letters* **10**(4):345-348.

Sleeper, H. P., Jr., 1970, *Bi-stable Oscillation Modes of the Sun and Long Range Prediction of Solar Activity*, Huntsville, Ala.: Northrop Services, Inc., Technical Report TR-709.

Sleeper, H. P., Jr., 1972, *Planetary Resonances, Bi-stable Oscillation Modes, and Solar Activity Cycles*, Washington, D.C.: National Aeronautics and Space Administration, NASA Contractor Report NASA CR-2035.

Sleeper, H. P., Jr., 1975, *Solar Activity Prediction Methods*, Huntsville, Ala.: Northrop Services.

Slutz, R. J.; Gray, T. B.; West, M. L.; Stewart, F. G.; and Loftin, M., 1970, *Solar activity prediction*, Boulder, Colo.: National Oceanographic and Atmospheric Administration, Final Report, NSASA-H-54409A.

Smith, E. J., 1964, Interplanetary magnetic fields, in D. P. LeGalley and A. Rosen (eds.), *Space Physics*, New York: John Wiley & Sons, pp. 350-396.

Smith, L. B., 1968, An observation of strong thermospheric winds during a geomagnetic storm, *Jour. Geophys. Research* **73**(15):4959-4963.

Smith, P. J., 1967, The intensity of the ancient geomagnetic field: A review and analysis, *Royal Astron. Soc. Geophys. Jour.* **12:**321-362.

Smock, J. C., 1888, *The Climate of New Jersey*, Trenton: New Jersey State Geologist Report, 416p.

Smythe, C. M., and Eddy, J. A., 1976, Low ebb for solar tides (abstract), *Am. Astron. Soc. Bull.* **8:**346.

Smythe, C. M., and Eddy, J. A., 1977, Planetary tides during the Maunder Sunspot Minimum, *Nature* **266:**434-435.

Snodgrass, H. B., and Howard, R., 1985, Torsional oscillations of the Sun, *Science* **228:**945-952.

Sofia, S.; O'Keefe, J.; Lesh, J. R.; and Endal, A. S., 1979, Solar constant: Constraints on possible variations derived from solar diameter measurements, *Science* **204:**1306-1308.

Somerville, R. C. J.; Quick, W. J.; Hansen, J. E.; Lacis, A. A.; and Stone, P. H., 1976, A search for short-term meteorological effects of solar variability in an atmospheric circulation model, *Jour. Geophys. Research* **81:**1572-1576.

Sonett, C. P., 1983, Is the sunspot spectrum really so complicated?, in B. M. McCormac (ed.), *Weather and Climate Response to Solar Variation*, Boulder: Colorado Associated University Press, pp. 607-613.

Sonett, C. P., 1984, Very long solar periods and the radiocarbon record, *Rev. Geophysics and Space Physics* **22**(3):239-254.

Sonett, C. P., and Suess, H. E., 1984, Correlation of bristlecone pine ring widths with atmospheric ^{14}C variations: A climate-Sun relation, *Nature* **307:**141-143.

Sonett, C. P., and Williams, G. E., 1985, Solar periodicities expressed in varves from glacial Skilak Lake, southern Alaska, *Jour. Geophys. Research* **90**(A-12):12,019-12,026.

Soutar, A., and Crill, P. A., 1977, Sedimentation and climate patterns in the Santa Barbara Basin during the 19th and 20th centuries, *Geol. Soc. America Bull.* **88:**1161-1172.

Southall, H., 1905, Long period rainfall averages, *Symons' Meteorol. Mag. (London)* **40:**50-51.

Southward, A. J.; Butler, E. I.; and Pennycuick, L., 1975, Recent cyclic changes in climate and in abundance of marine life, *Nature* **253:**714-717.

Sparkes, J. R., 1974, Sunspots and the business cycle, *Nature* **252:**520.

Spence, E. J., 1956, Weather and health, *Weather* **11**(10):335.

Spencer-Jones, H., 1955, *Sunspot and Geomagnetic Storm Data*, Royal Greenwich Observatory.

Spiegel, E. A., and Weiss, N. O., 1980, Magnetic activity and variations in solar luminosity, *Nature* **287:**616-617.

Spielmann, C., 1905, Die harten Winter 1783-84 und 1784-1785, *Nassovia (Wiesbaden)* **6:**6-8, 22-24.

Spillman, W. J., 1915, *On a Theory of Gravitation and Related Phenomena*, Washington, D.C.

Springer, B. D., 1983, Solar variability and quasi-stationary planetary wave behavior, in B. M. McCormac (ed.), *Weather and Climate Response to Solar Variation*, Boulder: Colorado Associated University Press, pp. 381-394.

Stacey, C. M., 1963, Cyclical measures: Some tidal aspects concerning equinoctical years, *New York Acad. Sci. Annals* **105:**421-460.

Stacey, C. M., 1967*a*, Earth motions, in R. W. Fairbridge (ed.), *The Encyclopedia of Atmospheric Sciences and Astrogeology*, Encyclopedia of Earth Sciences, vol. 2, New York: Reinhold Publishing Corporation, pp. 335-340.

Stacey, C. M., 1967*b*, Planetary intervals—The Titius-Bode Rule and modification, in R. W. Fairbridge (ed.), *The Encyclopedia of Atmospheric Sciences and Astrogeology*, Encyclopedia of Earth Sciences, vol. 2, New York: Reinhold Publishing Corporation, pp. 745-746.

Stacey, C. M., 1967*c*, Time and astronomic cycles, in R. W. Fairbridge (ed.), *The Encyclopedia of Atmospheric Sciences and Astrogeology*, Encyclopedia of Earth Sciences, vol. 2, New York: Reinhold Publishing Corporation, pp. 999-1003.

Stagg, J. M., 1931, Atmospheric pressure and the state of the Earth's magnetism, *Nature* **127:**402.

Stahlman, W. D., and Gingerich, O., 1963, *Solar and Planetary Longitudes for the Years -2500 to 2000*, Madison: University of Wisconsin Press, 566p.

Starkel, L., 1966, Post-glacial climate and the moulding of European relief, in J. S. Sawyer (ed.), *World Climate from 8000-0* B.C., Proceedings International Symposium, Imperial College, London, April 18-19, 1966, London: Royal Meteorological Society, pp. 15-33.

Starr, V. P., and Oort, A. H. 1973, Five-year climatic trend for the northern hemisphere, *Nature* **242:**310-313.

Steenbeck, M., and Krause, F., 1969, Zur Dynamotheorie stellarer und planetarer Magnetfelder. I. *Astron. Nach.* **291:**49-84.

Stenflo, J. O., 1970, The polar magnetic fields of July and August 1968, *Solar Physics* **13:**42-56.

Sternberg, R. S., and Damon, P. E., 1979, Re-evaluation of possible historical relationship between magnetic intensity and climate, *Nature* **278:**36-38.

Stetson, H. T., 1938, The Sun and the atmosphere, in *Smithsonian Institution Annual Report*, Washington, D.C., pp. 149-174.

Stetson, H. T., 1937, *Sunspots and Their Effects*, New York: McGraw-Hill, 201p.

Stetson, H. T., 1947, *Sunspots in Action*, New York: Ronald Press, 252p.

Stevens, R., 1985, Glaciomarine varves in late-Pleistocene clays near Göteborg, southwestern Sweden, *Boreas* **14**(2):127-132.

Stewart, F. G., and Ostrow, S. M., 1971, Improved version of the McNish-Lincoln method for prediction of solar activity, *I.T.U. Jour.* **37**(V).

Stockman, W. B., 1905, Periodic variation of rainfall in the arid region, *U.S. Weather Bur. Bull. N*, 15p.

Stockton, C. W.; Boggess, W. R.; and Meko, D. M., 1985, Climate and tree rings, in A. Hecht (ed.), *Paleoclimate Analysis and Modeling*, New York: Wiley-Interscience, pp. 71-150.

Stockwell, J. N., 1872, Memoir on the secular variations of the elements of the orbits of the eight principal planets, Mercury, Venus, the Earth, Mars, Jupiter, Saturn, Uranus, and Neptune; with tables of the same; together with the obliquity of the ecliptic and the precession of the equinoxes in both longitude and right ascension, *Smithsonian Inst. Contr. Knowledge Pub. 232* **18**(pt 3).

Stoiber, R. B., and Jepsen, A., 1973, Sulfur dioxide contributions to the atmosphere by volcanoes, *Science* **189:**577-578.

Stolov, H. L., and Cameron, A. G. W., 1964, Variations of geomagnetic activity with lunar phase, *Jour. Geophys. Research* **69**(23):4975-4982.

Stolov, H. L., and Shapiro, R., 1974, Investigation of the responses of the general circulation at 700 mbar to solar geomagnetic disturbance, *Jour. Geophys. Research* **79:**2161-2170.

Stommel, H., and Stommel, E., 1979, The year without a summer, *Sci. American* **240:**176-186.

Stommel, H., and Stommel, E., 1983, *Volcano Weather. The Story of 1816, The Year without a Summer,* Newport, R.I.: Seven Seas Press, 178p.

Stone, R. G., 1937, On the causes of deaths from heat at Detroit, July, 1936, *Am. Meteorol. Soc. Bull.* **18**(6):233-236.

Stothers, R., 1979, Solar activity cycle during classical antiquity, *Astronomy and Astrophysics* **77:**121-127.

Stothers, R., 1980, Giant solar flares in Antarctic ice, *Nature* **287:**365.

Stothers, R. B., 1984, The great Tambora eruption of 1815 and its aftermath, *Science* **224:**1191-1198.

Strangway, D. W., 1970, *History of the Earth's Magnetic Field,* New York: McGraw-Hill, 168p.

Stratton, F. J. M., 1911, On a possible phase relation between the planets and sunspot phenomena, *Royal Astron. Soc. Monthly Notices* **72**(9).

Streiff, A., 1926, On the investigation of cycles and the relation of the Bruckner and solar cycle, *Monthly Weather Rev.* **54**(7):289-296.

Streiff, A., 1927, Sunspots and rainfall, *Monthly Weather Rev.* **55**(2):69-71.

Streiff, A., 1929, The practical importance of climatic cycles in engineering, *Monthly Weather Rev.* **57**(10):405-411.

Streiff, M., 1979, Cyclic formation of coastal deposits and their indication of vertical sea-level change, *Oceanis* **5:**303-306.

Strickland, W. W., 1910, The effect of the Moon upon the weather, *Knowledge (London)* **7:**269-270.

Stringfellow, M. F., 1974, Lightning incidence in Britain and the solar cycle, *Nature* **249:**332-333.

Stromberg, B., 1985, Revision of the late glacial Swedish varve chronology, *Boreas* **14**(2):101-106.

Stromberg, G., 1913, Harmonic analysis of the air temperature in Stockholm 1894-1911 based on the periods of movement of the Sun and the Moon, *Svenska Hydrogr.-Biol. Komm. (Goteborg) Skr.,* pp. 1-9.

Stuiver, M., 1961, Variations in radiocarbon concentration and sunspot activity, *Jour. Geophys. Research* **66:**273-276.

Stuiver, M., 1978, Radiocarbon timescale tested against magnetic and other dating methods, *Nature* **273:**271-274.

Stuiver, M., 1980, Solar variability and climatic change during the current millenium, *Nature* **285:**868-871.

Stuiver, M., 1982, A high-precision calibration of the A.D. radiocarbon time scale, *Radiocarbon* **24:**1-26.

Stuiver, M., and Quay, P. D., 1980*a*, Changes in atmospheric carbon-14 attributed to a variable Sun, *Science* **207:**11-19.

Stuiver, M., and Quay, P. D., 1980*b*, Patterns of atmospheric 14C changes, *Radiocarbon* **22:**166-176.

Stuiver, M., and Quay, P. D., 1981, A 1600-year long record of solar change derived from atmospheric 14C levels, *Solar Physics* **74:**479-481.

Stummer, E., 1913, Mond und Wetter, *Geog. Anzeiger* **14:**183-185.

Stumpff, K., 1939, *Tafeln und Aufgaben zur harmonischen Analyse und Periodogrammrechnung,* Berlin.

Sucksdorff, E., 1956, The influence of the Moon and the inner planets on the geomagnetic activity, *Geophysica* **5**(2):95-106.

Suda, T., 1962, Some statistical aspects of solar-activity indices, *Meteorol. Soc. Japan Jour.* **40:**287-299.

Suda, T., 1976, Effect of solar activity on polar vortex and a hypothesis on linking mechanism, *Geophys. Mag.* **37**(4):361-379.

Suess, H. E., 1965, Secular variations of the cosmic-ray-produced carbon 14 in the atmosphere and their interpretations, *Jour. Geophys. Research* **70**(23):5937-5952.

Suess, H. E., 1970*a*, Bristlecone-pine calibration of the radiocarbon time scale 5200 B.C., to the present, in I. U. Olsson (ed.), *Radiocarbon Variations and Absolute Chronology: Proceedings of the Twelfth Nobel Symposium,* New York: John Wiley & Sons, pp. 303-309.

Suess, H. E., 1970*b*, The three causes of the ^{14}C fluctuations, their amplitudes and time constant, in I. U. Olsson (ed.), *Radiocarbon Variations and Absolute Chronology: Proceedings of the Twelfth Nobel Symposium,* New York: John Wiley & Sons, pp. 595-604.

Suess, H. E., 1974, Natural radiocarbon evidence bearing on climatic changes, in J. Labeyrie (ed.), *Les methodes quantitatives d'etude des variations du climat au cours du Pleistocene,* Paris: Editions du Centre National de la Recherche Scientifique, pp. 311-317.

Suess, H. E., 1978, LaJolla measurements of radiocarbon in tree-ring dated wood, *Radiocarbon* **20:**1-18.

Suess, H. E., 1980, Radiocarbon geophysics, *Endeavour* (new series) **4**(3):113-117.

Suess, H. E., 1981, Solar activity, cosmic-ray produced carbon-14, and the terrestrial climate, in *Sun and Climate: Proceedings of the Conference on Sun and Climate,* Toulouse: Centre National d'Etudes Spatiale, pp. 307-310.

Sugiura, M., 1976, Quasi-biennial geomagnetic variation caused by the Sun, *Geophys. Research Letters* **3**(11):643-646.

Sugiura, M., and Poros, D. J., 1977, Solar generated

quasi-biennial geomagnetic variations, *Jour. Geophys. Research* **82:**5621-5628.

Sutherland, W., 1908, Solar magnetic fields and the cause of terrestrial magnetism, *Terr. Magnetism and Atmos. Electricity* **13:**155-158.

Sutton, J. R., 1904, An introduction to the study of South African rainfall, *South African Philos. Soc. Trans. for 1903,* pp. 119-143.

Sutton, J. R., 1907*a,* Variability of temperature in South Africa, *South African Assoc. for Advance. Sci. Rept. for 1906,* pp. 135-142.

Sutton, J. R., 1907*b,* On the lunar cloud-period, *South African Philos. Soc. Trans.* **18:**313-320.

Sutton, J. R., 1911, Planetary rainfall, *Symons' Meteorol. Mag. (London)* **46:**161.

Svalgaard, L., and Wilcox, J. M., 1974, The spiral interplanetary magnetic field: A polarity and sunspot cycle variation, *Science* **186:**51-53.

Svenonius, B., and Olausson, E., 1977, Solar activity and weather conditions in Sweden for the period 1756-1975, *Jour. Interdisc. Cycle Research* **8:**222-225.

Svenonius, B., and Olausson, E., 1979, Cycles in solar activity, especially of long periods, and certain terrestrial relationships, *Palaeogeography, Palaeoclimatology, and Palaeoecology* **26:**89-97.

Swindells, Rev. B. G., 1923, Comparison of sunspot areas and terrestrial magnetic horizontal force ranges 1911-21, *Royal Astron. Soc. Monthly Notices* **83**(3):215-217.

Taira, K., 1980, Holocene events in Japan; palaeo-oceanography, volcanism and relative sea-level oscillation, *Palaeogeography, Palaeoclimatology, and Palaeoecology* **32:**69-77.

Talman, C. F., 1934, The Santa Ana, *Am. Meteorol. Soc. Bull.*

Tanaka, M.; Weare, B. C.; Navato, A. R.; and Newell, R. E., 1975, Recent African rainfall patterns, *Nature* **255:**201-203.

Tannehill, I. R., 1947, *Drought. Its Causes and Effects,* Princeton, N.J.: Princeton University Press, 264p.

Tavakol, R. K., 1978, Is the Sun almost-intransitive (sic)?, *Nature* **276:**802-803.

Thaler, J. S., 1979, West Point 152 years of weather records, *Weatherwise,* June 1979, pp. 112-115.

Thaler, J. S., 1983, A cyclical pattern in 19th and 20th century Hudson Valley winter temperatures, in B. M. McCormac (ed.), *Weather and Climate Response to Solar Variation,* Boulder: Colorado Associated University Press, pp. 189-196.

Thellier, E., and Thellier, O., 1959, Sur l'intensité du champ magnetique terrestre dans le passe historique et geologique, *Annales Géophysique* **15:**285-376.

Thomas, C., 1880, 7-yr cycle, *Illinois Dept. Agriculture Trans. 1880,* pp. 47-59.

Thompson, L. G., 1981, On the trail of the "Jupiter Effect," *Sky and Telescope* **63**(3):220-221.

Thompson, L. M., 1973, Cyclical weather patterns in the middle latitudes, *Jour. Soil and Water Conservation* **28**(1):87-89.

Thompson, L. M., 1975, Weather variability, climatic change, and grain production, *Science* **188:**535-541.

Thompson, S. L., and Schneider, S. H., 1979, A seasonal zonal energy balance climate model with interactive lower layer, *Jour. Geophys. Research* **84**(C5):2401-2414.

Thomson, A., 1936, Sunspots and weather forecasting in Canada, *Royal Astron. Soc. Canada Jour.* **30:**215-232.

Thorarinsson, S., 1939, *Grossenschwankungen der Gletscher in Island,* Washington, D.C.: International Commission of Snow and of Glaciers.

Thorarinsson, S., 1943, Oscillations of the Iceland glaciers in the last 250 years, *Geog. Annaler* **25:**1-54.

Thornthwaite, C. W., 1931, The climates of North America according to a new classification, *Geog. Rev.* **21**(4):633-655.

Tilms, R. A., 1980, *Judgement of Jupiter,* London: New English Library.

Tizard, T. H., 1907, Dr. Otto Pettersson on the influence of ice-melting on oceanic circulation, *Geog. Jour.* **30:**339-344.

Tizard, T. H., 1908, On the influence of ice melting on oceanic circulation, *Geog. Jour.* **31:**226.

Todd, Sir C., 1905, Coldest spring on record in South Australia, *Symons' Meteorol. Mag. (London)* **40:**219-221.

Tooley, M. J., 1974, Sea-level changes during the last 9000 years in northwest England, *Geog. Jour.* **140:**18-42.

Tooley, M. J., 1978, *Sea-Level Changes: North-west England during the Flandrian Stage,* Oxford Research Studies in Geography, Oxford: Clarendon Press, 232p.

Tooley, M. J., 1982, Sea-level changes in northern England, *Geologists' Assoc. London Proc.* **93:**43-51.

Toomre, A., 1966, On the coupling of the Earth's core and mantle during the 26,000-year precession, in B. G. Marsden and A. G. W. Cameron (eds.), *The Earth-Moon System,* New York: Plenum Press, pp. 33-45.

Toon, O. B.; Pollack, J. B.; Ward, W.; Burns, J. A.; and Bilski, K., 1980, The astronomical theory of climate change on Mars, *Icarus* **44:**552-607.

Topolansky, M., 1905, Einige Resultate der 20 jahrigen Registrierungen des Regenfalles in Wien, *Zeitschr. Meteorologische* **22:**113-119.

Trellis, M., 1967, Influence of the Sun's motion towards the apex and origins of centers of activity, *Astrophys. Letters* **1:**57-58.

Trenberth, K. E., 1975, A quasi-biennial standing wave in the Southern Hemisphere and interrelations with sea-surface temperature, *Royal Meteorol. Soc. Quart. Jour.* **101:**55-74.

Trenberth, K. E., 1980, Atmospheric quasi-biennial oscillations, *Monthly Weather Rev.* **108:**1370-1377.

Trewartha, G. T., 1961, *The Earth's Problem Climates,* Madison: University of Wisconsin Press, 334p.

Trewartha, G. T., 1968, *An Introduction to Climate,* 4th ed., New York: McGraw-Hill, 408p.

Tripp, W. B., 1912, Cycles of the Sun and weather and the impacts, *English Mechanic (London),* April 26.

Tromp, S. W., 1980, *Biometeorology. The Impact of the Weather and Climate on Humans and Their Environment (Animals and Plants),* London: Heyden, 346p.

Troup, A. J., 1965, The "Southern oscillation," *Royal Meteorol. Soc. Quart. Jour.* **91:**490-506.

Tucker, G. B., 1964, Solar influences on the weather, *Weather* **19:**302-311.

Tucker, G. B., 1979, The observed zonal wind cycle in the southern hemisphere stratosphere, *Royal Meteorol. Soc. Quart. Jour.* **105:**263-273.

Tucker, G. B., 1981, *The CO_2 Climate Connection,* Canberra: Australia Academy of Science, 54p.

Tuckerman, B., 1964, Planetary, lunar, and solar positions at five-day and ten-day intervals, AD 2 to AD 1649, *Am. Philos. Soc. Mem. 59,* 842p.

Turner, H. H., 1911, What can we learn from rainfall records?, *Royal Meteorol. Soc. Quart. Jour.* **37:**209-220.

Turner, H. H., 1914, Fourier series, *Royal Astron. Soc. Monthly Notices* **73:**549 and 714.

Turner, H. H., 1919*a,* On the 15-month periodicity in earthquake phenomena, *Royal Astron. Soc. Monthly Notices* **79**(6):461.

Turner, H. H., 1919*b,* On a long period (about 240 years) in Chinese earthquake records, *Royal Astron. Soc. Monthly Notices* **79**(7):531.

Turner, H. H., 1920, Note on the 240-year period in Chinese earthquakes in light of Dr. Fotheringham's paper, *Royal Astron. Soc. Monthly Notices* **80**(6):617-620.

Turner, H. H., 1925, Note on the alteration of the 11-year solar cycle, *Royal Astron. Soc. Monthly Notices* **85**(5):467.

Turner, H. H., 1926*a,* On a period of approximately 9.2 years in the Greenwich observations of magnetic declination and horizontal force, *Royal Astron. Soc. Monthly Notices* **86**(3):108-118.

Turner, H. H., 1926*b,* On an unsuccessful search for the 9.2 magnetic period in sunspot records, with a new analysis of those records back to 1610, *Royal Astron. Soc. Monthly Notices* **86**(3):119-130.

Tyler, W. F., 1907, The psycho-physical aspect of climate, with a theory concerning intensities of sensation, *Jour. Tropical Medicine* **10:**130-149.

Tyson, P. D., and Dyer, T. G. J., 1978, The predicted above normal rainfall for the seventies and the likelihood of droughts in the eighties in South Africa, *South Africa Jour. Sci.* **74:**372-377.

U.S. Weather Bureau, 1939, Reports on critical studies of methods of long-range weather forecasting, *Monthly Weather Rev.* (supp. No. 39).

Vail, O. E., 1917, Lithologic evidence of climatic pulsations, *Science* (new series) **46:**90-93.

Valnicek, B., 1952, Les eruptions chromospheriques et le temps, *Astron. Inst. Czechoslovakia Bull.* **3**(5):71-76 (also, 1953*a,* **4**(2):48-51; 1953*b,* **4**(4):97-99; 1953*c,* **4**(6):179-181).

van de Plassche, O., 1980, Holocene water level changes in the Rhine-Meuse delta as a function of changes in relative sea level, local tide range and river gradient, *Geologie en Mijnbouw* **59:**343-351.

van Loon, H., and Jenne, R. L., 1969, The half-yearly oscillations in the tropics of the Southern Hemisphere, *Jour. Atmos. Research* **26:**218-232.

van Loon, H., and Jenne, R. L., 1970, On the half-yearly oscillations in the tropics, *Tellus* **22**(4):391-398.

van Woerkom, A. J. J., 1953, The astronomical theory of climate changes, in H. Shapley (ed.), *Climatic Change: Evidence, Causes, and Effects,* Cambridge, Mass.: Harvard University Press, pp. 147-157.

Veeh, H. H., and Chappell, J., 1970, Astronomical theory of climatic change: Support from New Guinea, *Science* **167:**862-865.

Vegard, L., 1911, Studies of magnetic disturbances, *Nature* **85:**743-744.

Venne, D. E.; Nastrom, G. D.; and Belmont, A. G., 1983, Comment on "Evidence for a solar cycle signal in tropospheric winds" by G. D. Nastrom and A. G. Belmont, *Jour. Geophys. Research* **88:**11025-11030.

Ventosa, V., 1902, The Moon and thunderstorms, *Symons' Meteorol. Mag. (London)* **37:**161.

Vernekar, A. D., 1968, *Long-Period Global Variations of Incoming Radiation (Research on the Theory of Climate),* Hartford, Conn. The Travelers Research Center, 289p.

Vernekar, A. D., 1972, Long period global variations of incoming solar radiation, *Am. Meteorol. Soc. Meteorol. Mons. No. 12,* pp. 1-10.

Vernekar, A. D., 1977, Variations in the insolation caused by changes in orbital elements of the Earth, in O. R. White (ed.), *The Solar Output and Its Variation,* Boulder: Colorado Associated University Press, pp. 117-130.

Vestine, E. H., 1953, On variations of the geomagnetic field, fluid motions and the rate of the Earth's rotation, *Jour. Geophys. Research* **58:**127.

Vestine, E. H., and Kahle, A., 1968, The westward drift and geomagnetic secular change, *Geophys. Jour.* **15:**29-37.

Vestine, E. H.; Lanage, I.; LaPorte, L.; and Scott, W. E., 1947, The geomagnetic field, its description and analysis, Carnegie Inst. *Washington Pub.* 580.

Vestine, E. H.; LaPorte, L.; Lange, I.; Cooper, C.; and

Hendrix, W. C., 1947, Description of the Earth's main magnetic field and its secular change, 1905-1945, *Carnegie Inst. Washington Pub. 578,* 532p.

Vincent, C. E.; Davies, T. D.; and Beresford, A. K. C., 1979, Recent changes of level of Lake Naivasha, Kenya, as an indicator of equatorial westerlies over East Africa, *Clim. Change* **2:**175-189.

Visagie, P. J., 1966, Precipitation in South Africa and lunar phase, *Jour. Geophys. Research* **71**(14):3354-3350.

Visser, S. W., 1950, On Easton's period of 89 years, *Koninkl. Nederlandse Akad. Wetensch. Proc.* **53:**172-175.

Visser, S. W., 1959, On the connections between the 11-year sunspot period and the periods of about 3 and 7 years in world weather, *Geofisica Pura e Applicata* **43:**302-318.

Vitinski, Y. I., 1969, Solar cycles, *Solar System Research* **3:**99-110.

Volland, H., 1977, Can sunspots influence our weather?, *Nature* **269:**400-401.

Volland, H., and Schaefer, J., 1979, Cause and effect in some types of Sun-weather relationships, *Geophys. Research Letters* **6:**17-20.

von Humboldt, A., 1971, *Kosmos,* vol. 4 (trans. by E. C. Otte and B. H. Paul), London: Bell, 603p.

von Nobbe, 1903, Inwieweit beeinflussen Mond und Sonne das Wetter? *Deutsche. Naturf. U. Aerzte Gesell. Verh. (Leipzig)* **75:**148-150.

von Rudloff, H., 1967, Die Schwankungen und Pendelungen des Klimas in Europe seit dem Beginn der regelmassigen Instrumenten-Beobachtungen, (1670), *Die Wissenschaft* **122.**

von Schubert, O., 1927, Die 3-jahrige Luftdruckwelle, *Univ. Leipzig Geophys. Inst. Veroffentliche, no. 6.*

Waddington, C. J., 1967, Paleomagnetic field reversals and cosmic radiation, *Science* **158:**913-915.

Wagner, A., 1924, Eine bemerkenswerte 16 jahrige Klimaschwankung, mit einem Anhang: Mogliche Perioden verschiedener Wellenlange, *Akad. Wiss. Wien Sitz. Math.-Naturw. Kl.* **133**(ser. II):169-224.

Wagner, A., 1929, Untersuchung der Schwankung der allgemeinen Zirkulation, *Geog. Annaler* **11:**33-88.

Wagner, A., 1940, Klimaanderungen und Klimaschwankungen, *Die Wissenschaft* **92.**

Wagner, G., 1912, Der Einfluss des Mondes auf das Wetter, *Umschau* **17:**371-373.

Wagner, G., 1913, Der Einfluss des Mondes auf das Wetter, *Beitr. Geophysik* **12:**277-328, 528-587.

Wahl, E. W., and Bryson, R. A., 1975, Recent changes in Atlantic surface temperatures, *Nature* **54:**45-46.

Waldmeier, M., 1955, *Ergebnisse und Probleme der Sonnenforschung, Zweiter Auflage,* Leipzig: Geest and Portrig.

Waldmeier, M., 1959, Geomagnetic and solar data: Final relative sunspot numbers for 1958, *Jour. Geophys. Research* **64**(9):1347-1349.

Waldmeier, M., 1961, *The Sunspot-Activity in the Years 1610-1960,* Zurich: Schulthess, 171p.

Waldmeier, M., 1975, The beginning of a new cycle of solar activity, *Nature* **253:**419.

Waldmeier, M., 1976, The sunspot activity in the years 1961-1975, *Astron. Mitt.* **346:**1-19.

Walker, G. B., and O'Dea, P. L., 1925, Geomagnetic secular-change impulses, *EOS (Am. Geophys. Union Trans.)* **33**(61):797.

Walker, G. T., 1909, Correlation in seasonal variation of climate, *Indian Meteorol. Mem. No. 20* **6:**117-124.

Walker, G. T., 1915, Correlation in seasonal variation of weather, IV. Sunspots and rainfall, *Indian Meteorol. Mem. No. 21.*

Walker, G. T., 1925, On periodicity, *Royal Meteorol. Soc. Quart. Jour.* **51**(216):337-346.

Walker, G. W., 1909, Magnetic storms, *Nature* **82:**69.

Walker, G. H., 1910, Note on terrestrial magnetism, *Phys. Soc. [London] Proc.* **21:**890-891.

Wallace, H. A., 1920, Mathematical inquiry into the effect of weather on corn yield in the eight corn belt states, *Monthly Weather Rev.* **48:**429-456.

Wallace, J. M., 1967*a*, On the role of mean meridional motions in the biennial wind oscillation, *Royal Meteorol. Soc. Quart. Jour.* **93:**176-185.

Wallace, J. M., 1967*b*, A note on the role of radiation in the biennial oscillation, *Jour. Atmos. Sci.* **24:**598-599.

Wallace, J. M., 1973, General circulation of the tropical lower stratosphere, *Rev. Geophysics and Space Physics* **11:**191-272.

Wallace, J. M., and Holton, J. R., 1968, A diagnostic numerical model of the quasi-biennial oscillation, *Jour. Atmos. Sci.* **25:**280-292.

Wallenhorst, S. G., 1982, Sunspot numbers and solar cycles, *Sky and Telescope* **64**(3):234-236.

Wallis, W. F., 1901, Sun-Spots and magnetic storms [review of Sidgreaves, On the connection between solar spots and earth-magnetic storms], *Terr. Magnetism and Atmos. Electricity* **6:**95-96.

Walter, H., 1936, Die Periodicitat von Trocken- und Regenzeiten in Deutsch-Sudwestafrika auf Grund von Jahresringmessungen am Baumen, *Deutsche. Botan. Gesell. Ber.* **54.**

Wang, P.-K., 1980, On the relationship between winter thunder and the climatic change in China in the past 2200 years, *Clim. Change* **3:**37-46.

Wang, S. W., and Zhao, Z. C., 1979, The 36-year wetness oscillation in China and its mechanism, *Acta Meteorol. Sinica* **37:**64-73.

Ward, F., 1966, Determination of the solar rotation rate from the motion of identifiable features, *Astrophys. Jour.* **145:**416-425.

Ward, F., and Shapiro, R., 1962, Decomposition and comparison of time series of indices of solar activity, *Jour. Geophys. Research* **67**(2):541-554.

Ward, R. deC., 1908, *Climate, Considered Espe-*

cially in Relation to Man, New York: Putnam; and London: John Murray, 372p.

Ward, R. deC., 1910, Climate in some of its relations to man, *Popular Sci. Monthly* **76:**246-268.

Ward, W. R., 1973, Large-scale variations in the obliquity of Mars, *Science* **181:**260-262.

Ward, W. R.; Burns, J. A.; and Toon, O. B., 1979, Past obliquity oscillations of Mars: The role of the Tharsis uplift, *Jour. Geophys. Research* **84:**243-259.

Ward, W. T., and Russell, J. S., 1980, Winds in southeast Queensland and rain in Australia and their possible long-term relationship with sunspot number, *Clim. Change* **3:**89-104.

Warwick, J. W. 1959, Some remarks on the interaction of solar plasma and the geomagnetic field, *Jour. Geophys. Research* **64**(4):389-396.

Wasserfall, K. F., 1935, The 27-day period in temperature data, *Am. Geophys. Union Trans. for 1935*, pp. 161-162.

Waters, G. S., and Francis, P. D., 1958, A nuclear precession magnetometer, *Jour. Sci. Instruments* **35:**88.

Watkins, N. D., and Goodell, H. G., 1967, Geomagnetic polarity change and faunal extinction in the Southern Ocean, *Science* **156:**1083-1087.

Watson, A. E., 1901, A review of past severe winters in England, with deductions therefrom, *Royal Meteorol. Soc. Quart. Jour.* **27:**141-150.

Weakly, H. E., 1962, History of drought in Nebraska, *Jour. Soil and Water Conservation* **17:**271-275.

Webster, P. J., and Keller, J. L., 1975, Atmospheric variations: Vascillations and index cycles, *Jour. Atmos. Sci.* **32:**1283-1300.

Wedderburn, E. M., 1909*a*, Dr. O. Pettersson's observations on deep water oscillations, *Royal Soc. Edinburgh Proc.* **29:**602-606.

Wedderburn, E. M., 1909*b*, Temperature oscillations in lakes and in the ocean, *Scottish Geog. Mag.* **25:**591-597.

Wegener, A., 1924, *The Origin of Continents and Oceans*, 3rd ed. (English trans. by J. G. A. Skerl), London: Methuen and Company, 212p.

Weightman, R. H., 1940, *Forecasting from Synoptic Weather Charts*, Washington, D.C.

Weinbeck, R. S., and Yonger, D. N., 1978, Relationship of atmospheric ozone profiles to solar magnetic activity, *Geofisica Pura e Applicata* **116:**32-43.

Weinbeck, R. S., and Yonger, D. N., 1980, Sunspot numbers and area-weighted United States monthly temperature and precipitation, *Geofisica Pura e Applicata* **119:**209-222.

Weinstein, D. H., and Keeney, J., 1973, Apparent loss of angular momentum in the Earth-Moon system, *Nature* **244:**83-84.

Wells, N. C., and Puri, K., 1979*a*, Atmospheric feedback in a coupled ocean-atmosphere model, *Jour. Geophys. Research* **84**(C8):4971-4984.

Wells, N. C., and Puri, K., 1979*b*, The effect on a tropical sea surface temperature anomaly in a coupled ocean-atmosphere model, *Jour. Geophys. Research* **84**(C8):4985-4997.

Wetherald, R. T., and Manabe, S., 1975, Effect of changing the solar constant on the climate of a general circulation model, *Jour. Atmos. Sci.* **32:**2044-2059.

Wexler, H., 1937, Formation of polar anticyclones, *Monthly Weather Rev.* **65**(6):229-236.

Wexler, H., 1956*a*, Variations in insolation, general circulation and climate, *Tellus* **8:**480-494.

Wexler, H., 1956*b*, A look at some suggested solar-weather relationships, in J. W. Chapman and H. Wexler (eds.), *The Sun's Effects on the Earth's Atmosphere*, Climax, Colo.: High Altitude Observatory, Institute for Solar-Terrestrial Research, Technical Report No. 2, pp. 21-30.

Wexler, H., and Namias, J., 1938, Mean monthly isotropic charts and their relation to departures of summer rainfall, *Am. Geophys. Union Trans.* (19th annual meeting), pp. 164-170.

Weyl, P. K., 1968, The role of the oceans in climatic change: A theory of the ice ages, *Meteorol. Mon.* **8:**37-62.

Wheeler, W. H., 1906, The weather and March high-tides, *Naturalist*, pp. 110-111.

White, O. R., ed., 1977, *The Solar Output and Its Variation*, Boulder: Colorado Associated University Press, 526p.

Whiteley, J., 1904*a*, Table of rainfall in Halifax for twelve years, *Halifax Naturalist and Record of the Sci. Soc.* **8:**100.

Whiteley, J., 1904*b*, Table of adopted mean temperature in Halifax for twelve years, *Halifax Naturalist and Record of the Sci. Soc.* **8:**101.

Whitney, J. D., 1880-1882, The climate changes of later geological times, *Harvard College Mus. Comp. Zoology Mem.* **7:**1-120, 121-264, 265-394.

Wigley, T. M. L., 1980, Sun-climate links, *Nature* **288:**317-318.

Wigley, T. M. L., 1983, Climate and paleoclimate; What can we learn about solar luminosity variations?, *Solar Physics* **74:**435-471.

Wigley, T. M. L.; Ingram, M. J.; and Farmer, G., eds., 1981, *Climate and History: Studies in Past Climates and Their Impact on Man*, Cambridge: Cambridge University Press, 530p.

Wilcox, J. M., 1975, Solar activity and the weather, *Jour. Atmos. and Terr. Physics* **37:**237-256.

Wilcox, J. M., 1976, Solar structure and terrestrial weather, *Science* **192:**745-748.

Wilcox, J. M., 1979*a*, Tropospheric circulation and interplanetary magnetic sector boundaries followed by MeV proton streams, *Nature* **278:**840-841.

Wilcox, J. M., 1979*b*, Influence of the solar magnetic field on tropospheric circulation, in B. M. McCormac and T. A. Seliga (eds.), *Solar-Terrestrial Influences* *on Weather and Climate*, Dordrecht, The Netherlands: D. Reidel, pp. 149-159.

Wilcox, J. M., and Scherrer, P. H., 1979, Variation with time of a Sun-weather effect, *Nature* **280:**845-846 (discussion by R. G. Williams).

Wilcox, J. M., and Scherrer, P. H., 1980, *On the Nature of the Apparent Response of the Vorticity Area Index to the Solar Magnetic Field*, Stanford, Calif.: Stanford University, Institute for Plasma Research, SUIPR Report No. 802.

Wilcox, J. M.; Scherrer, P. H.; and Hoeksema, J. T., 1983, Interplanetary magnetic field and tropospheric circulation, in B. M. McCormac (ed.), *Weather and Climate Response to Solar Variation*, Boulder: Colorado Associated University Press, pp. 365-379.

Wilcox, J. M.; Scherrer, P. H.; Svalgaard, L.; Roberts, W. O.; Olson, R. H., 1973*a*, Solar magnetic structure: Influence on stratospheric circulation, *Science* **180:**185-186.

Wilcox, J. M.; Scherrer, P. H.; Svalgaard, L.; Roberts, W. O.; Olson, R. H.; and Jenne, R. L., 1973*b*, Influence of solar magnetic sector structure on terrestrial atmospheric vorticity, Stanford, Calif.: Stanford University, Institute for Plasma Research, SUIPR Report No. 530.

Wilcox, J. M.; Scherrer, P. H.; Svalgaard, L.; Roberts, W. O.; Olson, R. H.; and Jenne, R. L., 1974, Influence of solar magnetic sector structure on terrestrial atmospheric vorticity, *Jour. Atmos. Sci.* **31:**581-588.

Wilcox, J. M.; Severny, A.; and Colburn, D. S., 1969, Solar source of interplanetary magnetic fields, *Nature* **224:**353-354.

Wilcox, J. M.; Svalgaard, L.; and Scherrer, P. H., 1975, Seasonal variation and magnitude of the solar sector structure-atmospheric vorticity effect, *Nature* **255:**539-540.

Wilcox, J. M.; Svalgaard, L.; and Scherrer, P. H., 1976, On the reality of a Sun-weather effect, *Jour. Atmos. Sci.* **33:**1113-1116.

Willett, H. C., 1949*a*, Patterns of world weather changes, *EOS (Am. Geophys. Union Trans.)* **29**(6):803-809.

Willett, H. C., 1949*b*, Long-period fluctuations of the general circulation of the atmosphere, *Jour. Meteorology* **6**(1):34-50.

Willett, H. C., 1949*c*, Solar variability as a factor in the fluctuations of climate during geological time, *Geog. Annaler* **31**(1-4):295-315.

Willett, H. C., 1950, Temperature trends of the past century, *Royal Meteorol. Soc. Centennial Proc.*, pp. 195-206.

Willett, H. C., 1961, The pattern of solar-climatic relationships, *New York Acad. Sci. Annals* **95**(Article 1):89-106.

Willett, H. C., 1962, The relationship of total atmospheric ozone to the sunspot cycle, *Jour. Geophys. Research* **67:**661-670.

Willett, H. C., 1965, Solar-climatic relationships in the light of standardized climatic data, *Jour. Atmos. Sci.* **22:**120-136.

Willett, H. C., 1974, Recent statistical evidence in support of the predictive significance of solar-climatic cycles, *Monthly Weather Rev.* **102**(10):679-686.

Willett, H. C., 1980, Solar prediction of climatic change, *Phys. Geography* **1**(2):95-117.

Willett, J. C., 1979, Solar modulation of the supply current for atmospheric electricity?, *Jour. Geophys. Research* **84**(C8):4999-5002.

Williams, D., 1961, Sunspot cycle correlations, *New York Acad. Sci. Annals* **95**(Article 1):78-88.

Williams, G. E., 1981, Sunspot periods in the late Precambrian glacial climate and solar-planetary relations, *Nature* **291:**624-628.

Williams, G. E., 1983, Precambrian varves and sunspot cycles, in B. M. McCormac (ed.), *Weather and Climate Response to Solar Variation*, Boulder: Colorado Associated University Press, pp. 517-533.

Williams, G. E., and Sonett, C. P., 1985, Solar signature in sedimentary cycles from the late Precambrian Elatina Formation, Australia, *Nature* **318:**523-527.

Williams, J., 1978, The use of numerical models in studying climatic change, in J. R. Gribbin (ed.), *Climatic Change*, Cambridge: Cambridge University Press, pp. 178-190.

Williams, J., 1978, Spectral analysis of seasonal precipitation data from North America and Europe, *Monthly Weather Rev.* **106:**898-900.

Williams, J.; Barry, R. G.; and Washington, W. M., 1974, Simulation of the atmospheric circulation using the NCAR global circulation model with ice age boundary conditions, *Jour. Appl. Meteorology* **13:**305-317.

Williams, L. D.; Wigley, T. M. L.; and Kelly, P. M., 1981, Climatic trends at high northern latitudes during the last 4000 years compared with ^{14}C fluctuations, in *Sun and Climate, Proceedings of Conference on Sun and Climate*, Toulouse Centre National d'Etudes Spatiale, pp. 11-20.

Williams, R. G., 1978, A study of the energetics of a particular Sun-weather relation, *Geophys. Research Letters* **5:**519-522.

Williams, R. G., 1979, Williams replies, *Nature* **280:**846.

Williams, R. G., and Gerety, E. J., 1978, Does the troposphere respond to day-to-day changes in the solar magnetic field?, *Nature* **275:**200-201.

Williams, S. R., 1914, Possible factors in the variations of the Earth's magnetic field, *Science* (new series) **40:**606-607.

Willis, D. M.; Easterbrook, M. G.; and Stephenson, F. R., 1980, Seasonal variation of oriental sunspot sightings, *Nature* **287:**617-619.

Willis, E. H.; Tauber, H.; and Munnich, K. O., 1960, Variations in the atmospheric radiocarbon concentration over the past 1300 years, *Am. Jour. Sci.*, Radiocarbon Supp. No. 2, pp. 1-2.

Willson, R. C.; Duncan, C. H.; and Geist, J., 1980, Direct measurement of solar luminosity variations, *Science* **207:**177-179.

Willson, R. C.; Gulkis, S.; Jansen, M.; Hudson, H. S.; and Chapman, G. A., 1981, Observations of solar irradiance variability, *Science* **211:**700-702.

Willson, R. C.; Hudson, H.; and Woodward, M., 1984, The inconstant solar constant, *Sky and Telescope* **67**(6):500-503.

Wilsing, J., 1888, *Ableitung der Rotationsbewegung der Sonne aus Positionsbestimmungen von Fackeln,* Astrophysikalischen Observatoriums zu Potsdam.

Wilson, A. T., 1966, Variation in solar insolation to the south polar region as a trigger which induces instability in the Antarctic ice sheet, *Nature* **210:**477-478.

Wilson, C. R., and Haubrich, R. A., 1974, Atmospheric excitation of the Earth's wobble (abstract), *EOS (Am. Geophys. Union Trans.)* **55:**220.

Wilson, C. R., and Haubrich, R. A., 1977, Earthquakes, weather and wobble, *Geophys. Research Letters* **4:**283-284.

Wilson, O. C., 1978, Chromospheric variations in main-sequence stars, *Astrophys. Jour.* **226:**379-396.

Wilson, O. C.; Vaughan, A. H.; and Mihalas, D., 1981, The activity cycles of stars, *Sci. American* **244**(2):104-107, 111-114, 116, 118-119.

Wilson, W. M., 1902, Climate and man: With special regard to climate and climatic elements as curative or causative agencies of disease, in *Proceedings of the Convention of Weather Bureau Officials,* vol. 2, Washington, D.C.: Weather Bureau Officials, pp. 104-107.

Winkless, N., III, and Browning, I., 1975, *Climate and the Affairs of Men,* New York: Harper's Magazine Press, 228p.

Winston, J. S., 1955, Physical aspects of rapid cyclogenesis over the Gulf of Alaska, *Tellus* **7:**481-500.

Winter, T. C., and Wright, H. E., Jr., 1977, Paleohydrologic phenomena recorded by lake sediments, *EOS (Am. Geophys. Union Trans.)* **58:**188-196.

Wittman, A., 1978, The sunspot cycle before the Maunder Minimum, *Astronomy and Astrophysics* **66:**93-97.

Woillard, G., 1978, Grande Pile peat bog: A continuous pollen record for the last 140,000 years, *Quat. Research* **9**(1):1-21.

Wolf, R., 1858, Mittheilungen ueber die Sonnenflecken, *Astron. Mitt.* **6:**127-143. (Also, 1862, ibid.)

Wolfer, A., 1902, Revisions of Wolf's sunspot relative-numbers, *Monthly Weather Rev.* **30:**171-176.

Wolfer, A., 1904, Sonnenfleckenhaufigkeit 1902; Magnetische Variationen, *Naturf. Gesell. Zurich Viert.* **48:**376-429.

Wolff, C., 1976, Timing of solar cycles by rigid internal rotations, *Astrophys. Jour.* **205:**612-621.

Wolff, L. C., 1912, Die sakulare Anderung unseres Klimas, *Prometheus* **23:**305-311.

Wollin, G.; Ericson, D. B.; Ryan, W. B. F.; and Foster, J. H., 1971, Magnetism of the Earth and climatic changes, *Earth and Planetary Sci. Letters* **12:**173-183.

Wollin, G.; Ericson, D. B.; and Wollin, J., 1974, Geomagnetic variations and climatic changes 2,000,000 B.C.-1970 A.D., in J. Labeyrie (ed.), *Les methodes quantitatives d'etude des variations du climat au cours du Pleistocene,* Colloques internationaux du Centre national de la Recherche scientifique, No. 219, Gif-sur-Yvette, 5-9 juin 1973, Paris: Editions du Centre National de la Recherche scientifique, pp. 273-287.

Wollin, G.; Kukla, G. J.; Ericson, D. B.; Ryan, W. B. F.; and Wollin, J., 1973, Magnetic intensity and climatic changes 1925-1970, *Nature* **242:**34-37.

Wollin, G.; Ryan, W. B. F.; and Ericson, D. B., 1978, Climatic changes, magnetic intensity variations and fluctuations of the eccentricity of the Earth's orbit during the past 2,000,000 years and a mechanism which may be responsible for the relationship, *Earth and Planetary Sci. Letters* **41:**395-397.

Wollin, G.; Ryan, W. B. F.; and Ericson, D. B., 1981, Relationship between annual variations in the rate of change of magnetic intensity and those of surface air temperature, *Jour. Geomagnetism and Geoelectricity* **33:**545-567.

Wood, C. A., 1977, Preliminary chronology of Ethiopian droughts, in D. Dalby, R. J. Harrison-Church, and F. Bezzaz (eds.), *Drought in Africa,* vol. 2, London: International African Institute, pp. 68-73.

Wood, C. A., and Lovett, R. R., 1974, Rainfall, drought and the solar cycle, *Nature* **251:**594-596.

Wood, F. J., 1978, *The Strategic Role of Perigean Spring Tides in Nautical History and North American Coastal Flooding, 1635-1976,* Washington, D.C.: U.S. Government Printing Office, 538p.

Wood, F. J., 1985, *Tidal Dynamics, Coastal Flooding and Cycles of Gravitational Force,* Dordrecht, The Netherlands: D. Reidel, 712p.

Wood, K. D., 1972, Sunspots and planets, *Nature* **240:**91-93.

Wood, R. M., 1975, Comparison of sunspot periods with planetary synodic period resonances, *Nature* **255:**312-313.

Wood, R. M., and Wood, K. D., 1965, Solar motion and sunspot comparison, *Nature* **208:**129-131.

Wood, S. M., 1946, The planetary cycles, *The Illinois Engineer* **22**(2).

Wood, S. M., 1949, Long-term weather cycles and their causes, *The Illinois Engineer* **25**(3).

Woodbridge, D. D.; Macdonald, N. J.; and Pohrte, T. W., 1959, An apparent relationship between geomagnetic disturbances and changes in atmospheric circulation at 300 millibars, *Jour. Geophys. Research* **64**(3):331-341.

Woods, J. D., 1981, The memory of the ocean, in A. Berger (ed.), *Climatic Variations and Variability: Facts and Theories,* Dordrecht, The Netherlands: D. Reidel, pp. 63-83.

Woodward, C. M., 1904, The planetary equinoxes—An examination of Mr. Tice's theory, in E. B. Garriott (ed.), *Long-Range Weather Forecasts*, Washington, D.C.: U.S. Department of Agriculture, Weather Bureau Bulletin No. 35, pp. 11-31.

Wright, P. B., 1968, Wine harvests in Luxemburg and the biennial oscillation in European summers, *Weather* **23:**300-304.

Wright, P. B., 1975, *An Index of the Southern Oscillation*, Norwich, England: University of East Anglia, Climatic Research Unit, Research Publication No. 4.

Wright, P. B., 1977, *The Southern Oscillation—Patterns and Mechanisms of the Teleconnections and the Persistence*, Honolulu: Hawaiian Institute of Geophysics, HIG 77-13.

Wright, P. B., 1978, The Southern Oscillation, in A. B. Pittock, and others (eds.), *Climatic Change and Variability. A Southern Perspective*, Cambridge: Cambridge University Press, pp. 180-184.

Wright, P. B., 1985, The Southern Oscillation: An ocean-atmosphere feedback system?, *Am. Meteorol. Soc. Bull.* **66**(4):398-412.

Wulf, O. R., 1945, On the relation between geomagnetism and the circulation motions of the air in the atmosphere, *Terr. Magnetism and Atmos. Electricity* **50**(3):185-197.

Wulff, G., 1910, Influence de la pression de la lumiere solaire sur le pression barometrique de l'atmosphere terrestre, *Soc. Phys. et Histoire Nat. de Génève Compte Rendu* **27:**70-74.

Wundt, W., 1944, Die Ursache der Eiszeiten, *Geol. Rundschau* **34**(7/8):713-747.

Wunsch, C., 1967, The long-period tides, *Rev. Geophysics* **5**(4):447-475.

Wunsch, C., 1974, Dynamics of the pole tide and the damping of the Chandler wobble, *Geophys. Jour.* **39:**539-550.

Wyatt, T., 1984, Periodic fluctuations in marine fish populations, *Environ. Education and Information* **3**(2):137-162.

Wyrtki, K., 1975, El Niño—The dynamic response of the equatorial Pacific Ocean to atmospheric forcing, *Jour. Phys. Oceanography* **5**(4):572-582.

Wyrtki, K., 1977, Sea level during the 1972 El Niño, *Jour. Phys. Oceanography* **7**(6):779-787.

Xanthakis, J., 1959, L'expression del'activite solaire en fonction du temps d'Ascension, *Annales Astrophys.* **22:**855-876.

Xanthakis, J., 1960, The sunspot areas and the relative sunspot numbers, *Geofisica Pura e Applicata* **46:**11-22.

Xanthakis, J., ed., 1973*a*, *Solar Activity and Related Interplanetary and Terrestrial Phenomena*, Berlin: Springer-Verlag, 195p.

Xanthakis, J., 1973*b*, Solar activity and precipitation, in J. Xanthakis (ed.), *Solar Activity and Related Interplanetary and Terrestrial Phenomena*, Berlin: Springer-Verlag, pp. 20-47.

Xanthakis, J., 1975, *Solar Activity and Global Survey of Precipitation*, Research Centre for Astronomy and Applied Mathematics, Academy of Athens, Report.

Xanthakis, J., 1979, Reply to Pittock: Possible Sun-weather correlation, *Nature* **280:**254-255.

Yamamato, A., 1976, Paleoprecipitational change estimated from the grain size variations in the 200-m long core from Lake Biwa, in S. Horie (ed.), *Paleolimnology of Lake Biwa and the Japanese Pleistocene*, vol. 4, Otsu, Japan: Otsu Hydrobiological Station, pp. 179-203.

Yamamato, T., 1952, Secular variation of summer rainfall in the Far East, *Kagaku* **22**(2):96.

Yao, C. S., 1982, A statistical approach to historical records of flood and drought, *Jour. Appl. Meteorology* **21:**588-594.

Yapp, C. J., and Epstein, S., 1977, Climatic implications of D/H ratios of meteoric water over North America (9500-22,000 B.P.) as inferred from ancient wood cellulose C-H hydrogen, *Earth and Planetary Sci. Letters* **34:**333-350.

Yoshimura, H., 1978, Nonlinear astrophysical dynamo: Multiple-period dynamo wave oscillations and long-term modulations of the 22-year solar cycle, *Astrophys. Jour.* **226:**706-719.

Yukutake, T., 1965, The solar cycle contribution to the secular change in the geomagnetic field, *Jour. Geomagnetism and Geoelectricity* **17:**287-309.

Yukutake, T., 1972, The effect of change in the geomagnetic dipole moment on the rate of the Earth's rotation, *Jour. Geomagnetism and Geoelectricity* **24:**19-47.

Yukutake, T., 1973, Fluctuations in the Earth's rate of rotation related to changes in the geomagnetic dipole field, *Jour. Geomagnetism and Geoelectricity* **25:**195-212.

Yukutake, T., and Tachinaka, H., 1969, Separation of the Earth's magnetic field into the drifting and the standing parts, *Earthquake Research Inst. Tokyo Bull.* **47:**65.

Zatopek, A., and Krivsky, L., 1974, On the correlation between meteorological microseisms and solar activity, *Astron. Inst. Czechoslovakia Bull.* **25:**257.

Zehnder, L., 1923, *Die zyklische Sonnenbahn als Ursache der Sonnenfleckenperiode*, Halle a. d. Salle.

Zeller, E. J., and Parker, B. C., 1981, Nitrate ion in Antarctic firn as a marker for solar activity, *Geophys. Research Letters* **8:**895-898.

Zenger, K. W., 1905, *Die Luni-solare (Saros). Periode des Wetters und der grossen Erdstorungen 1887-1905*, Prague: Selbstverlag, 175p.

Zerefos, C. S., and Crutzen, P. J., 1975, Stratospheric thickness variations over the northern hemisphere and their possible relation to solar activity, *Jour. Geophys. Research* **80:**5041-5043.

Zeuner, F. E., 1938*a*, The Pleistocene chronology of Europe, *Serbian Royal Acad. Pub. 177*, First Section 87, pp. 127-204.

Zeuner, F. E., 1938*b*, The chronology of the Pleistocene sea levels, *Annals and Mag. Nat. History* **1**(Series 11)(4):389-405.

Zeuner, F. E., 1938*c*, Die Gleiderung des Pleistozans und des Palaolithikums in Palastina, *Geol. Rundschau* **29**(6):514-517.

Zeuner, F. E., 1939, Schwankungen der Sonnenstrahlung und des Klimas im Mittelmeergebiet wahrend des Quartars, *Geol. Rundschau* **30**(6):650-658.

Zeuner, F. E., 1945, *The Pleistocene Period*, London: Ray Society, 322p.

Zeuner, F. E., 1958, *Dating the Past*, 4th ed, London: Methuen and Company, 516p.

Zeuner, F. E., 1959, *The Pleistocene Period. Its Climate, Chronology and Faunal Succession*, 2nd ed., London: Hutchinson Scientific and Technical Company, 447p.

Zhu, K.-Z., 1973, A preliminary study on the climate fluctuations during the last 5000 years in China, *Sci. Sinica* **16**(2):226-256.

Zhukov, L. V., and Muzalevskii, Y. S., 1969, A correlation spectral analysis of the periodicities in solar analysis, *Soviet Astronomy* **13**:473-479.

Zirin, H., 1966, *The Solar Atmosphere*, Waltham, Mass.: Blaisdell Publishing Company, 501p.

Zirm, R. R., 1964, Variations in decay rate of satellites 1963-21, *Jour. Geophys. Research* **69**(21):4696-4697.

Zlotnick, B., and Rozwoda, W., 1976, The influence of solar activity on the temperature of the stratosphere, *Artificial Earth Satellites* **2**:55-70.

Author Citation Index

Subject Index